Models in Biology:

MATHEMATICS, STATISTICS AND COMPUTING

Models in Biology:

MATHEMATICS, STATISTICS AND COMPUTING

D. BROWN
The Babraham Institute, Cambridge, UK

and

P. ROTHERY
Natural Environment Research Council, British Antarctic Survey, Cambridge, UK

JOHN WILEY & SONS
Chichester · New York · Brisbane · Toronto · Singapore

Reprinted July 1994

Other Wiley Editorial Offices

John Wiley & Sons, Inc., 605 Third Avenue,
New York, NY 10158-0012, USA

Jacaranda Wiley Ltd, 33 Park Road, Milton,
Queensland 4064, Australia

John Wiley & Sons (Canada) Ltd, 22 Worcester Road,
Rexdale, Ontario M9W 1L1, Canada

John Wiley & Sons (SEA) Pte Ltd, 37 Jalan Pemimpin #05-04,
Block B, Union Industrial Building, Singapore 2057

Library of Congress Cataloging-in-Publication Data

Brown, D. (David)
Models in biology: mathematics, statistics, and computing / D.
Brown and P. Rothery.
p. cm.
Includes bibliographical references and index.
ISBN 0 471 93322 8
1. Biology—Mathematical models. 2. Biometry. I. Rothery, P.
(Peter) II. Title.
QH323.5.B756 1993
574'.01'5118—dc20 92-41229
CIP

British Library Cataloguing in Publication Data

A catalogue record for this book is available from the British Library

ISBN 0 471 93322 8

Typeset in 10/12pt Times by Mathematical Composition Setters Ltd, Salisbury, Wiltshire
Printed and bound in Great Britain by The Bath Press, Avon

Contents

Preface

Over the last few decades, biology has developed from a largely observational discipline of description and classification into a fully fledged branch of science involving complex theories and mathematical models. Today's research biologists increasingly need to formulate, and to study the behaviour of, a wide range of mathematical models. These models are sometimes stochastic, i.e. involving a random component, and sometimes deterministic. They can be descriptions of a cross-section of a population at a fixed time, or describe dynamic processes. They might be simply phenomenonological descriptions, or involve detailed mechanisms and sub-mechanism, even down to the level of the genetic code itself.

As well as formulating theories in precise mathematical form, and studying them—deducing consequences and analysing behaviour—there is an important requirement to put them to empirical test. This involves making observations and using the resulting data either to estimate aspects of the mathematical model, or to test the theory. Even though many biological models are at present deterministic, biological phenomena differ from their physical counterparts by being subject—at least apparently—to random influences. Thus the process of matching or comparing data with model predictions usually involves statistical methods.

Biologists therefore need to understand and to use methods of formulating deterministic and stochastic models, and of fitting and testing them against real data. Traditionally the relevant subjects, biomathematics and biostatistics, have been taught in separate courses, with little attempt to bring them together. One obstacle to the promotion of an integrated course is the lack of a suitable text book. Books on deterministic modelling in biology, and a larger number on statistical methods and probabilistic models for biologists, are already available, some of them excellent. But hardly any texts have been written which cover both areas adequately, and which attempt to integrate the two approaches. This book was written to fill the gap.

The book deals with deterministic and stochastic mathematical models, statistical methods for fitting and testing them, and computer programs for simulation and statistical analysis. Our intention is to foster a working understanding rather than merely mechanical skill in using techniques. We therefore do not just give recipes, although incidentally many techniques are operationally defined. Emphasis is given to learning about models by computer simulation as well as using formal mathematics. Models are treated as experimental subjects upon which experiments can be performed, and their behaviour studied and documented. There are many graphical illustrations, and most of the models are set in a biological context using examples involving real data, drawn from many biological fields.

This book starts with elementary methods of formulating deterministic and stochastic models of single populations or processes, of fitting and testing them, and of designing experimental and survey investigations. It assumes only mathematics taught in school up to about age 18, and some of these ideas are revised in the early chapters. The coverage extends from this introductory material to more complex dynamic and other models for interacting populations with applications in ecology, epidemiology, biochemistry and physiology. Techniques for analysing pattern and sequence, including DNA sequences, for fitting complex descriptive, multiple regression models, for modelling processes in space and time and neural networks are included in a more advanced section.

In attempting to cover such a wide field, there will inevitably be important areas which are covered only very sketchily or not at all; and many specialist books will cover their own areas more thoroughly than we could possibly do. Some subjects were excluded because we did not feel they were of wide enough interest, but some important areas had to be excluded because of our limited expertise, or for reasons of space. Some areas of biology, such as recent developments in genetics, or areas of statistics such as multivariate analysis, have very good treatments suitable for biologists elsewhere, and this was sometimes a further reason for exclusion. Other material, such as recent methods of formulating and studying dynamic stochastic models, are technically advanced and therefore probably more appropriate for mathematical specialists.

Computing techniques for implementing many of the methods using a number of packages for statistical analysis and model simulation are given in a separately printed computing examples supplement. The reasons for this are twofold. First of all, while it is very important to understand the basis of computing techniques, and to use some computing packages, no individual package or range of packages is essential to a course of study. Secondly, the details of implementation in any specific package are likely to be more ephemeral than the techniques or concepts, and the computing material is therefore likely to require earlier updating than the main text itself.

We are indebted to biologists, statisticians, mathematicians and computer scientists in numerous ways. The original idea for this book arose during the revision of a course in quantitative methods for first year biologists at Cambridge University, and we are grateful to those involved in this project for many useful ideas, in particular Chris Gilligan who was originally a co-author of this book, but had to pull out because of pressure of other work. Many biologists have stimulated us to get involved in their areas, and have provided much background detail and data which we have used in illustration. In particular we would like to thank Steve Albon, John Bicknell, Willie Close, John Coadwell, Joy Dauncey, Barbara Gorick, Nevin Hughes Jones, Doug Ingram, Mike Harris, Fred Harrison, Andrew Laurie, Gareth Leng, Ron Lewis-Smith, Reza Morovat, Robert Moss, Laurence Mount, Malcolm Mountford, Ian Newton, John Paterson, Dominique Poulain, Mike Smith, Jon Wakerley, Eurof Walters, Adam Watson, Terry Wells and Baige Zhao for providing us with the problems which we discuss or for giving us access to unpublished data. We have received help from the originators and distributors of the Genstat, Phaseplane, Minitab and SAS packages. Robert Marrs, Grenville Foster, Chris Greenwood at the Babraham Institute, Andy Wood at the British Antarctic Survey and other AFRC, NERC and Cambridge University staff have been of considerable help with computing problems. Michael Dixon has been a particularly understanding and supportive editor, and we would like to thank him and Nicola Sawyer, Lesley Winchester, Liz Warden and other staff at Wiley for their help. We are grateful to Jonathan Foweraker, Martin Major and Michael Usher who read parts of the manuscript at various stages, and particularly to Sam Pattenden who read and carefully criticised many chapters and provided help with computing problems. To all these and many others whose names we cannot mention individually, we extend our sincere thanks.

Finally we would like to acknowledge the debt we owe to our former colleagues from the Nature Conservancy Biometrics Section under the leadership of the late John Skellam. Could we offer better general advice to modellers in biology than the following extracts from one of Mr. Skellam's later papers entitled 'The Formulation and Interpretation of Mathematical Models of Diffusionary Processes in Biology' (Skellam,1973)?

> 'Roughly speaking, a model is a peculiar blend of fact and fantasy, of truth, half-truth and falsehood. In some ways a model may be reliable, in other ways only helpful and at times and in some respects thoroughly misleading. The fashionable dogma that hypothetical schemes can be tested in their totality in some absolute sense, is hardly conducive to creative thinking. It is indeed, just as great a mistake to take the imperfections of our models too seriously as it is to ignore them altogether. ...'

> 'Mathematical model-making in Population Dynamics looks easy, and so it is if we do not care about the uses and abuses to which our models may be put. In theory all that we have to do is to incorporate those things that matter and to ignore those things that do not matter, if we can, or if we can't, treat them in any arbitrary way that suits our mathematical convenience. By following the reassuring principle that conclusions hold more often than do the detailed premises, we can always entertain the prospect that results of wider applicability will emerge.
>
> The snag is that it is not given to anyone to know beforehand which features matter and which do not, and in a real life context all sorts of things matter in varying degrees. The best we can do is to begin with schemes that may not be entirely realistic but which are at least meaningful to others as well as to ourselves.
>
> It is important to scrutinise the models carefully and to explore numerous variants in order to allay fears and doubts created by the many artefacts we invariably introduce and glorify in the name of abstraction.
>
> Finally we must seek empirical verification wherever and whenever possible, and in particular determine by empirical means the true range of applicability of the theoretical schemes and speculations that we have been bold enough to propose.'

Introduction

1 TYPES OF MODEL

This book is about biological models which can be expressed in mathematical form. There are many kinds of model. They might be representations in two dimensions of varying degrees of realism; for example, a highly simplified abstract line drawing of a bull by Picasso, or, at the other extreme, solidly rendered cattle such as you might find in the glowing paintings of the seventeenth-century Dutch artist Aelbert Cuyp. Or they might be nearer reality by actually being three-dimensional, e.g. the models of ships and houses that amateur modellers and architects build—these are miniature versions of the original or projected object. In biological research, one species is often said to be a model for another, e.g. in the early stages of medical research, monkeys sometimes serve as models for man. Another interpretation of the word model is that of an ideal or a standard, something to be imitated. In the sense in which we shall use it, the term model means a simplified representation, which is designed to facilitate prediction and calculation, and which can be expressed in symbolic or mathematical form.

The simplest mathematical models take the form of equations showing how the magnitude of one variable can be calculated from others. Differential and difference equations predict rates of change as functions of the conditions or the components of the system. Random variation is an important element in most biological systems, and so stochastic models, which involve probabilities, occur frequently; and these often necessitate statistical methods for fitting and testing of the models. Finally, there are computer models which involve generating on a computer a simulation of the modelled process. In this book, we make extensive use of computer simulation for illustrating models of many kinds, and so most of the models also become computer models at some stage of the discussion.

One purpose of this Introduction is to introduce the different kinds of model, their main components, how they might be devised and manipulated, and how they might be related to data.

1.1 Deterministic and stochastic models

As an example, consider the relationship between the area, A, of a leaf of a particular species and an area index, x, formed from the product of the length, l, and width, w. This problem arose in a study of the factors affecting flowering in a particular population of the bee orchid, *Ophrys apifera*. We might expect that the underlying relationship between A and x is a straight line, in this case through the origin, since we would expect that if the index were close to zero, then the area would also be close to zero. We would say in symbols

$$A = bx.$$

If the leaves were of constant shape, we might also expect that the straight line relationship would be exact, and then the model is said to be deterministic. The term deterministic is used since the area, A, can be determined from the value of x for an individual leaf exactly. In practice of course for biological phenomena, things are usually more complicated than this. Suppose we have collected some data to attempt to determine this relationship in a particular case, in other words to fix the value of the unknown parameter b. Figure 1 shows data gathered from a population of bee orchids, *Ophrys apifera*, on $x = lw$ and A. The growth patterns of individual leaves are subject to many influences, some large in individual cases and many small ones, which cause the leaf to vary in shape and size. In highly controlled laboratory conditions, there would be fewer such influences, and it might be possible to account for all of them in a detailed model. In field work it would be impossible. For one thing we could not allow for all the genetic variation between different plants which involves a substantial random element. In any case, it is usually not necessary to do so. A good method of allowing for these influences is to add a further component to the model which represents the total effect of all the random influences on each individual leaf area. Because we cannot predict these with any certainty in any individual case, we say that they are random or stochastic. Stochastic comes from the Greek word, *stokhastikos*, which means 'capable of guessing'. We would not model these influences in detail, just possibly assess the average extent to which they blur the deterministic picture. How we do this is dealt with in detail in Chapter 3. For the moment we add a random element to the current model as follows

$$A = bx + \varepsilon.$$

Here ε, the Greek letter epsilon, represents the random departures from the underlying straight line relationship with x which we observe with measurements on leaves selected at random from the population. In any estimation we are not interested in the individual values of ε which vary from plant to plant, but they can be thought of as the difference between the predictions from the straight line, and the observed values

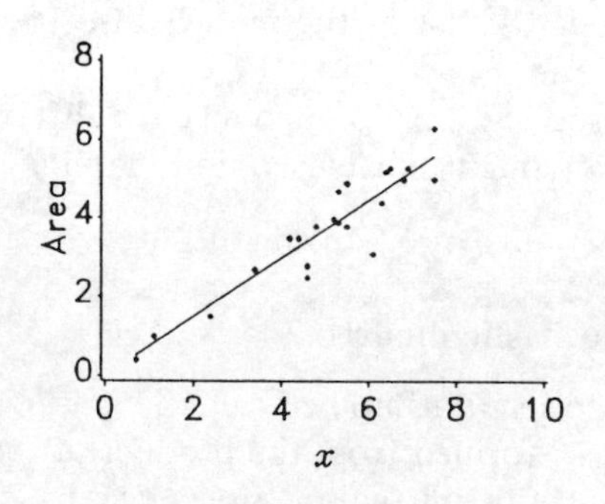

Figure 1 Leaf area plotted against leaf area index for the bee orchid data, together with a fitted straight line through the origin (data kindly provided by Terry Wells).

of leaf area. We might be interested in the distribution of ε—the range of variation, and the relative frequencies in different parts of the range. How to specify this is covered in Chapter 3.

1.2 Parameters and form

Now consider the way we think about the model itself. There are usually some aspects which we consider as being fixed. For example, in considering the relationship between the area and the index, the straight line model—whether correct or not—imposes a straitjacket on our interpretation of the relationship between A and x. That straitjacket takes the form of the following *fixed* element: the underlying relationship is a straight line, as opposed to a curved one. Even though, for a small sample, the straight line element of the model might not be distinguishable from a slight curve, it is still there and still constrains our thinking and analysis; and, if the straight line part of the model is correct and we were to take a very large sample of values and average the values at each x-value, we should see a straight line relationship.

But the model as given above does not state what the value of b is. b is termed a *parameter* and might be allowed to take values over a wide range. Which value it has will depend on the particular circumstances; and the value we select, or, as we say, *estimate* for it, depends on which method of estimation we use. Parameters are often referred to as changeable constants. In the real world, they are constants; but, as we do not know them, in our mental world they cannot be said to be fixed, and are therefore, in a sense, changeable.

By methods which we introduce in Chapter 5, we can fit a straight line to these data, its equation (Figure 1) being

$$A = 0.745x.$$

Thus we have succeeded in fixing what was previously unknown, although we should keep in mind that the value we have obtained for the parameter b is not exact, only an estimate which will be in error to some extent. In other cases, we might be able to fix parameter values by recourse to some theory—which would be possible if the leaves were a simple, fixed shape, e.g. elliptical with the same eccentricity—or to someone else's independent evaluations.

Thus there are two quite different ways in which we can speak of fixed parts of models. These two ways belong to our *descriptions* of two different worlds: the external world described in our models, and the mental world of the investigator. In the former we discriminate between deterministic and stochastic parts, stochastic parts having a randomly varying component. In the latter we discriminate between the fixed parts of our model, for example that the relationship is linear, and the parts about which there is more uncertainty, and are changeable, which are usually referred to as parameters. There might be some elements in our model which are fixed in our pictures of both the external world and in our mental world. For example, the acceleration due to gravity, or the atomic weight of sodium. And some quantities might change their status from being parameters at one stage of an investigation when their values have not been fixed precisely, to fixed constants at a later stage when—as a result of an estimation exercise, or some theoretical argument—these parameters might be given fixed values.

1.3 Descriptive and mechanistic models

The above model of leaf area is a simple descriptive model predicting leaf area from two simply made measurements, which is used in the case of the bee orchid in a non-destructive method of estimating leaf area, the eventual purpose being to relate the probability of flowering to leaf area. We expect leaf area to be fairly closely related to the leaf area index and devise an approximate model which we hope *describes* this relationship adequately for our purpose. We say describe firstly because we do not justify the systematic or deterministic straight line part of the model, other than to say that for leaves of a constant shape, we would expect a straight line. We might have gone on to consider the ways in which the leaf could change its shape as it grows, and discover that some other relationship would be more appropriate. Secondly, it could be said to be descriptive because we have not specified in detail the mechanisms causing the departures from the smooth relationship: we have lumped their effects together into a stochastic component of the model. However, for some purposes, the simple descriptive straight line model with random variation about it is quite adequate.

However, we could go on from this and devise a more detailed mechanistic model of the leaf structure and shape. Lindenmayer (1975) gives an example of a mechanistic model of leaf growth (discussed in more detail in Chapter 1). Leaves grow at the margins, and new cells originate from the cells at the outer boundary of the leaf, on the leaf margin. Therefore a model of a growing leaf need only specify what happens at the margin. This is probably the simplest case, but it is interesting to see how complex leaf structures can be derived from very simple replacement rules. The following example illustrates the potential.

Suppose a leaf consists of an assemblage of cells in various states. The symbols standing for cellular states in this simple model are a, b, c, d and k. We imagine the growth of a leaf as occurring in a number of discrete stages, and at each stage, we use replacement rules—rules which relate the state of the cells at the margins of the leaf at stage i to the states of the marginal cells at stage $i-1$. The replacement rules in this model are

$$a \to cbc, \quad b \to dad, \quad c \to k, \quad d \to a, \quad \text{and} \quad k \to k.$$

If we start from a, we obtain the following sequence.

Generation	Outcome
1	*a*
2	*cbc*
3	*kdadk*
4	*kacbcak*
5	*kcbckdadkcbck*
6	*kkdadkkacbcakkdadkk*
7	*kkacbcakkcbckdadkcbckkacbcakk*
8	*kkcbckdadkcbckkkdadkkacbcakkdadkkkcbckdadkcbckk*
...	...

These formulae can be interpreted in terms of leaf shapes by identifying the symbols as follows: a and b are sharp projecting leaf tips, c and d the lateral margins of lobes,

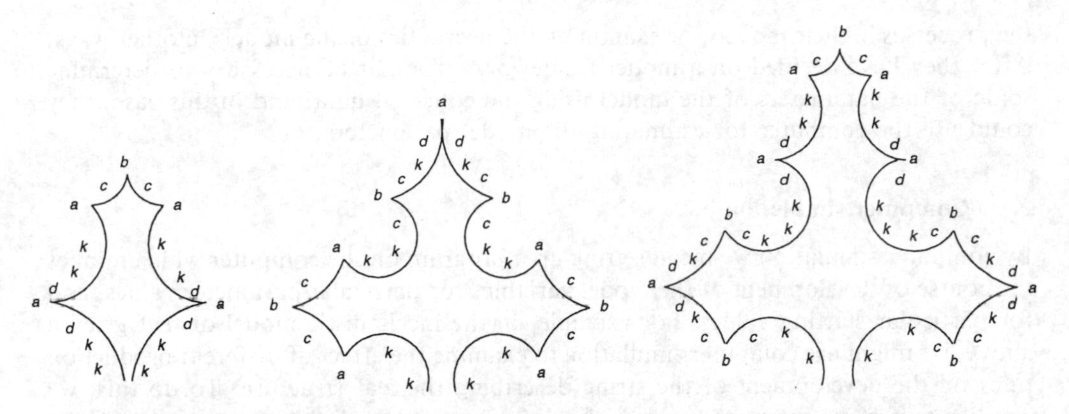

Figure 2 The development of the model leaf from generation 6 to generation 8, as discussed in the text.

and k's are identified with notches, the older the notch the more k's. So we obtain the leaf shapes in Figure 2 representing the strings in generations 6, 7 and 8 above. This, and the formulae derived in Chapter 1, shows how the centre lobe repeats the whole structure of the leaf two generations back, whereas the side lobes represent the whole structure three generations back.

1.4 Dynamic and non-dynamic models

A further distinction is between dynamic models which describe processes unfolding in time and models which do not involve time. Non-dynamic models include static models which describe a relationship at a fixed time, e.g. at the equilibrium of a process. They also include models which provide a cross-sectional picture of a population of individuals each of which is involved in some temporal process. In the above, the Lindenmayer model is dynamic, and the straight line model describes a cross-section of a population at a fixed time. There is not a hard and fast division between dynamic and static models, but rather a continuum. A useful framework for thinking about dynamic models based on relaxation time was given by Rapp (1980). Relaxation time can be defined informally as the time required to return to equilibrium after a small disturbance. The scales with examples and typical times are: electromechanical (neural firing, milliseconds), metabolic (enzyme catalysed reactions, seconds to minutes), epigenetic (short-term regulation of enzyme concentration, minutes to hours), developmental (differentiation, hours to years) and evolutionary (movement of genetic material through a population, months to years). Many models have events on more than one of these time scales, and in many cases, it might be simpler and also be an adequate approximation to think of the slower events as fixed, or to model the time scales separately.

2 COMPUTING

Modellers in biology use computers for two main purposes. The first use is simulating the processes in their models, or examining the properties of the models in other ways. After they have decided on a model framework, it might be necessary to determine some of the parameters of the model using the collected data, and in this case, they could use the computer for estimating the model parameters.

2.1 Computer simulation

By computer simulation we mean setting up a program on the computer which mimics the course of development of the model variables for particular parameter values, and for particular starting values. For example, in the mechanistic model of leaf growth above, we might use computer simulation to examine the effect of different production rules on the development of the string describing the leaf structure. To do this, we would set up a computer program which carries out the substitution rules automatically for any rules and for any starting values. Then it is merely a matter of feeding into the computer the rules, and the starting values and observing the development unfolding. We now outline a number of different ways in which simulation is used.

Instruction, problem clarification and informal experimentation

Simulation is useful for demonstrating mathematical models without getting too bogged down in the technical details of more formal mathematical methods. For those learning about models for the first time, it is particularly useful because it can demonstrate all the peculiarities of model behaviour which are sometimes not apparent in a mathematical analysis. Many mathematicians experiment with individual numerical cases, before embarking on setting down a general theory. So computer simulation is useful for experienced and inexperienced modellers to obtain a more intuitive understanding of a model.

Preliminary testing

Considering how you might simulate a model is also a good first test of a model, as often at this stage it becomes clear that the model is not completely defined in operational terms, and attention must be given to unambiguous specification of certain aspects.

Model verification

For models which are too complex for investigation of their properties by formal mathematics, simulation is an important tool for verifying that the model performs as intended. We are thus using computer simulation to perform mathematics which we cannot do directly. There is a very important difference however. A formal mathematical solution to a problem is in a sense universal; it applies to every case within a specified class. Computer simulation might be said to be *casewise mathematics*: it produces the solutions for cases specified by the initial conditions and particular parameter values. Skill must be exercised in selecting the cases to simulate, and, equally importantly, care is required in stating the range of applicability of the results of the simulation exercise.

Simulation may be used for this purpose with both deterministic and stochastic models. There is one major difference between the two uses: for a single case, i.e. a single set of starting values and parameter values, one simulation is adequate for a deterministic model, whereas many simulations—each one corresponding to an independently selected random choice for the stochastic elements—are usually necessary for a stochastic model.

Sensitivity analysis and robustness studies

A further aspect of a mathematical model is its robustness to errors in cases when we do not use exactly the right values of the parameters. Quite often, we do not know exactly what the value of a parameter is, and if the error we make in estimating that parameter affects the outcome of the model very much, then the model is of little practical value. This can be checked by mathematics, or by simulation. The method is quite simple: adjust the parameter in question slightly and assess whether the outcome of simulating the model changes only slightly. If it does change very little, we know that the margin of uncertainty surrounding that parameter is not materially affecting the results; if this is not the case, the model might need to be reconsidered. To carry out a proper sensitivity analysis is quite a task for models which are at all complex, as we must adjust the parameters singly, two at a time, three at a time, etc., as it might be that the parameter only becomes critical at a restricted set of values of the other parameters.

An important issue is the degree of complexity of the mathematical model which can be handled by simulation methods. Although it would appear that simulation frees us of the need to consider simple models, caution should be exercised in adding further complexity to a model. If a model is very complex, it is easier to come to the wrong conclusions in simulation exercises, mainly because the properties of models of a high degree of complexity are inadequately understood, and incidental features of a model formulation could critically affect the conclusions. This is a problem unless it is known *a priori* that the model is robust.

Methods for computer simulation and examples of its use occur throughout the book: particularly in Chapter 1 for deterministic models of growth and decline, Chapters 3 and 4 for stochastic models, Chapter 8 for models of biological interaction, Chapters 12 and 13 for spatio-temporal models and Chapter 15 for neural networks.

2.2 Fitting and testing models

The other area in which computers are very useful for biological modellers is parameter estimation, and testing of models or hypotheses against external data, in other words in statistical methods. Quite often a considerable amount of computation is required when using data on the observed outcome of a process to estimate the parameters, and this is usually best done on a computer. General methods of fitting models are dealt with in Chapters 5 and 11 in particular, but also in most of the other chapters for particular kinds of models: e.g. for patterns in space and time in Chapter 13, for empirically modelling a response as a function of many explanatory variables in Chapter 14, and for data from surveys and comparative experiments in Chapter 7. It is also frequently necessary to test a model against external data. This is an important part of the scientific method: to subject a hypothesis to the possibility of falsification, by comparing its predictions with freshly gathered data. Methods for doing this are again given in many chapters, but Chapters 6, 14 and 15 have more substantial treatments of some aspects of this topic. Computer simulation is also particularly useful for examining the properties of our statistical methods, and examples of its use are given in Chapters 5, 11 and 14.

3 PLAN OF THE BOOK

Part I is basic material essentially covering the construction of models for single populations and processes. The first two chapters cover the construction of deterministic models of growth and decline of populations and individuals and genetics. The next two chapters cover stochastic models, Chapter 3 mainly dealing with models in which the stochastic component is the largest part, mostly describing random variation about a single mean, and Chapter 4 dealing with models in which there are substantial stochastic and deterministic components—what we have termed *structured stochastic models*. The final chapter in this part is an introduction to statistical methods for fitting models, covering some but not all of the models dealt with in Chapters 1 to 4.

Part II's main emphasis is on biological comparisons, and design issues. Chapter 6 outlines techniques for comparing two or more populations or processes, and for comparing a population or process with a theoretical model, based on a sample of data. Chapter 7's focus is mainly on design: how to collect data when sampling from populations and processes, or when carrying out experiments in various contexts. There are also descriptions of the methods of analysing data from some of these designs.

Part III is about biological interactions of various kinds. Chapter 8 gives an introduction to the mathematical framework for models of interaction, mainly systems of differential equations, but also touching on difference equations and stochastic models. This material, together with some chapters from earlier parts, provides mathematical underpinning for both Chapters 9 and 10 which deal with applications in two broad areas of biology: Biochemistry and Physiology in Chapter 9, and the larger scale processes and populations of Ecology and Epidemiology in Chapter 10. Finally in this part, there is a chapter on more advanced model fitting, which covers methods of fitting models particularly from this section, but also some of the more advanced models from Part I; in particular, methods for fitting generalised linear models, non-linear statistical models and dynamic stochastic models.

Part IV covers more advanced topics. The chapters in this part are largely self-contained, in the sense that they do not refer to other chapters within Part IV, although most make use of material from earlier more general chapters. Chapter 12 deals with transport and diffusion. Chapter 13 covers models and statistical analysis of patterns in time, in space, and in sequences, particularly DNA sequences. Chapter 14 covers the general area of multiple regression, i.e. of methods for the statistical analysis of descriptive models in which a single response variable is a function of several explanatory variables. Chapter 15 deals with neural networks.

Courses could be constructed to use most of the material from Part I, although the chapter on genetics and the more advanced parts of Chapter 4 on structured stochastic models could be omitted. Chapters 1, 3 and 5 form core material in this part. Part II is mainly about statistical methods and design. Chapter 7 builds on the material in Chapter 6. Part III has two general chapters, Chapters 8 and 11, and these could be taken with either of the other two chapters in this part depending on the interests of the student. The chapters in Part IV have little interdependence and can be selected from as necessary to balance up a course.

3.1 Computing supplement

We feel that it is important for students to experiment with models and with fitting and testing procedures. In general, the technical details of the computing techniques have not been covered in the present text, but have been left to a computing supplement which will be bound separately. Apart from a general introduction, this supplement provides instructions for simulating, fitting and testing some of the models in the first three parts of the present text, using various simulation and statistical packages, and is cross-referenced in detail to the sections and examples of the present text. There are no references to the computing supplement in the present text.

Part I

MODELS FOR SINGLE POPULATIONS AND PROCESSES

1 Deterministic Models of Growth and Decline

1.1 INTRODUCTION

This chapter is about simple models for describing patterns of growth and decline or decay in biological populations and individual organisms. An example of individual growth taken from our own species is given in Figure 1.1(a) which shows observations from a pioneering study made by Gueneau de Montbeillard on the height of his first son beginning at birth in 1759 and continued for nearly eighteen years until 1777 at approximately six-monthly intervals (Scammon, 1927). Figure 1.1(b) gives an example of population growth showing the increase in the size of the population of the United States recorded during 1780–1940 from censuses carried out every ten years or so (Pearl *et al.*, 1940). In both cases the observed growth patterns are a result of complex underlying processes. Growth of the infant proceeds through cell division which needs to be fuelled by regular nourishment. There is a period of rapid growth during infancy and early childhood, a middle period extending from three to nearly thirteen in which growth is slow but constant; a marked spurt of growth prior to puberty, and a slower rate thereafter. Changes in the size of the US population are the net effect of increases from births and pioneering immigrants minus decreases from deaths and emigration. These demographic variables will in turn be affected by other factors. For example, births by family planning measures; deaths by outbreaks of disease; immigration and emigration by conditions prevailing in other countries. Despite the complexity of the processes involved the resulting growth appears fairly regular with the observations falling close to a smooth curve. In this chapter we present mathematical models for describing such patterns of growth or decline and the underlying processes involved.

In building mathematical models a useful broad distinction is between *mechanistic models* and *empirical* or *descriptive models*. Mechanistic models are more concerned with the 'nuts and bolts' of biological processes and the way in which the component parts fit together. They attempt to describe the observations in terms of fundamental postulates about the biological processes. By contrast empirical models provide quantitative descriptions of the patterns in the observations without attempting to describe the underlying processes or mechanisms involved. For example, an empirical model for the growth of the population of the United States could be a smooth curve drawn through the census data. A more mechanistic approach would be to build a model in terms of the separate components, such as the birth and death rates, which contribute to the overall population change. In a sense, all models are descriptive differing only in their degree of resolution and level of complexity. Nevertheless, the broad distinction between mechanistic and empirical models is a useful one.

This chapter deals with *deterministic models*, i.e. those in which a given input to the model always produces the same output. For example, in a deterministic model for population growth the input may be the size of the population at the start of some

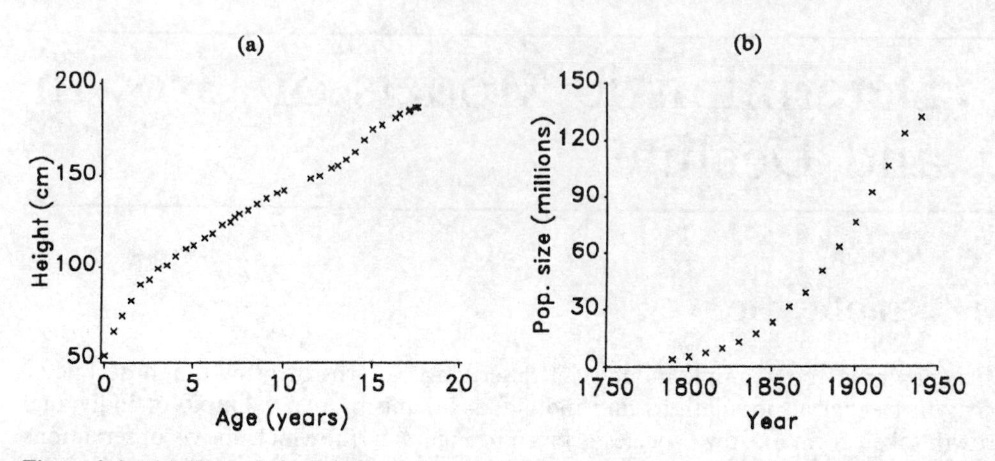

Figure 1.1 (a) Growth in height of Montbeillard's son; (b) growth of the population of the United States.

period and the output could be a single precise value for the predicted population size at some future time. Deterministic models contrast with *stochastic models* in which there is a random element such that for a given input to the model the outcome is not uniquely determined but takes a range of possible values. Real populations and individuals are subject to random influences which will affect their growth but we begin by simplifying the situation and ignore these complications. Despite the simplicity of the models we shall see that they can provide reasonable descriptions of real data on a wide range of growth patterns of populations and individuals. Later, in Chapter 3, we describe some simple stochastic models for random variation and in Chapter 4 we link deterministic and stochastic components to build more realistic models. We deal with the problem of how to fit the models to data in Chapter 5.

1.2 EXPONENTIAL GROWTH AND DECLINE

Exponential growth and decline are basic models for describing the change in size of biological populations. Essentially, this is because all organisms have an innate capacity to increase, or decrease, at a rate directly proportional to their numbers. This fundamental property is illustrated by a simple model proposed by Linnaeus in 1740 for population growth of an annual plant. He showed that if each plant produces only two seeds and if the two offspring in the next generation were to do the same then the numbers in successive generations would follow a geometric progression $2, 4, 8, 16, \ldots$, and so on: twenty years would suffice to produce a million seeds. This simple model captures the essence of exponential growth in which the population increases in direct proportion to its size.

In this section we extend the numerical model of Linnaeus by expressing the idea of exponential growth and decline more generally as mathematical models. We develop models for populations, such as annual plants, in which population change is studied over discrete times, and models for populations, such as our own species, in which change is a continuous process.

1.2.1 Exponential growth and decline in continuous time

In many situations growth, or decline, appears as a continuous process. For example, a colony of the bacterium *Escherichia coli* grows through asynchronous cell division, or binary fission, and may contain millions of cells. The addition of a single cell makes such a small change that the population size may be thought of as a variable changing continuously with time. The mathematical tools for describing the dynamics of continuous growth through infinitesimal changes are differential equations. We begin by developing a differential equation for exponential growth which we will then solve to find the mathematical equation for the size of the population at some specified time.

To fix ideas consider a bacterial colony. The successive population sizes over a small increment of time δt are related by

Colony size at time $t+\delta t$ = Colony size at time t
+ Growth increment in time $(t, t+\delta t)$

or, in symbols

$$N(t+\delta t) = N(t) + \delta N$$

The key property of exponential growth is that the growth increment is directly proportional to the size of the population. This is expressed mathematically by writing the growth increment in time $(t, t+\delta t)$ as

$$\delta N = rN\delta t$$

where the parameter r is referred to as the *rate constant of growth*. To make the expressions look simpler we have omitted the argument t but it is implicit that the population size N is a function of time. Rearranging this expression gives $r = \delta N / N\delta t$ as the fractional change in size per unit time. This is measured in units of reciprocal time—for example, 0.6 per hour or, equivalently, 0.01 per minute. For this reason, r is sometimes referred to as the *fractional growth rate*, the *relative growth rate* or the *specific growth rate*.

Dividing both sides of the above expression by δt gives the rate of change in size per unit time over the small time interval as

$$\frac{\delta N}{\delta t} = rN.$$

By considering an infinitesimally small time interval, or in mathematical jargon letting $\delta t \to 0$—in words, letting delta t tend to zero—the rate of change of population size is seen to satisfy the differential equation

$$\frac{\mathrm{d}N}{\mathrm{d}t} = rN.$$

The steps for solving the differential equation are as follows:

- Separate the variables and integrate so that

$$\int \frac{\mathrm{d}N}{N} = r \int \mathrm{d}t.$$

- Evaluate the integrals to give

$$\log_e N(t) = rt + c$$

where c is a constant of integration.
- Impose the condition $N(t) = N(0)$ at time $t = 0$ so that

$$\log_e N(t) = \log_e N(0) + rt.$$

- Take exponentials of both sides to obtain

$$N(t) = N(0)\, e^{rt}.$$

The population growth is called exponential from the exponential function which describes its form. An important property of exponential growth is that the logarithm of population size increases as a straight line with slope equal to the rate constant r (using logarithms to base e).

Figure 1.2 shows some patterns of exponential growth for a range of values of the rate constant and different initial population sizes. Note that for those populations with the same rate constant the increase in the logarithm of population size is a series of parallel lines separated by an amount depending only on their initial sizes.

For exponential growth the time taken to produce a doubling of population size is called the *doubling time*. A characteristic property of exponential growth is that the doubling time does not depend on the initial population size but is determined solely by the rate constant of growth. By definition, the doubling time T_2 satisfies the equation $2N = N\, e^{rT_2}$, for any value of N. Cancelling out the N so that $2 = e^{rT_2}$ and taking logarithms of both sides gives the doubling time as $T_2 = 0.6931/r$.

Doubling time for continuous exponential growth

$$T_2 = \frac{\log_e 2}{r} = \frac{0.6931}{r}$$

where r is the rate constant of growth in units of reciprocal time.

Example 1.1 Exponential growth of *Escherichia coli* in culture

The bacterium *Escherichia coli*, an inhabitant of the human intestine, is capable of very rapid growth. This growth can be recorded in culture by measuring the turbidity, through the scatter of incident light, to estimate the number of bacteria present. Figure 1.3 shows an example in a synthetic culture observed at about 20-minute intervals over about four hours (Stent, 1963). The population size shows a curvilinear increase and within half an hour it contains over one million cells ml^{-1}. On a logarithmic scale the increase is approximately linear indicating exponential growth. The rate constant of growth is estimated by the slope of the fitted line as 0.014 per minute or 0.84 per hour. Applying the formula for doubling time shows that the time for the population to double in size is $0.6931/0.014 = 49.5$ min. For the bacterium which reproduces through binary fission the doubling time can be interpreted as a generation time, or time between the moment when a parent cell has just been produced to the moment when it divides to produce a pair of daughter cells.

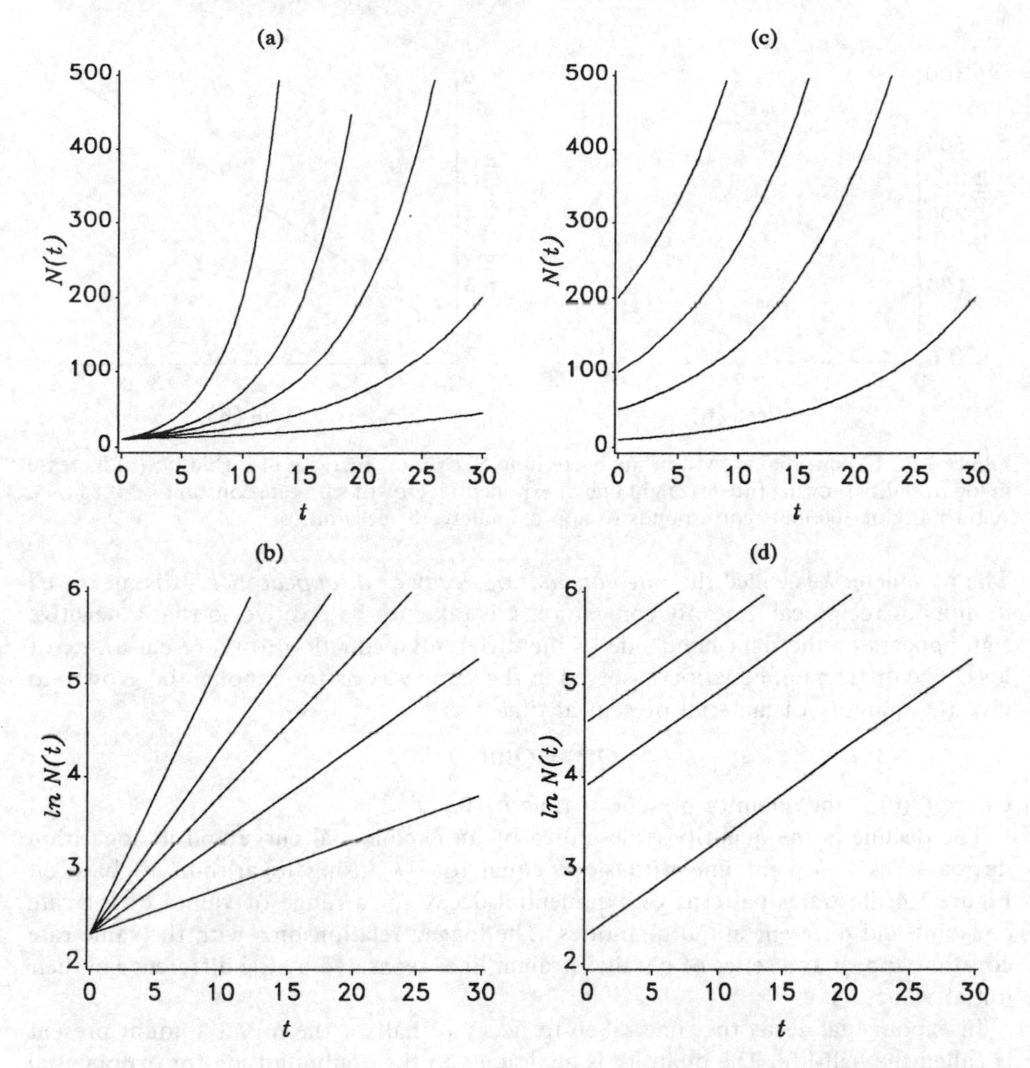

Figure 1.2 Examples of exponential growth in continuous time for different values of the rate constant, r, and initial population size, $N(0)$. (a) $N(0) = 10$, $r = 0.05$ (bottom), 0.10, 0.15, 0.20 and 0.25 (top); (b) logarithm of values in (a); (c) $r = 0.10$, $N(0) = 10$, 50, 100 and 200; (d) logarithm of values in (c).

Exponential decay

In the model of exponential growth the rate constant is positive but the model can also have a negative rate constant. This case corresponds to a rate of loss or decay in direct proportion to size or amount of material present. When the quantity of material present at time t can be described by a continuous variable $Q(t)$ the rate of change of the quantity at time t is given by the differential equation

$$\frac{dQ}{dt} = -kQ.$$

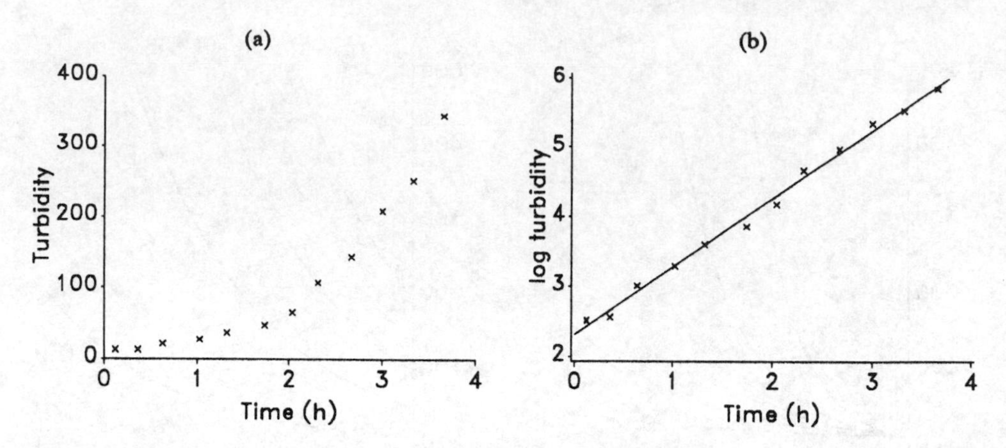

Figure 1.3 Exponential growth in the bacterium *E. coli.* (a) Increase in turbidity; (b) increase in log turbidity showing fitted straight line of exponential growth with rate constant $r = 0.84\ \text{h}^{-1}$. A turbidity of 100 units corresponds to approximately 10^8 cells/ml.

The parameter k is called the *rate constant of decay or disappearance*. It is measured in units of reciprocal time. By convention k is taken to be positive so that a negative sign appears on the right-hand side of the differential equation to represent a rate of loss. The differential equation is solved in the same way as for exponential growth to give the quantity of material present at time t as

$$Q(t) = Q(0)\ \text{e}^{-kt}$$

where $Q(0)$ is the quantity present at time $t = 0$.

The decline in the quantity is described by an exponential curve and its logarithm decreases as a straight line with slope equal to $-k$ (using logarithms to base e). Figure 1.4 illustrates patterns of exponential decay for a range of values of the rate constant and different initial quantities. The logged relationships with the same rate constant appear as a series of parallel straight lines separated by the difference in their initial values.

In exponential decay the time taken to decay to half of the initial amount present is called the half-life. The quantity is analogous to the doubling time for exponential growth. It does not depend on the initial amount present and is determined solely by the rate constant of decay. By definition, the half-life $T_{1/2}$ satisfies the equation $Q/2 = Q\ \text{e}^{-kT_{1/2}}$. Cancelling Q and taking logarithms of both sides gives the half-life as $T_{1/2} = 0.6931/k$.

Half-life for continuous exponential decay

$$T_{1/2} = -\frac{\log_e 1/2}{k} = \frac{0.6931}{k}$$

where k is the rate constant of decay in units of reciprocal time.

Example 1.2 Measurement of blood flow

Mathematical models are often used to estimate quantities which are difficult to

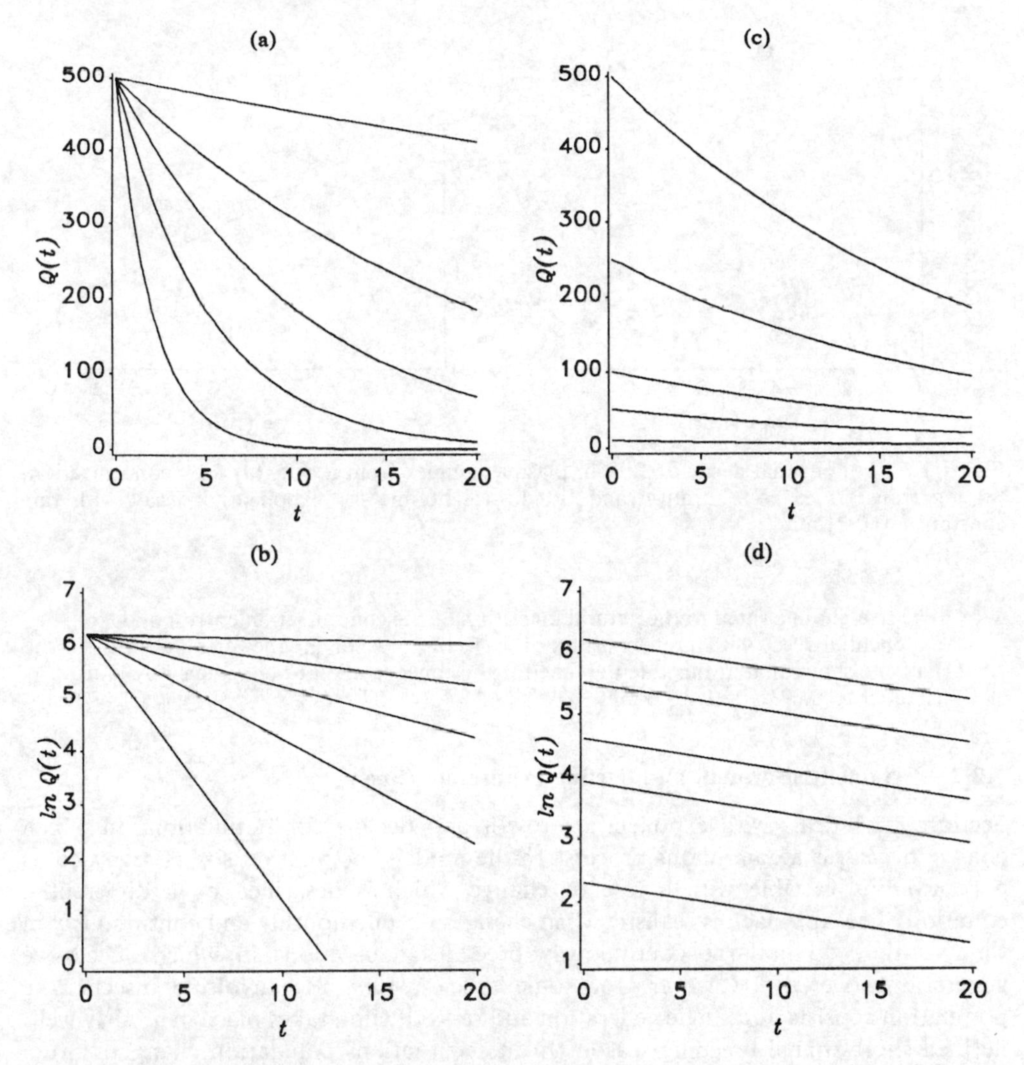

Figure 1.4 Examples of exponential decay in continuous time for different values of the rate constant, k, and initial quantities, $Q(0)$. (a) $Q(0) = 500$, $k = 0.01$ (top), 0.05, 0.10, 0.20 and 0.50 (bottom); (b) logarithm of values in (a); (c) $k = 0.05$, $Q(0) = 10$, 50, 100, 250 and 500; (d) logarithm of values in (c).

measure directly but which are of interest in a wider context. An example comes from compartmental analysis in which physiologists attempt to model the animal body as an assemblage of pools, or compartments, connected by the flow of body fluids. This requires ascertaining the various pathways in the system and measuring the flow rates. To do this, tracers are introduced into the system and their concentrations followed in time. The dye is well mixed into the animal's system so that the change in concentration declines smoothly and continuously. The technique is to use the rate of decline in concentration to measure the flow rate for a particular pool. Figure 1.5 shows some data on the decline in concentration of the sulphobromophthalein measured at 1 minute intervals in the jugular vein of a calf following injection of a standard dose (Harrison *et al.*, 1986). On a logarithmic scale the decline is approximately linear indicating a constant fractional loss of tracer.

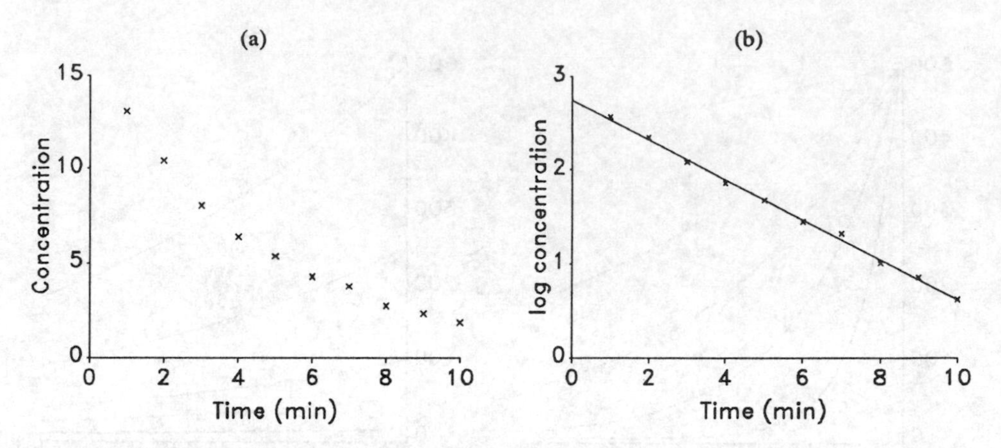

Figure 1.5 Exponential decay of BSP in plasma samples from a calf. (a) Raw concentration; (b) log transformed concentration and fitted straight line for exponential decay with rate constant $k = 0.21 \text{ min}^{-1}$.

The straight line fitted to the natural logarithm of the concentration corresponds to exponential decay with a rate constant of 0.21 min^{-1}. Applying the formula for the half-life shows that at this rate the time for the concentration to be reduced by half is equal to $0.6931/0.21 = 3.3$ min.

1.2.2 Exponential growth and decline in discrete time

Section 1.2.1 deals with exponential growth and decline for populations in which change occurs as a continuous process. In the models, population size is regarded as a continuous variable with a rate of change which is described by a differential equation. This approach is realistic when changes occur smoothly and continuously in time as they do in large continuously breeding populations in which successive generations overlap. However, for some species, such as univoltine insects, the population consists of a single generation and reproduction takes place in a fairly well-defined short annual breeding season. In these situations population change is often measured over discrete time periods from generation to generation or from year to year. To describe such changes we use difference equations in which the size of the population in one year is related to its size in the previous year.

Consider a simple model for an insect with a single annual breeding season and adult life span of one year. Suppose that each female produces a fixed number R of female offspring which survive to breed. The number of females in year $i + 1$ is then related to the number of females in year i by the difference equation

$$N_{i+1} = RN_i.$$

Starting with an initial number N_0 the size of the first generation is given by $N_1 = RN_0$, the size of the second generation by $N_2 = RN_1 = R^2N_0$, and so on. This illustrates the general solution of the difference equation in which the number of females in the ith year is given by

$$N_i = R^iN_0.$$

The parameter R is called the *geometric growth rate* or the *finite rate of increase*. It is sometimes denoted by the Greek letter λ. The sizes of successive generations form a geometric progression which increases when R exceeds one and declines when R is less than one. An analogy for geometric population growth is the capital accrued on a sum of money invested at a compound interest rate of $100 \times (R-1)\%$ per annum. Taking the logarithm of the population size gives

$$\log_e N_i = \log_e N_0 + i \log_e R$$

and shows a straight line increase with slope equal to $\log_e R$. The magnitude of the slope depends on the base of the logarithm but the increase is linear irrespective of the base.

By writing $R = e^r$ or $r = \log_e R$ the size of the population at the ith time is

$$N_i = N_0 \, e^{ir}$$

so that successive population sizes fall on an exponential curve at equally spaced intervals of time. This result means that the examples of exponential growth in continuous time shown in Figure 1.2 for different values of the rate constant r can be related to the patterns of exponential increase in discrete time with values of the geometric rate of increase given by $R = e^r$. For example, population sizes at unit time intervals for a geometric rate of increase of $R = 1.10$ per unit time fall on an exponential curve with rate constant $r = \log_e(1.10) = 0.0953$ per unit time.

Note that the correspondence between exponential growth in continuous time and discrete time applies only to the values of population size at selected equally spaced times. In between these times there is a fundamental difference between the models. In the continuous time model all population sizes fall on a smooth exponential curve whereas in the discrete time model this is not generally so. For example, consider an insect with a short annual breeding which dies after laying its eggs. When the eggs hatch there is a pulse in population size which may be followed by periods of varying survival rates before the pupae emerge as adults.

Example 1.3 Exponential growth of the population of collared doves in Britain and Ireland 1955–64

Following a dramatic expansion of its range across Europe from the Balkans to the North Sea in under twenty years the collared dove, *Streptopelia decaocto*, bred in Norfolk, England, in 1955 (Hudson, 1965). The subsequent increase in numbers over the next ten years is summarised in Figure 1.6. By 1964 the species had reached most areas of Britain and the population size, measured as the number of adults and juveniles at the end of the annual breeding season, had increased from 4 to an estimated 18 855. There is an approximately linear increase in the logarithm of the population size with a fitted straight line of slope equal to 0.98 which corresponds to exponential growth with a geometric rate of increase equal to $e^{0.98} = 2.66$ per annum. It is remarkable that this observed pattern of unlimited growth is similar to that for *E. coli* despite a substantial difference in the breeding cycles (Example 1.1). Note, however, that for the collared dove there is no implied pattern of smooth regular increase in time. Indeed, after the end of the breeding season the population will decline in size due to mortality until the influx of new recruits in the next season.

To explain the observed exponential increase in population size, we now take a more mechanistic approach and build a simple model which incorporates

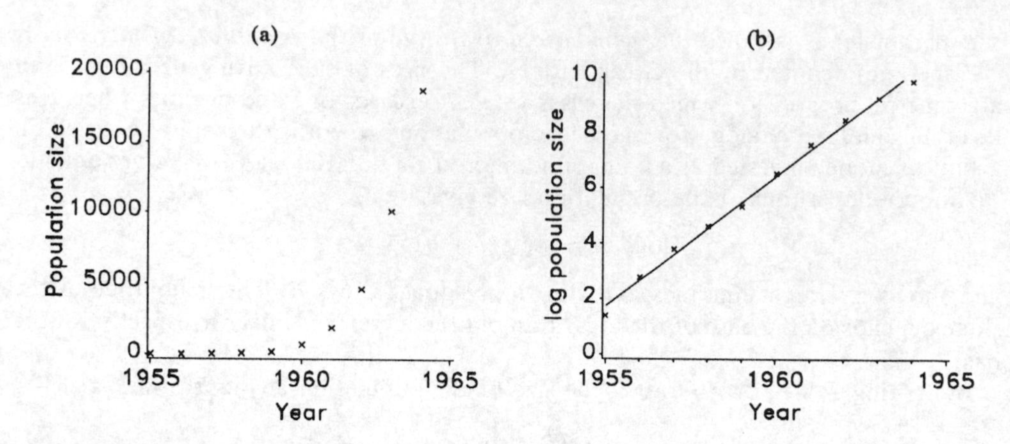

Figure 1.6 Exponential growth in a population of collared doves. (a) Increase in the number of adults plus young at end of breeding season; (b) approximately linear increase of log numbers with fitted straight line for exponential growth and geometric rate of increase 2.66 per annum.

information on survival and reproduction. Each year a number of breeding pairs rear young which are capable of breeding as adults in their first year after fledging. The number of adult females in the breeding population in a particular year is made up of the surviving adults from the previous year plus the number of their offspring which survive to be recruited. A word model for the population change is then

Number of adult females in breeding population year $(i+1)$
= Number of adult females in breeding population year i
× Proportion of adults surviving to year $(i+1)$
+ Number of young females produced in year i
× Proportion of young surviving to year $(i+1)$.

To develop this general equation for population change we consider the particular simple case in which survival rates and birth rates do not vary from year to year. A constant proportion s_a of adult females breeding in one year survive to breed in the next and a constant proportion s_y of young birds survive to become adults and breed at the end of their first year. Each female produces a constant number b of young of which half are assumed to be females. The above word model can now be written as a difference equation for the number of adult females (N_i) in the breeding population of year i given by

$$N_{i+1} = N_i s_a + \frac{N_i b s_y}{2}$$

or

$$N_{i+1} = \left(s_a + \frac{b s_y}{2}\right) N_i.$$

The formulation reduces to the simple model for geometric growth given by $N_{i+1} = RN_i$ and shows how the geometric rate of increase is related to the birth and death rates by $R = s_a + bs_y/2$.

Hofstetter (1954) studied collared doves in Germany and estimated average survival rates of 86% for adults and 60% for juveniles in their first year. Collared doves lay clutches of two eggs, although most pairs lay two clutches and some lay more so that each pair has the potential to rear 4 young. For a model population in which $s_a = 0.86$, $s_y = 0.60$ and $b = 4$, the calculated geometric rate of increase is given by $0.86 + 4 \times 0.60/2 = 2.06$ per annum. This is less than the observed value of 2.66 per annum for the population in Britain and Ireland and could be the result of increased survival, better breeding or a combination of both. For example, increasing the number of young per female to 6 gives a geometric rate of increase of 2.66.

The model illustrates that exponential growth is not restricted to simple situations akin to cell division in which all individuals respond identically. Despite the difference in survival rates of young and adult birds the model population increases exponentially at a steady rate.

Exponential decline

The simple difference equation $N_{i+1} = RN_i$ gives rise to exponential population growth when the geometric rate of increase R exceeds 1. When R is less than 1 successive population sizes are reduced by a constant fraction and the population size declines exponentially so that the logarithm decreases as a straight line

$$\log_e N_i = \log_e N_0 + i \log_e R$$

with slope $\log_e R < 0$. Since we can write this as $N_i = N_0 \, e^{i \log_e R}$, the series of population sizes fall at equally spaced times on an exponential curve with decay constant $k = -\log_e R$.

Example 1.4 Survival in the robin

Populations decline through the death of individuals but measuring death rates in wild populations is difficult because of the problem of monitoring the fate of individuals. In a population study of the robin *Erithacus rubecula* the ornithologist Lack (1965) met this problem by placing rings on the legs of adult birds and then retrapped them over subsequent breeding seasons. It was not possible to record the precise age of death of an individual but survival could be monitored as the proportion of birds surviving from year to year. Figure 1.7(a) shows the survival pattern of 129 adults as the percentage of birds alive 1, 2, 3 and 4 years after being ringed: 49 survived for at least 1 year, 20 for more than two years, 8 for 3 years or more and 2 for at least 4 years. So, the proportion of birds surviving successive years

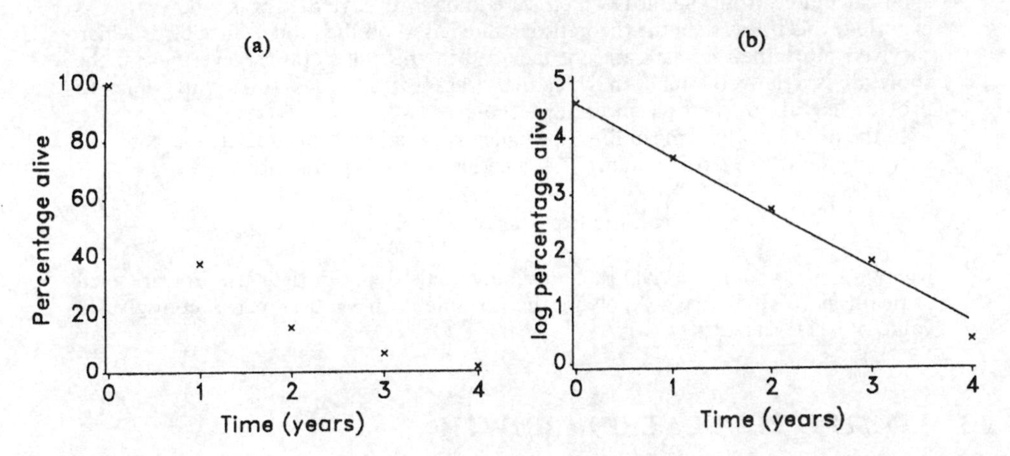

Figure 1.7 Pattern of survival in a cohort of 129 adult robins over 4 years after ringing. (a) Percentage of survivors; (b) log percentage of survivors with fitted straight line for exponential decline and constant annual survival rate of 0.38.

was as follows: year 1, 49/129 = 0.38; year 2, 20/49 = 0.41; year 3, 8/20 = 0.40; year 4, 2/8 = 0.25; with an overall annual survival rate of 79/206 = 0.38. The apparently low figure in the final year is based on only 8 birds alive at the start of the year and is therefore sensitive to just a single bird surviving but not being recorded. More generally, we might think of the observed survival rates as estimates of an average pattern of survival in a large population of adult birds. The analysis suggests a simple model with constant survival rate in which the percentage surviving over successive years declines exponentially according to

$$N_i = 100 \times 0.38^i$$

or

$$\log_e N_i = 4.605 - 0.968i.$$

Figure 1.7(b) shows that the values of the logarithm of the percentage of birds surviving from year to year are close to this straight line.

In his study, Lack aimed to find out how long robins lived. This question is less straightforward to answer than it is for human beings since births and deaths of robins are not so easy to document. Ideally, a large number of birds would be monitored, their ages at death recorded and the average calculated to give the mean life expectancy. In the absence of such detailed data we use the model to simulate the observations on a large number of birds and calculate the proportion dying in a particular year. This is simply the proportion alive at the start of the year multiplied by the mortality rate for the year. So, for example, the proportion dying in their 3rd year is $0.38^2 \times 0.62$. A contribution to the mean life expectancy is then calculated by forming the product of age of death and the corresponding proportion dying. Since time of death is not known precisely within the year we follow the convention of assuming that all deaths occur half-way through the year at ages 0.5, 1.5, etc. Summing the contributions to the life expectancy for the first ten years gives

$$0.5 \times 0.62 + 1.5 \times 0.38 \times 0.62 + 2.5 \times 0.38^2 \times 0.62 + \cdots + 9.5 \times 0.38^9 \times 0.62 = 1.11 \text{ years.}$$

The contribution from birds living longer than 10 years is negligible. It may come as a surprise to find that mean life expectancy in the adult robin is only just over a year. Note that for the data, which covers 4 years after ringing, the average life-span calculated from the birds which were known to have died was 1.06 years. As expected, this is less than the theoretical value based on the model since birds which survived more than 4 years are not included in the calculation. Nevertheless, the shortfall is relatively small showing that Lack's study was sufficiently long to provide useful information on life expectancy.

In the above analysis mean life expectancy is calculated for a particular constant survival rate of 0.38 per annum. The general formula is given by

$$\text{Mean life expectancy} = \frac{1}{(1-s)} - 0.5$$

where s is the annual survival rate and individuals dying in their ith year are aged at death as $(i - 0.5)$ years. Applying the formula with $s = 0.38$ per annum gives a value of 1.11 years.

1.3 LIMITED POPULATION GROWTH

In 1798 T.R. Malthus pointed out that although populations can increase exponentially their food supplies usually increase arithmetically so that population growth will eventually be limited. This important idea was later used by Charles Darwin (1859) to develop his theory of Evolution through Natural Selection. Darwin argued that most organisms produce many more offspring than can survive to breed so that competition for limited resources would lead to a struggle for existence and the survival of the fittest. He illustrated the point by applying the model for exponential growth to calculate the potential population increase in the elephant, a rather slow breeding species. He showed that after a period of about 750 years there would be nearly nineteen million elephants descended from a single pair.

For plants and animals there are many ways in which population growth might be limited. Shortage of food, lack of space or the accumulation of toxic substances in the environment may lead to increased mortality rates or reduced rates of reproduction and thereby prevent further population increase. These changes may involve increased competition for resources, either directly through overt interactions between individuals, or indirectly via a more passive depletion of the environment. Whatever the underlying processes, the result is limited population growth in which the relative growth rate decreases with increasing population size. The idea of limited population growth also applies at the molecular and cellular level. Chemical reactions can involve an increase in concentration of a product substance which is limited by the amount of reactant present and the nutrient uptake by a cell may be limited by the concentration of the surrounding medium. Often such processes lead to limited population growth in which the growth rate declines steadily with population size.

In this section we develop some mathematical models for limited population growth which apply to a wide range of biological processes and mechanisms. We begin with situations in which change occurs as a continuous process so that rates of changes can be described by differential equations. Then we consider changes measured over discrete times for which difference equations are appropriate.

1.3.1 Monomolecular model for limited growth in continuous time

A basic pattern of limited population growth corresponds to a growth rate which declines steadily towards zero. A particular case is the monomolecular growth curve which has been used extensively to describe the increase of the concentration of a product formed in a first-order chemical reaction. Another example is nutrient uptake by a cell in a constant medium. To derive the monomolecular model we consider the first-order chemical reaction represented by

$$A \rightarrow P$$

where A is the reactant and P is the product. The rate of the reaction at any instant is proportional to the concentration of reactant remaining at that time. So if $R(t)$ is the concentration of the reactant at time t then the rate of decline in concentration is given by the differential equation

$$\frac{\mathrm{d}R}{\mathrm{d}t} = -kR$$

where k is the reaction rate constant measured in units of concentration per unit time.

This is the differential equation for exponential decline (Section 1.2) with solution

$$R(t) = A\,\mathrm{e}^{-kt}$$

where A denotes the amount of reactant available at time $t = 0$.

If $C(t)$ is the concentration of the product at time t then, since $R(t) + C(t) = A$, we have

$$C(t) = A(1 - \mathrm{e}^{-kt}).$$

This is the *monomolecular* growth curve. Before describing its properties we derive it in a different way by working with the concentration of the product. When the rate of increase in concentration of the product at time t is proportional to the concentration of the reactant $R(t) = A - C(t)$, it satisfies the differential equation

$$\frac{\mathrm{d}C}{\mathrm{d}t} = k(A - C).$$

The equation can be solved using analytical methods as follows.

- Separate the variables and integrate so that

$$\int \frac{\mathrm{d}C}{A - C} = \int k\,\mathrm{d}t.$$

- Evaluate the integrals to give

$$-\log_e(A - C) = kt + c$$

  where c is a constant of integration.
- Impose the initial condition $C(t) = 0$ when $t = 0$ so that

$$-\log_e(A - C) = kt - \log_e A.$$

- Take exponentials of both sides and rearrange to obtain

$$C(t) = A(1 - \mathrm{e}^{-kt}).$$

This is the form of the *monomolecular* growth curve which was derived above from the exponential decline in the concentration of the reactant. The advantage of the alternative derivation is that it shows that monomolecular growth applies more generally to situations in which the growth rate declines linearly with the population size.

Figure 1.8 shows some monomolecular growth curves for a range of values of the rate constant k. The parameter A represents an upper limit, or asymptote, which theoretically is attained after an infinite amount of time. For practical purposes, however, by time $t = 5/k$ the concentration is equal to $A(1 - \mathrm{e}^{-5}) = 0.993$ and very close to the upper limit.

Example 1.5 Change in concentrations of *p*-chloroacetanilide and *N*-chloroacetanilide in a first-order chemical reaction

Figure 1.9 shows some data on the conversion of *N*-chloroacetanilide into *p*-chloroacetanilide (Blanksma, 1902). The fitted monomolecular model which closely

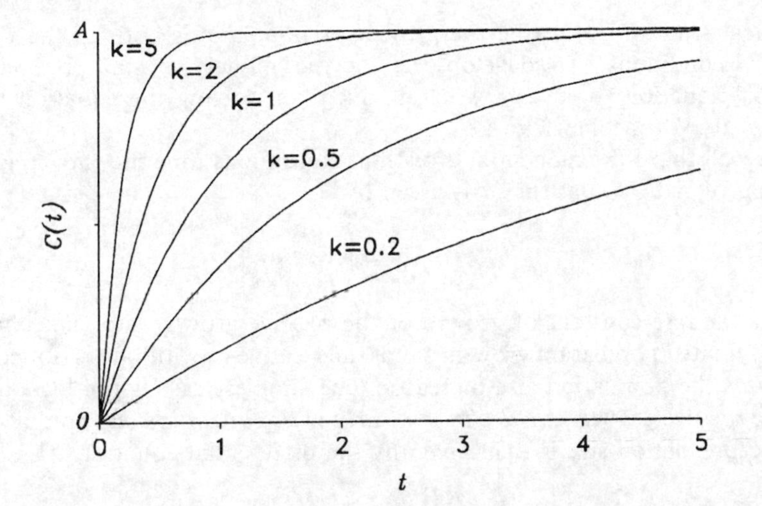

Figure 1.8 Examples of monomolecular growth curves for a range of values of the rate constant k.

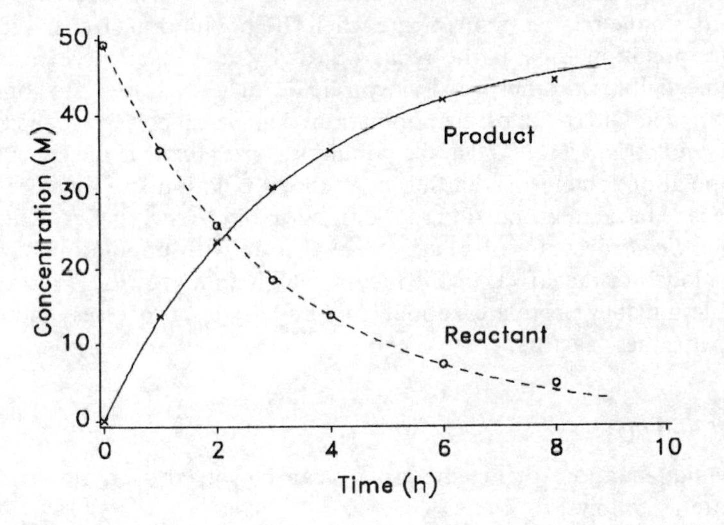

Figure 1.9 Observed increase in concentration of the product *p*-chloroacetanilide in reaction with the reactant *N*-chloroacetanilide. Solid line shows fitted monomolecular model; broken line, exponential decay.

describes the increase in concentration of *p*-chloroacetanilide has an upper limit of $A = 49.3$ M, the initial concentration of the reactant *N*-chloroacetanilide, and rate constant $k = 0.32\ \mathrm{M\,h^{-1}}$. Figure 1.9 also shows that the decline in the concentration of *N*-chloroacetanilide is close to exponential decay with rate constant $k = 0.32\ \mathrm{M\,h^{-1}}$.

1.3.2 Logistic model for limited population growth in continuous time

A basic pattern of limited population growth is a sigmoidal or S-shaped curve. A particular case is the logistic model which has been widely used in theoretical work and in empirical studies to describe the growth of populations both in the field and in laboratory conditions. To develop the logistic model we start by setting up a differential equation for the growth rate and then solving the equation to find the pattern of growth in time.

First, recall that for exponential growth in continuous time the rate of growth of a population of size $N(t)$ at time t is given by

$$\frac{\mathrm{d}N}{\mathrm{d}t} = rN$$

where r is the rate constant of growth or the relative growth rate. The characteristic feature of limited population growth is that the relative growth rate is not constant but decreases as the population size increases. The simplest case is the logistic model in which *the relative growth rate declines linearly with population size* so that the rate of change of population size is then given by the differential equation

$$\frac{\mathrm{d}N}{\mathrm{d}t} = rN\left(1 - \frac{N}{K}\right).$$

The parameter r is referred to as the *initial or intrinsic relative growth rate*. In the formulation, r is the relative growth rate when the population size is zero but a more tangible interpretation for r is the relative growth rate when the effects of limiting factors are negligible. In other words, r can be thought of as the rate constant for the *potential* exponential growth of the population. The parameter K is called the *carrying capacity or equilibrium level*. When the population size equals K the growth rate is zero and the population remains at that level. As the population size increases the relative growth rate is reduced in proportion to the fraction of the carrying capacity remaining. The assumed *linear* decrease of relative growth rate with population size is a purely descriptive model for the effects of the factors which limit growth. A more mechanistic model would involve postulates about the underlying processes and mechanisms associated with these factors.

Analytical solution

The differential equation for logistic growth can be solved using analytical methods. The steps are as follows.

- Separate the variables and integrate so that

$$\int \frac{K\,\mathrm{d}N}{N(K-N)} = \int r\,\mathrm{d}t.$$

- Express the integrand as partial fractions to give

$$\int \left[\frac{1}{N} + \frac{1}{K-N}\right] \mathrm{d}N = \int r\,\mathrm{d}t.$$

- Evaluate the integrals to get

$$\log_e N - \log_e(K-N) = rt + c$$

where c is a constant of integration.

- Impose the condition $N(t) = N(0)$ when $t = 0$ to give

$$\log_e N - \log_e(K - N) = rt + \log_e N(0) - \log_e\{K - N(0)\}.$$

- Take exponentials of both sides so that

$$\frac{N}{K - N} = \frac{N(0)\,e^{rt}}{K - N(0)}.$$

- Rearrange to obtain

$$N(t) = \frac{K}{1 + \left[\dfrac{K}{N(0)} - 1\right] e^{-rt}}.$$

This is the logistic equation for population growth.* The equation is sometimes written in the alternative equivalent form

$$N(t) = \frac{K}{1 + e^{-r(t-h)}}$$

by substituting $K/(N(0)) - 1 = e^{hr}$. The parameter h is the time at which the population reaches half the carrying capacity, i.e. $N(h) = K/2$. This form assumes that the initial population size is less than the carrying capacity K since the exponential function is always positive. We discuss the case when the initial population size exceeds K later.

Properties of the logistic growth curve

Figure 1.10 illustrates the following main properties of logistic growth.

(i) The increase of population size with time follows a sigmoidal or S-shaped curve (Figure 1.10(a));
(ii) The growth rate increases with size to a peak rate of $rK/4$ when the population reaches half the carrying capacity. This occurs at a point of inflexion on the growth curve where the slope is greatest (Figure 1.10(b));
(iii) The logarithm of population size increases at a decreasing rate (Figure 1.10(c));
(iv) The relative growth rate decreases linearly with population size with slope $-r/K$ (Figure 1.10(d)).

For logistic population growth, the carrying capacity K is reached when $N(t) = K$ or $1 + e^{-r(t-h)} = 1$, where h is the time to reach $K/2$. This implies $e^{-rt} = 0$ and therefore t is infinity. However, by time $t = h + 5/r$ the population size is equal to $K/(1 + e^{-5}) = K/1.0067 = 0.993K$ which is very close to K. Similarly, at time $h - 5/r$ the population size is $0.007K$. So for all practical purposes the complete growth curve covers the time span $h \pm 5/r$. Figure 1.11 illustrates this and shows how the shape of the curve changes as the rate constant r increases to steepen the central part of the curve.

---

* The logistic equation is sometimes called the Verhulst–Pearl equation after the mathematician Pierre Francois Verhulst (1804–49), who first derived the curve, and Raymond Pearl, who later in 1920 applied the curve to population growth of the United States.

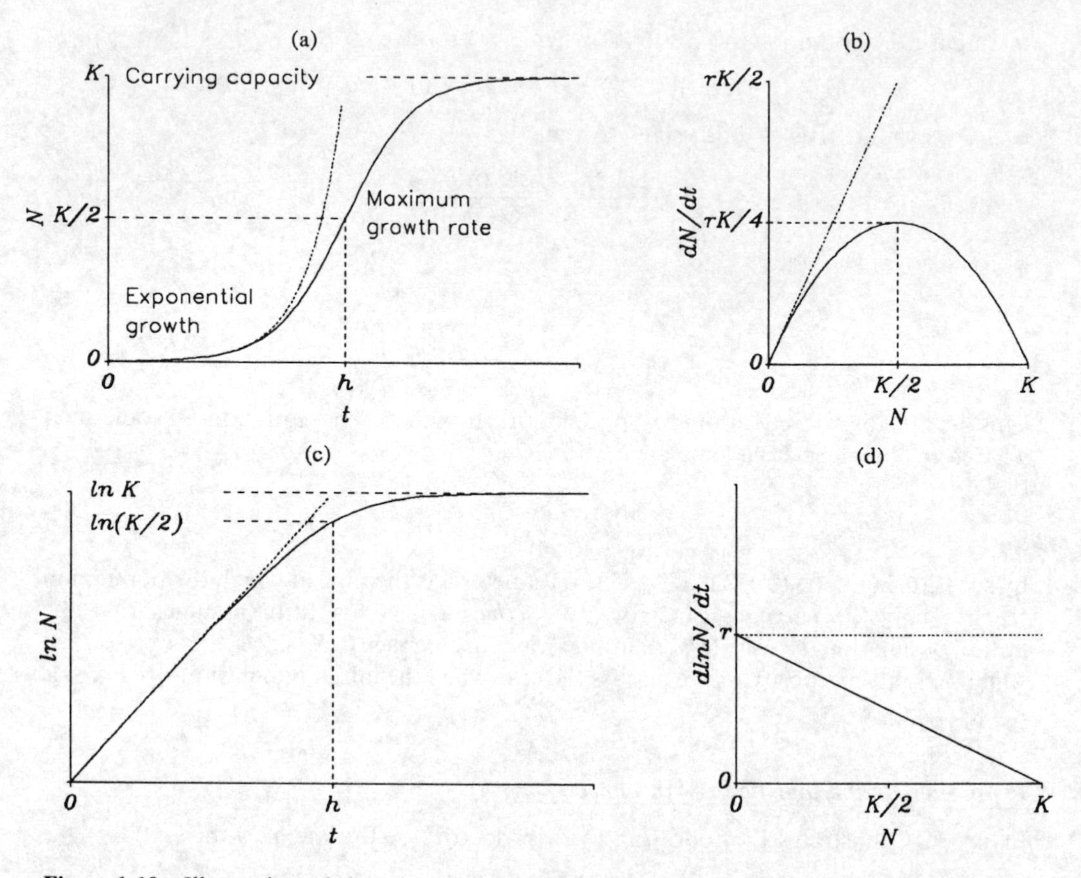

Figure 1.10 Illustration of the properties of logistic population growth. See text for details.

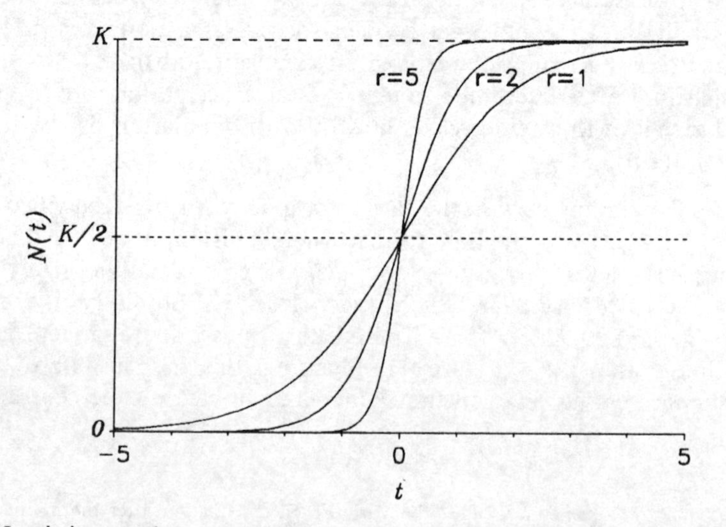

Figure 1.11 Logistic growth curves with the same carrying capacity but different values of the rate constant r.

Initial population size greater than the carrying capacity

The logistic curve depicts a population which is increasing steadily towards its carrying capacity K. Starting from any value on the curve the population size approaches the carrying capacity but never exceeds it. In theory, however, we can choose an initial population size which exceeds the carrying capacity. In this case the population *declines* steadily and returns to the carrying capacity (Figure 1.12). For this reason we say that the logistic model has a *stable equilibrium* at K because the population returns there from any displacement either above or below.

Example 1.6 Growth of a population of yeast cells

One of the earliest experimental studies of population growth was made by Carlson (1913) using yeast cells. In the experiment a measured amount of wort, which nourishes the plants, was seeded with a few cells and the amount of yeast was then measured at hourly intervals for 18 hours. Figure 1.13 shows the observed growth as a characteristic sigmoid, or S-shaped, pattern with an initial period in which the population is increasing at an increasing rate followed by a period in which the growth rate declines and the population settles down to a relatively constant size. Growth occurs as a continuous process and is clearly limited. The fitted model which describes growth rather well is a logistic curve with initial relative growth rate of 0.54 per hour and a carrying capacity of 665 units of yeast. If the population was growing exponentially at the above rate then it would double every 1.28 hours and starting from the initial amount of 9.6 units at time zero the time to reach 665 units would be 7.85 hours. With logistic growth the population size at this time is about half the carrying capacity.

1.3.3 Numerical solution of differential equations

In the previous section we derived the logistic growth curve as the solution of a differential equation by using analytical methods. This gave the size of the population at any particular time as an explicit formula involving the initial population size, the intrinsic relative growth rate (r) and the carrying capacity (K). Many problems,

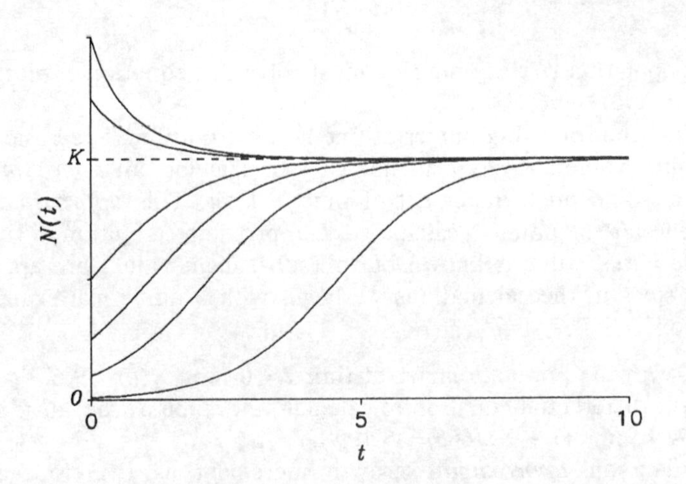

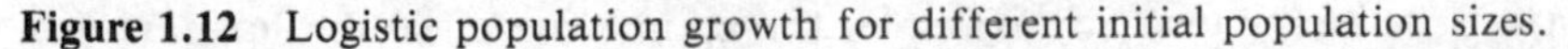
Figure 1.12 Logistic population growth for different initial population sizes.

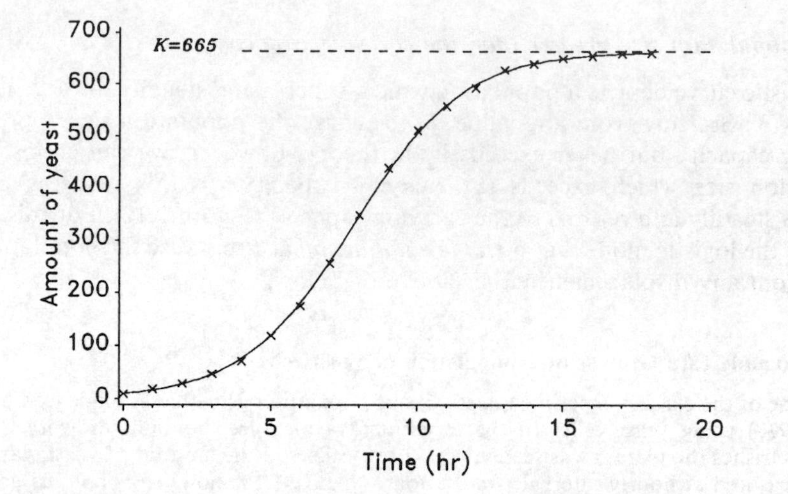

Figure 1.13 Growth of a population of yeast cells together with fitted logistic model, $K = 665$, $r = 0.54$ per hour.

however, involve differential equations which cannot be solved analytically and must be solved using numerical methods. There are many numerical techniques for solving differential equations which may be implemented using a wide range of computer programs. In this section we introduce the principle of the method (Chapter 8 gives more technical details of the numerical solution of differential equations). We use the logistic model for illustration so that we can compare the solution obtained by using the numerical method with the known analytical solution.

Consider the logistic model for the growth of the population of yeast cells with intrinsic relative growth rate $r = 0.54\ \mathrm{h}^{-1}$, carrying capacity $K = 665$ units and an initial population size of 9.6 units at time $t = 0$. The growth rate of the population at time t is given by the differential equation

$$\frac{\mathrm{d}N}{\mathrm{d}t} = 0.54N\left(1 - \frac{N}{665}\right).$$

To find the population size at time t we must solve this equation subject to the initial condition that $N(0) = 9.6$.

Now consider the following numerical method of solution. The basic idea is to use the known initial value and slope at time $t = 0$ to calculate an *approximate* value for the solution at some small increment of time δt later. This *approximate* value then becomes a new starting point to calculate the approximate solution at time $2\delta t$ and so on. Thus, the initial value is known but all subsequent values are approximate. To illustrate the steps in the calculations we begin with a numerical example using the above equation.

Step 1. Start with the population size at time $t = 0$, i.e. $N(0) = 9.6$.
Step 2. Use the differential equation to calculate the slope of the curve $\mathrm{d}N/\mathrm{d}t$ at time $t = 0$, i.e. $0.54 \times 9.6 \times (1 - 9.6/665) = 5.109$ units h^{-1}.
Step 3. Calculate an *approximate* growth increment as the slope of the curve multiplied by the time increment. For example, the approximate growth increment in

10 min is $5.109 \times 10/60 = 0.8515$ units. The divisor of 60 is to allow for the fact that the rate constant has units h^{-1}.

Step 4. Calculate the *approximate* size of the population at time $t = 10$ min as the population size at time $t = 0$ plus the *approximate* increment, i.e. $9.6 + 0.8515$ units. Figure 1.14 illustrates this part in the method.

Step 5. Repeat steps 1–4 starting with the calculated *approximate* population size at time t reached at the end of step 4 (i.e. 10 min first time round). Note that this will involve an *approximation* to the slope of the curve at time 10 min and at all subsequent times. Table 1.1 shows the results of iterating the process six times.

The basis of the numerical solution of a differential equation is a difference equation relating successive values of the solution at closely spaced times with general form given by

Numerical solution of a differential equation

General form of differential equation given by

$$\frac{dN}{dt} = f(N(t), t),$$

with initial value $N(0)$ at time $t = 0$.
Numerical solution uses the difference equation

$$\tilde{N}(t + \delta t) = \tilde{N}(t) + \frac{d\tilde{N}}{dt}\,\delta t$$

where $\tilde{N}(t)$ is a calculated value which is an *approximation* to the true value $N(t)$. The difference equation is iterated starting with the known value $N(0)$ at time $t = 0$ to produce the numerical solution $\tilde{N}(\delta t)$, $\tilde{N}(2\delta t)$ and so on. In practice, the time increment δt is made small to achieve a satisfactory approximation (see Tables 1.1 and 1.2).

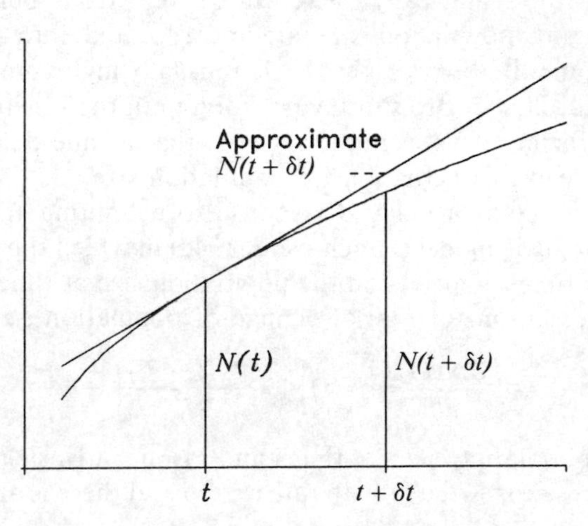

Figure 1.14 Illustration of basis of numerical method for solving differential equations by approximating the value of a function at time $t + \delta t$ using the known value and slope at time t.

Table 1.1 Numerical example to illustrate the steps 1–5 in the calculations for the numerical solution of a differential equation

Time (min)	Numerical solution		
	Calculated value	Calculated slope (h^{-1})	Calculated increment
0	9.6	5.109	0.8515
10	10.452	5.555	0.9258
20	11.378	6.039	1.0065
30	12.384	6.563	1.0938
40	13.478	7.131	1.1884
50	14.666	7.745	1.2908
60	15.957	8.410	1.4017

The numerical method produces an *approximate* solution to the differential equation which consists of a series of values at time intervals determined by the choice of the time increment. By making the time increment sufficiently small we aim to achieve a good approximation. This idea is illustrated in Table 1.2 which shows numerical solutions of the differential equation for the growth of the yeast population for a range of values of the time increment or step length. By decreasing the step length a closer approximation to the analytical solution is achieved. Figure 1.15 illustrates the effect.

1.3.4 Logistic model with time delay

In the logistic model the relative growth rate at a given instant decreases with the population size at that time. This implies that the factors which limit population growth are assumed to act instantaneously on changes in population size so that there is no delay, or lag. More realistically, there will be some delay before the effects of a factor due to changing population size are incorporated into the dynamics. For example, a population of herbivores may reach unusually high numbers by overgrazing its pasture and then decline due to shortage of food until the vegetation recovers from the effects of overgrazing. This example suggests that a time delay in response to a limiting factor may lead to fluctuations in population size.

To illustrate the effect of a delayed response to a limiting factor we consider a modification to the logistic model (Hutchinson, 1948) in which the relative growth rate of the population at time t depends on the population size at time $(t - T)$. This gives a delay-differential equation for rate of change of population size as

$$\frac{\mathrm{d}N(t)}{\mathrm{d}t} = rN(t)\left\{1 - \frac{N(t-T)}{K}\right\}$$

where $N(t)$ denotes population size★ at time t and r is the intrinsic relative growth rate. When the population size is K the growth rate is zero and there is an equilibrium value.

★ We write population size explicitly as a function of time to emphasise the lag in the equation.

Table 1.2 Illustration of numerical method for solving a differential equation using logistic growth in a population of yeast cells. Values of the solution are shown only at hourly intervals although intermediate values are calculated at intervals determined by the step length

Hours of growth	Numerical solution using different time steps δt			Analytical solution
	0.1667 h (10 min)	0.0833 h (5 min)	0.0333 h (2 min)	Infinitesimally small
0	9.6	9.6	9.6	9.6
1	16.0	16.1	16.2	16.3
2	26.4	26.9	27.3	27.5
3	43.1	44.4	45.3	45.8
4	69.5	72.1	73.8	74.9
5	109.4	114.0	117.0	119.0
6	166.3	173.4	177.9	181.0
7	240.5	250.1	256.0	259.6
8	327.2	337.8	344.3	348.5
9	415.7	425.4	431.1	434.9
10	493.9	501.3	505.5	508.3
11	554.6	559.3	562.0	563.7
12	597.1	599.6	601.0	602.0
13	624.5	625.6	626.3	626.7
14	641.3	641.7	642.0	642.2
15	651.3	651.4	651.5	651.5
16	657.1	657.1	657.1	657.1
17	660.5	660.4	660.4	660.4
18	662.4	662.4	662.3	662.3

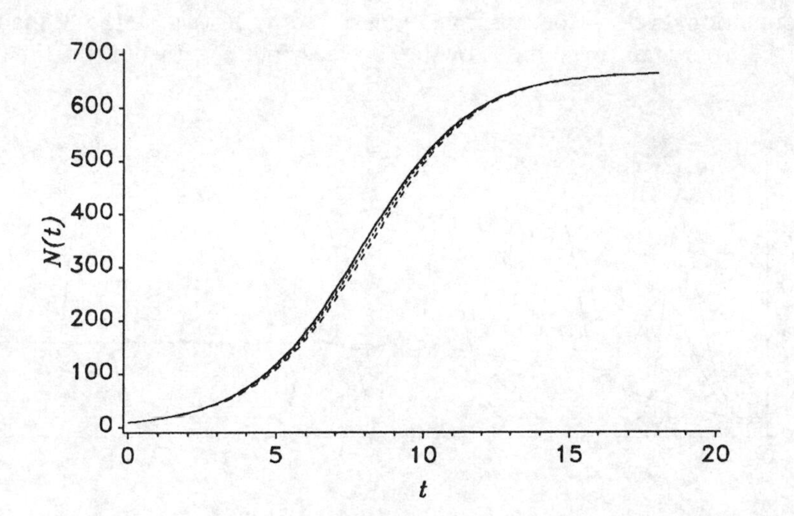

Figure 1.15 Illustration of the numerical solution of the logistic differential equation for growth in a population of yeast cells starting with $N(0) = 9.6$ and steplength δt. Solid line, analytical solution; dashed line, $\delta t = 10$ min; dotted line, $\delta t = 5$ min (see Table 1.2).

To explore the nature of this equilibrium and the dynamics of the model we use the numerical method to solve the differential equation and to simulate data for different values of the parameters. Figure 1.16 shows three series corresponding to a time lag of $T = 1$, $K = 10$ and values of $r = 0.3$, 1.2 and 2.0, starting from a population size of 2.0 over the previous time period of length $T = 1$. For a fixed length of time lag and increasing r the pattern of population change ranges from a steady increase to an equilibrium K, to damped oscillations converging on K and eventually to sustained oscillations called limit cycles.

The fluctuations in population size can occur because the time delay allows the population successively to overshoot, and then fall below, its equilibrium before a change in the growth rate stems the increase or checks the decline. Using theory we can show that the dynamic behaviour of the model depends on the relative magnitude of the intrinsic rate of increase and the time delay through the product rT as follows: $0 < rT < e^{-1}$—a steady increase or decrease to equilibrium; $e^{-1} < rT < \pi/2$—damped oscillations to equilibrium; $\pi/2 < rT$—sustained regular oscillations or limit cycles with amplitude and frequency depending on rT (Table 1.3).

The behaviour of the model is sometimes expressed in terms of the so-called *characteristic return time* $T_R = 1/r$. If the time lag is long relative to the characteristic return time the population will tend to overshoot leading to cyclical fluctuations.

Characteristic return time

$$T_R = \frac{1}{r}$$

The time taken for a population to increase in size by a factor of $e = 2.718$ growing exponentially with rate constant r.

Some natural populations show rather regular fluctuations in size which are called quasi-population cycles—for example, voles, red grouse and the Canadian lynx together with its main prey the snowshoe hare. Models with built-in time lags can

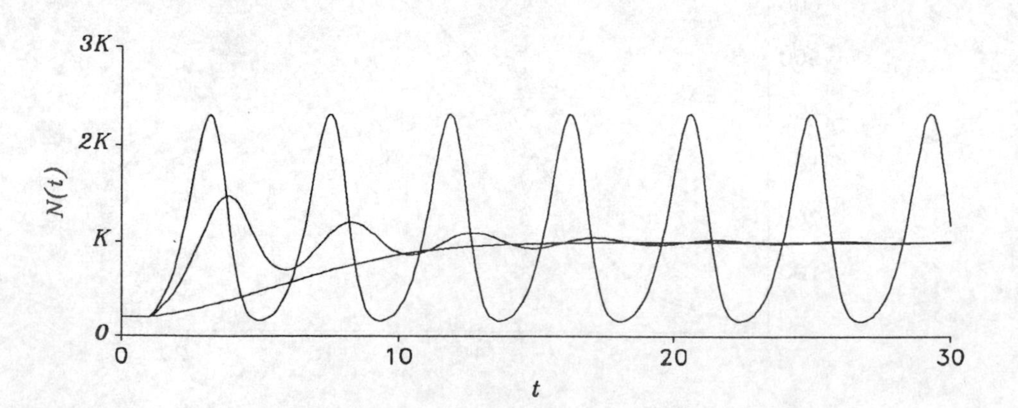

Figure 1.16 Illustration of the dynamic behaviour of the logistic model with a time delay for different values of rT showing a steady approach to an equilibrium ($rT = 0.30$), a damped oscillation ($rT = 1.2$) and a stable limit cycle ($rT = 1.8$).

Table 1.3 Amplitude (N_{max}/N_{min}) and period of the time-delay logistic model where r is the intrinsic rate of increase and T is the time lag. After May (1976a)

rT	Amplitude	Period
1.6	2.6	4.03
1.8	11.6	4.18
2.0	42.3	4.40
2.5	2930	5.36

mimic this type of fluctuation but in many cases the underlying biological mechanisms are poorly understood. In Chapter 13 we analyse a series of data showing quasi-cycles in red grouse and discuss a mechanistic model based on territorial competition. In Chapters 8 and 10 we develop models for the interaction between a predator and its prey which can also produce quasi-cycles.

1.3.5 Discrete time logistic model

In Sections 1.3.1 and 1.3.2 we presented models for limited population growth in which change occurs as a continuous process. In these models the population size is a continuous variable and differential equations are used to describe the rate of change of size with time. In some situations, however, the idea of a population which is changing smoothly and continuously in time is not very realistic. For example, in many insects the population consists of a single generation and breeding is confined to a short season. For these cases population change is often measured over discrete time periods from generation to generation or from one year to the next. Mathematical models to describe such changes use difference equations to relate population sizes at successive times. In Section 1.2 we showed that the essential character of exponential growth and decline is the same for models in continuous and discrete time, i.e. a constant relative rate of change, an exponential increase, or decrease, in population size and a linear increase, or decrease, in the logarithm of population size. For models of limited growth, however, the dynamic behaviour of continuous and discrete time models can be very different.

Consider a simple model for an insect with an annual life-cycle in which each adult female produces a constant number of female offspring R which survive to breed the next year (Section 1.2.2). In this model the number of females in year $i + 1$ is related to the number in year i by the difference equation

$$N_{i+1} = RN_i.$$

When the net reproductive rate R exceeds 1 this model leads to exponential growth with geometric rate of increase R. Now suppose that the growth is limited in some way so that as the population size increases the net reproductive rate declines. This could result from increased competition for a limited food supply leading to lower fecundity or increased mortality in the juvenile stages. Whatever the causes of the limited growth one simple way to model the effects is to express the reproductive rate as a decreasing function of population size. For illustration, consider the model in which the relative rate of increase declines linearly with population size. The difference equation relating the number of females in successive years is then

$$N_{i+1} = RN_i\left(1 - \frac{N_i}{K}\right).$$

The parameter K is used to measure the effect of the factors which limit growth. For a very large value of K growth is effectively unlimited, at least initially, and the population increases exponentially at rate R. To ensure that the population size never becomes negative each value of N_i must be less than K, i.e. $RN_i(1 - N_i/K) < K$. This condition implies that R must be less than 4 because the quadratic function $x(1 - x)$ has a maximum value of 1/4 when $x = 1/2$. Often, the model is written in terms of the scaled population size $X_i = N_i/K$, so that

$$X_{i+1} = RX_i(1 - X_i).$$

In this form the model is sometimes referred to as the quadratic map, because of the quadratic relationship between successive population sizes, or the logistic map because of its similar form to the logistic equation (May, 1987).

The corresponding equation for population change $\Delta X_i = X_{i+1} - X_i$ is

$$\Delta X_i = RX_i(X_E - X_i)$$

where $X_E = (R - 1)/R$.

The parameter X_E is called the *equilibrium* because it corresponds to zero change. When the population size is below the equilibrium the subsequent change is positive and the population increases towards the equilibrium and similarly when the population is above the equilibrium the population decreases towards it. From this it might be thought that the population would settle down to the equilibrium but this is not necessarily so. Large values of R can lead to the population overshooting the equilibrium and then falling below it leading to sustained fluctuations. Figure 1.17 gives a diagrammatic representation of the model which is useful for exploring the dynamics of difference equations for limited population growth.

Figure 1.17 Diagrammatic representation of the dynamics of the discrete time logistic model for limited population growth. The solid curved line in the figures on the left-hand side depicts the quadratic relationship between the population size in year i and $i + 1$. The intersection between this curve and the 45° line gives the equilibrium value. The dashed line represents the trajectory of a population starting from the initial value of $X_0 = 0.10$ and proceeding as follows: draw a vertical line to the curve to obtain X_1; draw a horizontal to the 45° line and then another vertical to the curve to find X_2 and so on. The right-hand side of the diagram shows the corresponding population changes in time. In (a) the dashed line has a staircase appearance as the population approaches the equilibrium steadily from below as in (b). In (c) the initial steps are followed by a spiral path as the population displays a damped oscillation about the equilibrium value as in (d). In (e) the dashed line wanders in an irregular manner and the population fluctuates about the equilibrium with no apparent tendency to settle down as in (f)—the effect is called *chaos*. This type of diagram can be used to explore the dynamics of models involving other relationships between successive population sizes. The behaviour of the system in the neighbourhood of the equilibrium point is determined by the slope of the curve at the intersection with the 45° line. If the slope is less steep than -1 the population will settle on its equilibrium. For steeper slopes sustained fluctuations occur.

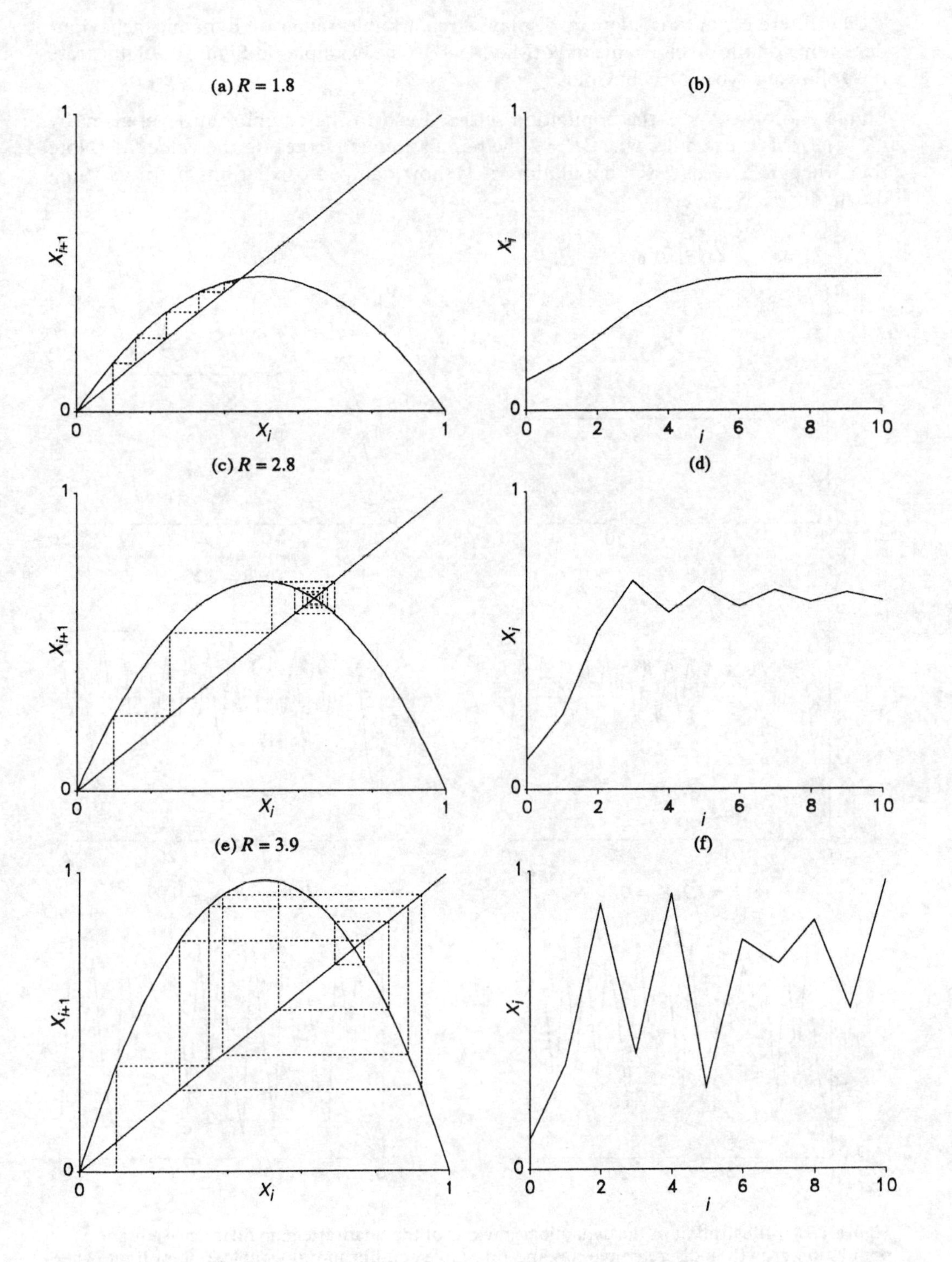

(a) $R = 1.8$
(b)
(c) $R = 2.8$
(d)
(e) $R = 3.9$
(f)
X_{i+1}
X_i
i
0
1
2
4
6
8
10

The discrete time logistic can display a remarkable range of dynamic behaviour depending on the precise value of R (May, 1987). The examples in Figure 1.18 illustrate the following types of behaviour.

Stable. For $1 < R < 3$ the population settles down to the equilibrium level given by $(R-1)/R$. For example, when $R = 2$ the population converges on the value 0.5. Note that when R exceeds 2 the population may show damped oscillations before settling down.

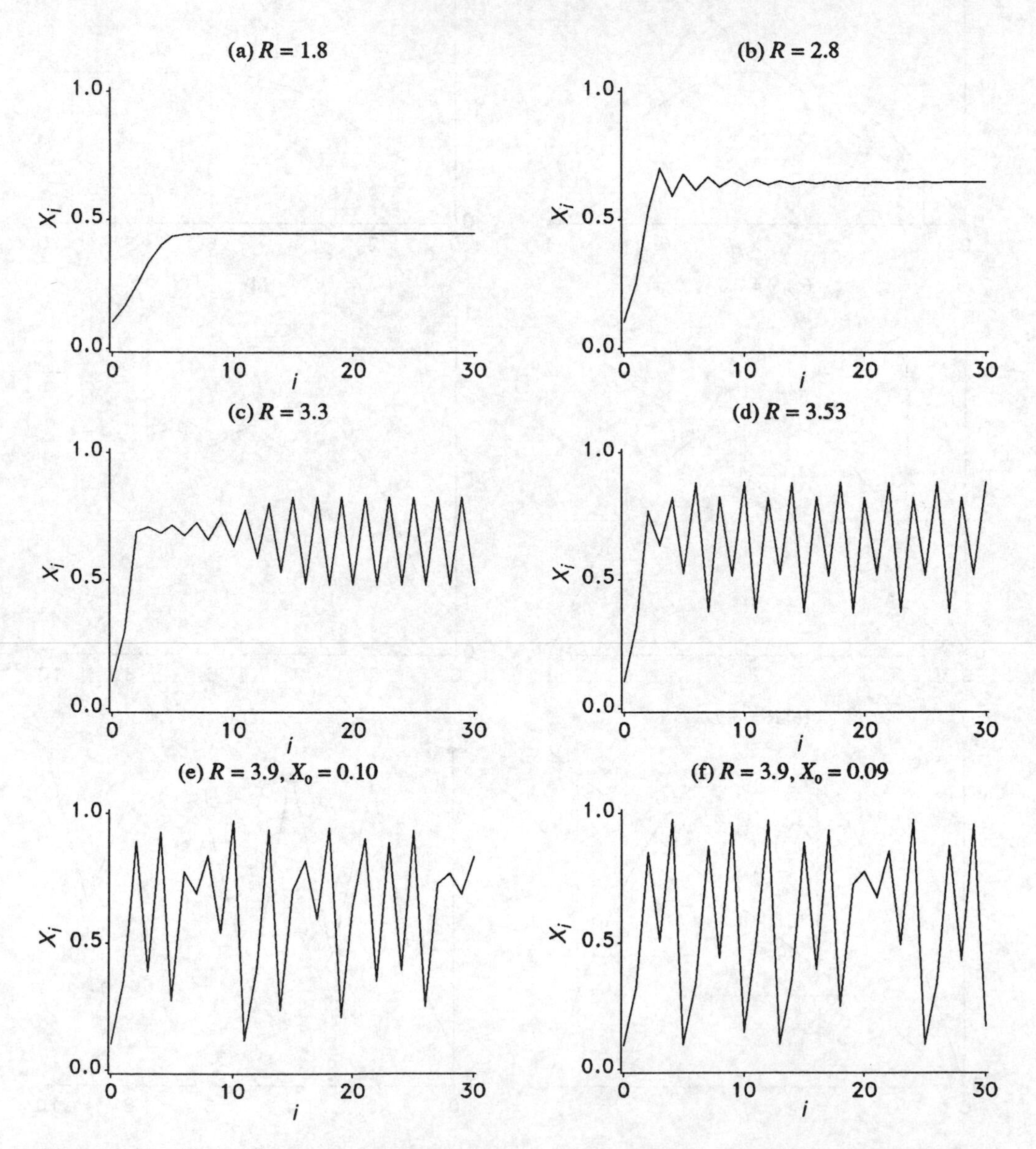

Figure 1.18 Illustration of the dynamic behaviour of the quadratic map difference equation for population growth in discrete time. (a) and (b) Stable equilibrium; (c) stable 2-point limit cycle; (d) stable 4-point limit cycle; (e) and (f) chaos. See text for details.

Limit cycles. For $3 < R < 3.45$, the population will eventually oscillate between two values in a so-called *stable two-point limit cycle*. For a slightly larger value of R this becomes a *stable four-point cycle* in which the population fluctuation is repeated every four years. As R gets progressively larger the cycle length increases to 8, 16, 32 and so on in a process called *period doubling*.

Chaos. For $3.57 < R < 4$ the population fluctuations are erratic and apparently random. The behaviour is called *chaos*.

A characteristic property of chaos is the sensitivity of the population size to the initial conditions. The two series of chaotic fluctuations shown in Figure 1.18(e) and (f) correspond to the same value of $R = 3.90$ with initial population sizes of 0.10 and 0.09—after only 6 years the population sizes are 0.77 and 0.34 respectively. This sensitivity to initial conditions is illustrated in another way in Figure 1.19 which shows the relationship between the initial population density and densities after 1, 5, 10 and 15 iterations over a narrow range of initial population density 0.100–0.110.

The above conditions for stable limit cycles and chaos in terms of the value of the single parameter R are specific to the particular model. More generally, the dynamic behaviour will depend on the form of the relationship between successive population sizes.

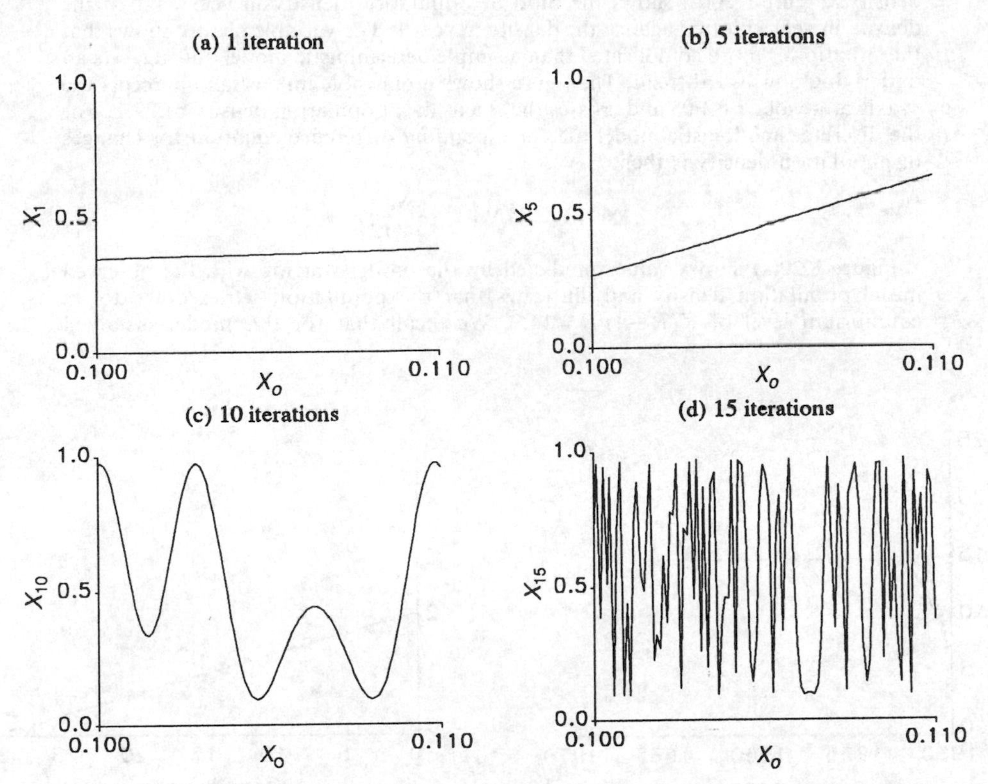

Figure 1.19 Illustration of the sensitivity to initial conditions of the chaotic dynamics of the quadratic map with $R = 3.9$. Figures show population density after 1, 5, 10 and 15 iterations for initial population densities in the range 0.10–0.11. By iteration 15, close initial values can be very different.

Example 1.7 Fluctuations in population size of the winter moth in Wytham Wood, 1950–68

The winter moth *Operophtera brumata* is an insect with an annual life cycle which is especially abundant on oaks *Quercus robur*. Adults emerge from the soil under oak trees in November and December and the females climb the trunks after mating to lay their eggs in the canopy and then die over winter. The eggs hatch when the oak buds are bursting in spring and the larvae crawl into the buds and feed on the leaves. When the larvae are fully grown in May they drop to the ground on silken threads and pupate in the earth where they remain until emerging as adult moths next winter. Varley & Gradwell (1968) estimated the size of a population of winter moths in Wytham Wood by placing traps around the trunks of oak trees to catch the ascending females. Figure 1.20(a) shows how the density of adult females changed over the 19-year period, 1950–68. Successive population sizes are joined by lines to guide the eye in assessing annual changes and are not meant to indicate the size of the population at intermediate times during the year. The observed population size fluctuates from year to year partly because of sampling errors but mainly because of real changes in the actual size of the population. Also, the population shows no apparent long-term trend and appears to fluctuate within limits. This raises the intriguing question of whether the dynamics could be described by a simple difference equation model for limited population growth in which the relative change from year to year decreases with increasing population density. Figure 1.20(b) shows the ratio of population density in year $i+1$ to the density in year i plotted against the density in year i. The variable picture shows that the situation is more complicated than a simple deterministic model but suggests an underlying downward trend. The figure shows a plausible line which intercepts the y-axis at a value of 2.05 and crosses the x-axis at a population density of 29.1. For the discrete time logistic model the corresponding difference equation for changes in population density is then

$$N_{i+1} = 2.05 N_i \left(1 - \frac{N_i}{29.1}\right).$$

Figure 1.20(a) shows values predicted by the model starting with the observed initial population density and illustrates that the population settles down to an equilibrium level of $K(R-1)/R = 14.9$. We recall that for this model sustained

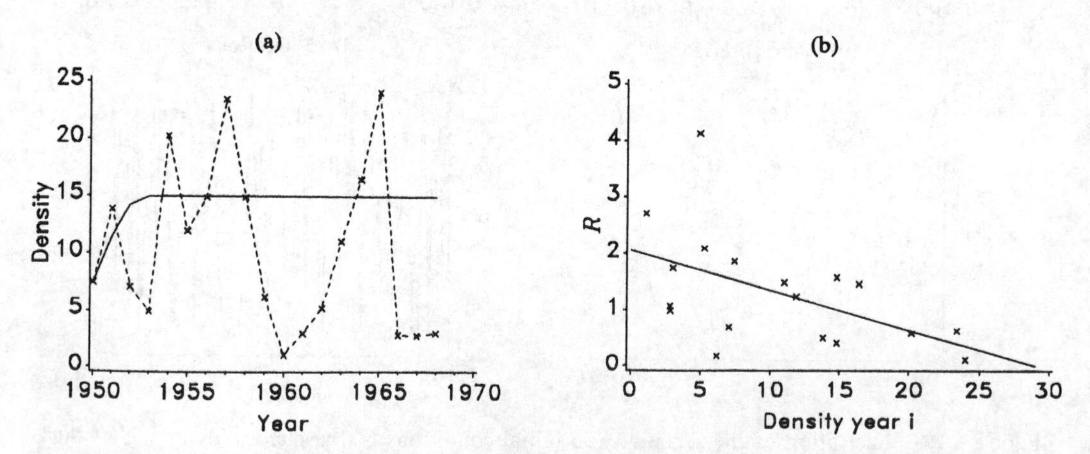

Figure 1.20 (a) Fluctuations in density of a population of winter moths. Solid line shows trajectory of a simple discrete logistic model; (b) year to year changes in density against density in previous year with fitted straight line decrease. Data from Manly (1990).

deterministic fluctuations require a value of R greater than 3. From visual inspection of Figure 1.20(b) this would appear to be rather high but the variability in the observations raises problems for estimating the underlying relationship.

The analysis addresses the rather naive question of whether the changes in size of the winter moth population can be described by a simple deterministic model for limited population growth. Varley and Gradwell carried out a more detailed analysis of the mortality in the different life stages. They found that: (a) much of the variation in the year to year changes was due to variation in the overwinter mortality of the larvae; (b) overwinter survival was not related to population size but since it is likely to be affected by various external factors such as weather and predation a simple explanation of the population dynamics is rather unlikely; (c) pupal mortality rate through predation increased with the density of the previous adult population and concluded that pupal predation could limit population growth. In Chapter 4 we present a simple model which illustrates how the different factors affect changes in population density.

1.3.6 General model for limited population growth in discrete time

For the discrete time logistic model there is a quadratic relationship between successive population sizes. Hassell (1975) proposed a more general model in which the size of the population at time $i + 1$ is related to the size at time i by the difference equation

$$N_{i+1} = \frac{RN_i}{(1 + aN_i)^b}$$

where N_i is the population density at time i. The model parameters R, a and b have the following interpretation. R is the intrinsic geometric rate of increase, i.e. at low densities the population increases exponentially with approximate rate equal to R. a is a scale parameter which depends on the units used to measure density. It is related to the equilibrium population density N_E, which occurs when $N_{i+1} = N_i = N_E$, by $a = (R^{1/b} - 1)/N_E$. b measures the effect of population density on the subsequent relative change and determines the form of the relationship between successive population sizes (Figure 1.21). For example, $b = 1$ gives a steady increase to an upper limit, or asymptote, of $R/a = RN_E/(R - 1)$. For $b > 1$, the curve increases to a peak and then declines to zero: with large values of R the dynamic behaviour is sometimes referred to as 'boom and bust' as rapid increases are followed by sharp declines. These two types of curve are sometimes used as models for the extremes of so-called 'contest' and 'scramble' competition. In contest competition some animals are successful at the expense of others which fail to survive or reproduce. In scramble competition resources are partitioned equally so that all or no individuals survive or breed. Hassell's model can be used to represent situations between the two extremes of contest and scramble competition.

The model displays a similar range of dynamic behaviour to the discrete logistic including monotonic damping, damped oscillations, stable limit cycles and chaos. The boundaries between the different types depend on combinations of the two parameters R and b. A brief summary is as follows: (a) $b < 1$, the population increases steadily to its equilibrium value whatever the value of R; (b) $1 < b < 2$, damped oscillations occur when R exceeds 4; (c) $2 < b$, stable limit cycles for $R > 10$ and eventually chaos for larger values of b and R.

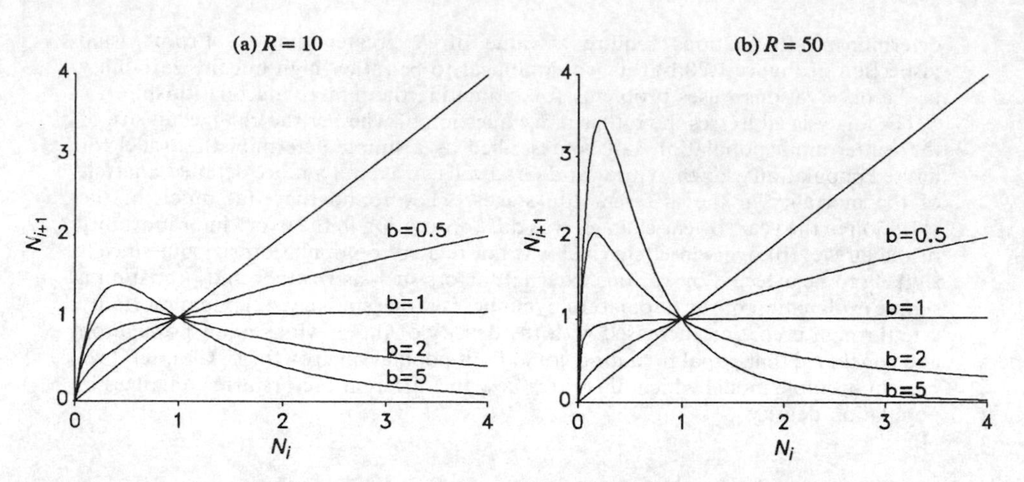

Figure 1.21 Illustration of properties of a general model for limited population growth in discrete time. The equilibrium value ($N_E = 1$, in each case) occurs at the intersection of the curve with the 45° line. When the absolute value of the slope of the curve at the intersection is less than one the equilibrium is stable.

The model includes the special case where successive population sizes are related by the difference equation

$$N_{i+1} = RN_i \, e^{-\mu N_i}.$$

This follows by first writing $a = \mu / b$ to give $N_{i+1} = RN_i / (1 + \mu N_i / b)^b$ and then allowing b to approach infinity so that $1/(1 + \mu N_i / b)^b$ approaches $e^{-\mu N_i}$. This model is often written in the equivalent form

$$N_{i+1} = N_i \, e^{r(1 - N_i / K)}$$

where $r = \log_e R$. The parameter K is the equilibrium population density because $N_i = K$ implies that $N_{i+1} = N_i$ and a zero change in density. This model is sometimes referred to as the discrete logistic but the relationship between successive population sizes is different from the discrete logistic model discussed in Section 1.3.5. May (1976a) discusses the dynamic behaviour of the model which ranges from monotonic damping, to damped oscillations, to stable limit cycles and eventually to chaos as the value of the parameter r increases.

1.4 GROWTH OF INDIVIDUAL ORGANISMS

So far we have dealt mainly with models to describe growth and decline of populations. In this section we consider the growth of the individual organism. Although an organism can be regarded as a coordinated heterogeneous population of cells, the substantial coordination makes the behaviour sometimes very different from that of a population of cells *in vitro*. One of the earliest models for organic growth was given by Ludwig von Bertalanffy (1938). His comments are worth quoting since they illustrate some of the difficulties in modelling biological phenomena:

> growth is, of course, extremely heterogeneous and complex in its descriptive as well as in its analytical aspects. Even the chemical side of the process, the synthesis of building materials specific to the body, is obscure. Mitosis is as yet not to be explained in physico-chemical terms. But even if we possessed the knowledge of these matters, we should be very remote from an understanding of growth in higher organisms. For the growth capacity of an organism does not depend on the single cells but on the organism as a whole; obviously the latter determines whether, where and when cells divide, increase, differentiate, how metaplasms and intercellular substances are formed, etc. Further, growth is widely dependent on external factors such as nutrition, temperature, life space, as well as such internal factors as hormones, progressive differentiation, change of water content, age, etc.

Von Bertalanffy noted the apparent hopelessness of providing a rational mathematical treatment and accepted the scepticism of critical investigators. Undaunted, however, he emphasised examples from physics and physiology in which complex phenomena can sometimes be described by the statistical laws and suggested that such procedures could apply to organic growth.

1.4.1 Von Bertalanffy's mechanistic growth model

Von Bertalanffy adopted a mechanistic approach by expressing growth as the net result of material gained through anabolic processes, less material lost through catabolism. Then he postulated that gain in weight is proportional to the surface area of the organism. A rate constant H is used to measure the average effect of the anabolic processes so that for an organism of surface area $S(t)$ at time t, the weight gain in a small time interval δt is then $HS(t)\delta t$. Loss in weight is assumed to be proportional to the weight of the organism. A rate constant C measures the average effect of the catabolic processes so that for an organism of weight $W(t)$ at time t, the weight loss in a small time interval is $CW(t)\delta t$. The resulting growth increment is then

$$\delta W(t) = HS(t)\delta t - CW(t)\delta t$$

so that the rate of growth in weight can be described by the differential equation*

$$\frac{\mathrm{d}W}{\mathrm{d}t} = HS - CW.$$

The model is not yet complete because it does not specify the relationship between surface area and weight. To do this von Bertalanffy considered animals which maintain the same shape while growing so that the surface area and the weight, or volume, are related to the square and cube of a linear dimension. For example, a sphere of radius r has surface area $4\pi r^2$ and volume $4\pi r^3/3$. More generally, we write the relationships of surface area and weight to length (L) as follows:

$$S = a_2 L^2 \quad \text{and} \quad W = a_3 L^3$$

where a_2 and a_3 are model parameters which will depend on the particular application.

---

* To simplify this expression and the ones which follow the functions $W(t)$ and $S(t)$ are shown without the argument t.

Substituting the above expressions into the differential equation for the rate of change of weight with time gives

$$\frac{\mathrm{d}(a_3 L^3)}{\mathrm{d}t} = Ha_2 L^2 - Ca_3 L^3.$$

Differentiating L^3 with respect to time gives $3L^2\,\mathrm{d}L/\mathrm{d}t$. Then, dividing both sides by $3a_3L^2$ we obtain

$$\frac{\mathrm{d}L}{\mathrm{d}t} = \frac{Ha_2}{3a_3} - \frac{CL}{3}.$$

In words, the length increases at a rate which decreases with increasing length. Growth stops (i.e. $\mathrm{d}L/\mathrm{d}t = 0$) when the maximum length of $L_{max} = Ha_2/Ca_3$ is reached. Using the maximum length as a parameter in the model the differential equation takes the form

$$\frac{\mathrm{d}L}{\mathrm{d}t} = k(L_{max} - L)$$

where the parameter $k = C/3$ or one-third of the catabolic rate constant. The steps for solving the differential equation are as follows.

- Separate the variables and integrate so that

$$\int \frac{\mathrm{d}L}{L_{max} - L} = \int k\ \mathrm{d}t.$$

- Evaluate the integrals to give

$$-\log_e(L_{max} - L) = kt + c$$

  where c is a constant of integration.
- Impose the condition $L = L(0)$ when $t = 0$ so that

$$-\log_e(L_{max} - L) = kt - \log_e[L_{max} - L(0)].$$

- Take exponentials of both sides and rearrange to obtain

$$L = L_{max} - [L_{max} - L(0)]\ \mathrm{e}^{-kt}.$$

  The equation is often written as

$$L(t) = L_{max}(1 - A\ \mathrm{e}^{-kt})$$

  where $A = 1 - L(0)/L_{max}$ is the proportion of the maximum length remaining for growth at time zero. Since the weight is assumed to be related to its length by $W = a_3L^3$, the weight at time t is given by

$$W(t) = W_{max}(1 - A\ \mathrm{e}^{-kt})^3$$

  where W_{max} is the maximum weight attained. This gives the equations of growth in length and weight derived by von Bertalanffy from the fundamental postulate that weight growth increment is the net result of weight gain through anabolism and weight loss via catabolic processes.

Figure 1.22 illustrates the form of the von Bertalanffy growth curves. Length

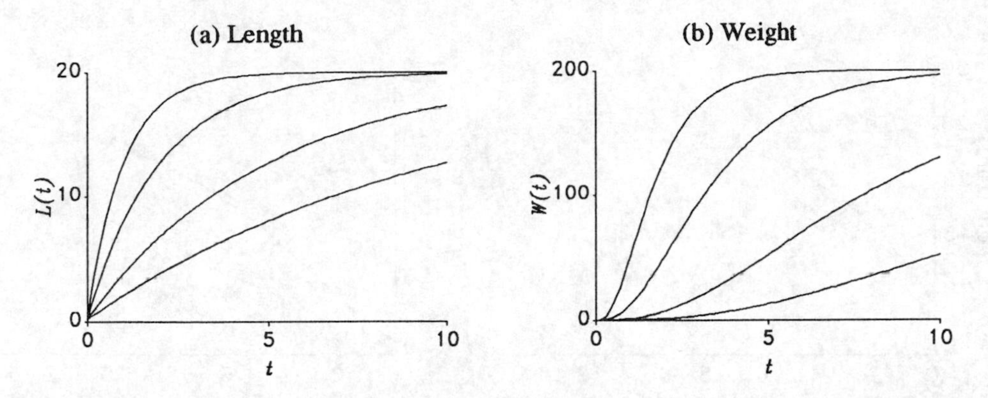

Figure 1.22 Examples of von Bertalanffy growth curves with $L_{\max} = 20$, $A = 0.99$, $W_{\max} = 200$ and $k = 0.1$ (bottom), 0.2, 0.5 and 1.0 (top).

increases at a decreasing rate whereas growth in weight shows a sigmoidal pattern with an initial increase in growth rate followed by a decrease. For weight there is a maximum growth rate corresponding to a point of inflexion on the growth curve which occurs when the organism reaches a weight of $8W_{\max}/27 = 0.296W_{\max}$, obtained by setting the second derivative of the weight growth curve equal to zero.

In the above derivation of the von Bertalanffy growth curves, we obtained a differential equation for the rate of growth in length. A corresponding differential equation for the rate of weight growth can be obtained as follows. First, weight and length are related by $W = a_3 L^3$ so that the rate of change of weight with time is related to the rate of change of length by $\mathrm{d}W/\mathrm{d}t = 3a_3 L^2\, \mathrm{d}L/\mathrm{d}t$. Substituting this expression into the differential equation for length gives the following differential equation for the rate of growth of weight as

$$\frac{\mathrm{d}W}{\mathrm{d}t} = 3kW\left[\left(\frac{W_{\max}}{W}\right)^{1/3} - 1\right].$$

It may be verified that the von Bertalanffy growth curve for weight at time t satisfies this differential equation.

Example 1.8 Application of the von Bertalanffy model to growth of the fish *Lebistes reticulatus*

Von Bertalanffy applied his model to describe the growth in length and weight of the fish *Lebistes reticulatus*. Newborn fish which emerge roughly in the shape of the adults were kept under relatively constant conditions and their length and weight were measured every week on a sample of about 20 fish taken on each occasion. Figure 1.23 shows the observed pattern of growth in mean length and weight. The fitted von Bertalanffy growth curves have a rate constant $k = 0.24$ per week, a maximum length $L_{\max} = 26.6$ mm and a maximum weight $W_{\max} = 175.6$ mg. It is seen that the models mimic the main features of the growth very well. Some of the deviations from the fitted models may be due to differences in the growth patterns of individual fish.

The fitted curves illustrate the properties of the model that the growth rate of length decreases steadily with age, whereas for weight there is an initial increase in the growth rate followed by a decline so that at some intermediate weight there is

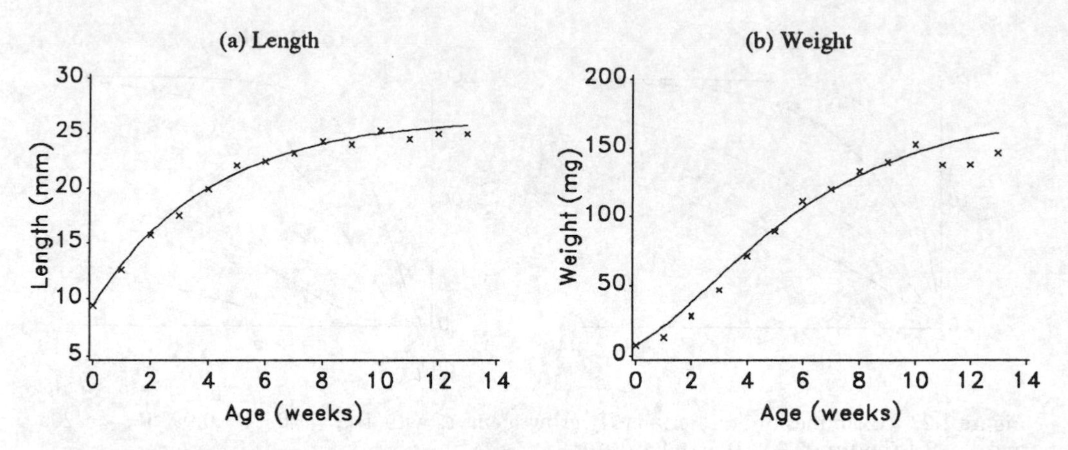

Figure 1.23 Observed growth in length and weight of the fish *Lebistes reticulatus* with fitted von Bertalanffy growth curves.

a maximum growth rate. This occurs at the point of inflexion on the S-shaped curve and can be obtained by setting the second derivative of $W(t)$ equal to zero. For the von Bertalanffy model the maximum rate of growth in weight occurs at a weight of $8W_{\max}/27 = 0.296W_{\max}$ and is equal to $4kW_{\max}/9 = 0.444kW_{\max}$. In other words, when the organism reaches about 30% of its final weight the growth rate is greatest and depends on the rate constant k and the final weight. For *Lebistes reticulatus* the maximum growth rate in weight is estimated as 18.7 mg per week and occurs when the fish weighs 52.0 mg.

1.4.2 Richards' family of growth models

Richards (1959) proposed a flexible empirical model for describing different patterns of organic growth in which the rate of growth in weight is given by the differential equation

$$\frac{dW}{dt} = \frac{kW}{(1-m)}\left[\left(\frac{W_{\max}}{W}\right)^{1-m} - 1\right]$$

where $W_{\max}$ is the maximum attainable size and k is a rate constant. Flexibility is achieved by varying the parameter m.

It can be verified that the solution to the differential equation is the growth curve given by

$$W(t) = W_{\max}(1 - A\,e^{-kt})^{1/(1-m)}.$$

Richards' family includes the following widely used models as special cases.

Monomolecular (case $m = 0$). In this model the growth rate decreases linearly with size and so the growth curve has no point of inflexion. An example occurs in von Bertalanffy's model for the growth in length.

von Bertalanffy (case $m = 2/3$). In this model the relative growth rate decreases linearly with the size raised to the power $1/3$. The growth curve has a point of inflexion at $W(t) = 0.296W_{\max}$ where the growth rate is greatest.

Logistic or **autocatalytic** (case $m = 2$). For this model the relative growth rate declines linearly with size. The growth curve is symmetrical about its point of inflexion at $W(t) = 0.5W_{max}$ where the growth rate is a maximum. In Section 1.3.2 we presented the logistic curve as a model for population growth but it is also used as a model for the growth of an organism.

Gompertz (case $m = 1$). At first sight this case appears impossible because it implies division by zero. It is, however, a limiting case and follows from the mathematical result that as x approaches zero then $(y^x - 1)/x$ approaches $\log_e y$. This model has the property that the relative growth rate declines linearly with the logarithm of size. The growth curve has a point of inflexion when $W(t) = 0.368W_{max}$ where the growth rate is maximum.

Table 1.4 summarises the main properties of these growth curves.

Table 1.4 Summary of the properties of some special cases of the Richards' family of growth models. $W(t)$ is size at time t, W_{max} is the maximum attainable size, k is a rate constant, A is a parameter which depends on the ratio $W(0)/W_{max}$ and reflects the position of the time origin. The maximum growth rate occurs at the point of inflexion on the growth curve and is obtained by setting the derivative of the growth rate equal to zero

Growth model	m	Equation of growth	Growth rate	Point of inflexion
Monomolecular	0	$W(t) = W_{max}(1 - A\,e^{-kt})$	$k[W_{max} - W(t)]$	None
von Bertalanffy	2/3	$W(t) = W_{max}(1 - A\,e^{-kt})^3$	$kW(t)\left[\left\{\frac{W_{max}}{W(t)}\right\}^{1/3} - 1\right]$	$0.296W_{max}$
Gompertz	1	$W(t) = W_{max}\,e^{-A\,e^{-kt}}$	$kW(t)\log_e\left\{\frac{W_{max}}{W(t)}\right\}$	$0.368W_{max}$
Logistic or autocatalytic	2	$W(t) = \frac{W_{max}}{(1 + A\,e^{-kt})}$	$kW(t)\left[1 - \frac{W(t)}{W_{max}}\right]$	$0.5W_{max}$

Figure 1.24 shows the general form of the monomolecular, von Bertalanffy, Gompertz and logistic models. For comparison, the curves are standardised so as to have the same maximum growth rate occurring at the same time. The Gompertz and the von Bertalanffy appear similar in that they lack the rotational symmetry of the logistic and show a more prolonged period of growth in the later stages of development. Note that the different curves can appear rather similar over restricted ranges of size so that in practice it might be difficult to choose between them. In an extensive study involving data from over 300 species of mammals, Zullinger *et al.* (1984) found that the Gompertz and the von Bertalanffy described the observed patterns equally well. Other applications of the Gompertz model have been made to describe the growth of solid tumours (Laird, 1965) and nestling birds (Ricklefs, 1968).

Example 1.9 Application of Richards' model to growth of melon seedlings

Richards applied his family of growth curves to growth in height of melon seedlings grown at different temperatures 20, 25, 30, 35 and 37.5 °C (Pearl *et al.*, 1934).

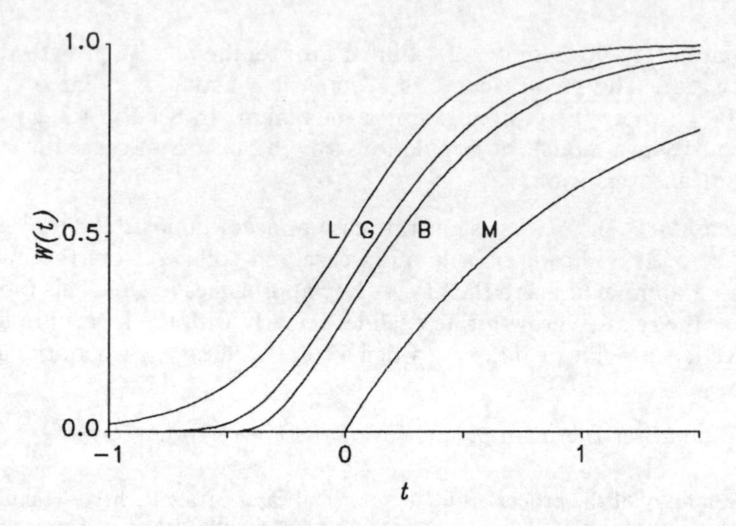

Figure 1.24 Particular members of the Richards' family of growth curves: monomolecular (M), von Bertalanffy (B), Gompertz (G) and logistic (L) standardised to have the same maximum growth at the same time.

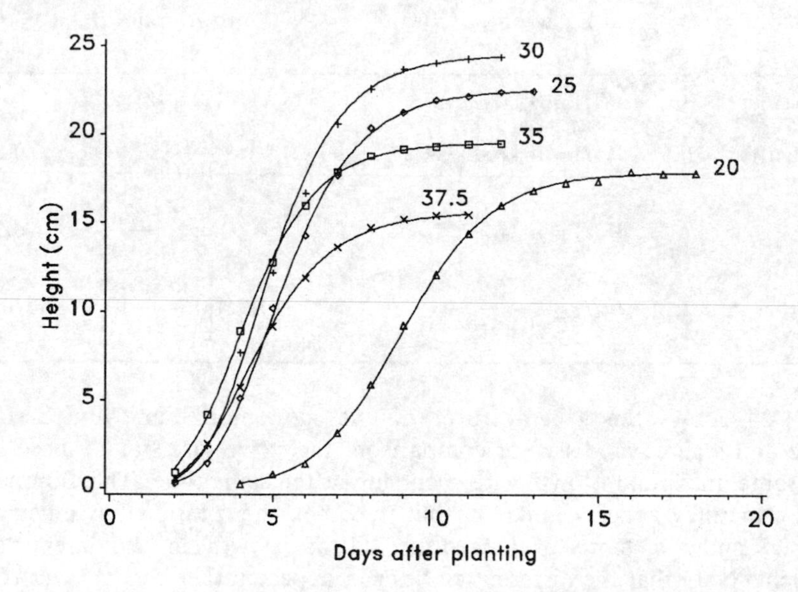

Figure 1.25 Richards' growth curves fitted to data on growth in height of melon seedlings at different temperatures (see Table 1.5).

Figure 1.25 shows that the model describes the observed patterns of growth rather well. The parameters of the fitted models are given in Table 1.5. Over the range 25–37.5 °C the calculated value of m is close to 1 which corresponds to a Gompertz growth curve. For the lower temperature of 20 °C the value is closer to 2 indicating a growth curve more like the logistic. This shows that changes in the temperature not only affect growth rates and heights attained but also the shape of the growth curve. In terms of the ultimate height attained the optimum temperature for growth lies between 25 °C and 35 °C.

Table 1.5 Parameter values of Richards' growth curves fitted to height of melon seedlings grown at different temperatures

Parameter	Temperature (°C)				
	20	25	30	35	37.5
W_{max} (cm)	17.8	22.5	24.5	19.5	15.6
k (day^{-1})	0.671	0.681	0.721	0.739	0.672
m	1.7	1.2	1.2	1.2	1.1

1.4.3 Allometric growth

Allometry is the study of the relative size of different parts of an organism and how the parts grow in relation to each other. One of the earliest workers in this field was Huxley (1932) who proposed a simple model for allometric growth in which the ratio of the relative growth rates of the two components remains constant. If we denote the sizes of the two components by Y and X then the model can be written as the differential equation

$$\frac{1}{Y}\frac{\mathrm{d}Y}{\mathrm{d}t}=\frac{b}{X}\frac{\mathrm{d}X}{\mathrm{d}t}$$

or

$$\frac{\mathrm{d}\log_e Y}{\mathrm{d}t}=b\frac{\mathrm{d}\log_e X}{\mathrm{d}t}$$

where the parameter b is the ratio of the relative growth rates of the variables Y and X. Integrating both sides gives

$$\log_e Y=\log_e A+b\log_e X$$

or

$$Y=AX^b.$$

This is called the allometric equation. It is a power function in which the size of one component is proportional to the size of the other component raised to some power b. The logarithms of the sizes are related by a straight line with slope b. The parameter b is a constant with no units of measurements whereas A measures the size of Y when X is of unit size and is therefore affected by the units of measurement. Figure 1.26 illustrates the form of the allometric equation for a range of values of the index b.

The allometric equation is not restricted to describing the relationship between different components such as the weights of two organs or the weight and length of an organism. For example, it is often found that the average metabolic rate* of an animal is related to its body weight by the allometric equation (Brody & Proctor,

* The average daily metabolic rate is the average metabolic rate of an active animal over a 24-hour period, at a temperature that the animal would normally experience in its natural environment, and when given access to food and water.

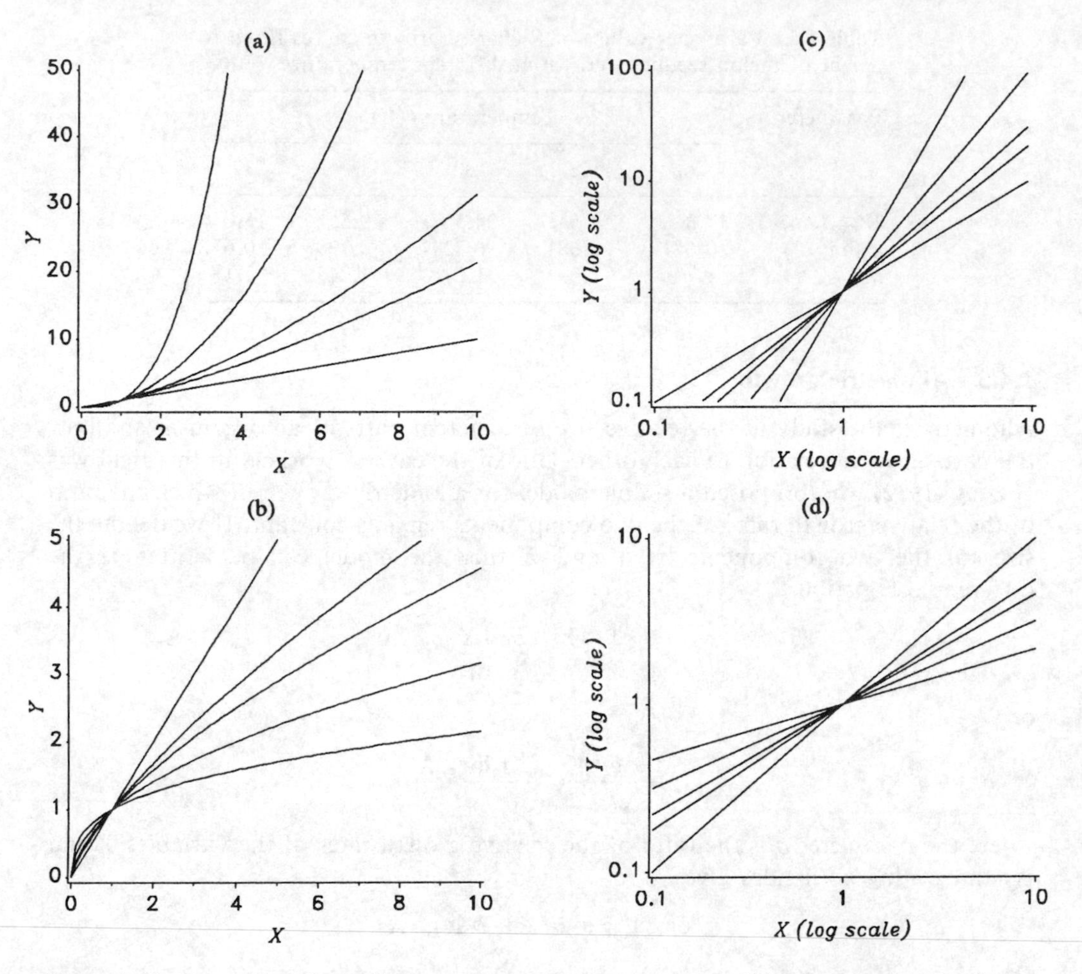

Figure 1.26 Examples of the allometric equation for $A = 1$ and different values of the index parameter b. (a) $b = 1$ (bottom), 1.33, 1.50, 2 and 3 (top); (b) $b = 0.33$ (bottom), 0.5, 0.66, 0.75 and 1 (top); (c) and (d) values on a log-log scale.

1932). In this case the index b varies from species to species but usually lies in the range 0.50–0.75.

Example 1.10 Allometric growth in the fiddler crab

Huxley (1932) applied the allometric equation to the growth of the large claw (chela) of the male fiddler crab *Uca pugnax* in relation to its body weight. The data on specimens from as small as 60 mg to those weighing over 30 times as much are shown in Figure 1.27(a). As the crab grows the relative size of the chela, expressed as a percentage of the body weight, increases from about 9% in the smallest crabs to 62% in the largest. Figure 1.27(b) shows the relationship between log chela weight and log of body weight. For animals with body weight less than about 0.75 g the points fall close to a straight line consistent with the allometric equation: the fitted line has slope $b = 1.62$. For the observations on the larger crabs the slope is less and equal to 1.25. Huxley noted that this change is associated with the onset of sexual

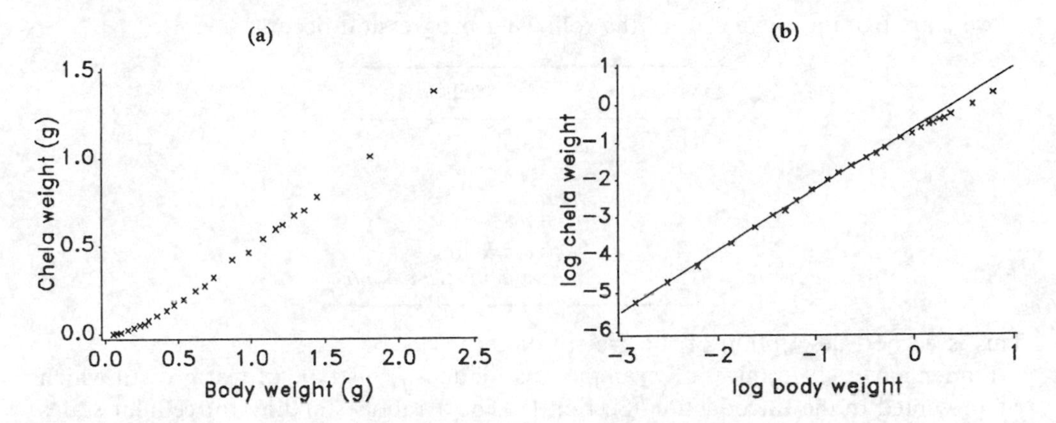

Figure 1.27 Allometric growth in the fiddler crab (see text).

maturity. So, a single allometric equation is not sufficient to describe the relationship between the weight of the chela and the body over the whole observed range of sizes but the model has two uses. First, it provides a quantitative description of the allometric growth in two periods. Second, it highlights the unexpected feature that the relative growth rate of the chela (a secondary sexual character) is reduced, rather than increased, relative to the relative growth rate in weight with the onset of sexual maturity.

1.5 DEVELOPMENT OF STRUCTURE

In Section 1.4 we presented some models for describing the growth of individual organisms using some measure of size such as length, height or weight. In this section we consider the growth and development of shape and structure such as the leaves on a plant or tree and present the Lindenmayer, or L systems, invented by Aristid Lindenmayer (Lindenmayer, 1968, 1975). His purpose was to provide a logical structure or language of plant growth, and initially his concern was with the topology of plants.

1.5.1 Lindenmayer systems

A Lindenmayer or L-system is a set of simple rules for growth and development. The general principle is to replace each element of an array by a string of elements according to the specified rule for the particular case, at each time point. Consider the simple example of the growth of a filament of the blue-green alga *Anabaena catenula*, which consists of short and long cells in a linear array, which we describe by a and b. Each such cell has a polarity, which is indicated by a subscript, l for left, and r for right. The L-system description of the growth of the bacteria is as follows for the four states:

$$a_r \rightarrow a_l b_r \qquad a_l \rightarrow b_l a_r \qquad b_r \rightarrow a_r \qquad b_l \rightarrow a_l.$$

If we start from a single cell a_r, the following progression occurs:

Division	Cells
0	a_r
1	$a_l b_r$
2	$b_l a_r a_r$
3	$a_l a_l b_r a_l b_r$
4	$b_l a_r b_l a_r a_r b_l a_r a_r$
5	$a_l a_l b_r a_l a_l b_r a_l b_r a_l a_l b_r a_l b_r$

This is a good description of the growth pattern.

Lindenmayer (1975) gives an example of a model of this type of leaf growth which we presented in the Introduction (Section 1). The symbols standing for cellular states are a, b, c, d and k and the production rules are

$$a \rightarrow cbc, \quad b \rightarrow dad, \quad c \rightarrow k, \quad d \rightarrow a, \quad \text{and} \quad k \rightarrow k.$$

Section 1 in the Introduction gives the sequences over 8 generations and shows how the formulae can be interpreted in terms of leaf shape by identifying the symbols as follows: a and b are sharp projecting leaf tips, c and d the lateral margins of lobes, and k's are identified with notches, the older the notch the more k's. Figure 2 in the Introduction shows the corresponding leaf shape.

Inspection of the sequences shows that some strings repeat themselves in the time course, e.g. *kcbck* appears in generation 5 and in generation 7, *kdadk* first appears in generation 3 and then reappears in generation 5 and 7 in the centre, and in generation 6 at the sides. We see that from generation 4 onwards the string at generation n, S_n, has the following relations to the strings 2 and 3 generations previously, S_{n-2} and S_{n-3}

$$S_n = kS_{n-3}S_{n-2}S_{n-3}k.$$

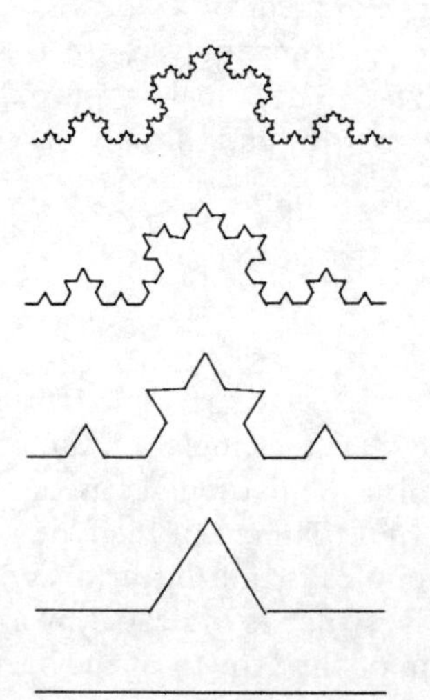

Figure 1.28 The development of part of the Koch snowflake fractal pattern using a L-system model (see text).

Lindenmayer models have a quite close relationship to models for generating fractals. The L-system model for the Koch snowflake involves replacement at each step of each straight line segment by the two outer thirds of the segment, joined by two sides of an equilateral triangle which has the missing middle third as its base (as in Figure 1.28). Thus we see that an equilateral triangle is thus converted—after an infinite sequence of applications of this substitution rule—into the Koch snowflake, devised by Helge von Koch in 1904. This is a curve, but it is nowhere smooth, having infinitely many indentations at every level. It is said to be self-similar in that the boundary is the same on every scale. The partial self-similarity which occurs with Lindenmayer systems, seen in the above example of leaf shape, is closely related to the self-similarity which we see in fractals.

Only very simple L-system models have been presented in this section. The models have been developed substantially since their invention in 1968, including models of trees, inflorescences and phyllotaxis. Prusinkiewicz & Lindenmayer (1990) give an elegantly presented summary of recent developments and applications.

1.6 MORE REALISTIC MODELS FOR POPULATION GROWTH AND DECLINE

This chapter deals with deterministic models of growth and decline for a single large homogeneous population of identical individuals in a constant environment. This is a gross simplification of real populations in which: (a) there is variation between individuals arising, for example, from age, size and genetic differences; (b) there may be a time-varying environment due to systematic variation such as seasonal and periodic changes or a stochastic environment with random variation in time; (c) there is a spatial distribution of individuals. These topics are beyond the scope of this chapter and the reader is referred to references below and to Table 1.6 which gives pointers to other parts of the book.

Age structure

For populations in which survival and fecundity vary with age, e.g. our own species, a useful framework is the *Leslie matrix* model (Leslie, 1945, 1948). The approach is to divide the population at a given time into distinct age classes and to model the

Table 1.6 Topics on modelling population growth and decline in other chapters

Topic	Location and Models
Stochastic effects	*Chapter* 4 Incorporating demographic and environmental stochasticity into dynamic models
	Chapter 13 Model for grouse population cycles
Population growth in time and space	*Chapter* 12 Combined diffusion and population growth

change of each age class. Changes in the total population are obtained by combining the constituent age classes. More generally, individuals can be classified by some other criterion such as size or stage in their life cycle. For a detailed discussion of matrix population models see Caswell (1989).

EXERCISES

1.1 Calculate the doubling time and time to achieve a 10-fold increase for populations growing exponentially in continuous time with intrinsic rates of increase: $r = 1$ per hour, 0.5 per day, 0.2 per week and 0.1 per year.

How do these times relate to exponential decay in continuous time with the above values for the rate constant of decay?

1.2 Consider a colony of bacteria which is growing exponentially with intrinsic rate of increase r per unit time and where there is immigration at a constant rate c per unit time.

(a) Write down a differential equation to describe the dynamics of the growth of the colony.

(b) Solve the equation to obtain an expression for the size of the colony at time t starting with a population of size N_0 at time zero.

(c) Show that after a long time the effect of immigration is to increase the size of the colony by a factor $[1 + c/(rN_0)]$, but that the effect on the relative growth rate is negligible.

1.3 Examine the following data on the increase in size of the USA population during 1790–1940 (Figure 1.1(b)) for evidence of: (i) exponential growth; (ii) limited growth.

Year	Population (millions)	Year	Population (millions)
1790	3.929	1870	38.558
1800	5.308	1880	50.156
1810	7.240	1890	62.948
1820	9.638	1900	75.995
1830	12.866	1910	91.972
1840	17.069	1920	105.711
1850	23.192	1930	122.775
1860	31.443	1940	131.410

1.4 A community is made up of two independent, continuously varying populations of size N_1 and N_2 with constant relative growth rates r_1 and r_2, respectively, i.e. $dN_1/dt = r_1 N_1$ and $dN_2/dt = r_2 N_2$.

(a) Show that the relative growth rate r of the total population of the community, $N = N_1 + N_2$, increases with time and that $dr/dt = N_1 N_2 (r_1 - r_2)^2 / N^2$. What is the eventual relative growth rate?

(b) Illustrate the dynamics by simulating data for two populations with $r_1 = 0.10$, $r_2 = 0.01$, for $t = 0$ to 50, using three different sets of initial population sizes: $(N_1, N_2) = (1, 1), (0.2, 1.8), (1.8, 0.2)$. Plot the logarithm of the total population size and comment on the differences in the three trajectories.

(c) For a community of M populations, of which the mth has population size N_m and constant relative growth rate r_m, the relative growth rate r of the total population of the community increases with time and $dr/dt = \Sigma p_m (r_m - \bar{r})^2$, where $p_m = N_m/N$ and $\bar{r} = \Sigma p_m r_m$. Discuss the validity and implications of the result for the world human population.

1.5 Suppose that the survival of an organism can be described by a model in continuous time for which the *instantaneous mortality rate* at age t is given by $m(t) = kt^a$, where $\alpha > 0$.

(a) Write down a differential equation for the number of surviving organisms in a large cohort of individuals of the same age and show that the number of survivors of age t can be written as

$$S(t) = S_0 \, e^{-(t/c)^b},$$

where S_0 is the initial size of the cohort. Find b and c in terms of a and k.

(b) Sketch the general shape of the survivorship curve showing the logarithm of the number of survivors against age for the three cases: $b < 1, b = 1, b > 1$.

(c) Compare the model survivorship curve for values of $b = 5$ and $c = 10.77$ with the following observed pattern of survival in a sample of female Svalbard reindeer after their first year (data provided by Dr Nick Tyler). Use the model to estimate the age to which 50% of the animals are expected to survive after their first year.

Age	Survival (%)
1	100
2	100
3	100
4	100
5	98.1
6	97.1
7	95.2
8	92.4
9	81.9
10	67.6
11	49.5
12	28.6
13	17.1
14	8.6
15	2.9
16	0.9
17	0.0

1.6 In a simple model for an epidemic in a population of size n, the number of infected individuals is given by

$$\frac{dY}{dt} = \beta Y(n - Y)$$

where β is a positive constant which measures the contact rate between infected individuals and susceptibles.

(a) Obtain an expression for the number of infected individuals at time t.

(b) For a population of size $n = 1000$, contact rate of $\beta = 0.001$ per day and an initial number of infected individuals $Y_0 = 10$, find the time until (i) 50% and (ii) 90% of the population is infected. Repeat the calculations for a population of size $n = 2000$ and comment on the results.

1.7 In the logistic model the relative growth rate of the population decreases linearly with population density. Smith (1963) tested this assumption for the water flea *Daphnia magna* and found that the growth rate declined with density but according to a concave curve and not a straight line. He proposed an alternative model to describe the rate of change of biomass in which

$$\frac{dM}{dt} = \frac{rM(K - M)}{[K + (r/c)M]}.$$

From a series of experiments, estimates of the parameters were as follows: $r = 0.44$ per day; $K = 15.0$ mg per 100 ml; $c = 0.127$ per day.

(a) Plot the relative growth rate against biomass and compare it with the logistic model having the same carrying capacity (i.e. $K = 15.0$ mg per 100 ml).

(b) Repeat (a) but using the absolute growth rate to illustrate the general property that the relationship is asymmetrical with a maximum at $M < K/2$.

(c) Simulate growth in a model population starting with biomass 0.75 mg per 100 ml and compare the pattern with that of the logistic model having the same carrying capacity.

1.8 Consider a model population with logistic growth which is subject to harvesting at a rate $h(t)$ so that

$$\frac{dN}{dt} = rN\left\{1 - \frac{N}{K}\right\} - h(t).$$

(a) Show that when the harvest rate is constant, i.e. $h(t) = h$, then a *maximum sustainable yield* of $h_{max} = rK/4$ occurs when the population is at half its carrying capacity.

(b) Now suppose that the harvesting rate is proportional to population size, i.e. $h(t) = qN(t)$, where q is a constant. Determine $N(t)$ explicitly and show that if $q < r$ the harvested population has a carrying capacity equal to $K(1 - q/r)$. What happens to the population when $q > r$?

1.9 Many studies of growth in plant monocultures have shown a decreasing relationship between the mean yield per plant and the density at harvest. The effect is attributed to intraspecific competition for the resources such as nutrients and light. A simple model to describe this yield–density relationship is

$$W = \frac{W_{max}}{(1 + aD)^b}$$

where W is the mean plant weight at harvest, D is the plant density at harvest and W_{max} is the mean plant weight in absence of competition and a and b are constants. For spring wheat, typical values for the parameters are: $W_{max} = 50$ g, $b = 0.60$, $a = 0.24$ m^2.

(a) Plot the yield–density relationship for the spring wheat data over a 100-fold range of plant densities, 10–1000 plants m^{-2} and show that at high densities the logarithm of plant weight declines approximately linearly with the logarithm of density.

(b) Examine the form of the relationship between total yield (i.e. $W \times D$) and plant density for the cases: $b < 1, b = 1, b > 1$.

(c) If there is an allometric relationship between the seed output per plant and mean plant weight (i.e. $S = AW^k$) show that the relationship between seed output per plant and density has the above form but with parameter values which can be expressed in terms of W_{max}, a, b, A and k.

1.10 In plant monocultures, self-thinning reduces the number of plants which survive until harvest. A model to describe mortality induced through intraspecific competition is

$$D = \frac{D_0}{(1 + mD_0)}$$

where D is the density at harvest, D_0 is the initial density and the parameter m measures the effect of density-dependent mortality—for spring wheat, a plausible value is $m = 0.00065$ m^2.

(a) Combine this model with that relating mean plant weight and harvest density (Exercise 1.9) to obtain a relationship between total yield and initial planting density.

(b) Examine the reduction in total yield in spring wheat due to the effect of mortality over the 100-fold range of plant densities 10–1000 plants m^{-2}.

1.11 In what circumstances does the discrete time logistic equation

$$N_{i+1} = N_i[1 - b(N_i - N_e)]$$

provide a model for the dynamics of a population?

Investigate the stability of the two equilibria of this equation in the cases: (i) $0 < bN_e < 1$; (ii) $1 < bN_e < 2$; (iii) $bN_e > 2$.

In the case $bN_e = 1$, find an exact solution by first writing $N_i = N_e + x_i$ and then solve the resulting equation for x_i. Hence show that when $bN_e = 1$, the equation at $N_i = N_e$ is stable to perturbations $N_i = N_e + \varepsilon$ provided that $|\varepsilon| < N_e$. Are perturbations with $|\varepsilon| > N_e$ relevant to population dynamics?
(Cambridge, NST, Part IA, Biological Mathematics, 1986.)

1.12 Growth in human height cannot be adequately described by a sigmoidal curve partly because of the spurt of growth at puberty. A model was proposed by Preece & Baines (1978) in which height is related to age by

$$H(t) = A - \frac{2(A - B)}{\exp[C(t - E)] + \exp[(D(t - E)]}$$

where $H(t)$ is height at age t. Model parameters are: A, adult height; B, height at age E; C and D, relative growth rate coefficients; E, approximate age at which the pubertal growth spurt occurs.

(a) Find an expression for the growth rate as a function of t and show that the relative growth rate at age E is given by $(A - B)(C + D)/2B$.

(b) Plot the model growth curve for the parameter values: $A = 187.0$ cm; $B = 173.9$ cm; $C = 0.1042$ per year; $D = 1.416$ per year; $E = 15.08$ years; and compare with the following measurements made by Montbeillard on the growth of his son (Figure 1.1(a)). Plot the growth rate and the relative growth rate to illustrate that the model mimics the spurt in growth around the age of puberty.

Age	Height (cm)	Age	Height (cm)
Birth	51.4	9.00	137.0
0.50	65.0	9.61	140.1
1.00	73.1	10.00	141.9
1.50	81.2	11.50	148.6
2.00	90.0	12.00	149.9
2.50	92.8	12.67	154.1
3.00	98.8	13.00	155.3
3.50	100.4	13.50	158.6
4.00	105.2	14.00	162.9
4.58	109.5	14.53	169.2
5.00	111.7	15.01	175.0
5.58	115.5	15.52	177.5
6.00	117.8	16.27	181.4
6.55	122.9	16.52	183.3
7.00	124.3	17.05	184.6
7.25	127.0	17.11	185.4
7.50	128.9	17.43	186.5
8.00	130.8	17.59	186.8
8.50	134.3		

1.13 In many species the vital rates of survival and reproduction vary with age. For age-structured populations, Leslie (1945) proposed a matrix model for population change. The population is divided into discrete age classes and the number of females in the ith age class at time t is denoted by N_{it}. The fecundity of a female in the ith age class (F_i) is defined as the number of daughters which survive to join the population in the youngest age class. The survival rate of females of age class i (s_i) is the proportion which survive to age class ($i + 1$). Equations relating the number of females in the different age classes at time t to the number at time $(t + 1)$ are then represented in matrix form

$$\begin{pmatrix} N_{0,t+1} \\ N_{1,t+1} \\ \cdot \\ \cdot \\ \cdot \\ N_{m,t+1} \end{pmatrix} = \begin{pmatrix} F_0 & F_1 & \cdots & F_{m-1} & F_m \\ s_0 & 0 & \cdots & \cdot & 0 \\ 0 & s_1 & \cdots & \cdot & 0 \\ \cdot & \cdot & \cdots & \cdot & \cdot \\ \cdot & \cdot & \cdots & \cdot & \cdot \\ 0 & 0 & \cdots & s_{m-1} & s_m \end{pmatrix} \begin{pmatrix} N_{0,t} \\ N_{1,t} \\ \cdot \\ \cdot \\ \cdot \\ N_{m,t} \end{pmatrix}$$

where age class m refers to individuals of age m or older. The equations for the changes in numbers in each age class follow from the rules for multiplying a column vector by the so-called *projection matrix* or *Leslie matrix*. For the Leslie matrix model, the population eventually increases, or decreases, exponentially with rate equal to the dominant eigenvalue of the projection matrix until a stable age distribution ensues proportional to the corresponding eigenvector.

In a study of the British grey seal, Harwood (1981) gives the following vital rates for seven age classes of females ranging from pups (age 0) to adults (age 5 and older).

Age (i)	Fecundity (F_i)	Survival (s_i)
0	0	0.657
1	0	0.930
2	0	0.930
3	0	0.930
4	0.08	0.935
5	0.28	0.935
5+	0.42	0.935

(a) Write down the projection matrix and the equations of change for the seven age classes.

(b) Starting with a density of 10 adult females, project the population forward by 20 years and monitor the changing age structure.

(c) Use the numerical simulations to show that the potential geometric rate of increase is about 7.6% per annum and that the proportions in the different age classes eventually settle down to: 0.179 (0); 0.109 (1); 0.095 (2); 0.082 (3); 0.070 (4); 0.061 (5); 0.404 (5+). Confirm that these values correspond to an eigenvalue and its eigenvector.

2 Deterministic Genetic Models

2.1 INTRODUCTION

The make-up and behaviour of any organism is determined to a large extent by its genetic constitution. Aspects such as blood group and eye colour are easily categorised variables which are almost completely determined by genetics. Skeletal growth often reflects past nutrition as well as genetics, while the dominating influence in more labile aspects such as metabolic rate would usually be the immediate environment. So to varying degrees genetics plays a part in most aspects of individual make-up and behaviour.

Similarly, most aspects of any biological population are affected by its genetic make-up. In a sense genetic constitution is more a characteristic of a population than of any individual. In a single individual there are always aspects of the genotype which do not find expression in the phenotype such as recessive alleles in a heterozygote. Correspondingly an individual's genotype can often only be elucidated by studying the appearance of a character in its relatives. This is one reason why the study of population genetics is of importance. Secondly, it is often easier to make predictions of the time course of genetic change in large populations than in the descendants of individuals: there is usually much more uncertainty about the outcome in any individual case than for the population as a whole. A third reason is that it is often the course of change in populations rather than individuals that is of interest, especially in studies of natural populations.

This chapter outlines a few important mathematical models which are of use in the study of the genetics of populations. All the models discussed in this chapter are *deterministic* in the sense that the predictions of the models are precise: if two modellers were to simulate a process using a deterministic model they would end up with the same outcome, unlike the case with stochastic models. Although the mechanism at the heart of genetics, genetic recombination, is one of the fundamental *stochastic* mechanisms in biology, it is possible to achieve some very useful results which are applicable to fairly large populations by considering models which are deterministic. The first section is a brief treatment, with examples, of Mendel's Laws. Material from two main areas of population genetics is then covered. The first is about gene frequencies in populations in which the individuals are mating completely at random, and there is no selection operating (i.e. no differences in fitness between the genotypes resulting), and we consider examples concerning the inheritance of blood groups. Secondly, we outline a very simple model of selection and apply it to the spread of the black-winged *carbonaria* form of the peppered moth, *Biston betularia*, following the industrial revolution. Moving on from this to the study of balanced polymorphisms, we encounter an example which is of practical importance: the continued prevalence of genes in a human population which confer immunity to malaria but simultaneously greater susceptibility to sickle cell anaemia.

We defer to the next two chapters some of the special problems of the inheritance of quantitative characters such as height, weight and yield (as opposed to qualitative ones such as eye colour) which are of greater interest to plant and animal breeders. We also defer to the next chapters predictions for small samples of individuals in which the random variability, for example in Mendel's Laws, is of greater importance.

Before we begin we should clarify what we mean by population in this context. The important thing which defines a population here is the potentiality for members to exchange genes in the course of sexual reproduction. There are various mating strategies which are possible within populations, but an important aspect in general is that there is exchange of genetic material over time between the members of a population. If there is no exchange then the two or more fractions between which there is no exchange form separate populations. We shall see that most of the results which we discuss in the necessarily elementary treatment outlined here are for populations in which mating is completely at random between the members of the population. This is usually only an approximation but it is adequate for many purposes.

2.1.1 Terminology

This section gives brief informal definitions of some technical terms in genetics.

Allele Any of the various forms in which a gene can occur

Chromosomes Microscopic rod-shaped structures, consisting of nucleoprotein arranged into genes responsible for the transmission of hereditary characteristics

Diploid Having paired homologous chromosomes so that twice the haploid number is present

Dominant Determining the phenotype when paired with a different allele

Gamete A haploid germ cell such as sperm or egg that fuses with another germ cell during fertilisation

Gene A unit of heredity composed of DNA occupying a fixed position on a chromosome

Genotype The genetic constitution of an organism, usually with respect to the genes under consideration

Haploid Having a single set of unpaired chromosomes, found in cells such as gametes

Heterozygous (genotype) A genotype having dissimilar alleles at a specific locus

Homologous chromosomes Two chromosomes one of maternal origin and the other of paternal origin that pair during meiosis and contain the same gene loci

Homozygous (genotype) Having identical alleles at a specific locus

Locus The position of a particular gene on a chromosome

Meiosis A type of cell division in which the nucleus divides into four daughter nuclei (rather than the usual two), each containing half the chromosome number of the parent nucleus, occurring in sexual reproduction

Mitosis Normal cell division in which the nucleus divides into two daughter nuclei, each containing the same chromosome number as the parent nucleus

Panmixia Random mating

Phenotype The observed characteristic of an organism as determined by the interaction of its genetic constitution and the environment

Recessive Only determining the phenotype when present in both alleles

Recombination The process by which genetic material of different origins (most typically from two sets of parental chromosomes) becomes combined during production of germ cells

Segregation The separation at meiosis of the two members of any pair of alleles into separate gametes

Zygote The diploid cell resulting from the union of the sperm and the ovum

2.2 MENDELIAN GENETICS

In a sexually reproducing population, the genes are scrambled each generation by segregation and recombination, the genotypes constantly changing as individuals die and new ones are born. The basic mechanisms by which genes are passed on from one generation to the next are quite simply stated and were discovered by Gregor Mendel in 1865 but remained relatively unknown until the turn of the century.

Mendel's Laws were derived from experiments on the inheritance of characters such as the colour and shape of the seeds of pea plants, and are as follows.

Mendel's Laws

Law of segregation. Every somatic (i.e. ordinary) cell of an individual carries a pair of genes for each character; the members of each pair separate during meiosis so that each gamete carries one gene of each pair, and gametes pair at random to form the genotype of the offspring.

Law of independent assortment. Each member of any pair of alleles is equally likely to be combined with either member of another pair of alleles, since they associate independently.

What do these laws mean? Figure 2.1 illustrates the first law. Mendel took seed and pollen from the F1 generation, i.e. the first generation cross of pure strains so that all

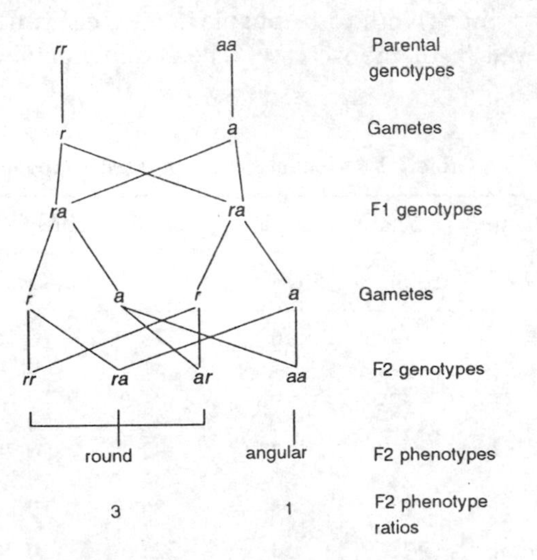

Figure 2.1 Illustration of Mendel's First Law (see Section 2.2).

the plants participating were heterozygous—having two different alleles at the locus in question let us say *r*, for round, and *a*, for angular seeds—which is then said to have genotype *ra*. Note that we do not differentiate between *ra* and *ar*. It is known that the *r* allele is dominant (i.e. if present in the genotype, it will determine the phenotype); so that the two genotypes *rr* and *ra* will have round seeds, and only the genotype *aa* will have angular seeds. Crossing two individuals both of genotype *ra*, there would be equal numbers of *r* and *a* gametes from both parents. They thus both contribute either one *r* or one *a* to each individual of the next generation and they do so in roughly equal proportions, so that we would expect equal numbers of *rr*, *ar*, *ra* and *aa* genotypes among the offspring (where just for the moment we regard the first of these as having come from parent 1 and the second as having come from parent 2). Since there is no discrimination between *ra* and *ar*, we expect the numbers of the genotypes—if we now ignore the parent of origin of each allele—(*rr*), (*ra* or *ar*) and (*aa*) to be in the proportions 1:2:1. However, as *r* is dominant and therefore if present determines the phenotype, we expect the numbers of round and angular seed to be in the ratio 3:1.

2.2.1 Examples

Some of Mendel's results are given in the Tables 2.1 and 2.2. The proportion of round seeds in Table 2.1 is 0.76 which is close to the value expected from Mendel's Laws of 0.75. There is variability of the proportions between the individual plants but they clearly cluster round 0.75. Mendel obtained data on other characters in his experiments, and his second law is derived from experiments in which data on two characters were collected for each seed. Some such data are given in Table 2.2. The allele for yellow is dominant to that for green, and so we would expect the number of yellow and green seeds to be in the ratio 3:1 as for shape. As the alleles segregate independently, we would expect three quarters of the round seeds also to be yellow, i.e. 9/16ths of the total to be round and yellow, and also one quarter of the round seeds to be angular, i.e. 3/16ths of the total. Similarly we expect 3/16ths of the total to be round and green, and only 1/16th to be angular and green, and this is approximately what has been achieved in the above case. The actually achieved percentages of the

Table 2.1 Mendel's data on seed shape

Plant	Shape of seed		Total	Proportion round
	round	angular		
a	19	10	29	0.66
b	24	7	31	0.77
c	26	6	32	0.81
d	22	10	32	0.69
e	25	7	32	0.78
f	28	6	34	0.82
g	27	8	35	0.77
Total	171	54	225	0.76

Table 2.2 Some of Mendel's data for the dihybrid cross (i.e. a cross between F1 hybrids) on the inheritance of the two characters, shape and colour

Colour	Shape	
	round	angular
Yellow	315	101
Green	108	32

total with the expected percentages in brackets were 57% (56%), 18% (19%), 19% (19%) and 6% (6%)—a very close agreement. We postpone to Chapter 6 discussion of how we would test whether data such as these conform sufficiently closely to the frequencies we would expect assuming the truth of the laws.

2.2.2 Gene or allele (relative) frequency

The gene or allele (relative) frequency of a particular allele A in a population is the proportion of the alleles at the specified locus over all the members of the population which are of the type in question. In other words it is the number of A alleles at this locus totalled over all the members of the population divided by the total number of alleles at the locus (i.e. $2n$ where n is the population size). It is usually denoted as p_A with a lower case p. It is also quite usual in genetics to leave out the word *relative*: whenever geneticists speak of gene or genotype frequency they mean relative frequency or proportion. In general we will not follow this convention so as to avoid confusion with other parts of the book.

A specification of the alleles and their proportions is always a briefer description of the genetic structure of a population than a similar specification of the genotypes, and for many purposes is quite adequate. If there are three allelic forms at a locus, A, B and C, then the number of possible genotypes from this one locus would be every possible pair from these three, of which there are 6: AA, AB, AC, BB, BC, CC. More generally there are $m(m+1)/2$ genotypes from a single locus when there are m possible alleles, and therefore it would usually be a much lengthier business specifying a list of the genotypes with their relative frequencies. When this calculation is performed for just a few loci each with 3 alleles, it soon becomes obvious that merely producing a complete list of genotypes is a quite impossible task; for 2 loci with 3 alleles, the number is just 6^2, rising to 6^{10} (about 60 million!) for 10 such loci.

The use of this simple model of the *gene pool*—that the genetic structure of a population can be expressed adequately by a list of genes and their relative frequencies—has been termed *beanbag genetics*. The genes in the population are likened to beans in a bag, the combinations in which they are found in the genotypes being just those which would arise by chance drawing of random samples of beans from the bag. J.B.S. Haldane gives an account (Haldane, 1964) of some of the consequences of this approach which were unravelled largely by the three great figures of population genetics in the first half of this century: R.A. Fisher, Sewall Wright and

Haldane himself. In this chapter, we cover only a small part of this material, giving a few of the elementary results which they obtained.

2.3 RANDOMLY MATING POPULATIONS: THE HARDY–WEINBERG PRINCIPLE

2.3.1 Two or more alleles at a single locus

Extensive data has been collected on blood groups of humans, which are known to be almost completely determined genetically unlike most phenotypic characters (Example 2.1).

Example 2.1 The M–N blood group

Race and Sanger (1954) collected the following data on the frequencies of the M–N blood group from a sample of 1279 English people. From this data we can easily infer that the relative frequency of the M allele by counting is $(2 \times 363 + 634)/(2 \times 363 + 2 \times 634 + 2 \times 282) = 1360/2558 = 0.532$, and that of the N allele is therefore $1 - 0.532 = 0.468$.

	Blood group		
	MM	MN	NN
Observed number	363	634	282
%	28.38%	49.57%	22.05%
Expected % (using Hardy–Weinberg principle)	28.27%	49.80%	21.93%

Gene relative frequencies: $p_M = 0.5317$ $p_N = 0.4683$.

Hardy and Weinberg independently in about 1908 discovered the law which bears both their names about the relationship expected between the allele frequencies and genotype frequencies after just one generation of random mating (i.e. independently of the alleles at the locus involved).

Hardy–Weinberg principle

If p_A and p_B are the relative frequencies of two alternative alleles, A and B, at the same locus, then *after one generation* of random mating the relative frequencies of the genotypes of the offspring, AA, AB, BB are $p_A^2, 2p_Ap_B, p_B^2$ respectively.

We derive the principle using the simple example above with two alleles M and N. Let P, $2Q$ and R be the relative frequencies of the MM, MN and NN genotypes respectively in the parent population. The various possible matings, their proportions, and the proportions of the various genotypes among the progeny are given in Table 2.3. The gene relative frequencies are then $p_M = P + Q$ and $p_N = Q + R$ respectively for the M and N alleles respectively. The table gives for each possible pairing of parental genotypes, the proportion of that pairing we expect, and the proportions of the various possible genotypes for the progeny of the pairing, which are

Table 2.3 Table illustrating the Hardy–Weinberg principle

Type of mating (male × female)		Proportion of matings	Proportion of progeny		
			MM	*MN*	*NN*
MM	*MM*	P^2	P^2		
MM	*MN*	$2PQ$	PQ	PQ	
MM	*NN*	PR		PR	
MN	*MM*	$2PQ$	PQ	PQ	
MN	*MN*	$4Q^2$	Q^2	$2Q^2$	Q^2
MN	*NN*	$2QR$		QR	QR
NN	*MM*	PR		PR	
NN	*MN*	$2QR$		QR	QR
NN	*NN*	R^2			R^2
Totals		1	$(P+Q)^2$ p_M^2	$2(P+Q)(Q+R)$ $2p_Mp_N$	$(Q+R)^2$ p_N^2

totalled at the base of the column, and shown to fall in the proportions $p_M^2, 2p_Mp_N, p_N^2$ for the genotypes *MM*, *MN* and *NN* respectively.

The Hardy–Weinberg principle enormously simplifies the study of natural diploid populations. It is only necessary to follow any changes in the gene frequencies as most populations mate approximately randomly, and hence it is possible to work out the zygotic frequencies from the gene frequencies. The extension to multiple alleles is straightforward and the formulae are very similar to those for two alleles. If there are n alleles, with relative frequencies $p_1, p_2, \ldots, p_n$, then the frequency of a particular homozygote, say ii, is just the square of the gene relative frequency involved p_i^2, and the relative frequency of the heterozygote ij is $2p_ip_j$. Hardy–Weinberg relative frequencies are attained in one generation of random mating as in the two allele case.

2.3.2 Two or more loci

One would think that a similar extension would apply to the case of multiple loci, but this is not the case. An equilibrium similar to that above, termed a *gametic equilibrium*, is eventually reached but not in one generation of random mating, and there are various quite frequently occurring obstacles to a swift transition to equilibrium. Consider the case of two loci each with two alleles. The relative frequencies of the different zygotes in terms of the gametic relative frequencies are given in Table 2.4. We follow the notation used by many workers that lower case p is used for gene relative frequencies and upper case for chromosome or gamete relative frequencies. It is also conventional that upper case P is used for genotype relative frequencies, and we also follow this convention below. The appropriate interpretation should be clear from the context.

We show below that an equilibrium is eventually established in which $P_{AB} = p_Ap_B$. But linkage (i.e. a tendency for alleles at positions close to each other on the same chromosome to be transferred together to the offspring) between closely positioned genes on the same chromosomes slows the progress. Note that when linkage occurs Mendel's Second Law no longer holds; Mendel was fortunate in that the characters he

Table 2.4 Relative frequencies of different zygotes arising from different combinations of parental genotypes

Homozygotes		Single heterozygotes		Double heterozygotes	
AB/AB	P_{AB}^2	Ab/AB	$2P_{Ab}P_{AB}$	AB/ab	$2P_{AB}P_{ab}$
Ab/Ab	P_{Ab}^2	aB/AB	$2P_{aB}P_{AB}$	aB/Ab	$2P_{aB}P_{Ab}$
aB/aB	P_{aB}^2	Ab/ab	$2P_{Ab}P_{ab}$		
ab/ab	P_{ab}^2	ab/aB	$2P_{ab}P_{aB}$		

studied were not sufficiently close on the same chromosome for there to be any appreciable linkage between them.

Also it will be seen from the table that when recombination occurs it only affects the gametic relative frequencies in the cases where the input gametes are double heterozygotes; in other cases the output gametes are the same as the input gametes. This slows progress towards equilibrium even when the genes are not linked at all, for example if they are on different chromosomes. In this case if we wish to model the approach to equilibrium, we are concerned with the speed at which $P_{AB} - p_A p_B$ approaches zero, and clearly this is dependent on the extent of linkage between the two loci. A model for the progress to equilibrium can be specified in terms of the gene relative frequencies p_A and p_B which we do not expect to change and the proportion of matings in which the gamete in question will be recombinant, r. For each pair of loci, there will be a probability that the genes at these loci are inherited from different parents, and this is what the quantity r is specifying. If the genes are on different chromosomes, there is no linkage, and so there is a chance of 0.5 that they are not inherited from the same parent; so for such unlinked genes, $r = 0.5$. If they are very close together on the same chromosome, there might be much less chance that they split up, and so in this case, r would approach zero.

Thus the AB gamete in the $(n+1)$th generation can arise in two ways: it can be present in the nth generation and recombination does not take place, or it is not present in the previous generation and therefore would be the result of recombination. We use a second subscript to denote the generation, e.g. $P_{AB,n}$. In the first case, the contribution to $P_{AB,n+1}$ is just $P_{AB,n}(1-r)$. In the second case, the contribution is r times the proportion of matings in which the alleles were present in the $(n-1)$th generation, but on different gametes. The proportion of matings in which the alleles were present on different gametes in the $(n-1)$th generation is just $p_A p_B$, since we are assuming that mating is at random. So we obtain, by adding up these two proportions (since there is no overlap between the two cases),

$$P_{AB,n+1} = (1-r)P_{AB,n} + rp_A p_B$$

and this can be rearranged in terms of the departure from the eventual equilibrium state

$$P_{AB,n+1} - p_A p_B = (1-r)(P_{AB,n} - p_A p_B).$$

This can be applied again and again, relating generation $n+1$ to generation n, and then n to $n-1$ and so on until we eventually obtain

$$P_{AB,n+1} - p_A p_B = (1-r)^n (P_{AB,1} - p_A p_B).$$

Thus we see that the progress towards equilibrium can be expressed in terms of a kind of negative compound interest in which the interest rate is recombination rate, r: the departure from equilibrium is reduced by a factor r each generation. The same relative rate applies to all alleles at each locus. This model is essentially the same as the model for exponential decline in discrete time discussed in Chapter 1.

The value of this expression in practice is that for closely linked genes it takes some time for equilibrium to be reached and we could in theory therefore obtain an estimate of the time at which populations originally fused from the present gene and gamete relative frequencies. We saw above that Hardy–Weinberg equilibrium at a single locus is achieved in one generation of random mating, and so the allele relative frequencies at a single locus are of no value for this purpose: the population forgets, as it were, in just one generation. But the 'memory' of past gamete frequencies is eroded only slowly if the recombination rate is low, and at most by 50% each generation, and hence this might be of use in elucidating the breeding history of a population a few generations back.

2.4 SELECTION

The above discussion assumes that individuals mate at random and that all genotypes survive and reproduce equally well, which is approximately true in many populations for some gene systems such as those for blood groups. For many other characters, it is clearly not true. Variation in these characters results in some individuals which are better fitted to survive individually and which have better reproductive performance, and hence contribute more surviving offspring to the next generation. In human populations for example, many diseases are known to be inherited, and in others the propensity to contract the disease has a genetic component. In animal populations, usually some individuals are better at finding a mate and reproducing than others, and this is often partially under genetic control. Therefore any realistic model of the change in the relative frequency of genes for these traits must make some allowance for selection.

Selection is of interest in a number of areas. Natural selection—the selection exercised by nature on wild populations—is important in evolutionary studies. Mutation is an important element in evolution, but once a favourable mutation has taken place, natural selection is a mechanism which ensures that the mutant gene becomes more abundant. The example of the peppered moth later in this section is an illustration of the swift growth of a mutant advantageous gene after substantial changes in the environment. The other major context in which selection is important is in plant or animal breeding in which those offspring with the best performance in one generation are used as stock for further selection by the breeder in the next. The question then is: given a particular selection regime, how quickly can a breed be improved?

Before we consider the more general formulation for one locus and two alleles we introduce a simple illustrative example. Suppose we have a population of 1000 individuals, in which the relative frequency of the allele A is 0.4. Then the relative frequency of the alternative allele, a, would be $1 - 0.4 = 0.6$. As we are assuming that these individuals are the result of random mating, the relative frequencies of the genotypes will satisfy the conditions for Hardy–Weinberg equilibrium, and will therefore be 0.16, 0.48, and 0.36 for the *AA*, *Aa* and *aa* genotypes respectively, i.e. 160, 480 and 360 individuals. Note that this means that we are counting the numbers of each genotype at the zygote stage, i.e. just after the ovum has been fertilised and before any selective forces have had time to act. We also ignore any random fluctuations, and attempt to model just the trend, for the case in which generations do not overlap as in many annual species. The fitness is defined as half the number of offspring resulting from each individual of the genotype measured at the zygote stage. This fitness measure is thus the net effect of the survival from the zygote stage to birth and sexual maturity, and then conception of offspring. There is a convention that this is divided by two so that if the average fitness of the whole population is 1, the total population size will not increase. Suppose for the moment that the *A* allele is dominant, and that the fitness of the *AA* and *Aa* zygotes both take the value 2, whereas the recessive homozygote *aa* has only fitness 1. These figures mean that for each zygote in generation n, there are on average 2 for *AA* and *Aa*, and 1 for *aa*, zygotes in generation $n + 1$. The average number of *A* alleles in the next generation per individual of the *AA* genotype in this would then be 2×2, for the *Aa* genotype 2×1 as there is only one *A* allele in the genotype, and 0 for the *aa* genotype, as there are no *A* alleles in this genotype. Thus the proportion of the *A* allele in the next generation is the sum of the products of these figures and the genotype frequencies: $160 \times 2 \times 2 + 480 \times 2 \times 1 + 360 \times 1 \times 0$ divided by the total number of alleles which is $160 \times 2 \times 2 + 480 \times 2 \times 2 + 360 \times 1 \times 2$ thus giving a proportion of $1600/3280 = 0.488$. Thus in one generation, the proportion of the *A* allele has increased substantially from 0.4 to 0.488 as a result of the selective advantage of the *AA* and *Aa* genotypes.

2.4.1 A general difference equation model

It is of interest to model changes over a number of generations in a more general situation, and also to examine how changes in the relative frequency of an allele are related to the starting point, and to the selective advantage or disadvantage the allele enjoys. In the model we outline here, we first use a general formulation for the case of one locus and two alleles, *A* and *a*, with relative frequencies in the nth generation, p_n and $q_n = 1 - p_n$. As above, the fitness is defined as half the number of surviving offspring per individual of the particular genotype. Thus when we say the fitness of the genotype *AA* is f_{AA} we mean that for each zygote of this genotype, there would on average be f_{AA} zygotes resulting in the next generation. Let the fitnesses of the *Aa* and *aa* genotypes be similarly f_{Aa} and f_{aa} respectively. The calculations are summarised in Table 2.5.

Thus the proportion of allele *A* in the $(n + 1)$th generation is given by

$$p_{n+1} = \frac{p_n^2 f_{AA} + p_n q_n f_{Aa}}{p_n^2 f_{AA} + 2p_n q_n f_{Aa} + q_n^2 f_{aa}} \tag{2.1}$$

2.4.2 A coefficient of selection

The above is a completely general expression which we will take up later in the chapter. For the moment we consider a restricted formulation which was first put forward by

Table 2.5 The calculations of the change in allele relative frequency between successive generations

Genotype	AA	Aa	aa
Relative frequency	p_n^2	$2p_nq_n$	q_n^2
Fitness	f_{AA}	f_{Aa}	f_{aa}
Contribution to gene pool in next generation	$p_n^2 f_{AA}$	$2p_nq_nf_{Aa}$	$q_n^2 f_{aa}$
Contribution allele A to gene pool in next generation	$p_n^2 f_{AA}$	$p_nq_nf_{Aa}$	0
Total contribution of all alleles to gene pool in next generation from all genotypes		$p_n^2 f_{AA} + 2p_nq_nf_{Aa} + q_n^2 f_{aa}$	
Contribution allele A to gene pool in next generation from all genotypes		$p_n^2 f_{AA} + p_nq_nf_{Aa}$	

Wright. The fitness of the AA genotype is written as f without any subscript, and this is considered as the favoured genotype in any selection which occurs. The fitnesses of the other genotypes relative to this are as follows: for the aa genotype, $f(1-s)$, i.e. a smaller value as s is positive; for the heterozygote, $f(1-hs)$, which allows for the effects of dominance. When $h=0$, the fitness of the heterozygote is the same as that of the AA zygote, and A is then the dominant allele. When $h=1$, the fitness of the heterozygote is the same as the aa genotype, and A is then recessive to a. s is termed the *coefficient of selection* by some authors. $s=0$ corresponds to no selection, but typical values obtained in laboratory experiments for genes concerned with metric characters are of the order of 0.2.

In this case, the above expression becomes

$$p_{n+1} = \frac{p_n^2 f + p_nq_nf(1-hs)}{p_n^2 f + 2p_nq_nf(1-hs) + q_n^2 f(1-s)} = \frac{p_n(1-q_nhs)}{1-sq_n(q_n+2p_nh)}$$

from which it follows that the change per generation is given by

$$\Delta p_{n+1} = p_{n+1} - p_n = \frac{sp_nq_n[q_n + h(p_n - q_n)]}{1 - sq_n(q_n + 2hp_n)}.$$

Example 2.2 Increase of *Biston betularia carbonaria* in Manchester

The first recorded instance of the dark *carbonaria* form of the peppered moth, *Biston betularia*, occurred near Manchester in 1848. Presumably the mutation responsible for the change had taken place at some time since the beginning of the industrial revolution during which time air pollution had increased very substantially. Tree trunks previously covered by lichen against which the *typica* form was almost invisible to birds, had lost their lichen covering and become blackened by soot. Selective forces against the *typica* form during this time had allowed the new form to establish itself and become sufficiently locally abundant for it to be noticed by collectors. Recent experiments have demonstrated that differential selection pressure is exerted by predation by birds in the polluted and unpolluted situations. In 1848 the *carbonaria* form constituted less than one per cent of the population, but by 1898 it accounted for over 95%. What selection pressure would be necessary to ensure such a rapid change?

By trial and error using the above equations, starting with $p_1 = 0.01$, using various possible values for s, and setting $h = 0$ as the *carbonaria* form is dominant to the *typica* form, we can determine what range of values of s could be consistent with the observed change in the relative frequency of the *carbonaria* form.

Table 2.6 gives the outcome for various possible values of s.
Hence we see that a value of s of approximately 0.48 is adequate to explain the observed change in relative frequency. This is quite strong selection pressure, but it is feasible that selective predation could achieve such a value, as experiments with the two phenotypes of moth against various backgrounds with some species of birds as predators have confirmed.

Table 2.6 The proportion of the *carbonaria* form in 1898 predicted as a result of selection of strength, s, starting from $p = 0.01$ in 1848

s	p in 1898	s	p in 1898
0.00	0.01	0.32	0.91
0.04	0.07	0.36	0.925
0.08	0.29	0.40	0.935
0.12	0.57	0.44	0.943
0.16	0.74	0.48	0.95
0.20	0.82	0.52	0.954
0.24	0.86	0.56	0.96
0.28	0.89	0.60	0.965

2.4.3 An approximate differential equation model for weak selection

In the case when s is small, corresponding to weak selection, the above expression can be simplified as follows. The denominator takes the value 1 approximately, and so we get

$$\Delta p_{n+1} = sp_n q_n[q_n + h(p_n - q_n)] = sp_n(1 - p_n)[1 - p_n + h(2p_n - 1)]$$

by substitution of $1 - p_n$ for q_n. If we assume that the time between generations is small, taking the value Δt, and we regard the process as operating in continuous time and therefore drop the subscripts, we obtain

$$\frac{\Delta p}{\Delta t} = sp(1 - p)[1 - p + h(2p - 1)]$$

which can be written in the more familiar differential equation form if we let Δt become very small (i.e. $\Delta t \to 0$)

$$\frac{\mathrm{d}p}{\mathrm{d}t} = sp(1 - p)[1 - p + h(2p - 1)].$$

This differential equation can be solved by bringing all the terms in p to the left-hand side, and the lone $\mathrm{d}t$ to the right, and then using partial fractions to split the left-hand side into integrable functions.

The differential equation is considerably simplified in the case of semi-dominance, i.e. when $h = 0.5$, as then the third factor on the right of the differential equation reduces to a constant 0.5, and we are left with the same equation as we had in Chapter 1 for logistic growth:

$$\frac{dp}{dt} = \frac{sp(1-p)}{2},$$

with solution

$$p = \frac{e^{s(t-t_0)/2}}{1 + e^{s(t-t_0)/2}},$$

where t_0 is the time at which $p = \frac{1}{2}$.

2.4.4 Polymorphism

In the above examples of models of selection at a single locus, one of the homozygotes is at an advantage over the other genotypes, and so the eventual outcome is that all but one of the alleles has been reduced to a very low relative frequency. The particular locus would have become monomorphic. In the models, this outcome occurs eventually, if only slowly, even if the fitness is only slightly higher for the favoured homozygote. Yet in the genotypes of many real populations there are substantial numbers of loci at which each of a number of alleles are quite abundant. There are a number of possible reasons for this. Many populations do not live in a stable environment, so that selection pressures would vary from year to year, and the direction of selection might well vary in parallel. A second explanation advanced by Fisher is that if the heterozygote is at an advantage rather than either of the homozygotes (in the case of two alleles at a locus), both alleles can continue in the population for some time. Indeed if the conditions remain constant for long enough, a stable mixture of all three genotypes results.

We adapt the general equation (2.1) to this situation by writing the fitnesses of the various genotypes as follows. The fittest is now the heterozygote so that f_{Aa} is written as f, with $f_{AA} = f(1-s)$ and $f_{aa} = f(1-r)$. Thus s and r are the proportions by which the two homozygotes are less fit than the heterozygote. By substitution of these values into equation (2.1) we obtain

$$p_{n+1} = \frac{p_n^2 f(1-s) + p_n q_n f}{p_n^2 f(1-s) + 2p_n q_n f + q_n^2 f(1-r)}$$

$$= \frac{p_n(1-sp_n)}{1 - sp_n^2 - rq_n^2}.$$

The equilibrium solution is obtained by putting $p_{n+1} = p_n$ and solving the resulting equation. Remembering also that $q_n = 1 - p_n$, the equilibrium solution is

$$p = \frac{r}{s+r}, \qquad q = \frac{s}{s+r}.$$

Thus an equilibrium is obtained in which both alleles are present, and depending on the values of s and r, the equilibrium proportions can take any value between 0 and 1.

Example 2.3 Sickle cell anaemia and malaria

The classical example of heterozygous advantage at a single locus resulting in a polymorphism is that of sickle cell anaemia which is a serious, usually fatal, disease afflicting children in some parts of Africa. This disease occurs in the Hb^SHb^S genotype, and as most cases die in childhood, the fitness is near zero. Susceptibility to heavy infection with parasites responsible for subtertian malaria is much less among the heterozygotes Hb^AHb^S than among the remaining homozygotes Hb^AHb^A. The relative frequency of heterozygotes is generally much higher in adults than either of the homozygotes. Thus the heterozygotes have a much higher fitness than both the homozygotes. This situation only pertains in those regions in which malaria is endemic, or was so until recently. When mass movement occurred to other regions where malaria was not so prevalent, as in the movement of Africans to America, the incidence of the HB^S allele has declined.

The relative frequency of the sickle cell allele (allele a in the above formulation) is about 0.1 in those parts of Africa where malaria is endemic. Using the above model, if we assume that $r = 1$, since sickle cell anaemia is usually fatal, we can calculate that $s = 1/9$, so that malaria has been an important influence reducing the fitness of the population.

2.5 OTHER GENETIC MODELS

This chapter only treats the simplest deterministic genetic models. Models and methods in the remainder of the book which are relevant in genetics are listed in Table 2.7.

Table 2.7 Genetic topics in other chapters

Topic	Location
Probability distribution of Mendelian proportions	*Chapter* 3 Binomial distribution
Continuous Normal distribution of multifactorial traits	*Chapter* 3 Normal distribution resulting from genetic mechanism
Relation between parental and offspring traits	*Chapter* 4 Offspring and mid-parent height in humans
Testing goodness of fit of predictions from Mendel's Laws to data	*Chapter* 6 Comparison of binomial proportions with theoretical expectation
Models for random genetic variation	*Chapter* 7 Components of variance
DNA sequences	*Chapter* 13 Comparison of DNA sequences

EXERCISES

2.1 (a) The relative frequency of the condition of albinism in man is about 1 in 20 000. Assuming that this is due to a single recessive autosomal gene, determine the relative frequency of the albino gene and the relative frequency of the carriers of the gene (i.e. of the heterozygote).

(b) The frequencies of three colour types in a British natural population of snails, *Cepea nemoralis*, were given by Cain (1960) as follows: brown 88, pink 83, yellow 42. The alleles responsible are B, b', and b respectively, in order of dominance. If p, q, and r are the allele relative frequencies, write down the expected relative frequencies of the three phenotypes in terms of p, q and r. Hence estimate the allele relative frequencies.

2.2 The frequencies of presence and absence of two characters, A and B, each thought to be under the control of a single dominant gene, in a diploid flowering plant are as follows:

A+ B+	A+ B−	A− B+	A− B−
83	21	24	13

What ratios would you expect and why? What sequence of hypotheses about the genetic control of A and B do you think it would be reasonable to test using these data?

2.3 One model for the ABO blood groups in man is that they are determined by three alleles, A, B and O of relative frequencies p, q and r. The following table shows the genotype of each blood group and their relative frequencies. Assuming Hardy–Weinberg equilibrium (derived from a sample of 190 177 airmen, from Race & Sanger, 1954), determine the allelic relative frequencies.

Blood group	A	B	O	AB
Genotype	AA AO	BB BO	OO	AB
Relative frequency (%)	41.716	8.560	46.684	3.040

Is there any scope in these data for testing the agreement with Hardy–Weinberg equilibrium?

2.4 Draw a sketch of the relationships between the relative frequencies of the genotypes at Hardy–Weinberg equilibrium, for the case where there are two alleles, relative frequencies p and $q = 1 - p$, and the relative frequency, p, of one of the alleles. Over which ranges of p would each of the genotypes be the most frequent of the three? If one particular *genotype* (which we do not know) occurs in the population with a relative frequency of 0.90, can we say anything from this graph about the relative frequencies of the alleles? If a particular *phenotype* occurs in the population with a relative frequency of 0.90, what can we then say?

2.5 *Sex-linked genes* are those which occur in the X chromosome. Therefore the (heterogametic, i.e. genotype XY) males have only half the complement of sex related genes as the (homogametic, i.e. genotype XX) females. In a particular population, the relative frequencies of the alleles and hence genotypes A_1 and A_2 of a sex related gene in the males are R and S, and of the genotypes A_1A_1, A_1A_2, A_2A_2 in the females are P, H and Q.

(a) Show that the relative frequency of the A_1 allele in the whole population is $(2P + H + R)/3$. What is the relative frequency of the A_2 allele?

(b) If the relative frequency of allele A_1 in the males is $p_{m,i}$ by year i, and in the females in the same year is $p_{f,i}$, then show that

$$p_{f,i} - p_{m,i} = -\tfrac{1}{2}(p_{f,i-1} - p_{m,i-1}).$$

(c) Describe the progress to equilibrium of the gene relative frequencies in the males and females. (See Falconer, 1977, pp. 17–19 for further discussion.)

2.6 By plotting out Δp, the change in frequency of the allele A per generation, against the allele frequency p in the cases of dominance, semi-dominance and zero dominance, demonstrate the relative importance of p, and the degree of dominance for the rate of change of allele relative frequency, in the case when $s = 0.1$, where s is the coefficient of selection (i.e. the fitness of the AA genotype is f, and the fitness of the aa genotype is $f(1-s)$).

2.7 The differential equation for the progress of weak selection (from Section 2.4.3), where s is the coefficient of selection, is

$$\frac{dp}{dt} = ps(1-p)[1-h+p(2h-1)] .$$

Solve the differential equation by collecting terms in p together on the left hand side, and writing the resulting expression in terms of partial fractions.

2.8 *Selection balanced by mutation.* In Section 2.4.2, the relationship

$$\Delta p_{n+1} = p_{n+1} - p_n = \frac{sp_n q_n[q_n + h(p_n - q_n)]}{1 - sq_n(q_n + 2hp_n)} .$$

is derived between the change in relative frequency of allele A between the n and $(n+1)$th generation, where s is the coefficient of selection against the a allele, and h takes different values depending on whether A is completely or partially dominant to a or not.

(a) If allele A mutates to allele a at rate μ, derive the modified form of this relationship.

(b) Find the equation satisfied by the equilibrium solution.

(c) Find the solution in the cases: (a) recessive mutant, i.e. $h = 0$, and (b) semidominance, $h = 0.5$. (See Crow, 1986, pp. 86–87 for further discussion.)

3 Simple Stochastic Models of Genetic, Environmental and Sampling Variation

3.1 INTRODUCTION

This and the following chapter are about models which include a random element. Random phenomena are also often known as stochastic, from the Greek *stokhastikos*, which means 'capable of guessing'. The characteristic feature of a random or stochastic model is that when we simulate the process on a computer, starting from a fixed starting point, repeated simulations will be different. We cannot predict with any certainty which particular outcome will occur in any individual simulation. However, if we carry out a very large number of simulations, then a stable pattern will eventually emerge: we usually see what is the range of possibilities, and what is the proportion (i.e. the *probability*) amongst the totality of outcomes, of each possible outcome. This chapter is about how we can model such systems which are irregular or uncertain, with a view to uncovering the potential shapes which would have emerged had we collected more and more data, but which are often obscured by the scatter resulting from restrictions on time or resources, or even imposed by the meagreness of the material which nature provides.

We confine our attention in this chapter to models which entirely consist of random components, and which we term simple or homogeneous stochastic models. Stochastic dynamic models, i.e. models of random processes in time, and models which also include systematic relationships such as, for example, a trend towards increased growth as concentration of nutrients increases, are dealt with in the next chapter on *Structured Stochastic Models*.

Sources of variation

Random variation can arise in many ways in biological systems, and we now briefly consider some of them.

1 *Internal sources: genetics, physiological fluctuations*. Many biological processes vary in such a haphazard way that it is natural that we should think of them as inherently random, i.e. their course, even over the next few seconds, cannot be predicted with certainty even if we have a complete and accurate knowledge of the state of the process and all the variables which affect it up to the present. We have seen in the previous chapter some of the deterministic laws of genetics applicable to the mass behaviour of genetic mechanisms. Yet genetic recombination, perhaps the most fundamental of biological processes, involves randomness, the effects of which are often immediately observable. The colour of the seeds from a particular cross of two

pea varieties such as those used by Mendel cannot be predicted other than down to a set of possibilities. Other qualitative characteristics of offspring which are under partial or virtually complete genetic control and hence unpredictable are sex, eye colour and blood group.

2 *External sources: the physical and biotic environment*. Organisms are also affected by their physical environment, many elements of which exhibit so little systematic structure that they can be regarded as almost purely stochastic. A prime example of unpredictability is the weather in England, but there is random variation in weather on some time scale in almost all countries. Often there are also substantial chance fluctuations in the biotic environment. This applies to multi- and single-species communities, individual organisms, parts of organisms from whole organs or compartments such as a gland or the blood stream, right down to cells, molecules and ions taking part in smaller scale interactions.

3 *Measurement error*. Whenever an observation is made on a continuous variable, some error is involved in the measurement. Errors in counts of clearly defined objects may be infrequent when the count is low, but occur more often as the count gets higher. In most fields of science, considerable efforts are made to keep errors as small as possible, and often the errors are so small that they can be ignored. This is also true in some, but not all, areas of biology. When measurements involve many stages or utilise biological material, it is quite common for the errors to be appreciable. Two components can be identified, a consistent systematic error or bias and a randomly varying component. A common way of reducing measurement error is to average out many independent measurements. This can reduce the random part to an acceptable level, but will not remove any bias. Often though, it is necessary to accept that random errors will occur, and to tailor the statistical treatment to accommodate them. It is impossible—without some additional information which might be used for adjustments—to use statistical methods to accommodate bias.

4 *Artificial sources: random sampling*. In many circumstances, it is impossible or very difficult to carry out a total census of a large population. Often a complete census or description is unnecessary anyway, as sufficiently accurate information can be obtained from a sample. Methods for obtaining an adequate sample and for utilising sample information are covered in Chapter 7, but it is appropriate to mention here the value of random sampling. First, we should say that systematic samples are often the best for studying the structure of a population or mechanism. But in cases where the major interest is in the population mean, a random sample is usually more useful. A strength of random sampling and statistical methods is that whatever the form of the population—and often we are in relative ignorance about it—the use of a random element in the selection of units to include in the sample will enable an estimate of the population mean of known accuracy to be obtained.

Reasons for measuring random variation

There are two views which can be taken about the random variation affecting the phenomenon under study, or the measurements we make of it: briefly stated, either it is a nuisance or it is interesting.

1 *Random variability as a nuisance*. In many studies in applied biology, and in much of classical statistics as applied, for example, to agriculture, the overall trend is

what is of interest, and the variability is a nuisance which somehow has to be filtered out of the data so that the underlying systematic pattern can be revealed. In many agricultural experiments, the object is to find treatments which optimise the total yield of a crop, first of all over the area of a field of an experimental farm, and then later in further trials over a whole region, or a whole country. The variability in the yield from plot to plot, from field to field, or from season to season is generally not of primary interest in its own right. It is measured only because it is required to determine whether a particular treatment is in reality better than another, or whether it is just a fluke of the sampling; and secondly to determine the likely margin of error on the predictions of the average yield under the optimum treatment (see Example 3.9).

2 *Random variability of primary interest*. The view that variability is a nuisance is quite common, but for a biologist rather shortsighted since the variation is often of great interest (see Examples 3.1, 3.2, 3.3 and 3.10). This view probably arises in the physical sciences, where variation is usually due to measurement error, and not an integral feature of the phenomenon studied. Statistical terminology can also be misleading: terms such as 'error variance' and 'error distribution' are often used when actual measurement error constitutes only a minor part of the variation. Sometimes it seems that the statisticians are saying that the biological material is in error for being so variable! In other areas such as genetics and breeding studies, the random variability is known to be very important in its own right.

3.2 SUMMARISING AND PRESENTING STOCHASTIC DATA

Before we discuss models for random variation using examples, we need to outline methods by which we can present the examples involving sample data. In this section, we consider two small samples of data: counts of the number of cells found in the fields of a haemocytometer, and the adult heights of a sample of people. It is not so clear as it is for data on patterns of growth and decline how best to present sample data. We could present both sets of data in the form of lists of values as obtained by the experimenter. Suppose we obtained the following data, in the two cases.

Haemocytometer counts:	3, 5, 0, 2, 2, 2, 1, 4, 2, 1, 0, 0, 3, 3, 1, 2, 5, 1, 0, 2
Heights:	124, 142, 181, 192, 129, 141, 173, 157, 105, 133, 127, 196, 164, 163, 164, 179, 148, 143, 150, 169, 165, 188, 93, 172, 152, 121, 201, 187, 167, 161, 152, 177, 164, 174, 195, 166, 169, 200, 181, 138.

3.2.1 Tables of grouped data and histograms

A first step in presenting these data, without any loss of information except the order in which the numbers occurred, is to group them in approximately increasing order, and to present all those within particular ranges on one line, or in one column, as in Tables 3.1 and 3.2(a).

These are complete listings of the data, but they are much more informative than the first ones. One can see at a glance where most of the data lie, and whether there

Table 3.1 Grouped haemocytometer data

0 0 0 0	
1 1 1 1	
2 2 2 2 2	2
3 3 3	
4	
5 5	

Table 3.2 Height data

(a) Grouped height data		(b) Stem-and-leaf plot of height data		
		Stem	Leaf	
93		90	3	
105		100	5	
.		110	.	
124 129 127 121		120	1 4 7 9	
133 138		130	3 8	
142 141 148 143		140	1 2 3 8	
157 150 152 152		150	0 2 2 7	
164 163 164 169 165	167 161 164 166 169	160	1 3 4 4 4	5 6 7 9 9
173 179 172 177 174		170	2 3 4 7 9	
181 188 187 181		180	1 1 7 8	
192 196 195		190	2 5 6	
201 202		200	0 1	

are any exceptionally large or small values and, as the complete data are given, readers can perform their own analysis if they wish. These tables are closely related to *stem-and-leaf plots*: the stem coincides with the baseline of each column (90, 100, . . . in the second example), and the leaves are the extra digits added to each baseline number (3, 5 etc.), as illustrated in Table 3.2(b).

A more conventional simple presentation of data is a *histogram*. For the haemocytometer data, we first get a list of the values which the count, R, has taken, and alongside each particular value, we also list the *frequency* among the whole list of counts which took the value r. For example for the list above, we would get the results in Table 3.3.

We can then plot out these numbers as in Figure 3.1(a), a series of columns of width one centred on the discrete values 0, 1, 2, 3, 4, 5 and heights equal to the frequencies in the data of the values 0, 1, 2 etc.

Table 3.3 Counts of frequency of each value for constructing a histogram for the subsample of haemocytometer data

Observation, r	0	1	2	3	4	5	6 and higher
Frequency	4	4	6	3	1	2	0
Proportion	0.20	0.20	0.30	0.15	0.05	0.10	0.00

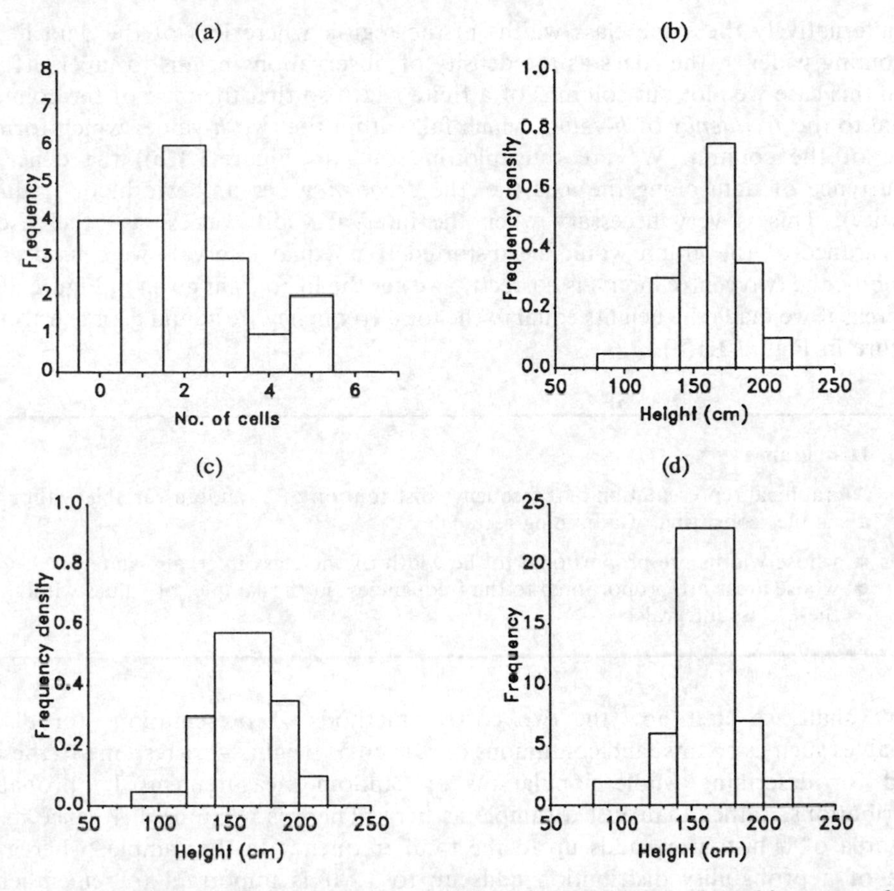

Figure 3.1 (a) A histogram of the haemocytometer data in Table 3.1. (b) A histogram of the height data in Table 3.2. (c) A histogram of the height data with the centre intervals grouped. (d) An illustration of the error due to making the column lengths, rather than column areas, equal to frequency when using unequally spaced class divisions.

For the data on heights something different is required as, supposing that height is measured to the nearest *cm*, the list of possible heights would be very long, and, unless we have very many observations, most of the cells in a table like that above would only have one or zero observations in them. So in this case, we group adjacent cells so that we might obtain a table such as Table 3.4. There is nothing special in this particular grouping; however, it is common to have uniform class widths over the whole range,

Table 3.4 Counts of frequencies of values within the ranges given for constructing a histogram of the sample of height data

Measurements, h	80–119.9	120–139.9	140–159.9	160–179.9	180–199.9	200–220
Frequency	2	6	8	15	7	2
Histogram height	2/40	6/20	8/20	15/20	7/20	2/20
Proportions	0.05	0.15	0.20	0.375	0.175	0.05

or alternatively the same class widths in the region where lots of the data lie, but becoming wider at the edges as the density of observations begins to taper off.

In this case we plot out columns of a finite width so that the *area* of the column is equal to the *frequency* of h-values which fall within the two h-values which form the base of the column. We are thus plotting out (in Figure 3.1(b)) the density of occurrence of data along the axis (i.e. the *frequency* per unit width, or frequency density). This is very necessary when the interval width varies, as otherwise the appearance of the graph would be distorted if unequal intervals were used. If we combine the two centre intervals correctly, we get the histogram given in Figure 3.1(c), whereas if we made the heights equal to the total frequency we would get the erroneous picture in Figure 3.1(d).

Histogram

A graphical representation of a frequency distribution of a random variable within a sample, consisting of adjoining rectangles

- whose widths are proportional to the width of the class intervals, and
- whose areas are proportional to the frequencies, in the sample, of values within these class intervals.

We shall see later how the two correct methods of presentation—for discrete variables such as counts, and continuous ones such as heights—correspond to the ways used for describing whole populations of random measurements by probability distributions, rather than just a sample as here. There is a minor difference in that the area of a histogram adds up to the total frequency in the sample, whereas the area of a probability distribution adds up to 1. It is important to remember the distinction between for example a histogram derived from a sample and the specification of the whole population or of a mechanism. We discuss this more fully in Chapter 5.

3.2.2 Sample cumulative frequency and distribution functions

A further very convenient method of presenting data such as these, which does not require decisions about grouping intervals is to plot out a cumulative frequency function of the data; in this case because it is a *sample* of data, we use the term *sample cumulative frequency function*. The curve amounts to a running tally of the number of cases encountered as we move from left to right along the axis, i.e. from lower to higher values of the random variable. Consider the sample of haemocytometer data: 4 zeros, 4 ones, 6 twos, 3 threes, 1 four and 2 fives, twenty numbers in all. Then the sample cumulative frequency function is obtained by plotting the value of r along the x-axis, and the cumulative frequency up to and including the value of r along the y-axis. So the figure consists of a number of steps: the first of height 4 at $r = 0$, the second of additional height 4 at $r = 1$ taking us up to 8, the third of further height 6 at $r = 2$ making 14, and so on as in Figure 3.2.

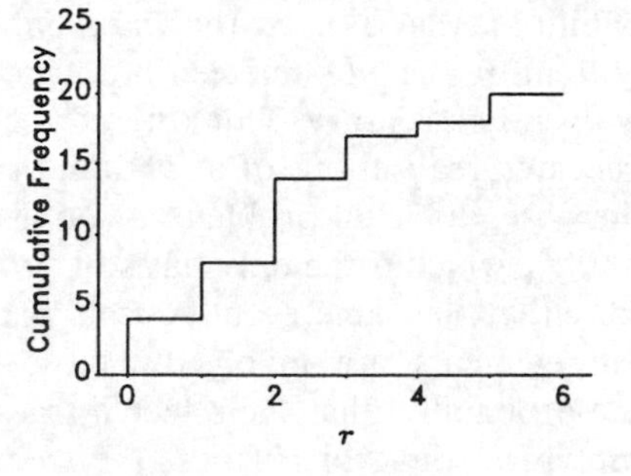

Figure 3.2 Cumulative frequency function for the sample of haemocytometer data.

Sample cumulative frequency function of r

A step diagram displaying the number of cases in a sample (plotted in the y-direction) which have r-values up to and including the value given along the x-axis. The height of each step is the number of values within the sample *at* each r-value.

The *sample cumulative distribution function* is identical to the sample cumulative frequency function except that the y-scale runs from 0 to 1, rather than 0 to n, the total sample size, and the steps are correspondingly in multiples of $1/n$ rather than 1. This function is also called the *empirical cumulative distribution function*.

3.3 PROBABILITY AND RANDOM VARIABLES

In the classic experiments of Mendel on peas, he produced various series of F2 crosses of peas with round and angular seeds. In one plant he obtained 26 round and 6 angular seeds; over a large number of his experiments he obtained 5474 round and 1850 angular seeds. His theory predicted that round and angular seeds would occur in the proportions 3:1. But he would not have expected exact agreement between his theoretical predictions and the results of his trials, especially in as small a sample size as 32, which he obtained in one plant. The proportion of round, and hence the proportion of angular seeds obtained by subtraction, are random variables with variation from plant to plant, and batch to batch of plants. How can we quantify such variation? We need to do this in order to be able to say whether the results of any particular experiment agree with a theory or are in conflict with it. We said above that the expected proportion of round seeds would be 0.75. That is to say that in a very large number of F2 crosses, say of the order of many millions, the proportion of round seeds will be 0.75, and of angular seeds 0.25. Another way of saying this is that the probability of obtaining a round seed is 0.75, and the probability of an angular one is 0.25.

Probability

The long-run *relative* frequency or proportion in a very large number of 'identical' repetitions of the *random* process or trial.

This definition of probability is that used by the *Frequentist* school of statisticians. It was first formulated by Venn in the late nineteenth century, and clarified by Fisher and von Mises in the early twentieth century. Thinking of probability by using a model of repeated sampling or repeated realisations of a random process in this way clarifies many confusing and otherwise difficult problems. von Mises (1931) states: 'The rational concept of probability, which is the only basis of probability calculus, applies only to problems in which either the same event repeats itself again and again, or a great number of uniform elements are involved at the same time.' Thus it is meaningless to say that the probability that the selected tree is an oak is 0.5, without also saying or at least implying from which forest the tree is selected, and without saying how we select a tree from the forest in a random manner.

There are thus three important aspects of probability which are outlined in Table 3.5: the *trial* (sometimes termed *simple experiment*—which need have nothing to do with a real experiment), the *sample space* and the *event*. The *trial* is the process we use for repeatedly obtaining outcomes, which includes selecting a unit from a population and making a defined measurement on it. The *sample space* is the total list of possible outcomes of the trials, a subset of which have the characteristic of interest and therefore constitute the *event* whose probability we are calculating. von Mises's insistence that the elements or trials be uniform seems guaranteed to produce a predictable outcome on which the use of the idea of probability is wasted. However, there is an implied *apart from a random element*, which injects some 'controlled' variability in the process. The beginner in probability and statistics usually has an uneasy feeling that there is something contradictory about the twin elements of this definition: the control (e.g. the probability of a head is 0.5) and the random variation (the outcome of any finite number of tosses of a coin cannot be exactly predicted). A leap of the imagination is required before the frequentist or any other notion of probability and randomness can be used. The main justification is that probability has been very fruitfully applied, especially in areas such as biology where random variation is especially common. The leap therefore needs to be made, but identifying the

Table 3.5 Trials, sample space, and event: definition and examples

Trial The process we use for repeatedly obtaining outcomes	*Sample space* The set of *all* possible outcomes.	*Event* The subset of the sample space which we are concerned with.
Examples of trials	*Examples of sample spaces*	*Examples of events*
(1) The species of a tree selected at random from Epping Forest on a particular date.	(1) {Oak, ash, beech, . . .}	(1) The tree is an oak
(2) The genotype of a randomly selected seed from a cross between two plants randomly selected from two pure homozygous genotypes, AA and aa	(2) {AA, Aa, aa}	(2) The genotype of the seed is Aa.

elements in Table 3.5 in any individual case can help by giving the notion a more mechanistic interpretation. The rules for handling probabilities are given in an appendix in Section 3.13.1 for those who need a firmer grounding or merely more experience before going on. A further preliminary which is required at this point is to introduce the term random variable and some conventions about its use.

Random variable

A variable which is subject to chance fluctuations and which can take a range of values.

Random variables are often written as capital letters, and in this book usually italicised, e.g. X, R. The values taken by the random variable if they are not specified numerical values are usually written in lower case, e.g. x, r. The x and r are here standing in for numbers such as 2.5 and 9.

3.4 THE BINOMIAL DISTRIBUTION

3.4.1 An example from Mendelian genetics

If we have a small number, n, of seeds, what are the probabilities that $0, 1, 2, \ldots, n$ of them are round, as opposed to angular? For the sake of definiteness, let us use the value that Mendel obtained in one plant, $n = 32$. To put the question in another way, if we repeated this whole process of obtaining batches of 32 seeds (perhaps from one plant each time, perhaps from a number of plants, but by crossing in exactly the same way under identical conditions) a very large number of times, in what proportion of repetitions would we obtain $0, 1, \ldots, 32$ round seeds? We would not usually actually perform a whole series of trials of size 32, only one, but whenever we speak of probabilities of outcomes of a trial we are—in our imagination—embedding this trial in an infinitely large series of hypothetical repetitions of the trial, and the probabilities of the various outcomes are simply the proportions of those outcomes in this very long series.

Example 3.1 Mendel's genetic data on peas

The importance of Mendel's experiments is that they demonstrated that inheritance of some characters follows a very simple mechanism. We saw in the last chapter that the overall proportion of angular seeds out of the total was approximately what was predicted by his theory. We can also check the variability of the proportion from plant to plant for consistency with a mechanism of random donation of alleles from each of the parents. The following data (in Table 3.6) on the proportion of round seeds, were obtained from three plants all with the same total number of seeds, $n = 32$. The overall proportion of round seeds is 0.76 which is quite close to the expected value under Mendel's Laws of 0.75. There is variability of the proportions from plant to plant, but they cluster round 0.75. If Mendel's theory is correct and the same mechanism operated in all the plants, the variability in the proportion round would follow a particular distribution, which we examine in this section.

Table 3.6 Mendel's F1 crosses concerned with shape of seed

Plant	Shape of seed		Total	Proportion round
	round	angular		
a	26	6	32	0.81
b	22	10	32	0.69
c	25	7	32	0.78
Total	73	23	96	0.76

The trial might consist of just one seed, in which case the outcome is either 0 or 1 round seed; when we said above that the probability of a round seed is 0.75, we meant that the proportion of a long run of trials (i.e. single seeds) in which the outcome is a round seed is 0.75. In order eventually to answer the question where $n = 32$, let us first consider $n = 2$. Table 3.7(a) shows the calculation of the probabilities that 0, 1 or 2 seeds will be round. In the left hand column we list the events which are possible outcomes, i.e. the sample space: 2R, 1R1A, 2A, and then in column 2 we give the associated values of the random variable, the number of round seeds, r. In column 3, we list the possible ways in which we can obtain the event in the first column. For two round seeds there is only one way. Column 4 gives the probability of this happening: it equals (0.75)(0.75), since at each stage there is a probability of 0.75 of a round seed occurring. For two angular seeds there is similarly only one way; and the probability is (0.25)(0.25), since the probability of an angular seed at each stage of the trial is 0.25. But for 1R1A, there are two possible ways: we could draw the round seed first or the angular seed first; and the probabilities of these are (0.75)(0.25) and (0.25)(0.75) respectively. In the final column these are collected together, to give 0.75^2 for $r = 2$, (2)(0.75)(0.25) for $r = 1$ and 0.25^2 for $r = 0$.

If now we take four seeds in each trial in a similar way, the calculations are as in Table 3.7(b). As for trials with $n = 2$, in column 3 we list the possible specific ordered outcomes which result in the r-value given in column 2, and then in column 4 give the probabilities of each of the ordered outcomes. You will again notice that the probabilities associated with any given r-value are all the same—the factors 0.75 and 0.25 occur in a different order in each line of the block, but the resulting individual probabilities (products made up of the individual factors, 0.25 and 0.75) are the same. Take a particular block of the table, that for $r = 3$. All of the probabilities of ordered outcomes take the same value, which is $(0.75)(0.75)(0.75)(0.25) = 0.1055$. In the final column we have this product, multiplied by a further term, which is the number of different orders of occurrence of the event 3R1A which is 4. This calculation of the number of different orders is easy in the case of $r = 3$: there are just 4 positions in the ordering for the single angular seed. For $r = 2$, it is more complex, and is dealt with in an appendix to this chapter (Section 3.13.2). But you will see that the number given in column 5 corresponds to the number of orders listed in column 3.

3.4.2 General formula for the binomial distribution

In order to answer the question we posed at the beginning of the previous section about the number of round seeds in Mendel's trials consisting of $n = 32$ seeds, in this section

Table 3.7 (a) A demonstration of the calculations of the probabilities of a binomial distribution with parameters $n = 2$ and $p = 0.75$

1. Event	2. Value of random variable R	3. Event in terms of specific ordered outcomes	4. Probabilities of ordered outcomes	5. Sum of probabilities of ordered outcomes
2R	$r = 2$	RR	(0.75)(0.75)	$0.75^2 = 0.5625$
1R1A	$r = 1$	RA	(0.75)(0.25)	$(2)(0.75)(0.25) = 0.375$
		AR	(0.25)(0.75)	
2A	$r = 0$	AA	(0.25)(0.25)	$0.25^2 = 0.0625$

(b) A demonstration of the calculations of the probabilities of a binomial distribution with parameters $n = 4$ and $p = 0.75$

1. Event	2. Value of random variable R	3. Event in terms of specific ordered outcomes	4. Probabilities of ordered outcomes	5. Sum of probabilities of ordered outcomes
4R	$r = 4$	RRRR	(0.75)(0.75)(0.75)(0.75)	$0.75^4 = 0.3164$
3R1A	$r = 3$	RRRA	(0.75)(0.75)(0.75)(0.25)	$(4)(0.75^3)(0.25) = 0.4219$
		RRAR	(0.75)(0.75)(0.25)(0.75)	
		RARR	(0.75)(0.25)(0.75)(0.75)	
		ARRR	(0.25)(0.75)(0.75)(0.75)	
2R2A	$r = 2$	RRAA	(0.75)(0.75)(0.25)(0.25)	$(6)(0.75^2)(0.25^2) = 0.2109$
		RARA	(0.75)(0.25)(0.75)(0.25)	
		RAAR	(0.75)(0.25)(0.25)(0.75)	
		ARRA	(0.25)(0.75)(0.75)(0.25)	
		ARAR	(0.25)(0.75)(0.25)(0.75)	
		AARR	(0.25)(0.25)(0.75)(0.75)	
1R3A	$r = 1$	RAAA	(0.75)(0.25)(0.25)(0.25)	$(4)(0.75)(0.25^3) = 0.0469$
		ARAA	(0.25)(0.75)(0.25)(0.25)	
		AARA	(0.25)(0.25)(0.75)(0.25)	
		AAAR	(0.25)(0.25)(0.25)(0.75)	
4A	$r = 0$	AAAA	(0.25)(0.25)(0.25)(0.25)	$0.25^4 = 0.0039$

we discuss a general formula—called the *Binomial distribution*—which we can use for any n. Following the logic we outlined above justifying a specific entry of Table 3.7, we might guess at the general formula, using the notation that $p = P(\text{round})$ which in this example is 0.75, and $1 - p = P(\text{angular}) = 0.25$:

$$P(r \text{ round seeds in } n \text{ trials}) = \begin{pmatrix}\text{no. of different ways} \\ \text{of selecting } r \text{ from } n\end{pmatrix} p^r(1-p)^{n-r}$$

The first quantity—the number of different ways of selecting r different objects from a list of n—is written as ${}^nC_r$ or $\binom{n}{r}$ and is also referred to as the number of combinations of r from n (Section 3.13.2).

So we have arrived at the following general definition. For completeness, we quote below the probability distribution the mean and variance which are explained in Section 3.7.

Binomial distribution

$$R \sim \text{binomial}(n, p)$$

Probability distribution $\quad P(R = r) = {}^nC_r p^r (1-p)^{(n-r)} \quad$ for $r = 1, 2, \ldots, n$

Mean and Variance $\quad E[R] = np \qquad \text{var}[R] = np(1-p)$

Before moving on, we need to define probability distribution in more general terms.

Probability distribution of a discrete random variable

A specification of the values which the discrete random variable can take, together with a list of the probabilities associated with those outcomes.

The distribution can be specified by a list of numbers, but more usually it is specified by a mathematical equation. This is just a general shorthand for the two lists of numbers: the outcomes and their probabilities.

3.4.3 Properties of the binomial distribution

Figure 3.3 gives visual representations of the distribution for various values of n and p. As r only takes integer values, the probabilities are drawn as spikes of length equal to the probability placed above the relevant integer value. The total length of the spikes is therefore 1.

The ranges of the distributions vary, as the maximum value in each case is n. But within those ranges, the shape of the distribution depends on the value of p. For $p = 0.5$, the distribution is symmetric for all values of n. For $p = 0.1$, the distributions are positively skewed, i.e. they tend to have high probabilities at low values of r, and lower probabilities at higher values. Another way of saying this is that the *mode* of the distribution—the value of r which has the highest probability associated with it—is not in the centre of the distribution but to the left. For $p = 0.5$, the mode is in the centre, and for $p = 0.75$, the mode is nearer to the right. It will be noticed though that for the larger n values, the distribution is more symmetrical, although for $p = 0.1$, the distribution is still quite asymmetrical for $n = 12$.

The binomial distribution is widely used in many areas of application throughout biology and statistics. An example of its application in the theory of random walks is given in Chapter 12.

Example 3.2 Yule's pea data

The following data (Table 3.8), collected by Yule, give the frequency of occurrence of double dominant YR peas in each of 269 four-seeded pods. Figure 3.4 demonstrates that the binomial distribution is a good fit to Yule's data.

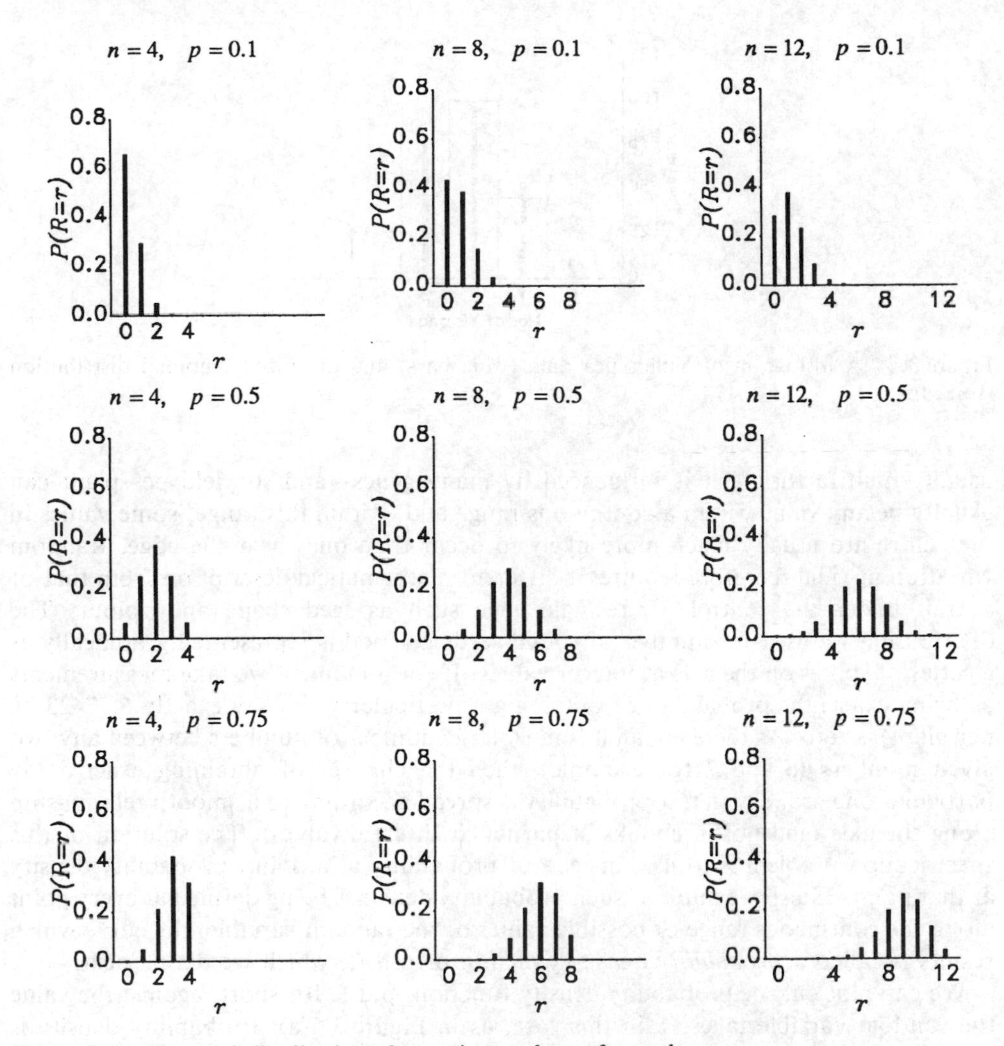

Figure 3.3 Binomial distributions for various values of n and p.

Table 3.8 Yule's data on frequency of YR peas in four-seeded pods

No. YR peas per pod	0	1	2	3	4	Total
No. of pods	16	45	100	82	26	269

3.5 PROBABILITY DENSITY FUNCTION OF CONTINUOUS RANDOM VARIABLES

In the previous section, we outlined a model for the distribution of a discrete, integer-valued random variable such as the number of YR peas in a four-seeded pod. In many breeding experiments, we are interested in the inheritance of continuous variates such as yield per plant. Yield differs from seed colour and shape in that its inheritance is

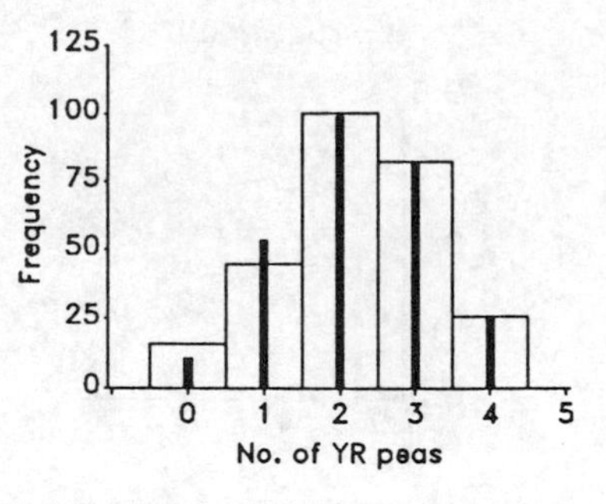

Figure 3.4 A histogram of Yule's pea data (wide bars) and the fitted binomial distribution (spikes).

usually multifactorial—it is influenced by many genes—and so yield per plant can usually be any value within a continuous range and, within this range, some values in the centre are usually much more likely to occur than ones near the edge. Random variation in yield therefore requires a different mathematical description from that of a trait under the control of a single gene such as seed shape and colour. The distribution we used for number of round seeds per pod is represented graphically as a series of spikes on the axis at integer values. If for a moment we take measurements as being exact, the probability of obtaining a particular yield, say exactly 6.77423 kg per plant, is zero. As there are an infinitely large number of numbers between any two given numbers (6 and 7 for example), then the chances of obtaining exactly any particular one are very slim if probability is spread according to a smooth relationship along the axis (and not in chunks at particular discrete values). The solution of this specification problem is to use, instead of probability at a point, probability density at that point. If also we think of such probability densities being defined at every point along the continuous range of possible values of the random variable, the curve which results is called a *probability density function* or p.d.f., which we define next.

We can plot out the probability density function, p.d.f. for short, against the value the random variable takes, x in this case, as in Figure 3.5(a). Probability density is

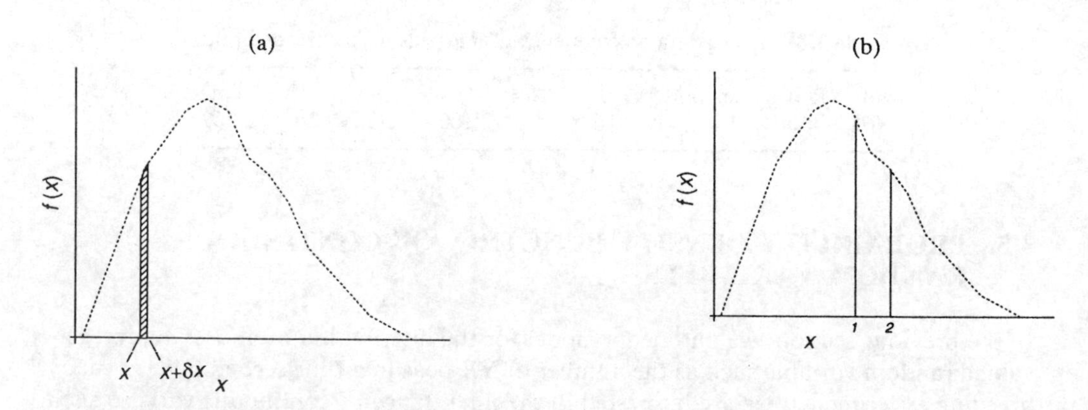

Figure 3.5 Demonstration of the idea of probability density function.

calculated at any point x by taking a very narrow strip of width δx starting at x, and dividing the area under the p.d.f. curve between these two values by the strip width δx.

Probability density function of a continuous random variable (p.d.f.)

A function $f(x)$ giving the density of probability, i.e.

$$f(x) = \frac{P(x < X < x + \delta x)}{\delta x} \quad \text{for small } \delta x.$$

As a consequence of this definition, we can obtain probabilities from the curve in two ways. If we wish to determine the probability that X lies in a very narrow interval, we can obtain an approximate answer by multiplying the value of the p.d.f. at the start, or better the mid-point, of the interval by the interval width. Secondly, we can calculate the probability that the random variable, X, lies between any two values of x, e.g. $x = 1$ and $x = 2$, as in Figure 3.5(b), by determining the area under the curve which lies between the two values. The total area under the curve equals 1, as the probability that X lies somewhere along the axis is 1. The two requirements for a curve to represent a valid probability density function are that the total area under the curve is 1, and that it is never less than zero. For a *probability distribution* of a *discrete* variable (such as the binomial distributions in Figure 3.3), there is similarly a minimum of zero, but also a maximum of 1. Probability density on the other hand can take values larger than 1, as only areas under the curve represent probabilities, and the p.d.f. can take values much greater than 1 over finite ranges, and yet the total area can still add up to 1. For an example of this, consider the p.d.f. of a negative exponential distribution with mean = 0.5 in Figure 3.14(a). You will notice from this discussion that even though we specify the way in which probability is distributed along the axis for continuous random variables by a probability density function, the term distribution is still used.

3.6 THE NORMAL OR GAUSSIAN DISTRIBUTION

3.6.1 Definition

We said above that the random variation in yield can be described by a probability density function, without specifying the specific form that the probability density function takes. In this section we discuss the *Normal* or *Gaussian distribution*, a very widely used continuous distribution, which describes the variation in yield in breeding experiments, and many other continuous random variables. Example 3.3 gives an application of the distribution to human height, and Sections 3.12.1 and 3.12.2 discuss other situations in which the Normal distribution is applicable. It is a bell-shaped distribution with the mathematical form

$$\frac{1}{\sigma\sqrt{2\pi}} \exp\left[-\frac{1}{2}\left(\frac{x-\mu}{\sigma}\right)^2\right]$$

The expression inside the square brackets is a quadratic, concave downwards function of x with its maximum at $x = \mu$ (see Figure 3.6(a)). Taking the exponential function of

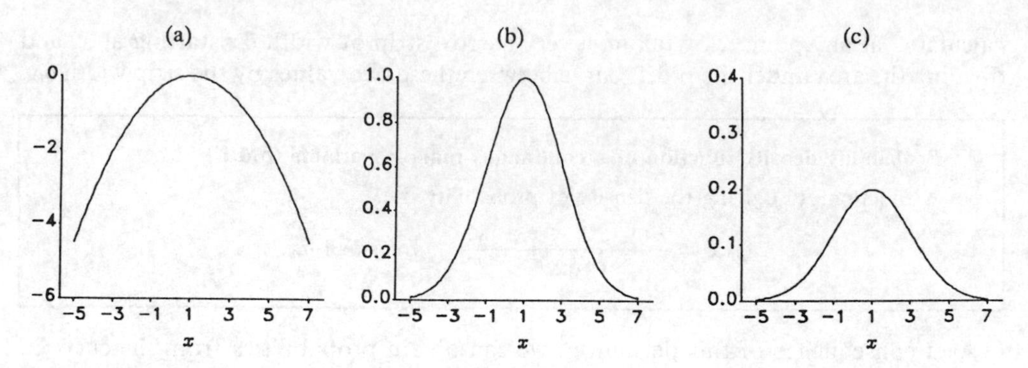

Figure 3.6 (a) the function $-\frac{1}{2}(x-\mu)^2/\sigma^2$, in the case when $\mu = 1$ and $\sigma = 2$; (b) the exponential transform of this, $\exp[-\frac{1}{2}(x-\mu)^2/\sigma^2]$; (c) the function scaled by the $1/\sqrt{2\pi}\,\sigma$ so that the area under the curve equals 1.

this expression produces a positive bell-shaped function, as in Figure 3.6(b). In order for the total area under the curve to be 1, we need to multiply by the constant $1/(\sqrt{2\pi}\sigma)$ so that the density function becomes as given above, and as illustrated in Figure 3.6(c).

Normal distribution

$$X \sim N(\mu, \sigma^2)$$

Probability density function (p.d.f.)

$$f(x) = \frac{1}{\sqrt{(2\pi)}\sigma} \exp\left(-\frac{1}{2}\frac{(x-\mu)^2}{\sigma^2}\right), \qquad -\infty < x < \infty$$

Mean and Variance $\quad E[X] = \mu \quad \text{var}[X] = \sigma^2$

Standardised Normal distribution

$$Z \sim N(0, 1)$$

$$Z = \frac{(X-\mu)}{\sigma}$$

Probability density function (p.d.f.)

$$f(z) = \frac{1}{\sqrt{2\pi}} \exp(-\tfrac{1}{2}z^2), \qquad -\infty < z < \infty$$

Mean and Variance $\quad E[Z] = 0 \quad \text{var}[Z] = 1$

3.6.2 The shape of the Normal distribution

Figure 3.7 illustrates a series of Normal distributions for different values of μ and σ. You will see that μ (pronounced mew) fixes the position of the *mode*—the point with the highest probability density—of the distribution on the x-axis, and σ (pronounced sigma) is related to the spread, the higher the value of σ the more spread out the distribution is. As the distribution is symmetric the mode of the distribution is also the

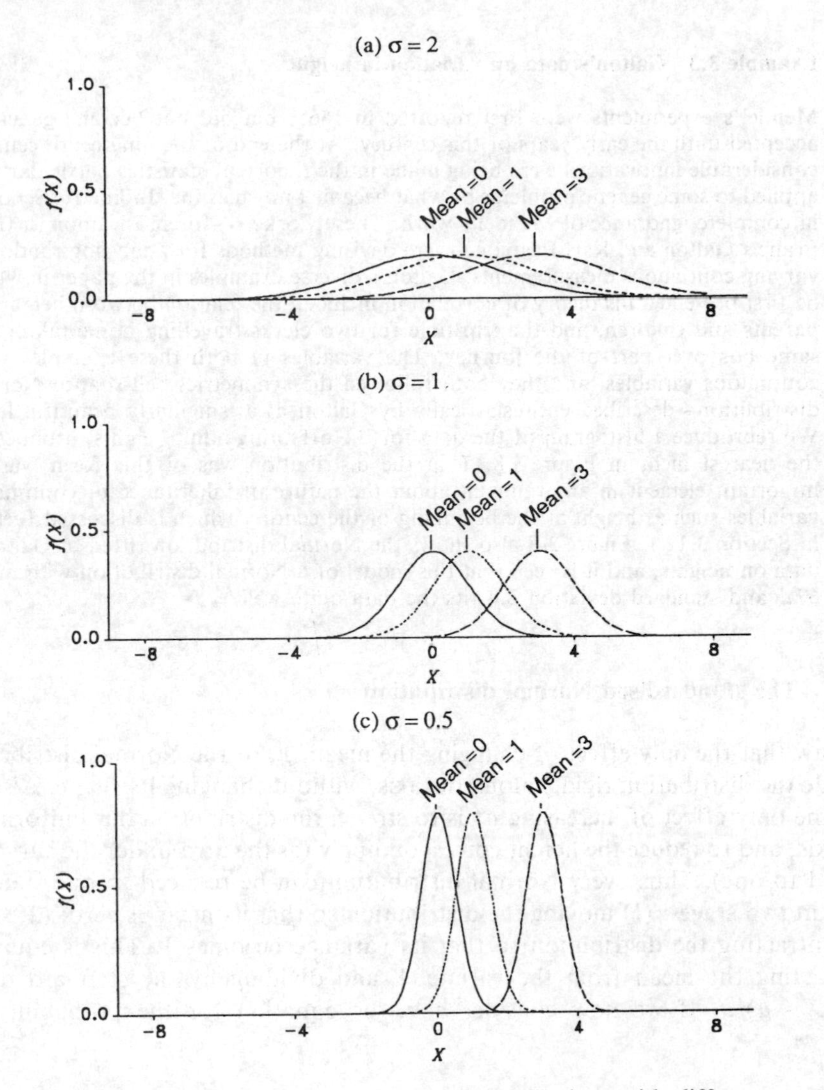

Figure 3.7 Probability density functions of Normal distributions with different means and variances.

mean. σ^2 is known as the *variance*, and σ the *standard deviation* (as defined in Section 3.7). σ could be thought for the moment as a linear measure of the spread: doubling σ results in the distribution covering twice as much of the axis. Figure 3.7 indicates that the Normal distribution is very flexible. However, what is consistent about the distribution—whatever the values of μ and σ are—is that the amount of probability between $\mu - \sigma$ and $\mu + \sigma$ is constant (and happens to equal 0.683), and that between $\mu - 2\sigma$ and $\mu + 2\sigma$ is also constant (and equal to 0.954). In fact, the probability between any two multiples of σ on either side of the mean is constant, no matter what μ and σ are. The range which includes 95% of the total probability is $\mu \pm 1.96\sigma$. This latter property is worth remembering as it is widely used in applications of the Normal distribution.

Example 3.3 Galton's data on variation in height

Mendel's experiments were first reported in 1865, but did not become generally accepted until the early years of this century. At the end of the nineteenth century considerable innovations were being made in the theory of statistics particularly as applied to some genetic problems by what became known as the 'Biometric School', in complete ignorance of Mendel's work. These workers—foremost amongst them Francis Galton and Karl Pearson—were devising methods for analysing randomly varying continuous measurements. Galton's diverse examples in the paper in which he first presented his theory of correlation included the relation between heights of parents and children, and the trip time for two clerks travelling home taking the same bus over part of the journey. The variables in both these examples were continuous variables, and they both followed the symmetric, bell-shaped Normal distribution—described enthusiastically by Galton as a 'singularly beautiful law'. We reproduce a histogram of the data for the offspring adult heights, grouped to the nearest inch, in Figure 3.8. That the distribution was of this form was an important element in an argument about the nature of inheritance of continuous variables such as height at the beginning of the century which is discussed further in Section 3.12.1. Figure 3.8 also shows the Normal distribution fitted to Galton's data on heights, and it is seen that this model of a Normal distribution with mean 67.5 and standard deviation 2.4 fits the data quite well.

3.6.3 The standardised Normal distribution

We saw that the only effect of changing the mean, μ, of the Normal distribution was to slide the distribution rigidly along the axis, without changing its shape. Also we saw that the only effect of increasing σ^2 is to stretch the distribution out uniformly along the axis, and to reduce the height correspondingly (as the area under the curve still has to add to one). Thus every Normal distribution can be reduced to one standardised form in two stages: (1) moving the distribution so that its mean is zero; (2) stretching or contracting the distribution so that its variance becomes 1. This is equivalent to subtracting the mean from the variate X and dividing by the standard deviation: $Z = (X - \mu)/\sigma$. If we now consider how the equation for the probability density

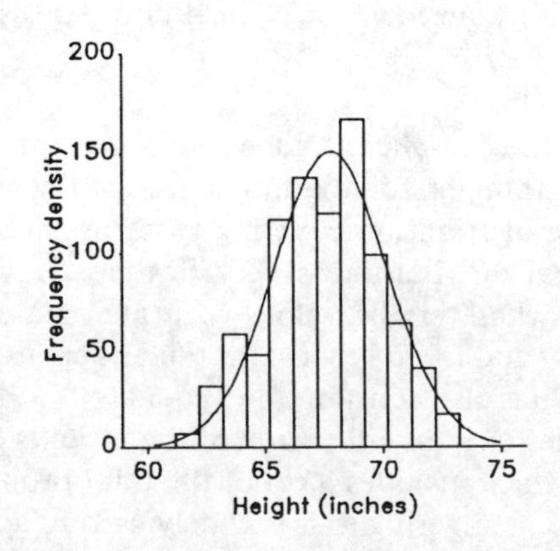

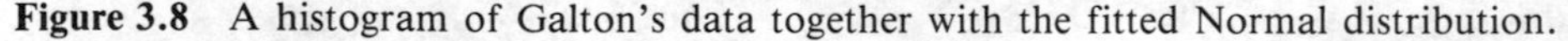

Figure 3.8 A histogram of Galton's data together with the fitted Normal distribution.

function changes, first of all the quantity inside the exponential function changes as follows

$$\frac{-\frac{1}{2}(x-\mu)^2}{\sigma^2} \rightarrow -\tfrac{1}{2}z^2$$

and since the standard deviation of the distribution becomes 1, the multiplier of the exponential becomes $1/\sqrt{2\pi}$, so that the new probability density function becomes as in the lower part of the box in Section 3.6.1. This is very useful because it means that calculations involving any Normal distribution can be easily transformed into terms of the standardised Normal, and only a single table of the standardised Normal distribution (Statistical Table S1) need be used (see Example 3.6).

3.7 MEASURES OF LOCATION AND SPREAD

Before we move on, we must consider some other general aspects of the two types of distribution, discrete and continuous, we have so far encountered. Once we have a mathematical description of a distribution, we are able to calculate its mean, or the position along the x-axis of its centre of gravity, and its variance or measure of spread.

3.7.1 Measures of location

First the mean, which is calculated exactly as the centre of gravity of a rigid body would be in mechanics. Its value could also be obtained by calculating the arithmetic average of an extremely large (infinitely large) random sample from the distribution.

Mean of a distribution

The position along the x- or r-axis of the centre of gravity of the distribution

$$\mathrm{E}[R] = \mu = \sum_{\text{all } r} rP(R=r) = \begin{bmatrix}\text{Sum over}\\ \text{all discrete}\\ \text{values of } r\end{bmatrix} r \times \text{Probability}(R=r)$$

for discrete distributions

$$\mathrm{E}[X] = \mu = \int_{-\infty}^{+\infty} xf(x)\,\mathrm{d}x = \begin{bmatrix}\text{Integral over}\\ \text{the whole}\\ \text{range of } x\end{bmatrix} x \times \text{Probability}(x < X \leqslant x + \mathrm{d}x)$$

for continuous distributions

Both of the above definitions are *weighted averages*, the weights being probabilities in the discrete case, and probability densities in the continuous (or probabilities of very thin slices, of thickness δx, through the continuous distribution). The reason why a weighted rather than an unweighted average is taken can be seen by considering the following example. Suppose the random variable R can only take 2 values, 1 and 2, with probabilities 0.9 and 0.1 respectively. The unweighted average of the 2 values is of course 1.5, whereas the weighted average is $(0.9)(1) + (0.1)(2)$, i.e. 1.1. If we take a large sample of say 1000 values from this distribution we would expect 900 of the values to be 1 and 100 to be 2. The average of these 1000 values would be $[(1)(900)+(2)(100)]/1000$, i.e. 1.1. So in this example, the mean calculated by the above rule would give the same value as the average we would expect in a very large sample, and the same is true for any population. Another term used for the mean, more in statistical texts, is *expectation*, written in symbols $\mathrm{E}[R]$, in words 'the expectation of R', or 'the expected value of R'. Other measures of location are useful in some circumstances, particularly for asymmetrical distributions. The main alternative to the mean is the median, but the mode is also sometimes useful.

Other location parameters

Median The value separating the top and bottom halves of the distribution. For a continuous distribution, 50% of the values are below the median, and 50% above it. For a discrete distribution, the percentage of values which are less than the median is below 50%, and similarly for values above the median (since the median itself has a finite probability of occurrence).

Mode The most frequently occurring value.

3.7.2 Variance

The variance is a measure of the extent to which the distribution spreads about the mean. There are various ways a measure of spread could be calculated, but the one which is most convenient is the mean squared deviation about the mean (again weighted by probability or probability density) as in the following definition.

Variance

Mean squared deviation about the mean

$$\mathrm{var}[R] = \sigma^2 = \sum_{\text{all } r} (r-\mu)^2 P(R=r) \qquad \text{for discrete distribution}$$

$$\mathrm{var}[X] = \sigma^2 = \int_{-\infty}^{+\infty} (x-\mu)^2 f(x)\,\mathrm{d}x \qquad \text{for continuous distributions}$$

Like the mean itself, the variance is an average, weighted by probability or probability density; this time of the squared deviation about the mean. The units of this measure would be in the square of the original scale of measurement of x or r—that is why it is referred to as σ^2. The square root of this quantity, or just σ, is known as the standard deviation of the distribution, σ, which has units which are the same as the original scale. This is the *population standard deviation*, rather than the sample standard deviation which we will encounter in Chapter 5.

It might seem strange to work with variances. We calculate the deviations from the mean, square them, add them up weighting by probabilities or probability densities, and then at the end have the bother of taking the square roots to get something we can work with. So the reader might well ask: what is so special about variances; why, as was said above, are they convenient? Why not use something which is both easier to think about and to calculate? The mean deviation about the mean (if we give the deviations of values below the mean negative signs, and those above positive signs, which are their proper signs) would be zero and is therefore of no use, but we could take the mean absolute deviation, i.e. mean deviation about the mean ignoring the sign. This is one alternative, but is not much used except in special circumstances.

The first part of the answer to the question—why are variances and standard deviations so convenient—is that the standard deviation certainly measures the scale of the distribution in the following sense. If we were to stretch out the r-scale of the probability distribution uniformly by a factor of 2, the standard deviation would increase by a multiple of 2 and the variance by a multiple of 4. The same would happen if we performed the same operation on the x-scale of a continuous probability density function (although in this case this would also involve dividing the p.d.f. by 2, in order to keep the area under the curve integrating to 1). For a direct illustration of this, see the way in which the Normal distribution changes with standard deviation in Figure 3.7. So standard deviation appears to measure the spread of the distribution. The main reason, however, for using standard deviation and variance is that variances are additive (which is described further in the next section). This phenomenon is important because it reflects the property of random variables that when you add or average a large number of them, the variability relative to the mean tends to get smoothed out, the *relative spread* is reduced. For example, large samples are taken in opinion polls so as to get a better estimate of programme or party ratings. The use of variance calculated in the above way makes these calculations easy.

3.7.3 Additivity of means and variances

We demonstrate the additive properties of means and variances in Example 3.4. First of all, let us state in symbols what we mean, in the following box.

Mean and variances of sums of independent random variables

If $X_1, X_2, \ldots, X_n$ are random variables with

(a) means denoted by $\mu_1, \mu_2, \ldots, \mu_n$, and
(b) variances denoted by $\sigma_1^2, \sigma_2^2, \ldots, \sigma_n^2$, and
(c) they are *independent* of each other (i.e. they do not tend to go up and down together or in opposition; they are effectively totally unrelated), then

$$\text{mean}[X_1 + X_2 + \cdots + X_n] = \mu_1 + \mu_2 + \cdots + \mu_n$$
$$\text{var}[X_1 + X_2 + \cdots + X_n] = \sigma_1^2 + \sigma_2^2 + \cdots + \sigma_n^2$$

Example 3.4 Adding individual weights to obtain the total weight

We discuss the relationship between the variance of the total weight of cartons of 6 apples and the variance of individual apple weight at a fruit farm. Suppose that the distribution of individual apple weight in a particular season is Normal with mean 120 g, and standard deviation 10 g, in other words with variance 100 g^2. The mean of the distribution of total carton weight is just the sum of the means of the individual fruit, i.e. $120 + 120 + 120 + 120 + 120 + 120 = 720$. The special thing about variances is that they can be added in the same way; the variance of total weight equals 100 + 100 + 100 + 100 + 100 + 100 = 600, provided that the apples are selected independently from the whole crop. So the total weight is Normally distributed with mean 720 g, and variance 600 g^2, or 24.5^2 g^2.

So although the total weight has increased by a multiple of 6, the standard deviation has only increased by a multiple of 2.45, which is the square root of 6. Thus the relative spread of the total weight is less than the relative spread of the individual apple weights, and this is what is observed in practice. If for example we had added up the standard deviations as we did the means, then the relative spread would have remained the same, so this essential feature of real variability would not be correctly modelled; and similar problems would have occurred with other measures of variability such as mean absolute deviation.

3.7.4 Cumulative distribution function

We had to use different methods, for discrete and continuous random variables, of apportioning the total probability among the possible values of the random variable:

(a) $\sigma = 2$

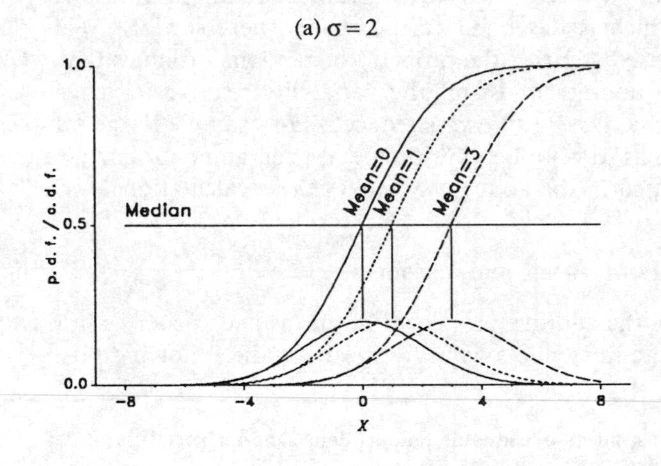

(b) $\sigma = 1$

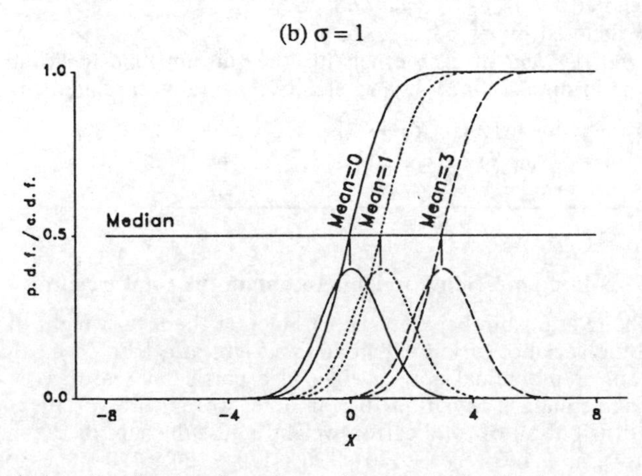

Figure 3.9 Cumulative distribution functions of Normal distributions.

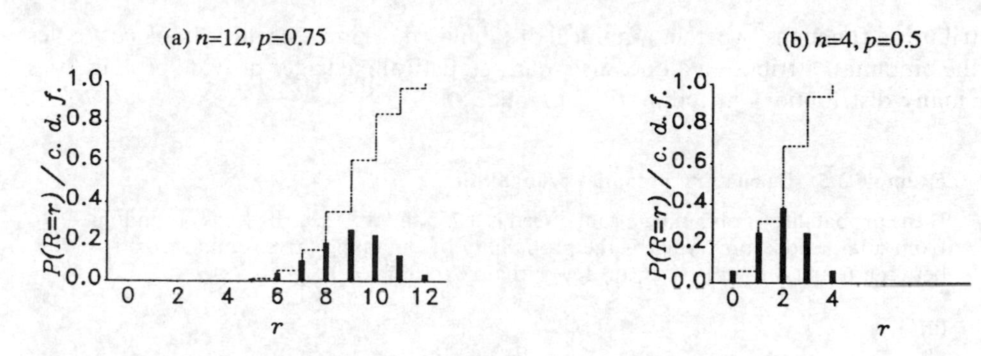

Figure 3.10 Cumulative distribution function of a binomial distribution.

in terms of probability density functions for continuous variables, and probability distributions for discrete ones. A further very convenient concept which does not require different definitions for continuous and discrete random variables is the cumulative distribution function. This, as its name implies, cumulates probability as one moves along the axis, from lower to higher values of the random variable, x or r. In other words, the cumulative distribution function, c.d.f. for short, gives the probability that the random variable takes any value less than or equal to the value being considered.

Cumulative distribution function (c.d.f.)

Probability that the random variable takes any values less than or equal to the value being considered

— for a discrete random variable, R

$$F(r) = P(R \leqslant r) = \sum_{s=0}^{r} P_s$$

— for a continuous random variable, X

$$F(x) = P(X \leqslant x) = \int_{-\infty}^{x} f(u)\,\mathrm{d}u$$

Note that s and u in the expressions for c.d.f. are dummy variables, they could equally well have been any other symbol. The c.d.f.'s of some Normal distributions and a binomial distribution are given in Figures 3.9 and 3.10. An application of the Normal c.d.f. curve in the calculation of probabilities is given in Example 3.6.

3.8 CALCULATING PROBABILITIES

3.8.1 Binomial probabilities

For the binomial distribution, the individual probabilities can be calculated directly, and the probability that r lies within a specified range can be calculated by summation of the individual probabilities. Alternatively they can be calculated using cumulative distribution functions, whether graphical or tabulated. There are whole books of tables of the binomial distribution alone; also many statistical packages provide probabilities for many distributions including the binomial.

Example 3.5 Calculating binomial probabilities

If the probability of obtaining a round seed is 0.75, and we select 10 seeds at random from a large collection, what is the probability of obtaining (a) 8 round seeds; (b) between 6 and 8 round seeds; (c) fewer than 7 round seeds?

(a)

$$P(R=8) = \binom{10}{8} p^8 (1-p)^{10-8} = \frac{10.9.8.7.6.5.4.3}{8.7.6.5.4.3.2.1.} 0.75^8 0.25^2$$

$$= \frac{10.9}{2.1} 0.75^8 0.25^2 = 0.2816$$

(b) If the question means 'between 6 and 8 seeds inclusive' then we need to obtain the probability of observing 6 or 7 or 8

$$= P_6 + P_7 + P_8.$$

P_8 was obtained immediately above.

$$P_7 = \binom{10}{7} p^7 (1-p)^3 = \frac{10.9.8}{3.2.1} 0.75^7 0.25^3 = 0.2503$$

and

$$P_6 = \binom{10}{6} 0.75^6 0.25^4 = \frac{10.9.8.7}{4.3.2.1} 0.75^6 0.25^4 = 0.1460$$

So $P_6 + P_7 + P_8 = 0.2816 + 0.2503 + 0.1460 = 0.6779$.

(c) The probability that we obtain fewer than 7 round seeds can be written

$$P(R<7) = 1 - P(R \geqslant 7) = 1 - [P(R=7) + P(R=8) + P(R=9) + P(R=10)]$$

P_7 and P_8 were calculated above and P_9 and P_{10} can be calculated similarly:

$$P(R=9) = \binom{10}{9} 0.75^9 0.25^1 = (10)(0.75^9)(0.25^1) = 0.1877$$

and

$$P(R=10) = \binom{10}{10} 0.75^{10} 0.25^0 = (1)(0.75^{10})(1) = 0.0563.$$

(Note also that $P(R=0) = \binom{n}{0} 0.25^{10} 0.75^0 = (1) 0.25^{10} = 0.0$. The convention is that $\binom{n}{0} = \binom{n}{n}$ is always 1.)

So $P(R<7) = 1 - [0.2503 + 0.2816 + 0.1877 + 0.0563] = 0.2241$.

3.8.2 Normal probabilities

The probability density of the Normal distribution can be calculated directly, but there is no simple way to obtain the probability that x or z lie within a specified range. The equivalent of adding probabilities is the determination of the area under the p.d.f. curve over the given range as the distribution is continuous, but the p.d.f. of the Normal distribution cannot be integrated analytically. It can only be done by numerical means—the computer equivalent of counting the squares under a graph of the probability density function drawn for a specific case. The good thing about the Normal distribution is that any specific case can be reduced to a problem about a standardised Normal, and tables of the cumulative distribution function of this for a wide range of values of z have been tabulated (see Statistical Table S1). This is usually referred to as $\Phi(z)$. Alternatively rather than looking up the tables we could read the value of $\Phi(z)$ off the graph of the c.d.f. of the standardised Normal (as given in Figure 3.9), but this is usually not accurate enough for most purposes. An example illustrates the method.

Example 3.6 Calculating Normal probabilities

If yield per plot, Y, follows a Normal distribution with mean = 20 kg m^{-2}, and variance 2^2 (kg m^{-2})^2, what are the values of the following probabilities:

$(a)\ P(Y \leqslant 25)\quad (b)\ P(18 < Y < 20)\quad (c)\ P(Y > 15)$?

(a) $$P(Y \leqslant 25) = P\left(\frac{Y-20}{2} \leqslant \frac{25-20}{2}\right) = P\left(Z \leqslant \frac{25-20}{2}\right) = P(Z \leqslant 2.5).$$

The tables of the Normal distribution give $P(Z < 2.5) = \Phi(2.5) = 0.9938$ which is the required probability. Note that $P(Z \leqslant 2.5) = P(Z < 2.5)$ because $P(Z = 2.5) = 0$ as Z is a continuous random variable and the probability that a continuous random variable takes a specific value is always zero.

(b) $$P(18 < Y < 20) = P\left(\frac{18-20}{2} < Z < \frac{20-20}{2}\right) = P(-1.0 < Z \leqslant 0)$$
$$= P(Z < 0.0) - P(Z < -1.0) = \Phi(0.0) - \Phi(-1.0)$$
$$= 0.5000 - 0.1587 = 0.3413.$$

(c) $$P(Y > 15) = 1 - P(Y \leqslant 15) = 1 - \Phi(-2.5) = 1 - 0.0062 = 0.9938.$$

3.8.3 Computer simulation of samples from distributions

In earlier chapters, we found that computer simulation was a useful tool for the investigation of a mathematical model or of methods for relating a model to collected data. The same is true for models of random mechanisms. We can use simulation to check calculations, for example of probabilities, made in other ways, or to do calculations which cannot be done by analytical means (i.e. by explicit algebra and calculus). Simulation is also useful for getting an intuitive insight into a new problem and for demonstrating the properties of distributions or of certain statistical methods to users who possibly do not have the time or the expertise to work through these properties in a more theoretical way. In physics, one would be able to derive the relationship between the time it takes for a freely falling apple to reach the ground and the initial height above the ground from Newton's Laws of motion. This would be straightforward algebra. However, an alternative is to set up an experiment and actually measure the time taken to fall various heights, and from these to either determine the relationship in the particular case, or to verify that the relationship derived by a more theoretically inclined person is correct. Computer simulation is similarly a way of conducting experiments or demonstrations in probability and statistics.

The principle is illustrated by two simple exercises. In these exercises, we do not take the details down to the most elementary level. It is not necessary for the user to do this anyway. In most statistical packages there are facilities for generating 'uniformly distributed random numbers', or samples of any size from a wide variety of distributions. Usually a short command is all that is necessary. When writing one's own program rather than using a package, libraries of subroutines for the same purpose which can be called from the user's program are widely available.

We should briefly state what we mean by a *uniformly distributed random number*. A uniformly distributed random number, or uniform random deviate as it is more commonly known, on the range 0 to 1 is one generated by a chance mechanism such that every number between 0 and 1 is equally likely to be generated; no part of the range is preferred. We discuss uniform random deviates in more detail in Section 3.11.1. The reader might be interested to know how the computer generates random numbers. In fact, the random numbers generated by a computer are not truly random, but are generated using a fixed mathematical rule, so that if you know the rule and the last few numbers you are able to predict the rest. They are therefore known as *pseudo-random numbers*. Nevertheless they behave for many purposes as if they were random.

Example 3.7 Normal probability calculations by computer simulation

Suppose that the yield distribution from a plot is Normally distributed with mean 20 kg m^{-2} and standard deviation 2 kg m^{-2}, determine by simulation the answers to the questions in Example 3.6.

Method

Generate 1000 random numbers from a Normal distribution with mean 20 and standard deviation 2. Determine the number out of 1000 which are (a) less than or equal to 25, (b) between 18 and 20, (c) over 15. Express these numbers as proportions of the total number of random numbers generated. As a supplementary to the above plot out a histogram of the 1000 numbers just to check that the distribution looks Normal.

Results

Histogram of 1000 samples.

Midpoint	Count	Each * = 10
13.5	0	.
14.5	4	.
15.5	20	**
16.5	48	*****
17.5	70	*******
18.5	145	***************
19.5	196	********************
20.5	188	*******************
21.5	169	*****************
22.5	99	*********
23.5	37	****
24.5	17	**
25.5	5	*
26.5	2	.

(a) The number less than 25 is $1000 - 7 = 993$. Hence estimated $P(Y \leqslant 25) = 0.993$.

(b) The number between 18 and 20 is $145 + 196 = 341$. Hence estimated $P(18 < Y < 20) = 0.341$.

(c) The number over 15 is $1000 - 4 = 996$. Hence estimated $P(Y > 15) = 0.996$.

Example 3.8 Binomial probability calculations by computer simulation

Let us now do the same for the binomial example above. We estimate the probabilities of obtaining:

(a) 8 round seeds.
(b) between 6 and 8 round seeds.
(c) fewer than 7 round seeds, if the number of round seeds follows a binomial distribution with $n = 10$, and $p = 0.75$.

This is done by taking 1000 samples from the binomial distribution and determining the proportion of samples which have exactly 8 round seeds, between 6 and 8 round seeds, and fewer than 7. The number of samples with each number of seeds in one simulation involving 1000 samples is given in Table 3.9 from which we obtain:

Estimated probability (8 round seeds) = 0.282.
Estimated probability (between 6 and 8 round seeds)
$$= 0.152 + 0.257 + 0.282 = 0.691.$$
Estimated probability (fewer than 7 round seeds)
$$= 1 - 0.063 - 0.181 - 0.282 - 0.257 = 0.217.$$

Table 3.9 Results of simulation of a binomial distribution, $n = 10$, $p = 0.75$

No. of round seeds	0	1	2	3	4	5	6	7	8	9	10
No. of samples	0	0	4	3	7	51	152	257	282	181	63

3.9 THE POISSON DISTRIBUTION

3.9.1 The Poisson process

We saw above how the number of successes (i.e. round seeds), R, in a fixed number of trials (seeds of any shape), n, was distributed according to a binomial distribution, with parameters n and p. We now consider a second discrete distribution, this time usually applicable to the rate per unit time, or per unit of space. Whereas for the binomial, the maximum value of R is n, there is, theoretically at least, no maximum value of R for the Poisson distribution. This is the classical distribution for describing point events occurring completely at random and with a uniform rate in time or space. Consider its application, in the case of a process in time, to nerve cell firing. The process itself, as opposed to the count distribution, is called a *Poisson Process*, and two things are necessary for it to occur. In each small interval of time, δt, the probability of a spike occurring is constant, $\lambda\delta t$; this is what is meant by a uniform rate, and the rate would then be given by λ. Secondly, the occurrence of a spike in any small interval is totally independent of whether any spikes have occurred in any other interval. The second of these requirements is only likely to be approximately satisfied

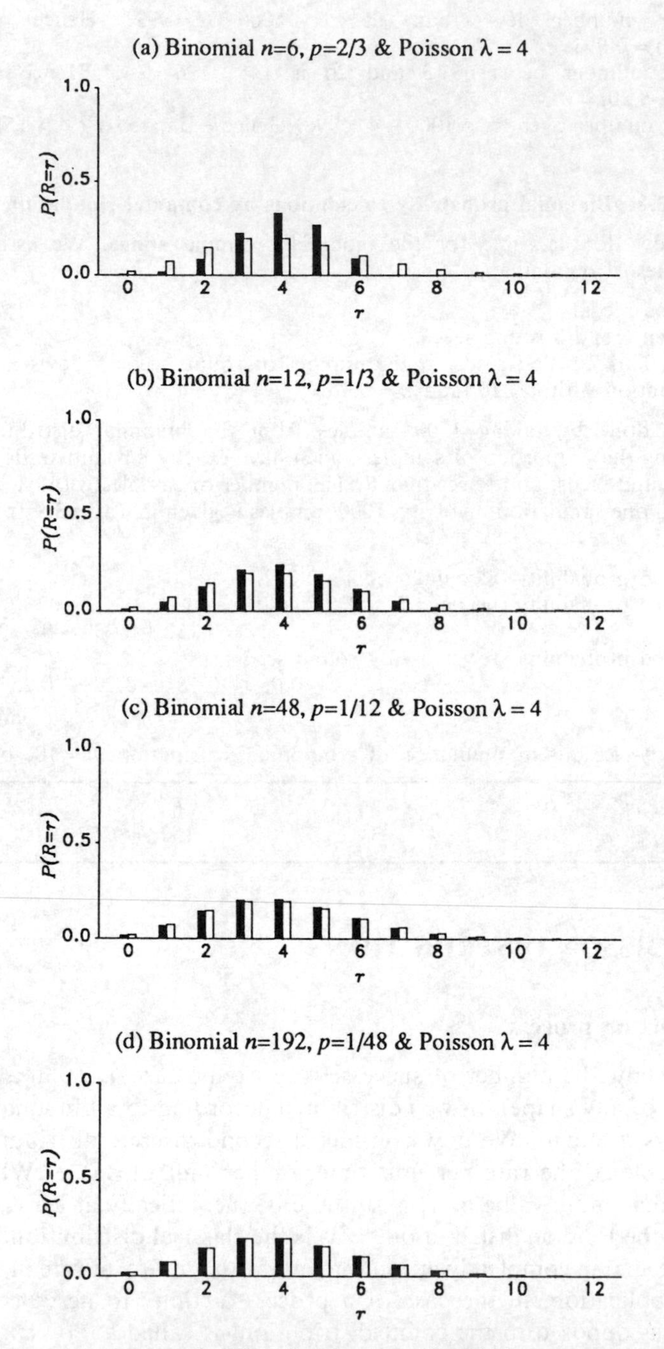

Figure 3.11 Various binomial probability distributions all with the same mean, plotted together with a Poisson distribution with the same mean, showing the convergence of the binomial to the Poisson as n is increased. The open bars give the probability distribution of the Poisson with mean 4 in each case. The solid bars give binomial probability distributions with parameters (a) $n = 6$, $p = 2/3$, (b) $n = 12$, $p = 1/3$, (c) $n = 48$, $p = 1/12$, (d) $n = 192$, $p = 1/48$.

in the case of neurone firing patterns, because there is usually a brief refractory period after a spike, during which no further spikes are possible. However, in many cases this is so short compared with the average interval between spikes that it can be ignored. Because of this last property, the Poisson process is often said to have no memory.

There are 2- and 3-dimensional spatial versions of a Poisson process, in which the probability of an event in a small unit of space, δA or δV, is $\lambda\delta A$ or $\lambda\delta V$ respectively, together with the independence requirement applied to occurrence in all such small units of space (see Chapter 13). This is the reason why the Poisson count distribution, as described in the next section, is an adequate model for the distribution of the yeast cell counts which Student investigated (see Example 3.9).

3.9.2 The Poisson distribution (of counts)

In order to determine the distribution of the number of spikes per second, we start from the binomial distribution. Divide the second under consideration into n intervals of length $1/n$, where n is so large, and therefore $1/n$ so small, that the chances of two spikes occurring in such a short interval are virtually zero. The probability of obtaining a spike in one interval is, from the above definition, $\lambda\delta t = \lambda/n$, so the probability of obtaining r spikes in n intervals is given using the binomial distribution by

$$P(r) = \binom{n}{r} p^r (1-p)^{(n-r)}$$

$$= \binom{n}{r} \left(\lambda \frac{1}{n}\right)^r \left(1 - \lambda \frac{1}{n}\right)^{(n-r)}$$

which can be simplified, as we let n become very large to $e^{-\lambda}\lambda^r/r!$; or in mathematical language, as $n \to \infty$, $P(r) \to e^{-\lambda}\lambda^r/r!$. For those readers who prefer a graphical demonstration, Figure 3.11 shows a series of binomial distributions, all with same mean but increasing n and decreasing p values, getting closer and closer to the Poisson.

Poisson distribution

$$R \sim \text{Poisson}(\lambda)$$

Probability distribution $\quad P(R = r) = e^{-\lambda}\dfrac{\lambda^r}{r!} \quad r = 0, 1, 2 \ldots$

Mean and Variance $\quad E[R] = \lambda \quad \text{var}[R] = \lambda$

The Poisson distribution is illustrated for various values of λ in Figure 3.12.

Example 3.9 Data on Student's haemocytometer counts

W.S. Gossett who in the early part of this century worked for the Guinness firm of brewers and adopted the pseudonym of 'Student' obtained the following data on counts of yeast cells in 400 randomly positioned squares of a haemocytometer, given in Table 3.10 and Figure 3.13.

Student's haemocytometer data was well modelled by the Poisson distribution, and his data plus the fitted Poisson is given in Figure 3.13. We might ask what his

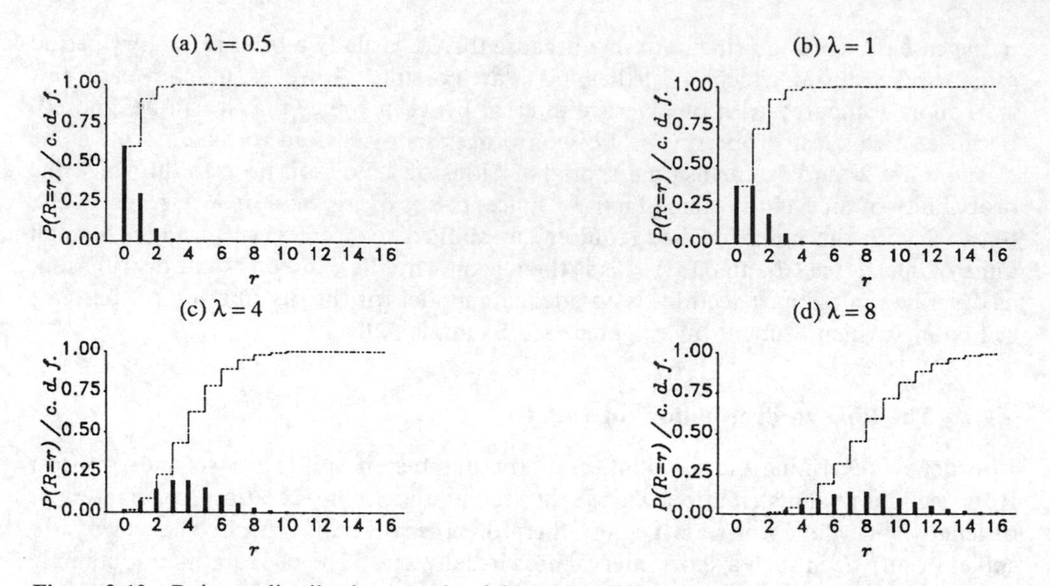

Figure 3.12 Poisson distributions (and c.d.f.'s) with different means.

Table 3.10 Counts of yeast cells in Student's haemocytometer data

No. of cells	0	1	2	3	4	5	beyond
Observed frequency	213	128	37	18	3	1	0

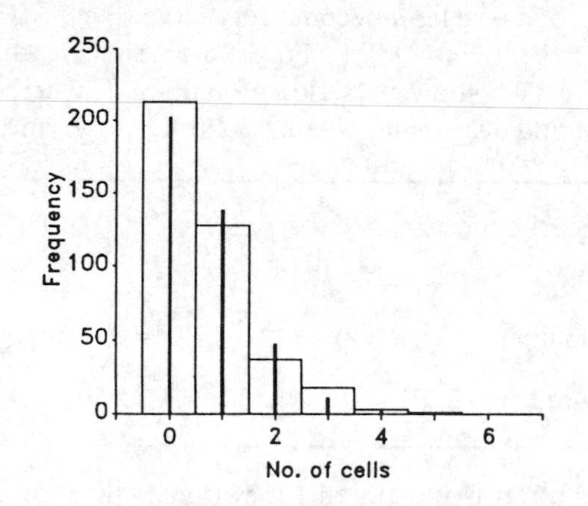

Figure 3.13 Student's haemocytometer data and the fitted Poisson distribution.

interest was in studying these data? There was the theoretical interest in seeing that a particularly simple model could be used for data of this type. More importantly scientists who were using this method to obtain an estimate of the density of cells in the medium could use the Poisson model to assess how many fields they needed to take in order to achieve a specified accuracy of estimation.

We see above that the variance of a Poisson distribution is equal to the mean. This provides a rule for the accuracy of a Poisson count. The standard deviation and coefficient of variation (the standard deviation of the count as a proportion of the average count) are then given by

$$\text{s.d.}[R] = \sqrt{\text{var}[R]} = \sqrt{\lambda} \qquad \text{c.v.}[R] = \frac{\sqrt{\text{var}[R]}}{\text{mean}[R]} = \frac{\sqrt{\lambda}}{\lambda} = \frac{1}{\sqrt{\lambda}}.$$

This leads to the very convenient rule for random events that the standard deviation of a count—the count estimating the rate of incidence of the events of a Poisson process—can be estimated by merely taking the square root of that count. A widely used version of this result in any work involving radioactive tracers is that the standard deviation of the number of particles emitted, N, is just $\sqrt{N}$, and expressed as a proportion of the total number, $1/\sqrt{N}$.

3.9.3 The negative exponential distribution (of intervals)

Data can be collected from a point process such as the firing of neurones in two main ways: first, the rate (per second or over some other time interval) can be calculated; secondly, the intervals between spikes can be obtained and recorded. We have determined the distribution of the firing rate for a Poisson process in the previous section. We now determine the distribution of the interval between events, T. This is a continuous distribution, so we will calculate the probability density function. Use of the definition of probability density from Section 3.5 gives us the first line below.

$$\begin{aligned}
&P(T \text{ lies between } t \text{ and } t+\delta t)\\
&\quad = P(t < T < t+\delta t) = f(t)\delta t\\
&\quad = P(\text{no spike in } [0,t]) \times P(1 \text{ spike in } [t, t+\delta t])\\
&\quad = P(\text{Poisson count (mean } \lambda t) = 0) \times P(\text{Poisson count (mean } \lambda\delta t) = 1)\\
&\quad = e^{-\lambda t} \times e^{-\lambda\delta t}\lambda\delta t\\
&\quad \cong e^{-\lambda t}\lambda\delta t \qquad \text{as } e^{-\lambda\delta t} \cong 1.
\end{aligned}$$

From this we obtain the definition overleaf.

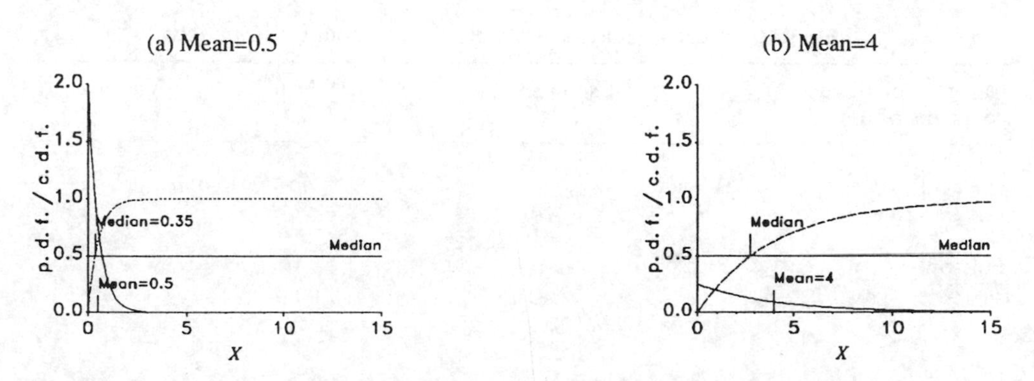

Figure 3.14 Negative exponential probability density functions and cumulative distribution functions for two different values of λ: (a) $\lambda = 2$, mean $= 0.5$, (b) $\lambda = 0.25$, mean $= 4$.

Negative exponential distribution

$$T \sim \text{exponential}(\lambda)$$

Probability density function $\quad f(t) = \lambda\, e^{-\lambda t} \qquad 0 \leqslant t < \infty$

Mean and Variance $\quad E[T] = 1/\lambda \qquad \text{var}[T] = 1/\lambda^2$

Figure 3.14 illustrates the distribution for two values of λ.

3.10 COUNT DISTRIBUTIONS OTHER THAN THE BINOMIAL AND POISSON

The Poisson distribution is used as a count distribution of events in a completely random point process. A characteristic of this distribution among count distributions is that the variance and the mean are equal. Table 3.11 gives some count distributions which are often used as alternatives to the Poisson, together with their means and variances. The main categorisation into over-dispersed and under-dispersed distributions is dependent on whether the variance (of the count of the process) is greater or less than the mean. These distributions and this distinction are used in plant ecology when describing the abundance of plants in randomly selected quadrats, although underdispersed populations are fairly uncommon, the most usual case being over-dispersion. Another area is in modelling point processes in time, such as nerve cell firing; here underdispersion does occur. If a very large sample of data is available, or there is appropriate background knowledge, then there is a very extensive catalogue of distributions which can be used as alternatives to the positive and negative binomial, but in many practical instances the distributions given here can act as quite adequate descriptive models. The negative binomial arises in various ways but there is one which should be mentioned specifically. It is sometimes known as the 'binomial waiting time distribution', the reason being that this is the distribution of the number of failures it is necessary to observe before encountering the kth success, where the probability of success in a single trial is p.

Table 3.11 A categorisation of simple discrete count distributions

Category of discrete count distributions	Examples	Mean (m)	Variance
Under-dispersed var $[R]$ < mean $[R]$	Binomial $P(R=r) = {}^nC_r p^r (1-p)^{(n-r)}$	np	npq or $m - \dfrac{m^2}{n}$
Poisson var $[R]$ = mean $[R]$	Poisson $P(R=r) = e^{-\lambda} \dfrac{\lambda^r}{r!}$	λ	λ or m
Over-dispersed var $[R]$ > mean $[R]$	Negative binomial $P(R=r) = {}^{k+r-1}C_r p^k (1-p)^r$	$\dfrac{kq}{p}$	$\dfrac{kq}{p^2}$ or $m + \dfrac{m^2}{k}$

3.10.1 Examples of fitted models to point process data

Models for the count distribution and the intervals between events are discussed for a neurophysiological example below.

Example 3.10 Data on the firing patterns of neurones

Cells in the supraoptic nucleus (next to the pituitary gland at the base of the brain) are of two main types: oxytocin (OXT) cells which have a major function during the processes of parturition and lactation in females, and vasopressin (AVP) cells whose main role is in maintaining the water balance of the body. Both of these cells secrete their respective hormones as a response to continued firing, and have been extensively studied in the rat. When subjected to very low stimulation from other parts of the body, the two types of cell behave similarly and the pattern can often not be discriminated from a completely random pattern of firing, but as the stimulation increases, the behaviour of the cells diverge. The AVP cells fire phasically—i.e. at high rates of up to 30 spikes/second for periods of up to a minute or so, and then fall silent for a similar period. One possible reason for this which

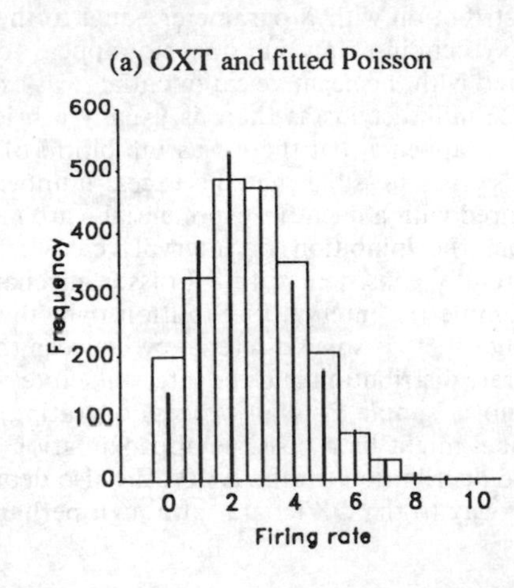

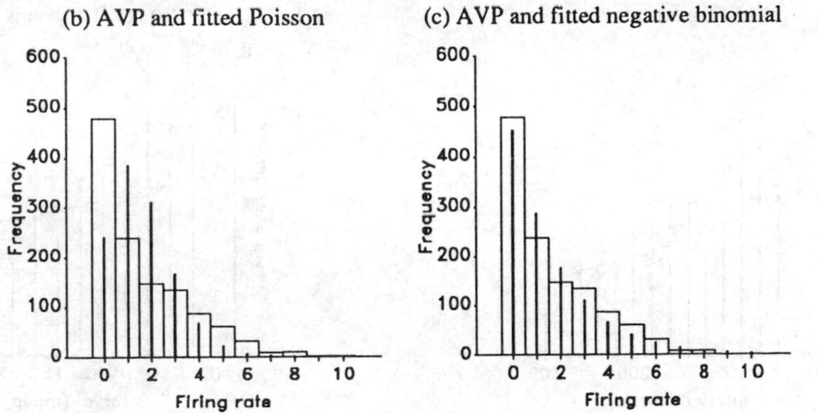

Figure 3.15 Vasopressin (AVP) and oxytocin (OXT) firing rate histograms and fitted distributions.

has been confirmed by further experiments is that the AVP cell becomes fatigued after a period of prolonged stimulation, and requires some time to recover. The background activity of the OXT cells on the other hand becomes more uniform than would be the case with a completely random pattern of activity. Analysing the nature of the variability in firing rate over time provides clues as to the mechanism of action of these cells. We consider one example of the distribution of firing rates at a low firing rate from each type of cell (from Poulain *et al.*, 1988), which are exhibited in Figure 3.15; and of the intervals between spikes in Figure 3.16.

We attempted to fit the Poisson and negative binomial distributions to the two sets of firing rate data. The histograms of the firing rate data and the fitted distributions are given in Figure 3.15. The OXT firing rate data was reasonably well approximated by a Poisson distribution, but not so the AVP data. The variance/mean ratios of the firing rates were 1.00 and 2.15 respectively, and so we might expect that the OXT but not the AVP data would be well fitted by Poisson distributions. In the latter case, a negative binomial distribution fitted much better, though still not completely satisfactorily.

If the Poisson process model were really adequate for the oxytocin data we would expect that the interspike intervals would follow a negative exponential distribution. Data on the intervals were only available up to 300 ms for the OXT data and up to 240 ms for the AVP data, and the distributions are given in Figure 3.16. The negative exponential distribution with λ parameter equal to the mean firing rate is also displayed for the oxytocin intervals and does not appear to be a very good fit. A deficiency, as compared with the negative exponential, was expected in the region near zero up to about 20 milliseconds as there is usually a brief refractory period after a spike. However, it appears that there was inhibition of longer intervals up to about 60 milliseconds. It is possible that the excess number of intervals in the range 60 to 180—compared with a negative exponential—are merely these intervals deferred. So apart from the inhibition of intervals below 60 milliseconds, the distribution could be broadly consistent with a Poisson mechanism. It would also have been better to examine the interval distribution over the complete range of intervals. Thus we see that there is some discrepancy between the results of the two approaches—via firing rate distribution and the interspike interval distribution. The conclusion is that it is not a simple Poisson process operating, although for some purposes a Poisson model might be a possible approximation.

The interspike interval distribution for the AVP cells also departs from a negative exponential in a similar way to the OXT data, although perhaps not as extremely.

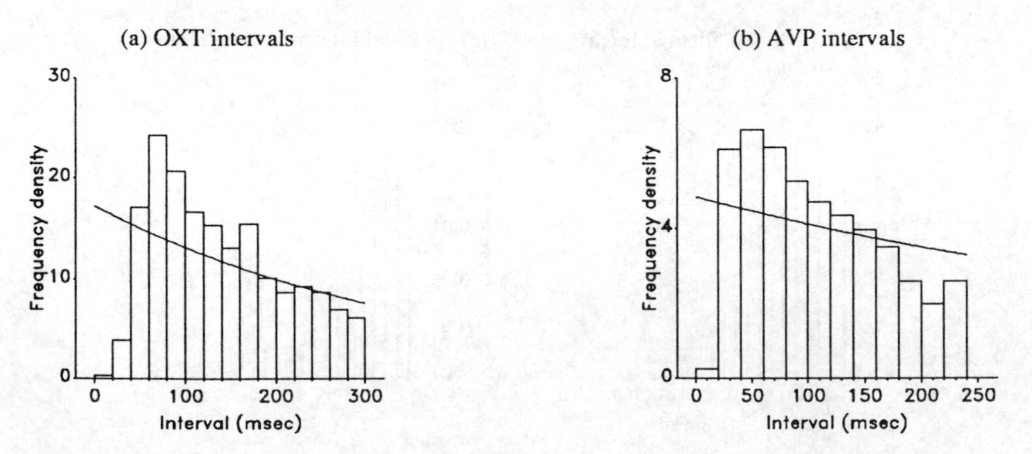

Figure 3.16 Vasopressin (AVP) and oxytocin (OXT) interspike interval histograms and fitted distributions.

The line plotted in this case is approximately the best fitting negative exponential to the interval data. The suggestion from these and similar analyses of many more cells is that the interspike interval distributions for the two types of cell are broadly similar (although in general there are more small intervals for a given mean level of activity in the AVP cells), but that a different mechanism of ordering the intervals in the two types of cell ensures a different firing rate distribution. Further analysis indicated that the small intervals in the AVP cells are much more clustered than in the oxytocin cells. This might be responsible for the fact that the negative binomial is a better fitting model for the AVP firing rate data.

3.11 OTHER CONTINUOUS DISTRIBUTIONS

3.11.1 The uniform distribution

We next consider two alternatives to the two continuous distributions we have encountered so far. The uniform distribution is one in which the probability density is constant over some particular range of x-values, between $x = a$ and $x = b$. In order for the total area under the curve to be 1, the probability density therefore has to be as given below.

Uniform distribution

$$X \sim \text{uniform}[a, b]$$

Probability density function (p.d.f.)

$$f(x) = \frac{1}{(b-a)} \quad \text{for} \quad a \leqslant x \leqslant b$$

Mean and Variance $E[X] = (a+b)/2 \qquad \text{var}[X] = \dfrac{(b-a)^2}{12}$

One use of this distribution is to describe rounding error. In reading an apparatus to say 1 place of decimals, assuming that the rounding is done correctly, the error would be distributed uniformly on the range -0.05 to $+0.05$. Another use, which crops up in the theory of some statistical methods, is of estimated probabilities which have to lie in the range 0 to 1. A third use is in random number generation which we mentioned briefly above.

3.11.2 The lognormal distribution

Some measurements such as weight or length can never be negative, and in cases where the mean is low compared with the standard deviation, often such distributions tend to be more squashed up near the origin, and with a long tail to the right; they are said to be *positively skew*. The Normal distribution can have negative values for low values of μ relative to σ, and is symmetric, and is therefore not entirely appropriate for modelling these measurements. In these cases, an adequate model is often obtained by considering that the logarithm of the weight is Normally distributed. Then the weight would be said to be following a lognormal distribution. The simplest way of approaching this is merely to transform the data by taking logarithms and then to work with a Normal distribution.

Hastings & Peacock (1974) is a source of information on other distributions, as well as a more complete list of the properties of the distributions we have dealt with here, in a very convenient format.

3.12 ORIGINS OF NORMALITY OF DISTRIBUTION

Normal distributions are both very convenient to work with and arise in many contexts. This section deals with two cases in which Normal distributions arise as a result of other processes.

3.12.1 Normally distributed height measurements as a result of the *discrete* actions of several genes

At the turn of the century a very heated debate about the nature of inheritance of continuously variable traits such as adult height as opposed to qualitative ones such as eye colour and blood group took place between those of the Biometric School who had been working with the genetics of continuous traits, and the proponents of Mendelian theory. It appeared that there was incompatibility between the two approaches. One of the issues was how the variation in continuous variables could be explained by a discrete Mendelian mechanism. G.U. Yule (1902) showed that there was no incompatibility between the two approaches.

Yule quoted Francis Galton (1889):

> human stature is not a simple element, but a sum of the accumulated lengths or thicknesses of more than a hundred bodily parts, each so distinct from the rest as to have earned a name by which it can be specified. The list includes about fifty separate bones, situated in the skull, the spine, the pelvis, the two legs, and in the two ankles and in the feet.

Yule continued as follows:

> Surely it would be a very moderate estimate that the number of units could not be less than 50? Yet this would suffice to give, on the simplest Mendelian assumption that each unit can only exhibit two types, not some mere ten thousand different values of stature, the run of which would be indistinguishable from strictly continuous variation, but *over a thousand-million million different types*! Even if the variations of 'units' do take place by discrete steps *only* (which is unproven), discontinuous variation must merge insensibly into continuous variation simply owing to the compound nature of the majority of characters with which one deals.

We can demonstrate his argument—and go further to show what distributional form of height results—by a simple simulation program. Suppose that the lengths of each of the n elements is determined by a single locus, and that the genes at the n loci involved segregate independently, to produce a limb length or thickness for each genotype as in the following table. The genotype relative frequencies are also given.

Table 3.10 Notation for the genotypes, their relative frequencies, and the effect on the phenotype (the limb length) resulting for each

Genotype at ith locus	A_iA_i	a_ia_i	A_ia_i
Genotype relative frequency	p_i^2	$(1-p_i)^2$	$2p_i(1-p_i)$
Length of ith 'limb'	l_i	$l_i(1-s_i)$	$l_i(1-h_is_i)$

Table 3.11 Parameter values used in the simulation of Section 3.12.1

Gene	1	2	3	4	5	6	7	8	9	10	11	12	13	14	15	16
p_i	0.2	0.3	0.1	0.5	0.6	0.9	0.4	0.3	0.5	0.1	0.2	0.3	0.1	0.5	0.6	0.9
s_i	0.2	0.3	0.4	0.2	0.1	0.4	0.2	0.3	0.4	0.2	0.2	0.3	0.4	0.2	0.1	0.4

For simplicity, although it is not at all necessary, these relative frequencies are those obtaining at Hardy–Weinberg equilibrium.

The parameters, l_i, s_i and h_i can be adjusted to reflect A_i being dominant to a_i ($h_i = 0$) or alternatively a semi-dominance situation ($h_i = 0.5$). Then we obtain total height by adding up these lengths or thicknesses.

Figure 3.17 gives histograms for total length (or height) for 500 realisations of this model, for different values of n, and for sets of (l_i, h_i, s_i, p_i) as follows. In each case, $l_i = 1$ and $h_i = 0$ (i.e. the case of dominance, the limb length being unity for the dominant phenotype), and p_i and s_i are taken in sequence from Table 3.12. Only a small number of genes are needed for the distribution to be approximately Normal.

3.12.2 The widespread applicability of the Normal distribution: the central limit theorem

We saw above how the distribution of a continuous trait would be approximately Normal even if it were determined entirely genetically, provided that more than a very

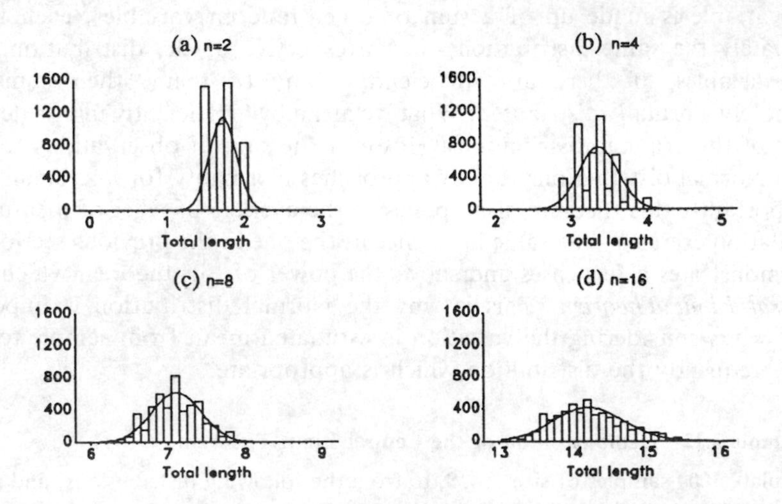

Figure 3.17 Distributions of height resulting from simulated genetic mechanisms (Section 3.12.1). Normal distributions with the same means and variances are overplotted.

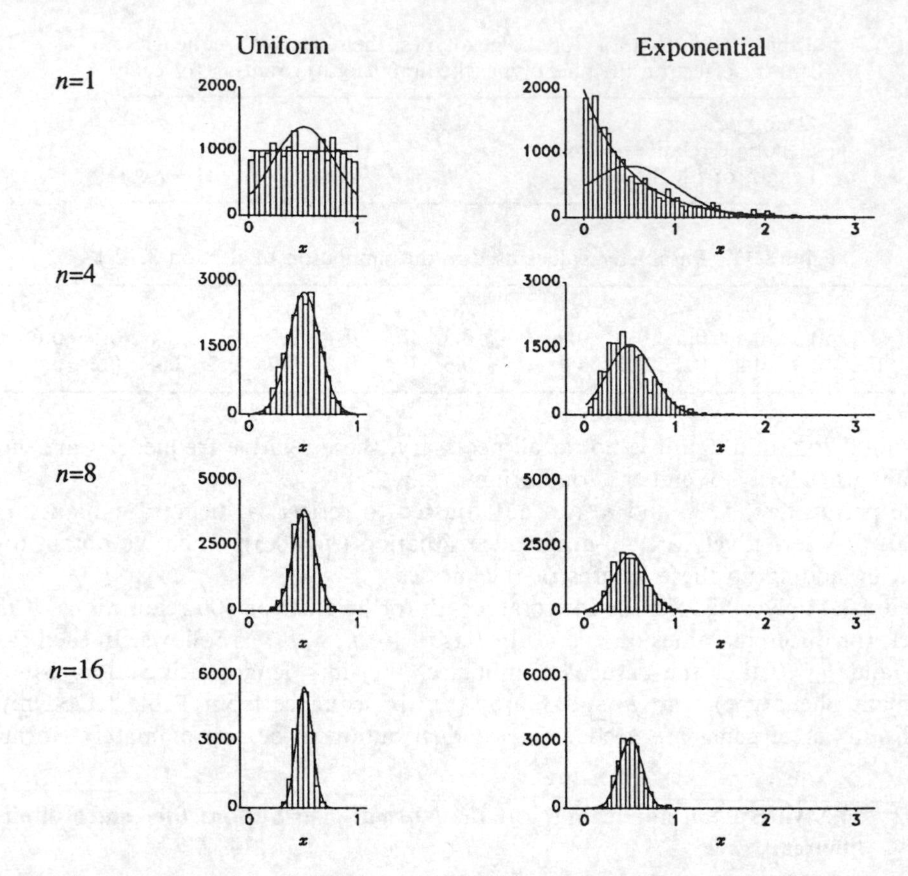

Figure 3.18 Demonstration of the Central Limit Theorem (see Example 3.12).

few genes were involved. This phenomenon applies also to other influences. If a random variable is made up of a sum of other random variables, each following approximately the same distribution, then irrespective of the distribution of those random variables, if there are sufficiently many of them, their sum will be approximately Normally distributed. What we mean by 'sufficiently many' depends on the shape of the original distributions. However the sum of observations from some quite non-Normal distributions rapidly approaches Normality for n as small as 6–20. More importantly this theorem also applies to the average of a set of measurements. The simulation exercise—the same in essence as the one in the previous section—given below demonstrates a few cases and shows the power of this theorem which is called the *Central Limit Theorem*. This is why the Normal distribution is important in statistics: when considering the variation in estimated mean from sample to sample, this is so frequently the distribution which is appropriate.

Example 3.12 Demonstration of the Central Limit Theorem

Simulate 1000 samples of size 1, 4, 8, 16 from the following distributions, and plot the histograms of the sample averages:

(a) Uniform [0, 1];

(b) Negative exponential, with rate parameter $\lambda = 2$, i.e. mean = 0.5.

The histograms are given in Figure 3.18, overplotted with Normal distributions with the same mean and variance, and, for the samples of size 1, the parent distribution.

3.13 APPENDICES

3.13.1 Rules for calculating probabilities

This appendix outlines the main rules for calculating probabilities using an immunological example.

Example 3.13 A population of T cells

Suppose that a restricted population of T cells has receptors for some related ligands A_1, A_2 and A_3, and two other ligands which occur in just one form, B and C respectively. The cells have either A_1, A_2, A_3 receptors or none (A_0) at the A site but no cell has more than one A receptor. Suppose that the proportions (and hence probabilities) overall in the population are $P(A_1) = 0.10$, $P(A_2) = 0.20$, $P(A_3) = 0.30$, $P(A_4) = 0.40$, where P stands for probability. Suppose also that the B and C receptors are equally abundant and $P(B) = P(C) = 0.70$, so the probabilities of these two receptors not being present are $P(B_0) = P(C_0) = 0.30$.

In this population, the A and B receptors are at different positions on the cell, and there is no relation between them, i.e. they occur totally *independently*. We should emphasise what independence means here. In order to demonstrate this, first split the population of cells down into four subpopulations: those which have the A_1 receptor, those which have the A_2 receptor, those with the A_3 receptor and those with no A receptor, i.e. A_0. Independence between A and B means that the probability of obtaining a B receptor in each of these subpopulations is the same. Knowledge of which A receptor is on a cell does not help us to know whether there is a B receptor on the cell or not. So we get the probabilities of occurrence of every combination of receptor of types A and B in Table 3.12(a). In the body of the table are the probabilities of each combination of the two receptor classes. Along the bottom margin are the overall probabilities of each form of the A receptor. In the right hand margins are the overall probabilities of B being present and absent. The probability in the lower right corner is just the total probability, which is 1, indicating that something must happen. In Table 3.12(b), these have been converted into expected frequencies out of 1000 cells. Quite often, this is the easiest way to resolve lack of clarity in how probabilities can be combined to work with expected numbers.

It is important to understand the nature of the trials over which the probability is being calculated, and the sample space. In these examples, the probabilities refer to the presence of each receptor type on a randomly selected cell from a specific population. The *trial* consists of selecting a cell at random from the population, and noting what receptors *of the types under consideration* it contains. The *sample space* is the set of possible cells with each possible combination of receptors of the types under consideration. We considered probability calculations involving the receptors A and B above. We next consider calculations involving the receptors A and C. Considering the two situations separately, the sample spaces for calculations involving A and B is the set of possibilities in Table 3.12; whereas the sample space for calculation involving A and C is the set of possibilities in Table 3.13. Since we are given no information about the relationship between B and C, we cannot calculate probabilities for the complete sample space.

Table 3.12

(a) Proportions (i.e. probabilities) of A and B

	A_1	A_2	A_3	A_0	Margin
B	0.07	0.14	0.21	0.28	0.70
B_0	0.03	0.06	0.09	0.12	0.30
Margin	0.10	0.20	0.30	0.40	1.00

(b) Expected numbers out of a total of 1000

	A_1	A_2	A_3	A_0	Margin
B	70	140	210	280	700
B_0	30	60	90	120	300
Margin	100	200	300	400	1000

Table 3.13

(a) Proportions (i.e. probabilities) of A and C

	A_1	A_2	A_3	A_0	Margin
C	0.10	0.20	0.27	0.13	0.70
C_0	0.00	0.00	0.03	0.27	0.30
Margin	0.10	0.20	0.30	0.40	1.00

(b) Expected numbers out of a total of 1000

	A_1	A_2	A_3	A_0	Margin
C	100	200	270	130	700
C_0	0	0	30	270	300
Margin	100	200	300	400	1000

A and C are not independent, in that C is more likely to occur if one of the forms of A is present (unlike A and B considered above). You can see from Table 3.13 that the dependence of C on A is very strong. If A_1 or A_2 occurs, then C is certain to occur; and the occurrence of A_3 makes it more likely that C occurs than if A does not occur at all (i.e. in the case A_0). Table 3.13(a) and (b) gives the probabilities in the same format as Table 3.12. Note that the marginal probabilities and frequencies of the various forms of A are the same as previously, as they must be. And since the two receptors B and C are equally abundant, they have the same marginal values also. Only the probabilities and expected numbers in the body of Table 3.13 are different from those in Table 3.12.

(1) *Conditional probability*. We use the notation $P(X \mid Y)$ to mean the probability of event X given that Y has occurred or is present (sometimes described as 'conditional on the event Y').

For example, $P(A_1 \mid B)$ is the probability that receptor type A_1 is present given that B is present; in other words it is the proportion of those cells which have B which also have A_1. It can be calculated as (No. of cells with B and A_1)/(No. of cells with B), which equals $70/700 = 0.10$ in the above example.

(2) *Independence*. If events X and Y are independent, then

$$P(X \mid Y) = P(X) \quad \text{and} \quad P(Y \mid X) = P(Y).$$

for example, the A and B receptor types are *independent*, i.e. the presence of a B receptor on a cell does not tell us anything about whether an A receptor is present on that cell nor which type of A receptor is present. This can be written in terms of conditional probability as $P(A_i \mid B) = P(A_i)$. In particular we see that $P(A_1) = 0.10$ which equals $P(A_1 \mid B)$ which we calculated in the previous paragraph.

On the other hand, A and C are interrelated. $P(C \mid A_1) = P(C \mid A_2) = 1$, $P(C \mid A_3) = 0.90$, and $P(C \mid A_0) = 0.325$, whereas $P(C) = 0.70$.

Simultaneous occurrences or intersection of events

Venn diagrams depicting subsets of the sample space are useful for thinking about probability calculations. The proportion of the total sample space occupied by a set in the diagram gives the probability of the set of events corresponding to the set. Simultaneous occurrences of events can be depicted as intersections of sets in a Venn diagram, as in Figure 3.19.

(3) *Product rule for probability of simultaneous occurrence of independent events.* If two events are independent, then the probability of their joint occurrence is given by the product of their individual probabilities of occurring.

i.e. $$P(X \text{ and } Y) = P(X)P(Y) \quad \text{if } X, Y \text{ independent.}$$

e.g. $$P(A_1 \text{ and } B) = P(A_1)P(B) = 0.1 \times 0.7 = 0.07,$$

$$P(A_2 \text{ and } B_0) = P(A_2)P(B_0) = 0.2 \times 0.3 = 0.06.$$

(4) *Probability of simultaneous occurrence of non-independent events.* In the general case, when two events are not necessarily independent, then the probability of their joint occurrence is given by the product of the probability of one event multiplied by the probability of the second event *conditional on the first having occurred.*

i.e. $$P(X \text{ and } Y) = P(X)P(Y \mid X) = P(Y)P(X \mid Y)$$

if X, Y are non-independent.

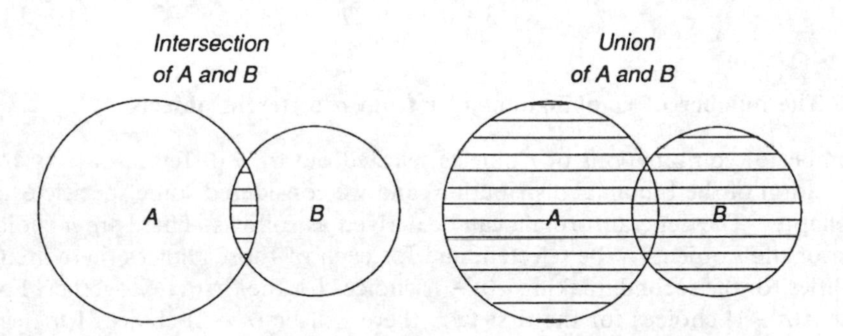

Figure 3.19 Venn diagrams illustrating the simultaneous occurrence of two events as the intersection of the two sets in the sample space which correspond to the two events, and alternative occurrences as the union of the two sets.

For example, we see from the second table that $P(A_1 \text{ and } C) = 0.1$ and this can also be obtained from the formulae

$$P(A_1 \text{ and } C) = P(C)P(A_1 \mid C) = 0.7 \times (1/7) = 0.1$$

or as

$$P(A_1)P(C \mid A_1) = 0.1 \times 1 = 0.1.$$

Alternative occurrences or union of events

Alternative occurrences of events can be interpreted as the union of sets in a Venn diagram (Figure 3.19).

(5) *Summation rule for probability of union of disjoint events.* If two events X and Y *cannot* occur simultaneously, then the probability of one *or* the other occurring is the sum of their separate probabilities,

$$P(X \text{ or } Y) = P(X) + P(Y).$$

e.g.

$$P(A_1 \text{ or } A_2) = P(A_1) + P(A_2) = 0.1 + 0.2 = 0.3.$$

(6) *Probability of union of non-disjoint events.* If two events X and Y *can* occur simultaneously, then the probability of one *or* the other occurring *or* both is the sum of their separate probabilities minus the probability of their joint occurrence,

$$P(X \text{ or } Y) = P(X) + P(Y) - P(X \text{ and } Y).$$

The reason for the formulae can be seen from the Venn diagram; in $P(X) + P(Y)$ the intersection of the two sets has been counted twice, and therefore has to be subtracted.

e.g.

$$\begin{aligned} P(A_1 \text{ or } B) &= P(A_1) + P(B) - P(A_1 \text{ and } B) \\ &= 0.1 + 0.7 - 0.07 = 0.73. \end{aligned}$$

A second example is

$$\begin{aligned} P(A_3 \text{ or } C) &= P(A_3) + P(C) - P(A_3 \text{ and } C) \\ &= 0.3 + 0.7 - 0.27 = 0.73. \end{aligned}$$

3.13.2 The number of combinations of r from n different objects

The number of combinations of r objects selected out of n different objects arises in the derivation of the binomial distribution, and we considered some specific examples in the chapter. The general formula can be derived as follows. There are n choices for the first of the r objects to be selected, and for each of those choices there are $(n-1)$ possibilities for the second, making $n(n-1)$ choices for the first two together. For each of these $n(n-1)$ choices for the first two, there will be $(n-2)$ choices for the third, and so on, until by the time we get to the rth choice, there are only $(n-r+1)$ objects left. So the total number of possible choices is $n(n-1)(n-2)\ldots(n-r+2)(n-r+1)$. This is not quite the end of the matter though. Let us consider the number of combinations of 2 from 4 different objects {A, B, C, D} as a specific example. There are 4×3 choices by the above argument, and they are as follows.

First choice	A	B	C	D
2nd choices	B, C, D	A, C, D	A, B, D	A, B, C

The list resulting is thus AB, AC, AD, BA, BC, BD, CA, CB, CD, DA, DB, DC and this includes every pair twice but in a different order, e.g. AB and BA. If the order mattered, then this calculation would be all right. This list is known as the number of permutations of $r = 2$ from $n = 4$, and the corresponding general formula is:

$$\text{No. of permutations of } r \text{ from } n = {}^nP_r = n(n-1)\ldots(n-r+1)$$

$$= \frac{n(n-1)(n-2)\ldots 1}{(n-r)(n-r-1)\ldots 2.1}$$

$$= \frac{n!}{(n-r)!}$$

When order does not matter, we can make the correction by dividing by two, or in the general case of ${}^nC_r$, we divide by the number of ways of rearranging r objects in a different order (from the above argument this is just the number of r objects selected from r, i.e. $r(r-1)\ldots 2.1$). So the general expression can be written

$$\text{No. of combinations of } r \text{ from } n = {}^nC_r = \frac{n(n-1)\ldots(n-r+1)}{r(r-1)\ldots 2.1}$$

$$= \frac{n(n-1)(n-2)\ldots 1}{r(r-1)\ldots 2.1\ (n-r)(n-r-1)\ldots 2.1}$$

$$= \frac{n!}{r!(n-r)!}$$

EXERCISES

3.1 Plant A gives 20 seeds, of which 8 would produce a plant with white flowers. Twelve would produce red flowers. Plant B (of the same species) gives 70 seeds of which 42 would produce white flowers and 28 red. Every seed will germinate if sown.

(a) One plant is chosen by the spin of a fair coin and one seed is selected from it at random. What is the probability that it will produce a red flower?

(b) Alternatively, the 90 seeds are thoroughly mixed and one randomly chosen seed is sown; what is the probability of a red flower?
(Cambridge, NST, Part 1A, Biological Mathematics, 1981.)

3.2 In a large series of breeding experiments, 50% of plants had red flowers, 30% had white flowers and 20% had blue flowers. 65% had round leaves, and 35% pointed leaves. The two characteristics flower colour and leaf shape are completely genetically determined and are inherited independently.

(a) What is the probability of obtaining a plant with red flowers and pointed leaves?

(b) What is the probability of obtaining a plant with red or blue flowers?

(c) What is the probability of obtaining a plant with red or blue flowers and round leaves?

Additionally there is a relationship between flower colour and flowering date. The probability that a red flowering plant flowers early is 0.8 (and hence of flowering late is 0.2). The probability that a blue or white flowering plant flowers late is 0.3. Flowering date is independent of leaf shape.

(d) What is the probability of flowering late given that a plant has blue flowers?

(e) What is the probability that a randomly selected plant flowers late?

(f) What is the probability that a plant with round leaves flowers late?

3.3 Discuss possible definitions of trial, sample space and event associated with the following probability statements.

(a) The probability of obtaining a green seed (g), in a series of crosses of pure strains of g and y genotypes, is 1/4.

(b) The probability of obtaining a yield of more than 5 kg of tomatoes per pot in a glasshouse uniformity trial is 0.4.

(c) The probability of observing 6 failures before the first success is q^6p.

3.4 The proportion of cells in a blood sample with a binding site for antigen A is $p_A = 0.05$. Assuming cells are randomly mixed and further sampled at random, what are the probabilities of the following events:

(a) observing no cells with the antigen binding site in a sample of 10 cells;
(b) observing 2 cells with the site in a sample of 10 cells;
(c) observing 9 or more cells with the site in a sample of 10 cells;
(d) observing more than 10% of the cells with the binding site in a sample of 100 cells (using a Normal approximation).

Using a simulation program, demonstrate the effectiveness of the Normal approximation in (d). Also assess how effective a Normal approximation would be in (a)–(c), also using simulation.

3.5 Two similar but non-competing bird species have different breeding patterns. A simplified model is as follows. The long-term clutch size distributions for each species (where p_r is the probability that the number of eggs laid to a randomly chosen breeding pair in any year is r) are: for species X, $p_2 = 1.00$, and $p_r = 0$ for $r \neq 2$; for species Y, $p_1 = 0.80$, $p_2 = 0.10$, $p_3 = 0.10$ and $p_r = 0$ otherwise.

In a good year (a mild spring) *all* the offspring of both species survive until the following breeding season; in a poor year, because the amount of food per average territory is only sufficient to rear one offspring, only those offspring hatched in clutches of *one* survive, and all other offspring die.

(a) What is the expected number of surviving offspring from an average breeding pair of species X and an average breeding pair of species Y in a good year?

(b) What is the expected number of surviving offspring from an average breeding pair of species X and an average pair of species Y in a poor year?

(c) In northern Scotland, the proportions of good and poor years are 0.4 and 0.6, while in southern England they are 0.7 and 0.3. Assuming good and poor years occur in a random sequence, what are the expected number of surviving offspring per breeding pair in the two locations for the two species?

(d) Both species breed for two years starting at one year of age and then die or leave the breeding population. Comment on the likely success of the two species in the two areas. (Cambridge, NST, Part 1A, Biological Mathematics, 1988.)

3.6 The number of eggs deposited in a 24-hour period by the female of a certain species of toad is approximately a Normal variable with mean 310 and standard deviation 20.

(a) Determine the approximate probability that a given female deposits fewer than 200 eggs in 24 hours.

(b) The numbers deposited in successive 24-hour periods can be regarded as statistically independent. Determine the approximate probability that in the total 120-hour period in which eggs are deposited the total number deposited by one female is less than 1450.

(c) Determine also the approximate probability that the total numbers deposited by 2 females differ by more than 150.

(Cambridge, NST, Part 1A, Biological Mathematics, 1984.)

3.7 An exhaustive study is made of two species X and Y in an area A. It is found that the weights of the females and males of each species were Normally distributed with means and standard deviations (kg) as follows:

	Mean	Standard deviation
Females species X	100	10
Males species X	125	10
Females species Y	120	12
Males species Y	160	12

You should assume that the study was so large that these means and standard deviations are known exactly.

Calculate the probabilities that:

(i) the weight of a randomly chosen male of species X is greater than 144 kg;

(ii) the weight of a randomly chosen male of species X is more than 10 kg heavier than a randomly chosen male of species Y;

(iii) the average weight of the female and the male in a randomly selected mating pair is greater for species X than for Y, assuming that mating is at random within species.

(Cambridge, NST, Part 1A, Biological Mathematics, 1987.)

3.8 Insects arrive at a light trap independently, and with probability $\lambda\delta t$ of an arrival in any interval of duration δt. Show that the probability density function of T_1, the time interval between arrivals, is $f(t_1) = \lambda\, \mathrm{e}^{-\lambda t_1}$.

[You may assume that $\lim_{n\to\infty} (1 + x/n)^n = \mathrm{e}^x$, for all x.]

Find the mean and variance of T_1. Hence or otherwise, obtain the mean and variance of T_n, the time from an arrival to the nth arrival thereafter.

(Cambridge, NST, Part 1A, Biological Mathematics, 1977.)

3.9 Diagnoses that a patient needs renal dialysis occur at random at a more or less constant rate throughout the year. Ignoring complications about weekends and holidays, what distribution would you expect the number of patients, R, needing renal dialysis on a randomly selected day to follow? If μ is the mean rate of occurrence per day, write down the form of the probability distribution of R. Show that

$$\frac{P(R = r+1)}{P(R = r)} = \frac{\mu}{r+1}.$$

A hospital has three kidney machines, each of which can only be used by one patient per day. If $\mu = 2$, calculate the probability that:

(a) no kidney machine will be needed on a given day;

(b) some patients cannot be dialysed on a given day;

(c) there will be exactly one day in a seven-day period when some patients cannot be dialysed.

(Cambridge, NST, Part 1A, Biological Mathematics, 1982.)

3.10 Explain how the Poisson distribution may be considered as a special case of the binomial distribution, illustrating your argument by a suitable example. Hence obtain the Poisson probability formula. You may quote

$$\lim_{n\to\infty} \left(1 + \frac{x}{n}\right)^n = \mathrm{e}^x, \text{ for all } x.$$

If the mean of a Poisson variable is $N > 10$, obtain approximate limits within which 90% of observations of the variable may be expected to lie.
(Cambridge NST, Part 1A, Biological Mathematics, 1978.)

3.11 Evaluate $P(R = 0)$, $P(R = 1)$ and $P(R > 1)$ for the following distributions: (a) binomial $n = 8$, $p = 1/12$; (b) binomial, $n = 16$, $p = 1/24$; (c) binomial, $n = 32$, $p = 1/48$; (d) Poisson, $\lambda = 2/3$, and thus demonstrate the convergence which you are asked to prove in Exercise 3.10.

3.12 (a) A random variable has a uniform distribution over the interval a, b. State its probability density function and show that

$$E[X] = \tfrac{1}{2}(a + b), \text{var}[X] = \tfrac{1}{12}(b - a)^2.$$

(b) The wing spans of 45 birds are measured, and the measurements are rounded to the nearest whole cm. The rounding errors may be regarded as independent random variables from a uniform distribution on $[-0.5, 0.5]$. If the observed mean of the rounded wing spans is $\bar{X}$, find the probability that the difference between $\bar{X}$ and the mean of the actual values is less than 0.05.
(Cambridge, NST, Part 1A, Biological Mathematics, 1982.)

3.13 By simulation verify the answers to Exercise 3.12. In (a), it is sufficient to consider the case $a = 0, b = 1$.
(Hint: if $Y \sim Uniform[0, 1]$, and $X = a + bY$, then $X \sim Uniform[a, b]$ and hence

$$E[X] = a + b\text{E}[Y], \qquad \text{var}[X] = b^2 \text{ var}[Y].)$$

3.14 (a) Assuming that insemination of dairy cattle are statistically independent trials with the same chance p of success (i.e. conception), show that the probability that exactly r inseminations are needed to establish pregnancies in each member of a herd of k cows is

$$P(R = r) = \binom{r-1}{k-1} p^k (1 - p)^{r-k} \qquad (r = k, k + 1, \ldots).$$

(b) The distribution of the number of attempts until the single cow is successfully inseminated follows a geometric distribution. Derive the form of this distribution from the above result. Suggest other biological situations in which the result might be useful.

4 Structured Stochastic Models: Models with Both Deterministic and Stochastic Elements

4.1 INTRODUCTION

In previous chapters we have considered models which were entirely deterministic and others which were entirely stochastic. These models can be rather artificial when applied to many areas of biology. Random variability is a characteristic feature of biology as opposed to the physical sciences, and therefore such variability needs to be built into any realistic models we construct. On the other hand, there are parts of biological systems which follow the laws of chemistry and physics or the systematic laws underlying biological mechanisms such as those of population genetics. They change in some systematic way with time or in response to variation in the conditions. Therefore it is sometimes equally necessary to involve in a biological model a substantial deterministic element similar to those we have introduced in Chapters 1 and 2.

In this chapter we build models which have both stochastic and deterministic components. We consider two different problems. First, how do we model situations in which there is a broad trend in response as the level of some explanatory variable changes, but also some chance variability about that trend? We do this by incorporating a deterministic component—reflecting the trend—into the more descriptive stochastic models we have encountered in Chapter 3. Secondly, we approach the problem from the other end, and see how the mechanistic deterministic models we have dealt with in Chapters 1 and 2 change when the simplifying assumptions which we needed to make there—for example, that the population considered was large or that the environment was invariable—no longer hold. In particular, we present *dynamic stochastic models* and show the different ways in which a random component can be incorporated into a deterministic dynamic model. In both types of model we link the deterministic and stochastic components to produce *structured stochastic models* as more realistic descriptions of biological phenomena.

4.2 DESCRIPTIVE MODELS FOR CONTINUOUS MEASUREMENTS

In this section we present descriptive models for continuous measurements which link together a deterministic systematic relationship and a random component. We begin with the simplest case of a straight line and then extend the approach to curvilinear relationships.

4.2.1 Straight line-Normal model or linear regression

There are many situations in which the relationship between a response variable and an explanatory variable appears to be a straight line except for some random variability. This section shows how we can utilise the Normal distribution model of Chapter 3 to depict the random variation component of this relationship as the straight line-Normal model.

Example 4.1 Growth rate of rats against dose of vitamin B_2

Figure 4.1 illustrates data from an experiment in which male weanling rats were dosed with different rates of riboflavin or vitamin B_2 (Clarke *et al.*, 1940). Growth tends to increase with the level of the vitamin but the relationship is a variable one and appears as a scatter of points about some underlying trend. How might we model this relationship, so as to express in numerical form the response of growth rate to dose of the vitamin? The experiment was also carried out on female rats, and there was interest in comparison of the sexes as well as in characterising the response of the sexes separately.

If there was no random variation, we could model the relationship between growth rate, y, and $x = \log(\text{dose})$ by a straight line model

$$y = \alpha + \beta x,$$

where β is the slope of the line, i.e. the change in growth rate for a unit change in log(dose), and α is the intercept, i.e. the value of the growth rate when $x = \log(\text{dose}) = 0$. In the standard stochastic model for straight line relationships, it is assumed that the response variable (e.g. growth rate), but not the explanatory variable (e.g. log(dose)), has an additional random component ε. To indicate that the response is a random variable we use a capital, Y, and write the model as

$$Y = \alpha + \beta x + \varepsilon.$$

The next thing which needs to be decided is what distribution the ε follow, since they are random variables. The simplest model and the one which crops up most often in

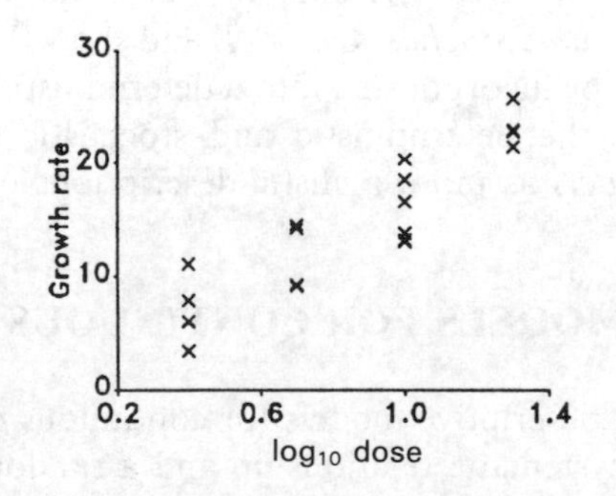

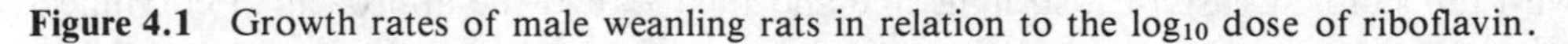
Figure 4.1 Growth rates of male weanling rats in relation to the $\log_{10}$ dose of riboflavin.

applications is that ε follows the symmetric, bell-shaped Normal distribution. There is no need for the distribution of the ε to have a mean other than zero, since they are measuring *departures* from the straight line (in the Y direction). It is also usually assumed, unless we have evidence to the contrary, that the variance of the distribution of ε is constant. In particular, we assume that there is no relationship between var$[\varepsilon]$ and x: speaking loosely, the random scatter about the line is of the same width throughout the range of x. So we say $\varepsilon \sim N(0, \sigma^2)$.

The model is represented schematically in Figure 4.2 with the distribution of the ε at each value of x depicted by small Normal density functions.

Now let us consider a particular case with intercept $\alpha = -0.53$, and the slope $\beta = 17.6$—these seem at least plausible values for these data (Figure 5.10). The aspect of the model which remains to be given a numerical value is the variance of the random component σ^2, which we set at 7 for the moment (these values for the parameters were obtained by using the methods described in Chapter 5). As with previous models, to see how it works out in practice we can use the model to simulate some data. We use exactly the same set of x—or log(dose)—values as in the real data. Figure 4.3 displays two different realisations. The deterministic or systematic part of the model is the same in each case, only the random part varies from realisation to realisation.

That the Normal distribution is an appropriate model for the variation about the line is an assumption which it is difficult to check empirically with such small data

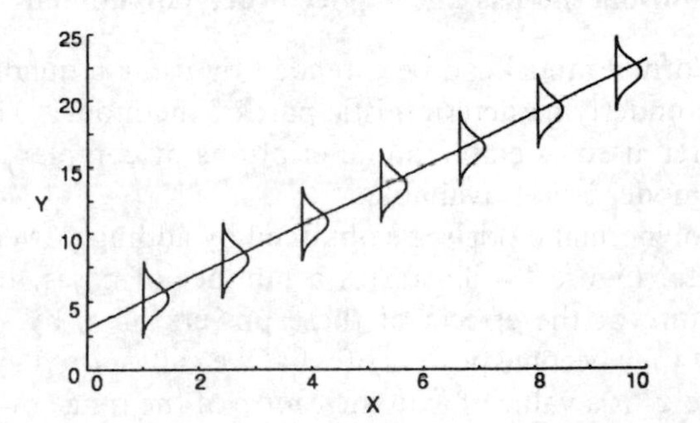

Figure 4.2 Schematic representation of the straight line-Normal model.

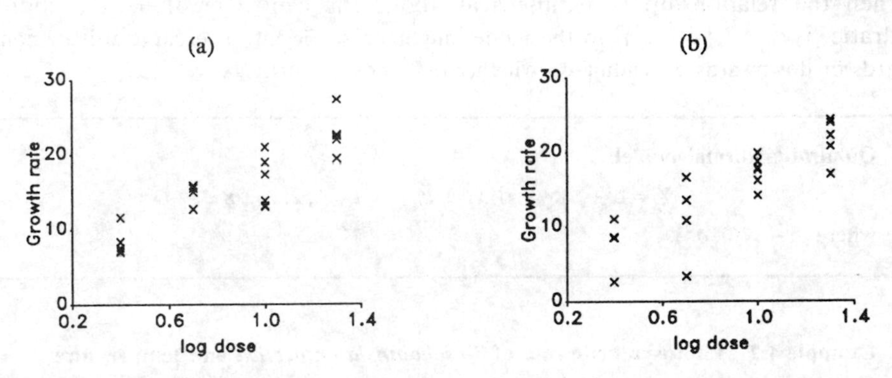

Figure 4.3 Realisations of the straight line-Normal model using plausible parameter values for the rat growth data.

sets. In this case it appears to be a reasonable assumption, as growth over a long enough time period is likely to consist of increments in many body components and, by the arguments presented in the previous chapter, the variation could be expected to be approximately Normally distributed. In practice, we might also have many previous sets of data which justify the assumption.

The general straight line-Normal model

Finally, we write this model in a form which is commonly used when a set of data is available and it is desired to fit the model to the data. In this form, each data point is identified with a subscript i. The object of this is that it makes explicit which components of the model vary from one data point to another.

Straight line-Normal model or linear regression

$$Y_i = \alpha + \beta x_i + \varepsilon_i, \qquad i = 1, 2, \ldots, n$$

where $\varepsilon_i \sim N(0, \sigma^2)$.

4.2.2 Quadratic-Normal models and higher order polynomials

The straight line-Normal model can be extended by using a quadratic or higher order polynomial for the underlying deterministic part of the model. These models are very flexible and are often used as empirical descriptions of complex relationships when a more mechanistic model is not available.

Quadratic and polynomial models are obtained by adding powers of the explanatory variable, x_i^2, x_i^3, etc. Figure 4.4 illustrates a number of cases. It is usually easier to understand the nature of the effects of these powers of x_i by slightly changing the formulation so that they become powers of what we call *centred explanatory variables*, $x_{ci} = x_i - x^*$, where x^* is a value of x in the *centre* of the range of x values used. Often x^* is the average of the x values in the data, and in that case there are some advantages over the uncentred explanatory variable when it comes to fitting the model.

When the relationship is symmetrical about the centring value, x^*, adding a quadratic $\beta_2 x_{ci}^2 = \beta_2 (x_i - x^*)^2$ to the model might be sufficient. The curve will be concave upwards or downwards depending on whether β_2 is positive or negative.

Quadratic-Normal model

$$Y_i = \alpha + \beta_1 x_i + \beta_2 x_i^2 + \varepsilon_i, \qquad i = 1, 2, \ldots, n$$

where $\varepsilon_i \sim N(0, \sigma^2)$.

Example 4.2 Photosynthetic rate of *Deschampsia antarctica* and temperature

The rate of photosynthesis of this Antarctic species of grass was determined for a series of environmental temperatures (Table 4.1). One objective was to determine

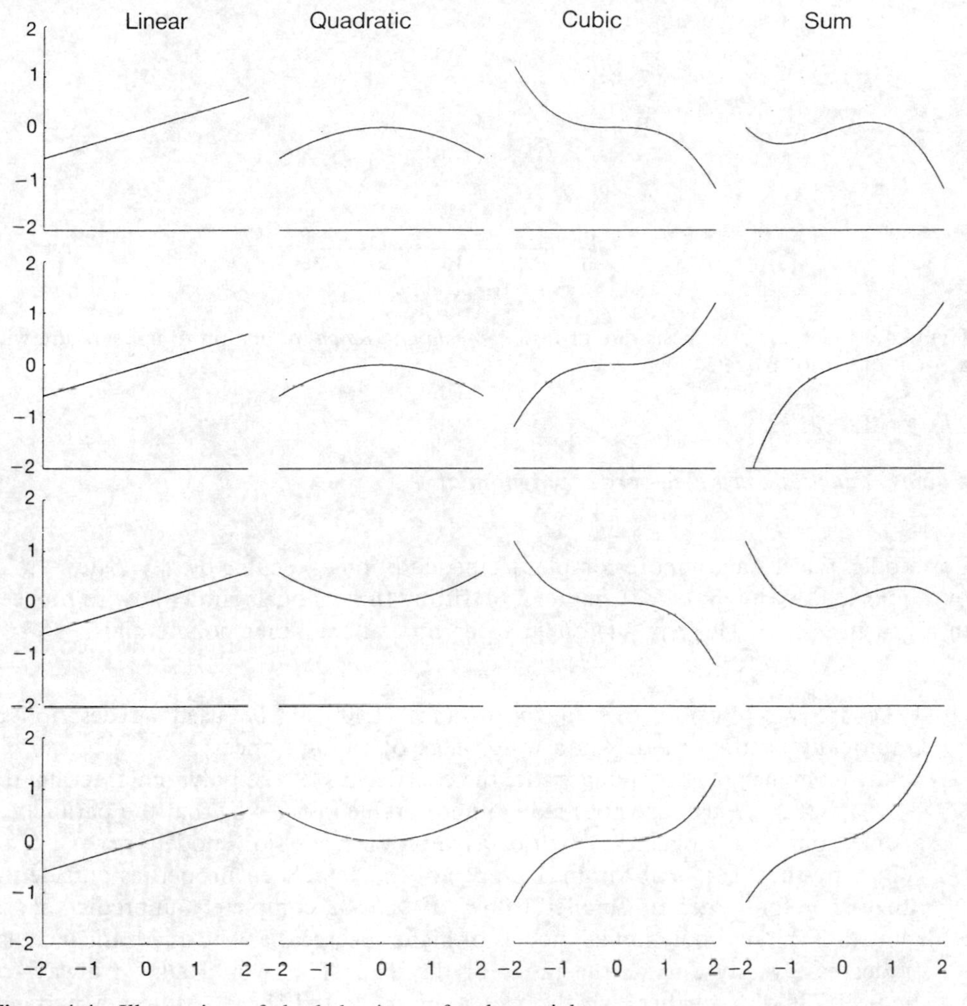

Figure 4.4 Illustration of the behaviour of polynomials.

Table 4.1 Net photosynthetic rate (P%) of *Deschampsia antarctica* at different temperatures (T °C). Data kindly provided by Dr. Ron Lewis-Smith

T (°C)	−1.5	0	2.5	5	7	10	12	15	17	20	22	25	27	30
P (%)	33	46	55	80	87	93	95	91	89	77	72	54	46	34

the temperature at which photosynthesis is maximum. Figure 4.5 shows the data and a fitted quadratic model with equation

$$y = 46.37 + 6.77x - 0.249x^2$$

(see Section 5.2.6). Although not fitting the data exactly, it follows the main trend for these data. The maximum value of a quadratic occurs when $\beta_1 + 2\beta_2 x_i = 0$. In this case this gives a temperature of 13.6 °C, when the fitted photosynthesis is 92.4%.

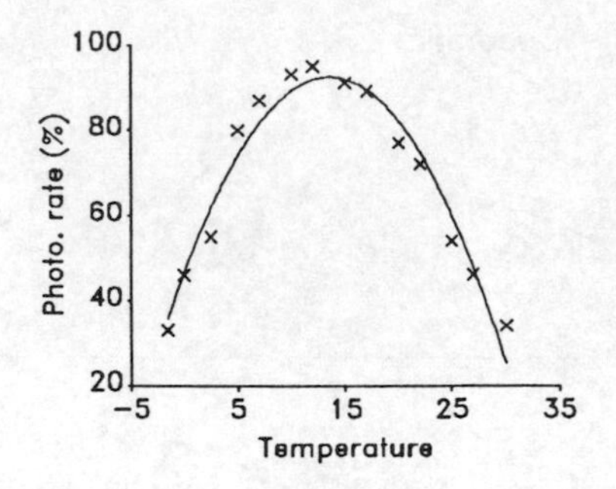

Figure 4.5 Net photosynthesis rate of *Deschampsia antarctica* in relation to temperature with a fitted quadratic model.

Cubics, quartics and higher order polynomials

For models which have a more complex shape, addition of a cubic $\beta_3 x_{ci}^3 = \beta_3(x_i - x^*)^3$ or a quartic $\beta_4 x_{ci}^4$ might help. Chapter 5 on fitting these models shows how to proceed in a practical case. Three points ought to be made here about polynomials:

(1) They are very flexible over a fixed range, and can be used as descriptive, empirically justified models in a wide range of circumstances;
(2) Apart from the linear and quadratic, the coefficients of the polynomial terms, the $\beta_1, \beta_2, \beta_3, \beta_4, \ldots$, etc. are not readily interpretable in the way that the parameters of an exponential, logistic growth or asymptotic regression model are;
(3) While they might fit well within the range of the data, their properties outside the range, i.e. for larger or smaller values of x_i, are completely unpredictable in general. Typically, polynomial models of higher degree than the quadratic increase or decrease wildly outside the range of the data. This would almost always be inconsistent with further data were it to be collected there. Although it is not generally sensible to extrapolate a model outside the range of the data, this should usually only be done with models based on the biology of the situation such as those in Section 4.2.3 which might be expected to have more appropriate behaviour.

4.2.3 Non-linear Normal models

We can greatly increase our catalogue of models by using any of the deterministic models in Chapter 1 and adding a stochastic component as in the straight line-Normal model above. These curves can be fitted to data from a 'Growth and Decline' context, in which case the justification of the systematic part of the model will be as given in that chapter, or they can be used as simple descriptive models. Two widely used models are as follows.

Exponential-Normal model

$$Y_i = \alpha + \beta\, e^{\gamma x_i} + \varepsilon_i.$$

Logistic-Normal model

$$Y_i = \alpha + \beta \frac{e^{\gamma(x_i - m)}}{1 + e^{\gamma(x_i - m)}} + \varepsilon_i,$$

where $\varepsilon_i \sim N(0, \sigma^2)$, $i = 1, 2, \ldots, n$.

Example 4.3 Application of logistic-Normal model to microvillus growth data

As cells move up villi—tongue-like projections from the intestine wall—the height of microvilli on them grow, thus enlarging the active area of the absorptive surface of the intestine. The change in microvilli height with distance along the villus has been modelled by a logistic model (Figure 4.6). This is a routine method of *approximately* describing data such as these and for quantifying growth by relating functions of the parameters of the logistic model to the growth rate of the organism concerned. Problems occur in that there is often rather little data at the lower end of the curve, and at the upper end it is often necessary to exclude points because microvilli appear to shrink as cells senesce prior to being sloughed off the tip of the villus. Most probably a fully appropriate model would be more complex—a more flexible, asymmetric S-shaped curve, with a further decline phase at the top—but generally speaking the data are not adequate to fix the parameters of more complex models precisely. The logistic model follows the general pattern of growth, once the period of senescence has been excluded, sufficiently well for useful estimates to be obtained. The fitted model is given by

$$h = 0.7557 + 1.325 \frac{e^{0.01429(d-103)}}{1 + e^{0.01429(d-103)}}.$$

This has a lower asymptote at $h = 0.7557$ an upper asymptote at $h = 0.7557 + 1.325 = 2.081$. When $d = 103$, the value of h is halfway between the upper and lower limits.

4.2.4 Complications

There are various other generalisations of these models which we will not deal with in this chapter. However, an important restriction on the above simple models for

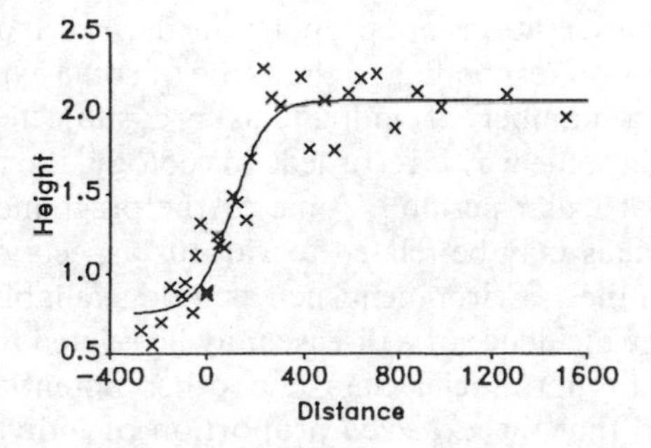

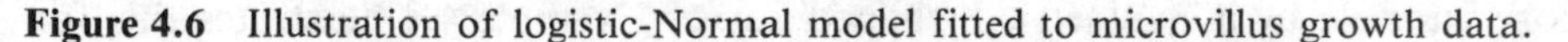

Figure 4.6 Illustration of logistic-Normal model fitted to microvillus growth data.

continuous response variables is that the variance of the random component, σ^2, is the same throughout the range of the model. In many cases *non-constancy of variance* at different values of x will occur, and checks should be made on this assumption if the data are sufficient. Alternative model formulations are possible in which σ^2 is different for different values of the explanatory variables, especially when the variability is systematically related to the mean or the value of the explanatory variable. In some situations it may be preferable to transform the observations, for example by working with logarithms. This, of course, affects the form of the systematic part of the model but it also affects the variance. In Chapter 11 we discuss the implications of these complications for fitting models to data and give some examples.

A further restriction is that the random component of the measurements is independent from one experimental unit to another. An example of *non-independence* occurs when data are collected from litters of animals. Because animals from the same litter are in general likely to have a more similar response than animals in different litters as they have some genetic and environmental influences in common, this assumption of independence will often not be justified. In some cases it might have little impact on the modelling, but in others it could reduce the effective replication very considerably. A second commonly occurring example of non-independence is when *repeated observations* are made on the same individual. Here the stochastic elements of observations close together in time are more closely related than those from observations far apart in time. The analysis of such studies can be quite difficult and requires special methods (Chapters 7 and 11).

Other distributional assumptions could be made for the random components. We have in this section presented the most commonly used one for continuous measurements—the Normal distribution. In many cases the main interest is in the deterministic part, and provided the distribution of the variation about the curve is not very different from a Normal distribution, the effect of incorrectly assuming a Normal distribution on the estimates of the deterministic part will be small. The next section deals with a model which is widely applicable for data which are proportions derived from independent binary responses.

4.3 DESCRIPTIVE MODELS FOR PROPORTIONS

4.3.1 Introduction

In many situations the observations are proportions derived from binary data and we wish to relate the proportion responding to the value of some explanatory variable. For example, in bioassay a number of individuals are subjected to a trial and the proportion responding in some way is recorded. In ecology, the relationship of survival of individuals over a particular period of time or the pregnancy rate of females in a population of wild animals may be related to various aspects of the individuals such as size, condition and of their environment such as food availability, weather variables, etc. In epidemiology, the incidence of a disease may be related to weight, diet, whether the individual smokes or not, social class and other potential indicators. In these situations we often find that the *expected* proportion of individuals is related to the explanatory variable by a symmetric S-shaped curve. Example 4.4 illustrates this for the proportion of individuals with coronary heart disease status in relation to age in a sample of 100 subjects. Typically, however, there is some variability about the curve arising from variation in the responses of a small number of individuals sampled from a larger population. In this section we present a model to describe the systematic and random components for relating an observed proportion to the value of an explanatory variable.

4.3.2 Logistic response curve

The first component of the model is the mathematical expression we use for the expected proportion p responding. This must lie between 0 and 1, and it quite often happens in practical examples that there is a range over which p changes slowly in response to an increase in the explanatory variable x, after which p increases more quickly for intermediate values of p, and then for higher values of p a lower increase again. In other words, p forms an S-shaped curve against x. One formulation which satisfies these requirements and has certain technical advantages, as well as advantages of interpretation, is the logistic model:

$$P(\text{reaction on the } i\text{th subject}) = p_i = \frac{e^{\alpha+\beta x_i}}{1+e^{\alpha+\beta x_i}}$$

where x_i is the value of the explanatory variable which may be log(dose), age, etc.

Figure 4.7 illustrates the following properties of the model:

(1) p lies between 0 and 1, as required;
(2) It forms a symmetric curve in that the lower asymptote is a mirror image of the upper one (twice reflected);
(3) The mid-point of the curve, and the point where the slope is steepest, is given by

$$x = -\alpha/\beta;$$

(4) To get a rough idea of the extent and slope of the curve, it can be well approximated by a straight line with slope $\beta/6$ over an appropriate range (for $\alpha + \beta x$ between -3 and $+3$), and outside this range taking the value 0 or 1.

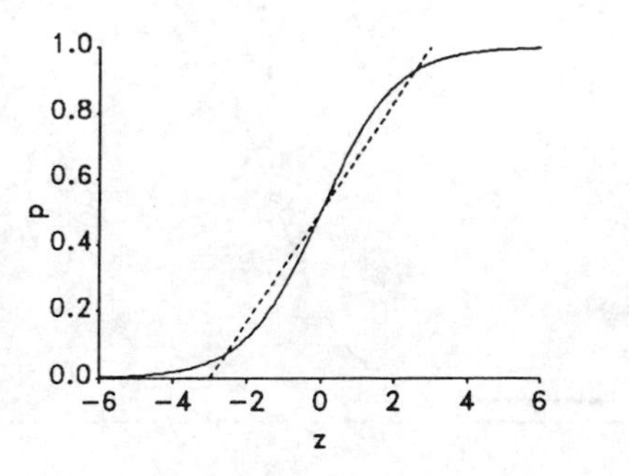

Figure 4.7 Illustration of properties of the logistic response curve relating the probability of a reaction to the value of a standardised explanatory variable $z = \alpha + \beta x$.

4.3.3 Binomial model of the random component

There is still one aspect of this model which we need to specify. The model specifies the *expected* proportion for each x_i value when α and β are given numerical values, but says nothing about the variation in the responses. In the model for continuous variables, we postulated a Normal distribution for the variability about the line or curve but what distribution for the variability about the expected proportion would it be appropriate to specify here? If we consider just one value of x, for a moment, the answer will be clear. Suppose that the model predicts that the expected proportion would be 0.7, and that there are 10 individuals in all. If each individual responds *independently* of the others then the *number* responding follows a binomial distribution with $n = 10$ and $p = 0.7$ (Chapter 3, Section 4). Similarly, a binomial distribution would operate at any other value of x each with its own n, and $p =$ the expected proportion calculated from the logistic function.

4.3.4 The logistic-binomial model

Combining the systematic component of the logistic curve and the binomial model for the random variation gives

Logistic-binomial model for proportions

$$P(\text{reaction on the } i\text{th subject}) = p_i = \frac{e^{\alpha + \beta x_i}}{1 + e^{\alpha + \beta x_i}}$$

where the number reacting, R_i, out of n_i subjects for each x_i value, follows the binomial distribution:

$$R_i \sim \text{binomial}(n_i, p_i)$$

The logistic-binomial model is represented diagrammatically in Figure 4.8. Note how the *shape* of the distribution changes as the probability of reaction changes. This contrasts with the logistic-Normal model for a continuous variate in which the distributions about the curve are all the same—just their means are different—so that

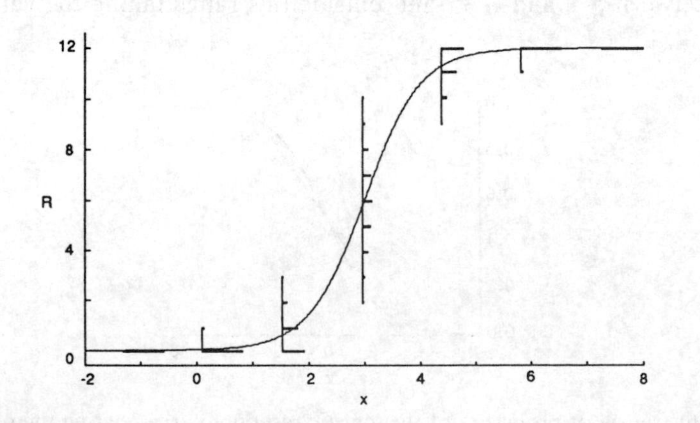

Figure 4.8 Schematic representation of the logistic-binomial model.

the shapes are the same. This change of shape in the distribution and, in particular, the non-constant variance is the reason why the methods used for fitting models involving Normal variation about a line cannot be applied without modification to this case. The method for fitting the logistic-binomial model is given in Chapter 11.

One further cautionary remark needs to be made about circumstances in which this model is not applicable. It is possible to obtain data in the form of proportions from data other than counts. An example is when the hormone level as a result of a treatment is expressed as a proportion of the hormone level during a control period, or when the amount of the yield in a mixed cropping situation which is of one of the components is expressed as a proportion of total yield. Such proportions are obtained as the ratio of two continuous measurements, and not a ratio of counts as is required for the above theory. It follows that this model is *not* then applicable. The logistic expression could be used, but the distribution of the observed proportion would not be given by the binomial distribution.

Example 4.4 Logistic-binomial model relating the incidence of coronary heart disease status to age

Table 4.2 summarises some data on the incidence of coronary heart disease status (CHD) in relation to age in 100 subjects participating in a study (Hosmer & Lemeshow, 1989). Figure 4.9 shows that the proportion of individuals with CHD increases with age as an S-shaped curve and that the underlying relationship is well described by the fitted logistic curve

$$p = \frac{\exp[-5.31 + 0.111\,age]}{1 + \exp[-5.31 + 0.111\,age]}$$

Table 4.2 Frequency of CHD against age class in a sample of 100 patients

	Age class							
	20–29	30–34	35–39	40–44	45–49	50–54	55–59	60–69
Number (n)	10	15	12	15	13	8	17	10
With CHD (r)	1	2	3	5	6	5	13	8
Proportion (p)	0.10	0.13	0.25	0.33	0.46	0.63	0.76	0.80

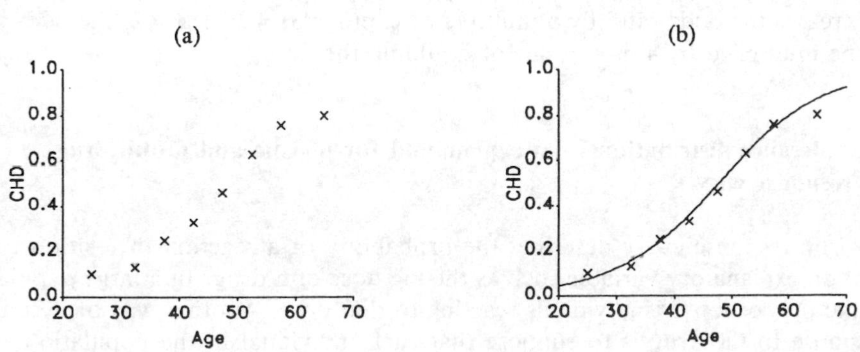

Figure 4.9 (a) Proportion of patients with CHD against age in a sample of 100 patients; (b) data with fitted logistic curve.

4.3.5 Log odds form of the logistic

One very useful property of the logistic model is that by applying a transformation to both sides of the equation, it becomes linear. To derive this we use the odds of an event given by $p/(1-p)$, i.e. the ratio of the probability of the event happening to the probability of it not happening. For example, if an event occurs with probability 2/3, the odds of it happening are 2 to 1. Conversely, if the odds of an event happening are 2 to 1 then it occurs with probability 2/3. For the logistic curve the odds of a reaction for x_i are calculated as follows

$$p_i = \frac{e^{\alpha+\beta x_i}}{1+e^{\alpha+\beta x_i}}$$

$$1 - p_i = 1 - \frac{e^{\alpha+\beta x_i}}{1+e^{\alpha+\beta x_i}} = \frac{1}{1+e^{\alpha+\beta x_i}}$$

with odds

$$\frac{p_i}{1-p_i} = e^{\alpha+\beta x_i}.$$

Taking $\log_e$'s on both sides, we obtain

Log odds form of the logistic

$$\log_e\left(\frac{p_i}{1-p_i}\right) = \text{logit}(p_i) = \alpha + \beta x_i$$

In toxicological studies the relationship is often a function of $x = \log(\text{dose})$ so that the effect of doubling the dose is to add $\log_e(2) = 0.693$ to $\log_e(\text{dose})$. Thus, no matter what the initial dose, each time the dose is doubled the log (odds ratio) increases by 0.693β, i.e. the odds ratio is multiplied by $\exp(0.693\beta)$. Thus if $\beta = 1$, doubling the dose increases the odds ratio by a multiple of $\exp(0.693) = 2$; if $\beta = 2$, the odds ratio would be multiplied by 4 as a result of doubling the dose.

4.3.6 Tolerance distributions—an explanation for logistic and probit drug-response curves

The logistic response curve describes the probability of a reaction in relation to the value of an explanatory variable such as the log dose of a drug. In a large population this is the proportion of individuals reacting to that dose. Another way of describing the response to the drug is to suppose that each individual in the population has a susceptibility or tolerance to drug action and if the susceptibility of the individual is less than the received dose then the individual will react. For this model there is a distribution of tolerances, or tolerance distribution, which describes the variation in susceptibility of the individuals in the population. Denote the cumulative distribution function of tolerances by $F(t)$, i.e. the proportion of individuals in the population with tolerance less than or equal to t. For an individual drawn at random the probability of a reaction is related to the cumulative distribution of tolerances by

$$P(\text{reaction to dose } d) = P(\text{tolerance} \leqslant d) = F(d).$$

A particular model is when the tolerances follow a lognormal distribution, i.e. the distribution of the logarithm of the tolerances is Normal. So the proportion reacting will be related to the logarithm of the dose by the cumulative Normal distribution. This curve is the basis of the statistical method called *Probit Analysis*. Often, however, we use the logistic curve to describe the probability of a reaction. This model has a shape very similar to the cumulative Normal distribution but is preferred because it has more convenient statistical properties and is simpler to interpret.

4.4 DYNAMIC STOCHASTIC MODELS

In Chapter 1 we developed dynamic deterministic models for the growth and decline of a single population and, in Chapter 2, for changes in the genetic structure over successive generations. The characteristic feature of these deterministic models is that, for specified values of the parameters and given initial conditions, subsequent values calculated from the model do not vary, i.e. a given input always produces the same outcome. This contrasts with the behaviour of a stochastic model in which the outcome varies randomly from realisation to realisation of the same model with the same input. Sometimes the actual process being modelled by a deterministic model could be stochastic, as for example, in genetics, but when the population, or sample, is large the chance variability is ironed out, and useful results are obtained by regarding the process as deterministic. In other situations, however, a random element can make a substantial difference. In this section we discuss some of the deficiencies of deterministic dynamic models and present some ways of incorporating a random element to form dynamic stochastic models.

First, we deal with *demographic stochasticity* which arises because populations contain a finite integer number of individuals subject to chance events. Demographic stochasticity can be very important for *small populations at risk of extinction* where chance variation could be of great practical significance. Second, we consider *sampling variability* in situations where only a *small sample* from a larger population is available. For example, changes in the frequency of a gene which is subject to selection could be recorded using small samples of individuals from a large population. Thirdly, we introduce *environmental stochasticity* due to random temporal variation in the environment. For example, the survival and breeding success of a bird species may be affected by weather. Environmental stochasticity leads to fluctuations at the population level, as distinct from demographic stochasticity which refers to chance variation between individuals in the *same* environment. We deal mainly with dynamic models which use difference equations for changes in discrete time rather than differential equations in continuous time. The reason is that it is easier to think about models in discrete time and it is also much simpler to simulate data from them. With dynamic stochastic models in continuous time there are difficult problems associated with how long a random influence lasts: these models are beyond the scope of this book.

4.4.1 Demographic stochasticity

Demographic stochasticity arises because populations consist of a finite and integer number of individuals subject to chance events. The effect leads to chance fluctuations which are most marked in small populations with few individuals. In a large population the effects of demographic stochasticity are ironed out. To illustrate this idea we begin with a very simple example using the binomial distribution. Then we present a more detailed example showing how demographic stochasticity can be incorporated into a deterministic dynamic model and how it can affect the dynamics.

Example 4.5 Effects of demographic stochasticity in different sized populations
Consider a population in which each individual survives until the following year with probability 0.5, independently of the others. Starting with N individuals the number of individuals which survive follows a binomial distribution with mean $0.5N$ and standard deviation $0.5\sqrt{N}$ (Chapter 3). The relative magnitude of the random variation in the number surviving, measured by the ratio of the standard deviation to the mean, decreases with population size as $1/\sqrt{N}$. For example, in a small population with $N = 5$ the ratio is 45% compared with only 2% for a larger population with $N = 1000$. Furthermore, in the smaller population there is a non-negligible probability of $1/32 = 0.03$ that all the individuals will die and the population will become extinct; for the larger population the probability of extinction is infinitesimally small. In other words the effects of demographic stochasticity are negligible for large populations.

4.4.2 A general stochastic difference equation formulation

To illustrate the way in which demographic stochasticity can be incorporated into deterministic dynamic models we consider the general form of the difference equation for exponential growth in discrete time (Chapter 1) given by

$$N_{i+1} = R_i N_i$$

or

$$N_{i+1} - N_i = (R_i - 1)N_i$$

where $(R_i - 1)$ is the relative growth rate of the population from year i to year $i + 1$.

To make the transition from a deterministic difference equation to a dynamic stochastic model, we replace the single integer value, N_i, at time i, by a probability distribution $P_i(N_i)$, which specifies the range of possible *integer* values that N_i can take, and the probability of each value occurring. Then the process is defined by rules specifying the probability distribution of the number in the population at time $i + 1$ in terms of the number in the population at time i. The example below illustrates the approach.

4.4.3 Exponential growth in discrete time with demographic stochasticity—extinction of small populations

Consider a model of an annual species, say an insect. In a deterministic model of this situation, each individual would give birth to a fixed number of offspring which survive to reproduce themselves, and would then die. Then numbers in successive years are related by the difference equation

$$N_{i+1} = RN_i$$

with solution

$$N_i = R^i N_0,$$

i.e. the population would increase exponentially. The model can be rewritten as

$$N_{i+1} = RN_i = \sum_{j=1}^{N_i} R.$$

This form is used to emphasise that the number in each year is the sum of the surviving offspring of the parents of the previous year and that each of the N_i parents produces R offspring. The stochastic version replaces each R in this last form by a random variable, with mean R, but which varies from individual to individual according to some probability distribution. One of the simplest models which we can take is for the number of offspring of each parent to follow a Poisson distribution with mean R. So the model becomes

$$N_{i+1} = \sum_{j=1}^{N_i} R_j$$

where $R_j \sim$ Poisson (mean R).

Consider a specific example. Suppose that we start with 1 individual at time 0, and let the value of $R = 1.7$. Then the number of offspring of this individual will follow a Poisson distribution with mean 1.7. Then suppose that the outcome of this Poisson distribution is 3, so that at the end of the first year there are 3 individuals in the population. Then each of these individuals will give rise to a number of offspring determined by a Poisson distribution, so that the number after two years would follow a distribution which would be given by the sum of three Poisson distributions. And so the process would continue.

Figure 4.10 demonstrates eight outcomes of this process as far as 20 seasons. One way in which this differs from the deterministic version is that there is a very real possibility of extinction. In some years no offspring are born to *every* individual and so the population dies out. As the population size increases extinction becomes less and less likely. Also, the stochastic simulations which survive the initial period show a fairly steady relative growth rate, equal to that of the deterministic model, illustrating that the effect of demographic stochasticity diminishes with population size. These stochastic simulations also maintain later numbers which are consistently higher, or lower, than the deterministic population trajectory with larger values corresponding to those simulations which quite by chance achieved high growth rates at the start.

The above general framework applies to more complex situations. For example, we could incorporate demographic stochasticity into a model of limited population growth

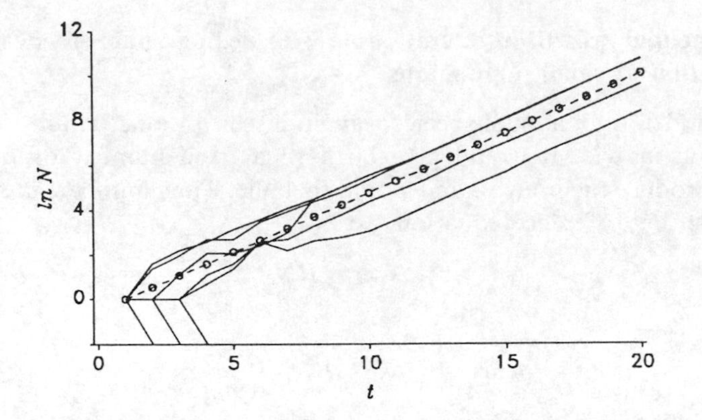

Figure 4.10 Simulated data showing eight realisations of a simple model of exponential growth in discrete time with demographic stochasticity. ○--○ = outcome of equivalent deterministic model.

by allowing the relative growth rate to vary with population density. One model based on the discrete logistic difference equation is to allow the number of daughters produced by each female to follow a Poisson distribution with mean $e^{r(1-N_i/K)}$, where N_i denotes population density. Such a model population would vary randomly about a mean level but with reduced variation in a larger population.

4.4.4 Sampling variability and measurement error

In the genetic models considered in Chapter 2 we assumed that we had such large populations that demographic variability could be ignored and also that we were observing the whole population. It is often the case, however, that we can only sample a population, and the sample might be so small that the resulting sampling variability affects the appearance of the sample trajectory very much. To illustrate this we consider the model to describe the process of selection against an allele of *Drosophila* which is lethal when homozygous and is responsible for stubble bristle when heterozygous. The difference equation for changes in gene frequency from one generation to the next is

$$p_{n+1} = \frac{p_n(1 - q_n hs)}{1 - sq_n(q_n + 2p_n h)}.$$

where $q_n = 1 - p_n$ is the gene frequency in the nth generation, s is the selection coefficient and h measures the relative fitness of the heterozygote (Chapter 2). If we can only observe a small sample of say size 20 on each occasion, then the estimate of p_n derived from the sample, written $\hat{p}_n$ to emphasise that it is an estimate, will not follow the course of the above process exactly, but will randomly fluctuate about it. If we sample 20 individuals at random on each occasion and these individuals are independently drawn on each occasion, then the number showing the trait out of the 20 sampled will follow a binomial distribution with $n = 20$ and $p = p_n$. Thus in simulating data from the model we need to allow for this extra variability.

Figure 4.11 demonstrates the appearance of two sample curves for a sample size $n = 20$ with $s = 1$ and $h = 0.12$, and the trajectory of the deterministic model is given

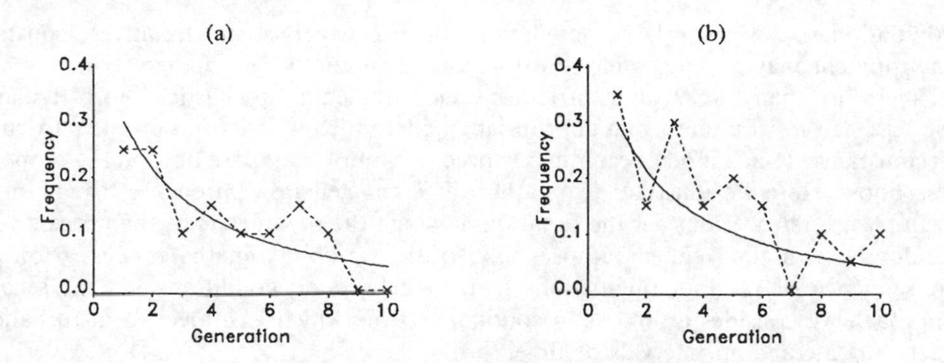

Figure 4.11 Two sample curves based on samples of size $n = 20$ for estimating the frequency of a gene in a population (broken line). Solid line shows decline in gene frequency in the population.

for comparison. It is seen that the appearance of the sample curves is much more jagged. The character of these simulations differs from those used to illustrate the effect of demographic stochasticity in two fundamental ways. First, with demographic stochasticity and a small population there is a chance of extinction and several trajectories petered out. Here this does not occur because there is no uncertainty in the process, just in our sample from it. Even if a sample we take does not contain any individual with the gene, it does not mean that the gene has died out in the population. Secondly, with demographic stochasticity once the curve had wandered away from the deterministic trajectory it occasionally took some time to get back on to expectation. In other words, there is some continuity *in the departures* from the deterministic model. In the present case, as the samples are completely independent on each occasion there is no such continuity except for an apparent continuity occurring sometimes by chance.

4.4.5 Environmental stochasticity

A type of random variation which is quite different in nature from either demographic stochasticity or sampling variability involves temporal variation in the environment of the population, sometimes called *environmental stochasticity*. Some examples and their potential effects on the dynamics of a population are: (a) seasonal variation in conditions such as weather, with associated differences in survival at different times of the year; (b) year-to-year fluctuations in food affecting the average number of young produced in a particular year; (c) shocks to a system such as electrical impulses in the case of neurones or catastrophic events causing large scale mortality in a population.

An important characteristic of environmental stochasticity is that its effects lead to fluctuations at the population level in both large and small populations. For example, in a 'bad' year the response of all individuals in the population will be affected. This contrasts with the effects of demographic stochasticity which are relatively small in large populations as individual random differences between individuals in a given year are ironed out. Because of this, environmental stochasticity is often more important than demographic stochasticity for both short-term and long-term changes in population size. Even a large population which can persist in a relatively constant environment may be vulnerable to extinction when conditions change.

There are many ways of incorporating environmental stochasticity into dynamic models, and in formulating an appropriate model much will depend on the particular circumstances. Such models can quickly become complex and are beyond the scope of this book. Here we illustrate two rather different general approaches using simple examples. First, we consider the situation in which the parameters of the process vary randomly with time either because of intrinsic variability in the process, or as a consequence of random fluctuations in the background conditions. In the second approach we consider a process in continuous time which is subject to disturbances occurring at random intervals of time.

Exponential growth in discrete time with environmental stochasticity—large populations

Consider the general form of the difference equation model for exponential growth in discrete time in which successive population sizes are given by

$$N_{i+1} = R_i N_i$$

where R_i is the geometric rate of increase from year i to year $i+1$. We take the particular case of an annual species in which individuals live for at most one year, reproduce and then die so that R_i is also the net reproductive rate.

For the simple deterministic model the growth rate is a constant R. When there is demographic stochasticity the net reproductive rate varies between individuals so that R_i is an average value for N_i individuals. Now consider a *large* population so that the effect of demographic stochasticity can be ignored. To incorporate environmental stochasticity we let the net reproductive rate R_i vary randomly from year to year and write

$$R_i = R\delta_i$$

where the mean net reproductive rate R is an average value over a hypothetically infinitely long run of years and δ_i is a positive random term reflecting variation for the ith year. This is the simplest formulation. More generally, the net reproductive rate may be related to the value of some *driving variable* or to population size. Also, there may be non-randomness in the form of runs of 'good' or 'bad' years. Substituting in the difference equation for change in population size gives

$$N_{i+1} = R\delta_i N_i$$

or

$$\log_e N_{i+1} = \log_e R + \log_e N_i + \log_e \delta_i.$$

Starting from some initial log population size the next log population size is obtained by *adding* a constant and a random quantity. This is repeated a number of times to generate a series. The model is sometimes referred to as a random walk model in one dimension because the log population size after n years can be represented as a point on the x-axis reached by taking n steps, either to the left or to right of the origin, with lengths sampled independently and at random from some distribution. For example, a random walk starting at $x = 0$ with five steps 3, -2, 1, 0 and 4 moves 3 to the right, 2 to the left, 1 to the right, stays where it is and moves 4 to the right, and ends up at $x = 3 - 2 + 1 + 0 + 4 = 6$. More generally, the model is written as

Random walk model for population growth

$$\log_e N_{i+1} = r + \log_e N_i + \varepsilon_i$$

where N_i is the population density in year i and r is the mean change in the logarithm of population size. The ε_i are *independent* random variables from a Normal distribution with mean zero and variance σ^2, i.e. $\varepsilon_i \sim N(0, \sigma^2)$, sometimes called *white noise*.

Figure 4.12 shows eight outcomes of the model over 20 years for $R = 1.7$ ($r = 0.531$) and values of $\sigma^2 = 0.10$ and 0.40. Individual trajectories fan out, following irregular paths some of which appear to diverge from the trajectory of the deterministic model. Observations in a particular realisation tend to lie on the same side of the deterministic

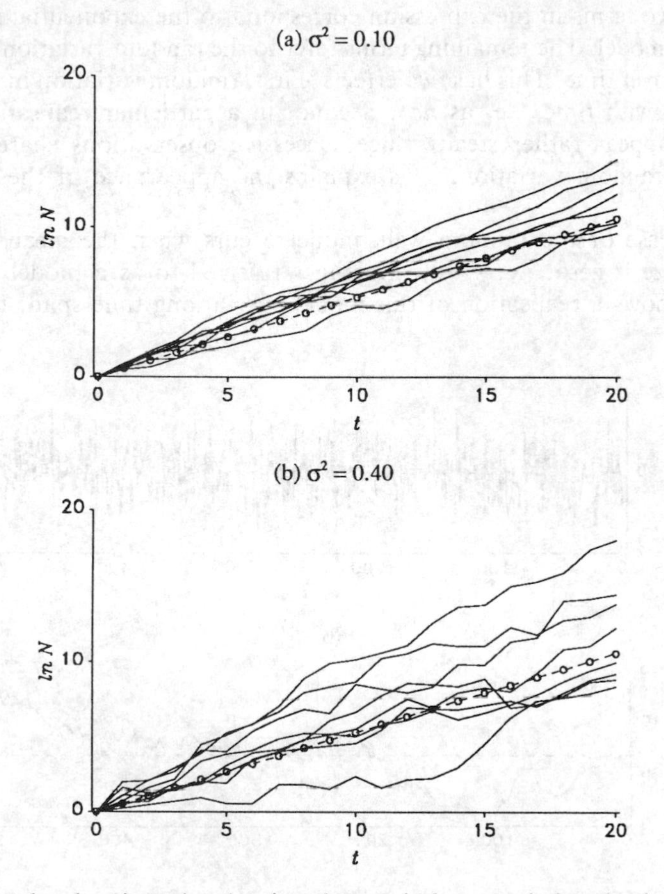

Figure 4.12 Simulated trajectories showing changes in log population density for a random walk model with different values for the variance of year-to-year change in log population size. ○–○ = outcome of equivalent deterministic model.

line rather than being scattered about it. Increasing the variance of the population change leads to more variation within trajectories and a greater spread between trajectories. The environmental stochasticity produces rather different looking trajectories from those when only demographic stochasticity is involved (Figure 4.10). With demographic stochasticity, variation in the relative growth rate become negligible as the population increases in size whereas with environmental stochasticity there is variation both in the short-term changes and the long-term trends.

The effects of environmental stochasticity in the random walk model can be explained as follows. Successively iterating the difference equation gives

$$\log_e N_1 = r + \log_e N_0 + \varepsilon_0$$

and

$$\log_e N_2 = r + \log_e N_1 + \varepsilon_1 = 2r + \log_e N_0 + \varepsilon_1 + \varepsilon_0,$$

so that in general

$$\log_e N_n = nr + \log_e N_0 + \varepsilon_{n-1} + \varepsilon_{n-2} + \cdots + \varepsilon_0.$$

The first two terms in the expression correspond to the exponential increase of the deterministic model. The remaining terms refer to the random variation which is being accumulated over time. This has two effects. First, random variation in log population size increases with time, i.e. as $n\sigma^2$. Second, in a particular realisation population growth will appear rather steady since successive observations share much of the accumulated random variation. This explains the appearance of the trajectories in Figure 4.12.

A special case of the random walk model occurs when the mean change in log population size is zero, i.e. $r = 0$, sometimes referred to as a model with no *drift*. Figure 4.13 shows a realisation of this model over a long time span. It is interesting

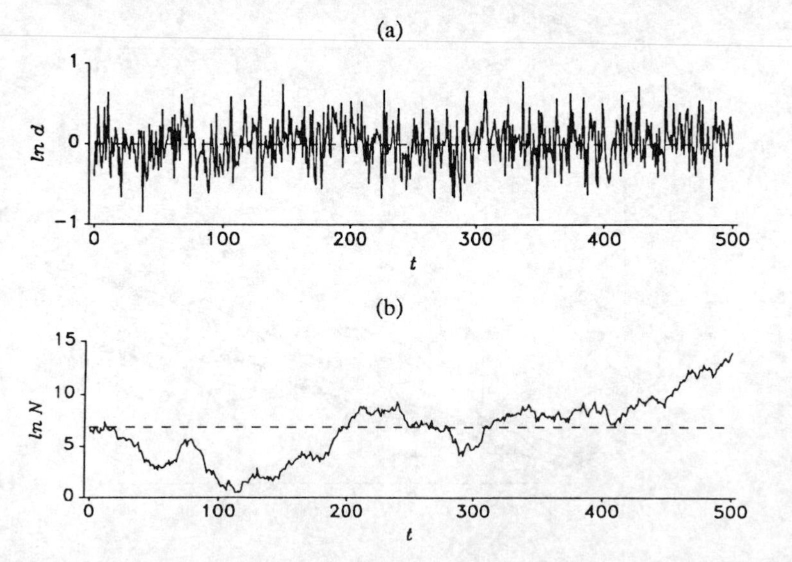

Figure 4.13 Realisation of a random walk model in which the changes in log population size are independently and Normally distributed with mean zero and variance = 0.10. (a) Changes in log population size; (b) log population size.

to see how the phenomenon of increased variability with time can give the false impression of a trend and how this can vary depending which section of the series we happen to look at. The random walk model is unrealistic for most real populations since the increased variation with time implies eventual extinction or indefinite increase. Nevertheless, over short periods population change may be dominated by random variation and Figure 4.13(b) shows that such a population could in theory persist for long periods.

Limited population growth in discrete time with environmental stochasticity

In the model for exponential growth with environmental stochasticity the growth rate varies randomly from one time to the next. This illustrates a general approach of incorporating environmental stochasticity in a deterministic dynamic model by allowing the parameters to vary randomly. This method extends to other models with more parameters although these models quickly become very complex and outside the scope of this book. Even with only two parameters there are several different possibilities. For example, consider the deterministic discrete time model for limited population growth in which successive population densities are given by the difference equation

$$N_{i+1} = N_i \, e^{-r(1-N_i/K)}$$

where r is the intrinsic relative growth rate and K is the equilibrium population density. A model with environmental stochasticity could have: (a) variable intrinsic relative growth rate r with constant K; (b) constant r, variable K; (c) r and K varying but independently of each other; (d) r and K covarying—for example, in 'good' years breeding success may be high and the habitat may be able to support a large population, increasing both r and K, and in 'poor years' both r and K may be reduced. In formulating an appropriate model much will depend on particular circumstances. Example 4.6 shows an application of a simple model for limited growth in discrete time with environmental stochasticity to the winter moth *Operophtera brumata*.

Example 4.6 Population fluctuations of the winter moth in Wytham Wood, 1950–68

Many annual species of insect pass through various stages in their life cycle from egg to adult during which they are subject to different forces of mortality. In a long-term study of the winter moth *Operophtera brumata* Varley & Gradwell (1968) identified six sequential stages: (1) egg to full grown larvae; (2) parasitism of larvae by the *Cyzensis albicans*; (3) parasitism by non-specific Diptera and Hymenoptera; (4) parasitism by *Plistophora operophterae*; (5) pupal predation; (6) adults. The changes in size of successive generations can be expressed quite generally by the difference equation

$$N_{i+1} = E_i s_{1i} s_{2i} s_{3i} s_{4i} s_{5i} s_{6i} N_i,$$

where E_i is the egg output from adults in year i and $s_{1i}, s_{2i}, \ldots, s_{6i}$ are the survival rates in the different stages. Survival rates in the different stages vary from year to year due to factors such as weather, food and abundance of parasites, and changes in population size. A basic problem is to analyse the contribution of the variation in the different survival rates to changes in population density. One approach called *key-factor* analysis is to calculate *k*-values (i.e. $-\ln s$) for each stage and to plot them: (a) in time; and (b) against the logarithm of the density of individuals entering the stage. This can indicate which survival rates vary most and which may be limiting population growth. Figure 4.14 shows untransformed survival rates plotted against the logarithm of the density entering each stage. Some survival rates show substantial year-to-year variation: for example, survival in stage 1 varies approximately 22-fold. The survival rate in stage 6 also appears quite variable but the relative variation, which is important for population change, is much less and approximately 5-fold. The survival rate in stage 5 shows a decline with the logarithm of the density entering that stage. A plausible model of this relationship is given by

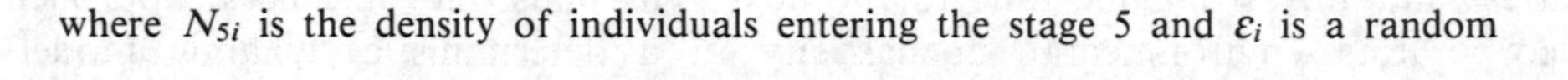

$$s_{5i} = \frac{1}{1 + \exp[-0.662 + 0.495 \ln N_{5i} + \varepsilon_i]}$$

where N_{5i} is the density of individuals entering the stage 5 and ε_i is a random

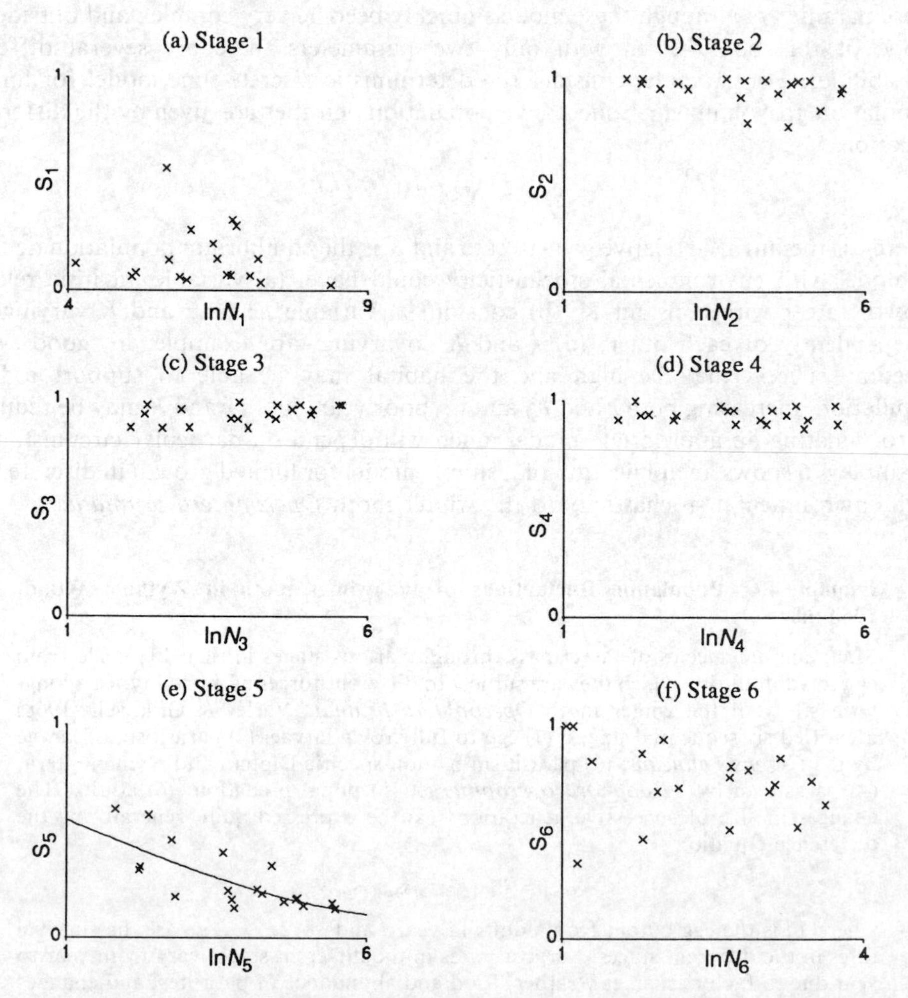

Figure 4.14 Survival rates in different life stages of the winter moth plotted against log density entering each stage. Data from Manly (1990).

deviation. Substituting in the general model gives the following difference equation model for change in density from year to year

$$N_{i+1} = \frac{E_i s_{1i} s_{2i} s_{3i} s_{4i} s_{6i} N_i}{1 + \exp[-0.662 + 0.495 \ln(E_i s_{1i} s_{2i} s_{3i} s_{4i} N_i) + \varepsilon_i]}.$$

This is a dynamic stochastic model with: (a) a systematic component for the relationship between survival in the stage 5 and the density entering that stage with an added random element; (b) random variation in survival rates for other stages. Figure 4.15(a)–(d) gives four scenarios to illustrate the contribution of the systematic and random components to the changes in population density. In each case the egg output is constant at 26.25. Figure 4.15(a) shows changes simulated using the purely deterministic model in which the survival rates in all stages except stage 5 are held at their observed average values. The population increases to a stable level. In Figure 4.15(b) the effect of limited growth has been removed by holding the survival rate in stage 5 at its average value and simulating changes by using the observed pattern of survival in the other stages. In this case the population fluctuations show a very large amplitude compared with that observed illustrating the role of the relationship between survival in stage 5 and density for containing the size of the population. In Figure 4.15(c) changes have been simulated by adding to the systematic component the observed pattern of survival in stage 1 while keeping the other survival rates at their average values. This simple model mimics the changes in population density rather well. Adding the observed pattern of

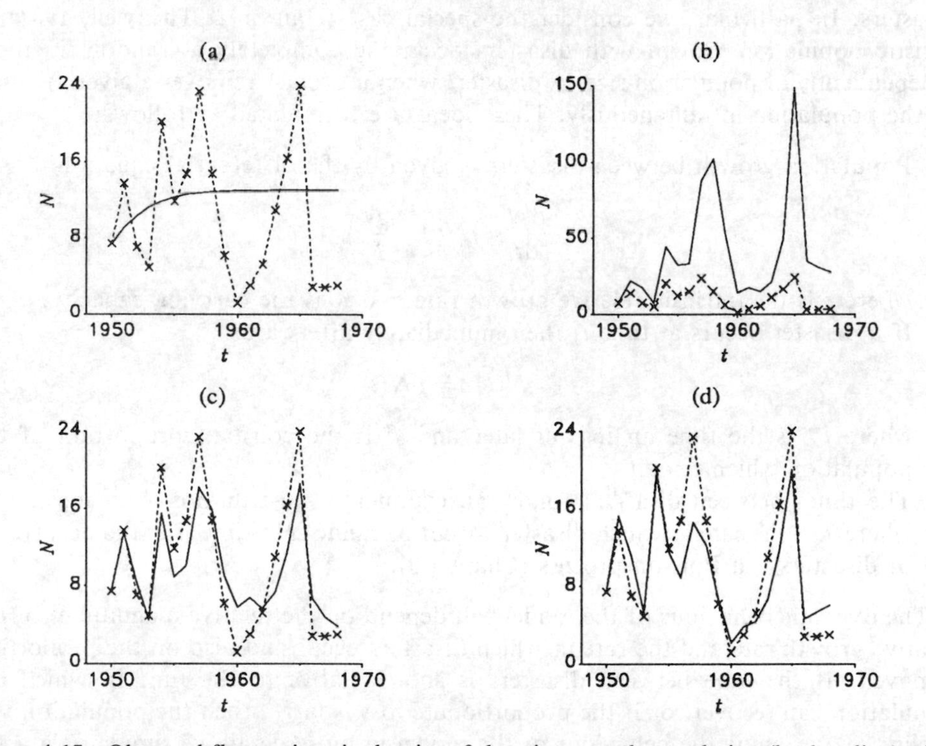

Figure 4.15 Observed fluctuations in density of the winter moth population (broken line) with simulated data (solid line) for models assuming different scenarios for mortality in the life stages: (a) deterministic model with pupal survival decreasing with pupal density; (b) observed pattern of mortality but with constant pupal mortality; (c) deterministic model with observed pattern of pupal mortality; (d) same as (c) with added observed pattern of mortality for remaining stages.

survival rates in the other stages changes the detailed fit of the model but not the general features of change (Figure 4.15(d)). The analysis suggests that the changes in the size of the population results can be largely accounted for by the interplay of two factors: (a) random variation in survival from egg to larvae leading to fluctuations in population density; (b) survival in the pupal stage declining with pupal density which limits growth and contains the magnitude of fluctuations due to the environmental stochasticity. Manly (1990) describes a theoretical approach to these data which also identifies these two factors as important for population change.

Logistic population growth in continuous time with random disturbances

So far we have illustrated the effects of environmental stochasticity using difference equation models for growth in discrete time. In these models variability was incorporated by allowing the parameters of the model to vary from one time to the next, i.e. random changes occur at fixed times. In this section we consider a different situation where population growth is a continuous process which is subject to disturbances occurring at irregular intervals of time. For example, an intertidal population which may be periodically depleted by severe storm conditions. In the model, population growth is assumed to be deterministic between randomly occurring disasters. In particular, we consider the special case (Hanson & Tuckwell, 1981) of logistic population growth with disasters occurring completely at random in time, independently of population size. A disaster, when it occurs, removes a given fraction of the population instantaneously. These ideas are formulated as follows:

(1) Population growth between disasters is given by the differential equation

$$\frac{dN}{dt} = rN\left(1 - \frac{N}{K}\right)$$

where r is the intrinsic relative growth rate and K is the carrying capacity.

(2) If a disaster occurs at time t, then immediately afterwards

$$N(t^+) = pN(t)$$

where t^+ is the time an instant later and p is the constant proportion of the population which remains.

(3) The times between disasters follow an exponential distribution with mean $1/\lambda$, where λ is the rate at which disasters occur at random in time, i.e. the occurrence of disasters is a Poisson process (Chapter 3).

The dynamic behaviour of the model will depend on the relative magnitudes of the relative growth rate and the rate at which disasters occur and also on the proportion removed. If the time between disasters is short relative to the time in which the population can recover, or if the proportionate loss is large, then the population will decline. To examine the behaviour of the model in more detail we simulate data for given values of the parameters. The steps in the simulation are as follows:

Step 1. Start from $N(t) = N(t_0)$ at time $t = t_0$.

Step 2. Sample an observation from an exponential distribution with mean $1/\lambda$, say T_1, for the time *interval* to the first disaster.

Step 3. Calculate population growth during the time interval $(t_0, t_0 + T_1)$ up to the first disaster from the logistic model, i.e.

$$N(t) = \frac{KN(t_0)}{\{N(t_0) + [K - N(t_0)]\ e^{-r(t-t_0)}\}}.$$

Step 4. Calculate population size immediately after the disaster as $pN(t_0 + T_1)$.
Step 5. Repeat steps 1–4 with the new population size at time $t = t_0 + T_1$.

Table 4.3 gives a numerical example of the method using $r = 1$, $K = 1$, $p = 1/2$ and $\lambda = 0.5$, i.e. mean time between disasters of 2 time units. Figure 4.16(a) shows the resulting fluctuations in population size in a longer series. Figures 4.16(b) and (c) illustrate the effect of increasing the disaster rate using $\lambda = 1$ and 2. For $\lambda = 1$ the population declines but then following a run at low levels starts to increase again. For $\lambda = 2$ the population is effectively driven to extinction. To explain these results we consider the behaviour of the corresponding deterministic model in which disasters occur at regular time intervals equal to the mean time between random events. For this model the population size is reduced by a proportion p every $1/\lambda$ units of time. If the population is to offset this reduction before the next disaster then it must increase to $N(t)$ in time $1/\lambda$ starting from $pN(t)$. In other words the deterministic model avoids a decline when

$$N(t + 1/\lambda) = \frac{KpN(t)}{\{pN(t) + [K - pN(t)]\ e^{-r/\lambda}\}} > N(t)$$

or, rearranging,

$$\frac{N(t)}{K} < \frac{(e^{r/\lambda} - 1/p)}{(e^{r/\lambda} - 1)}.$$

Thus the condition for the population to recover from a disaster depends on the ratio of the population size to its carrying capacity, the proportion removed and the ratio of the intrinsic relative growth rate to the rate at which the disasters occur. If the

Table 4.3 Numerical example of method to simulate data from the logistic model with random disasters using: intrinsic relative growth rate $r = 1$, carrying capacity $K = 1$, mean time between disasters $1/\lambda = 2$, proportion of population remaining after disaster $p = 1/2$, initial population size $N(0) = 0.5$

Time	Time interval to disaster	Population size:	
		just before disaster	just after disaster
0	2.729	0.939	0.469
2.729	0.135	0.503	0.252
2.865	0.412	0.337	0.168
3.277	0.625	0.274	0.137
3.901	1.492	0.414	0.207
5.393	2.443	0.750	0.375

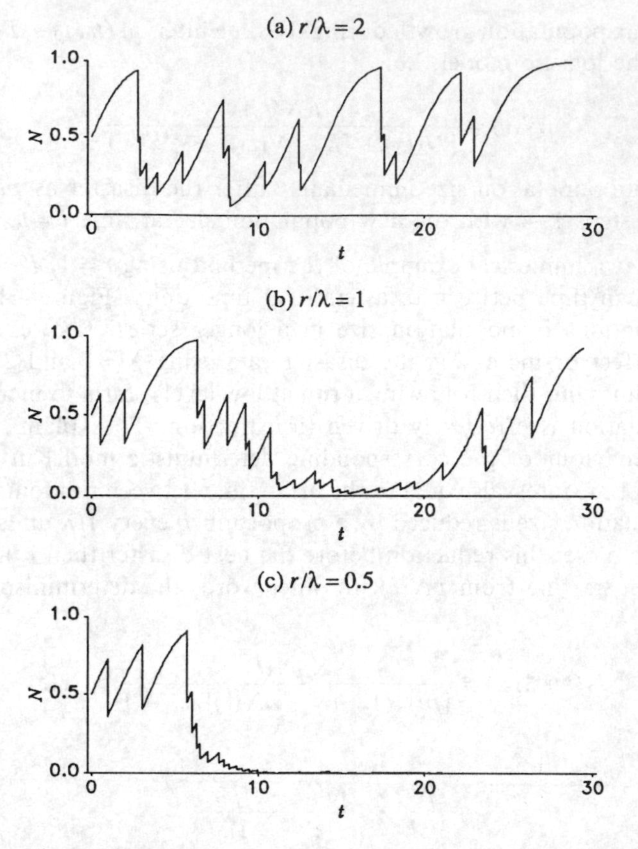

Figure 4.16 Illustration of logistic population growth with random disasters for $r = 1$, $K = 1$, proportion remaining, $p = 1/2$ and rate of disasters $\lambda = 0.5$ (a), 1.0 (b), 2.0 (c).

right-hand side of the above expression is negative, i.e. $r/\lambda < -\log_e p$, the population will never recover. For example, when $p = 1/2$ so that $-\log_e p = 0.6931$ and when $r/\lambda = 0.5$ the population never recovers. This is essentially what eventually happened in the run of the stochastic model (Figure 4.16(c)) although, by chance, the population increased initially. When $r/\lambda = 1$ and $p = 1/2$ the deterministic model population recovers from a disaster when $N(t)/K < 0.418$. For the stochastic model, however, this is not necessarily the case and Figure 4.16(b) shows a prolonged period where the population is below this level. This can occur in a stochastic model when the times between disasters are, by chance, less than the mean time between disasters. This happens in about 63% of the times between disasters because of the skewness in the distribution of times between disasters. In this sense the stochastic model is more prone to declines than the deterministic model with the same disaster rate.

4.5 MODELS FOR CORRELATED RANDOM VARIATION

In Section 4.2.1 we presented the straight line-Normal model for relating a response variable Y to the value of an explanatory variable x. For a given value of x, values of Y follow a Normal distribution with mean $\alpha + \beta x$ and constant variance. As an example we used the relationship between growth rate of weanling rats and the dose of vitamin B_2 where the values of the explanatory variable are *fixed* by the experimenter and the response variable, growth rate, is subject to the random variation arising from differences between individual rats. There are other situations, however, where we wish to describe the relationship between two variables *both* of which vary randomly. An example described below is the relationship between mid-parent heights (i.e. the averages of the heights of the two parents) and the heights of their offspring in a large population. In this section we present two models for this kind of correlated random variation. In the first approach we consider the variation in one of the variables for fixed values of the other—a so-called *conditional model*. Perhaps surprisingly we shall see that the descriptive model which results is very similar to the straight line-Normal model formulated for growth in relation to some experimental treatment. In the second approach we include a random element in both variables and present a *joint distribution model*, the bivariate Normal distribution.

4.5.1 Offspring and mid-parent height in humans

Extensive data have been collected for many characters and species on the relationship between the value of the character measured on the offspring and the mean of the two parents. Table 4.4 illustrates the classic set of such data on adult height in humans

Table 4.4 Relationship between offspring height and mid-parent height

Height child − 0.2 (in)	Mid-parent height − 0.5 (in)											
	<63.5	64	65	66	67	68	69	70	71	72	>72.5	n
>73.5	—	—	—	—	—	—	5	3	2	4	—	14
73	—	—	—	—	—	3	4	3	2	2	3	17
72	—	—	1	—	4	4	11	4	9	7	1	41
71	—	—	2	—	11	18	20	7	4	2	—	64
70	—	—	5	4	19	21	25	14	10	1	—	99
69	1	2	7	13	38	48	33	18	5	2	—	167
68	1	—	7	14	28	34	20	12	3	1	—	120
67	2	5	11	17	38	31	27	3	4	—	—	138
66	2	5	11	17	36	25	17	1	3	—	—	117
65	1	1	7	2	15	16	4	1	1	—	—	48
64	4	4	5	5	14	11	16	—	—	—	—	59
63	2	4	9	3	5	7	1	1	—	—	—	32
62	—	1	—	3	3	—	—	—	—	—	—	7
<61.5	1	1	1	—	—	1	—	1	—	—	—	5
Totals	14	23	66	78	211	219	183	68	43	19	4	928
Mean	65.1	65.2	66.5	66.8	67.4	67.8	68.5	69.4	69.9	71.7	72.7	
s.d.	2.13	2.10	2.41	1.95	2.15	2.24	2.46	2.18	2.22	1.65	0.50	

compiled by Galton (1889). The question is how strongly is the character transmitted from parents to offspring? If there were perfect replication of the parental value in the offspring, the relationship would be a straight line with slope 1, but this is not the case. How can we reduce such data so that meaningful comparisons can be made between characters and species? The first step in such reduction is to devise a model which can be used to describe the data. In comparing the phenomenon between characters and species, we would then compare the parameters of the model fitted to each data set. The problem of fitting the model is postponed to the next chapter; here we consider only how we might draw up the model in the first place.

We might first suspect that we could theoretically work out what form of mathematical model would be most appropriate for this relationship, but as we have a large collection of data, we could attempt to discover from the data alone what an appropriate model would be. We notice from inspection of Table 4.4 that if we were to plot the offspring height on the y-axis against mid-parent height on the x-axis, we should get an elliptical cloud of points with a greater density of points towards the centre. The table presents columns of data which correspond to sets of points which would fall within thin vertical sections which we could cut through such a graph. Each column gives the heights of all the offspring whose mid-parent height falls within a specific inch wide band. By calculating the means and variances or standard deviations of these restricted sets of offspring heights we can get an idea of the broad trend in mean, and in the magnitude of the variability about that mean, of offspring height for a given mid-parent height. The means and standard deviations are given at the foot of the table and show that the mean increases approximately as a straight line with mid-parent height, and the standard deviation about the line is approximately constant. If we plot out histograms of the variation of offspring height for fixed mid-parent heights (we do not include the two outer bands of mid-parent height because the sample sizes are very small) we see also that a Normal distribution of the departures about the line might not be a bad approximation (Figure 4.17). Thus the variation in height of offspring from a given value of mid-parent height can be described by the straight line-

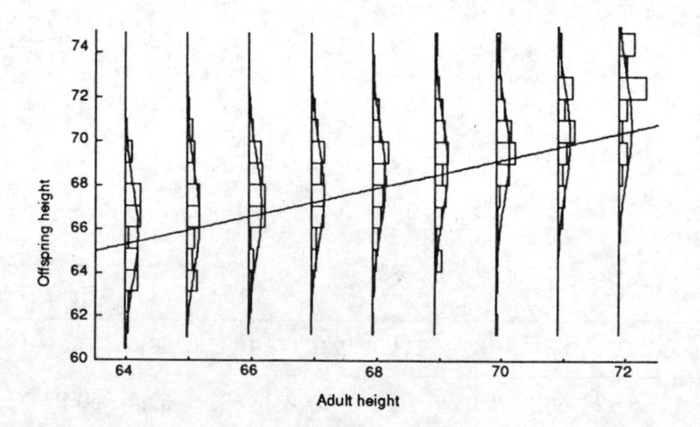

Figure 4.17 Histograms of offspring height data for different values of mid-parent heights.

Normal model in much the same way that growth of rats is related to different doses of a nutritional supplement, despite the different original situation.

It is also seen that the slope of the line is less than 1, so that the *expected* departure of the offspring height from the mean is less than the departure of the mid-parent height from the mean. This is the phenomenon of *regression towards the mean*, first reported by Galton, and from which the statistical method for fitting such models—regression analysis—derives its name. We said above that, if there were perfect replication of the parental phenotype in the offspring, we would expect a slope of 1, i.e. no regression towards the mean. A slope of less than one can be explained by the fact that height is subject to environmental variation as well as having a genetic component.

4.5.2 Correlated Normal random variables and the bivariate Normal distribution

Galton's data illustrates the idea of correlated Normal random variation where two variables each follow a Normal distribution and where the distribution of one variable for a fixed value of the other also follows a Normal distribution. The theoretical distribution which has these properties is called the bivariate Normal distribution. In this section we briefly present the main properties of this distribution and show how to simulate data from it.

Figure 4.18 shows simulated data for some bivariate Normal distributions as scatter diagrams of one variable, Y, against the other variable, X—the method for simulating the data is described below. Each plot shows an elliptical shaped scatter of points but the 'strength of the relationship' varies between sets of data. The appearance of the scatter plots is related to the values of the parameters in the bivariate distribution. These parameters and the main properties of the distribution are described briefly below.

In the bivariate Normal distribution of two random variables X and Y: (a) both X and Y are Normally distributed; (b) for a given value of X, Y follows the straight line-Normal model in X, and similarly for X given Y. The model has five parameters. These are the mean and variance of the Normal distributions of X and Y, and the so-called *covariance* between X and Y. The covariance is the mean value of the cross-product of the deviations of X and Y from their mean values, i.e. it is the average value of the quantity $(X-\mu_X)(Y-\mu_Y)$ in a hypothetical infinite number of randomly selected pairs of (X, Y) values. If the deviations from the mean tend to be similar in sign the covariance will be positive; and conversely when the deviations tend to be of opposite sign the covariance will be negative. A related quantity is the *correlation coefficient*, ρ, which is the covariance scaled by dividing by the product of the two standard deviations; it has the property that it lies between $(-1, 1)$.*

* A correlation coefficient can be calculated from samples of observations selected from the population but the distinction between the population parameter and the sample values which vary from sample to sample is important.

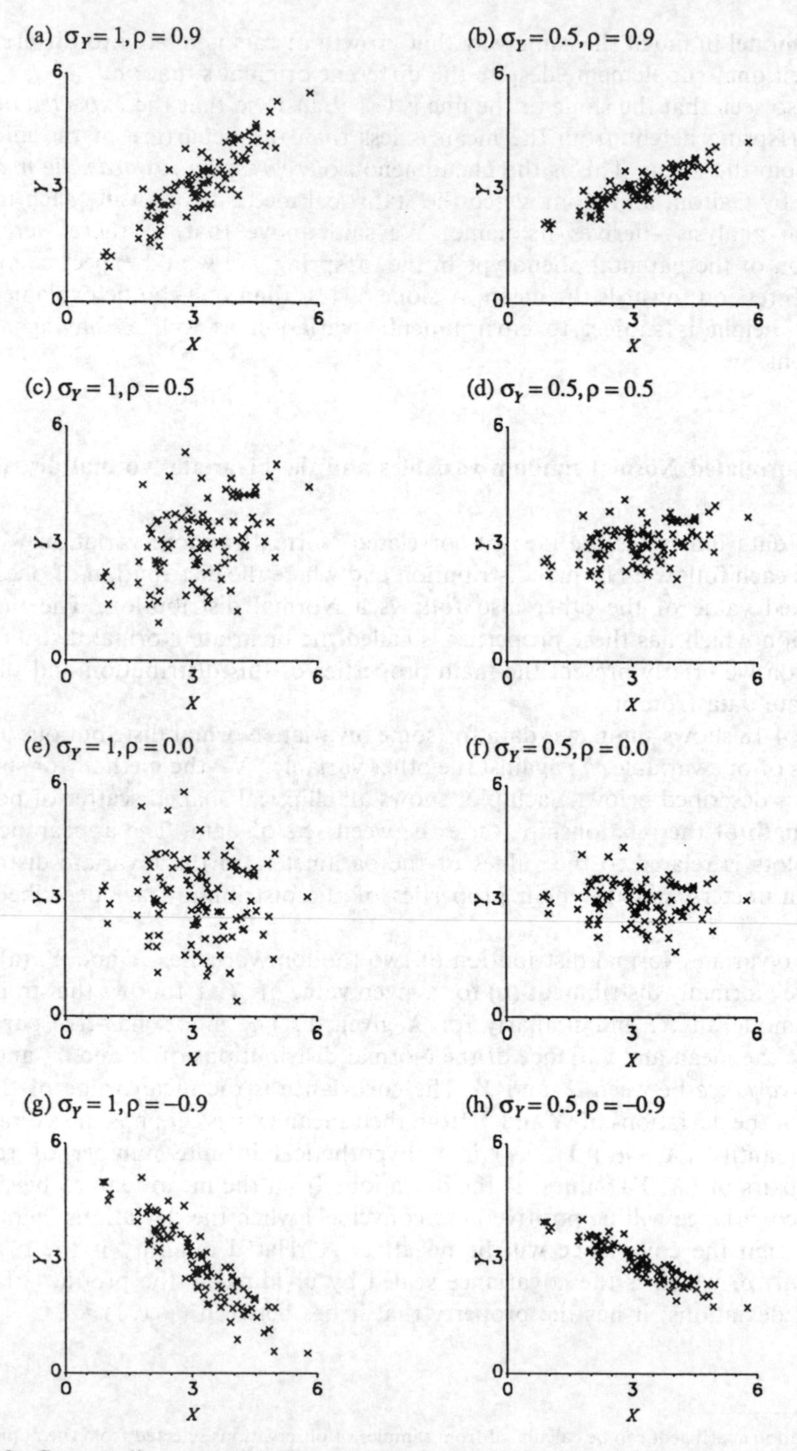

Figure 4.18 Scatter diagrams showing simulated correlated Normal variates X and Y for different values of the correlation coefficient and standard deviation of Y. Each diagram shows 100 pairs of points derived from the same set of standardised Normal variates.

Bivariate Normal distribution: correlated Normal random variation

$$X \sim N(\mu_X, \sigma_X^2); \qquad Y \sim N(\mu_Y, \sigma_Y^2)$$

Distribution of Y given X: $\quad N\left(\mu_Y + \frac{\sigma_{XY}}{\sigma_X}(X - \mu_X), \sigma_Y^2(1 - \rho^2)\right)$

(similarly for X given Y)

Covariance: $\quad \text{cov}[X, Y] = \text{mean}[(X - \mu_X)(Y - \mu_Y)] = \sigma_{XY}$

Correlation: $\quad \rho = \sigma_{XY}/\sigma_X\sigma_Y$

Figure 4.18 illustrates that as the absolute value of the correlation coefficient increases the elliptical shaped scatter of points becomes tighter. In the straight line-Normal model describing the variation in Y for given X the slope of the line is $\sigma_{XY}/\sigma_X = \rho\sigma_Y/\sigma_X$ and the variance about it is $\sigma_Y^2(1 - \rho^2)$. Figure 4.18 illustrates these results by showing how the slope of the line and the scatter about it vary as ρ and σ_Y are changed.

Covariance and correlation

Correlated random variation is not restricted to the Normal distribution but a discussion of other models is beyond the scope of this book. However, the covariance between a pair of random variables, i.e. the mean of the cross-product of the deviations from their means, is used quite generally for any random variables whatever the form of their distribution, and similarly for the correlation coefficient. Note, however, that a *sample* correlation coefficient (r) is often calculated for a set of data values $(x_1, y_1), (x_2, y_2), \ldots, (x_n, y_n)$.

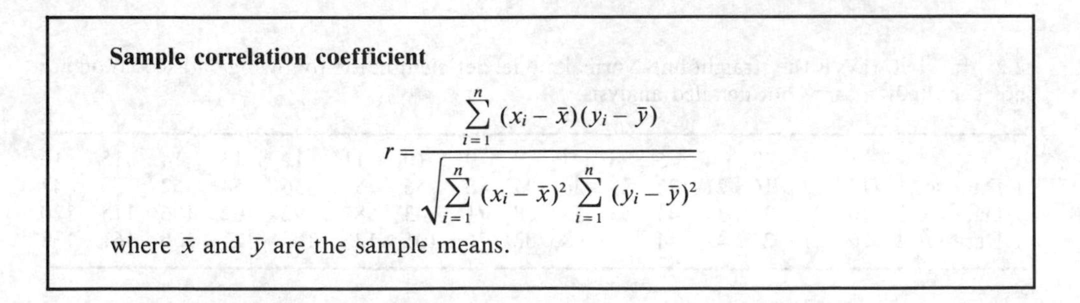

Sample correlation coefficient

$$r = \frac{\sum_{i=1}^{n} (x_i - \bar{x})(y_i - \bar{y})}{\sqrt{\sum_{i=1}^{n} (x_i - \bar{x})^2 \sum_{i=1}^{n} (y_i - \bar{y})^2}}$$

where $\bar{x}$ and $\bar{y}$ are the sample means.

If the observations are a random sample from some distribution of pairs of values then r can be regarded as an *estimate* of the population value which would vary from sample to sample and therefore be subject to sampling error. More often, r is used simply as a measure of the strength of the linear relationship in the sample values (the sample correlation coefficient between two variables which are functionally related can be zero). In many applications the values of X, such as dose of vitamin B_2 or some other explanatory variable, are deliberately chosen by the experimenter so that a model in which both X and Y are random is not relevant.

Simulating correlated Normal random variables

There are times when we might want to simulate correlated Normal random variation, i.e. sample observations from a bivariate Normal distribution. For example, to vary the parameters in a dynamic stochastic model to mimic environmental variation from year to year (Section 4.4.5). One way to do this is first to sample X from the specified Normal distribution and then sample Y from the Normal distribution with mean determined by the value of X.

Simulating data from a bivariate Normal distribution

$$X = \mu_X + \sigma_X Z_1$$

$$Y = \mu_Y + \rho\sigma_Y Z_1 + \sigma_Y\sqrt{1-\rho^2}Z_2$$

where Z_1 and Z_2 are independent standardised Normal random variables, μ and σ refer to mean and standard deviation and ρ is the *correlation coefficient* between X and Y.

EXERCISES

4.1 For each of the three data sets below, suggest a model to describe the underlying relationship between y and x. In each case, estimate by eye the parameters of the model and simulate a set of data by adding random variation from a Normal distribution with mean zero and standard deviation 0.8.

x	1	2	3	4	5	6	7	8	9	10	11	12
y Data set 1	3.7	4.5	3.7	6.8	6.4	7.8	8.5	9.5	11.4	12.9	12.6	13.4
y Data set 2	2.0	7.1	10.0	11.3	13.9	15.0	14.2	14.1	12.1	8.7	8.2	2.7
y Data set 3	4.2	8.6	10.3	10.7	12.7	12.1	12.4	12.8	12.5	14.8	16.4	20.9

4.2 In what ways is the straight line-Normal model deficient in the following datasets? You are not expected to carry out detailed analysis.

x	1	2	3	4	5	6	7	8	9	10	11	12	13	14	15	16
y Data set 1	15	13	16	21	22	29	36	45	46	56	53	56	54	52	50	47
y Data set 2	10	21	27	39	44	55	58	68	71	43	87	92	102	106	115	120
y Data set 3	10	21	27	42	44	53	68	82	76	101	125	103	121	156	161	138

4.3 Consider the following particular cases of the asymptotic exponential and the rectangular hyperbola models with additive random errors

$$Y = 3(1 - e^{-0.1386x}) + \varepsilon; \qquad Y = \frac{3x}{(5+x)} + \varepsilon$$

where $\varepsilon \sim N(0, (0.1)^2)$.

(a) Comment on the differences and similarities of the two underlying response curves.

(b) For each model, simulate one value of Y for each of $x = 1, 2, \ldots, 30$. Use your results to illustrate the problems of choosing between models and why, in a particular application, it might be important to have knowledge of the underlying mechanisms involved.

4.4 In what situations might models of the form

$$Y = 3(1 - e^{-0.1386x})\,\delta; \qquad Y = \frac{3x}{(5+x)}\,\delta$$

where $\ln \delta \sim N(0, \sigma^2)$, be more appropriate than those given in Exercise 4.3? Simulate some data to illustrate your answer.

4.5 Consider the following model relating the value of a positive continuous response variable, Y, to an explanatory variable, x, in which

$$Y = Ax^{\alpha}\, e^{-\beta x}\delta$$

where $\ln \delta \sim N(0, \sigma^2)$.

(a) Show that for $\alpha, \beta > 0$, the underlying relationship increases to a maximum at $x_0 = \alpha/\beta$ and then declines to zero. Find the maximum in terms of the parameters A, α and β.

(b) The data below were simulated from a model with $A = 3$, $\alpha = 1$, $\beta = 0.05$ and $\sigma = 0.10$. Simulate five more sets of data and display them for comparison with the model as: (i) y against x; (ii) $\ln y$ against x. Comment on the form of the random component for modelling this type of relationship.

x	5	5	10	10	15	15	20	20	25	25
y	10.7	13.4	18.4	18.9	21.3	21.4	24.3	18.9	21.1	22.0
x	30	30	35	35	40	40	45	45	50	50
y	20.6	19.5	17.5	16.3	15.5	16.3	15.1	12.6	13.7	14.8

(Note: Applications of the model include the relationship between photosynthesis rate and solar radiation, and plant yield and density.)

4.6 The table below gives some data on survival times of leukaemia patients (Feigl & Zelen, 1965). These are time to death, y, in weeks from diagnosis and $\log_{10}$ (initial white blood cell count), x, for 17 patients suffering from leukaemia.

x	3.36	2.88	3.63	3.41	3.78	4.02	4.00	4.23	3.73
y	65	156	100	134	16	108	121	4	39
x	3.85	3.97	4.51	4.54	5.00	5.00	4.72	5.00	
y	143	56	26	22	1	1	5	65	

(a) For describing the relationship, discuss the limitations of the straight line-Normal model and models of the form $Y = f(x) + \varepsilon$, where $\varepsilon \sim N(0, \sigma^2)$.

(b) Comment on the advantages of the model $Y = \beta_0\, e^{\beta_1 x}\delta$, where δ is a random term which follows an exponential distribution with mean equal to one.

(c) Compare the observed relationship with the model for which $\beta_0 = 4800$ and $\beta_1 = -1.109$. Simulate five sets of data for comparison with the observed random scatter of points. (For further discussion of this example see Cox & Snell (1981).)

4.7 Carry out a numerical illustration of the similarity of the logistic response curve and the cumulative Normal distribution as follows. To compare curves, find the logistic curve which is equal to the standardised cumulative Normal function at the values of the explanatory variable x corresponding to probability of a reaction equal to 0.10, 0.50 and 0.90. Tabulate the values of both functions over a suitable range of values for p.

What are the practical implications of the result for descriptive models of the relationship between p and x?

4.8 The flexibility of the logistic-binomial model can be increased by relating logit(p) to a curvilinear function of x. One example is a quadratic model in which

$$\text{logit}(p) = \alpha + \beta_1 x + \beta_2 x^2.$$

(a) Show that for $\beta_2 < 0$, the underlying proportion increases to a peak at $x_0 = -\beta_1/2\beta_2$ and find the maximum value in terms of α, β_1 and β_2.

(b) The table below contains data on survival in sparrowhawks of different ages (provided by Dr Ian Newton). Compare the observed pattern of survival with that for the logistic model with $\alpha = -0.854$, $\beta_1 = 0.837$, $\beta_2 = -0.103$.

Age	1	2	3	4	5	6	7	8	9
Number of birds	77	149	182	118	78	46	27	10	4
Number surviving	35	89	130	79	52	28	14	3	1

(c) For each age, simulate 100 proportions from the appropriate binomial distribution. Comment on how the distributions change with age and the problem of assessing the apparent reduced survival in 8 and 9 year olds.

(d) Examine the fitted model over a wider age range 1–20 years and comment on the use of the model for estimating the survival rate of a 12-year-old bird.

4.9 Explain the difference between the logistic-binomial model with logit(p) = $\alpha + \beta x$ and the model

$$\ln\left(\frac{Y}{1-Y}\right) = \alpha + \beta x + \varepsilon$$

where $\varepsilon \sim N(0, \sigma^2)$. Illustrate your answer by suggesting an application for each model.

For each of $x = -5, -4, \ldots, 5$, simulate 100 values of Y for the model with $\alpha = 0$, $\beta = 1$, $\sigma^2 = 1$ and use the results to illustrate: (i) how the distribution of Y changes with x; (ii) that the standard deviation of Y is *approximately* proportional to mean[Y] (1 − mean[Y]).

4.10 Consider a stochastic version of the allometric relationship between two variables X and Y

$$Y = AX^{\beta}\delta$$

where $\ln \delta \sim N(0, \sigma^2)$.

(a) Simulate 100 pairs of data values for the model with $A = 1.5, \beta = 0.60, \sigma^2 = 0.01$ and where the values of $\ln X$ are sampled at random from a Normal distribution with mean 2 and variance 1.

(b) Comment on the appearance of scatter diagrams of the simulated data showing $\ln Y$ against $\ln X$, and Y against X.

(c) Simulate data for a range of values of X and σ^2 to illustrate the result

$$\text{mean } [Y \text{ for given value of } X] = AX^{\beta}\, e^{\sigma^2/2}.$$

(d) Suppose that in a particular application, the allometric equation refers to the relationship between the volume of timber in a tree, Y, and tree diameter, X, at some height above ground. What are the implications of the result in (c) for estimating the total timber volume in a stand by applying the allometric equation to measurements of diameter made on a sample of trees?

4.11 In the southern elephant seal the cows come ashore to have their pups starting in early autumn. Each female stays ashore for about 30 days to give birth and wean the pup before returning to the sea. The table below shows a typical haul-out pattern for a breeding population from counts of cows ashore at times throughout the season.

Date	Count	Date	Count
23/9	24	23/10	983
25/9	26	25/10	1001
27/9	51	27/10	992
29/9	95	29/10	955
1/10	147	31/10	896
3/10	239	2/11	785
5/10	342	4/11	669
7/10	482	6/11	544
9/10	594	8/11	434
11/10	678	10/11	362
13/10	802	12/11	274
15/10	843	14/11	196
17/10	891	16/11	140
19/10	891	18/11	112
21/10	931	20/11	88

A simple model to describe this pattern is as follows. The times at which the females come ashore follow a Normal distribution with mean μ, measured in days from some convenient date, and standard deviation σ. Each female stays ashore for S days before returning to sea.

(a) Show that the expected proportion of the population ashore at time t is given by

$$p(t) = \Phi\left(\frac{t-\mu}{\sigma}\right) - \Phi\left(\frac{t-S-\mu}{\sigma}\right)$$

where Φ is the cumulative distribution function of the standardised Normal distribution.

(b) Plot the observed counts with the curve of expected values for a model with $\mu = 38.1$ days from the 1st September, $\sigma = 7.3$ days, $S = 30$ days and population size $N = 1030$.

(c) Use analytical methods to show that there is a peak in the curve at time $\mu + S/2$ and find an expression for the maximum expected proportion ashore in terms of the ratio S/σ.

(d) Comment on a stochastic version of the model in which the number of females ashore at time t follows a binomial distribution.

(e) How would you set up a simulation study to find the curve for a model in which the time spent ashore follows a Normal distribution with mean S standard deviation σ_S.

4.12 The polymerase chain reaction is utilised in a method in molecular biology for replicating as unworkable small number of copies of a region of DNA to obtain a number of copies (of the order of 10^7) such that standard techniques can be applied. In a simple model of replication, the number of copies increases from generation to generation according to binary fission in which a copy is replicated (or divides into two) with probability p. For this model, theory shows that the mean and variance of the number of copies in the nth generation satisfy the difference equations

$$\mu_{n+1} = (1+p)\mu_n; \qquad \sigma_{n+1}^2 = (1+p)^2\sigma_n^2 + p(1-p)\mu_n.$$

(a) Show that starting with a fixed number N_0 of copies, the mean and variance after n generations are given by

$$\mu_n = N_0(1+p)^n; \qquad \sigma_n^2 = N_0(1-p)(1+p)^{n-1}[(1+p)^n - 1].$$

(b) Assuming that for large n the distribution of the number of copies is approximately Normal, show that the number of generations required for the number of copies to exceed 10^7 with probability $1-\alpha$ satisfies the equation

$$10^7 = N_0(1+p)^n\{1 + z_\alpha\sqrt{(1-p)/[N_0(1+p)]}\}$$

where z_α is the 100α per cent point of the standardised Normal distribution.

(c) For $p = 0.75$ and $N_0 = 10$, find the value of n required to obtain 10^7 copies with probability 0.95.

4.13 Discuss the following two views about the role of stochastic variation in the dynamics of biological populations and processes.

View A: Variation is simply 'noise' which tends to obscure what is really happening.
View B: Variation is an integral part of the dynamics of the population or process.

In your answer, distinguish between the different types of stochastic variation and their potential effects.

4.14 In a deterministic model of grouse population cycles (Chapter 13) changes in population density are given by the difference equation

$$N_{t+1} = s_t N_t \{1 + \lambda \exp[-\exp(\alpha N_t + \beta R_t)]\}$$

where N_t is the density of breeding males in year t, s_t is the overwinter survival rate from year t to $t+1$, $R_t = N_t / (s_{t-1} N_{t-1}) - 1$, λ is the mean number of young males fledged, α and β are model parameters which determine the recruitment of young birds into the breeding population.

(a) Simulate data for a series of 100 years with $\lambda = 3$, $s = 0.5$, $\alpha = 0.0013$ and $\beta = -1.00$, starting with $N_1 = 600$, $N_2 = 800$ to illustrate population cycles of period approximately 8 years.

(b) Simulate data for the corresponding dynamic stochastic model in which survival varies randomly from year to year according to a Normal distribution with mean 0.5 and standard deviation 0.1 to illustrate population cycles of approximate period 8 years with varying amplitude and which drift out of phase.

4.15 Simulate 200 pairs of values from bivariate Normal distributions with means $\mu_X = \mu_Y = 0$, standard deviations $\sigma_X = \sigma_Y = 1$ and correlation coefficients $\rho = -0.9(0.1)0.9$ to illustrate the result that the variance of the sum of two correlated random variables is given by

$$\text{var}[X + Y] = \sigma_X^2 + \sigma_Y^2 + 2\rho\sigma_X\sigma_Y.$$

What is the corresponding result for the variance of the sum of three correlated random variables X, Y and Z? (Hint: $\text{cov}[X, Y + Z] = \text{cov}[X, Y] + \text{cov}[Y, Z]$.)

Apply the above result to Exercise 3.6 assuming that the number of eggs laid in successive days is positively correlated with serial correlation, $\rho = 0.2$. Repeat the exercise with serial correlation coefficient, $\rho = 0.5$. From these results describe the effects of increasing the serial correlation. (Hint: The sum of two correlated Normal random variables follows a Normal distribution.)

4.16 In a dynamic stochastic model for an annual insect species, population densities in successive years are given by the difference equation

$$N_{i+1} = E_i s_{1i} s_{2i} N_i$$

where N_i is the population density in year i, and where egg output (E_i), survival from egg to larva (s_{1i}) and survival from larva to adult (s_{2i}) vary randomly from year to year.

Comment on the relevance of the results of the first part of Exercise 4.15 to the variance of the year to year changes in the logarithm of population density.

5 Fitting Models: Constants, Straight Lines, Polynomials and Non-linear Models

5.1 INTRODUCTION

So far in this book, we have concentrated on formulating, simulating and analysing the behaviour of a range of mathematical models: models for population growth and genetic change, for random processes, and for processes which combine both random and substantial deterministic elements. The aim has been to present a catalogue of models, and to discuss how to formulate models for any biological situation. While engaged in this, we have deliberately sidestepped the methods of determining what the values of the parameters of the models are, assuming instead that the parameters are known. In some cases, the parameters are known from other studies, or do not need any special techniques for their determination. It is obvious, for example, that many species in temperate zones breed only once a year, and some species invariably have at most one offspring. In other cases, some of the parameters of the model are not amenable to a common-sense approach, and have not been separately determined. What do we do then? You might think that modelling in these circumstances would be less worthwhile, but the techniques of statistics come to our aid: we can often use observed data on the outcome of the process to determine the values of some if not all of the unknown parameters of the model. This estimation procedure is the subject of this chapter.

In a sense, estimation can be thought of as the inverse process to simulation. Simulation proceeds from a completely specified model of the population or process to generate possible data, whereas estimation uses real data to attempt to determine unknown aspects (i.e. the numerical values of the parameters) of the model.

In Chapters 1 and 2 we considered deterministic models of growth and decline of populations and also population genetic models, in which there is no random element. Fitting these models can sometimes be done without any special techniques. Many models can be transformed into straight lines. Then the transformed data can be plotted out, a straight line drawn through the resulting points, and from the slope and intercept of the line, the unknown parameters of the model determined. However, this can usually only be done for the very simplest models, and we therefore need some techniques for fitting more complex deterministic models. In any case, deterministic models rarely, if ever, describe the observations exactly. Usually there is some random variation which cannot be accounted for by the deterministic model. In Chapters 3 and 4 we discussed sources of variation which are common in biological systems, and how they might be incorporated into a mathematical model. An example is the straight line-Normal model for the relationship between two variables, in which there is Normally distributed random variation about an underlying straight line relationship. Plots of simulated values from such a model result in a scatter of points—as do plots of real data. What do we do then? A straight line can be drawn through the centre of the resulting cloud of points, but usually no two people would draw exactly the same line. In scientific research subjective procedures like this are best avoided. It is also difficult if not impossible to quantify the errors that might have been made, either as a result of lack of skill on the part of the drawer of the line, or more importantly because the cloud of points has by chance an untypical scatter about the underlying line. An objective and effective procedure with some built-in error monitoring is clearly desirable.

This chapter deals with three different but related aspects of the problem, which can be summarised by the following three questions and their answers.

What methods do we use to estimate the parameters of the underlying relationship? For example, in the case of a straight line-Normal model, how do we estimate the slope and intercept of the line? In Section 5.2, we outline the general procedure called least squares, which in this case finds those values of the slope and intercept which minimise the sum of squares of deviations between the data and the fitted line.

How do we estimate the random variation about the underlying relationship? The magnitude of the random variation about the deterministic part of the model is obtained as a by-product of the least squares estimation procedure. As well as estimates of the parameters of the deterministic part of the model, we also need to have some description of the random variation of the observations about the underlying relationship; in the straight line example, it is given by the variance of the random scatter about the line, and the least squares method gives an estimate as a by-product of estimating the parameters of the underlying deterministic relationship (Section 5.3).

How reliable are the estimates of the parameters of the deterministic and random parts of the model? In science, it is vital to know how reliable our procedures are and this applies no less to statistical methods for estimating parameters than to other methods, e.g. chemical analytical methods or methods of sectioning tissue. We outline methods of describing the uncertainty in our estimates, and of determining the uncertainty in any particular case. We use the notion of sampling distribution of an estimator, which is a specification of the variation in the estimate we would observe if we repeated our experiment or survey many times drawing separate random samples and producing fresh estimates each time. This can tell us in some cases how far from the true parameter value the estimate obtained in any particular sample *might* be.

5.2 ESTIMATING THE PARAMETERS OF THE DETERMINISTIC PART OF THE MODEL

It is simpler to illustrate the method of estimation of parameters first of all, using the example of fitting a straight line, before introducing any general principles.

5.2.1 Notation for estimates

In a particular case, we can never determine the parameters of a model exactly, except by pure fluke. There will be errors in our estimates. So we use a hat notation, which is useful for referring to estimates rather than the parameter values themselves, e.g. we say that $\hat{\beta}$ is an estimate of β. We also remind readers below of the convention for random variables and realised values of random variables.

Notation and terminology for estimates and random variables

Estimates of population or model parameters are indicated by hats or circumflexes placed above the parameter, e.g. $\hat{\mu}$ is an estimate of the population mean, μ; $\hat{\beta}$ is an estimate of the slope parameter, β, in the straight line-Normal model.

An estimator is a rule by which we calculate an estimate; *an estimate* is a specific case of an estimator calculated for a particular set of data.

Random variables are usually indicated by capital or (occasionally) Greek letters, e.g. Y, Y_i, ε.

Realised values of the random variables (i.e. data values) are usually indicated by lower case letters, e.g. y, y_i.

5.2.2 Fitting the straight line-Normal model

In Chapter 4, on structured stochastic models, we gave examples where the relationship between a response (Y) and an explanatory variable (x) could be represented as a straight line with some superimposed Normally distributed random variation. For example, the straight line-Normal model was shown to be a good description of the relationship between growth rate in rats and the log dose of vitamin B_2. We demonstrated this by simulating models for particular values of the parameters, without referring to how the parameters had been estimated, and showed that the simulations resembled the real data in broad outline. How might we have obtained these estimates? The following section outlines the use of the least squares method for estimating the parameters of a straight line model.

Suppose we have a sample of data (e.g. as in Figure 5.1) (x_1, y_1), $(x_2, y_2), \ldots, (x_n, y_n)$ to which we wish to fit the model

$$Y_i = \alpha + \beta x_i + \varepsilon_i$$

where $\varepsilon_i \sim N(0, \sigma^2)$. ε is the Greek lower case letter *epsilon*. We use the method of least squares to obtain the estimates of α and β.

5.2.3 The least squares principle

The method of Least Squares has a long history. It was discovered independently by the mathematicians Gauss and Legendre about two hundred years ago. Today it is widely used in both the theory of model fitting and in computer software for doing so. The general rationale, discussed using the example of the straight line model, is as follows.

When we fit a model we attempt to split each y-value in the data (each 'observed y') into two portions, a 'fitted y' reflecting the underlying straight line and a 'residual' reflecting the random departure from the straight line:

$$\text{observed } y = \text{fitted } y + \text{residual}.$$

The fitted values must lie on the line or curve, and the residual is obtained by subtracting the fitted from the observed value. In any modelling procedure we usually wish to characterise as much as possible of the overall variation in the y-variable as being due to the deterministic part. We therefore want to make the fitted values such that the residuals are as small as possible *on average*. We do not want to minimise the residual for each and every observation separately, or for any particular observation, just on average in some way. We therefore need to find some means of summing the residuals in a meaningful way. Taking the straightforward arithmetic sum

$$(y_1 - \text{fitted } y_1) + (y_2 - \text{fitted } y_2) + \cdots + (y_n - \text{fitted } y_n)$$

would not do because negative and positive residuals would cancel one another out. We could take mean absolute deviation* between the data and the fitted line,

$$|y_1 - \text{fitted } y_1| + |y_2 - \text{fitted } y_2| + \cdots + |y_n - \text{fitted } y_n|$$

but this is not so convenient theoretically, since it does not result in such simple formulae for estimates, as we shall see later for the straight line model; and so it is not so easy to implement in a computer program. The most useful method is to sum the squares of the differences

$$(y_1 - \text{fitted } y_1)^2 + (y_2 - \text{fitted } y_2)^2 + \cdots + (y_n - \text{fitted } y_n)^2.$$

This is known as the least squares criterion.

Least squares criterion for estimating parameters

The Least Squares estimates of model parameters are obtained by minimising the sum of the squared deviations of the observations from the fitted line or curve i.e. by minimising

$$S = \sum_{i=1}^{n} (\text{observed } y_i - \text{fitted } y_i)^2.$$

This procedure works quite well in general, and is simpler computationally than many alternative procedures. It can be applied to models with almost any deterministic component, but in its simplest form is restricted to models with random components which (i) have distributions which are reasonably near Normal in shape, i.e. are symmetric and have the characteristic bell-shaped form, and which (ii) have constant variance; in particular var$[\varepsilon_i]$ should not change as x varies.

Perhaps the most important reason for the widespread use of least squares is that the resulting estimators have good properties. In particular, they do not systematically mislead us about the true values of the parameters (the estimates are said to be unbiased); and secondly, for certain classes of model, they are the most accurate estimators. We discuss this in more detail in Section 5.4.2.

* The modulus or absolute value of x, denoted $|x|$, is the value of x without its sign, e.g. $|-3| = 3$, $|21| = 21$.

5.2.4 Least squares estimates for the straight line model

Returning to the straight line model, we find the values of $\hat{\alpha}$ and $\hat{\beta}$ by minimisation of the sum of squares of these deviations

$$S = \sum_{i=1}^{n} (\text{observed } y_i - \text{fitted } y_i)^2$$

$$= \sum_{i=1}^{n} (y_i - \hat{\alpha} - \hat{\beta} x_i)^2.$$

The estimated line which results is sometimes referred to as the least squares line.

The search for a line to fit the small sample of data

x	1	1	2	2	3	3	4	4	5	5
y	3.17	13.25	19.80	14.18	11.43	25.85	13.81	25.49	26.94	38.86

plotted in Figure 5.1 is illustrated in Table 5.1 below which shows values of the sum of squared deviations calculated for lines with a range of different slopes and intercepts. As we move away from the centre of the table the size of the sum of the squared deviations increases. If you think about fitting a line by eye, you can see how this comes about. For example, if the intercept is fixed and the slope is allowed to vary then some lines will be too steep and others too flat. If the slope is fixed and the line is moved up and down then some lines will be too high and others too low.

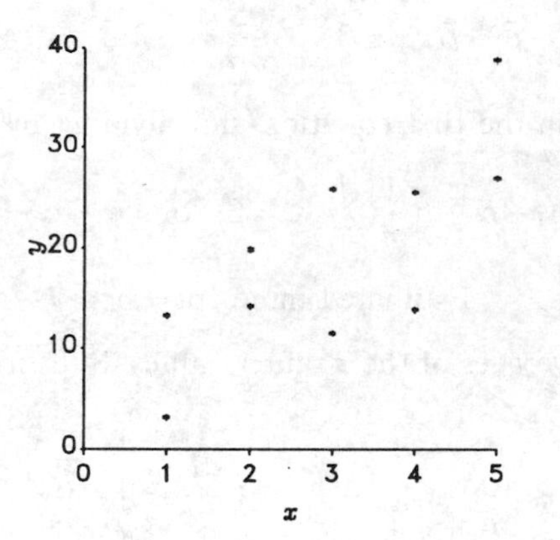

Figure 5.1 The data to which a straight line is fitted in Section 5.2.4.

Table 5.1 Sum of squared deviations calculated for lines with the slopes and intercepts given

		Value of estimated slope ($\hat{\beta}$)				
		5.0	5.1	5.2	5.3	5.4
Value of estimated intercept ($\hat{\alpha}$)	3.5	402.75	398.37	396.18	396.20	398.41
	3.6	401.29	397.51	395.93	396.54	399.36
	3.7	400.04	396.85	**395.87**	397.09	400.50
	3.8	398.98	396.40	396.01	397.83	401.85
	3.9	398.13	396.14	396.36	398.77	403.39

The values in the table can be visualised as points on a surface, called the *sum of squares surface*, which describes how the sum of the squared deviations changes as the slope and intercept of the fitted line vary. This surface has a bowl-shaped appearance with the lowest point corresponding to the location of the slope and intercept of the least squares line. The shape of the sum of squares surface is sometimes represented as a contour diagram with lines connecting points of equal height. For fitting a straight line the contours form concentric ellipses. For other models the shape of these contours varies with both the form of the model and the data. We do not dwell on this aspect here but note that this geometrical interpretation of least squares can be useful for thinking about procedures for fitting models and for assessing properties of the estimates.

Table 5.1 enables us to locate the slope and intercept of the least squares line to within ± 0.1. More accuracy could be obtained using a finer grid, or by a search method. In practice neither are necessary because the slope and intercept can be calculated from formulae which we now derive. For a set of n data points with coordinates $(x_i, y_i), i = 1, \ldots, n$ the *fitted* line $y = \hat{\alpha} + \hat{\beta}x$ is chosen so that the sum of squares

$$S = \sum_{i=1}^{n} (y_i - \hat{\alpha} - \hat{\beta}x_i)^2$$

is minimised. The necessary conditions for this are that the partial derivatives of S with respect to $\hat{\alpha}$ and $\hat{\beta}$ are equal to zero, or

$$\frac{\partial S}{\partial \hat{\alpha}} = -2 \sum_{i=1}^{n} (y_i - \hat{\alpha} - \hat{\beta}x_i) = 0 \qquad \frac{\partial S}{\partial \hat{\beta}} = -2 \sum_{i=1}^{n} x_i(y_i - \hat{\alpha} - \hat{\beta}x_i) = 0.$$

Cancelling out the -2 in the first equation and dividing by n gives us

$$\frac{1}{n}\sum_{i=1}^{n} y_i - \hat{\alpha} - \hat{\beta}\frac{1}{n}\sum_{i=1}^{n} x_i = 0 \quad \text{or} \quad \bar{y} - \hat{\alpha} - \hat{\beta}\bar{x} = 0$$

$$\text{i.e.} \qquad \text{Estimated intercept} = \hat{\alpha} = \bar{y} - \hat{\beta}\bar{x}$$

where $\bar{x}$ and $\bar{y}$ are the averages of the x and y values. Substituting for $\hat{\alpha}$ in the second equation gives

$$\sum_{i=1}^{n} x_i[y_i - \bar{y} - \hat{\beta}(x_i - \bar{x})] = 0$$

which can be rearranged to give

$$\sum_{i=1}^{n} x_i(y_i - \bar{y}) - \hat{\beta} \sum_{i=1}^{n} x_i(x_i - \bar{x}) = 0.$$

Using the fact that the sums of the deviations of the x and y values about their average values are zero (i.e. $\Sigma(x_i - \bar{x}) = \Sigma(y_i - \bar{y}) = 0$) and rearranging gives the following

$$\text{Estimated slope} = \hat{\beta} = \frac{\sum_{i=1}^{n} x_i(y_i - \bar{y})}{\sum_{i=1}^{n} x_i(x_i - \bar{x})} = \frac{\sum_{i=1}^{n} (x_i - \bar{x})(y_i - \bar{y})}{\sum_{i=1}^{n} (x_i - \bar{x})^2}.$$

The second expression for the slope shows that the estimated slope is only a function of the deviations of the observations from their means—the means $\bar{x}, \bar{y}$ do not otherwise enter into the estimation of β, as you would expect. The expression for the fitted line shows that it passes through the centroid, $(\bar{x}, \bar{y})$, of the scatter of points. The line is then rotated about the centroid until the sum of the squared deviations is least. Seen in this way it is apparent that points with values of x far away from the centroid will have a larger effect in determining the slope of the line because of their larger contribution to the sum of the squared deviations as the slope is changed. Nowadays, fitting a straight line by least squares is a simple exercise using one of a wide range of statistical packages (see Example 5.1). It can be done routinely without having to think at all about the details. However, some familiarity with the expressions for the estimates is helpful in understanding the properties of the method and its application to more complex problems.

Example 5.1 Fitting a straight line in Minitab

A straight line is fitted to the data from Figure 5.1 using the Minitab package. y has been read into column C2 and x into column C1. The command is shorthand for 'regress the column C2 on 1 other column C1'.

```
MTB > REGRESS C2 1 C1

The regression equation is
C2 = 3.67 + 5.20 C1

Predictor    Coef      Stdev     t-ratio     p
Constant     3.666     5.217     0.70        0.502
C1           5.204     1.573     3.31        0.011

s = 7.034       R-sq = 57.8%       R-sq(adj) = 52.5%

Analysis of Variance
SOURCE       DF     SS        MS        F         p
Regression    1     541.63    541.63    10.95     0.011
Error         8     395.87    49.48
Total         9     937.50
```

The estimated intercept and slope are 3.666 and 5.204 which agree well with the results of our numerical search of the least squares surface, 3.7 and 5.2. We will be returning to discussion of the columns headed 'Stdev t-ratio p' and the 'Analysis of Variance' later in Section 5.4.5.

5.2.5 Fitting a mean

A common problem in biology is to estimate the mean of a variable in a specified population. A microbiologist counts the cells on slides in replicate samples from a well-mixed culture to estimate the density of microbes. An entomologist counts wireworms in a sample of soil cores to estimate the density in a field population. Despite the marked differences in organisms, the scale of sampling and technique, both scientists have the same basic aim of estimating the mean density in a population, in the presence of random variation in their data.

The estimator to use in this situation is fairly obvious: the sample mean is a reasonable estimator of the population mean. Does the principle of least squares come up with the same estimator? Suppose we have sample of data $(y_1, y_2, \ldots, y_n)$ of size

n. Then a model which is sufficiently defined for the purposes of least squares estimation of the population mean is

$$Y_i = \mu + \varepsilon_i$$

where μ is the population mean. For the moment, we do not need to say what the distribution of ε_i is. The fitted value in this case is just $\hat{\mu}$. So to use the least squares criterion, the sums of squares of deviations between observed and fitted values is

$$S = \sum_{i=1}^{n} (y_i - \hat{\mu})^2.$$

To obtain the least squares estimate of μ, we therefore need to find the value, $\hat{\mu}$ which minimises S. Differentiating with respect to $\hat{\mu}$ we obtain

$$\frac{dS}{d\hat{\mu}} = -2\Sigma(y_i - \hat{\mu})$$

and equating this to zero

$$\sum_{i=1}^{n} (y_i - \hat{\mu}) = 0$$

and, dividing by n and rearranging, we obtain the estimate

$$\hat{\mu} = \frac{\sum_{i=1}^{n} y_i}{n} = \bar{y}.$$

So the sensible estimator is also the least squares estimator in this simple situation.

5.2.6 Fitting a quadratic (and polynomials)

In Chapter 4, we discussed quadratic models (and higher order polynomials) as models providing flexible empirical descriptions of the relationship between two variables, e.g. the relationship between photosynthesis and temperature for the plant *Deschampsia antarctica*. Although polynomials can contain many more parameters than the straight line, most of the results presented in Section 5.2.4 for fitting a straight line extend easily to fitting polynomials, and most of the estimation details are taken care of by statistical packages.

Model

The quadratic model in its usual form is

$$Y_i = \alpha + \beta_1 x_i + \beta_2 x_i^2 + \varepsilon_i, \qquad i = 1, 2, \ldots, n$$

where Y_i is the response of the ith individual with value of the explanatory variable x_i. The ε_i represent random variation in the responses about the curve, which are assumed to be distributed Normally with mean zero and unknown variance σ^2.

Fitting the quadratic model by least squares

Fitting the model by least squares involves finding the values of the parameters, α, β_1, β_2 so that the sum of the squared deviations about the fitted curve is least (and σ^2 which we will deal with in Section 5.3). For this model, there are three parameters and the method reduces to solving three simultaneous linear equations in the three unknowns. Higher order polynomials with more parameters lead to more equations but these can be solved in a similar way. In practice, the calculations are best carried out on the computer and many software packages are available for this purpose. The calculations* are broadly similar to those for fitting a straight line. That parameters can be estimated by a simple automatic procedure is a major reason for the widespread use of least squares as an estimation procedure for what are known as *linear models* (i.e. models which are linear in their parameters, see Section 5.2.7 for a further discussion).

Example 5.2 Photosynthesis and temperature in *Deschampsia antarctica*

To illustrate the fitting of a quadratic model we use the data relating photosynthesis to temperature in the plant *Deschampsia antarctica*. Referring to the general form of the model given in Section 4.2, y_i is the percentage of the maximum photosynthesis rate for the ith observation and x_i is the corresponding temperature. There are $n = 14$ pairs of observations. Using a computer package to fit the model by least squares would typically produce output in a similar form to that for fitting a straight line (see Section 5.2.4) containing parameter estimates with estimated standard errors and the analysis of variance table giving the breakdown of the total sums of squares and degrees of freedom. The essential information for the present purposes of fitting the model is summarised below (this output is from Genstat; temperature and temperature squared have been previously put into variates T and T2).

*** Estimates of regression coefficients ***

	Estimate	Standard error	t
Constant	46.37	2.80	16.54
T	6.77	0.47	14.41
T2	−0.249	0.016	−15.42

Figure 5.2 shows the fitted model with the data. Note that fitted values to the right of the peak are generally larger than the observations suggesting a possible departure from the symmetrical form of the quadratic model, but these differences are small compared with the residual standard deviation ($\sqrt{\text{residual mean square}} = \sqrt{26.6} = 5.2$

---

* One aspect of computation associated with polynomials is, however, worth mentioning. Problems with rounding error are eased by centring and scaling the explanatory variables before calculating powers. *Rounding errors* can occur because of the way large numbers are stored in the computer when values of x are squared, cubed, or raised to higher powers. When the differences of such large quantities are calculated in, for example, determining fitted values of polynomials, the relative numerical error in the result can sometimes be very substantial. Some software may allow for this but the user should be aware of the problem. Scaling and centring the calculated values, before squaring and cubing etc., so as to occupy some fixed range, such as $(-1, 1)$, may help. For example, if x takes the values 0, 10, 20, 30, 40, 50, 60, 70, 80, 90, 100, then it is better to transform by subtracting 50 and dividing by 50, thus transforming the x-values to $-1, -0.8, -0.6, -0.4, -0.2, 0.0, 0.2, 0.4, 0.6, 0.8, 1.0$. Transforming using the observed mean and standard deviation of the x values is another possibility:

$$x_s = \frac{(x - \bar{x})}{s_x}$$

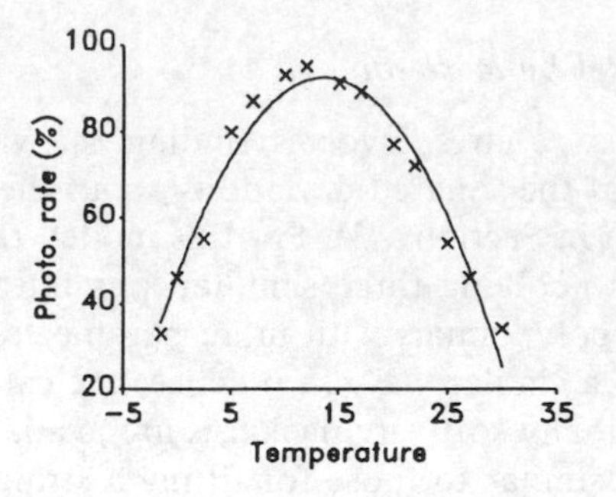

Figure 5.2 Photosynthesis of *Deschampsia antarctica* plotted against temperature, together with the fitted model.

which is a measure of the random variation about the fitted curve). We return to a discussion of other elements in this computer output, such as standard errors of estimates in Section 5.4.6, where we outline how to quantify the uncertainty in estimates of population parameters.

5.2.7 Fitting non-linear models

For many situations a straight line, or a polynomial model, is a poor description of the data. For example, the sigmoid-shaped pattern of limited growth with an upper asymptote cannot be mimicked by a polynomial and other models such as the logistic are more realistic (Chapter 1). In this section we discuss the more difficult problems of fitting models of this type referred to as non-linear models.

Non-linear models: models which are non-linear in the parameters

The essential difference between linear and non-linear models, as these terms are used in statistical theory and practice, can be illustrated using the following models, depicting very different relationships between Y and x:

(1) $Y = \alpha + \beta x + \varepsilon$ a straight line.
(2) $Y = \alpha + \beta x + \gamma x^2 + \varepsilon$ a quadratic.
(3) $Y = \alpha\, e^{\beta x} + \varepsilon$ an exponential increase.
(4) $Y = \alpha(1 - e^{-\beta x}) + \varepsilon$ an asymptotic exponential with upper limit α.

ε is a random variable with mean equal to zero. If, however, we think of Y in relation to the parameters in the model we see that models (1) and (2) both have the property that the mean response is a linear combination of the parameters. Both models are members of what statisticians call the family of *linear models* which can be written in the general form

$$Y = \alpha + \beta_1 \begin{pmatrix}\text{something}\\ \text{involving}\\ \text{just } x \text{ values}\end{pmatrix} + \beta_2 \begin{pmatrix}\text{2nd quantity}\\ \text{involving}\\ \text{just } x \text{ values}\end{pmatrix} + \cdots + \beta_k \begin{pmatrix}k\text{th quantity}\\ \text{involving}\\ \text{just } x \text{ values}\end{pmatrix} + \varepsilon$$

where the quantities in brackets are *known* values for any particular data set (i.e. numbers calculated from the explanatory variables, and not involving the model parameters). Any model which is not linear in this sense is said to be *non-linear* or, more precisely, *non-linear in its parameters*. Models (3) and (4) can be seen to be non-linear since the mean response is an exponential, rather than a linear, function of the parameter β. Non-linear models take many forms and there is no simple mathematical characterisation as there is for the linear model.

Some non-linear models can be converted to linear ones by transformation of Y or x or both. This is illustrated by the difference between models (3) and (4). In model (3) the *mean* response (i.e. $y = \alpha \exp(\beta x)$, leaving aside for the moment the random element, ε) can be transformed to a linear model by taking logarithms. In model (4), a transformation to linearity is not possible unless the value of the parameter α is known. However, these transformations also often cause distortion of the random component, ε, which requires special treatment, a discussion of which is postponed until Chapter 11.

5.2.8 Fitting a non-linear model by least squares

Fitting a non-linear model by least squares involves finding those values of the parameters for which the sum of the squared deviations of the observations from the fitted values is least. The criterion is the same as it is for fitting a linear model. Finding the parameter estimates is, however, less straightforward. The essential difference is that for a linear model the parameter estimates can be calculated from the data using fairly simple formulae—comparable to those for fitting a straight line—but for a non-linear model this is not generally possible.

To illustrate the problem of fitting a non-linear model by least squares, we fit a model (this 1-parameter model which does not arise much in practice is chosen as a simple illustration) for exponential decay

$$Y_i = e^{-\beta x_i} + \varepsilon_i$$

to the set of data given in Figure 5.3(a). The problem is to estimate the rate constant β. If we apply the least squares criterion then we need to find the value of the estimate

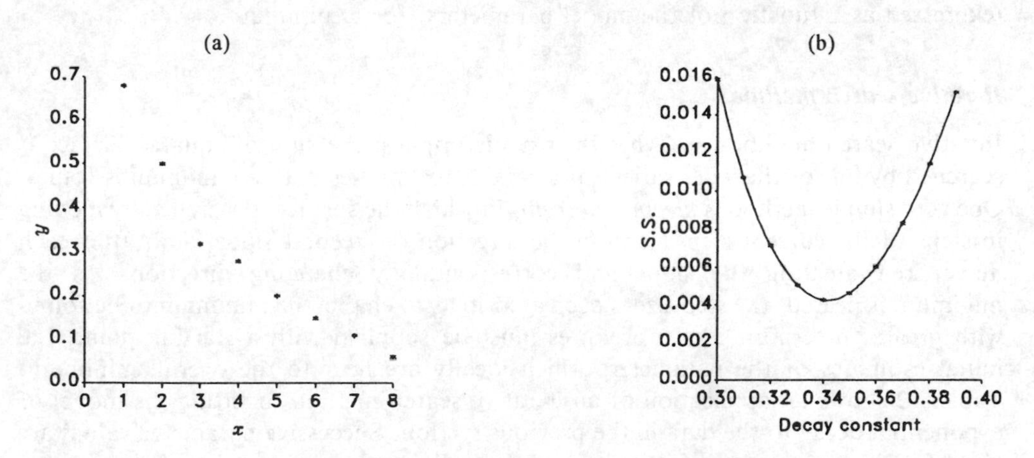

Figure 5.3 (a) Data for fitting the exponential model; (b) the curve of the sum of squares of deviations between the data and the fitted curve against possible values of the estimated parameter, $\hat{\beta}$.

$\hat{\beta}$ for which the sum of the squared deviations of the observations about the fitted curve is least. In other words we need to minimise

$$S = \sum_{i=1}^{8} (y_i - e^{-\hat{\beta}x_i})^2.$$

If we calculate this quantity S for a number of values of $\hat{\beta}$, we could plot S against $\hat{\beta}$, so as to get an idea of the relationship between S and $\hat{\beta}$. Figure 5.3(b) shows the concave upwards curve which relates S to $\hat{\beta}$ which has a minimum when $\hat{\beta}$ is about 0.35. The exact value occurs at the point on the curve where the slope is zero.

Can we find this value in any other way? By setting the derivative of S with respect to $\hat{\beta}$ equal to zero we find that the least squares estimate $\hat{\beta}$ must satisfy the equation

$$2 \sum_{i=1}^{8} x_i \, e^{-\hat{\beta}x_i}(y_i - e^{-\hat{\beta}x_i}) = 0.$$

This messy equation, which is seen (after the numerical values of x_i and y_i have been substituted in) to involve only $\hat{\beta}$, presents problems for even the best mathematicians as it is not possible to convert the equation into one of the form

$$\hat{\beta} = \text{(an expression involving only the } x\text{'s and the } y\text{'s)}.$$

Thus even with a single parameter and the simplest non-linear model there is no explicit formula which enables us to calculate the estimate of the parameter from the data. This is generally a problem when fitting non-linear models: finding the least squares estimate using analytical techniques is usually not possible and we must resort to numerical methods.

A simple and sometimes effective approach is to calculate values of the sum of squares surface over successively finer grids of points (in the parameter space, with as many dimensions as there are parameters) until the location of the minimum has been found to a sufficient accuracy (as in Figure 5.3(b) where we have used a 1-dimensional grid). This can work well for particular problems but for more general use we often employ iterative methods, using algorithms which search the sum of squares surface (expressed as a function of the model parameters) for a minimum.

Iterative search methods

Iterative search methods are what their name implies: the sum of squares surface is searched by one of the wide variety of available techniques until a minimum is found. One very simple method is *steepest descent*, in which the surface is searched by moving in steps of the current step length in the direction of steepest slope, and, after each move, re-evaluating the slope and correspondingly changing direction. As the minimum is neared, the step size is decreased so as to enable the minimum to be found with greater precision. The procedures must be supplied with a starting point, i.e. initial estimates of the parameters which ideally are near to the overall minimum. Table 5.2 shows an application of an iterative search method to fitting the model of exponential decay to the data in the previous section. Successive parameter values are given for three sequences of iterations with the different starting values for the decay constant of 0.0, 0.3 and 1.0. When two successive values are within some prespecified amount the process is said to have converged. For a starting value of 0.3 this happens

Table 5.2 Iterative searches employing different starting values for the slope parameter, β_0

Iteration	Starting value β_0		
	0.30	0.0	1.00
1	0.3430	0.1478	0.6940
2	0.3396	0.2720	0.5657
3	0.3430	0.3329	0.2524
4	0.3428	0.3427	0.3342
5		0.3428	0.3440
6		0.3428	0.3428

after four iterations when the value of 0.3428 is reached. For the more remote starting values of 0.0 and 1.0 the process converges but more iterations are required. This illustrates the general point that when using iterative procedures it is best to try to choose 'good' starting values, i.e. values relatively close to the required estimates.

Iterative fitting problems

In the above example, with a fairly simple model involving only a single parameter to be estimated, the iterative process converges quickly to the least squares estimate and it makes little difference which starting value we choose. In practice, however, when fitting more complex models with more parameters, convergence of the iterations to the required values is not guaranteed. Two problems which can arise are: (a) the process may not converge; (b) the process may converge but not to the least squares estimates. In the latter case there may be a point on the sum of squares surface which is lower than the immediately surrounding points but which is not the lowest point overall. Such a point is referred to as a *local minimum*. The iterative process may settle into this local indentation and fail to find the so-called *global minimum* of the surface which corresponds to the least squares estimates. The occurrence of these problems will in general depend on the model being fitted and the extent of the data, and we refer the reader to Chapter 11 for a more detailed discussion.

5.3 ESTIMATING THE RANDOM COMPONENT OF THE MODEL

We have seen above that least squares can be used for estimating the parameters of the deterministic component of a model. Can we use it to get an idea of the parameters of that part of the model which describes the random variation about the deterministic underlying relationship? The random component in almost all models is assumed to have a mean of zero. The next most important aspect of the random component is its variance—this gives a measure of its spread. In many cases, the situation is so complex that we cannot go beyond an estimate of the variance, and the main emphasis is generally on estimating the variance of the random component. We first consider this problem in the context of the simplest deterministic component of the model, just a single mean.

5.3.1 Estimating the population variance

In this section we find an estimator of the variance of the random component of the model from a random sample of observations, in the case when there is only a very simple deterministic component of the model: the population mean. The model is

$$Y_i = \mu + \varepsilon_i$$

where $\varepsilon_i \sim N(0, \sigma^2)$. We shall show in Section 5.4.2 that we need to estimate σ^2 in order to assess the errors in using the sample mean to estimate the population mean; then the variability is a nuisance. In other situations, however, estimating the population variance could be of central interest.

The population variance measures the spread of the population distribution about the mean: the mean squared deviation of the individual values in the population about the population mean. So, a plausible estimator is the sample mean squared deviation calculated as the sum of the squared deviations of the observations in the sample about the sample mean divided by the sample size. This can also be derived as a by-product of the least squares estimation of the parameters of the model. In Section 5.2.5, we set out to minimise the sum of squares, S, given by

$$S = \sum_{i=1}^{n} (y_i - \hat{\mu})^2$$

where y_i is a single observation. The least squares estimate of the population mean was also the common-sense one

$$\hat{\mu} = \bar{y} = (1/n) \sum_{i=1}^{n} y_i.$$

Substituting this into the formula for S, we get

$$S = \sum_{i=1}^{n} (y_i - \bar{y})^2.$$

This is the portion of the original variation which is left after fitting the mean. We need to divide this by n to obtain an estimate of the mean square deviation per observation, and so, unsurprisingly, we arrive at the same estimator as we obtained intuitively in the last paragraph.

The sample mean squared deviation, however, tends to *underestimate*⋆ the population variance. The effect is demonstrated in Table 5.3 which shows average values of the sample mean squared deviation in 1000 simulated samples of different sizes from a Normal distribution with a mean of zero and variance equal to one. The simulations illustrate the general result that the mean of the sampling distribution of the sample mean squared deviation is less than the population variance by a

⋆ This underestimation occurs because the same sample is used to estimate the mean as to calculate the sum of squares of deviations, and the sample mean tends to be slightly nearer the observations in its own sample than the true population mean would be.

Table 5.3 Estimation of variance using simple random samples. Results of a simulation experiment with samples of size n from $N(0, 1)$

Sample size n	$(n-1)/n$	Average value in 1000 simulated samples	
		Sample mean squared deviation	Sample variance, s^2
2	0.50	0.51	1.02
3	0.67	0.67	1.01
4	0.75	0.74	0.99
5	0.80	0.82	1.02
10	0.90	0.94	1.04
50	0.98	0.98	1.00

factor of $(n-1)/n$. To adjust for this bias we divide the sum of the squared deviations by $(n-1)$ rather than by n and calculate what is termed the *sample variance*, denoted by s^2:

$$\text{sample variance} = s^2 = \sum_{i=1}^{n} \frac{(y_i - \bar{y})^2}{(n-1)}.$$

The table illustrates the result that, in random samples from a population distribution, the mean of the sampling distribution of the sample variance is equal to the population variance, i.e. the sampling distribution of the sample variance is centred on (in the sense of having a mean which is equal to) the true population variance. In technical language, we say that the sample variance is an *unbiased* estimator of the population variance.

The divisor in the expression for sample variance, $n-1$, is known as the *degrees of freedom* of the estimate s^2. It is the number of independent deviations which have been squared and added to give the numerator of s^2. n deviations have been used in total, but after $n-1$ of them have been fixed, there is no freedom left to fix the last one, it is determined. For example, for three deviations, if the first two are -1 and -2, the third one must be $+3$, since the deviations within the sample about the sample average must sum to zero by definition. This degree of freedom (out of the original n, one corresponding to each of the independent observations in the sample) lost from s^2 has in a sense been used up in estimating the mean.

5.3.2 Estimating the variance about the regression line

Fitting a straight line-Normal model by least squares provides estimates of the two parameters of the deterministic part of the model, the slope and the intercept. A further parameter in the model, but this time of the random part, is the variance of the responses about the line. In this section we show how to estimate the variance.

The variance about the regression line is the mean squared deviation of the response from the mean response in the *population* for a given value of x. This is assumed to be the same for all values of x. How to check this is discussed in Section 5.3.4. A plausible estimate is the mean squared deviation about the fitted line for the sample, defined as the sum of the squared deviations about the fitted line divided by the number of observations. It is, however, easy to see that this estimate is not always satisfactory. For example, when there are only two data points the fitted line will always pass through them leaving a zero value for the estimated variance. More generally, the mean squared deviation about the fitted line tends to underestimate the variance of the responses about the line. This bias is demonstrated by the results of 1000 simulations of a straight line-Normal model, which will also be used later (see Section 5.4.5 for more details). For example, with a variance about the underlying line equal to 2, the average values of the mean squared deviations about the least squares lines for data with $n = 5$, 10 and 20 points are given in Table 5.4.

The general result is that mean squared deviation about the fitted line is, on average, less than the variance about the regression line by a factor $(n-2)/n$. To allow for this bias we divide the sum of the squared deviations about the fitted line by $(n-2)$. The effectiveness of this correction can be seen in the table.

We usually refer to the sum of the squared deviations about the regression line as the *residual sum of squares*. This is a general terminology and applies to a wide range of models. It stems from the idea of the *residual* which is the difference between the observed response and the value calculated from the fitted model: Residual $= (y_i - \hat{y}_i)$. The *residual sum of squares* is then the sum of the squared residuals. An important associated quantity is the *degrees of freedom for residual*. For fitting many models this is calculated as the number of observations minus the number of estimated parameters (of the deterministic part of the model) used in the calculation of the fitted values. For example, when fitting the straight line-Normal model there are two parameters, a slope and an intercept, used to calculate the fitted values, so if there are n data points the residual degrees of freedom are $n - 2$. The *residual mean square* is the residual sum of squares divided by the residual degrees of freedom.

Finally, we note that output from computer programs for fitting the linear regression model usually contains the residual sum of squares as a line in a so-called *analysis of variance table*, e.g. the following (taken from the analysis in Example 5.5)

Analysis of Variance

SOURCE	DF	SS	MS	F	p
Regression	1	697.25	697.25	97.85	0.000
Error	18	128.26	7.13		
Total	19	825.51			

Table 5.4 Estimation of residual variance in a straight line-Normal model. Simulation results for a model in which the true residual variance is 2

Sample size (n)	Average value in 1000 simulated samples	
	Sample mean squared deviation from fitted line	Sum of squared deviations divided by $n - 2$
5	1.21	2.02
10	1.59	1.99
20	1.79	1.99

The analysis of variance table can be regarded as a balance sheet of variation, a splitting up of the total variation into components reflecting the partition we have made of each observation. We have split each observation, y_i into two components, a fitted value from the regression, $\hat{y}_i$, and the residual or deviation between the individual observation and the fitted value, $y_i - \hat{y}_i$. The same is true about the deviation of each observation from the sample mean: $(y_i - \bar{y}) = (\hat{y}_i - \bar{y}) + (y_i - \hat{y}_i)$. It turns out that the same can be said for the sum of squares of these quantities:

$$\sum_{i=1}^{n} (y_i - \bar{y})^2 = \sum_{i=1}^{n} (\hat{y}_i - \bar{y})^2 + \sum_{i=1}^{n} (y_i - \hat{y}_i)^2$$

and this is reflected in the analysis of variance above. So we get the following splits:

(a) total sum of squares of the deviations of the observations about the overall mean $(\Sigma_{i=1}^{n}(y_i - \bar{y})^2 = 825.51)$ = Residual sum of squares $(\Sigma_{i=1}^{n}(y_i - \hat{y}_i)^2 = 128.26)$ + Regression, or Model, sum of squares $(\Sigma_{i=1}^{n}(\hat{y}_i - \bar{y})^2 = 697.25)$;
(b) A corresponding breakdown of the degrees of freedom as Total ($n - 1$ or 19) = Residual ($n - 2$ or 18) + Regression (1);
(c) Mean squares obtained by dividing the sums of squares by their degrees of freedom. The residual mean square is the required estimate, s^2, of σ^2.

In Section 5.4.5, this example is discussed further.

5.3.3 Estimating the residual variance for a quadratic, polynomial or non-linear model

The variance about the polynomial curve is estimated unbiasedly by the residual mean square. This is the sum of the squared deviations of the observations about the fitted values divided by the residual degrees of freedom. The latter are obtained from: number of observations – number of parameters in the model. For example, when fitting the quadratic the residual degrees of freedom are $(n - 3)$, more generally $(n - p)$, where p parameters of the deterministic part of the model are estimated.

5.3.4 Complications: Is the variance constant for all *x*?

The above methods for estimating variance are based on the assumption that the variance of the random part of the model, ε, is the same for all values of x. This is quite often not the case (see Figure 5.4 for an example). How can this be checked, and what can be done about it? One method of checking the assumption is by plotting out residuals (e.g. $\hat{\varepsilon}_i = y_i - \hat{\alpha} - \hat{\beta}x_i$ for the straight line model) against x_i. An increasing scatter of points as x_i increases suggests that the variance is increasing with x_i. If alternatively there are replicate measurements of y_i at each value of x_i, then a variance can be calculated for each x-value, and plotted against x_i. It is important to carry out such checks. Draper & Smith (1981) give methods for checking variance homogeneity. Chapter 11 outlines methods, such as weighted least squares, which can be used when there is a trend in variance with x. Chapter 14 gives other model-checking methods based on residuals.

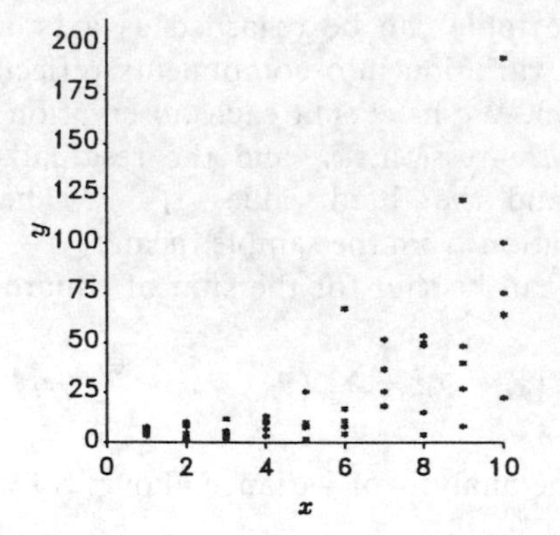

Figure 5.4 An example of inequality of variance of the random component for different values of x.

5.3.5 Fitting a distribution

In Chapter 3, we displayed distributions fitted to a number of sets of data. In some cases, these distributions can be fitted by estimating the mean and variance of the distribution. For a Normal distribution, estimates of the mean and variance are all that are required. For binomial, Poisson and negative exponential distributions, the mean is sufficient to fix the whole distribution. For the binomial, the number of trials, n, is also required, but that is usually known *a priori*. However, there are some distributions which have two unknown parameters, which do not correspond to the mean and variance. There are also other distributions which contain more than two unknown parameters. How do we fit these?

Method of moments

The simplest method to explain is the method of moments. The first two moments of a distribution are the familiar mean and the variance (higher moments measure the skewness, and the peakedness of a distribution). For many distributions in which there are exactly two unknown parameters, we can usually estimate the parameters, from a sample of data, by equating the expressions for the mean and variance in terms of the parameters to the sample mean and variance, and solving the resulting equations. Consider as an example the negative binomial distribution. The mean of the distribution is m, and the variance is $m + m^2/k$, where k is the other parameter of the negative binomial distribution. By equating these to the sample mean, $\bar{x}$, and variance, s^2,

$$\bar{x} = \hat{m}, \qquad s^2 = \hat{m} + \frac{\hat{m}^2}{\hat{k}}$$

estimates of the parameters can be obtained.

Maximum likelihood

A second more generally used method of fitting is maximum likelihood. We do not have space to discuss the method here (see Chapter 11 and Ross, 1990). It generally provides good estimates when large samples are available, and is also applicable to a wide variety of other estimation problems. Genstat has facilities for fitting a number of distributions by maximum likelihood.

Example 5.3 Fitting a negative binomial distribution

We illustrate the methods by fitting a negative binomial distribution to data on counts of Colorado potato beetles, *Leptinotarsa decemlineata*, by G. E. Beall in 1938 in a uniformity trial.

No. of beetles	0	1	2	3	4	5	6	7	8	9	10	11	12	13	>13
No. of samples	12	17	32	19	20	13	9	11	5	2	2	0	1	1	0

(1) *Fitting by moments.* The sample mean and variance are $\bar{x} = 3.59$, $s^2 = 6.73$, and hence the moment estimators of the parameters m and k are $\hat{m} = \bar{x} = 3.59$, and

$$\hat{k} = \frac{\hat{m}^2}{s^2 - \hat{m}} = \frac{\bar{x}^2}{s^2 - \bar{x}} = \frac{3.59^2}{6.73-3.59} = 4.10.$$

(2) *Maximum likelihood.* The estimates are: $\hat{m} = 3.60$ (s.e. = 0.22), $\hat{k} = 3.92$ (s.e. = 1.03).

These are quite close to the moment estimators in this case.

5.4 HOW ACCURATE ARE OUR ESTIMATES?

All the above methods are very useful, but it is important to supplement them with some means of assessing the accuracy of the estimates we have obtained. First of all, what do we mean by accuracy in this context? We do not mean inaccuracies due to errors of procedure, but the effects of *errors due to sampling* on the value of the estimate. We are estimating the parameters by drawing a *random sample* of data from some situation (see Chapter 7 for a discussion of the reasons for the use of a random sample). What we want to know is: had we obtained a different randomly drawn sample, are the resulting estimates likely to have been very different? We also would like to know whether our estimates are likely to be near to the unknown true values of the parameters. This section discusses how we can obtain answers to these questions.

It is probably helpful to mention at this point the kind of statement we are going to end up with. We know that our estimate (of the population parameter we are interested in) will almost certainly be in error. We shall show how to specify the range of possible error in the following way. A range will be calculated from our sample data within which we are quite sure that the true value will lie. In particular we shall specify a range within which the true value will lie with probability 0.95. So that if we envisaged ourselves repeating the experiment or survey many times taking a different random sample each time, we could expect that in 95% of those repetitions, the interval we calculate, called a *95% confidence interval*, would contain the true value. This seems a tall order, but for many models it is quite feasible, and usually involves plugging a few numbers into a simple formula. Readers who want to skip further theory could restrict their attention to the examples in this section.

One way to assess the properties of the estimators of model parameters we have discussed above is to carry out the sampling process on a computer. This might appear artificial and unrealistic, especially if little is known about the real population. We shall see, however, that by experimenting with a range of models useful results can be obtained.

5.4.1 Determining the accuracy of estimation of the population mean

An illustrative example—a single Normal distribution

We start with a model in which the variation in the population is described by a Normal distribution. A single observation selected at random from the population is then regarded as a random variable with a Normal probability distribution. This model has two parameters a mean μ and a variance σ^2. For a single observation Y we can write $Y \sim N(\mu, \sigma^2)$ or

$$Y = \mu + \varepsilon$$

where $\varepsilon \sim N(0, \sigma^2)$. The problem is to estimate the population mean μ. Note that the model also includes another parameter σ^2 which measures the variation in the population. This is, of course, central to the problem because without the variation in the observations they would all be the same and a single observation would suffice to estimate the population mean. Since the population mean is regarded as unknown then it seems reasonable to suppose that the variance is also not known. So, at first sight, the model appears to make the problem more difficult by introducing another parameter to be estimated. We have discussed estimating variances in Section 5.3.

To gain some familiarity with the model and to explore the problem, we start by simulating some observations from one particular Normally distributed population of mean 10 and the variance 4, using the methods of Chapter 3. Table 5.5 gives the results of taking six samples each of five observations from this distribution. The similarity of the sample means to the population value of 10 is striking, despite the relatively large spread of the observations in each sample. Nevertheless, the sample means vary—a value as low as 8.31 occurs. To describe precisely how the sample mean varies, we introduce the important notion of the sampling distribution of the sample mean.

Table 5.5 Six samples of size 5 and their sample means

Sample number	Sample observations					Sample mean
1	7.84	6.13	12.17	6.84	14.00	9.40
2	9.42	8.94	13.04	8.60	10.66	10.13
3	12.85	9.31	12.81	10.58	9.41	10.99
4	7.58	5.24	10.48	6.53	11.70	8.31
5	11.26	9.15	11.17	9.76	11.64	10.60
6	8.32	10.23	13.57	11.91	9.92	10.79

5.4.2 Sampling distribution of the sample mean

The sampling distribution of the sample mean describes the way in which the sample mean varies in repeated random samples of the same size from the same population. It is called a sampling distribution to emphasise that (i) it is a probability distribution, and (ii) it specifies the behaviour of an estimator in repeated drawings of a random sample from the population. We can get an idea of what the sampling distribution looks like by carrying out a large simulation exercise—drawing lots of samples of size n from a completely specified population, calculating the sample means for each such sample, and drawing up a histogram of the results. An example for the above case, in which we drew samples of size 5 from an $N(10, 2^2)$ distribution, is given in Figure 5.5. We have already calculated lots of sampling distributions, without specifically using the term sampling distribution. For example, in Chapter 3 we showed how the probability distribution of the sample mean converges to a Normal distribution by simulating random samples of different sizes from various populations. All of these distributions of sample means were sampling distributions.

We now discuss the main properties of the sample mean considered as an estimator* of the population mean. The properties of any estimator—in relation to the population parameter being estimated—are best described in terms of the sampling distribution of the estimator. There are three main questions that can be asked about any distribution. (i) What is its location? Whereabouts on the scale of the variable is it centred, i.e. what is its mean? (ii) How spread out is it, i.e. what is its variance? (iii) What shape is the distribution? All of these are important for specifying the distribution sufficiently to be able to use it, and the three following paragraphs deal with these points in turn.

(1) *The mean of the sampling distribution of the sample mean is equal to the population mean.* This says that when the sample mean is used to estimate the

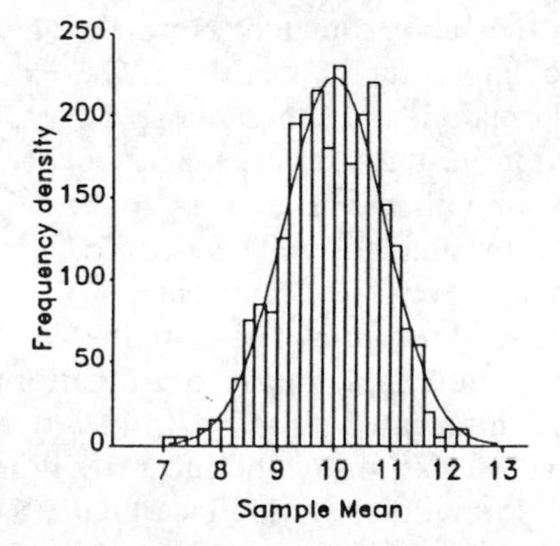

Figure 5.5 The sampling distribution of the sample mean, for sample of size 5 from an $N(10, 2^2)$ distribution, and a superimposed Normal distribution, with mean 10, and standard deviation $2/\sqrt{5}$.

* An *estimator* is a random variable (in repeated drawings of fresh random samples of size n) obtained from a calculation using the observations in each sample, with the purpose of estimating a population parameter. The estimate is the value obtained for one particular sample of size n.

population mean there is no systematic error or bias. Some means will be too large and others too small but in a long hypothetical sequence of sample means the deviations cancel each other out; errors of estimation are essentially due to random deviations from the population mean. The property is sometimes described by saying that the sample mean is an *unbiased* estimator of the population mean. You can see that this is approximately so in the example we gave above of a sampling distribution (Figure 5.5). In this case we had drawn repeated samples of size 5, and the histogram is centred approximately on the value 10—the same as the population mean. When we calculate the mean of the histogram we get 9.9 which is very close to 10. If the histogram had been based on an infinitely large number of random samples, it would have been exactly 10.

(2) *The variance of the sampling distribution of the sample mean is the population variance divided by the sample size, i.e.* σ^2/n. This says that the spread of the sampling distribution about its mean, and therefore about the population mean, decreases with increasing sample size so that in larger samples we expect smaller errors of estimation. We will see below that the probable magnitude of the random errors in the estimate of the population mean can be measured by using the *standard error of the sample mean*. This is defined as the square root of the variance of the sampling distribution of the sample mean. It is also equal to the population standard deviation divided by the square root of the sample size (i.e. $\sigma/\sqrt{n}$). Note that the terminology standard error of the sample mean is used to refer to the magnitude of the random variation, from sample to sample, of the sample mean. It is so called to avoid possible confusion with the *standard deviation* which usually refers to the variation of individual values of the random variable.

(3) *The shape of the sampling distribution of the sample mean is Normal.* If the distribution of the raw observations is Normal, as our model states, then the sampling distribution of the sample mean is exactly Normal in form. We see below (Section 5.4.4) that even if the distribution of the raw observations is not Normal, the sampling distribution is often approximately Normal. This is one reason why the Normal distribution is so important in statistics. The example above (Figure 5.5) appears approximately Normal. If we superimpose a Normal distribution with the same mean as the population mean, and variance σ^2/n, we obtain the smooth curve given in the figure. This smooth curve appears to be a good description of the histogram we have built up by simulation. This therefore gives a demonstration (not a proof) of the truth of the above three statements.

Taken together the above three properties can be used to measure the likely magnitude of error in using the sample mean to estimate the population mean. We make use of the result that in repeated sampling from *any* Normal distribution only 5% of the observations will deviate from the mean by more than ± 1.96 times the standard deviation of the Normal distribution (see Figure 5.6).

So, when the sample mean is used to estimate the population mean, in only about 5% of cases will the estimate be in error by more than ± 1.96 times the standard error. In other words, the error will be *within* $\pm 1.96\sigma/\sqrt{n}$ in approximately 95% of cases. For means based on samples of size 36 from a population with variance equal to one this gives errors within $\pm 1.96 \times 1/\sqrt{36} = \pm 0.327$ for 95% of the time. The idea that an estimate is very likely to be within approximately two standard errors of the true value is an important one which occurs in many estimation problems and we will see other

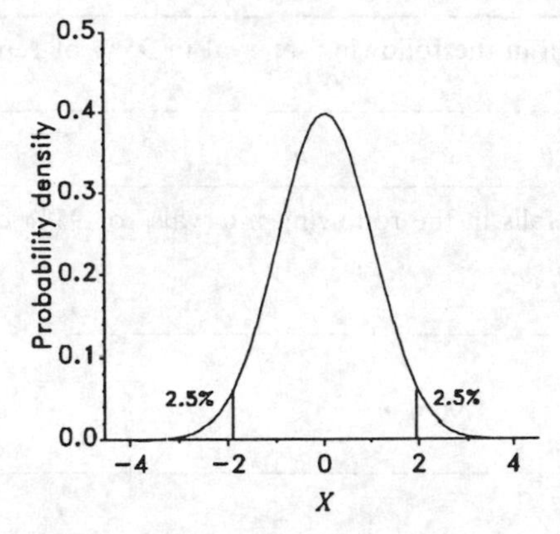

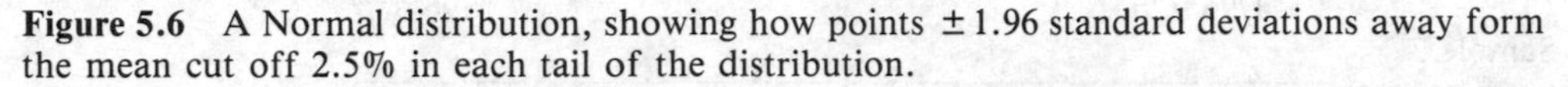

Figure 5.6 A Normal distribution, showing how points ± 1.96 standard deviations away form the mean cut off 2.5% in each tail of the distribution.

examples in this chapter. To calculate the likely magnitude of the error in estimating the population mean we need to know the population variance. In practice this is unknown so we proceed to estimate it from the sample.

5.4.3 Calculating a confidence interval for the population mean

The calculations when the population variance, σ^2, is known

In this section we use the above properties of the sampling distribution of the sample mean to demonstrate the notion of a confidence interval and show how to calculate such an interval from the data.

We start by reiterating the central result.

For 95% of repeated samples of size n from a Normal distribution with standard deviation σ, the sample mean, $\bar{Y}$, will fall within the interval $\mu \pm 1.96\sigma/\sqrt{n}$.

The result can be stated the other way round as follows.

For 95% of repeated samples of size n from a Normal distribution with standard deviation σ, the random interval calculated using the formula $\bar{Y} \pm 1.96\sigma/\sqrt{n}$ will contain the population mean.

The equivalence of the statements is illustrated in Figure 5.7. The upper interval is fixed and represents the range within which sample means fall with probability 0.95, in repeated drawings of samples of size n. The lower intervals calculated afresh for each sample move from side to side reflecting the variation in the sample mean. You will notice that whenever $\bar{y}$ falls in the upper interval, the lower intervals contains the population mean μ; also this occurs with probability 0.95 (actually, in this case, in *all* of the four samples illustrated above). For any particular sample, the interval either does or does not contain the population mean, but in the long run, i.e. for a large number of samples, 95% of the intervals will contain the population mean. In random samples of size from a Normal distribution with standard deviation σ the interval $\bar{y} \pm 1.96\sigma/\sqrt{n}$ is called a *95% confidence interval for the population mean*. When the

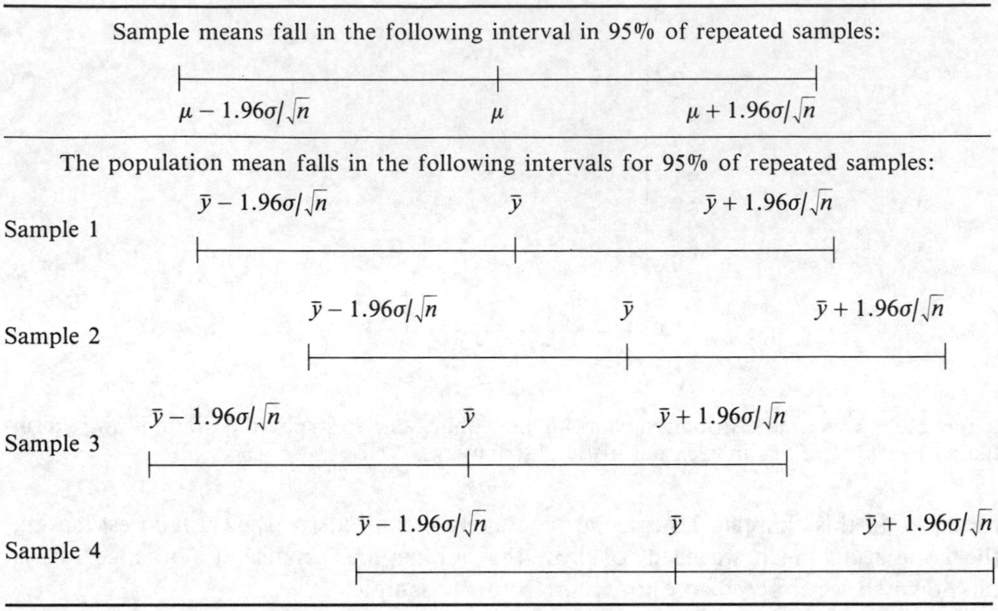

Sample means fall in the following interval in 95% of repeated samples:

$\mu - 1.96\sigma/\sqrt{n}$ μ $\mu + 1.96\sigma/\sqrt{n}$

The population mean falls in the following intervals for 95% of repeated samples:

Sample 1 $\bar{y} - 1.96\sigma/\sqrt{n}$ $\bar{y}$ $\bar{y} + 1.96\sigma/\sqrt{n}$

Sample 2 $\bar{y} - 1.96\sigma/\sqrt{n}$ $\bar{y}$ $\bar{y} + 1.96\sigma/\sqrt{n}$

Sample 3 $\bar{y} - 1.96\sigma/\sqrt{n}$ $\bar{y}$ $\bar{y} + 1.96\sigma/\sqrt{n}$

Sample 4 $\bar{y} - 1.96\sigma/\sqrt{n}$ $\bar{y}$ $\bar{y} + 1.96\sigma/\sqrt{n}$

Figure 5.7 A demonstration of the equivalence between the proportion of sample means which fall in an interval about the population mean, and the proportion of confidence intervals which enclose the population mean.

distribution of the observations is not Normal then approximately 95% of the intervals will contain the population mean provided the sample size is large enough. We discuss this further below in Section 5.4.4 on robustness.

But in general we do not know the population variance, σ^2

We can see from the above formula that to calculate a 95% confidence interval for the population mean we need the population variance. When the population variance and hence the population standard deviation, σ, are not known, we can estimate σ by using the sample standard deviation, s, as discussed in Section 5.3.1. The 95% confidence interval is then calculated by substituting the sample standard deviation, s, for σ in the above formula. But a further slight modification is also necessary. The value of 1.96 is a standardised Normal deviate which cuts off 2.5% of the area under the density function in each tail. This needs to be replaced by a quantity $t_{0.025,(n-1)}$, which is a similar value for a different distribution, known as Student's t-distribution with $(n-1)$ degrees of freedom. The effect of the use of $t_{0.025,(n-1)}$ rather than 1.96 is to widen the intervals slightly. It can be proved that this makes an appropriate allowance for the fact that we have estimated the population variance rather than knowing it exactly. Table S2 (Appendix) gives the values of $t_{0.025,(n-1)}$ for various sample sizes. You will see that for larger sample sizes, the value of t is hardly any larger than 1.96, but as sample size, n, gets below 10, t increases substantially, reflecting the fact that, as n gets smaller, the uncertainty in the estimation of σ gets greater. We still want to have the interval cover the true mean with probability 0.95 in repeated sampling. The only way this can be achieved is for the interval to widen.

Example 5.4 Calculation of a 95% confidence interval on the population mean

Sample of data: 7.84, 6.13, 12.17, 6.84, 14.00, 12.88, 5.92.

Sample mean: $\bar{y} = 9.40$
Sample variance: $s^2 = 12.13$
Sample standard deviation: $s = 3.48$
Sample size: $n = 7$
Estimated standard error of the sample mean: $s/\sqrt{n} = 1.32$
95% confidence interval:

$$\bar{y} \pm t_{0.025,(n-1)} s/\sqrt{n} = 9.40 \pm 2.45 \times 1.32 = 9.40 \pm 3.23 \text{ or } (6.17, 12.63).$$

This particular interval happens to contain the true population mean of 10 (which we know in this case, as we are in the very artificial situation that the data were obtained by simulation from a known model), but it is just one among many possible intervals, some of which would not. Figure 5.8 shows how calculated intervals vary from one sample to the next in 50 simulations from the same population. The intervals now vary in their length as well as their location, and they still vary in whether they contain the population mean. The only guarantee we have is that, in a much longer sequence, 95% of the intervals would contain the population mean.

The above method of calculating confidence intervals was developed for samples from a Normal distribution. When the distribution is not Normal we use the fact that the sampling distribution of the sample mean is approximately Normal, and that s^2 estimates σ^2, and calculate the confidence interval in the same way. The important proviso is that in repeated samples from the same distribution the calculated confidence interval will contain the population mean in *approximately* 95% of cases. We discuss this in more detail in Section 5.4.4.

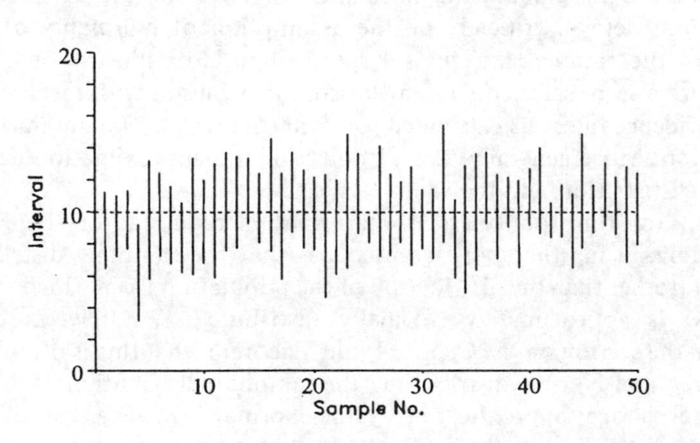

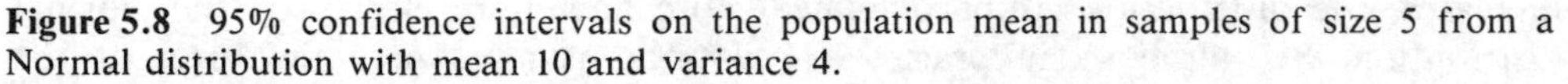
Figure 5.8 95% confidence intervals on the population mean in samples of size 5 from a Normal distribution with mean 10 and variance 4.

5.4.4 Robustness

One important step in any mathematical modelling or model-fitting exercise is checking which parts of the model are strictly needed for the conclusions you wish to draw. In some cases, it might be quite easy to produce an argument by which the conclusions can be derived from a given set of assumptions. But it might also be that the very same conclusions can be drawn from a smaller set of assumptions. Some of the assumptions in the original set are not *necessary*. More formally, suppose that the components or assumptions of a model can be written down separately as A_1, A_2, A_3, and one interesting facet of the behaviour of the model can be written down as a statement B. Quite often it is possible to say that if A_1, A_2 and A_3 hold, then you can draw the conclusion that B holds. But sometimes, not all of the three assumptions A_1, A_2 and A_3 are *necessary* for B to hold; or one, say A_3, might not need to hold exactly. We would then say that the behaviour, B, of the model is robust to departures from the assumption A_3.

The method outlined above for obtaining a confidence interval on the population mean assumes the following:

A_1: the individual units in the sample are drawn *from the same distribution*;
A_2: the individual units in the sample are drawn *at random* from the distribution;
A_3: the individual units in the sample are drawn *independently* from the distribution;
A_4: the distribution from which the individual units in the sample are drawn is *Normal*.

Assumptions A_1, A_2, A_3 are important for the argument by which we obtain 95% confidence intervals on the mean, but the fourth is not at all critical. Normal distributions do occur quite widely in natural data, but there are many important situations where they do not. For example, the Normal distribution is a model for continuous measurements. It would therefore be unrealistic for the whole class of data in the form of counts, especially when the counts are low and the discreteness of the distribution matters. For such cases, a Poisson or negative binomial distribution might be more appropriate.

One remarkable thing about the calculation of 95% confidence intervals above is that it does not depend critically on the assumption of Normality of distribution, provided that the sample size, n, is large enough. For illustration, 1000 random samples of size 5 were selected from a Poisson distribution with mean equal to 4. Of the 95% confidence intervals calculated for each sample 93.9% contained the mean of the distribution. More generally, the adequacy of the approximation depends on the form of the distribution and the sample size.

The reason for this robustness is that what is required for the calculation of confidence limits using the above argument is that the sampling distribution of the sample mean (rather than the distribution of the population from which we have drawn our samples) is approximately Normally distributed. We have already seen in Chapter 3, in the section on the Central Limit Theorem, that this is often so. For *small* samples from a non-Normal distribution, the sampling distribution of the mean might be only rather poorly approximated by the Normal. However, as the sample size increases, the distribution is approximated more and more closely by the Normal distribution. So, although in practice we do not usually know the form of the population distribution, this is not as serious as it seems, because the Normal distribution approximation to the sampling distribution of the sample mean can be achieved by taking a large sample. Chapter 3 gives a simulation method for determining how large the sample needs to be for some particular populations, together with some examples of this phenomenon. The main aspect of a population distribution to look out for which cause problems in this respect are extreme skewness, or distributions which have heavy tails compared with the bell-shaped Normal; i.e. those distributions for which there is a moderate chance of obtaining a sample which contains just one or two very large or very small values.

5.4.5 Sampling errors in fitting a straight line

The method of least squares is an objective procedure for fitting a straight line. In this section we examine the method in terms of the likely magnitude of the errors in the estimated slope and intercept of the fitted line.

The sampling distributions of $\hat{\beta}$ and $\hat{\alpha}$ are Normally distributed with means β and α

To illustrate the general results, we first simulate the process of drawing sample data from the particular model:

$$Y_i = 1 + 6x_i + \varepsilon_i, \qquad i = 1, \ldots, 5$$

where $\varepsilon_i \sim N(0, 4)$. Table 5.6 shows five simulations of the model together with the least squares estimates of the slope and intercept.

Estimates vary because of the stochastic element in the model. The variation in the parameter estimates is described by their sampling distributions which can be obtained by simulating a large number of samples. Figures 5.9(a) and 5.9(b) show histograms for 1000 simulations. What we see are bell-shaped distributions centred around the model parameter values. These illustrate the general results that: (a) the least squares estimates of both slope and intercept follow Normal distributions; (b) the means of the sampling distributions are equal to the model parameter values (i.e. 6 for the slope and 1 for the intercept). The estimates are therefore unbiased.

Table 5.6 Simulated samples from the linear model given in Section 5.4.5

x_i	$\mathrm{E}[Y_i] = 1 + 6x_i$	Five sets of simulated data				
1	7	8.34	6.38	6.39	9.23	8.88
2	13	11.52	15.87	8.39	7.17	9.72
3	19	15.33	21.88	18.53	14.62	20.75
4	25	28.10	24.13	22.62	23.65	24.77
5	31	32.49	29.99	32.58	32.96	30.59
Estimated intercept ($\hat{\alpha}$)		−0.31	3.01	−2.28	−1.66	1.40
Estimated slope ($\hat{\beta}$)		6.49	5.55	6.66	6.39	5.85

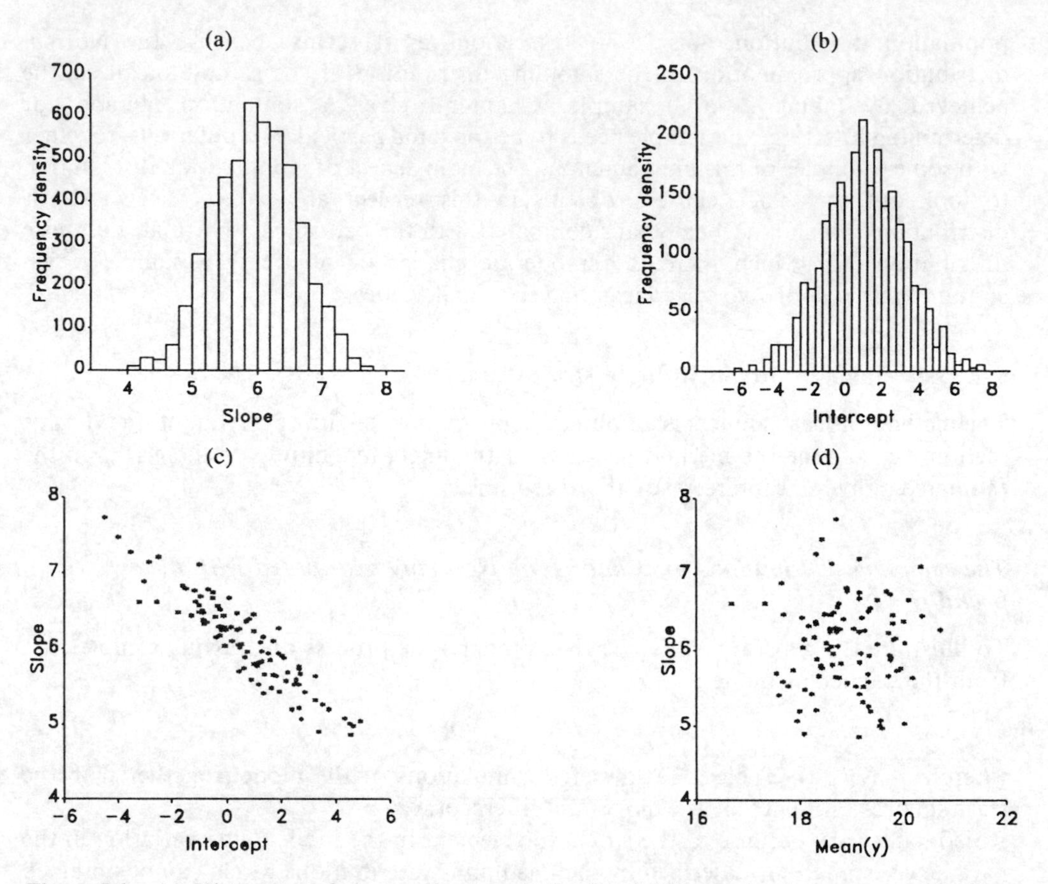

Figure 5.9 (a) Histogram of slope estimates obtained in 1000 samples from the straight line-Normal model given in Section 5.4.6; (b) histogram of intercept estimates in the same sampling experiment; (c) scatter-plot of the slope estimates plotted against the intercept estimates; (d) scatter-plot of the slope estimates plotted against $\bar{y}$.

The variances of the sampling distributions of $\hat{\beta}$ and $\hat{\alpha}$

A shorthand terminology for the variance of the sampling distribution of an estimator is the *sampling variance of the estimator*. As might be surmised, the sampling variance of the estimated slope depends on the number of observations, the range of the values of the explanatory variable x and the magnitude of the variance about the regression line. The relationship is illustrated in Table 5.7 which shows the sampling variance of the estimated slope in 1000 simulations of models involving different combinations of these factors.

The table shows that the sampling variance of the estimated slope decreases when: (a) the variance of y about the line decreases; (b) the range of the values of x is increased; and (c) the number of data points for each value of x is increased. The general result, where n here is the total sample size (i.e. 5, 10 or 20 in the above simulations), is

$$\text{Sampling variance of estimated slope} = \frac{\sigma^2}{\sum_{i=1}^{n} (x_i - \bar{x})^2}$$

Table 5.7 Relationship between sampling variance of the estimated slope for various designs, and values of residual variance, σ^2

Number of values of y for each x	Design 1 $x = 1, 2, 3, 4, 5$		Design 2 $x = 2, 2.5, 3, 3.5, 4$	
	$\sigma^2 = 2$	$\sigma^2 = 4$	$\sigma^2 = 2$	$\sigma^2 = 4$
1	0.20	0.41	0.77	1.62
2	0.10	0.20	0.44	0.82
4	0.05	0.10	0.20	0.41

$$\text{Estimated by } \frac{\text{residual mean square}}{\sum_{i=1}^{n} (x_i - \bar{x})^2}$$

For example, in the first design with a single value of y for each value of x, $\Sigma(x_i - \bar{x})^2 = 10$. For a variance about the regression line equal to 2 the formula for the sampling variance of $\hat{\beta}$ gives 0.2 which agrees with the simulated value.

The sampling variance of the estimated intercept could be examined in a similar way but here we simply quote the result and give an intuitive justification. First, note that the intercept is estimated as $\hat{\alpha} = \bar{y} - \hat{\beta}\bar{x}$. The implication of this is that the intercept is affected by variability in the raw data in two ways: (a) through the resulting errors in $\bar{y}$ (the average of the y values and the ordinate of the centroid of the data)—if $\bar{y}$ is by chance too high, then so is $\hat{\alpha}$, similarly if it is too low; and (b) through errors in $\hat{\beta}$—if the line is by chance too steep, then this will induce a similar error, weighted by $\bar{x}$, the distance in the x-direction between the position of the centroid and where the line intercepts the axis. Furthermore, estimates of the slope vary independently of $\bar{y}$ from sample to sample (see Figure 5.9(d)). These two properties can be combined to give the sampling variance of the estimated intercept as

$$\text{Sampling variance of } \bar{y} + \bar{x}^2 \times \text{sampling variance of estimated slope}$$

$$= \frac{\sigma^2}{n} + \frac{\bar{x}^2\sigma^2}{\sum_{i=1}^{n} (x_i - \bar{x})^2}$$

$$\text{Estimated by } \frac{\text{residual mean square}}{n} + \frac{\bar{x}^2 \times \text{residual mean square}}{\sum_{i=1}^{n} (x_i - \bar{x})^2}$$

The first term in this expression comes from the variation in $\bar{y}$ which is a mean of n observations each with variance equal to σ^2. The contribution from errors in the estimated slope increases as $\bar{x}$, the x-coordinate of the centroid, moves further from the origin.

Figure 5.9(c) shows a scatter plot of the estimated slope and intercept for the first 100 simulations. This illustrates the negative correlation between the estimates; so that when the estimated slope is high the estimated intercept tends to be low. This makes sense when one visualises the effect of rotating the line about the centroid of the data points. Figure 5.9(d) demonstrates that there is no correlation between the estimated slope and the average of the values of y, the ordinate of the centroid. This illustrates the general point that some aspects of a model might be estimated independently of each other, whereas for others estimates may be correlated. We discuss this point later in Chapter 11.

The importance of the above results is that they allow us to measure the likely magnitude of the errors in the estimates of the slope and the intercept as follows.

Confidence intervals for the slope and intercept

In this section we show how to calculate confidence intervals for the slope and intercept of the linear regression model. Calculating confidence intervals for the slope and intercept in the straight line-Normal model proceeds along similar lines to the calculations for a single mean (Section 5.4.3), using the above expressions This is because (a) sampling distributions of the estimates are Normal and (b) the standard errors can be estimated by using the residual mean square to estimate the variance about the regression line (Section 5.3.2). The slight modification is that the degrees of freedom in the value of Student's t are now $(n-2)$, the residual degrees of freedom. This allows for the fact that in fitting the straight line, we have estimated two parameters of the deterministic part of the model, α and β, rather than just one when we were considering the estimation of the population mean.

Summary. 95% Confidence interval for slope and intercept

Generally:

Estimate $\pm\, t_{0.025,(n-2)} \times$ standard error for the estimate

$$= \text{estimate} \pm t_{0.025,(n-2)}\sqrt{\text{sampling variance of the estimate}}$$

For the slope:

$$\text{Estimated slope} \pm t_{0.025,(n-2)}\sqrt{\frac{\text{residual mean square}}{\sum_{i=1}^{n}(x_i-\bar{x})^2}}$$

For the intercept:

$$\text{Estimated intercept} \pm t_{0.025,(n-2)}\sqrt{\left(\begin{matrix}\text{residual}\\ \text{mean}\\ \text{square}\end{matrix}\right)\left(\frac{1}{n}+\frac{\bar{x}^2}{\sum_{i=1}^{n}(x_i-\bar{x})^2}\right)}$$

Example 5.5 Growth rate of rats against dose of vitamin B_2

The data are discussed in Chapter 4. The model we use is:

$$Y_i = \alpha + \beta x_i + \varepsilon_i, \qquad i = 1, 2, \ldots, n$$
$$\varepsilon_i \sim N(0, \sigma^2)$$

where the growth rate and $\log_{10}$ (dose) of vitamin B_2 for the ith animal are denoted by Y_i and x_i respectively, and there are $n=20$ rats.

There is now a wide range of computer software for fitting linear regression models so we concentrate on the results of the analysis and their interpretation, rather than the details of the calculations. The table below gives a fairly typical presentation of the computer output.
The regression equation is

$y = -0.53 + 17.56x$

Predictor	Coef	Stdev	t-ratio	p
Constant	−0.527	1.647	−0.32	0.753
x	17.562	1.775	9.89	0.000

s = 2.669 R-sq = 84.5% R-sq(adj) = 83.6%

Analysis of Variance

SOURCE	DF	SS	MS	F	p
Regression	1	697.25	697.25	97.85	0.000
Error	18	128.26	7.13		
Total	19	825.51			

The first part of the table gives the estimated parameters of the fitted line together with their estimated standard errors. Note that 'Stdev' (short for standard deviation) is used in this package to refer to the estimated standard error, illustrating that terminology varies between packages. In our description so far, we have used standard error when we refer to random variation in an estimate to avoid confusion with the standard deviation, which often refers to variation of the individual observations. The fitted line has a slope of 17.56 (s.e. = 1.78) and intercept of −0.53 (s.e. = 1.65). Figure 5.10 shows the data together with the fitted line. Note that the model relates growth rate to log dose so that the effect of doubling the original dose level is to increase growth rate by $17.56 \times \log_{10} 2 = 5.29$.

The lower part of the output gives the analysis of variance table which we discussed in Section 5.3.2. For estimating the variance about the regression line and setting confidence intervals on the parameters we are concerned with the residual mean square and its degrees of freedom. The appropriate line in the table is labelled as an 'Error' sum of squares which is sometimes used as an alternative name for the residual sum of squares. The estimated variance about the line is 7.13 with an associated 18 degrees of freedom. Note that the value of s given above the table is the square root of this.

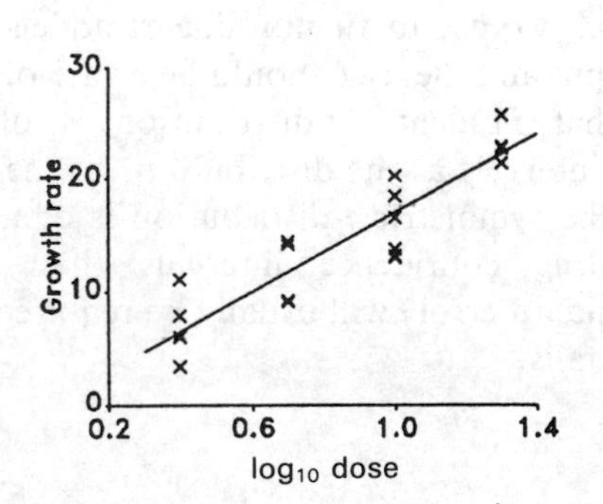

Figure 5.10 Growth rate of rats in relation to dose of vitamin B_2. Data and fitted straight line. See Example 5.5.

To calculate a 95% confidence interval for the slope (not given in the computer output) we need the two-tailed 5% point of Student's t-distribution with 18 degrees of freedom which is 2.10. This is then multiplied by the estimated standard error to give a 95% confidence for the estimated slope as 17.56 ± 3.73. Similarly, a 95% confidence interval for the intercept is 0.53 ± 3.46.

5.4.6 Sampling errors in fitting quadratics and polynomials

Fitting a polynomial by least squares produces estimates of the parameters β_0, β_1, $\beta_2, \ldots$, etc. which are unbiased and distributed Normally. Furthermore, the sampling variances of the estimates are equal to the residual variance (i.e. variance of deviations about the curve) multiplied by quantities which can be calculated from the values of x. The situation is analogous to that for a straight line where the sampling variance of the estimated slope is equal to $\sigma^2/\Sigma(x_i - \bar{x})^2$. The variance of the random element, σ^2, can be estimated by a similar analysis of variance to that employed for the straight line as discussed in Section 5.3.2. These results can be combined to calculate a 95% confidence interval for each parameter:

$$\text{Parameter estimate} \pm t_{0.025,(n-p)} \times \text{estimated standard error}$$

where $t_{0.025,(n-p)}$ is the two-tailed 5% point of Student's t-distribution with $(n-p)$ degrees of freedom.

The close correspondence between these results and those for fitting the linear regression model may seem surprising especially as polynomials can describe such a wide variety of curvilinear relationships between y and x (Chapter 4). The similarity arises because in both types of model the parameters occur as linear combinations in the model, i.e.

$$\beta_0 + \begin{pmatrix}\text{something}\\\text{involving } x\text{'s}\end{pmatrix} \times \beta_1 + \begin{pmatrix}\text{something}\\\text{involving } x\text{'s}\end{pmatrix} \times \beta_2 + \cdots$$

5.4.7 Sampling errors fitting non-linear models

For non-linear models the above results which were used to construct confidence intervals on model parameters do not apply exactly and the properties of the sampling distributions of the estimators can only be obtained approximately. In practice, the computer packages which we use to fit non-linear models usually produce standard errors as part of the output and the user should be aware of their approximate nature. A further aspect is that Student's t-distribution is often not appropriate for determining confidence intervals as the distributions of the estimates are often highly skewed, and therefore the symmetric t-distribution is of no use. More sophisticated approaches to calculating confidence intervals than simply using parameter estimate $\pm\, t_{0.025,\upsilon} \times$ (standard error) will usually be required. The reader is referred to Chapter 11 for more details.

EXERCISES

5.1 The following samples of data were collected in surveys of clutch size in each of two types of habitat, the first at the centre of the breeding range of the species, and the second at the margin of the breeding range.

Central habitat	2, 3, 1, 0, 4, 1, 2, 1, 3, 1, 3, 2
Marginal habitat	0, 1, 0, 5, 0, 0, 1, 2, 0

(a) Estimate mean clutch size, and the standard deviation of clutch size in the two habitats from these data.

(b) Obtain 95% confidence intervals on the means in the two habitats.

(c) Comment on any approximations which you need to make in the two cases.

5.2 *Confidence intervals on the population mean of non-Normal populations.* In Example 3.12, the rapid approach to Normality of the sampling distributions of the means of samples drawn from non-Normal populations is demonstrated. Using simulation, consider the impact of the non-Normality on the true confidence of nominal 95% confidence intervals, i.e. the percentage which contain the true mean, for samples of size $n = 4, 8, 16$ from (a) a uniform [0, 1] and (b) a negative exponential, with rate parameter $\lambda = 2$, mean $= 1/2$. (See also Exercise 5.3 for a related exercise using these same simulations.)

5.3 *Confidence intervals on the population median.* Confidence intervals on the median, rather than the mean of the population, using data from a random sample of size n, $x_1, x_2, \ldots, x_n$ can be calculated as follows:

(i) order the values in the sample in ascending order to obtain $x_{(1)}, x_{(2)}, \ldots, x_{(n)}$;
(ii) take the kth value from both ends of the ordered list, where k is as given in the following table. Exact 95% confidence intervals cannot be obtained for all sample sizes, and the exact confidence, c, is also given in the table.

n	6	7	8	9	10	12	14	16	20	25	30	40	60
k	1	1	1	2	2	3	3	4	6	8	10	14	22
c (%)	96.9	98.4	99.2	96.1	97.9	96.1	98.7	97.9	95.9	95.7	95.7	96.2	97.3

(a) Obtain approximate 95% confidence intervals on the population median clutch size for the two habitats in Exercise 5.1.

(b) Verify in the simulation exercises in Exercise 5.2 that this formula does result in confidence intervals with approximately the degree of confidence indicated. You should:

(i) obtain expressions for the medians of the populations for the two distributions;
(ii) repeat the simulations above (except for sample size $n = 4$ when no confidence interval of worthwhile size can be obtained on the median by this method) assessing, in each case, whether the confidence intervals calculated as above do include the true median, and determine the proportion which do.

5.4 In a study to determine the number of type A cells in blood samples, the number per field (a square field in a microscope graticule) is counted for a number of randomly selected fields. It is thought that type A cells are thoroughly mixed in the sample and that all the cells are about the same size and are quite small compared with the size of the field.

(a) What would you expect the distribution of the number of cells per randomly selected field to be? Justify your answer.

(b) How would you estimate m, the mean number of cells per field?

(c) What would be the probability that a field contains zero cells?

(d) In the case when m is quite low, how would you use the answer to (c) to obtain an alternative estimate of m?

(e) In practice, how might the above model fail, and would such failures be important in your estimation scheme?
(Cambridge, NST, Part 1A, Biological Mathematics, 1988.)

5.5 The data on the relationship between leaf area, A, and leaf area index, x, given in Figure 1 of the Introduction are as follows:

x	y	x	y
0.7	0.4	5.5	3.8
1.1	1.0	6.1	3.1
2.4	1.5	5.3	4.7
3.4	2.7	5.5	4.9
4.6	2.5	6.3	4.4
4.6	2.8	6.8	5.0
4.2	3.5	6.4	5.2
4.4	3.5	6.5	5.3
4.8	3.8	6.9	5.3
5.2	4.0	7.5	5.0
5.3	3.9	7.5	6.3

(a) Fit the ordinary straight line-Normal model to these data.
(b) Estimate the residual variance.
(c) Obtain 95% confidence intervals on the slope and the intercept.
(d) Does it seem plausible from these results that the true intercept is zero?
(e) Fit a straight line-Normal model which passes through the origin. The least squares estimate of the slope is given by $\hat{\beta} = (\Sigma\ xy)/(\Sigma\ x^2)$.

5.6 The following data relate heart weight (h) to body weight (w) in female cats (data due to H.G.O. Hock, quoted from Fisher, 1947).

w	2.3	3.0	2.9	2.4	2.3	2.0	2.2	2.1	2.3	2.1	2.1	2.2	2.0	2.3	2.2	2.3
h	9.6	10.6	9.9	8.7	10.1	7.0	11.0	8.2	9.0	7.3	8.5	9.7	7.4	7.3	7.1	9.0
w	2.1	2.0	2.9	2.7	2.6	2.3	2.6	2.1	2.1	2.7	2.2	2.3	2.1	2.4	2.7	2.3
h	7.6	9.5	10.1	10.2	10.1	9.5	8.7	7.2	9.8	10.8	9.1	11.2	8.1	10.2	8.5	10.1
w	2.1	2.2	2.3	2.1	2.9	3.0	2.2	2.4	2.4	2.5	2.5	2.3	2.3	2.6	2.3	
h	8.7	10.9	7.9	8.3	10.1	13.0	8.7	6.3	8.8	10.9	9.0	9.7	8.4	10.1	10.6	

Formulate and fit a model suitable for predicting heart weight from body weight. Estimate all the parameters including the residual variance, σ^2. Obtain some measure of the uncertainty of estimation of the parameters of the curve (apart from σ^2).

5.7 The straight line-Normal model through the origin is $Y_i = \beta x_i + \varepsilon_i$, where the ε_i are independently Normally distributed with mean zero and variance σ^2.
(a) Obtain the least squares estimate of β, given that a sample of data, (x_1, y_1), $(x_2, y_2), \ldots, (x_n, y_n)$, is available.
(b) Obtain an expression for the variance of the estimate.
(c) By writing $(y - \hat{\beta}x)^2 = [(y - \beta x) + (\beta - \hat{\beta})x]^2$, show that $1/(n-1)\ \Sigma\ (y - \hat{\beta}x)^2$ is an unbiased estimate of σ^2.
(d) Write down an expression for a 95% confidence interval on the slope.

5.8 The systolic blood pressures (mm Hg) of a random sample of women over 30 years of age were recorded and grouped into 10 year age classes. The mean blood pressure for each age class is given below:

Mean blood pressure, y	114	124	143	158	166
Midpoint of age class (years), x	35	45	55	65	75

(a) Given that $\Sigma\,(x-\bar{x})(y-\bar{y}) = 1380$, $\Sigma\,(x-\bar{x})^2 = 1000$, and $\Sigma\,(y-\bar{y})^2 = 1936$, estimate the slope and intercept of the straight line regression of mean blood pressure on age, and sketch the line and the points.

(b) Estimate the residual variance of the relationship, and test whether or not the slope is significantly different from zero.

(c) Comment on the grouping of data into age classes. What approximations does this involve? In your answer, you should comment on any inadequacies in the specification of the problem. For example, how many women are there in each age group?

5.9 *Estimating the mean of lognormal distributions.* For random variables which are known to be lognormally distributed, two possible methods are available for estimating the population mean from a random sample, $X_1, X_2, \ldots, X_n$:

Estimator 1. As above, use the arithmetic mean of the sample,

$$E_1 = \bar{X} = \sum_{i=1}^{n} X_i.$$

Estimator 2. Use the antilog of the arithmetic mean of the log transformed values, i.e. if $Y_i = \log_e(X_i)$ for $i = 1, \ldots, n$, use the estimator

$$E_2 = \exp(\bar{Y}) = \exp\left(\frac{1}{n}\sum_{i=1}^{n} Y_i\right).$$

Verify by taking repeated random samples from the lognormal distribution such that $\log_e(X) \sim N(0, 4)$ of sizes 1, 4 and 8 that in these cases:

(a) The mean of the lognormal distribution is $\exp(\mu + \sigma^2/2)$;

(b) $E_1 = \bar{X}$ is an unbiased estimator of the population mean;

(c) $E_2 = \exp(\bar{Y})$ is a biased estimator of the population mean, and that the mean of E_2 in repeated sampling is given by $\exp(\mu + \sigma^2/2n)$.

5.10 For the three sets of data given in Exercise 4.1, fit the most appropriate relationships between y and x, and assess the uncertainty in the fitted curves.

At least one of the fitted curves should be a polynomial. Sometimes fitting procedures work better when the explanatory variables are centred, i.e. adjusted so as to have a mean of approximately zero, before carrying out the fits. Investigate the effect of centring, by subtracting the mean of the explanatory variable, x, before calculating powers and estimating the polynomial.

5.11 The fish data of Example 1.8 illustrating the von Bertalanffy growth curve

$$\begin{aligned} L &= L_{\max} - [L_{\max} - L(0)]\,e^{-kt} \\ &= L_{\max}(1 - A\,e^{-kt}). \end{aligned}$$

where $A = 1 - L(0)/L_{\max}$ is the proportion of the maximum length remaining for growth at time zero, are given in the following table.

Age (weeks)	Length (mm)	Weight (mg)
0	9.4	7.6
1	12.6	13.0
2	15.7	28.0
3	17.5	46.7
4	19.9	71.4
5	22.1	89.3
6	22.5	111.2
7	23.2	120.0
8	24.3	133.3
9	24.0	140.0
10	25.3	152.5
11	24.5	137.5
12	25.0	137.5
13	25.0	146.0

These are mean weights and lengths of samples of fish independently drawn from the population on each occasion.

(a) Fit the model to length data assuming a length at $t = 0$ of 9.4.
(b) Fit the full three parameter model, thus estimating A from the data rather than using the initial value of L.
(c) Fit the full three parameter model for weight

$$W(t) = W_{\max}(1 - A\,e^{-kt})^3.$$

(d) Are the two fitted curves consistent with each other?

5.12 The data obtained by Carlson (1913) on the growth of yeast cells is as follows.

Time (h)	Yeast amount
0	9.6
1	18.3
2	29.0
3	47.2
4	71.1
5	119.1
6	174.6
7	257.3
8	350.7
9	441.0
10	513.3
11	559.7
12	594.8
13	629.8
14	640.8
15	651.1
16	655.9
17	659.6
18	661.8

Ignoring any problems about serial correlation in the observations since these are measurements on the same colony over time (this is a longitudinal study—see Section 7.4.3), fit the model

$$N(t) = \frac{K}{1 + \left[\dfrac{K}{N(0)} - 1\right] e^{-rt}}$$

by non-linear least squares using the known initial value $N(0) = 9.6$. Is there any evidence of serial correlation in the random departures from the curve? (*Hint*: calculate the correlation between successive residuals.)

Part II

BIOLOGICAL COMPARISONS AND DESIGN ISSUES

6 Statistical Methods for Comparing Biological Populations and Processes

6.1 INTRODUCTION

This chapter is about methods for obtaining an answer to the general question: is there a difference? There are many possible comparisons, but those we consider in this chapter are between two or more biological or statistical populations or realisations of random processes, or of single parameters of models fitted to them, e.g. means or slope coefficients. We also consider how to compare aspects of a population or process with their expectations according to a hypothetical model, which might serve as a test of the model. Biological comparisons are not always straightforward. With variable material, even if a big difference is observed between samples from two populations, we need to determine whether this reflects a real difference between the populations, or whether we just happen to have selected by chance individuals with a high response from the first population and those with a low response from the second, *identical* population.

In the next few sections we discuss some uses of comparisons, the aspects of a population or model which are often compared, and the types of data which can be used for making the comparisons. We then outline the general rationale of the *significance test*—the general method we propose for answering the question 'Is there a difference?'—considering the case of the comparison of the means of two populations. We then consider other methods for comparing the means of two populations for different types of data, before moving on to comparison of means of three or more populations. We also outline methods for comparing the spreads of populations, and slopes and intercepts in straight line models. We discuss situations in which it is inappropriate to ask the simple question: is there a difference? Finally we consider how these methods might be adapted for comparing a real population with a model.

We do not give detailed instructions for carrying out most of the tests, since many statistical packages are available to do this, and there is no need to burden the reader with many formulae and calculation techniques. Any reader requiring this detail is referred to Sokal and Rohlf (1969), where methods of calculation are given for many test procedures.

6.1.1 Uses of comparisons

It is frequently necessary to make comparisons in biological research or development. At the more applied end of the spectrum, there is often practical value in comparing a number of treatments of biological material to determine which has either optimal or least detrimental effect. A frequent aim in agricultural field trials is to establish which farming practices—e.g. addition of fertilisers, the use of different harvesting techniques, application of insectides or fungicides—produce a *discernible* effect on the yield. Such trials have resulted in greatly increased agricultural efficiency. In ecology, we might compare populations in different habitats, or at different times (e.g. of birds of prey on moorland and farmland, or before and after the introduction of organochlorine pesticides). In the last thirty years, substantial improvements in medical practice have occurred partly due to the increased use of clinical trials in which new treatments of some medical condition are compared with the current standard treatment.

Considering areas of basic research, it might be helpful, in revealing aspects of an underlying mechanism, to know if a controlled manipulation of the material produces a discernible effect up or down. Such manipulation is often of no practical or economic interest. An example occurring in the area of brain function is whether the response of a particular area of brain tissue is affected by the application of a drug which blocks the action of natural opiates. If there is a difference in response, then these natural opiates are likely to be implicated in the mechanism. This type of question often arises at the qualitative stage of research, when a predictive model has not been fully formulated, and there is no direct interest in the exact magnitude of the differences. It might not be possible to produce a quantitative predictive model because of the difficulty of measuring the critical components. There is also the possibility that the parameters will change, and so effort directed towards a precise estimation would be wasted. One example is unidirectional genetic change in a population as a result of selection, so that detailed estimates obtained at one time might not be relevant later.

6.1.2 Types of comparison

So far we have introduced many different models which might be appropriate for biological populations and processes. If we fit the same model to two or more populations or processes, we might wish to compare the whole model between the two or more situations, asking the question: is there evidence against the hypothesis that the models are identical? More often there are some critical parameters which need to be compared between the models. For the purely stochastic models of Chapter 3, the most common comparisons are of means, but spreads and the whole distributions are also frequently compared. For other more complex models, such as those in Chapter 4 (structured stochastic models), we might wish to compare the values of one or more parameters. An example we deal with in this chapter is comparison of straight line-Normal models, where we often wish to compare slopes, and possibly intercepts on the axes. Some types of comparison are too complex for this introductory chapter and finally we give pointers to more involved comparisons in other chapters.

6.1.3 Types of population and samples

Biological and statistical populations

We should emphasise here that the techniques given in this chapter are applicable to biological populations in the narrow sense, but are also applicable to a more general population type, which we term *statistical populations*. The distinction between the two types has been mentioned before (see Chapter 3), but it is worthwhile reiterating it here. The term statistical population is much more generalised, being applied to any collection of units which can be randomly sampled. It could be a population of responses of a single individual which we sample to determine the mean response rate of that individual over an experimental period during which different treatments might be applied. It could be a population of spatially distributed quadrats, which we sample to estimate the density of organisms per unit area.

Types of population

A *biological population* of organisms can be defined as a group of individuals of the same species inhabiting a given area.

A *statistical population* is the entire finite or infinite aggregate of units from which samples are drawn.

Real and model populations and processes

With increasing use of stochastic models in biology, there is also a need for a different type of comparison: a comparison of the predictions of a stochastic model with data obtained by sampling a real population or process. Before discussing this in more detail in Section 6.6, it is first necessary to deal with the methods for comparing models of real populations or processes on the basis of sample data from them.

Simple random sampling, randomisation and independence

For most statistical comparison techniques, it is important that a random element be involved in the selection of individuals to be included in the samples drawn from the populations being compared. The basic random selection procedure for biological populations is simple random sampling: sampling which involves a random element in the selection of the samples, and which gives each possible sample of the same size an equal chance of being selected. When comparing two treatments, an important safeguard which ensures that any differences can reasonably be attributed to differences in treatment is randomisation of the allocation of treatments to the experimental units. The equivalent of simple random sampling is a completely randomised design, in which every possible allocation of treatments to units (with the same replication) is equally likely. Either simple random sampling in the case of comparison of two populations, or a completely randomised design in the case of a comparison of two treatments, is assumed for all the techniques discussed in this chapter (see also Chapter 7).

A key aspect of both these procedures is that the units included in the sample from each population, or allocated to each treatment, respond independently of each other. In experiments, there should be no grouping factor within the replication of each treatment. For example, if two litters each consisting of 4 litter-mates were allocated to each of two treatments, we could not regard the responses of the 8 individuals allocated to each treatment as independent. Individuals within the same litter are more likely to respond similarly, since they have some genetic influences in common. Sampling designs in which some of the replication consists of repeated observations on the same individuals are also not adequate. More complex designs and methods of analysis are necessary in these cases (see Chapter 7).

6.1.4 Types of data

The procedure we use for a comparison, particularly for comparing means, depends on the type of data collected. This section briefly reviews the common types of data in ascending order of information content.

(1) *Categorical data, in particular binary data, and derived proportions*. Categorical data are derived by grouping individuals into classes or groups: for example, female or male; Irish, Scottish, Welsh, or English; pregnant or non-pregnant. Data in which there are just two possible outcomes are said to be binary. Categories are the simplest scale of measurement: they cannot be ordered, and distances between them cannot be defined. For example, we cannot put the nationalities of the United Kingdom in any sensible order, except on the basis of some other measurements. One numerical measurement derived from categorical data deserves special mention: the *proportion*, for example the proportion of girls in a class of schoolchildren, the proportion of heads in 10 tosses of a coin. We saw in Chapter 3 that the binomial distribution is an appropriate description for proportions derived from *binary data* (where there are only two categories), and data conforming to it have special methods of statistical analysis*.

(2) *Rank data or ordinal measurements*. The next most general class of measurement is the *rank*. These are categories, but have the additional characteristic that they can be put in order. A classic example of a naturally occurring ranking in biology is *pecking order*. The order is related to the dominance of the bird. Even though it is often impossible to measure dominance at all meaningfully, it is usually clear what the pecking order of a set of birds is. Ranks are also sometimes called *ordinal measurements*.

(3) *Continuous measurements on an interval scale*. These are measurements such as weights, densities, times, distances, growth rates, which could theoretically at least take any value. If we analyse what this involves in more detail, there are a number of component features. First, like rank data, they can be put in order. For example, if we have percentage growth rates such as 10.2, 15.4, 17.3, 8.4, these can easily be put in order 8.4, 10.2, 15.4, 17.1. Secondly, the intervals between them can be put in order. It is meaningful to say that the intervals 1.8, 5.2 and 1.7 can be ordered and that the second is about three times the other two. Such measurements are said to be on an *interval scale*.

6.1.5 Moving down the progression of data scales

One further feature of this catalogue of data types is that a datum at a higher level of

* Note that here we are not talking about proportions based on the ratio of two *continuous* measurements such as the proportion by weight of gold in a sample of ore. Such proportions could follow any distribution (between 0 and 1 of course) and need not be at all related to the binomial distribution.

information content can always be reduced to one at a lower level, by ignoring some of the information at the higher level. These might seem like rather academic manipulations, but many robust comparison procedures—which make few assumptions about the model underlying the data—function by moving down the progression of data scales. This trick is worth noting as it can be used in many situations and requires no special expertise.

(1) *Continuous measurements to ranks*. For example, the two sets of continuous data $A = \{1.1, 2.1, 0.5\}$, $B = \{0.3, -0.1, 1.4\}$ can be reduced to ranks $A = \{4, 6, 3\}$ and $B = \{2, 1, 5\}$ when they are ranked together. This is less informative in a sense, but it does mean that a statistical method which can be used for ranks can be applied to the original data set just by transforming to ranks.
(2) *Reduction further to binary data*. They can further be reduced to binary data by, for example, coding any value which is less than the joint median as 0 and any value above the joint median as 1. The joint median for the above data set would be calculated as the average of the 3rd and 4th value as there are an even number (6) of values, and so is 0.8, and so the binary equivalents are $\{1, 1, 0\}$ and $\{0, 0, 1\}$.
(3) *Expression as proportions*. These can then be reduced to proportions when we say that two thirds of the values in sample A are below the joint median of the two samples, and one third are above it.

6.2 THE SIGNIFICANCE TEST

6.2.1 Introduction

In this section we introduce the concept of the significance test which is the main general method used for determining whether there is a difference between two or more populations. In Chapter 5, we discussed how to estimate the mean of a single population, and to quantify the possible errors of estimation. If we had two populations whose means we wish to compare, we could similarly estimate the individual population means by the sample means, and obtain 95% confidence intervals on each of the population means. We could also easily estimate the difference of the means. By methods which we shall outline later in this chapter, we can obtain a 95% confidence interval on the difference of the two means: limits within which the difference of the true (i.e. population) means would lie with probability 0.95. If such an interval did not contain zero, we would say that there is strong evidence that the two population means do actually differ. But there is another method of answering the question: do the two population means differ? In this section we outline the rationale of the significance test for this purpose, using two simple examples: the comparison of two populations by a randomisation test, and by Student's *t*-test. In the following section, we summarise some alternative tests for comparison of the means of two populations.

Null hypothesis

We first need to define the hypothesis we are testing. The word *hypothesis* is used because it is about a situation which might or might not be true. The hypothesis is also usually framed in terms of one or more *parameters* of a model *for the population* or *populations* (e.g. the means of two populations, the regression slopes in a straight line-Normal model), although it could be just that the two populations are identical in every respect. Usually it is something like: the two means are the same, $\mu_1 = \mu_2$, or that the population mean takes a particular constant value, $\mu_1 = 5.5$, and so the general term which is given to it is the *null hypothesis*. The null hypothesis has to be *completely specific or exact* about the issue in question: the particular aspect or aspects of the populations which are being compared. The hypothesis that the mean is greater than zero, $\mu_1 > 0$, would not be sufficiently specific to be a null hypothesis. Statistical hypotheses are usually referred to by a capital H, with a subscript zero giving H_0 for the null hypothesis. So the null hypothesis in the particular example we are considering could be that the two population means are equal, i.e. in symbols, H_0: $\mu_1 = \mu_2$. One other feature about the null hypothesis is that although it has to be exact about the feature being compared, other features could be imprecisely specified or not specified at all. Another null hypothesis that is very commonly of interest when comparing two populations is that the distributions of the two populations are identical:

$$H_0\text{: Population distribution}_1 \equiv \text{Population distribution}_2.$$

Nothing else need be specified. The distributions could be of any shape, the variances and means could take any value, so long as they are the same for the two populations.

Test-statistic

The next element of a significance test is a *test-statistic*, T. This is a specified function of the observations contained in the sample, i.e. in any individual case a number which is calculated from the observations in the sample. In the test for comparing population means, a sensible test-statistic from which to start is the difference of the observed sample means $T = \bar{x}_1 - \bar{x}_2$. Another possibility is to use sample medians which are more useful if the population distributions are very skewed (strictly speaking we would then be testing the null hypothesis that the medians, rather than the arithmetic means, of the two populations are the same).

Alternative hypothesis

When we carry out a test that two populations are identical we usually have in mind some kind of departure from the null hypothesis which we particularly wish to detect, something which is of more significance for our thinking or the choice of practical actions we wish to take. The test-statistic we use reflects the particular aspect of the populations which we wish to compare. A frequent alternative hypothesis is that the population means are different. The test-statistic $T = \bar{x}_1 - \bar{x}_2$ is then the appropriate one. However, we could be interested in knowing whether the variances are different. This might occur in a comparison of two industrial processes for construction of electronic components where the consistency of the product is paramount. An appropriate test-statistic is then $T = s_1^2/s_2^2$, the ratio of the two sample variances. We take the ratio rather than a difference for technical statistical reasons; it is easier that way to conduct a good test, but this detail is of no concern to us here. The alternative hypothesis is discussed further in Section 6.2.3.

Table 6.1 Conventional categorisation of p-values

p-value	Strength of evidence	Description	Shorthand used	
$p > 0.10$	Reasonable consistency with H_0	Not significant	$p > 0.10$	n.s.
$0.05 < p \leqslant 0.10$	Slight evidence against H_0	Significant at 10%	$p < 0.10$	†
$0.01 < p \leqslant 0.05$	Moderate evidence against H_0	Significant at 5%	$p < 0.05$	*
$0.001 < p \leqslant 0.01$	Strong evidence against H_0	Significant at 1%	$p < 0.01$	**
$p \leqslant 0.001$	Very strong evidence against H_0	Significant at 0.1%	$p < 0.001$	***

Rationale

The rationale of the hypothesis test is based on asking the question:

How frequently would a value of the test-statistic arise as extreme as, or more extreme than, that observed if the null hypothesis were true?

The probability of such an event—a value of T as extreme or more extreme than that obtained, T_{obs}—is termed the *p-value*. If it turns out that the p-value is very small, then we are faced with two choices. We could say that the null hypothesis is true and a rare event has occurred. Alternatively, we could say that such an event as that we have observed is unlikely to have occurred if the null hypothesis is true, and so we reject the null hypothesis. The second choice seems to many people to be the more reasonable, and we follow this choice when we say that 'we reject the null hypothesis'. If on the other hand, the p-value is large, then we have no grounds for rejecting the null hypothesis. In determining what is more extreme we need to bear in mind the alternative hypothesis.

Significance level

What constitutes a small p-value is a matter of judgement and might vary from case to case, but there are certain conventions which have become accepted in biological research (see Table 6.1). The most commonly used threshold is 0.05. If the p-value is smaller than 0.05, then the result of the significance test is said to be '*statistically significant at 5%*', or we say that 'we reject the null hypothesis at 5%'. Other thresholds which are conventionally cited as *stronger* evidence against the null hypothesis are 0.01 or 1%, and 0.001 or 0.1%. These 5%, 1% and 0.1% threshold levels should not be regarded too rigidly. A result which is almost significant at 5% is not really any less important than one which just achieves significance at this level. However, journals reporting scientific research are sometimes strict about it, insisting on only reporting results which have reached significance at least at 5%. Some allow a 10% level as well.

6.2.2 Calculation of the p-value

Method 1. Student's t-test for comparing the means of two Normal populations

How do we calculate the p-value? There are a number of ways, and we now consider a common test for comparing two population means which is known as Student's t-test, so called after the *nom de plume*, 'Student', of the inventor of the test W. S. Gossett who worked for the Guinness Brewing Company in the early 1900s. Suppose that our two samples are those from the treatment T_1 and T_2 in Example 6.1. The data then are

T_1: 100, 108, 119, 127, 132, 135, 136, 164
T_2: 122, 130, 138, 142, 152, 154, 176.

This test starts with the difference of the two means, $T = \bar{x}_1 - \bar{x}_2$, as test statistic. In order to carry out the test of the null hypothesis

$$H_0: \mu_1 = \mu_2$$

we need to make further assumptions (we shall see later, in the section on robustness, that these need not be exactly true under certain circumstances) which are:

Further assumption 1: The two population variances are the same $\sigma_1^2 = \sigma_2^2 = \sigma^2$.
Further assumption 2: The two populations are Normally distributed.

We can work out from Chapter 3 that the sampling distribution of T (in repeated random samples from the two populations involved) is Normal with mean $(\mu_1 - \mu_2)$, and variance $\sigma^2(1/n_1 + 1/n_2)$, where μ_1, μ_2 and σ^2 are the true means and common variance of the two populations respectively. The problem is that σ^2 is not known. Student's contribution was to work out the distribution of the statistic—in repeated drawing of samples from the two populations—given by $\bar{x}_1 - \bar{x}_2$ divided by its *estimated* standard deviation $\sqrt{s^2(1/n_1 + 1/n_2)}$, i.e.

$$t = \frac{(\bar{x}_1 - \bar{x}_2)}{\sqrt{s^2\left(\frac{1}{n_1} + \frac{1}{n_2}\right)}}.$$

This t-statistic results from replacing σ^2 by an estimate, s^2, calculated from the data of the two samples, using the formula in the box below. The distribution of t is rather more spread out than the Normal distribution of the equivalent statistic in which the divisor is $\sqrt{\sigma^2(1/n_1 + 1/n_2)}$. The degree of spread is greater the fewer observations which enter into the estimation of σ^2. The practical effect of this result for the user is that t is compared with the distribution which is written t_ν, and referred to as Student's t on ν (Greek, pronounced 'nyu') degrees of freedom. The test then takes the following form.

Student's t-test for comparing two populations

Calculate

$$t = \frac{(\bar{x}_1 - \bar{x}_2)}{\sqrt{s^2\left(\frac{1}{n_1} + \frac{1}{n_2}\right)}}$$

where the pooled variance estimate,

$$s^2 = \frac{(n_1 - 1)s_1^2 + (n_2 - 1)s_2^2}{n_1 + n_2 - 2}$$

and compare t with the percentage points of Student's t-distribution on ($\nu = n_1 + n_2 - 2$) degrees of freedom (or obtain exact p-value in statistical package).

95% confidence interval on the difference between means

$$\bar{x}_1 - \bar{x}_2 \pm t_{0.025,\nu}\sqrt{s^2\left(\frac{1}{n_1} + \frac{1}{n_2}\right)}$$

Tables of the t-distribution

Student was able to tabulate the distribution of this t statistic, for different degrees of freedom (ν). Tables used to carry out the tests (contained in most books of statistical tables, and in Appendix Table S2) consist of the values of t which cut off 2.5%, 0.5% and 0.05% in the upper tail of the distribution. Since the distribution is symmetric about zero, there are similar symmetrically positioned points on the other side of zero which cut off the same percentage of the area in the lower tail (see Figure 6.1). Since the total areas cut off are 5%, 1% and 0.1% these are known as *2-sided 5%, 1% and 0.1% points of Student's t-distribution*. The absolute value is denoted $t_{0.025,\nu}$. By comparison of an observed value of t with these percentage points we can determine whether the p-value is less than 0.05, 0.01, or 0.001. If we are beyond the percentage point, then we declare the result as being significant at the appropriate level of significance. Some statistical packages give the actual tail probability (see Examples 6.2 and 6.3) so there is no need to look up further tables, and quite often p-values are quoted in reports of test results.

Example 6.1 Effect of fertiliser supplements on relative growth rate of plants

This simulated data is from an experiment in which three treatments—three different fertiliser supplements—are applied at random to a set of 23 experimental plants of similar initial weight, 8 receiving the first and third supplement, and 7 the second. The experimental measurement is relative growth rate in weight (multiplied by 1000 to give more convenient whole numbers).

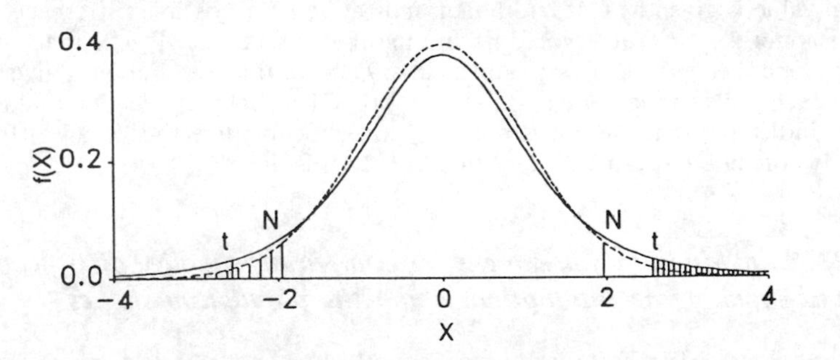

Figure 6.1 Student's t-distribution for 5 degrees of freedom, and a standardised Normal distribution for comparison. t_5 is the distribution with the lower value in the centre and greater weight in the tails, i.e. the least peaked. The 2-sided 5% points (i.e. cutting off 2.5% in each tail) of t_5 which equal ± 2.57 are indicated by t, and of the Normal distribution, which equal ± 1.96, by N.

										$\bar{x}$	s
T_1 (A)	100	119	127	132			164			127.63	19.60
	108			135							
				136							
T_2 (B)			122	130	142	152		176		144.86	17.81
				138		154					
T_3 (C)			121	132	141	152	168		201	151.38	24.61
					142	154					

Example 6.2 Comparison of two means using Student's *t*-test.

The data, for the first two treatments from Example 6.1, are:

T_1: 100, 108, 119, 127, 132, 135, 136, 164
T_2: 122, 130, 138, 142, 152, 154, 176.

The two samples of data have been read into columns C1 and C2 of the Minitab worksheet, before the following commands and output.

```
MTB  > twosample c1 c2;
SUBC > pooled.
   TWOSAMPLE T FOR C1 VS C2
        N      MEAN      STDEV     SE MEAN
C1      8      127.6     19.6        6.9
C2      7      144.9     17.8        6.7

95 PCT CI FOR MU C1 – MU C2: (–38.3, 3.8)

TTEST MU C1 = MU C2 (VS NE): T = –1.77 P = 0.10 DF = 13

POOLED STDEV = 18.8
```

The headings are fairly self-explanatory: N, MEAN, STDEV, SE MEAN are the individual sample sizes, n_1, n_2, means, $\bar{x}_1$, $\bar{x}_2$, standard deviations s_1, s_2, and standard errors of the means, $s_1/\sqrt{n_1}$, $s_2/\sqrt{n_2}$ respectively. The *t*-value = -1.77, and the *p*-value is given as 0.10. Although indicative of a possible difference, this *p*-value is not very low and would not normally be taken as evidence that there was a difference between the two treatments. A 95% confidence interval is given on the difference between the population means (MU C1 – MU C2). The interval is rather wide, indicating that the test is not very powerful for detecting small differences. Finally, the pooled standard deviation, $s = 18.8$ is given.

Method 2. A randomisation argument for comparing equality of two population distributions—making no assumptions about the population shapes

We now consider a further method of calculating the *p*-value which involves only minimal assumptions about the forms of the population distribution. Suppose that our two samples are those from the treatment T_1 and T_2 in Example 6.1, as used in the *t*-test above. The null hypothesis we wish to test is H_0: the two populations are identical. The test-statistic we use is $T = \bar{x}_1 - \bar{x}_2$, as we are mainly interested in differences between the locations of the two population distributions. The calculation now depends on the following statement:

If there were no true difference between the two populations (i.e. if there is no treatment effect), then every observation would be equally likely to have come from population 1 as from population 2.

After all, the units which received each treatment were selected at random from the 15 individuals; if there were no treatment effect, any set of 8 individuals could have been the individuals in sample 1. So, assuming the truth of this null hypothesis, each possible re-allocation of the 15 observations into a set of 8 in one sample and of 7 in the other is equally likely. We therefore split the complete set of observations (100, 108, 119, 127, 132, 135, 136, 164, 122, 130, 138, 142, 152, 154, 176) into two sets of size 8 and 7 respectively *at random*, and recalculate the value of the test-statistic, T. We repeat this a number of times and obtain different values of T, a process which is illustrated in Table 6.2.

A histogram of a very large number of values of T obtained using a Genstat program for carrying out a randomisation test is given in Figure 6.2. This gives a specification of the range of values of T which could have been observed and the relative frequencies of each of them, if the null hypothesis of no treatment effect is true. This histogram thus enables the p-value of any result to be calculated. The observed value of T, denoted $T_{obs} = 127.6 - 144.9 = -17.3$. There are 24 out of the 500 randomisations which are less than the observed value of the test statistic, T. So we can say that values as extreme or more extreme are likely to occur—if the null hypothesis is true—in 4.8% or less of repeated samples. Allowing for the possibility of results as extreme or more extreme in both directions away from the mean (since the alternative hypothesis is simply that the means of the two distributions are not equal in either direction, see also Section 6.2.3), we obtain a p-value of 0.096. The distribution given in Figure 6.2, is called the *randomisation distribution of T*. The test procedure is called an *observation randomisation test* (see Table 6.8).

Table 6.2 Building up the randomisation distribution of T

Total set: 100, 108, 119, 127, 132, 135, 136, 164, 122, 130, 138, 142, 152, 154, 176

Number of randomisation	Allocated to T_1	Mean T_1	Allocated to T_2	Mean T_2	Test statistic, T
1	154 130 127 136 108 132 122 138	130.9	100 164 119 135 142 152 176	141.1	−10.2
2	135 132 122 154 164 176 130 119	141.5	127 142 136 152 100 108 138	129.0	12.5
3	154 122 130 135 152 100 108 136	129.6	176 127 142 132 138 119 164	142.6	−13.0
4	154 138 132 119 142 108 130 176	137.4	136 152 100 127 135 122 164	133.7	3.7
etc.					

	$T<$	-22.5	9	*****
-22.5	$\leqslant T<$	-20.0	6	***
-20.0	$\leqslant T<$	-17.5	7	***
-17.5	$\leqslant T<$	-15.0	18	*********
-15.0	$\leqslant T<$	-12.5	18	*********
-12.5	$\leqslant T<$	-10.0	27	**************
-10.0	$\leqslant T<$	-7.5	30	***************
-7.5	$\leqslant T<$	-5.0	44	**********************
-5.0	$\leqslant T<$	-2.5	51	**************************
-2.5	$\leqslant T<$	0.0	48	************************
0.0	$\leqslant T<$	2.5	60	*******************************
2.5	$\leqslant T<$	5.0	37	*******************
5.0	$\leqslant T<$	7.5	33	*****************
7.5	$\leqslant T<$	10.0	26	*************
10.0	$\leqslant T<$	12.5	26	*************
12.5	$\leqslant T<$	15.0	23	************
15.0	$\leqslant T<$	17.5	18	*********
17.5	$\leqslant T<$	20.0	8	***
20.0	$\leqslant T$		11	******

Figure 6.2 Histogram of values of T obtained in randomisation test with 500 randomisations for comparison of treatments 1 and 2 of Example 6.1.

6.2.3 Power of significance tests

Alternative hypothesis

We stressed above the importance of the *alternative* as well as the null hypothesis. There are a number of reasons for this. First there are a number of possible alternative classes of hypothesis to some null hypotheses. As we discuss in the next section, it is only when we have specified what type of hypothesis we wish to discriminate from the null hypothesis, that the performance of any test procedure can even be considered. Secondly, some null hypotheses are not sufficiently specific to fix a test-statistic. The dependence of the test-statistic on the alternative hypothesis is discussed.

Alternative hypothesis and power of tests

In the above case of comparison of two means the null hypothesis is H_0: $\mu_1 = \mu_2$ while the *alternative hypothesis* is H_A: $\mu_1 \neq \mu_2$ or H_A: $\Delta = \mu_1 - \mu_2 \neq 0$. It is a class of hypotheses, rather than a single hypothesis, because there are many, in fact infinitely many, possible ways in which $\mu_1 \neq \mu_2$, and we wish our test to be able to detect as many of these as possible. It is commonly referred to as simply the alternative hypothesis, but it is in general less specific than the null hypothesis. This alternative class is not a very interesting one, but it enables us to specify the power of a test. The *power of a test* is the probability of declaring a significant result when a specified member of the alternative class of hypotheses is true. This applies to a test procedure at a fixed significance level, e.g. 0.05. The distance from the null hypothesis of any specific alternative in this case is measured by Δ. If we plot the power against Δ we get what is known as a *power curve* as in Figure 6.3. We would expect that with any reasonable test procedure, the power would be very little more than the significance level, 0.05,

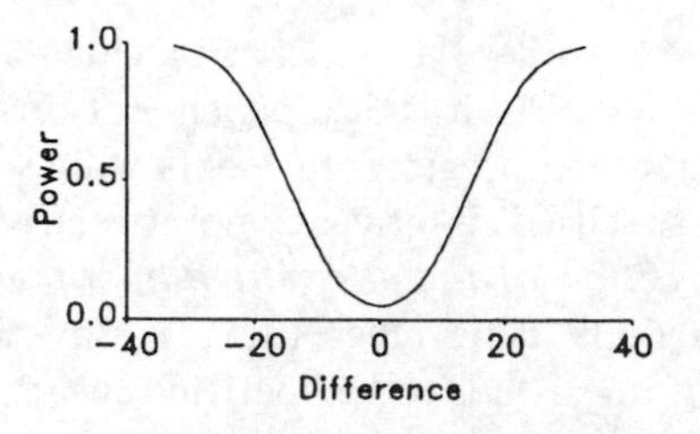

Figure 6.3 A typical power curve of a t-test. The curve gives the probability of declaring a significant result at 5%, i.e. of detecting the difference when the true difference between the means is Δ. The curve shape is dependent on n_1, n_2 and σ^2.

for alternatives which are very close to the null hypothesis, and the further we move away from the null hypothesis, the higher the power. The power curve, in this case for the t-test, goes through 0.05 when $\Delta = 0$, as the significance level is 0.05. The power increases as we move away from $\Delta = 0$, in both directions.

Implications of the alternative hypothesis for the test-statistic

There are usually many ways in which a null hypothesis can be false. For the randomisation test, the null hypothesis was

$$H_0\text{: Population distribution}_1 \equiv \text{Population distribution}_2.$$

We opted for a test-statistic $T = \bar{x}_1 - \bar{x}_2$. We could equally well have chosen $T = s_1^2/s_2^2$, the ratio of the two sample variances. The two statistics are focusing on different aspects of the null hypothesis, the means in one case, the spreads of the two distributions in the other. With a randomisation test, we are completely free to use whatever statistic we wish, and both these test-statistics are acceptable statistics for testing the null hypothesis that the two population distributions are identical. How does this fit in with our theory? The difference between them is that the first has high power against the alternative hypothesis that the means of the two populations are different; the second has high power against the alternative hypothesis that the variances of the two populations are different.

So with any null hypothesis, there are frequently many different test-statistics, and quite often they are all valid tests of the null hypothesis. The difference between them often is that they differ in the types of departure from the null hypothesis which they can pick up.

The critical region—one or two-tailed tests?

Any result of a test based on a single test-statistic, T, could be plotted on an axis through the origin, the distance from the origin on the axis indicating the value of the statistic. Part of this axis could be labelled a *critical region* in such a way that if T falls in the critical region we declare a significant result; if it falls elsewhere we declare non-significant. So for Student's t-test (testing at 5%) discussed above, the critical region is in two parts as in Figure 6.1, the boundaries being given by $\pm t_{0.025,\nu}$. The critical region is partially determined by the sampling distribution of the test-statistic under the null-hypothesis, since the area under the distribution calculated over the critical region has to equal 0.05 for a 5% test. But it is also dependent on the alternative hypothesis, in the following sense. The critical region represents values of T which are as extreme or more extreme than the specified significance point relative to expectation under the null hypothesis, *in the direction of the alternative hypothesis*. As we saw above, the alternative hypothesis partially determines the test-statistic used, but it has other effects: the alternative hypothesis has to be specified in order to know what we mean by more extreme in the above statement. When the alternative is H_A: $\mu_1 \neq \mu_2$, then a critical region in two equal parts (i.e. cutting off equal *areas* in the two tails) is appropriate, since we wish to declare as significant extreme results in both directions. If, however, the alternative is H_A: $\mu_1 > \mu_2$, the critical region would only consist of one tail—the right-hand one if $T = \bar{x}_1 - \bar{x}_2$—of area 0.05 of the sampling distribution. One-tailed tests of this kind could occur when, for example, in comparing results with an existing standard method, we are only interested in testing for treatments which are better than the standard. By focusing our interest in this way, higher power of detection of alternatives of interest can be obtained, without incurring other penalties.

Non-significant results, power and confidence intervals

Power curves are not normally constructed each time a test is carried out, but considerations of power affect the choice of test procedure for comparing two populations in any particular case. However, it is particularly important to think about the power of the test when we get a non-significant result. Non-significant results could arise for a number of reasons: (a) the null hypothesis is true; (b) the test has very low power for detecting departures from the null hypothesis of a magnitude which are likely to occur; (c) the departure from the null hypothesis which has actually occurred is small (this is saying that we are quite close to the null hypothesis). One useful rule which helps to overcome this difficulty is: wherever possible, calculate a confidence interval on the difference after carrying out a test. The width of the confidence interval tells us what differences we could expect to discriminate. If the interval is narrow, then we can more readily interpret a non-significant result as indicating no difference; if wide, then the conclusion ought to be that the test has little power. An example is given in Example 6.2. There the 95% confidence interval is quite wide, indicating that the test has little power.

A further point is relevant in this discussion. In order that we declare statistical significance and reject the null hypothesis, we typically have to have observed a result which would only occur 5% of the time or less if the null hypothesis were true. Thus, in a way, the odds are loaded against the alternative hypothesis, and so we need to be doubly cautious in interpreting a non-significant result. With the odds in favour of non-significance, it seems more appropriate to interpret a non-significant result rather cautiously unless we know that the test has high power to detect differences of a size which interests us.

6.2.4 Further aspects of significance tests

Statistical and biological significance

Statistical significance should not be confused with biological significance. The former is solely concerned with the answer to the question: is there evidence of a difference in the data? It might well happen that even though a difference can be demonstrated quite clearly, the difference is so small in magnitude that it is of no consequence either practically or theoretically or both. Equally it might happen that though the difference between two treatments is not statistically significant, the true difference between the populations—which we have not been able to discern with our current experiment—could be of profound biological significance. Also just because one difference is statistically significant at 0.1%, whereas a second is barely significant at 5%, it does not mean that the second is of less importance. Other things besides the actual magnitude of the difference are involved in determining whether a difference is statistically significant: the variability of the material, the sample sizes employed are two important influences.

Robustness of tests

The assumptions which underly the test procedure should also be borne in mind when choosing a test. For example, Student's t-test assumes that the distributions of the two populations from which the samples are drawn are Normal. How reliable will the test be if the populations are not exactly Normally distributed, but only approximately so? We say that the test procedure is *robust* to departures from the assumption of Normality if it is not substantially affected by such departures. It turns out that if the sample sizes are large enough (at least 6, but preferably larger) the exact Normality of the parent distribution is not crucial for the t-test. The larger the sample size, the more robust the test is to departures from Normality, i.e. it can accommodate greater departures without being substantially affected. Another assumption of Student's t-test is equality of variance, $\sigma_1^2 = \sigma_2^2$, which is perhaps the more important of the two assumptions if the two sample sizes are not the same. If the sample sizes are equal or nearly equal, then the test can also accommodate quite large departures from equality of variances.

Terminology for error rates

Errors of two types occur in significance testing:

(1) *Type 1 errors* or *false positives*, when a significant result is declared but H_0 is true. The rate is equal to the significance level.
(2) *Type 2 errors* or *false negatives*, when a significant result is *not* declared but H_0 is not true. The rate is given by (1-power). This rate would take a different value for each alternative.

6.3 COMPARISON OF MEANS AND MEDIANS OF TWO POPULATIONS

In this section, we outline some further tests for comparing the means and medians of two populations, using data from *independently* collected samples from each population (i.e. the samples are not paired or blocked in any way, see Chapter 7).

There is insufficient space here for much detail, and most readers do not need to know the details of the calculations, since statistical packages include facilities for performing them. The Appendix gives some practical recommendations and a tabular summary (Table 6.8) of the main tests which are routinely used for comparison of means of two or more populations.

6.3.1 Continuous measurements or ranks

The Welch test for continuous measurements: an unpooled-variance version of the t-test for comparing means

Student's t-test is the commonly used test for comparison of two means, but this test assumes that the variance within the two populations is the same, and another test—which does not make this assumption—is generally to be preferred: the *Welch test* named after its inventor. This is a variant of Student's t-test, but the estimates of variance within each population, s_1^2, s_2^2 are not pooled. The degrees of freedom of the t-distribution (to which the calculated value of t is referred) are determined according to a special formula, given below. Unless sample sizes are very small, or we definitely know that the variance within the two populations are the same, it is probably better to use the Welch test all the time.

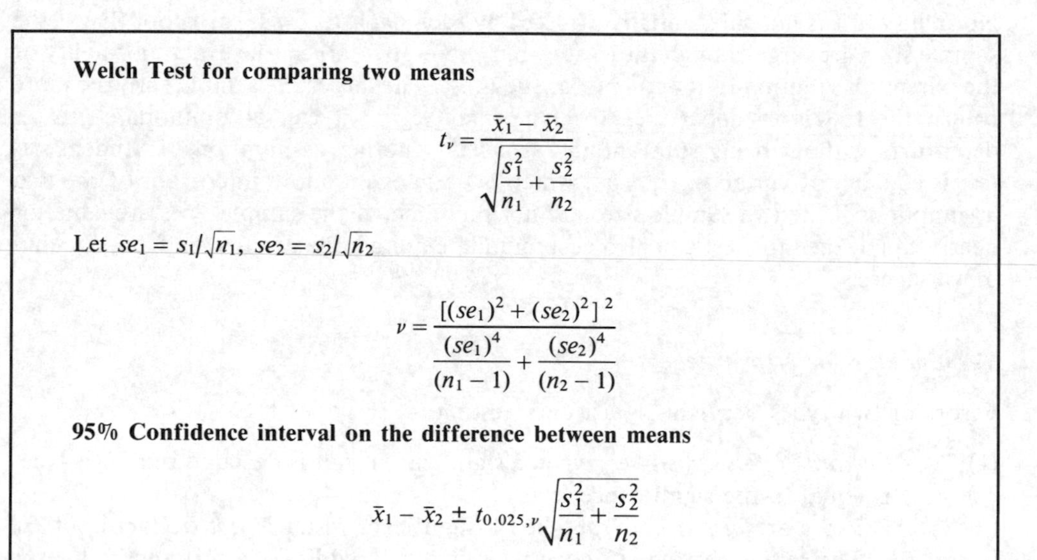

Welch Test for comparing two means

$$t_\nu = \frac{\bar{x}_1 - \bar{x}_2}{\sqrt{\dfrac{s_1^2}{n_1} + \dfrac{s_2^2}{n_2}}}$$

Let $se_1 = s_1/\sqrt{n_1}$, $se_2 = s_2/\sqrt{n_2}$

$$\nu = \frac{[(se_1)^2 + (se_2)^2]^2}{\dfrac{(se_1)^4}{(n_1 - 1)} + \dfrac{(se_2)^4}{(n_2 - 1)}}$$

95% Confidence interval on the difference between means

$$\bar{x}_1 - \bar{x}_2 \pm t_{0.025,\nu}\sqrt{\frac{s_1^2}{n_1} + \frac{s_2^2}{n_2}}$$

Robust tests for continuous measurements or ranks: the Mann–Whitney U-test for ranks

The observation randomisation test outlined in Section 6.2.2 is a robust test for comparing two populations with high power against differences of means. This test is also available for rank data (or continuous data after they have been ranked) and it can be performed in exactly the same way. Being applied to ranks, it is called the rank randomisation test, or Mann–Whitney U-test after its originators and the test-statistic

U (equivalent to the difference of the means of the ranks, but calculated by merely counting inversions of ranks) which they used. It can be used for data in which the ranking is inherent, or for continuous data in which case it is necessary to replace the observations with the ranks in the combined set consisting of the two samples. Before the days of high-speed computers and modern statistical packages, the Mann–Whitney *U*-test had the advantage that the determination of the probability of the observed value of the test-statistic, *U*, or a more extreme value, could be done without the use of a computer. It only involved very simple calculations using the sample values, and reference to tables of significance points of the *U*-statistic. Related to this test, approximately 95% confidence limits can be obtained on the difference between the population medians. This is useful in interpreting the results of the significance test.

A further advantage of the test is that the power of the test is maintained when a small number of outliers (i.e. values very far from the rest of the data) occur. It quite often happens, when working with data which is highly non-Normal, or when occasional spuriously high or low readings occur, that the Mann–Whitney *U* test is able to detect differences which are non-significant according to a *t*- or Welch test. The reason is that, by replacing the raw data by ranks, we sometimes reduce the importance of values a long way from the sample mean.

Example 6.3 The Welch and Mann–Whitney U-test applied to the fertiliser data

The data are:

T_1: 100, 108, 119, 127, 132, 135, 136, 164
T_2: 122, 130, 138, 142, 152, 154, 176.

The two samples of data have been read into columns C1 and C2 of the Minitab worksheet, before the following commands and output.

```
MTB > twosample c1 c2

TWOSAMPLE T FOR C1 VS C2

        N     MEAN     STDEV     SE MEAN
C1      8    127.6      19.6         6.9
C2      7    144.9      17.8         6.7

95 PCT CI FOR MU C1 - MU C2: (-38.3, 3.8)

TTEST MU C1 = MU C2 (VS NE): T = -1.78 P=0.10 DF = 12

MTB > mann c1 c2

Mann-Whitney Confidence Interval and Test

C1     N = 8     Median = 129.50
C2     N = 7     Median = 142.00
Point estimate for ETA1 - ETA2 is  -17.50
95.7 pct c.i. for ETA1 - ETA2 is (-40.99, 5.01)
W = 49.0
Test of ETA1 = ETA2 vs. ETA1 n.e. ETA2 is significant at 0.0933
Cannot reject at alpha = 0.05
```

In Minitab, the Welch test is carried out as an unpooled version of the TWOSAMPLE command. The effective degrees of freedom are just 1 less than for the t-test, because the standard deviations are quite similar. This is not always the case. The p-value is the same to 2 significant figures, demonstrating in this case that little power is lost by carrying out the more widely applicable Welch test. The Mann–Whitney U-test gives a very similar result. Note that the confidence intervals on the differences of the means and the medians are rather similar in width. Both indicate the degree of sensitivity of the test procedure: loosely speaking, it is unlikely to resolve differences less than about 23 (at a 5% level of significance).

6.3.2 Binary data and derived proportions

Contingency tables, Pearson X^2 and G-tests.

Quite different test procedures are necessary for proportions which are based on counts of discrete objects. The null hypothesis is H_0: $P_A = P_B$ and the alternative hypothesis is H_A: $P_A \neq P_B$ where P_A and P_B are the proportions in the two populations being compared. The test uses a test statistic based on the difference in the sample proportions although the calculations are more conveniently laid out in the form of what is known as a *contingency table*. The method is most easily explained using an example.

Example 6.4 Testing for a difference in two proportions—comparisons of the incidences of multiple births

Random samples of pregnancies from two flocks of sheep were assessed for the incidence of single or multiple births. Out of 30 pregnancies in area A, 20 resulted in single births, and of 37 in area B, 31 resulted in single births. Is the proportion of single births the same between the two areas? This data can be expressed in a contingency table (Table 6.3(a)) of observed frequencies.

Calculation of expected frequencies

The first step in the method involves working out the frequencies which would be expected in the body of the table if there were no difference between the proportions

Table 6.3 Observed and expected frequencies, and contributions to the chi-squared statistics

	Area A	Area B	Total
(a) Observed frequencies			
Single	20	31	51
Multiple	10	6	16
Total	30	37	67
(b) Expected Frequencies			
Single	22.84	28.16	
Multiple	7.16	8.84	
(c) Contribution to G			
Single	−5.31	5.96	
Multiple	6.68	−4.65	
Contribution to Pearson X^2			
Single	0.35	0.29	
Multiple	1.13	0.91	

in the two populations, and these are given in Table 6.3(b). They are easily derived. The proportion of single births in the two areas combined is $51/67 = 0.761$, and therefore the proportion of multiple births is 0.239. Applying these proportions to each area separately gives us the required expected frequencies:

Single births, Area A: $30 \times 0.761 = 22.84$;
Multiple births, Area A: $30 \times 0.239 = 7.16$;
Single births, Area B: $37 \times 0.761 = 28.16$.
Multiple births, Area B: $37 \times 0.239 = 8.84$.

Calculation of χ^2 test-statistics

Next we need some measure of discrepancy between the observed frequencies and those expected on the null hypothesis, which would effectively discriminate between the null hypothesis and the alternative hypothesis.

One measure, based on the difference $(f_o - f_e)$ between observed frequencies and those expected assuming the null hypothesis (this statistic can also be shown to be a function of the difference in the two proportions), which has been extensively used before the era of the computer is known as *Pearson chi-squared:*

$$\text{Pearson } X^2 = \sum_{\text{all cells in table}} \frac{(f_o - f_e)^2}{f_e}$$

(the symbol X is the capital of the greek letter 'chi', pronounced 'kie', but written like a capital X). An alternative measure which uses the ratio f_o/f_e is known as the *deviance chi-squared*, or *G-statistic*:

$$G = \sum_{\text{all cells in table}} 2f_o \ln\left(\frac{f_o}{f_e}\right).$$

Both of these statistics are measures of the discrepancy between the observed and expected which take low values when the observed and expected are similar. The first is a weighted sum of squares of differences between observed and expected (see Chapter 11 for further discussion of weighted least squares). The second is more difficult to motivate intuitively, but takes rather similar numerical values in most cases, and is more convenient for complex comparisons. Tests based on deviance chi-squared are known as *G-tests* by some authors.

Distribution of Pearson X^2 and G under H_0

Both of these measures are approximately distributed as χ_1^2, chi-squared on 1 degree of freedom, and therefore can be compared with the percentage points of chi-squared. In the above case, the contributions to chi-squared are given in the right hand table, and the total in this example is 2.68 for deviance chi-squared and 2.67 for Pearson chi-squared. The tabulated percentage points of the χ_1^2 distribution are 3.84 at 5%, 6.63 at 1% and 10.83 at 0.1%, and so the result is not significant at 5%. Therefore we cannot reject the null hypothesis of equality of proportions between the two areas.

Validity of χ^2 approximation

This test relies on an approximation which is only valid for large expected frequencies. A *continuity correction* can be made to the Pearson X^2 statistic which improves the approximation. This involves reducing the absolute magnitude of the difference between f_o and f_e by 0.5 before squaring and dividing by f_e. An approximate guide to the adequacy of the approximation is that the test for a 2×2

table is only valid if all expected frequencies are greater than 5. For tables with lower expected frequencies, it is better to apply *Fisher's exact test for 2 × 2 contingency tables* (see Siegel, 1956 for further details).

6.4 COMPARISON OF MEANS AND MEDIANS OF MORE THAN TWO POPULATIONS

In situations with more than two populations, pairs of populations can of course be compared using the techniques we have outlined so far. But the means of all the populations can also be simultaneously compared, and this section deals with methods for doing this. Table 6.8 lists tests for comparing all the means in one test, which can then be followed up by tests comparing pairs of means, either all possible pairs, or perhaps just a selected set (e.g. comparison of treatments with the control and not with each other).

6.4.1 Continuous measurements: the *F*-test and the analysis of variance

With three or more populations, it is sometimes necessary to compare all the population means in one test. It is then a test of a null hypothesis of equality of all the means, and the alternative hypothesis is that at least one of the means is different. With continuous Normally-distributed variables, the test which is used is the *F-test*, and the preliminary analysis of the data before carrying out such a test is called a *One Way Analysis of Variance*. There are two essential aspects to the *F*-test. First it is a portmanteau test—a simultaneous comparison of all the means. Secondly, just as in Student's *t*-test, there is a pooling of estimates of population variances.

Consider Example 6.1, in which three fertiliser supplements T_1, T_2, and T_3 were given to samples of experimental plants, and the relative growth rates measured. We have already shown how to compare two of these treatments. We now answer the question: are the three treatments equal in their effect on relative growth rate, or alternatively is there evidence that at least one of the supplements has a different effect. The means and standard deviations of the treatments are as follows, together with the overall mean and pooled standard deviation (see below for method of calculation).

Treatment	Sample size	Sample mean	Sample standard deviation
A	8	127.63	19.60
B	7	144.86	17.81
C	8	151.38	24.61
		Overall mean	Pooled standard deviation
		141.1	21.0

When comparing two treatments, we started with a test-statistic which was just the difference of the two sample means. In this situation we need a test-statistic which in some way simultaneously takes account of the differences between all the means. The statistic used is the sum of squares of the differences of the individual means from the overall mean. The overall mean in the above case is 141.1. This is the mean of all

23 observations which is slightly different from the mean of the three means above, because the replication of the three treatments is slightly different. The test-statistic is

$$SS_{\text{treatment}} = 8 \times (127.63 - 141.13)^2 + 7 \times (144.86 - 141.13)^2 + 8 \times (151.38 - 141.13)^2 = 2396.0.$$

The treatments which have higher replication figure more prominently in this statistic (the first number in each term is the replication of the treatment) which seems sensible as there is in a sense more evidence the higher the replication. We call this statistic $SS_{\text{treatment}}$ to stand for sum of squares, because it is a sum of squares weighted by the treatment replication, and we use the subscript *treatment* because it is due to difference in the effects of the treatments. The more nearly equal the treatment means are, the smaller is $SS_{\text{treatment}}$, and when all the treatment sample means are the same, $SS_{\text{treatment}}$ takes the value zero. So large values will indicate a substantial departure from the null hypothesis of equal population means.

A further step in the construction of the test-statistic is to make an allowance for the number of means which enter into the calculation. It seems sensible that the greater the number of means, the higher $SS_{\text{treatment}}$ could be expected to be. The appropriate correction for this effect is to divide by (number of means – 1). Another word for this divisor is the *degrees of freedom* on which the statistic $SS_{\text{treatment}}$ is based.* The statistic then changes its name to become $MS_{\text{treatment}}$. How do we then determine whether such a large value as that observed, $2396/2 = 1198$, is likely to have arisen by chance? How do we determine the probability of observing a value as large as or greater than this, if the null hypothesis is true, i.e. there are no differences between all the population means? For Student's t-test, we divided the difference between the means by the standard error of the difference and so it seems sensible here to divide by a quantity which measures, as near as is possible with sample data, what we would expect $MS_{\text{treatment}}$ to be if there were no effects of treatment. In this case, we use a pooled estimate of the variances within treatments, calculated in a natural extension of the way we calculated the pooled variance for the t-test:

$$s^2_{\text{pooled}} = \frac{(n_1 - 1)s_1^2 + (n_2 - 1)s_2^2 + (n_3 - 1)s_3^2}{(n_1 - 1) + (n_2 - 1) + (n_3 - 1)}$$

$$= \frac{7 \times 19.6^2 + 6 \times 17.81^2 + 7 \times 24.61^2}{7 + 6 + 7}$$

$$= \frac{8832.6}{20} = 441.6$$

$$\left[= \frac{SS_{\text{residual}}}{df_{\text{residual}}} = MS_{\text{residual}} \right].$$

Dimensional considerations indicate that we need to divide $SS_{\text{treatment}}$ by a quantity which has the same units, i.e. which are in the square of the units of measurement of the raw data, and the quantity, MS_{residual}, or pooled residual variance, is in just such units, and it turns out that this is the appropriate divisor.

---

* See Chapter 7 for further discussion of degrees of freedom; in general for treatment sums of squares in the analysis of variance, the degrees of freedom are calculated as number of treatment levels being compared minus 1.

The technique is usually part of an analysis very widely used in statistics called *analysis of variance*, sometimes abbreviated to ANOVA. In this case, we can lay out the calculations in a convenient analysis of variance table.

Source	*df*	*SS*	*MS*	*F*
Treatments	2	2396.0	1198.0	2.71
Residual	20	8832.6	441.6	
Total	22	11228.6		

We here see all the quantities we have motivated above fitting together in a neat table. There is a further line and a further column. The further column is just the calculation of the final test-statistic, which is called the *F-ratio*, or *variance ratio*. This quantity has the property that if there are no treatment effects, i.e. the means of the populations from which the three samples are thought of as having come are equal, it has an expectation of 1. Sampling error would of course displace it from 1 both up and down, but on average in a very large number of repetitions of the experiment (only if there are no treatment effects) it would average to 1. Large values indicate that there is excess variation between the treatment means as compared with MS_{residual}, i.e. they reveal the presence of real treatment effects. The distribution which this ratio follows under the null hypothesis of no treatment effects is the *F-distribution* on 2 and 20 degrees of freeedom. The numbers 2 and 20, the 'degrees of freedom' quoted are those on which the numerator and denominator mean squares which make up the *F*-ratio are based. The value of 2.71 can then be compared with tables of the *F*-distribution on 2 and 20 degrees of freedom. The tabulated percentage points are 5% 3.49, 1% 5.85, and 0.1% 9.95. The value achieved does not reach the tabulated value at 5%, and so we say that in this case the difference between the means is not significant at 5%.

The further line in the above table, labelled total, gives us an estimate of the variation between all 23 observations, if there are no treatment effects at all. The figure in the *SS* column is merely the numerator which you would calculate in a sample variance estimate between all 23 observations, and the figure in the *df* column is the degrees of freedom on which this would be based, and which is therefore 22 (i.e. 23 − 1). You will notice that the values above the solid line in each column add up to the value below the solid line (in the *Total* line of the table). This is a convenient feature which enables the calculations to be checked. It is also the motivation behind the title *Analysis of Variance Table*. The table produces an analysis—or splitting up—of the total *SS* into a part due to variation between treatments and a part which is due to residual variation within treatments. The *df* are similarly split up. The above is the simplest example of analysis of variance. It can be applied to much more complex experimental situations in which a number of different types of treatments are applied to the experimental units. This will be developed further in the next chapter.

To summarise then, the *F*-test of the equality of more than two means involves two aspects. First of all, it is a simultaneous comparison of all the means, and this, as well as statistical convenience, dictates the form of the numerator of the *F*-statistic: a sum of squares of differences of treatment means from the overall mean divided by the degrees of freedom. The second element is that the variance within populations is pooled just as in Student's t-test, except that in this case there are three or more within population variances to pool. The ratio of these two quantities has an expectation of 1 if there are no differences between the means of the populations, and we can carry out a significance test by comparing it with the tabulated percentage points of the F-distribution, high values indicating significance. The whole calculation can be conveniently tabulated in an Analysis of Variance table.

The F-test assumes that the observations are samples from Normally distributed populations with the same variance, but possibly different means. This model, sometimes called the *Normal Model* (justifying the Analysis of Variance), can be written in symbols

$$Y_{1j} = \mu_1 + \varepsilon_{1j}, \qquad j = 1, \ldots, n_1$$
$$Y_{2j} = \mu_2 + \varepsilon_{2j}, \qquad j = 1, \ldots, n_2$$
$$Y_{3j} = \mu_3 + \varepsilon_{3j}, \qquad j = 1, \ldots, n_3$$

where the means of the populations are μ_1, μ_2, μ_3, the levels of replication of the treatments are n_1, n_2, n_3 respectively, and the ε's are Normally distributed with mean zero and common variance for the three treatments, σ^2.

The really interesting part of the analysis of an experiment such as this is yet to come; the analysis of variance is just a preliminary sieve to check whether there is evidence of any differences at all between the means.

6.4.2 The nature of the heterogeneity—comparisons between means

It is important when comparing the means of more than two populations not to feel that a complete analysis consists of an F-test, Kruskal–Wallis (see Section 6.4.5) or similar test. If a significant result is obtained, it is then necessary to examine the population means or medians to see how the heterogeneity occurs. This is probably the most important and interesting part of the analysis. The initial comparison of all the means is just a preliminary sieve to assess whether there is any evidence of differences between treatments or populations.

Many computer packages which perform an analysis of variance, also give tables of means of treatments, or populations or other groupings, together with a measure of the uncertainty (i.e. a standard error) attached to the means. As an example, we consider the data of Example 6.1. In this case, the initial analysis of variance did not produce a significant F-test result, and so we would not normally proceed further. However, these data illustrate some interesting relationships between the overall F-test and comparisons of the individual pairs of means, and it can serve anyway to show what we would do had we initally obtained a significant F-test result. Genstat produces a table of means and standard errors of differences between means as follows.

Example 6.5 Genstat output for comparison of means in data from Example 6.1

```
***** Tables of Means *****

Grand mean 141.1

    Treat          A        B        C
               127.6    144.9    151.4
    rep.           8        7        8
```

```
*** Standard errors of differences of means ***
Table        TREAT
rep.         unequal
s.e.d.       11.23  X  min.rep
             10.88     max-min
             10.51     max.rep
(No comparisons in categories where s.e.d. marked with an X)
```

The overall mean is 141.1 and the treatment means are 127.6 for A (with a replication of 8), 144.9 for B (replication 7), and 151.4 for C (replication 8). The standard errors of differences between means are coded for those *pairs* of treatment means both of which have maximum replication (i.e. 8), those pairs with minimum replication (i.e. 7, of which there are none) and those pairs one of which has minimum and the other of which has maximum replication (i.e. 7 and 8). They are calculated using the formula

$$s.e.d. = \sqrt{(\text{residual mean square})\left(\frac{1}{n_1}+\frac{1}{n_2}\right)}$$

$$= \sqrt{441.6\left(\frac{1}{8}+\frac{1}{7}\right)} = 10.88$$

to give just one example.

In order to compare means we can calculate 95% confidence intervals—using the appropriate s.e.d.—as follows:

$$\text{B} - \text{A} \quad \text{difference} \pm t_{0.025}, \nu \times s.e.d.$$

$$= 144.9 - 127.6 \pm 2.09 \times 10.88$$

$$= 17.3 \pm 22.7$$

and so we obtain for all the pairwise comparisons the table of differences and 95% confidence intervals, given in Figure 6.4

This analysis indicates that the only difference between the means which is significantly different from zero is that between A and C, but that there are wide margins of uncertainty for all the differences between the means.

6.4.3 The multiple comparison problem—getting significant results by chance

There are problems with such an analysis which become apparent if we think of how many pairwise comparisons could be made between 10 treatment means: $10 \times 9/2 = 45$ (the number of pairs selected from 10). The guarantee that we provide with significance

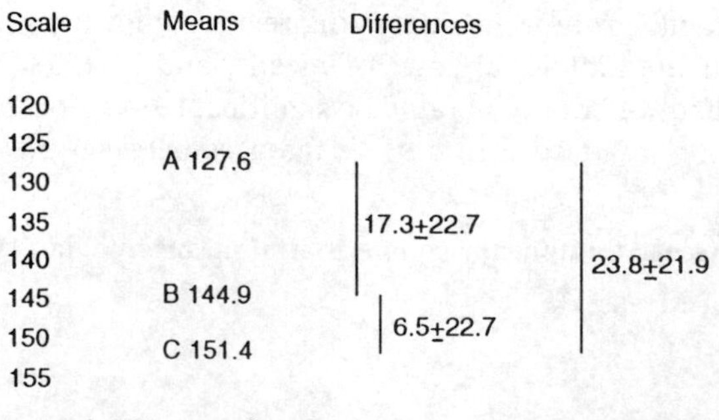

Scale	Means	Differences	
120			
125	A 127.6		
130			
135		17.3±22.7	
140			23.8±21.9
145	B 144.9		
150	C 151.4	6.5±22.7	
155			

Figure 6.4 The means, and difference between means, with 95% confidence intervals, for the data of Example 6.1.

testing at 5% is that such significant differences would arise by chance if there were no true difference between the means on average 5% of the time. This is what testing at 5% means. So, even if there were no true differences between the treatment means, we would expect differences to show up as significant 5% of the time, i.e. in this case in $0.05 \times 45 = 2.25$ comparisons. So we expect that about two comparisons will give a significant result even if there are no true differences due to treatments. This is unsatisfactory, and the remedy that some workers apply is effectively to reduce the significance levels of the individual comparisons (known as the comparison-wise error rate) so that the chances of one or more significant results in the whole battery of comparisons in the experiment or survey is 5%. This 5% is known as the experiment-wise error rate. One approximate rule is to divide the desired experiment-wise error rate by the number of comparisons being made, and then use this as the comparison-wise error rate. This rule is derived from what is known as the Bonferroni Inequality.* In the example above, none of the comparisons would be significant using the rule.

There are very many rules which can be used for *multiple comparisons between means* (the interested reader is referred to Sokal & Rohlf (1969)). We do not discuss these procedures further here, as we feel they are of marginal importance, and that, whichever rule is adopted, there will be some criticisms which can be made of it. So long as the scientist is aware of the multiple comparison problem, and adjusts the significance levels to suit the number of comparisons made, that will in general be sufficient. The other advice which follows from this is that in general it is not a good thing to 'trawl' amongst a large number of means for significance in this way. If it is done properly taking due account of all the comparisons which are being made, then each individual comparison will be so insensitive that some differences of importance might be missed; and if it is not done properly there is a much greater likelihood of some differences being declared significant by chance.

What could be done in an experiment in which a large number of treatments are being compared is that the differences which are important should be nominated in advance, and provided there are not too many of them, this should not cause any problems with falsely significant differences. An alternative is to try to construct a model of the differences, and this is taken up in a very simple manner in the next section.

6.4.4 The inappropriateness of testing for heterogeneity

We have now seen a number of examples of testing whether means differ between populations. The reader might think that it is a good initial method of examining any set of data from a number of populations. However, it should be stressed that, although there are many circumstances in which it is entirely appropriate, there are also many where it is not. Consider as an example of the latter the data set in Example 6.1. Here we were comparing the means of three populations, and the F-test in Section 6.4.1 indicates that there are no significant differences between the means of the three populations. Individual comparisons between the treatments indicate that the two most

---

* The argument for this inequality is as follows.

Prob(at least one significant result in n independent tests (each at α) =

$$1-(1-\alpha)^n = 1-(1-n\alpha+n(n-1)/2\alpha^2\ldots) > 1-(1-n\alpha) = n\alpha.$$

extreme are significantly different, except that when an adjustment is made for multiple comparisons this difference becomes non-significant.

Suppose now we were given the extra information that the main difference between the fertiliser supplements is that they include different doses of a chemical: T_1 does not contain the chemical, T_2 has 5 units of the chemical and T_3 has 10 units. Then a plausible analysis is to fit a straight line-Normal model (as in Chapter 5) between the relative growth rate (RGR) and the dose of the chemical, d. The fitted model (with standard errors and other statistics associated with the goodness of fit) is:

$$RGR = 129.3 + 2.38d$$

$$(\text{s.e. } 6.73)(\text{s.e. } 1.03)$$

$$r = 0.45, \text{ residual s.d.} = 20.7 \text{ on } 21\ df.$$

The slope of the line is significantly different from zero at 5%. Thus we have demonstrated a difference between the means of the populations by showing that they are systematically related to an explanatory variable. This example demonstrates that it is often better to seek a systematic relationship with some other quantity rather than ask the question: is there a difference? In some cases no appropriate potential explanatory variable suggests itself, and then we are forced to search for differences in a less directed manner, but in general it is not the best strategy of analysis.

6.4.5 The observation and rank randomisation tests

These are performed exactly as in the observation randomisation test of two populations, except that the test-statistic is a sum of squares of treatment means about the overall mean as in the F-test. The observed value of the test-statistic is compared with those obtained by randomly allocating all the observations to the treatment or population groups in the same proportions as in the observed sample, and assessing whether the observed value of the statistic or a more extreme one has a low probability or not. If this probability is lower than 0.05, we say the result is significant at 5%, and similarly for 1% and 0.1%.

For rank data, there is a rank randomisation test, called the Kruskal–Wallis test, after its two inventors. Just as the F-test is the many population generalisation of the t-test, so the Kruskal–Wallis test is the many population generalisation of the Mann–Whitney U-test. The rationale of the test is the same as in the observation randomisation test, and it can be performed in the same way, but there are shortcuts resulting from the fact that the distribution of the test-statistic for rank data can be calculated for a general case, as was the case with the Mann–Whitney U-test.

6.4.6 Proportions and categorical variables

We now consider the extensions of the two sample test for comparing proportions to the case of independent samples from more than two treatments. The extension is rather similar in principle to that we made in the case of the F-test. The calculations take the same form as those for comparing two proportions, being based on calculating the expected frequencies assuming equality of proportions, and then the same χ^2 statistics, over the larger contingency table which results.

Table 6.4

	Area A	Area B	Area C	Total
(a) Observed frequencies				
Single	20	31	15	66
Multiple	10	6	6	22
Total	30	37	21	88
(b) Expected frequencies				
Single	22.50	27.75	15.75	
Multiple	7.50	9.25	5.25	
Total	30	37	21	
(c) Contribution to deviance X^2				
Single	−4.71	6.87	−1.46	
Multiple	5.75	−5.19	1.60	

So total deviance $X_2^2 = 2.852$; not significant ($P > 0.05$).

Example 6.6 Comparison between more than two proportions

Suppose that instead of two flocks of sheep in Example 6.4, we had observed 3 flocks and the results were as in the contingency table (Table 6.4).

The expected frequencies are calculated in exactly the same way as previously, by multiplying the total frequency in the column by the observed proportion over all three columns, e.g. for column 1 $30 \times 66/88 = 22.50$; $30 \times 22/88 = 7.5$. The deviance X^2 is calculated in the same way also, giving a total of 2.85, which is not significant, indicating that there is no difference in the proportions between the 3 flocks.

More than two categories

The above chi-squared method is completely general in that it can be applied to data in which there are more than two categories. Just as we have extended the original method to cope with more than two populations, so also can we extend it to more than two categories.

6.5 MORE COMPLEX COMPARISONS

6.5.1 Continuous measurements—comparisons of scale

It is often of interest to compare the spreads of populations. For example—some treatments might enhance the normal level of functioning of some individuals in a population whereas the functioning of others might be depressed, or not change at all. The data in Example 6.7 below is a possible case of this.

Example 6.7 Hormone release from the pituitary

A study was carried out to investigate how the amount of hormone secreted by the pituitary gland of the rat under electrical stimulation is affected by the presence or absence of various compounds in the surrounding fluid. The experimental measurement was the ratio of the amount of hormone released under electrical stimulation when the gland is bathed in the experimental substance to that released when bathed in a neutral saline solution. The results for two compounds are presented, saline (S) and yohimbine (Y). The data were as follows. There is unequal replication of the two treatments since some samples were lost from the Y treatment.

S	1.0	1.3	1.5	1.7	1.9	2.1						
	1.1	1.3		1.7								
	1.1											
Y			1.4		1.8	2.0	2.3	2.6		3.7		4.2

It appears that the response of the yohimbine treated animals is much more variable than the control animals. Is the difference statistically significant? The standard method for comparing two variances is the *F*-test.

F-test for comparing variances

$$\frac{s_1^2}{s_2^2} \sim F_{\nu_1, \nu_2}$$

where $\nu_1 = n_1 - 1$, and $\nu_2 = n_2 - 1$ are the degrees of freedom on which s_1^2, and s_2^2 are estimated.

The standard deviation of the control animals is 0.371 on 9 *df* whereas that of the yohimbine treated ones is 1.024 on 6 *df*. So $F_{6,9} = 1.049/0.138 = 7.60$ which is significant at 5% (two-sided 5% point of $F_{6,9} = 4.32$). This suggests that the spreads of the control and yohimbine treated group are different.

However, the *F*-test for comparing two variances is rather sensitive to non-Normality, and the two populations here might be skewed, so unless a robust test could be found the result must be rather suspect. Few general techniques exist for comparing the spreads of non-Normal populations, and we have no space to deal with them here.

6.5.2 Comparison of slopes

In Chapter 14, we consider the fitting of complex models with a number of explanatory variables, and in that chapter construct a model for the heat loss of experimental pigs as a function of environmental temperature and energy intake. Data was collected on 8 pigs at each of the temperatures 10, 15, 20, 25°C and 6 at 30 °C. Within each temperature the pigs were kept on a wide span of energy intakes. As there are observations on 8 animals at all but one of the temperatures, and each of these sets covers a wide range of energy intakes, one strategy of model construction is to fit separate models of heat loss against energy intake for each temperature separately and then to assess what aspects of the relationships with energy intake vary between temperatures. As an example of how to proceed, let us consider the data for two temperatures 15 and 20 °C. The separate linear models relating heat loss to energy intake are given in Table 6.5.

There are various ways we could test whether the relationship is the same within the two groups. The usual first step is to test for equality of the slopes, and only if this is satisfied, then to go on to test for equality of the intercepts. If the slopes are the same, then the distance in the *y*-direction between the two lines is the same for all *x*.

Table 6.5 Linear models relating heat loss to energy intake at 15 and 20 °C

15 °C					20 °C				
$h = 489$ +	$0.2099e$				$h = 293.4$ +	$0.3443e$			
(s.e. 31.3)	(s.e. 0.026)				(s.e. 42.0)	(s.e. 0.035)			
ANOVA					ANOVA				
Source	*df*	*SS*	*MS*	*F*	Source	*df*	*SS*	*MS*	*F*
Regression	1	71 963	71 963	64.52	Regression	1	189 126	189 126	98.85
Residual	6	6 692	1 115		Residual	6	11 479	1 913	
Total	7	78 655			Total	7	200 605		

If the slopes are different then the distance varies depending on the value of x. A test of equality of intercept is in effect a test of equality of the distance between the two lines in the y-direction only in the case when the slopes are the same.

The regression slopes could be tested first using an approximate t-test. The test-statistic is the difference of the slopes divided by the standard error of the difference, i.e.

$$t = \frac{\hat{\beta}_1 - \hat{\beta}_2}{\sqrt{(\text{s.e.}(\hat{\beta}_1))^2 + (\text{s.e.}(\hat{\beta}_2))^2}}$$

which in this case is

$$\frac{0.2099 - 0.3443}{\sqrt{0.0260^2 + 0.0346^2}} = -\frac{0.1344}{0.04328} = -3.10$$

Like the Welch test for comparing two means without assuming that the variance within the two populations is the same, the degrees of freedom of the t-statistic here can be calculated by a rather complex formula. Usually it is unnecessary, as the effective degrees of freedom are between the smaller of the two degrees of freedom participating and their sum, 6 and 12 in this case, and the difference is clearly significant at 5% but not at 1%.

An alternative method which can be easily implemented in many computer packages is to fit the following combined models to the two groups:

Model (a) One straight line with the same slope and the same intercept;
Model (b) Two straight lines with the same slope, but different intercepts;
Model (c) Two straight lines with different slopes and different intercepts.

This method makes the additional assumption that the residual variances after fitting the straight lines are the same between the two groups. When this is reasonable, the analysis below might be more powerful; when there is evidence of substantial variance heterogeneity, then it could be misleading.

An ANOVA for the sequence of fits can be assembled as follows.

Source	*df*	*SS*	*MS*	*F*
Fitting a common straight line	1	245 502	245 502	162.1
Difference in intercept (same slope)	1	7 929	7 929	5.2
Difference in slope	1	14 589	14 589	9.63
Residual	12	18 171	1 514.3	
Total	15	286 191		

Thus the portion of the total sum of squares which is accounted for by fitting a single straight line model is very considerable. Allowing for a difference between the two slopes makes a substantial improvement in the fit. The difference between the slopes is highly significant as the observed F-value is higher than the 1% point of the F-distribution (which is 9.33). Assuming a common slope, differences in intercept make a significant improvement, but this is an academic exercise as we have already found a difference between the slopes (in most cases* it makes no sense to have a common intercept with different slopes), so that then practically all the variation (i.e. the sum of squares) is accounted for. Thus a single straight line model is not tenable for the two groups; different slopes and therefore intercepts are required.

6.6 COMPARISON WITH THEORETICAL EXPECTATION—TESTS OF THE FIT OF A MODEL

The methods we have dealt with in this chapter can be adapted to deal with the problem of comparison of some aspect of a real population from which we have drawn a sample with the expectations derived from a theoretical model.

6.6.1 Comparison of a single mean with a theoretical prediction—one sample *t*-test

Student's t-test can be easily modified to test the agreement between the mean of a population with a theoretically predicted mean, using a sample of data from the population. Suppose we wish to test the null hypothesis that the mean of the population is 100. So if μ_x is the population mean of x, the null and alternative hypotheses are

$$H_0: \mu_x = 100 \qquad H_A: \mu_x \neq 100.$$

The test procedure based on the t-distribution, given a sample $x_1, x_2, \ldots, x_n$, is as follows. Calculate $t_{n-1} = (\bar{x} - 100)/(s/\sqrt{n})$ and compare t_{n-1} with tables of Student's t on $n - 1$ d.f.

* The exceptional case is when a number of lines are known *a priori* to pass through one point, for example the origin.

Example 6.8 Test of agreement with a hypothesised mean

Read data from treatment 1 of Example 6.1 into column C1.

```
MTB > TTEST 100 C1

TEST OF MU = 100.000 VS MU N.E. 100.000

      N   MEAN     STDEV    SE MEAN    T      P VALUE
C1    8   127.625  19.603   6.931      3.99   0.0054
```

Thus there is strong evidence for a real difference from the hypothesised value of 100. A 95% confidence interval on the mean is 127.6 ± 16.4.

6.6.2 Comparison of a binomial proportion with theoretical expectation

In Chapter 3 we gave some of Mendel's data from his breeding experiments on peas, in particular data from a series of dihybrid crosses on the proportion of round and wrinkled seeds. Suppose we had obtained data from one of his experiments, and wished to test whether the data conformed with expectation derived from his Laws. Example 6.9 shows how to carry out the test.

Example 6.9 Comparison of observed and expected frequencies of round and wrinkled peas in Mendel's breeding experiments

Table 6.6 Observed and expected frequencies and contributions to χ^2-statistics

	(a) Observed frequencies	(b) Expected frequencies	(c) Contribution to G	Contribution to Pearson X^2
Round	19	21.75	−5.14	0.23
Wrinkled	10	7.25	6.43	0.70
Total	29	29	1.29	0.93

Calculation of expected frequencies

The first step in the method involves working out the frequencies which would be expected in the body of the table if there were no difference between the observed proportions and those expected on the basis of the theory. We saw in Chapter 2 that the expected proportions are 0.75:0.25, and hence the expected frequencies are $29 \times 0.75 = 21.75$ and $29 \times 0.25 = 7.25$ given in Table 6.6(b).

Calculation of χ^2 test-statistics

We use the same measures of discrepancy between the observed frequencies and those expected on the null hypothesis as in the comparison of two sample proportions:

Pearson chi-squared:

$$\text{Pearson } X^2 = \sum_{\text{all cells in table}} \frac{(|f_o - f_e| - 0.5)^2}{f_e}$$

(with continuity correction)

Deviance chi-squared, or G-statistic:

$$G = \sum_{\text{all cells in table}} 2f_o \ln\left(\frac{f_o}{f_e}\right).$$

Both statistics are approximately distributed as χ_1^2, under H_0. The contributions to chi-squared are given in the right hand columns, and the totals are 1.29 for deviance chi-squared and 0.93 for Pearson chi-squared with continuity correction. The tabulated upper 5% point of the χ_1^2 distribution is 3.84, and so the result is not significant at 5%. Therefore we cannot reject the null hypothesis that the proportion of round seeds was 0.75.

6.6.3 Comparing the whole distribution with a theoretical model

Take as an example the vasopressin cell firing rate data which we found in Chapter 3 was not well fitted by a Poisson distribution. How can we carry out a formal goodness-of-fit test of the whole distribution? The empirical cumulative distribution function from the sample of data is compared with the theoretical cumulative distribution function, and $\sqrt{n}$ times the maximum distance in a vertical direction

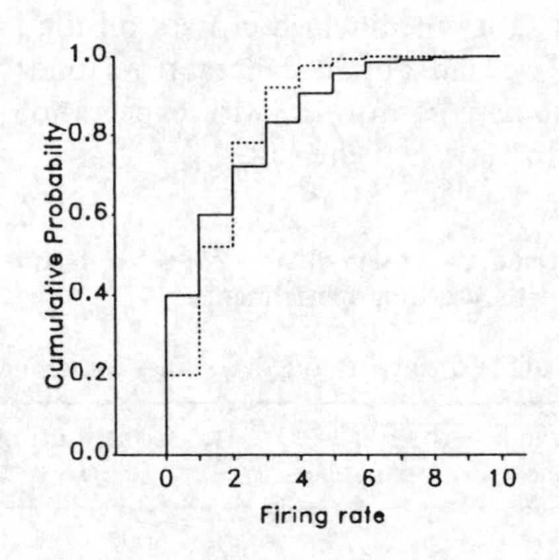

Figure 6.5 The *e.c.d.f.* of the vasopressin firing rate data (from Chapter 3, solid line), and the *c.d.f* of the fitted Poisson distribution (dotted line).

Table 6.7

Topic	Location and techniques
Comparison of sequences of models with many parameters	*Chapter 14* Multiple regression models
Assessment of incidental assumptions: distributional form, serial independence	*Chapter 14* Test of residuals
Comparison of models involving space and time	*Chapter 13* Building and testing series of models of increasing complexity for time series and point processes
Comparison of sequences	*Chapter 13* Comparison of DNA and protein sequences
Comparison in the presence of heterogeneity and trends	*Chapter 7* Design and analysis for stratified and structured populations
Model checking based on transformation to a simpler model	*Chapter 11* Linearisation and its consequences

compared with tabulated percentage points of the distribution given in Table 23 of Lindley & Scott (1984), where n is the sample size. The *e.c.d.f.* for these data and the cumulative distribution function for a Poisson distribution with the same mean $(= 1.609)$ are given in Figure 6.5. The maximum vertical displacement between the two is $0.3998 - 0.2000 = 0.1998$ (over the range 0 to 1). $\sqrt{1198}$ times this far exceeds the tabulated value at 0.1% for this sample size ($n = 1198$) of 1.949. There is very strong evidence therefore that the Poisson distribution is a poor fit to these data.

6.7 Other comparison methods

The methods of this chapter cover only the simplest comparison methods. Pointers to treatments of related topics in other parts of the book are given in Table 6.7.

6.8 APPENDIX: THE CHOICE OF TEST FOR COMPARING MEANS AND MEDIANS OF TWO OR MORE POPULATIONS

Table 6.8 Suggested tests for comparison of means/medians. Reference should also be made to the text (see section number in bottom right hand corner in square brackets) for possible limitations. The sample sizes in this table are given only as approximate guidance; the suitability of the tests will depend on the individual circumstances

Form of data	Two populations		More than two populations
	Large samples ($n > 12$)	Small samples ($5 \leqslant n < 12$)	
Continuous Normal (variances within groups equal	(1) t-test [6.2.2]	(1) t-test [6.2.2]	(1) One way analysis of variance [6.4.1]
Continuous Normal (variances within groups not necessarily equal)	(2) Welch test [6.3.1]	(2) Welch test [6.3.1]	
Continuous—not necessarily Normal	(3) Observation randomisation (also (1) and (2) if non-Normality not very great) [6.2.2]	(3) Observation randomisation (also (1) or (2) if only very slight non-Normality) [6.2.2]	(3) Observation randomisation (also (1) above if only slight non-Normality) [6.4.5]
Continuous-not necessarily Normal and rank data	(4) Mann–Whitney U (rank randomisation) [6.3.1]	(4) Mann–Whitney U (rank randomisation) [6.3.1]	(4) Kruskal–Wallis Test (rank randomisation) [6.4.5]
Frequencies and proportions (assuming independently responding individuals)	*All expected frequencies* $\geqslant 5$ Chi-square contingency table [6.3.2]	*Some expected frequencies* < 5 Fisher's exact test [6.3.2]	*All expected frequencies* $\geqslant 5$ Chi-square contingency table *Some expected frequencies* < 5 Fisher's exact test

EXERCISES

6.1 The following are the results of an experiment into the diffusion rates of carbon dioxide through soil of different porosity:

Fine soil	20	30	18	23	23	28	23	26	27	26	12	17	25
Coarse soil	19	30	32	28	15	26	35	18	25	27	35	34	

(a) Is there a statistically significant difference between the rates in fine and coarse soil?
(b) Calculate a 95% confidence interval for the difference between the mean diffusion rates in the two types of soil.
(Cambridge, NST, Part 1A, Biological Mathematics, 1981.)

6.2 Ten samples of a cell suspension were used in an experiment to compare the effects of two media (*A* and *B*) on the survival times (hours) of the cells at $-20°C$. Each sample was divided in half and one randomly chosen half was diluted in medium *A*, and the other in medium *B*. The survival times were as follows:

A	19	17	34	16	22	19	24	18	12	26
B	8	9	19	11	9	9	12	8	3	15

Explain under what statistical conditions a *t*-test could validly be used to test whether the mean survival times in the two media differ. Carry out the appropriate *t*-test, and obtain 95% confidence limits for the difference between the two mean survival times.
(Cambridge NST, Part 1A, Biological Mathematics, 1978.)

6.3 Samples *A* and *B* are drawn from Normally distributed populations with the same variance. Perform a suitable test to examine the hypothesis that the mean of a population *A* is twice the mean of population *B*.

A	9	24	15	12	11	19	13	15	21	6
B	4	7	5	9	10	15	10	2		

Also examine the assumption that *A* and *B* have the same population variance.
(Cambridge, NST, Part II Special Subject Statistics, 1969.)

6.4 What is the Mann–Whitney *U*-test and when should it be used? Illustrate your answer using a hypothetical set of data.

6.5 In an experiment to compare the growth rate of rats given two different feeding treatments, A and B, in carefully controlled environmental conditions, the following growth increments (g) from day 10 to day 30 were obtained:

A	4.3	5.2	4.6	4.8	4.9	5.3	4.8	5.1	4.9	
B	5.3	5.2	5.9	5.3	6.2	6.1	5.7	5.4	5.1	5.8

In a separate experiment on rats kept in a more variable environment over a longer period, the following results were obtained:

A	7.3	7.2	8.9	7.3	7.2	6.9	10.7	7.4	
C	8.6	10.5	9.2	8.1	9.8	10.6	9.6	10.3	9.8

What evidence is there of a difference in growth between treatment A and B, and between A and C? Comment on the biological and statistical significance of any differences.

6.6 In order to compare the effectiveness of a traditional remedy (R) and a placebo (P) in reducing temperature, a fixed dose of the traditional remedy and a similar dose of a placebo were each given in a double-blind trial to 30 patients, and the number of patients whose temperature fell by more than 0.2 degree within 2 hours of administration of the treatment, and the number whose temperature stayed the same or rose were noted. The numbers were as follows:

	Placebo	Traditional remedy
Number whose temperature fell by 0.2 degree	17	12
Number whose temperature did not fall by 0.2 degree	13	18

Test whether the placebo and the traditional remedy differed in their effectiveness.

6.7 How would you perform an observation randomisation test to compare the two treatments in Example 6.1? What are the advantages, if any, of this test over a two sample t-test, or similar parametric test?

6.8 Over a period, 1406 cases of stomach cancer were recorded at certain hospitals in the north of England. Blood group classification showed these to be 603 group O; 625 A; 107 B; 44 AB. For each case, a patient not having stomach cancer was selected at random in the same hospital; the blood groups of these were 713, 552, 96, 45 of O, A, B, AB respectively.

By means of an appropriate test, determine whether lack of association between cancer and blood group is an acceptable hypothesis.
(Cambridge, NST, Part 1A, Biological Mathematics, 1982.)

6.9 Assuming that inseminations of dairy cattle are statistically independent trials with the same chance p of success (i.e. conception), the probability that exactly r inseminations are needed to establish pregnancies in each member of a herd of k cows is

$$P(R=r)=\binom{r-1}{k-1}p^k(1-p)^{r-k} \qquad (r=k,k+1,\ldots)$$

(the negative binomial distribution as derived in Exercise 3.14).
Records for 40 herds all of size $k=10$ gave the following frequencies for $r=10, 11, \ldots$.

r	10	11	12	13	14	15	16	17	18	19	20
Frequency	0	5	8	7	4	2	4	4	2	3	1

Determine whether these are an acceptable fit to the distribution obtained when $p=0.7$.
(Cambridge, NST, Part 1A, Biological Mathematics, 1984.)

6.10 Test the hypotheses you derived in Exercise 2.2 about the genetic control of the characters A and B using the data given there.

6.11 Two hundred rice plants were examined for the incidence of an insect pest. The distribution of insects in the sample of 200 plants is given in the following table.

Number of insects per plant	0	1	2	3	4	5	$\geqslant 6$
Number of plants	69	40	38	20	16	8	9

Mean of observations = 1.735; variance of observations = 3.744.

By fitting the negative binomial distribution with the same mean and variance to these data, and then using the χ^2 or some other test, determine whether it is reasonable to conclude that the negative binomial adequately describes the incidence of the pest.
(Cambridge, NST, Part 1A, Biological Mathematics, 1975.)
Hint: in the χ^2 test, expected frequencies are calculated from the fitted distribution, and then one of the usual chi-square test statistics, as discussed in the chapter, is used to test the discrepancy between the observed and expected frequencies, by comparing the observed value of this test statistic with the distribution of χ^2 on $n - p - 1$ degrees of freedom, where n is the number of frequencies used in the test, and p is the number of parameters estimated when fitting the model.

6.12 A sample of adult females was randomly selected from each of two natural populations of red deer in 1980. The individuals contained in these samples were marked and then released back into the populations. Data were collected on the numbers of female and male offspring born to these females in the two years 1981 and 1982. Red deer normally only produce single births. The numbers were as follows:

	1981		1982	
	Females	Males	Females	Males
Population A	102	85	87	97
Population B	47	39	43	41

(a) Test the null hypothesis that the sex ratio at birth in population A in 1981 was 50:50.
(b) Test the null hypothesis that the sex ratio at birth in the two populations was the same in 1981?
(c) Is it appropriate to carry out the same type of test as you used in (b) if you wish to compare the 1981 and 1982 sex ratios in population A? Give reasons for your answer.
(Cambridge, NST, Part 1A, Biological Mathematics, 1987.)

6.13 The following data, treatment means ($\bar{x}$), standard deviations (s), and standard errors of the means (s.e.), were obtained in a completely randomised design for the comparison of three treatments

Treatment	Observations						$\bar{x}$	s	s.e.
A	100	98	105	103	102	107	102.5	3.27	1.33
B	108	114	116	113	117	119	114.5	3.83	1.57
C	112	108	100	84	129	122	109.2	16.03	6.54

Carry out the following three analyses:

(a) an F-test of the differences between A, B, and C;
(b) Three separate t-tests of A vs B, B vs C, C vs A including in each test only the data on the two treatments in question;
(c) Three t-tests, as in (b), but using the residual mean square from the analysis of variance table to estimate the standard errors of the differences between the means.

Comment on the usefulness of these analyses for comparing the three treatments, and explain the inconsistencies between the results. You may assume that the populations from which these data are samples are Normally distributed. What are the advantages and disadvantages of these three procedures in general? How would you resolve the inconsistencies between approaches (a) and (c) when the number of treatments is large? Give a brief rationale of the proposed method.
(Adapted from Cambridge, NST, Part II General, Special Subject Statistics, 1981.)

6.14 (a) Define Type I and Type II errors when testing the null hypothesis H_0: $\mu = 10$ against the alternative H_A: $\mu > 10$, where μ is the mean of the population, using a random sample of size n from the population.

(b) Explain the concept of power function of a test.

(c) Why is it important to consider power when interpreting the results of a significance test?

(d) What other errors apart from Type I and Type II errors might be made in testing this hypothesis?

In your discussion you should refer to the following two hypothetical samples of data collected (in completely separate situations) to test the null hypothesis given above.

Sample 1: 31.3, 2.7, 19.2, 12.1, 20.9, 7.3, 25.0, 30.6, 5.8
Sample 2: 13.3, 12.1, 14.0, 11.8, 9.5, 17.0, 14.2, 13.5, 14.1

7 Sampling, Controlling and Measuring the Random and Systematic Variation: Design of Experiments and Surveys

7.1 INTRODUCTION

In Chapters 1–4 we presented a range of models for describing biological populations and processes including deterministic models for growth and decline, and genetics, simple stochastic models and more complex models with both deterministic and random elements. Then, in Chapter 5 we dealt with methods for fitting these models to data and, in Chapter 6, statistical techniques to compare populations and processes using samples of observations. In this chapter we deal with the *design* of the biological investigation and, in particular, how we take observations on populations and processes to obtain data for fitting models and making comparisons. We begin by drawing the important distinction between surveys and experiments.

7.1.1 Surveys and experiments

In a survey we are interested in *describing* the properties of some population or system, usually from samples of material selected from it. In an experiment we *manipulate* the material or system to answer specific questions, and to establish causal links between some factor and a measured response. For example, an epidemiological survey may be performed to estimate the frequency of a disease in a population or to study how it varies with risk factors such as obesity, smoking habit and blood pressures. A physiological experiment may be set up to investigate the effect of food intake and temperature on the metabolism of thyroid hormones in pigs. The survey may be able to demonstrate correlations of incidence of disease with several factors but it cannot establish a causal link. In the experiment, however, we can, in principle, disentangle the effects of the different factors by manipulation of the material and control of the conditions.

In a survey, an element of representativeness in selection of material is critical. Ideally, we would like the properties of the sample to be very similar to those of the population. In an experiment the requirement for representative material is less strong and often rather artificially uniform material is used, e.g. inbred lines. This is to reduce extraneous random variation which might obscure the true effects of the different factors. The approach can be particularly effective for studying biological mechanisms and how different factors operate. A disadvantage of working with artificially uniform material is that the validity of the conclusions may not extend to different but related material. This could be particularly important in applied work comparing, for example, different varieties of crops intended for large scale use. For some problems it is necessary to include material which is representative of this variation in any experiments.

Despite the differences in aims and interpretation of surveys and experiments, some fundamental principles of design apply to both. For example, in Section 7.2 we present the basic method of *simple random sampling*, used in surveys for selecting individuals from a population, and the method of *treatment randomisation*, used in experiments for allocating treatments to experimental material or subjects. Both methods deliberately involve a random element to avoid bias and to justify the application of statistical techniques for the analysis of the data.

7.1.2 Biological, statistical and treatment populations

In Chapter 6 we defined a *biological population* as a group of individuals of the same species inhabiting a given area, and a *statistical population* as the entire finite or infinite aggregate of units from which samples are drawn, sometimes called a *sampling frame*—for example, cores of mud in sampling chironomid larvae from a lake. In an experiment, treatments are applied to experimental units selected from some population and it is useful to think of a *treatment population*, i.e. the population of responses that would have been obtained by applying a particular treatment to the whole population from which the experimental units have been, or could be thought to have been, drawn. The treatment population is an abstraction because in a particular experiment each treatment is only allocated to a sample of experimental units. In the analysis, however, we regard the observed responses as a random sample from the corresponding treatment population which we assume to be infinitely large.*

To clarify what we mean by *treatment population*, consider Example 6.1 (from Chapter 6), also discussed in Section 7.2.2, in which the effects of three feeding supplements on the relative growth rates of plants were compared. The plants used in this experiment can be thought of as being sampled at random from some population. There would thus be three notional treatment populations: *treatment population 1*, the responses of all the plants in the population if they had received treatment 1; *treatment population 2*, the responses of all the plants in the population if they had received treatment 2; *treatment population 3*, the responses of all the plants in the population if they had received treatment 3. We shall see later that the usual experimental protocol—in the case where each treatment would have 8 replicates, for example—is to obtain a sample of 24 units from the population in the first place, in the ideal case at random, and then to allocate the three treatments at random to these 24 units, 8 units receiving each treatment. Thus we can think of the 8 units receiving treatment 1 as a sample of 8 from treatment population 1, and similarly for the units receiving treatments 2 and 3.

7.1.3 Plan of chapter

In many sections of this chapter methods for both surveys and experiments are

---

* This is the basis of the so-called Normal model and the method of Analysis of Variance which is used to compare treatments, as presented in the following sections and in Section 6.4.1. The Analysis of Variance can also be justified in a more complex manner without recourse to the Normal model (see Kempthorne, 1952).

presented. The chapter is divided according to the characteristics of the systematic and random variability of the biological material used in either the experiment or the survey. Table 7.1 contains a key to the chapter giving a classification of population types, aims of the investigation and designs considered. We start by presenting basic designs for estimating population parameters, such as the mean and variance, of a homogeneous population (i.e. where there is no obvious classification of the material into groups which differ in a systematic way). Designs are also given for utilising material from such a population in experiments for comparing applied treatments. Then we consider heterogeneous populations and discuss designs to control for substantial heterogeneity when estimating population parameters and comparing the means of applied treatments. Next we discuss designs for studying trends, patterns and relationships; a broad topic which we give a rather selective treatment, concentrating on a few key designs. Then the important idea of Factorial Structure is covered: a useful framework for the design and analysis of surveys and experiments in which there are several factors occurring in combination. After this we give a very brief account of designs for estimating variances in data where there are different components of random variation, for example between sires, dams and offspring in an animal breeding

Table 7.1 Plan of chapter showing different designs and applications

Section	Population classification	Aims	Survey design	Experimental design
7.2	Homogeneous random variation	Estimating and comparing population parameters	Simple random sampling	Completely randomised design
7.3	Heterogeneous with both systematic and random variation	Estimating and comparing means	Stratified random sampling	Randomised block design
7.4	Structured: trends, temporal or spatial pattern, autocorrelation	Analysis of pattern and process. Estimating relationships	Systematic sampling	• Systematic design • Response surface design • Repeated measures
7.5	Factorial structure	Estimating and comparing means for combinations of factors	Factorial designs for surveys and experiments	
7.6	Components of random variation	Estimating variance components Estimating and comparing means	Nested or hierarchical designs for surveys and experiments	
7.7	Model simulated data	Analysis of model behaviour		Factorial design

experiment. In the final section we show how design principles can be applied to study the behaviour of a dynamic stochastic model using simulated data. For some of the major designs, not discussed elsewhere in the book, we also outline methods of analysis in this chapter.

7.2 DESIGNS FOR HOMOGENEOUS POPULATIONS: ESTIMATING AND COMPARING POPULATION PARAMETERS AND THE EFFECTS OF APPLIED TREATMENTS

This section is about simple designs for estimating and comparing population parameters in homogeneous populations. We deal with designs for surveys where the aim is to *describe* some aspect of a population, and for experiments where the aim is to compare two or more treatments *applied* to the biological material. We present the two key ideas of *simple random sampling*, for selection of individuals from a population, and *randomisation*, for allocation of treatments to experimental subjects. In both cases a *random* element is deliberately introduced into the process of selection of individuals. An important practical reason for this is to avoid subjective biases on the part of the investigator. A second reason is so that we can apply a body of statistical theory to calculate the precision of our estimates and to carry out significance tests for real differences.

Both simple random sampling and randomisation apply particularly to populations for which we are *indifferent* about which units to include in the sample or use for each treatment. This is reasonable for homogeneous populations but if there is some clear and known heterogeneity, then it makes sense to constrain the random selection of units or allocation of treatments by using the methods of *stratification* and *blocking* (Section 7.3).

The designs described here for estimating and comparing parameters of the population distribution, such as the mean and variance, may be thought of as designs for fitting the simple stochastic models presented in Chapter 3. In other situations, however, we may be more interested in the population structure—such as patterns in time or space—in which case simple random sampling is inappropriate and a systematic sample with more regularly spaced observations is often better (Section 7.4).

7.2.1 Sampling a homogeneous population: simple random sampling

Simple random sampling is a basic method of sampling in which individual units are selected from the population: at random; independently of each other; and each with the same chance of selection.

Simple random sample

One selected by a *process* involving a chance mechanism which gives every possible sample of a given size the same chance of selection.

Simple random sampling is like a fair lottery, or dealing a hand of cards from a well shuffled pack. In practice, however, the usual method of drawing a simple random sample is to use tables of random numbers or to generate random selections on a computer. We emphasise that in simple random sampling individuals are selected independently of each other. So, for example, if we are sampling plots in a field, two plots which are far apart are just as likely to occur in the same sample as two neighbouring plots. This would not be so if a cluster of plots were sampled around a randomly located position. Simple random sampling is practically important for two reasons.

(1) The simple random sample is free of *selection bias*—no particular individual is any more or less likely than any other to be included in the sample. Thus, simple random sampling can be used to safeguard against potential biases inherent in subjective or *ad hoc* methods of selection.
(2) Simple random sampling provides estimates of population parameters with known properties, and a theory for measuring the potential errors of the estimates.

The properties of the simple random sample refer to what happens in the hypothetical *long-run* of a very large number of repeated samples from the population. Simple random sampling does not guarantee that a *particular* sample will produce a good estimate of, say, the population mean, but it does allow us to measure the likely magnitude of the chance deviations of the sample estimate from the corresponding population value.

Analysis of a simple random sample

A remarkable property of the simple random sample is that it can be used to estimate all aspects of the population *distribution* such as location (e.g. mean and median), spread (e.g. variance, standard deviation and inter-quartile range) and distributional form (e.g. Normal, lognormal). Information on other aspects such as spatial or temporal pattern is difficult to utilise (Section 7.4). Here we are concerned with simple random sampling as a procedure for estimating parameters of the population distribution, in particular the mean and variance.

In the basic model for simple random sampling we write the observations as

$$Y_i = \mu + \varepsilon_i$$

where μ is the population mean and the ε_i are random variables which vary *independently* with mean zero and variance σ^2. This is just another way of saying that the observations are drawn independently at random from a population distribution with mean μ and variance σ^2.

Chapter 5 deals with fitting this model, i.e. estimating the mean and variance. The main results are: (a) that the sample mean $\bar{Y}$ is an unbiased estimator of the population mean with variance, from sample to sample, σ^2/n; (b) the sample variance s^2 is an unbiased estimate of the population variance; (c) the sampling distribution of the sample mean is *approximately* Normal whatever the form of the population distribution; (d) an *approximate* 95% confidence interval for the population mean is $\bar{Y} \pm t_{0.025,(n-1)}s/\sqrt{n}$, where $t_{0.025,(n-1)}$ is the two-tailed 5% point of Student's t-distribution with $(n-1)$ degrees of freedom.

In the above model for simple random sampling there is no depletion of the population by the removal of individuals. At each selection the observation is assumed to come from the *same* distribution unaffected by previous selections, sometimes called an *infinite population model*. In practice, real populations contain finite numbers of individuals which are usually *sampled without replacement*, i.e. not returned to the population following selection. Surprisingly this requires only a minor modification to the results based on the infinite population model. The expression for $\text{var}[\bar{Y}]$ is changed slightly.

Variance of the sampling distribution of the sample mean in simple random samples without replacement

$$\text{var}[\bar{Y}] = \frac{\sigma^2}{n}\left(1 - \frac{n}{N}\right) = \frac{\sigma^2}{n}(1 - f) \approx \frac{\sigma^2}{n} \quad \text{if } f \text{ is small,}$$

where n is the sample size, N is the population size (i.e. the number of sampling units), σ^2 is the population variance (defined as the sum of squared deviations about the population mean divided by $N - 1$) and f is the *sampling fraction* (i.e. the proportion of units selected).

In practice, populations are often so large that the effect of the sampling fraction is negligible. To allow for the finite population size, calculate a 95% confidence interval as *sample mean* $\pm\, t_{0.025,(n-1)} s\sqrt{(1-f)/n}$, where $t_{0.025,(n-1)}$ is the two-tailed 5% point of Student's t-distribution with $(n - 1)$ degrees of freedom.

Example 7.1 Illustration of properties of simple random sampling from a finite population

In the previous section we outlined the properties of simple random sampling by assuming a model for observations selected at random from an infinite population. For a more concrete illustration we can take a population of known values and repeatedly select samples at random to analyse their properties. In this section we present an example using a population of plot timber volumes (Table 7.2) ignoring for the moment the subdivision into 6 compartments.

The population mean volume is 194 units and the variance is 4341 units^2. For a simple random sample of ten plots selected without replacement the timber volumes were: 153, 212, 83, 159, 306, 253, 277, 236, 218, 65. A 95% confidence interval for the population mean is calculated as follows: sample mean $\bar{Y} = 196.2$; sample variance $s^2 = 6377.5$; sample standard deviation $s = 79.9$; sampling fraction $f = 10/150 = 0.0667$; standard error of sample mean $= \sqrt{6377.5(1 - 0.0667)/10} = 24.40$; $t_{0.025,9} = 2.26$; 95% confidence interval $= 196 \pm 2.26 \times 24.40 = 196 \pm 55$.

In this case the confidence interval contains the population mean. To illustrate the sampling properties of the sample mean and sample variance, 5000 sets of simple random samples were selected of size 5, 10 and 30, sampling with and without replacement. Table 7.3 illustrates that both the population mean and variance are estimated unbiasedly by the sample values, and also that there is a slight reduction in the variance of the sample mean due to the effect of the sample fraction.

7.2.2 Experimental design utilising material from a homogeneous population: treatment randomisation and the completely randomised design

In designing an experiment to compare two or more treatments, we need to decide how to allocate treatments to the experimental subjects. A subjective assignment is unsatisfactory because it is affected by the judgement of the experimenter and may therefore be biased. One approach is to allocate treatments at random, i.e. to use randomisation.

Table 7.2 A population of timber volumes (1/10 m^3) of 150 plots (0.1 ha) in 6 compartments of an uneven-aged forest stand (Loetsch & Haller, 1964). [R] locations of 10 plots selected at random from the whole population

	1					2				
	130	153[R]	153	112	200	106	100	147	118	165
	124	106	136	130	165	141	194	212[R]	136	83[R]
A	177	165	136	124	171	106	82	177	147	165
	165	112	124	118	153	118	224	136	118	159[R]
	100	82	118	153	147	130	130	112	88	118
	224	247	217	230	130	259	277	100	147	171
	253	200	135	271	277	271	230	206	242	177
B	212	277	265	212	206	171	289	259	183	247
	224	283	247	300	100	318	277	306	177	200
	100	141	265	277	306[R]	165	253[R]	265	271	159
	277	330	253	218	177	353	330	253	171	194
	224	212	159	224	141	183	283	188	147	183
C	271	318	200	271	218	253	260	200	147	259
	277	277[R]	206	236[R]	230	230	294	165	294	212
	130	218[R]	65[R]	171	165	194	171	206	312	94

Table 7.3 Illustration of the properties of the sampling distribution of the sample mean and the sample variance using computer simulation of 5000 simple random samples of sizes 5, 10 and 30 from the population of plot timber volumes of size $N = 150$, mean $\mu = 194$, $\sigma^2 = 4341$ (Table 7.2). Theoretical values are given in parentheses

Sample size (n)	Sampling fraction (f)	Sampling without replacement			Sampling with replacement		
		Average of sample means	Variance of sample means	Average of sample variances	Average of sample means	Variance of sample means	Average of sample variances
5	0.033	194	839 (839)	4350	194	875 (868)	4334
10	0.067	194	405 (405)	4320	194	434 (434)	4316
30	0.200	194	116 (116)	4341	194	142 (145)	4326

Randomisation

An objective procedure in which treatments are assigned at random to experimental units.

We could randomise by using shuffled cards or cards drawn out of a hat but the main method is to use tables of random numbers or to generate random allocations of treatments on a computer.

Randomisation can be used to avoid biases which could arise with a systematic assignment. For example, if two treatments were applied alternately in time or space then an underlying trend would lead to an apparent treatment difference in the absence of an effect. A further benefit of randomisation is to conceal from the person involved which treatments are applied to which units: clinical trials are sometimes *double blind* with neither the patient nor doctor knowing the treatment. Cox (1958) discusses an interesting example of selection bias from subjective allocation in the Lanarkshire school milk experiment assessing the effect on child growth of receiving extra milk. Finally, randomisation provides a statistical basis for testing differences between treatments and for estimating the precision of estimated differences.

In a *completely randomised design,* treatments are assigned to units completely at random subject only to the condition that they occur the required number of times. This allocation is reasonable in situations where there are no obvious systematic differences between the experimental units. The design is simple and flexible. In particular, if for some reason, unconnected with the effect of the treatments, some of the observations are missing, either lost or due to damage, then the method of analysis is not much affected. The interpretation would, of course, be different if the missing observations were *caused* by one of the treatments.

To illustrate the completely randomised design, consider the example in Chapter 6 involving a comparison of the effect of three feeding supplements T_1, T_2 and T_3 on the relative growth rate of a plant. The experimental units are plants of similar initial weights so that a completely random assignment of treatments to plants seems reasonable. If there were large differences in the initial weights of the available plants, it would be better to group plants of similar weight so as to compare feeding supplements on plants of similar initial weight (Section 7.3.3).

Mechanics of randomisation

Suppose that in the above example we decide to use 8 plants for each feeding supplement, i.e. 8 replicates per treatment (Section 7.8.1 describes methods for determining the number of replicates). How do we assign the three different feeding supplements at random to the 24 plants?

First, number the plants 0, 1,. . ., 23. Then, take random digits in pairs from a table of random numbers. If a number exceeds 23 subtract a suitable multiple of 24. Reject any repeats. Allocate the plants for the first 8 numbers to T_1, for the second 8 numbers to T_2 and for the remaining numbers to T_3. Basically, we are arranging the numbers 0, 1, . . ., 23 in random order. Some statistical packages have procedures for generating random permutations which could be used to carry out the randomisation.

Analysis of the completely randomised design

The analysis of the completely randomised design for comparing means is described in Chapter 6 and illustrated there with a worked example. The main steps are as follows. A preliminary portmanteau test of treatment differences is carried out using a one-way analysis of variance. The residual mean square from the analysis of variance is used to estimate the amount of random variation in the responses. This is then used to estimate standard errors for treatment means, to calculate confidence intervals and to test for differences between pairs of treatment means. For the example in Chapter 6, there are only seven observations for one of the treatments but the method of analysis is the same for both equal and unequal replication of treatments. This flexibility contrasts with more structured designs in which data with unequally replicated treatments usually require special methods of analysis. The analysis assumes that the responses are Normally distributed with constant variance. Alternatively, an observation randomisation test of treatment differences can be used if the assumptions of Normality are in doubt. The completely randomised design is not confined to comparing means of the different treatments. Methods for comparing variances and the overall distributions of responses described in Chapter 6 also apply.

7.2.3 Lack of independence and pseudo-replication

An important property of simple random sampling and randomisation is that units are selected, or allocated to treatments, *independently* of each other. In some situations this may not be so. There may be clusters of sampling units, repeat measurements on the same material, observations close together in time or space, or treatments applied to related units such as animals from the same litter. In each case there is a *lack of independence*. *Pseudo-replication* occurs when non-independent observations are regarded as independent. It can lead to a false picture of the true amount of replication and spurious accuracy. Section 7.6 describes certain types of designs in which some of the observations share components of random variation and are therefore not independent of each other.

7.3 DESIGNS FOR HETEROGENEOUS POPULATIONS: CONTROLLING AND REDUCING RANDOM ERROR IN POPULATION ESTIMATION AND TREATMENT COMPARISON

In Section 7.2 we introduced some basic designs for estimating and comparing means and other parameters in homogeneous populations. These involved the procedures of simple random sampling and randomisation to avoid selection bias and to provide a statistical basis for analysis. In many situations, however, there is some heterogeneity in the population of interest—for example, in a mixed population, males may be heavier on average than females. In simple random sampling no attempt is made to allow for this heterogeneity due to systematic differences between groups of individuals; e.g. a sample could conceivably contain members of only one sex. Similarly, randomisation in a completely randomised design could result in all the members of one sex being assigned to one treatment. This would be unsatisfactory for comparing treatments if there was a substantial systematic difference in the measured response between males and females. In this section we consider ways of incorporating knowledge of the biological material into the structure of the design to allow for systematic differences between units. The aim is to control and reduce the amount of random error so as to increase the precision of estimation and comparison of means. Simple random sampling and randomisation remain as basic procedures *but* for selection *within* groups of units.

7.3.1 General approaches to the control and reduction of random error

For simple random sampling and randomisation a key result is that the standard error of the sample mean is equal to the population standard deviation divided by the square root of the sample size, i.e. $\sigma/\sqrt{n}$. Therefore, one way to reduce the standard error is to increase the sample size. In practice, however, this approach is usually limited by available resources—to *halve* the standard error requires a *fourfold* increase in sample size. A second way to increase precision is to reduce the amount of random variation which is measured by the population standard deviation. Some approaches are as follows.

Reducing measurement error. In most studies there is some error in taking measurements on the sampling or experimental units. For example, in recording weights of individual animals fed on different diets, or in estimating the density of a pest species using core sub-samples from individual field plots treated with various pesticides. Other random errors may arise due to inaccuracies in administering the treatments or failure to control experimental conditions. In many cases these errors can be reduced by refining the techniques and often they can be made small relative to the intrinsic variability of the biological material.

Use of uniform experimental material. Another way to reduce the amount of random variation is to work with artificially uniform material or in artificially uniform conditions—for example, using animals from an inbred line. A problem with this approach is that although precise comparisons can be made, conclusions might be restricted to a limited range of conditions.

Stratification and blocking. In sample surveys, stratification involves grouping the sampling units into strata which consist of units with similar responses. We then combine means of samples from relatively homogeneous strata to form a precise estimate of the population mean. In experiments, we use blocking to group the units into blocks made up of similar units and compare treatments within such blocks so as to reduce error. In observational studies, the similar technique of matching individuals by factors such as age and sex is sometimes used. Stratification, blocking and matching are basic aspects of design for the control and reduction of error which are discussed in this section.

Use of supplementary observations. This involves making supplementary measurements on quantities which are thought to be related to the response variable. The idea is to use such information in the analysis to compare groups. For example, in a clinical trial with random allocation of treatments we might find that the individuals in the control group are older than those in the treated group. If the measured response is thought to be related to age, a *post hoc* adjustment of the differences in response due to age between groups would be prudent. When there are several supplementary measurements the technique of multiple regression may be useful (Chapter 14).

7.3.2 Survey design for heterogeneous populations: stratified random sampling

In stratified random sampling the population is subdivided into *strata* and a simple random sample is taken from each stratum. The sample means for each stratum are then combined to estimate the population mean. The aim is to form strata which consist of similar sampling units to reduce the estimation error of each stratum sample mean and thereby increase the precision of the estimated population mean. A requirement for stratified random sampling is that the size of each stratum is known. We begin with the analysis and then illustrate the gains in precision from stratified random sampling.

Analysis of stratified random samples

A stratified random sample is just a set of simple random samples, so we can represent the observations by the model

$$Y_{ij} = \mu_i + \varepsilon_{ij}.$$

Y_{ij} is the jth observation from the ith stratum, μ_i is the mean for the ith stratum and ε_{ij} is a random term for the variability *within* the ith stratum which is distributed with zero mean and variance σ_i^2. The overall population mean is a *weighted* mean of the stratum means with weights proportional to the stratum size, i.e.

$$\mu = \frac{1}{N} \sum_{i=1}^{k} N_i \mu_i$$

where N_i is the number of sampling units in the ith stratum in the population. The sample mean for the ith stratum $\bar{Y}_i$ is an unbiased estimator of the population stratum mean so that an unbiased estimator of the population mean is a weighted average of the sample stratum means given by

$$\hat{\mu} = \frac{1}{N} \sum_{i=1}^{k} N_i \bar{Y}_i.$$

Using the formula for the variance of the sample mean of a simple random sample, and the fact that the stratum means are *independent*, the variance of the estimated population mean is given by

$$\text{var}[\hat{\mu}] = \frac{1}{N^2} \sum_{i=1}^{k} \frac{N_i^2 \sigma_i^2 (1 - f_i)}{n_i}$$

where n_i is the sample size for the ith stratum and $f_i = n_i/N_i$ is the sampling fraction. This variance is estimated unbiasedly by using the stratum sample variance, s_i^2, to estimate the stratum population variance σ_i^2. From the above expression we can see that, for fixed sample sizes, the error of the estimate is determined by the magnitude of the random variation *within* the strata σ_i^2. So, to increase precision, we usually aim to make units within strata as similar as possible.

Example 7.2 Stratified random sampling to estimate the mean density of a breeding population of puffins

These data are from a study to estimate the size of a breeding population of puffins. Counts were made of the number of occupied nesting burrows in quadrats of size 30 m^2. The colony was subdivided into four strata with different densities of burrows and a random sample of quadrats was taken within each stratum. Table 7.4 gives summary statistics for use in the analysis.

Table 7.4 Summary statistics for analysis of a stratified random sample of quadrats in the puffin breeding area

Population		Sample			
Stratum (i)	Stratum size (N_i)	Sample size (n_i)	Sampling fraction (f_i)	Mean ($\bar{Y}_i$)	Standard deviation (s_i)
1	409	10	10/409	4.70	4.62
2	493	13	13/493	4.23	5.00
3	728	17	17/728	11.88	7.35
4	823	16	16/823	15.06	9.54

The estimated mean and its standard error are calculated as follows.
Estimated population mean

$$= \frac{1}{N} \sum_{i=1}^{k} N_i \bar{Y}_i \quad \text{(general case)}$$

$$= (409 \times 4.70 + \cdots + 823 \times 15.06) / (409 + \cdots + 823)$$

$$= 25\,050.7/2453$$

$$= 10.21$$

Standard error of estimated population mean

$$= \sqrt{\frac{1}{N^2} \sum_{i=1}^{k} \frac{N_i^2 s_i^2 (1 - f_i)}{n_i}} \quad \text{(general case)}$$

$$= \sqrt{\frac{1}{2453^2} \left\{ \frac{409^2 \times 4.62^2 \times (1 - 10/409)}{10} + \cdots + \frac{823^2 \times 15.06^2 \times (1 - 16/823)}{16} \right\}}$$

$$= 1.02$$

The estimated overall density of 10.21 burrows per quadrat (s.e. = 1.02) corresponds to an estimated size of the breeding population of 25 045 pairs (s.e. = 2594).

Illustration of gains in precision from stratified random sampling

To illustrate how stratified random sampling can reduce the error in the estimated population mean we use the population of 150 plot timber volumes consisting of 6 compartments each with 25 plots (Section 7.2.1). It makes sense to use these compartments as strata because they contain trees of the same age and similar size. Observations randomly selected within strata will therefore be less variable than observations taken from the population as a whole. Table 7.5 shows that the standard error of the estimated mean using a stratified random sample with 2 plots taken at random from each of the 6 compartments is 14.7 compared with 18.2 for a simple random sample of 12 plots.

Choice of strata and sample size in each stratum

The choice of strata in stratified random sampling will depend on what is known about

Table 7.5 Illustration of gains in precision for estimating a population mean by stratified random sampling of a population of plot timber volumes

	No stratification Simple random sampling ($n = 12$)	
	$\mu = 194.2$ $\sigma = 65.9$	
	s.e. estimated mean **18.2**	
	Stratification by compartments 6 strata; 2 units per stratum	
	1	2
A	$\mu = 138.2$ $\sigma = 27.5$	$\mu = 136.5$ $\sigma = 37.7$
B	$\mu = 224.0$ $\sigma = 60.4$	$\mu = 224.8$ $\sigma = 56.8$
C	$\mu = 218.7$ $\sigma = 60.6$	$\mu = 223.0$ $\sigma = 64.1$
	s.e. estimated mean **14.7**	

the variability of the biological material. In the above example on estimating plot timber volume, forest compartments form useful strata because trees in the same compartment are known to be of the same age and similar size. The guiding principle is to subdivide the population into relatively homogeneous strata. For a given stratification, however, we can show from theory that for a fixed total number of sampling units the variance of the estimated population mean will be smallest when the sample size in each stratum is proportional to the product (*stratum size*) × (*stratum standard deviation*). This so-called *optimal* allocation constitutes a theoretical ideal because, in practice, the stratum standard deviations will not usually be known in advance. A practical compromise is to use *proportional* allocation in which the sample size is directly proportional to the size of the stratum. Another general scheme which is sometimes used is *equal* allocation of sample sizes to each stratum.

In the analysis of a stratified random sample the variability in each stratum is estimated by the stratum sample variance which requires that the stratum sample size must be at least two. However, since an estimated variance based on a random sample of size two will be rather variable it is usually best to make sure the sample size is at least six–ten. Finally, we stress that the possible gains in stratified random sampling by allocating different sample sizes to each stratum do *not* extend to experimental design and analysis where *equal replication* usually leads to a much simpler analysis and easier interpretation of the results.

7.3.3 Experimental design utilising material from a heterogeneous population: randomised blocks

The basic idea of *randomised blocks* is that of grouping the experimental units into sets, or *blocks*, so that the experimental units in a block are as alike as possible. Treatments are then allocated at random to units *within* each block. In the simplest case each treatment occurs once in each block. Treatment comparisons are based on *differences* between the observations *within* blocks of similar units. The aim is to increase the precision of the estimated treatment differences by eliminating systematic variation due to differences between blocks from the treatment comparisons.

Some of the ways in which units may be formed into blocks so as to reduce the experimental error are as follows.

Neighbouring plots. Measurements on experimental units which are close together in space or time are usually more similar than those on units which are far apart. In agricultural field trials, similarity of neighbouring plots may be due to heterogeneity on a larger scale or trends in soil fertility. To allow for this, blocks are often made up of compact groups of contiguous plots. The same principle may apply in time if there are systematic differences between samples processed on different days. This can be allowed for by arranging that treatments in a block be processed on the same day.

Animal litters. Sibs are usually more similar than unrelated individuals because of genetic or maternal effects, so that litters form natural blocks. One possible limitation for randomised blocks is that the litter size should be as large, or larger, than the number of treatments to be compared.

Supplementary measurements. Sometimes individuals may be grouped according to a measurement which is thought to be associated with the observed response. For example, in growth studies the gain in weight of an individual may be related to its weight at the start of the experiment. In this case, individuals of similar initial weights may be grouped into a block. A simple way to do this is to order the individuals by weight and then form groups from consecutive observations. In some metabolic studies there may be complex weight dependencies so that the simplest way is to keep weight as constant as possible.

Example 7.3 Randomised blocks to compare four varieties of wheat

Randomised blocks were developed by R. A. Fisher in the 1920s for agricultural field trials. Such trials involve, for example, the comparison of a number of varieties of a crop or different types of manurial treatment. Plots are grouped into blocks as a form of local control to allow for systematic differences in yield due to variation in soil fertility. Treatments are then allocated at random within blocks so that each treatment occurs once in each block. Randomisation is carried out separately and independently for each block so that there is no correspondence between the order of the treatments in different blocks. The design is illustrated below by a trial to compare four varieties of wheat A, B, C, D in five randomised blocks. In this case the blocks run at right-angles to a fertility gradient to reduce the variation between plots within blocks.

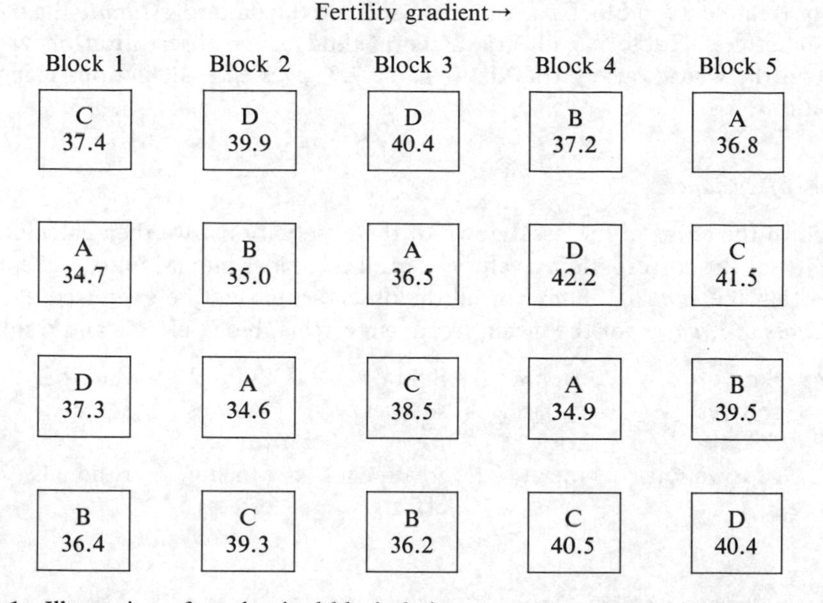

Figure 7.1 Illustration of randomised block design to compare the yields of four varieties of wheat. Numbers are yields per plot in pounds per 1/50th acre from harvesting a central portion of each plot allowing a buffer zone to avoid edge-effects.

7.3.4 Analysis of randomised blocks

The main aim of the analysis of randomised blocks is to estimate the magnitude of the differences between the treatments. The true treatment differences are obscured by random variation due to inherent differences between plots within a block. To compare treatments we therefore need to know the amount of random variation in the observations. This is usually estimated from the data using the technique of *analysis of variance*. In Chapter 6 we showed how this was done for a completely randomised design by using a one-way analysis of variance, in which the residual mean square is used to estimate the variance of the observations. To construct an analysis of variance for randomised blocks we need to allow for the effect of blocking, and we do this by using an *additive model* for the observations.

Additive model

In the *additive model* for a randomised blocks design each observation is represented as a sum of terms given by

Observation treatment i block j = Overall mean + Effect of treatment i + Effect of block j + Residual

or

$$Y_{ij} = m + t_i + b_j + \varepsilon_{ij}$$

where m, t_i (for $i = 1 \dots t$) and b_j (for $j = 1 \dots b$) are fixed quantities and ε_{ij} is a random variable with constant variance σ^2. The ε_{ij} tend to obscure the true value of the mean for treatment i in block j, but we can still use the data to *estimate* the block and treatment effects. Table 7.6 illustrates the method using observation on variety D, block 1 in the wheat variety trial data. Table 7.7 gives the calculations using all the observations.

Analysis of variance

For each of the terms in the breakdown of the observations, we then calculate a *sum of squares* of the corresponding values over all the experimental units. A remarkable result is that the *sum of squares* of all the observations can be expressed as the sum of the *sums of squares* for the mean, treatment effects, block effects and residual, i.e.

Sum of squares of observations = Sum of squares for mean + Sum of squares for treatment effects + Sum of squares for block effects + Sum of squares for residuals

or

$$SS_{\text{total}} = SS_{\text{mean}} + SS_{\text{treatments}} + SS_{\text{blocks}} + SS_{\text{residual}}$$

or

$$\Sigma\, y_{ij}^2 = \Sigma\, \bar{y}_{..}^2 + \Sigma\, (\bar{y}_{i.} - \bar{y}_{..})^2 + \Sigma\, (\bar{y}_{.j} - \bar{y}_{..})^2 + \Sigma\, (y_{ij} - \bar{y}_{i.} - \bar{y}_{.j} + \bar{y}_{..})^2$$

where y_{ij} is the observed value for variety i block j and a 'bar' denotes an average over values indicated by the 'dot' subscript, e.g. $\bar{y}_{i.}$ is the mean for variety i. This illustrates the general form of the *analysis of variance* for randomised blocks. The *sums of squares* measure the separate contributions to the total variation arising from treatment differences, block effects and random variation between the units within a block. Details of the calculations for the wheat variety trial are shown in Table 7.7 to

Table 7.6 Breakdown of a single observation in a randomised block design corresponding to an additive model in variety and block effects

Term	Method of calculation (Variety D, Block 1)									Result
Overall mean	Mean of all observations (37.96)		.		.		.	=		37.96
Variety effect	Mean for variety (40.04)	−	*Overall mean* (37.96)		.		.	=		2.08
Block effect	Mean for block (36.45)	−	*Overall mean* (37.96)		.		.	=		−1.51
Residual	Observation (37.3)	−	*Overall mean* (37.96)	−	*Variety effect* (2.08)	−	*Block effect* (−1.51)	=		−1.23
Observation = sum of the effects			.		.		.	=		37.30

Table 7.7 Illustration of the additive breakdown of the observations in a randomised block design for every observation and the partition of the total sum of squares in the analysis of variance using data from a wheat variety trial (Figure 7.1). Sum of squares (*SS*) on the right equals the sum of squares of the values in the body of the table. Observations which are free to take any value are shaded to illustrate the number of degrees of freedom

	Variety	Block 1	2	3	4	5	Mean	*SS*	*df*
Overall mean	A	37.96	37.96	37.96	37.96	37.96	37.96		
	B	37.96	37.96	37.96	37.96	37.96	37.96		
	C	37.96	37.96	37.96	37.96	37.96	37.96		
	D	37.96	37.96	37.96	37.96	37.96	37.96	28819.232	1
+	Mean	37.96	37.96	37.96	37.96	37.96	37.96	+	+
Variety effects	A	−2.46	−2.46	−2.46	−2.46	−2.46	−2.46		
	B	−1.10	−1.10	−1.10	−1.10	−1.10	−1.10		
	C	1.48	1.48	1.48	1.48	1.48	1.48		
	D	2.08	2.08	2.08	2.08	2.08	2.08	68.892	3
+	Mean	0.00	0.00	0.00	0.00	0.00	0.00	+	+
Block effects	A	−1.51	−0.76	−0.06	0.74	1.59	0.00		
	B	−1.51	−0.76	−0.06	0.74	1.59	0.00		
	C	−1.51	−0.76	−0.06	0.74	1.59	0.00		
	D	−1.51	−0.76	−0.06	0.74	1.59	0.00	23.748	4
+	Mean	−1.51	−0.76	−0.06	0.74	1.59	0.00	+	+
Residuals	A	0.71	−0.14	1.06	−1.34	−0.29	0.00		
	B	1.05	−1.10	−0.60	−0.40	1.05	0.00		
	C	−0.53	0.62	−0.88	0.32	0.47	0.00		
	D	−1.23	0.62	0.42	1.42	−1.23	0.00	14.828	12
	Mean	0.00	0.00	0.00	0.00	0.00	0.00	=	=
Observations	A	34.7	34.6	36.5	34.9	36.8	35.50		
	B	36.4	35.0	36.2	37.2	39.5	36.86		
	C	37.4	39.3	38.5	40.5	41.5	39.44		
	D	37.3	39.9	40.4	42.2	40.4	40.04	28926.7	20
	Mean	36.45	37.20	37.90	38.70	39.55	37.96		

illustrate the principle of the method. In practice, the *analysis of variance* for a randomised block design can be routinely carried out using a wide range of statistical packages. Example 7.4 shows output from the statistical package Genstat 5 which has excellent facilities for the analysis of designed experiments. In some other packages the same table is produced for a *two-way analysis of variance*. The use of the output for comparing varieties (treatments) is described in the next sections. These comparisons involve some assumptions about the data which are discussed in the section below headed 'Assumptions in the Analysis'.

Example 7.4 Output from the analysis of a wheat variety trial randomised block design using Genstat 5

					Authors' annotation
*****Analysis of variance*****					*General case*
Variate: Yield					*t treatments*
					b blocks
Source of variation	d.f.	s.s.	m.s.	v.r.	d.f.
Block stratum	4	23.748	5.937	4.80	$(b-1)$
Block.*Units* stratum					
Variety	3	68.892	22.964	18.58	$(t-1)$
Residual	12	14.828	1.236		$(b-1)(t-1)$
Total	19	107.468			$bt-1$

*****Tables of means*****
Variate: Yield
Grand mean 37.96

Variety	A	B	C	D
	35.50	36.86	39.44	40.04

*** Standard errors of differences of means ***

Table	Variety
rep.	5
s.e.d.	0.703

Interpretation of the analysis

Estimating the amount of random variation

When we speak of the random variation in the data from a randomised block design, we mean the variance of the observations not due to real treatment effects or systematic differences between blocks, in other words the variance of the ε_{ij}. To estimate this residual variance it therefore seems reasonable to use the average of the squared residuals since each residual is obtained by subtracting block and treatment effects from the observations. In fact, this intuitive approach tends to underestimate the variance and to correct for this we divide the *residual sum of squares* by the *residual degrees of freedom*. This gives the *residual mean square* which is an unbiased estimator of the error variance. The *residual degrees of freedom* is given by the number of *independent* residuals and is less than the total number of residuals because in each block, and for each treatment, the residuals must sum to zero. For the above example with 5 blocks and 4 varieties, 12 residuals comprising any 3 from each of any 4 blocks would determine the remaining 8 residuals. In this sense there are effectively only 12 *residual degrees of freedom*. This idea is illustrated in Table 7.7 where the groups of observations which are free to take any values are shaded.

$$MS_{\text{residual}} = \frac{SS_{\text{residual}}}{df_{\text{residual}}} = \frac{14.83}{12} = 1.24.$$

This gives a standard deviation of $\sqrt{1.24} = 1.11$ pounds per 1/50th acre.

Comparison of varieties (treatments)

True differences between treatments are estimated by the *observed* difference between the treatment means. The Genstat output gives the precision of these estimates calculated from

$$\text{standard error of estimated difference between two means} = \sqrt{\frac{2 \times MS_{\text{residual}}}{b}}$$

In the wheat variety trial the estimated difference in mean yield between variety C and variety B is equal to $39.44 - 36.86 = 2.58$ pounds per 1/50 acre with standard error $\sqrt{2 \times 1.24/5} = 0.703$. To calculate a 95% confidence interval for the true difference we multiply the standard error by the t-value with the degrees of freedom of the residual mean square. In this example, there are 12 degrees of freedom for residual so that the appropriate t-value is 2.18. A 95% confidence interval for the difference in mean yields between variety C and variety B is then 2.58 ± 1.53 pounds per acre.

A portmanteau F-test for treatment differences is given in the analysis of variance table as a variance ratio (v.r.) of the treatment mean square to the residual mean square (see Chapter 6). The significance level is obtained from tables of the F-distribution with $(t-1)$ and $(b-1)(t-1)$ degrees of freedom. For the wheat variety trial, the portmanteau F-test of the null hypothesis of no variety differences gives $F_{3,12} = 22.96/1.24 = 18.58$ ($p < 0.001$), i.e. there is strong evidence for variety differences.

Gains in precision by blocking

The purpose of blocking is to increase the precision of the estimated treatment differences by allowing for systematic differences between blocks. One way to assess the gain is to amalgamate block and residual sums of squares to estimate the error mean square in the absence of blocking (i.e. that which we would have obtained using a completely randomised design). In the wheat variety trial, an estimate of the error variance with random allocation of varieties to plots is $(23.75 + 14.83)/16 = 2.41$, compared with the residual mean square of 1.24 for the randomised block design. In this case, blocking has reduced the standard errors of the estimated variety differences by factor of about $\sqrt{1.24/2.41} = 0.72$. When calculating 95% confidence intervals, the standard errors are multiplied by t on the degrees of freedom of the residual mean square. In the randomised block design, $t = 2.18$ ($df = 12$) compared with $t = 2.12$ ($df = 16$) for the completely randomised design. So the slight loss of efficiency due to the reduced degrees of freedom is not sufficient to offset the substantial reduction in standard error achieved by blocking.

Assumptions in the analysis

The analysis of the randomised block design assumes that each observation is the *sum* of contributions from systematic effects due to treatments and blocks, and random uncontrolled variation which follows a Normal distribution with constant variance, i.e. an *additive-Normal model*. In this section we briefly discuss these assumptions and their effects on the analysis.

(1) *Additivity of treatment effects.* This is a key assumption which states that when a particular treatment is applied to a particular experimental unit the effect is to

add a *constant* amount to the response of that unit. A consequence of this is that the true difference between any pair of treatments is the same in every block. We say that in randomised blocks the (treatment × block) *interaction* is assumed to be zero (Section 7.5.3 gives a discussion of the idea of interaction). The additivity of treatment effects will, in general, depend on the chosen scale of measurement. For example, a pesticide to control slugs may affect individual survival and therefore effectively kill a fraction of those present on a plot. In this case, the treatment effect is additive on a logarithmic scale, sometimes called a *multiplicative* effect. Working with the logarithm of the response is one approach. The effect of a block × treatment interaction in the analysis of variance is to inflate the residual mean square so that the amount of random variation tends to be overestimated.

(2) *Constant variance*. The assumption that the variance of the random element is the same for each treatment and each block is the justification for using the residual mean square in the analysis of variance to estimate the variance. For estimating the standard error of a pair of estimated treatment differences, the analysis is robust to differences in variances between blocks because each pair occurs once in each block. This, however, does not apply when some treatments are more variable than others. When the variance of the treatment response is approximately related to the mean we can sometimes work with transformed data, e.g. logarithms (see Chapter 11).

(3) *Normality*. The assumption that the observations follow a Normal distribution is necessary to justify the theory for calculating confidence intervals for treatment differences, and *p*-values in significance tests. However, the design is fairly robust to departures from Normality because the analysis utilises sample means which are approximately Normally distributed whatever the distribution of the observations.

Other blocking arrangements and methods for controlling random error

Some other designs which use the principle of grouping units into similar blocks are as follows: (a) Latin squares, to allow for two sources of systematic variation, in which treatments occur only once in each row and once in each column; (b) incomplete blocks when there are more treatments than there are units in a block; (c) lattice squares, for two-way elimination of error, when there are more treatments than units in either rows or columns. For details, see Cox (1958) and Cochran & Cox (1957). A different approach to reducing random error in randomised field experiments was proposed by Papadakis (1937) in which plot values are adjusted by using the values on neighbouring plots. This method of *neighbour analysis* rests on the assumption that the environmental effect of a plot is related to the effects of its immediate neighbours, i.e. there is spatial autocorrelation (Chapter 13). For a discussion of the method and its application in the analysis of variety trials, see Kempton & Howes (1981).

7.4 DESIGNS FOR STRUCTURED POPULATIONS: TRENDS, PATTERNS IN SPACE AND TIME, AUTOCORRELATION

In Section 7.2 we discussed basic designs using simple random sampling and treatment randomisation for estimating and comparing parameters such as the mean and variance of homogeneous populations. Then, in Section 7.3, we described stratified random sampling and randomised block designs for reducing the error in estimating and comparing means in heterogeneous populations. This is done by grouping the units into relatively homogeneous strata or blocks, and sampling at random within strata, or allocating treatments at random within blocks. These designs are most useful when the major interest lies in the means, variances or some other parameter of the population distribution, in which case the *deliberately* imposed randomness avoids bias and leads to estimates of sampling errors. However, for analysing population structure the artificial source of variation from random sampling adds a further level of complexity which tends to obscure pattern and is an obstacle to fitting models. For example, relating a response at one point to responses at nearby points as a function of distance is very complex for randomly placed points and much simpler for a grid of points. Also, if we wish to describe the spatial distribution of an organism in some defined area, then a grid of sampling points is a sensible scheme since it *guarantees* an even coverage of sampling units which would help, for example, to construct a distribution map. By contrast, a random sample of points would produce an irregular arrangement of points, possibly leaving large areas unsampled with others containing relatively high densities.

In this section we discuss designs for studying the structure of biological populations and processes. This structure may refer to spatial or temporal pattern, or relationships between variables, so that in general there could be observations in space, time, or on more than one variable for individuals from some unstructured population. The designs are relevant when collecting data to fit a wide range of models including those found in: Structured Stochastic Models (Chapter 4); Models in Space and Time (Chapter 12); Pattern and Sequence (Chapter 13); and Complex Relationships (Chapter 14). We shall use examples from these chapters to illustrate the ideas but for more details on models and model fitting the reader is referred to the appropriate chapter. A common feature of the designs is that we avoid the haphazard and irregular configurations which can arise with random sampling by using by a more regular arrangement or *systematic design*.

7.4.1 Systematic sampling

In *systematic sampling* the observations are made according to some predetermined pattern, usually involving a regular spacing of units either in time, space or within some other ordering of the units in the population. The relative positions of the units within the sample are fixed so that a more even coverage of the population or process is achieved. Some examples are as follows.

(1) A systematic sample may involve taking every 10th unit in the population (more generally every kth unit). In cases where the units can be numbered serially to form a sampling frame, the first unit would be selected at random from the first ten units to determine the subsequent pattern, i.e. i, $i+10$, $i+20, \ldots$ etc. In other situations we may simply take the order from events as they occur naturally, for example every 10th feed delivered to chicks in a breeding colony of seabirds, where it is not possible to draw up a sampling frame beforehand. The advantages of this approach are that it is relatively simple to carry out and the units are distributed evenly

throughout the population. Such a scheme provides an alternative method to simple random sampling, or stratified random sampling, for estimating the population mean although in some cases the systematic sample can be very misleading (see below).

(2) In a systematic sample of a population or process in time, the observations are taken at equally spaced times. Sometimes the timing and frequency of observations will be determined by the nature of the biological processes. For example, in a study of the growth and decline of an annual bird species the size of the breeding population may be recorded at intervals of approximately one year. In other cases, little may be known about the underlying processes so that, at least initially, the sampling must be frequent enough to avoid missing any essential detail. Systematic sampling, with observations at regular time intervals, has considerable advantages for analysing patterns and building models for populations and processes. In Chapter 13 we present an example involving the analysis of population cycles in the red grouse from a regular series of annual grouse bags spanning over 100 years.

(3) Systematic samples are often used in studies of the spatial distribution of organisms and their relation to environmental factors. For example, in one-dimension a transect of regularly spaced quadrats may be used to relate the density of an intertidal invertebrate to distance from low water, or to the substrate particle size. In two-dimensions, a regular grid of sampling points is usually the most satisfactory way to map spatial pattern, except when a complete census of the area is available. Similarly, studies of animal movement often employ a systematic arrangement centred on the point of release of a group of individuals.

Systematic sampling for estimating the population mean: advantages and disadvantages

One of the main themes of this section is to stress the advantages of systematic sampling when interest lies in aspects of population structure such as temporal or spatial pattern, or relationships between variables. However, in some situations systematic sampling produces a more precise estimate of the population mean than either simple random sampling or stratified random sampling. The systematic sample is often advantageous in populations for which units close together are more similar than those which are further apart. This is because by spacing out the units in the sample, the configurations containing similar neighbouring units, which are possible with random sampling, are excluded, thus reducing the variation in the sample mean. A deficiency, however, is that the sampling variance of the sample mean from a single systematic sample cannot be estimated unbiasedly without making some assumptions about the nature of the variability in the population. This contrasts with simple random sampling and stratified random sampling where there is an unbiased estimator of the variance of the estimated population mean *whatever the population distribution*. A situation when this problem does not arise is when systematic sampling is used to subsample a sampling unit—for example, the vegetation on a plot in a grazing experiment—and where the error of the sample mean is estimated from the variation between the units, and not from the variation in the systematic sample within each unit.

Periodic populations

Systematic sampling can be very misleading if the spacing of the sample coincides with a periodicity in the population. In this case the sampling variance of the systematic sample mean would be large compared with that of a random sample mean (σ^2/n) whereas the sample variance (s^2) would underestimate the population variance (σ^2). So, s^2/n doubly underestimates the sampling variance of the systematic sample mean.

7.4.2 Sampling spatial point patterns

In studying the spatial distribution of an organism it is sometimes possible to draw up a distribution map showing the locations of all the individuals. Such data can be used to test hypotheses about the underlying processes which may have generated the pattern. In Chapter 13, we present an example involving the spatial distribution of several species of ground-nesting ducks, where interest lies in detecting any regularity of spacing as opposed to a completely random distribution. To answer this question satisfactorily we require a distribution map showing the precise location of each nest and, in particular, we need to be sure that all the nests have been recorded so that any apparent spacing is not due to nests being missed. In other situations a complete mapping may not be feasible and some method of sampling is needed. One approach is to count the number of individuals in small sub-areas called *quadrats*. The positioning of the quadrats may be random or systematic, although for analysis of pattern a systematic arrangement is preferred. Greig-Smith (1952) suggested a systematic division of the area into an array of contiguous quadrats, and a method for analysing the spatial pattern on different scales by combining counts from neighbouring quadrats. A second approach is to use *distance methods* and to measure distances of individuals either from selected points, from other individuals or a mixture of both. Such distance measurements can be used to test for non-randomness in the spatial distribution and, in some cases, to estimate density (Diggle, 1983).

7.4.3 Longitudinal data and cross-sectional data

When populations and processes are observed at different times it is useful to distinguish between two types of data:

Longitudinal data—a series of measurements in time made on the *same* individual, or sector of the population or total population;
Cross-sectional data—a series of measurements in time made on *different* individuals, or samples of individuals, at each time sampled from the same population.

An example of longitudinal data is the series of measurements on the height of Montbeillard's son recorded at about six-monthly intervals from birth to age 18 (Chapter 1). With longitudinal data we can describe the changes in size of an *individual* whereas cross-sectional data can only describe changes in the *average* size of all the individuals in a population. For example, in a cross-sectional study of child growth, Variot and Chaumet (1906) recorded changes in the mean height of male Parisians in separate samples of children drawn from Public Schools Dispensaries Out-Patient Departments (Scammon, 1927).

The above example of longitudinal data involves a series of measurements made on the same *individual*, but the idea can also apply when measurements are made on the same *population*, or more generally on the same biological unit. Both longitudinal data and cross-sectional data occur widely in studies on dynamic populations and processes, or where the emphasis is on estimating a *change* in time. An advantage of longitudinal data is that, by making measurements on the same unit, the random variation between different units which would tend to obscure systematic changes is removed, giving a more precise estimate of change. Sometimes measuring changes in time must involve different units—for example, when sampling is destructive or disruptive, as it would be when recording changes in vegetation by harvesting sample plots.

Finally, an essential point is that with longitudinal data the observations are serially dependent whereas with cross-sectional data they are statistically independent. This basic difference between the two types of data is important for fitting models (Chapter 11).

Repeated measures designs

In a *repeated measures design* a sequence of observations is made on one or more experimental units which have been allocated to one or more treatments. The aim is to investigate the response to a particular treatment with time, to estimate treatment differences and how they vary with time. Observations in time are the most common but the idea can apply to repeated measurements in space—for example, the sequence could be different locations on the unit, say on the lower, middle and upper leaves of a plant. In this section we briefly mention the main aspects of repeated measures designs. For some interesting examples, see Rowell & Walters (1976).

Two basic design questions are the choice of times to make the measurements and the allocation of treatments to units.

(1) *Time profile*. In general, the times at which observations are made and the length of the series will depend on the particular application. Sometimes there may be few observations on each unit spanning a relatively short time, whereas in other situations there may be many observations on each unit forming a long time series. In almost all cases it greatly simplifies the analysis if the observations on different units are taken at the same time and, in particular, that there are no, or at least very few, missing observations in each series. More generally, considerations of design for response curves are relevant (Section 7.4.4).

(2) *Allocation of treatments to experimental units*. Allocation of treatments could be completely random when there is no obvious heterogeneity, or a randomised block design may be preferable if there are systematic differences amongst the units.

Analysis of repeated measures designs

The method of analysing a repeated measures design will depend on the given situation and, in particular, must allow for the way in which treatments are allocated to the experimental units. Often, however, the main problem is how to compare treatments in terms of the complex response of a series of measurements on each unit. Some approaches are mentioned briefly below, but for a more detailed discussion see Crowder & Hand (1990).

First, we note the *apparently* similar structure of the repeated measures design to the so-called *split-plot, or split-unit, design* in which each unit is split and the levels of some factor are allocated at random to the split-units within units. Both designs involve several measurements on the same unit, but the fundamental difference is that the levels of the factor corresponding to the times at which the repeated measurements are made are not allocated at random within units. Some observations are always taken closer together than others, and so will tend to be more similar than those further apart. In the split-unit design, randomisation of the treatment factor levels within units makes sure that, on average, no pair of levels of the split-unit factor is closer than any other. Analysing the repeated measures design as a split-unit design can give misleading results.

Analysing measures of the repeated measures profile
In this approach the profile of the repeated measurements is summarised by using measures which are relevant to the particular problem. These are then calculated for each unit and used to compare treatments. There are many possibilities depending on the particular context. Some examples are the overall mean when there is no systematic trend in the profile, averages calculated over particular periods, e.g. early and late, or the area under the profile when a cumulative effect is of interest. In some cases it might be possible to describe the profile by fitting a model and using the estimated parameters to compare treatments—for example, the slope and intercept of a straight line or the initial relative growth rate (r) and upper limit (K) of the logistic growth curve. In these situations, the models in Chapters 1 and 4, and the methods of fitting models in Chapters 5 and 11 could apply.

Analysis of data at each time separately
In this approach treatments are compared using the observations at each time separately, allowing for any blocking structure. If the various assumptions of the analysis are approximately satisfied, then the method provides valid treatment comparisons at each time. The method is not recommended in general since: (a) it is likely to be relatively insensitive because individual observations are more variable than averages over periods of time or other summary measures; (b) it does not examine how treatment effects *change* with time, i.e. the treatment × time interaction; (c) with a relatively long time series there is the *multiple comparison problem*, with statistically significant results occurring by chance (Chapter 6); (d) the separate analyses are not *independent* of each other because they use the same experimental units—differences detected at later times may simply be reflecting earlier differences.

7.4.4 Designs for response curves and relationships

In Chapter 4, we presented models for describing the relationship between a response variable and an explanatory variable in which there is a random element in the measured response. Two important cases are the straight line-Normal model for continuous measurements and the logistic-binomial model for proportions derived from binary responses. In this section we deal with ways of collecting data to study this type of relationship, sometimes referred to as a *response curve*. For these problems, two basic questions of design are: what values of the explanatory variable to use, and how many replicates of each? Such questions need to be considered in the context of a particular case but it is useful to distinguish between designs for reducing the error in estimating the parameters in a known model and designs for exploring the form of the underlying relationship. We shall see that a design which is efficient for estimation may be very poor for exploration. The ideas presented here extend to relationships and models with two or more explanatory variables. We deal with these models in Chapter 14 and describe an example in which heat loss in the pig is related to environmental temperature and metabolisable energy intake.

What range for the explanatory variable?

Often a suitable range for the explanatory variable will be determined by the particular application. For example, in studying the relationship between a response and the dose of a drug, a realistic range of values could be known either from other studies or by carrying out a pilot investigation. If the response curve is to be used for prediction then the range of the explanatory variable should cover the range over which predictions are to be made. A similar point could be made about comparing two response curves where it makes sense to choose comparable ranges for the explanatory variables so that there is a large overlap: with little or no overlap the comparison rests on the assumed form of the underlying relationship.

What values for the explanatory variable and how many replicates?

A basic design which often works well in practice is to use equally spaced values of the explanatory variable, with equal numbers of observations at each value. In bioassay, the doses are often equally spaced on a log scale (Chapter 4). Equal spacing can be wasteful when there are parts of the response curve which change very little: ideally, the x-values should be closer together where the curve is changing most rapidly, but to implement this would require some prior knowledge of the curve. In general, the choice of x-values within the range and the number of replicates per value will depend on what is known, or assumed, about the curve. In some cases simulation studies could be set up to explore the relative merits of different schemes. Here we present a simple example using the straight line-Normal model to illustrate the trade-off between designs for efficient estimation of parameters and designs which are more suitable for model checking. Then we mention briefly some aspects of designs for non-linear models.

Straight line example

Suppose that we wish to fit a straight line over the range $(1, 5)$ and that we can manage 30 observations in all. What values of x and how many replicates of each value should we use? Different designs can be compared using the standard error of the estimated regression coefficient as a criterion, i.e. $\sigma/\sqrt{S_{xx}}$, where S_{xx} is the sum of squares of the deviations of the x-values about the mean value (Chapter 5). Table 7.8 gives calculated standard errors for: Design 1 with 15 observations at $x = 1$ and 5; Design 2 with 10 observations at each of $x = 1$, 3 and 5; Design 3 with 6 observations at each of $x = 1, 2, 3, 4$ and 5 (the amount of replication is rather unrealistic and is used

Table 7.8 Standard errors of estimated slope in the straight line-Normal model for three designs each with 30 observations. A residual variance of 120 is used to make the standard error for Design 1 equal to one

x_i	Number of observations		
	Design 1	Design 2	Design 3
1	15	10	6
2	0	0	6
3	0	10	6
4	0	0	6
5	15	10	6
S_{xx}	120	80	60
s.e. $\hat{\beta}$	1	1.225	1.414

simply to obtain integer values). It may come as a surprise to see that Design 1 with half the observations at each end gives the smallest standard error. This is followed by Design 2 and then Design 3. However, although Design 1 has high precision, it is useless for checking the assumption of a straight line model since all the points are at the ends of the range. For this purpose, either Design 2 or 3 is preferable. In practice, we would not usually opt for Design 1 unless we were interested solely in estimating the change over the full range. Instead, we would compromise by including observations at intermediate values of x to provide some check on the form of the model.

Non-linear models

The theory of designs for fitting non-linear models is more complicated than that for linear models and is beyond the scope of this book. Nevertheless, similar design objectives apply, i.e. (a) efficient estimation of parameters; (b) providing a check on the form of the model. In considering different designs, the calculation of *parameter loadings*, which measure the effect of changes in each observation on the estimated parameters, can be helpful in deciding where to place further observations in order to fix a parameter (Ross, 1990). In Chapter 11 we discuss some of the potential difficulties in fitting non-linear models which are relevant to the choice of design.

7.5 FACTORIAL STRUCTURES

7.5.1 Introduction

So far we have dealt with designs to assess the effect of a single factor on some measured response, say yields of varieties of wheat or plant growth in relation to feeding supplements. Often, however, we wish to examine the effect of several factors on some response. For example, an agricultural field trial to compare yields of different varieties of wheat grown with different dressings of a fertiliser; an epidemiological investigation to relate the incidence of some disease in different groups of people classified by age, sex and smoking habits. Designs involving two, or more, factors are said to have a *factorial structure*. *Factorial designs* occur widely in both observational and experimental studies and in this section we deal with their design and analysis.

In a *factorial structure* each *factor* occurs at a number of *levels*—for example, 2 sexes, 4 varieties of wheat, 3 diets—and responses are classified as combinations of one level from each factor. A *complete factorial structure* involves responses from all the possible factor combinations—4 varieties of wheat $\times$ 3 fertiliser dressings gives 12 possible factor combinations. Complete arrangements are more common in controlled experiments than in observational studies, in which some factor combinations may be absent or poorly represented in the population, and therefore difficult to include in the sample. In a factorial design, the aim is to examine how some response varies in relation to the various factor combinations. For example, are there differences in yield between different varieties of wheat and how do these differences vary with the amount of fertiliser dressing? If the incidence of a disease is related to smoking habit, is the relationship different for males and females? For these questions, an important general concept is that of *interaction* between two or more factors. Broadly speaking, *interaction* between two factors means that the difference between the responses at different levels of one factor depends on the particular level of the other. A more detailed discussion of *interaction* is given in Section 7.5.3. First, we present two examples of designs with *factorial structure*.

Example 7.5(a) The effect of temperature and food intake on the number of thyroid hormone receptors in pigs

This experiment was carried out to examine the effect of food intake and temperature on the metabolism of thyroid hormones in pigs (Dauncey *et al.*, 1988). The measured response was the number of thyroid hormone receptors in metabolically active tissue. Pigs were kept at either 10 or 35 °C and fed either a high or low food intake. The experimental treatments form a 2×2 factorial structure with the factor temperature at two levels and food intake also at two levels. The four possible factor combinations (treatments) were allocated at random within groups of litter-mate pigs in a randomised block design with 4 replicates per treatment (Table 7.9).

Example 7.5(b) Dieldrin levels in eggshells of sparrowhawks

The following example is based on an observational study to compare the dieldrin levels in the eggshells of two populations of sparrowhawks over two years. The design has a 2×2 factorial structure with two areas (moorland and arable land) and two years. Table 7.10 shows some simulated data for eggshell fragments in eight nests selected at random from each population in each year.

7.5.2 Types of factors

Before further discussing the main ideas of factorial structures we give a broad classification of the different types of factor to illustrate the wide range of situations

Table 7.9 Number of thyroid hormone receptors in metabolically active tissue of pigs in relation to temperature and food intake

Litter	10 °C		35 °C		
	Low	High	Low	High	Mean
1	43	120	126	157	111.5
2	121	236	246	470	268.25
3	43	46	149	210	112
4	18	32	41	81	43
Mean	56.25	108.5	140.5	229.5	133.6875

Table 7.10 Dieldrin levels as $\log_e$(p.p.m.) in samples of eggshells from two sparrowhawk populations in two years

Year	Area	Dieldrin levels	Mean
1	1	3.55 4.03 3.88 3.99 3.80 3.72 3.47 3.79	3.78
1	2	4.24 4.24 4.18 3.88 3.85 3.88 3.80 4.14	4.03
2	1	3.74 3.99 3.96 3.98 3.71 3.61 3.67 3.68	3.79
2	2	3.80 3.61 3.46 3.47 4.03 4.05 3.99 3.77	3.77

in which factorial structures apply. It is useful to distinguish between *quantitative* and *qualitative* factors.

Quantitative Factors

Quantitative factors are those for which the different levels correspond to some numerical value of a variable, sometimes referred to as a *carrier variable*. In the above pig experiment, temperature is a quantitative factor with corresponding carrier variable the ambient temperature (°C). The levels of the second factor, food intake, are given simply as High and Low but they do in fact correspond to specified amounts of food eaten. In situations with quantitative factors we are often interested in analysing the relationship between the response variable and the carrier variable, rather than with the response at particular levels of each factor. This is the general problem of *response surface analysis* (Chapter 14).

Qualitative Factors

Qualitative factors reflect specific characteristics such as variety, region, occupation or sex with no particular numerical values or order amongst the levels. Both qualitative and quantitative factors can occur in the same experiment or survey. For example, varieties of wheat may be grown with different amounts of some fertiliser dressing. With this mixture of qualitative and quantitative factors we would like to know if the relationship between the yield and the amount of fertiliser differs between varieties. *Ordered qualitative factors* arise when the levels of a qualitative factor can be ordered without specifying a numerical value—for example, it might be possible to order the roughness of treated surfaces without measuring specific values.

Fixed and random effects

In many situations each level of the qualitative factor is of interest in itself—for example, the yields of the *particular* varieties of wheat in a field trial. In these cases, we refer to *fixed effects*. In other situations, however, the particular levels of a qualitative factor are of little interest in themselves and are regarded as a sample from some larger population, i.e. as levels of a *sampled qualitative factor* or *random effects*. For example, if different varieties of wheat are grown in a trial spanning several years at different sites we may consider years and sites as samples from some population, i.e. we have *random effects* of years and sites and *fixed effects* of varieties. The *interaction* of two factors can also be considered as a random effect when the *difference* between two levels of a factor varies randomly. This would be relevant to the variety trial example where differences between varieties, and how these differences vary between years, are of interest.

Crossed and nested factors

An important distinction in both design and analysis is between crossed and nested factors. Two factors are said to be *crossed* when each level of one factor occurs across the levels of the other factor, i.e. the various possible combinations of factor levels can be arranged in a two-way table with rows and columns corresponding to the levels of each factor. If every level of one factor occurs with each level of the other, so that the two-way table is complete, then the factors are *completely crossed*, otherwise they are *partly crossed*. A different situation is when the levels of a factor A occur only *within* the levels of the other factor B in which case we say that the levels of A are *nested* within B, or A is nested within B. To illustrate the distinction between crossed and nested factors consider Example 7.5(b) on dieldrin in sparrowhawk eggs. In this design the factors area and years are crossed. Now consider just one area, area 1. Within each year there are observations on eggshell fragments taken from 8 nests in breeding territories selected at random and independently in each year. In this case, if we think of territory as a third factor with 8 levels, territories are nested within years. Now consider a different design in which the *same* 8 territories in area 1 were used in both years. In this case, the factors year and territory are crossed because each territory occurs in each year. Table 7.11 illustrates the difference.

In a particular design the factors may be crossed, nested or a combination of the two, and there are many possibilities. In this section, we deal with the simplest designs in which the factors are crossed, and where, for each combination of factor levels, the responses are measured on units which have been selected *independently* from some population. In Section 7.6, we discuss nested designs in which the factor levels are regarded as random effects and where observations contain several components of random variation.

Table 7.11 Illustration of crossed and nested factors

		Territory number							
Factors crossed	Year 1	1	2	3	4	5	6	7	8
	Year 2	1	2	3	4	5	6	7	8
Factors nested	Year 1	1	2	3	4	5	6	7	8
	Year 2	9	10	11	12	13	14	15	16

7.5.3 Two-factor designs

We begin with the simplest case of a factorial design with two factors each at two levels but the ideas apply also to designs with more than two levels and with more than two factors.

Main effects and interactions

In a factorial structure an important question is whether the pattern of responses for the different factor combinations can be summarised in a fairly concise and meaningful way. For quantitative factors, one approach is to describe the relationship between the response variable and the carrier variable by a mathematical function. This method, which is sometimes referred to as response surface analysis, is discussed in Chapter 14. For qualitative factors where there are no numerical values for the factor levels, we use a different approach based on the ideas of *main effects* and *interaction*. In broad terms, *main effects* measure differences between the levels of a factor when averaged over the levels of the other factor(s). *Interaction* measures how the differences between the levels of a particular factor vary with the levels of the other factor.

We shall describe the meaning of main effects and interaction in a 2×2 factorial design. For illustration we use fictitious data and assume that there is no experimental error so that the true mean responses are known exactly. In practice, the values of the true means will be obscured by random variation but we will return to this aspect later. Table 7.12 shows two sets of mean responses for the four combinations of factors. In

Table 7.12 Illustration of interaction in a 2×2 factorial structure using fictitious data for the *true* treatment mean responses

Absence of interaction

Difference between two levels of one factor is the *same* for both levels of the other factor

Factor *A*	Factor *B* Level 1	Factor *B* Level 2	Mean
Level 1	50	150	100
Level 2	100	200	150
Mean	75	175	125

Presence of interaction

Difference between two levels of one factor depends on the level of the other factor

Factor *A*	Factor *B* Level 1	Factor *B* Level 2	Mean
Level 1	50	150	100
Level 2	100	280	190
Mean	75	215	145

the table of means on the left-hand side, the difference between the two levels of A is the same for both levels of B. Similarly, the difference between the two levels of B is the same for both levels of A. When differences between the levels of one of the factors are the *same* for all levels of the other factor, we say that there is no *interaction*, or that the factors *do not interact*. For the data in the right-hand table, the difference between the two levels of A is 50 (B at level 1) and 130 (B at level 2). Similarly, the difference between the two levels of B is different for the two levels of A. When the difference between the two levels of one factor depends on the level of the other factor we say that there is *interaction* between the two factors, or that the two factors *interact*. Patterns of interaction can be represented graphically by choosing the levels of one factor for the x-axis and plotting a separate graph for each level of the other factor. For the above example, Figure 7.2 shows the mean response plotted against levels of A for each level of B. When there is no interaction the lines are parallel, and non-parallel lines indicate an interaction. In this case, the difference between the two levels of A is larger in the higher level of B, and the lines diverge. In other cases the lines may cross indicating a change in the order of the levels of one factor with the levels of the other. The graphical display can be particularly helpful for looking for patterns of interaction when there are more than two levels of each factor (Figure 7.3).

Analysis of a two-factor design

To illustrate the analysis of the two-factor design, we use the data from the study on the effect of food intake and temperature on thyroid hormone metabolism in pigs. The design is a 2 × 2 factorial structure with treatments arranged in randomised blocks of litter-mates. We could simply analyse the data as a randomised block design with four treatments using an overall test for treatment differences, and significance tests and confidence intervals for comparing pairs of treatment means (Section 7.3.4). This approach, however, ignores the factorial structure of the treatments, and does not directly address the questions which the factorial structure is designed to answer. For

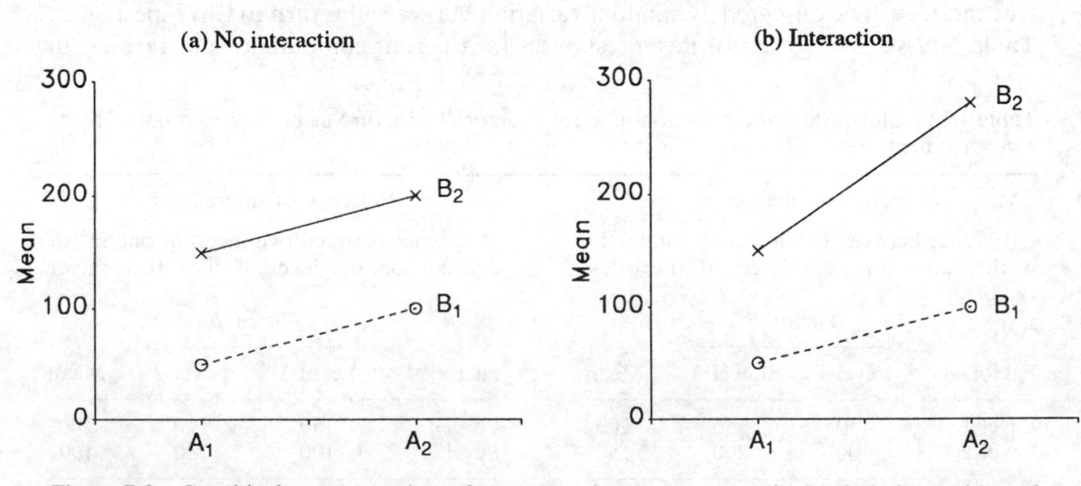

Figure 7.2 Graphical representation of patterns of mean response in 2 × 2 designs with and without interaction between the two factors (Table 7.12).

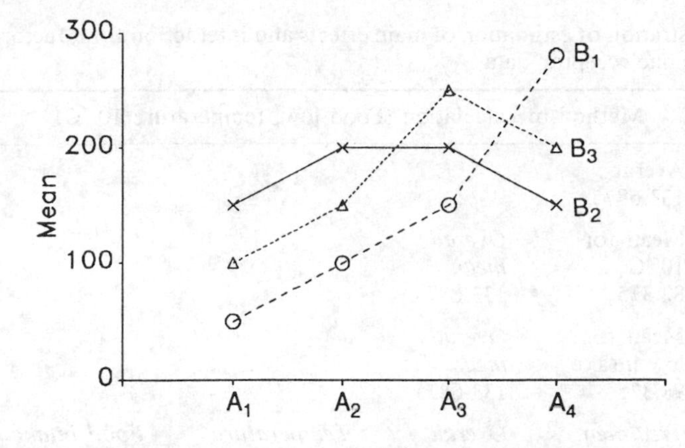

Figure 7.3 Illustration of a pattern of interaction in a 4 × 3 factorial structure.

example, does the effect of temperature vary with food intake, i.e. is there a (temperature × food intake) interaction? If so, what is the magnitude of the interaction? To answer such questions, we define more clearly the *main effects* and *interaction*. Then we show how the variation among the treatment means (i.e. the treatment sum of squares) can be partitioned into a sum of contributions due to main effects and interaction.

Model definition of main effects and interaction

In a two-factor structure the main effects measure differences between the levels of one factor averaged over levels of the other factor. Interaction measures how the differences between levels of one factor vary with levels of the other. We now define these effects using the pig example for illustration. First, express the *true* treatment mean as a sum of terms

Treatment mean Temperature (level i) Food (level j)	= Overall mean	+ *Main effect* Temperature (level i)	+ *Main effect* Food (level j)	+ *Interaction* Temperature (i) × Food (j)

or

$$m_{ij} = m + t_i + f_j + (tf)_{ij}$$

The *overall mean* is the average of all the treatment means. The *main effect* of temperature at a particular level is the average of the means for that level minus the overall mean. The *main effect* of food intake at a particular level is calculated in a similar way. The *interaction* term is then obtained as the treatment mean minus the sum of the overall mean and the main effects. These quantities are unknown model parameters which could be calculated *if* the treatment means were known.

Estimation of main effects and interaction

In practice, the true treatment means are obscured by the random variation so we *estimate* them using the *observed* means, which we then use to estimate the main effects and interaction. Table 7.13 shows the calculations for low food intake at 10 °C. Table 7.14 gives the complete calculations.

Table 7.13 Illustration of estimation of main effects and interaction for a factorial design using the thyroid hormone receptor data

Effect	Method of calculation (Food low, temperature 10 °C)								Result
Overall mean	Average 133.6875		.		.			=	133.6875
+ *Main effect Temperature*	Mean for 10 °C 82.375	−	*Overall mean* 133.6875		.		.	=	+ −51.3125
+ *Main effect Food intake*	Mean for low intake 98.375	−	*Overall mean* 133.6875		.			=	+ −35.3125
+ *Interaction effect*	*Treatment mean* 56.25	−	*Overall mean* 133.6875	−	*Temperature effect* −51.3125	−	*Food intake effect* −35.3125	=	+ 9.187
Treatment mean (Food low, temperature 10 °C)								=	56.25

Table 7.14 Mean effects and interaction, and the corresponding partition of the treatment sum of squares, in a two-factor design using data from a factorial experiment to study the effect of temperature and food intake on the number of thyroid receptors in pigs

	Food intake	Temperature (°C) 10	35	Mean	Sum of squares
Treatment means (4 replicates)	Low	56.25	140.50	98.375	
	High	108.50	229.50	169.000	349 387
	Mean	82.375	185.000	133.6875	−
Overall mean	Low	133.6875	133.6875	133.6875	
	High	133.6875	133.6875	133.6875	285 956
	Mean	133.6875	133.6875	133.6875	=
Temperature *Main effects*	Low	−51.3125	51.3125	0.000	
	High	−51.3125	51.3125	0.000	42 128
	Mean	−51.3125	51.3125	0.000	+
Food intake *Main effects*	Low	−35.3125	−35.3125	−35.3125	
	High	35.3125	35.3125	35.3125	19 952
	Mean	0.000	0.000	0.000	+
Temp. × Food *Interaction*	Low	9.1875	−9.1875	0.000	
	High	−9.1875	9.1875	0.000	1351
	Mean	0.000	0.000	0.000	

Analysis of variance: partitioning the treatment sum of squares into sums of squares for main effects and interaction

For the two-factor design, we calculate a treatment sum of squares for the different factor combinations from the sum of the squares of the deviations of each treatment mean from the overall mean. For the complete design, i.e. with all combinations of the factor levels, a remarkable result is that the treatment sum of squares can be expressed as a sum of the squares of the terms for main effects and interaction, i.e.

Treatment	=	Temperature	+	Food intake	+	Temp. × food
		Main effect		*Main effect*		*Interaction*
Sum of squares		*Sum of squares*		*Sum of squares*		*Sum of squares*

or

$$SS_{\text{treatments}} = SS_{\text{temp}} + SS_{\text{food}} + SS_{\text{temp} \times \text{food}}$$

The sum of squares for each effect is obtained by summing the squares of the corresponding terms in the breakdown of the treatment mean. For example, the temperature main effect sum of squares is given by $4((-51.3125)^2 + \cdots + (51.3125)^2) =$ 42 128 (Table 7.14). To make the sum balance, each term is counted once for each replicate, i.e. in this case four times. This breakdown of the treatment sum of squares into components due to main effects and interaction is analogous to the breakdown of the total sum of squares in the analysis of variance for the randomised block design (Section 7.3.4). For randomised blocks, there are sums of squares for treatments, blocks and residual so that the *interpretation* of the components is rather different. In numerical terms, however, the procedure for partitioning the sum of squares is the same. In fact, for *any* complete two-way table of numbers it is possible to partition the total sum of squares into variation between row means, column mean and (row × column) interaction effects.

The analysis of a complete two-factor design can be routinely carried out using a wide range of statistical packages. The output in Example 7.6 was produced using Genstat 5.

Example 7.6 Output from the analysis of a 2 × 2 factorial experiment in a randomised block design using Genstat 5

					Authors' Annotation
*****Analysis of variance*****					*General case*
					Treatment = a^*b
Variate: Recept					*r blocks*
Source of variation	d.f.	s.s.	m.s.	v.r.	d.f.
Litter stratum	3	109176	36392	10.98	$(r-1)$
Litter.*Units* stratum					
Temp	1	42128	42128	12.71	$(a-1)$
Food	1	19952	19952	6.02	$(b-1)$
Temp.Food	1	1351	1351	0.41	$(a-1)(b-1)$
Residual	9	29820	3313		
Total	15	202425			$rab-1$

```
*MESSAGE: the following units have large residuals
Litter 2   *units* 4   106.   s.e. 43.

*****Tables of means*****
Variate: Recept

Temp      10.00     35.00
           82       185
Food       Low      High
           98       169

Temp      Food      Low       High
10.00                56        109
35.00               141        230

***Standard errors of differences of means***
Table     Temp      Food      Temp
                              Food
rep.        8         8         4
s.e.d.    28.8      28.8      40.7
```

Interpretation of the analysis

In the analysis of variance table the partition of the treatment sums of squares provides portmanteau F-tests for the main effects of the two factors, and their interaction, as the ratio of the corresponding mean square to the residual mean square (v.r.). On the null hypothesis of no effect, the ratio follows an F-distribution with degrees of freedom for the effect and the residual. The F-ratio for the main effect of food is 6.02 which is compared with the F-distribution with 1 and 9 degrees of freedom with an upper 5% point of 5.12. We conclude that the main effect of food is statistically significant ($p < 0.05$). Similarly, the main effect of temperature is statistically significant ($F_{1,9} = 12.71, p < 0.01$). The F-ratio for the interaction effect is rather small, 0.41, and not statistically significant. So there is no evidence that the difference between the two levels of food is different for the two levels of temperature. A reasonable summary of the effect of food (High – Low) is the difference between the two overall means, i.e. $169 - 98 = 71$ (s.e. = 28.8, obtained as $\sqrt{3313 \times 2/8}$ since we are measuring the difference between two means each based on 8 observations with estimated variance 3313).

7.5.4 Interpretation of interactions

The interpretation of the analysis in Example 7.6 is relatively straightforward because the interaction effect is small and not statistically significant. Often this is not so and there is a statistically significant interaction, i.e. differences between the levels of one factor depend on the level of the other factor. When this happens the problem is to interpret the pattern of the interaction. A useful first step is to examine the two-way table, or a graph, showing the means of the different factor combinations. Adding standard errors helps to give some idea of differences which could be due to random variation, and the means can be compared by using a t-test—bearing in mind the problem of multiple comparisons if there are many combinations. In some cases, it can happen that the interaction arises from a particular subset of the levels of the two factors with no interaction among the remainder. In others, however, there may be no simple pattern. A rather special case is an interaction which can be removed by changing the scale of measurement, i.e. transforming the observations. For example, if the effect of a treatment is to multiply the response by a constant amount, i.e. a *multiplicative* treatment effect, then working with logarithms will remove the interaction effect.

7.5.5 Balance, orthogonality and non-orthogonality

The factorial designs considered here are *balanced* in the sense that each combination of factor levels occurs the same number of times. This balance has considerable advantages for analysis and interpretation of the results.

(1) The analysis is simple because the comparisons of the different factor combinations are based on averages over the appropriate levels. Also, the main effects of the two factors and their interaction are estimated *independently* of each other and the treatment sum of squares can be partitioned into separate contributions for the main effects and the interaction.
(2) For a balanced design the analysis is more robust to departures from the assumptions of the additive-Normal model.

In a factorial design, balance may be lost in two rather different ways.

(1) All factor combinations may be present but not with equal replication.
(2) Some factor combinations may be missing.

Case (2) is much more serious than case (1) because whole factors are not present so that some effects *cannot* be estimated. In both situations, the analysis is more complicated than the balanced case because the effects cannot be estimated independently using simple averages over the different factor combinations. We refer to this as *non-orthogonality*, which means that the estimates for one factor are affected by those of the other, i.e. they are not estimated independently. Chapter 14 discusses non-orthogonality in more detail and shows how it can affect the interpretation of the analysis. Non-orthogonality can be a major problem for survey data where it may be difficult, or impossible, to achieve a balanced design. Also, in a designed experiment some observations may be missing, for reasons unconnected with the applied treatment, leading to non-orthogonality. The statistical packages GENSTAT and SAS have facilities for the analysis of non-orthogonal designs so that the problem of analysis is not a technical one. However, it can mean that much more time has to be spent on analysis and, more particularly, on interpretation of the results.

7.6 DESIGNS FOR ESTIMATING COMPONENTS OF VARIANCE

In the designs for homogeneous populations (Section 7.2), the main interest was estimating and comparing parameters, such as the mean and variance, using simple random sampling. For heterogeneous populations (Section 7.3) we used stratification or blocking to increase precision for estimating and comparing means. Both cases involve only a single component of random variation, either in the population as a whole or within strata, or blocks of experimental units. Furthermore, in estimating and comparing means the random variation was regarded as a nuisance since it obscures the underlying values to be estimated. In other situations, there may be several components of random variation which are quantities of interest in their own right and which need to be estimated from the data. This type of problem arises in genetical studies where animal breeding experiments are used to analyse the variability in certain strains. A basic design involves a sample of sires, each of which is mated with a number of dams, with measurements, such as back-fat thickness or milk yield, recorded on the offspring. Figure 7.4 illustrates this for observations stemming from a single sire—in the actual experiment there would be several sires. When different dams are used for each sire the design has a *nested*, or *hierarchical structure*, of dams *within* sires, and offspring *within* dams *within* sires. For this structure, there are different components of variation: (a) between sires; (b) between dams within sires; (c) between the offspring for each sire/dam combination. The animal breeder is interested in estimating these components of variation and their relative contributions to the total variation.

In this design we can think of sires and dams as factors with levels of dams nested within levels of sires. We usually regard the animals as a sample from a wider population so that the levels are random effects (Section 7.5.2). There are more complex designs, used in genetics and other fields, with components of variation for both nested and crossed factors but these are beyond the scope of this book. Here we shall deal with the simplest case of a nested, or hierarchical, design with only *two* levels and *two* components of random variation.

7.6.1 Nested or hierarchical designs

Figure 7.4 illustrates a nested, or hierarchical, design for an animal breeding experiment. The structure is different from the designs with observations on units selected *independently* from some population. In the nested design, observations made on *different* sires are *independent* of each other, but those on dams from the *same* sire, or from the *same* dam, are not independent. The models for designs with a single random error term need to be augmented to reflect the nested structure of the random

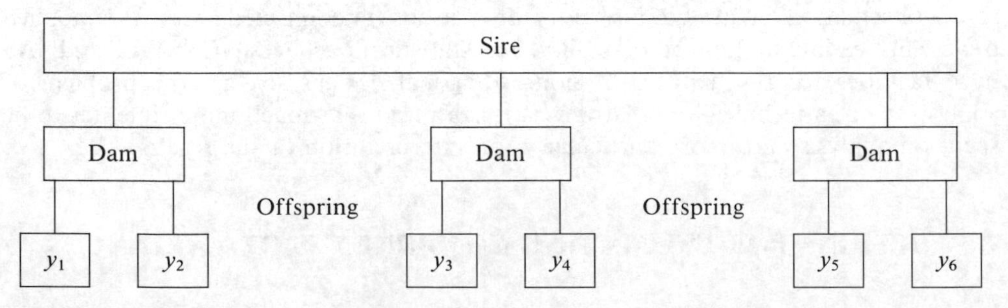

Figure 7.4 Illustration of a basic design for an animal breeding experiment with a nested, or hierarchical, structure. Observations (y_i) on each offspring, which could be milk yield or back-fat thickness, are shown only for a single sire.

variation. However, before dealing with models, and the analysis of nested designs, we present some more examples to illustrate the different ways a nested design can arise in practice.

Litters as experimental units or sampling units

In some experiments the responses are recorded on whole litters, or groups of experimental subjects—for example, treatments applied to female mice to study the effect on their offspring. Table 7.15 shows data on placenta weights in a control group of animals. Placenta weight varies between offspring in the same litter and also between litters. For comparison with other groups we need to consider both components of random variation. This is the simplest nested design with two levels, or strata, and two components of random variation. Section 7.6.2 describes the *one-way random effects model* which is used to analyse this design and which forms the basis of the analysis of designs with three or more levels.

Subsampling: primary and secondary units

In some situations it is too laborious to process all the material in a sampling unit—for example, a core may contain several hundred chironomid larvae. One way is to reduce the size of the sampling unit, but when this is not feasible an alternative approach is *subsampling*, sometimes called *two-stage sampling*. First, a sample of *primary units* is selected from the population and then a sample of *secondary units* is taken from each primary unit. Often the material within each primary unit is well-mixed, or homogenised, before subsampling. When the data are counts, this may lead to a Poisson distribution in subsamples. More generally, mixing increases precision by reducing the variation between replicate subsamples from each primary unit. Mixing, however, cannot reduce the variation between the primary units which is determined by the population distribution and the method of selection.

Split-plot experiments

A practical problem which can arise in a factorial experiment is that the size of experimental units which is suitable for one factor may not be convenient for another. For example, a spraying treatment may need to be applied over a relatively large area compared with the size of a plot required to grow a single variety. One approach is a

Table 7.15 Placenta weights (g) of three offspring from 10 litters of mice

Litter	Offspring weights	Litter	Offspring weights
1	1.01 0.95 1.08	6	0.71 0.90 0.49
2	0.79 0.57 0.66	7	0.96 0.71 1.03
3	0.72 0.76 0.62	8	1.03 0.96 0.98
4	0.82 1.05 0.85	9	0.80 0.80 0.79
5	0.91 0.80 0.62	10	0.69 0.90 0.71

split-plot design in which *whole plots* are subdivided into *split plots* to take levels of the factor which can be applied on a smaller scale. The whole plots could form a completely randomised design or they may be arranged in randomised blocks. The design has a nested structure with split plots nested within whole plots. For discussion of this design, see Cox (1958).

7.6.2 Analysis of nested designs

In a nested design, the components of random variation can either be a nuisance, when comparing means, or they can be quantities of interest in their own right. In either case we need to estimate the magnitude of the components of variation. The main method for doing this is to construct an analysis of variance which apportions the total variation in the observations into sums of squares for variation between and within the levels in the design. For illustration, we use two components of variation but the approach applies to nested designs with more than two levels.

One-way random effects model

The one-way random effects model is the basic model for estimating the components of variation in nested designs. It applies when there are two components of random variation in the observations, which we refer to as between and within group variation. For example, in the placenta weights of mice (Table 7.15) there is random variation in weights of offspring within and between litters. In the one-way random effects model each observation is expressed as a sum of terms given by

Observation	=	Population	+	Random effect	+	Random effect
group i		mean		group i		group i
individual j						individual j

or

$$Y_{ij} \quad = \quad \mu \quad + \quad B_i \quad + \quad \varepsilon_{ij}$$

where the random effects for groups and individuals are *independent* of each other and follow Normal distributions with zero mean and variances σ_b^2 and σ^2 respectively.

Estimating the components of variance

To estimate the variances in the one-way random effects model, we use the one-way analysis of variance (Chapter 6). We consider the case of equal numbers in each group but the method can be modified for unequal group sample sizes.

Table 7.16 gives the calculations for the mice placenta weight data with 3 offspring per litter, and also shows the quantities being estimated in the general case. The within group (residual or error) mean square estimates the within group variance σ^2. The between group mean square estimates the variance of a group mean, but multiplied by the number in each group, i.e. $\sigma^2 + 3\sigma_b^2$. This is because each squared deviation is counted three times in the analysis of variance so that the sums of squares add up to the total. This gives: $\widehat{\sigma^2} = 0.0144$; $\widehat{\sigma^2 + 3\sigma_b^2} = 0.0452$, and therefore $\widehat{\sigma_b^2} = (0.0452 - 0.0144)/3 = 0.0103$. A quantity which is sometimes calculated is the

Table 7.16 Illustration of estimating the components of variance in the one-way random effects model using mouse placenta weight data (Table 7.15) and for the general case of k litters with m mice per litter

Mouse placenta weight data				General case	
Source of variation	Sum of squares	*df*	Mean square	*df*	Quantities estimated by mean square
Between litters	0.4071	9	0.0452	$k-1$	$\sigma^2 + m\sigma_b^2$
Within litters	0.2878	20	0.0144	$k(m-1)$	σ^2
Total	0.6949	29		$km-1$	

repeatability—defined as the ratio of the between group component of variance to the total variance—which measures the relative similarity of the observations in the same group. For these data, the estimated repeatability is $0.0103/(0.0103 + 0.0144) = 0.42$. In genetic applications, repeatability gives an upper limit on the heritability of a trait.

The method can produce a negative estimate of the between group component whereas the actual component could not be negative. Negative estimates can occur by chance and are more likely when the variance component is small. One approach is to set the component to zero but this estimate is biased.

Variance of the sample mean

In some situations the main interest lies in estimating the population mean rather than the variance components. For these cases the random variation is a nuisance but the components of variance need to be estimated to assess the precision of the estimated mean. To do this we use the result that the variance of the mean of the observations in a single group of m individuals is equal to $\sigma_b^2 + \sigma^2/m$. This is because the random component of the mean of the observations in a particular group is the sum of the group random effect, which has variance σ_b^2, and the mean of m independent individual random effects, each with variance σ^2. The variance of the sample mean based on k independently selected groups, each with m individuals, is then obtained by dividing the variance of a single group mean by k, i.e.

$$\text{var}\,[\text{sample mean}] = \frac{\sigma_b^2}{k} + \frac{\sigma^2}{mk}.$$

The first term in the expression is due to variation between groups and decreases with the number of groups k. The second term is due to variation within groups and decreases with the total number of observations $n = km$. The variance can thus be reduced by increasing either the number of groups or the number of individuals in each group. Increasing the group size, however, only reduces the contribution from within group variation so that there is a lower limit determined by the number of groups. We stress the difference between the variance of the sample mean of n observations in a random sample of k litters, each with m mice, and that based on a simple random sample of n mice.

When the number of individuals in each group is the same, the variance of the sample mean is estimated as the between group mean square divided by (number of litters × litter size) = 0.0452/30 = 0.0015, i.e. a standard error of $\sqrt{0.0015} = 0.039$. For the case of *equal* group sizes we do not have to estimate the separate components of variance to obtain the standard error of the sample mean. Estimates of the separate components are useful, however, for assessing the relative precision of different designs by calculating standard errors for different combinations of litter size and number of litters. For example, a design with 10 litters of size 2 leads to a standard error of $\sqrt{\hat{\sigma}^2/20 + \hat{\sigma}_b^2/10} = \sqrt{0.0144/20 + 0.0103/10} = 0.042$, compared with 0.039 using three mice per litter. This small loss in precision can be offset by using 12 litters instead of 10. In practice, the choice of the number of groups and group size will involve other considerations but calculating the precision of different schemes can be useful.

Designs with more than two levels

For nested designs with more than two levels, the components of variance can be estimated by using the method for the one-way random effects model to partition the variance at each level into between and within components in a *nested analysis of variance*. Estimates of the components of variance are then obtained as linear combinations of the mean squares in the table. Table 7.17 illustrates the approach for a design with three levels in which the observations are described by the model

$$Y_{ijk} = \mu + A_i + B_{ij} + \varepsilon_{ijk}$$

where μ is the population mean, A_i is a random effect with mean zero and variance σ_a^2, B_{ij} is a random effect with mean zero and variance σ_b^2 and ε_{ijk} is a random effect with mean zero and variance σ^2.

Advantages of balance in nested designs

The above examples of nested designs are *balanced* because the group sizes within each level of the design are equal. This balance is the reason why the components of variance

Table 7.17 Illustration of using nested analysis of variance for estimating the components of variance in a three level nested design. Bar denotes average over levels of the factor(s) indicated by dot subscript. See text for model

Source of variation	df	Sum of squares	Quantities estimated by mean square
A	$a - 1$	$\Sigma(\bar{y}_{i..} - \bar{y}_{...})^2$	$\sigma^2 + r\sigma_b^2 + br\sigma_a^2$
B within A	$a(b - 1)$	$\Sigma(\bar{y}_{ij.} - \bar{y}_{i..})^2$	$\sigma^2 + r\sigma_b^2$
Residual	$ab(r - 1)$	$\Sigma(y_{ijk} - \bar{y}_{ij.})^2$	σ^2
Total	$abr - 1$	$\Sigma(y_{ijk} - \bar{y}_{...})^2$	

can be estimated simply from the mean squares in the nested analysis of variance table. When the group sizes vary within levels, the quantities estimated by each mean square are more complicated expressions involving the different group sizes. The analysis is much less straightforward than it is for the balanced design. One method, REML (Residual Maximum Likelihood), is available as a PC package and also as a procedure within the statistical package GENSTAT. This is a good general method which can be used to estimate the parameters in models with both systematic and random components of variation, using balanced or unbalanced data. For some examples see Robinson (1987).

7.7 DESIGN OF SIMULATION EXPERIMENTS

7.7.1 Introduction

Some biological models are so complex that the only method of assessing their properties presently available is computer simulation. While it is important to study simple special cases analytically, there is also a need for properly designing and analysing simulation experiments with a model sufficiently complex to exhibit behaviour resembling that of the real biological system. Although the statistical theory of experimental design has been devised for comparative experiments involving real biological material, some of the ideas are also useful for simulation experiments. However, in such experiments we usually know much more about the mechanism than in experiments with real biological material, and so standard experimental design procedures sometimes need modification.

7.7.2 Design issues

It is worthwhile itemising the issues which arise and some possible answers or points which should be taken into account in deciding each issue. We cannot give full answers, merely pointers to possibilities, as so much depends on the individual case.

Model behaviour to study

(1) What aspect of the behaviour of the model do we study? This is related both to the purpose of the modelling exercise, and to the ease of measuring the behaviour by statistical analysis of the simulated outcomes.

Choice of parameters and parameter value combinations

(2) All models include parameters whose exact values are usually unknown and therefore the behaviour needs to be assessed for a range of values of each of those parameters. How do we know which are the important parameters?

(a) First of all, some parameters are merely scaling parameters. They do not qualitatively affect the nature of the response, but set its scale. For example, in a model which describes the short-term growth of a plant, the time scale, the maximum growth achieved, the range of concentrations of a nutrient might all be parameters which do not affect the qualitative behaviour of the model, but just determine the scale in time, and space, and nutrient concentration over which it operates.

(b) Secondly, we need to attempt to identify *a priori* which parameters amongst the non-scaling parameters are the critical ones of most interest, and which ones we have to give values to in order to simulate the model at all, but which are unlikely substantially to affect the outcome of the simulations. We might need to change our minds after carrying out the simulations, but it is necessary beforehand to specify which parameters are initially of primary interest.

(3) Having decided which are the important parameters, what values do we give them? In general it seems sensible to give each important parameter a number of values spanning a range which either is likely to occur in nature, or where interesting transitions of model behaviour are likely to occur.

(4) Having decided what values we give to each parameter, how do we combine them? It seems sensible to adopt a factorial design, in which each value of parameter a occurs together with each value of parameter b, and so on. This makes interpretation of the results of the simulation study easier. However, in some circumstances a factorial design might not be appropriate. For example, some parameter combinations might be outside the physiological range. If they represent stresses imposed on a model organism, it often occurs that the organism would not survive a number of severe stresses, and so there would be no point in attempting to predict the model system's growth rate under the stresses, since it would never be observable.

(5) It is possible that such experiments using a factorial design could be very expensive, and some means of cutting down the effort is required. Fractional replicates of higher order factorial schemes—in which an ingeniously constructed fraction of the whole factorial experiment is performed with little loss of information about the major questions—could be considered (for details, see Cox, 1958).

Simulation protocol

(6) How long do we run the simulation for? The time scale of interest depends on the purpose of the modelling and the likely behaviour of the model: how long might it take to reach a stable equilibrium or limit cycle, how much random variability is there, how long do we wish to predict for? We also need to have a sufficiently long run to be able to estimate aspects of the behaviour we are interested in.

(7) What starting values should be used? It might be necessary to use a number of starting values for each parameter combination. If there are a number of attractors to which the system tends (see Chapter 8) then it might be sensible to use a systematic layout of starting values over the range of interest, and then an assessment of the relationship between eventual behaviour and initial starting value can be made.

(8) If it is a stochastic model, how many replications of each parameter set/starting values combination should be run? Two could perhaps be run initially in order to monitor the random variation in the outcomes, followed by further replicates if necessary.

(9) Should independent realisations of the random processes involved be used for each

parameter combination or should the same realisations be used over all parameter combinations? Such realisations could be regarded as a blocking factor. It is probably simpler to use different realisations initially, but in some circumstances greater precision and insight into the process could be obtained by using the same realisations over all parameter combinations. This is an example of a *variance-reduction technique* in stochastic simulation for improving precision and saving costs. For a discussion of other variance-reduction techniques and illustrations of their use, see Morgan (1984).

Monitoring, analysis and presentation of output

(10) Is it necessary to monitor the whole process at least graphically for each replicate of each parameter combination? Initially, detailed monitoring is desirable until an understanding of the dynamics is obtained. This also serves as a check that no untoward behaviour occurs as a result of inadequate numerical routines, or random number generators, or other inadequacy of simulation technique.

(11) How should the results be presented? Assuming that some statistic can be derived from the time courses of the simulations which represents the feature of interest, then tabular summaries involving means or medians, standard deviations or inter-quartile and total ranges for each parameter combination would be a good first step. If there are many parameters which have been varied, then statistical analysis of such tables might be needed, to assess the sensitivity to each parameter, and to assess what interactions occur in the results. It might be necessary to model the relation between the model behaviour and the parameters, using response surface methods.

Example 7.7 An investigation of the Fitzhugh–Nagumo model subject to random stimulation and a continuous current

In order to make the discussion more concrete, we now discuss a specific physiological model from the chapter on biological interaction: the Fitzhugh–Nagumo model of the firing of a neuronal cell. The model is expressed as two differential equations in the variables v, the departure of the membrane potential from its equilibrium, and w, a recovery variable representing the effects of changes in some voltage-dependent channel conductances which tend to bring the system back to equilibrium when disturbed:

$$\frac{dv}{dt} = -v(v - v_1)(v - v_2) - w + I$$

$$\frac{dw}{dt} = \varepsilon(v - d_w w).$$

This model has a parameter I which represents a continuous current applied to the neurone. One input which is not given explicitly in the differential equation framework is the stochastic, discrete, input due to the discharges of connected neurones. We add a single stochastic element to the model to represent these effects, which takes the form of discrete stimulatory perturbations of v, Δv such that:

(1) perturbations occur randomly in time according to a Poisson Process, rate λ;
(2) the magnitude of the perturbations follows a lognormal distribution with parameters μ and σ^2, i.e. $\ln \Delta v \sim N(\mu, \sigma^2)$.

In a real neurone, there will be inhibitory inputs as well, but for the moment we ignore them. A Poisson Process model for the input times is plausible for some cells, as they have many, relatively unco-ordinated inputs. A lognormal is used for the distribution of the magnitudes of the disturbances, as they would typically be of variable magnitude, but, if they are to be excitatory, are always positive.

We now consider the design issues under the headings outlined in the previous section.

Model behaviour to study

In Chapter 8, we discuss how, when the applied current, I, is zero, the neurone is in an excitable state, and disturbances, if large enough, will induce a spike in the cell, followed by a refractory period. However, as the applied current is increased (until the nullclines for the two variables cross on the central rising portion of the inverted N-shaped v-nullcline) the cell goes into continuous regular firing, whether or not discrete stimuli are applied. There is some interest in how the cell will respond to a mixture of inputs—both stochastic discrete inputs and the continuous current. What is the periodicity of firing of the cell when subject to stochastic input of a given rate λ, and how is this modified when a current of a given magnitude, I, is applied? How might we reasonably study this question?

The first thing is that the question is probably too complex for much headway with an analytical approach, and therefore simulation is necessary.

Choice of parameters and parameter value combinations

Let us initially consider the simple case with specific values for the parameters of $v_1 = 1, v_2 = 0.2, d_w = 2.5, \varepsilon = 0.01$ and also fix the parameters of the lognormal distribution of the simulations $\mu = -1.2, \sigma^2 = 0.05^2$. The two parameters of main interest are then λ and I. We first fix appropriate ranges of these parameters. For I, this is the range between $I = 0$ and the value necessary to move the intersection of the nullclines on to the central section of the v-nullcline. $I = 0.07$ does this. Suppose for simplicity we consider I taking the eight values 0, 0.01, 0.02, 0.03, 0.04, 0.05, 0.06, 0.07. Secondly, we need to fix the range of λ. If the stimuli are arriving very infrequently with the cell in its simple 'excitable' state, i.e. $I = 0$, then each supra-threshold stimulation will result in a discharge. So it seems worthwhile to consider a range for λ which starts comfortably in this region, call this λ_0, but also consider a number of multiples of this, e.g. 2, 4, 8, 16 and 32 times. So, for simplicity, we might arrive at a design which takes every possible combination of these values for the two parameters, an 8×6 factorial design.

Simulation protocol

As there is some randomness, we need replicate trials so that we can get some idea of the variability of the response, so initially we take two replicates for each parameter combination. We need to decide on a time scale for the simulations, and we could arrange it so that there are at least say between 5 and 10 successful stimulations in the least responsive case. Ideally we would have long simulations and many replicates for each parameter combination, but we need to keep the effort involved in the study reasonably within bounds. If $\lambda_0 = 0.5$, a time period of 10 time units should give of the order of 5 spikes in the least stimulated case, provided the refractory period is not too long.

Monitoring, analysis and presentation of results

The above restriction to a time period of 10 time units—which is sufficient for about 5 spikes—enables us to plot out the results easily and examine them, as well as assessing the firing frequency. This is important in the initial stages of an investigation—we should be able easily to monitor the details, so as both to check that we are doing sensible things, and to get a better understanding of the results. The determination of the spike frequency was done by counting spikes in this case, but could easily have been done automatically. Spikes were counted if they were above a certain magnitude and started from close to the baseline. The results of one simulation are given in Figure 7.5; this is the standard graphical output which we might use in the initial trials for every simulation. The numerical results are given in Table 7.18.

This is sufficient to give a good idea of what is happening. In general for the lower values of I, the response rate falls as stimulation rate falls, which is what we might expect. For higher values of I, certainly 0.06, this does not happen so markedly, and this appears to be the region in which a minimum response frequency occurs irrespective of stimulus frequency. The change between these two patterns occurs quite sharply, between $I = 0.05$ and $I = 0.06$.

A descriptive statistical analysis of the simulation results might be required both for understanding the results and for presenting them to others. For example, in this case, a simple response surface model might be first constructed of the relationship of spike frequency to I and λ, and then the parameters of this relationship could in turn be related to the other parameters d_w, ε, v_1, v_2, μ, σ. In other words, we might proceed much as we do with the analysis of a real experiment. However, the fact

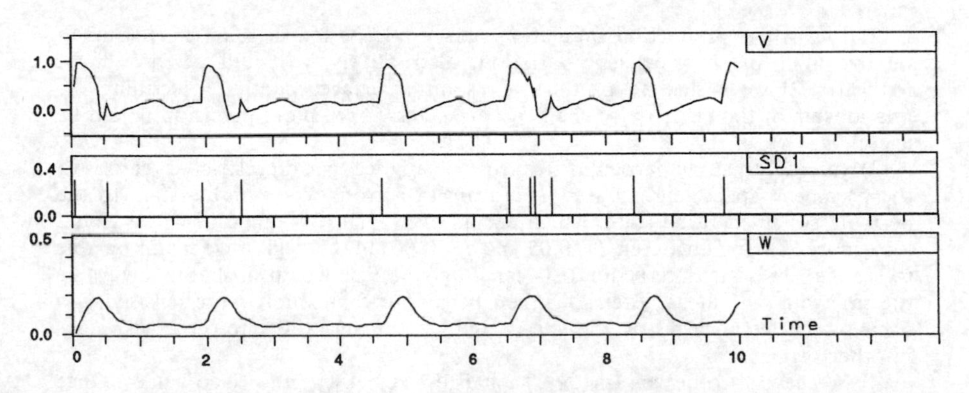

Figure 7.5 A simulation of the Fitzhugh–Nagumo model with random perturbations arriving at an average rate $\lambda = 1$ msec^{-1}, and subject to an applied current of $I = 0.05$. The remainder of the parameters are as in the text. Time scale of graph in msecs.

Table 7.18 Number of responses in 10 msec as a result of different frequencies of stimulation, λ, and input current strengths, I

	Frequency					
I	16	8	4	2	1	0.5
0.0	15 14	9 11	7 8	8 7	5 5	6 3
0.01	12 13	9 12	9 9	9 7	5 6	5 4
0.02	14 14	12 12	8 10	7 7	5 6	5 3
0.03	15 13	13 11	9 10	8 6	7 7	4 4
0.04	13 15	13 11	11 9	10 9	4 4	5 2
0.05	16 14	12 11	10 9	10 8	6 6	5 5
0.06	13 15	14 12	12 10	11 11	10 10	10 10
0.07	16 15	12 13	11 12	12 10	11 10	10 10

that we have a specification of the mechanism means that we should proceed more carefully, continually reassessing our proposed scheme to ensure that we cannot capitalise on the knowledge of the mechanism, and reduce the effort in some way.

This kind of experiment with a model might give an understanding of the type of response surface which might be obtained in a real experiment. Just as we obtained the logistic response curve as the solution of the logistic differential equation, we might obtain the form of a descriptive response surface for use in experiments with real neurones from a pilot study of this kind with model neurones. The difference between the logistic and the present model is that the present model cannot be solved analytically. Nevertheless, we might be able to get some understanding of the nature of the response surface from cheaper—and therefore possibly more extensive—experiments with the model. The experiments with the model also have the advantage that we know better what mechanistic elements are in the model, unlike the real neurone.

Further simulations

We do not have space to pursue this example in any more detail than this, but a few pointers to further work could be made.

(1) The ranges of these two parameters appear to cover the region of interest, although it is possible that if we were to widen the ranges, different patterns would emerge.

(2) Ideally our production simulation runs would be longer than these so as to remove more of the stochastic variation; alternatively we could perform more replicates. If we wished to produce a response surface equation describing the relationship of the response to the two parameters, I, λ, more precision would be useful.

(3) In order to study particular features, we might focus on restricted regions of the parameter space, e.g. to study the transition from the excitable state to the periodic sustained firing state, which occurs over a narrow range of I, we would look more closely at between $I = 0.05$ and $I = 0.06$. In this region we might be able to see if there is any interaction between I and λ. This would probably always be the case with non-linear systems, as their behaviour can exhibit acute sensitivity to some parameters over restricted regions, and be quite insensitive to the exact values in other regions.

(4) We have set other parameters, such as d_w, ε, v_1, v_2, μ, σ to fixed values. Once we have found a reasonable description of the relationship of the response to I and λ over a particular range of these parameters, we need to check how important the values of the other parameters are in this behaviour. We could carry out a higher order factorial experiment involving I, λ and some other parameters, e.g. ε, d_w, μ, probably with only a few levels of each. If these parameters are unimportant, then this should show up in zero interaction between the values of the extra parameters and the central parameters.

7.8 TECHNICAL APPENDIX

7.8.1 Determination of sample size

An important aspect of planning a survey or an experiment is the choice of sample size. Increasing the number of sampling, or experimental, units leads to more precise estimates but resources are usually limited so that in practice some compromise must be found. Sometimes the sample size is entirely determined by the practical circumstances in which case it is important to assess the precision that will result.

Estimating and comparing means

One criterion for determining the sample size when estimating a mean, or the difference between two means, is that a 95% confidence interval should be less than some specified level, E. This means

$$t_{v,0.025}\frac{s}{\sqrt{n}} < E,$$

$$n > \frac{(t_{v,0.025})^2 s^2}{E^2}.$$

If σ^2 is known, rather than estimated from the sample to be taken, then

$$n > \frac{(1.96)^2\sigma^2}{E^2} \approx \frac{4\sigma^2}{E^2}.$$

Example

Consider a very large population of timber plots and suppose that previous experience suggests that the volume of timber varies from plot to plot with a standard deviation of about 50 units. What size of simple random sample is required to estimate the population mean to within 30 units? Using the above formula the required sample size is $n = 4 \times 50^2/30^2 = 11$. This would give 10 degrees of freedom for estimating the population standard deviation and a t-value of 2.23 for calculating 95% confidence intervals. To allow for this we increase the sample size to $n = 2.23^2 \times 50^2/30^2 = 14$. A further adjustment to allow for the 13 degrees of freedom by using a t-value of 2.16 gives $n = 2.16^2 \times 50^2/30^2 = 13$.

Robustness considerations

The above calculations are concerned with determining the sample size to achieve a certain precision. Another consideration is the robustness of the analysis when the observations are non-Normally distributed. For this reason it is prudent to avoid very small sample sizes and use a minimum of 6, but more if possible.

Power considerations

The method for determining sample size is related to the idea of the power of a test for detecting a difference, described in Chapter 6. A useful rule of thumb is that for a t-test of size 5%, a true difference of three standard errors of the estimated difference is detected in about 80% of cases provided that the available degrees of freedom exceed 15.

EXERCISES

7.1 Discuss the principles and practice of: (i) simple random sampling and randomisation; (ii) stratification and blocking, in surveys and experiments for estimating and comparing population means.

7.2 The data in the table below are from an experiment to compare the potency of two plant viruses on leaves of tobacco plants. Because individual leaves vary in their susceptibility, the viruses were applied at random to the two halves of each of ten leaves in a *paired comparison design*. Potency was measured by the number of lesions appearing on the half leaf.

Leaf	1	2	3	4	5	6	7	8	9	10
Virus A	9	8	3	4	8	4	17	3	14	20
Virus B	19	8	13	5	16	8	17	6	19	17

Two ways to analyse a paired comparison design are: (i) as a randomised block design with two treatments per block; (ii) considering differences within pairs as a random sample of differences and using this sample to estimate the mean difference and calculate a confidence interval. Apply these two approaches to illustrate that: (a) the sample variance of the difference in response within pairs is equal to twice the residual mean square of the randomised blocks analysis of variance; (b) both methods result in the same 95% confidence interval on the difference in treatment means.

For the general case, let y_{ij} denote the observation on the jth treatment in the ith pair $(i = 1, \ldots, b)$, and $d_i = y_{i2} - y_{i1}$ the corresponding difference between treatments. Show that the residual sum of squares in the analysis of variance table is given by

$$SS_{\text{residual}} = \frac{1}{2} \sum_{i=1}^{b} (d_i - \bar{d})^2$$

where $\bar{d}$ is the mean difference.

Comment on the appropriateness of the above analyses to the virus data and suggest another approach.

7.3 Discuss the design of a large-scale experiment to investigate the effect of fluoride tablets on the condition of children's teeth. Previous evidence suggests that the effect is beneficial if the tablets are taken over a period of one year or more. Furthermore, a preliminary survey shows that a large number of primary schools in different parts of the country are willing to participate in the experiment provided that the number of children involved does not exceed the size of a single class.

7.4 In some immunological experiments, rectangular plastic plates containing 8 rows of 12 wells—small circular indentations usually with round bottoms—are used to assess whether biological material in liquid suspensions responds differently to a variety of treatments. Before carrying out a long series of experiments some uniformity trials were carried out, and the data below are the results of one such trial given as counts of radioactivity emitted, in thousands. Because of the way the material is harvested from the wells, it was thought that (i) there might be variation from column to column, and from row to row on the plate; also (ii) that the counts near the edge might be lower. Carry out a suitable analysis to test these possibilities separately. If you detect any differences, give some indication of their magnitude.

The experimenter wishes to lay out a 5×3 factorial experiment on a single plate. Bearing in mind the results of the uniformity trial, how would you suggest that he does it? Assuming that the results of the uniformity trial apply to the proposed experiment, for your suggested design estimate (i) the residual variance; (ii) the standard error of the difference between two treatment means.

		Column											
		1	2	3	4	5	6	7	8	9	10	11	12
	1	61	67	78	79	65	71	76	75	68	67	73	64
	2	60	66	79	80	73	74	86	83	84	71	74	65
	3	73	76	81	84	94	85	72	85	76	86	75	73
Row	4	66	65	80	82	83	86	82	82	79	77	80	71
	5	73	87	82	78	78	80	80	79	80	80	88	72
	6	75	80	82	78	78	79	73	74	80	82	75	69
	7	70	78	78	82	82	81	76	91	70	77	81	77
	8	70	61	68	75	75	76	73	74	72	77	65	76

(Cambridge, NST, Part II, Special Subject Statistics, 1981.)

7.5 By considering a stratified random sample with two strata, illustrate the general result that for a fixed number of sampling units the variance of the estimated population mean is least when the sample size for each stratum is proportional to the product of stratum size and stratum standard deviation, i.e. for the ith stratum $n_i \propto N_i\sigma_i$.

For the population of puffin burrows with four strata (Example 7.2, Table 7.4) show that the standard error of the estimated mean for the stratified random samples with equal, proportional and estimated optimal allocation of 56 sampling units are as in the table below. Given that the estimated standard deviation of the unstratified population is 8.72, find the estimated precision of the corresponding simple random sample.

Allocation		Stratum sample size (n_i)				s.e. Estimated mean
Equal	$n_i = n$	14	14	14	14	1.09
Proportional	$n_i \propto N_i$	9	11	17	19	0.98
Optimal	$n_i \propto N_i s_i$	6	8	17	25	0.95

7.6 Explain the meaning of: (i) main effects; (ii) interaction, in a factorial design with two factors.

The data below are from a 3×3 factorial experiment to compare the yields (lb per plot) of three varieties of beans (A, B, C) sown at 3 spacings (4 in, 8 in, 12 in) in a randomised block design with 4 blocks. Carry out an analysis of the data and discuss your findings.

Suggest a plausible statistical model to describe the systematic and random variation in the observations. How might such a model be used to compare the performance of the different varieties and the effect of spacing?

Block	1	1	1	1	1	1	1	1	1	2	2	2	2	2	2	2	2	2
Variety	A	A	A	B	B	B	C	C	C	A	A	A	B	B	B	C	C	C
Spacing	4	8	12	4	8	12	4	8	12	4	8	12	4	8	12	4	8	12
Yield	52	55	56	68	62	59	65	65	77	39	55	55	60	57	48	50	62	83
Block	3	3	3	3	3	3	3	3	3	4	4	4	4	4	4	4	4	4
Variety	A	A	A	B	B	B	C	C	C	A	A	A	B	B	B	C	C	C
Spacing	4	8	12	4	8	12	4	8	12	4	8	12	4	8	12	4	8	12
Yield	47	45	50	62	54	51	55	65	61	42	46	57	68	54	50	51	68	66

7.7 For a 2 × 2 factorial structure with factors A and B, give a pictorial representation of the six patterns in the true treatment means shown in the table below, where + indicates the presence of an effect and − indicates absence.

Pattern	1	2	3	4	5	6
Main effect of A	+	+	+	−	−	−
Main effect of B	+	+	−	+	−	−
Interaction	+	−	−	−	−	+

Analyse the data on dieldrin levels in eggshells from nests in random samples of territories in two years and two areas (Example 7.5(b), Table 7.10) as a 2 × 2 factorial design and discuss your findings.

How would you modify the analysis if within each area the same territories were used in each year? What advantages or disadvantages might there be in using the same territories as opposed to selecting them at random and independently in the two years?

7.8 Suppose that an experimenter suspects a straight line-Normal model for the relationship between Y and x, over the range $-1 < x < 1$, but also admits the possibility of an underlying quadratic response. Discuss the merits of the designs given below for fitting the straight line model and checking for a quadratic effect. Each design involves a total of 14 observations but with different numbers of observations n_i at different values of x_i. For a discussion of this exercise, see Draper & Smith (1981, Chapter 1).

Design 1		2		3		4		5		6		7	
x_i	n_i	x_i	n_i	x_i	n_i	x_i	n_i	x_i	n_i	x_i	n_i	x_i	n_i
−1	1	−1	2	−1	4	−1	5	−1	5	−1	6	−1	7
−11/13	1												
−9/13	1	−2/3	2										
−7/13	1			−1/2	2								
−5/13	1	−1/3	2			−1/3	2						
−3/13	1												
−1/13	1												
		0	2	0	2			0	4	0	2		
1/13	1												
3/13	1	1/3	2			1/3	2						
5/13	1												
7/13	1			1/2	2								
9/13	1	2/3	2										
11/13	1												
1	1	1	2	1	4	1	5	1	5	1	6	1	7

7.9 The table below shows body weights at 6 weeks of female mice in an animal breeding experiment involving 6 sires each mated with 3 dams with measurements on 2 offspring per litter, forming a nested, or hierarchical, design. Use the analysis described in Table 7.17 to estimate the components of variance due to sires, dams and offspring within a litter.

Sire	A	A	A	A	A	A	B	B	B	B	B	B
Dam	1	1	2	2	3	3	4	4	5	5	6	6
Body weight (g)	19.3	21.9	22.7	24.6	21.0	19.1	19.3	21.9	20.6	17.9	21.0	19.3
Sire	C	C	C	C	C	C	D	D	D	D	D	D
Dam	7	7	8	8	9	9	10	10	11	11	12	12
Body weight (g)	23.3	23.6	17.4	19.6	21.8	21.1	17.1	21.2	18.5	19.3	18.8	22.1
Sire	E	E	E	E	E	E	F	F	F	F	F	F
Dam	13	13	14	14	15	15	16	16	17	17	18	18
Body weight (g)	22.2	20.7	21.5	21.7	17.9	17.4	18.7	21.6	17.3	15.7	16.5	16.5

7.10 The data below are from an experiment to estimate the calcium concentration in turnip greens (Snedecor & Cochran, 1967). Four plants were selected at random, and then three leaves were randomly selected from each plant. Two 100 mg samples were taken from each leaf and the concentration of calcium determined by microchemical methods.

Plant	Leaf	Determination		Plant	Leaf	Determination	
1	1	3.28	3.09	3	1	2.77	2.66
	2	3.52	3.48		2	3.74	3.44
	3	2.88	2.80		3	2.55	2.55
2	1	2.46	2.44	4	1	3.78	3.87
	2	1.87	1.92		2	4.07	4.12
	3	2.19	2.19		3	3.31	3.31

(a) Write down a statistical model with terms for random variation between plants, between leaves within plants and between determinations on each leaf.

(b) Use a nested analysis of variance to estimate the three components of variance.

(c) For the general case, with a sample of a plants, b leaves per plant and r determinations per leaf, write down the expression for the standard error of the sample mean concentration of calcium.

(d) If $r = 1$ determination was made on each of $b = 2$ leaves for each of $a = 12$ plants, what is the standard error of the mean? With one determination on one leaf from each plant, estimate the number of plants required for a standard error of 0.15.

(e) Comment on the use of 4 plants as in the original design.

7.11 Consider the following dynamic stochastic model for an annual butterfly species in which population densities in successive years are related by

$$N_{t+1} = \frac{RKN_t}{(K + RN_t)}.$$

The intrinsic rate of increase, R, and the population ceiling, K, vary randomly from year to year, independently of N_t, such that $\ln R \sim N(\mu_R, \sigma_R^2)$, $\ln K \sim (\mu_K, \sigma_K^2)$ with correlation coefficient ρ between $\ln R$ and $\ln K$.

Discuss the design of a simulation study to examine how the mean and variance of population density are related to values of the parameters $(\mu_R, \sigma_R, \sigma_K, \rho)$ for a specified value of μ_K.

Part III

BIOLOGICAL INTERACTIONS

8 The Mathematics of Interaction

8.1 INTRODUCTION

This part of the book is about models of biological interaction, with two main themes. The first is the construction and exploration of the properties of dynamic models of interacting systems; the second is fitting these models and related models which quantify the results of biological interactions.

In Part III, we discuss ecological interactions between populations of whole organisms, for example of predator and prey or of competing species. On a smaller scale, we deal with chemical and similar interactions between populations of different types of cell, or of molecules or ions. Another interaction discussed is that between attributes of a single cell in mechanisms of nerve impulse generation: between the membrane potential and a recovery variable reflecting changes in ion channel conductances. Though these models are on very different scales both spatially and temporally, they do share certain features which define the scope of Part III.

(1) The models are of *changes in time*, whether discrete or continuous. In some studies we might only be interested in the eventual equilibrium which is reached, but in general we start with a dynamic model.
(2) There are *two or more participants*, often different species or different types of molecule or cell.
(3) The participants in the interaction are either spatially *homogeneous*, or can be considered so, or can even be considered as being at one point. We are not here concerned at all with the spatial distribution—this is dealt with in Chapters 12 (on transport and diffusion) and 13 (on spatial and temporal pattern and sequence).
(4) We initially deal with *deterministic* models of interactions, but also consider how they can be generalised to include a *random* element.

The second main theme of Part III is statistical techniques for fitting models arising out of biological interactions. Many are of the relationship between the interacting components at the eventual equilibrium, and most of these, particularly from Biochemistry and Physiology, fall within the categories of *non-linear* (in the statistical sense) and *generalised linear models*, which also have wider application. Fitting such models as well as dynamic models of interaction is the main subject of Chapter 11.

8.2 TYPES OF MODEL OF INTERACTION

The object of this chapter is to discuss methods of formulation of dynamic models of biological interaction, of displaying them, and of examining their important properties. We make use of five models. The first four are of differential equations, in which the rate of change of each component is in general a function of the magnitude of all the components, and the changes take place smoothly in time. The fifth is a difference equation, which is particularly appropriate for ecological models of interacting populations in which discrete generations and synchronous breeding occur.

Model 1. The 'linear' model in two variables, which can be used for interpreting tracer experiments in studies of turnover of drugs or other substances in the bloodstream, but also is useful when examining the properties of more complex non-linear models.
Model 2. The Lotka–Volterra, two component, 'non-linear' model which has been used both as a simple description of predation and of some simple chemical reactions.
Model 3. The Fitzhugh–Nagumo 'non-linear' model of nerve cell firing involving two variables, which exhibits excitability—quickly returning to equilibrium as a result of stimulations below the threshold, but substantially responding once the threshold is crossed—and fixed amplitude, fixed period oscillations.
Model 4. The Lorenz model in three variables, originally written down as a greatly simplified model of the meteorological phenomenon of cellular convection, but which is now more famous for its ability to display deterministic chaotic behaviour.
Model 5. The Nicholson–Bailey difference equation model of host–parasitoid interaction.

These models are all considerable over-simplifications, and most real situations would be affected to varying degrees by stochastic effects of the various kinds which we have already introduced in the single population context (in Chapter 4 on structured stochastic models). In any real situation, quite often there is also substantial heterogeneity of some kind, and the distributed model system that would be required to describe such heterogeneity might well behave differently. However, both in learning about models and in any particular modelling exercise it is necessary to start simply and to add greater complexity only when necessary.

8.3 SECOND ORDER DIFFERENTIAL EQUATIONS AND THE PENDULUM

8.3.1 The importance of the second order, homogeneous, linear differential equation

Before we can begin our examination of models for interaction, we must introduce some simple but important models of second order differential equations in one response variable which differ from those considered already. The new feature is that they involve accelerations as well as velocities, and hence second differentials with respect to time. We use as an example a physical model: that of a pendulum, first of all in the idealised frictionless case, then involving friction. Solutions of these models also provide us with solutions of any *linear* system of two differential equations, and through them, with the tools required to assess the behaviour near equilibrium points of more biologically interesting non-linear systems. Thus a thorough examination of this type of differential equation is an essential preliminary to studying at least Models 1 to 3 above. It is equally important for the reader to understand the main principles of the methods of solution as it is to be able to apply the methods to solve any particular example. As we shall see later in this chapter, Model 4, that involves three differential equations, involves new features which cannot be captured by the second order differential equation.

8.3.2 Constructing and solving the differential equation

The equation of motion for an ideal pendulum subject to no drag or friction is derived in the legend to Figure 8.1 (the details of the derivation are not important for our purposes in this section), and is

$$\frac{d^2x}{dt^2} = -cx$$

where x is the angular displacement of the pendulum, and c is a positive constant.

This is a second order differential equation because it involves the second differential with respect to time. In order to solve this, we need to look for something which when differentiated twice will give us $-c$ times what we started with. A glance through the tables of differentials suggests that the solution is either a sine function or cosine

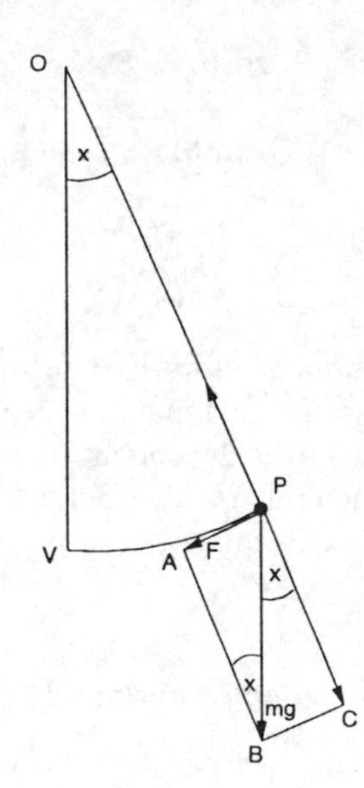

Figure 8.1 The force diagram for a pendulum consisting of a weight at P swinging at the end of a weightless rod of length R suspended from point O. Force mg vertically downwards has a component $mg \sin x$ in the direction of motion. This is equal to mass × acceleration and thus equals $-mR\, d^2x/dt^2$. Thus we have the equation of motion

$$-mR\frac{d^2x}{dt^2} = mg \sin x$$

i.e.

$$\frac{d^2x}{dt^2} + \frac{g}{R}x \cong 0 \quad \text{or} \quad \frac{d^2x}{dt^2} = -cx$$

as for small x, $\sin x \cong x$. The rod on which the mass at P swings is rigid, and therefore the components of gravity along the length of the rod and the tension in the rod balance exactly and therefore do not concern us.

function of $\sqrt{c}t$ multiplied by any arbitrary constant, i.e. $A\sin(\sqrt{c}t)$ or $B\cos(\sqrt{c}t)$, so we take as a general solution a sum of two such terms

$$x = A\sin(\sqrt{c}t) + B\cos(\sqrt{c}t).$$

To prove that this is a solution, we differentiate

$$\frac{\mathrm{d}x}{\mathrm{d}t} = A\sqrt{c}\cos(\sqrt{c}t) - B\sqrt{c}\sin(\sqrt{c}t)$$

and again

$$\frac{\mathrm{d}^2x}{\mathrm{d}t^2} = -Ac\sin(\sqrt{c}t) - Bc\cos(\sqrt{c}t)$$

$$= -cx.$$

Then we need to find the values of the arbitrary constants A and B. Whatever the values that A and B take, the above expression is still a solution. However, the values of A and B can be fixed by considering the initial conditions. Suppose that the pendulum starts from rest when $x = x_0$, then

$$x(0) = A\sin(0) + B\cos(0) = B = x_0 \quad \text{i.e.} \quad B = x_0$$

$$\frac{\mathrm{d}x}{\mathrm{d}t} = \sqrt{c}A\cos(0) - \sqrt{c}B\sin(0) = \sqrt{c}A = 0 \quad \text{i.e.} \quad A = 0.$$

Hence the solution is

$$x = x_0\cos(\sqrt{c}t).$$

This motion is called *simple harmonic motion*. The solution is a cosine wave which repeats itself every $2\pi/\sqrt{c}$, and so we say that the pendulum has period $2\pi/\sqrt{c}$, and amplitude x_0. Note that the amplitude is dependent on the initial condition; the further out you pull the pendulum and then release it (provided you stay within the region that the linear approximation $x \cong \sin x$ is valid) the larger the amplitude of the resulting oscillations.

8.3.3 Solution using a complex variable shorthand

Before we consider what happens when friction or some other source of drag introduces complications, we consider another method of solving the differential equation in the simple frictionless situation. We inferred from the form of the derivative of sine and cosine functions that the solution above was of the form $A\sin\sqrt{c}t$ or $A\cos\sqrt{c}t$. Suppose we had alternatively thought that a solution might be an exponential $E\,\mathrm{e}^{\lambda t}$. We use a different letter E since the arbitrary constant would in general be different from those above. Then substituting in the equation we obtain

$$E\lambda^2\,\mathrm{e}^{\lambda t} + Ec\,\mathrm{e}^{\lambda t} = E\,\mathrm{e}^{\lambda t}(\lambda^2 + c) = 0$$

and so $E\,\mathrm{e}^{\lambda t}$ is a solution provided that

$$\lambda^2 + c = 0 \quad \text{i.e.} \quad \lambda^2 = -c \qquad \text{or } \lambda = \pm i\sqrt{c}$$

where i is the square root of minus one. So, if for a moment we do not worry too much about having complex solutions, the general solution is $E\,e^{i\sqrt{c}t} + F\,e^{-i\sqrt{c}t}$. This can be rewritten by using the identity

$$e^{i\theta} = \cos\theta + i\sin\theta$$

as

$$x = (E + F)\cos(\sqrt{c}t) + i(E - F)\sin(\sqrt{c}t)$$

This looks more like the previous general solution we derived except that the arbitrary constants contain i. This is not really a problem however as the arbitrary constants in any solution are set to whatever values are necessary to make the solution satisfy the initial conditions. In this case, the initial conditions are that when $t = 0$, $x = x_0$ and $\mathrm{d}x/\mathrm{d}t = 0$. So

$$x(0) = (E + F) = x_0$$

and

$$\frac{\mathrm{d}x(0)}{\mathrm{d}t} = i(E - F)(\sqrt{c}) = 0$$

and therefore $E = F$. Hence $E = F = x_0/2$. The solution is therefore

$$x = x_0\cos(\sqrt{c}t)$$

which is exactly what we got previously. So these two methods are different ways of getting to the same result in the end. The second method is just a shorthand, in general more convenient, for the full calculation involving sine and cosine functions. The complex number shorthand is more powerful than this example indicates, as we shall see in the next section.

8.3.4 The equations of motion of a more realistic pendulum

Pendulums in the real world are not of the idealised kind we considered in the previous section, but suffer from drag or friction which would eventually cause them to stop swinging (in some clocks this is counteracted by weights or a spring which are rewound periodically). This drag increases with velocity, and is usually taken to be proportional to the velocity. So the equation of motion becomes

$$\frac{\mathrm{d}^2x}{\mathrm{d}t^2} = -cx - b\frac{\mathrm{d}x}{\mathrm{d}t}$$

i.e.

$$\frac{\mathrm{d}^2x}{\mathrm{d}t^2} + b\frac{\mathrm{d}x}{\mathrm{d}t} + cx = 0$$

where b is a positive constant.

8.3.5 Solution using complex variables

We use the second method from the previous section of obtaining the solutions of such an equation. Try a solution $x = A\,e^{\lambda t}$. Differentiating

$$\frac{dx}{dt} = \lambda A\,e^{\lambda t}, \qquad \frac{d^2x}{dt^2} = \lambda^2 A\,e^{\lambda t}$$

and substituting this into the differential equation we get

$$\lambda^2 A\,e^{\lambda t} + \lambda b A\,e^{\lambda t} + cA\,e^{\lambda t} = 0.$$

For this to be true for all t, which is what we require of a solution of the differential equation, then λ must satisfy the equation

$$\lambda^2 + b\lambda + c = 0$$

i.e.

$$\lambda_1, \lambda_2 = \frac{-b \pm \sqrt{b^2 - 4c}}{2}.$$

Suppose for the moment that $b^2 < 4c$, which is quite plausible since the drag forces are likely to be quite small compared with the gravitational forces in any well designed pendulum, then the two solutions will be of the form $\lambda_1, \lambda_2 = -b/2 \pm i\omega$, where $\omega^2 = c - b^2/4$. So we can choose either of these roots to insert in our exponential solution, and therefore we have two solutions

$$x = A\,e^{(-b/2 + i\omega)t} \quad \text{and} \quad x = B\,e^{(-b/2 - i\omega)t}$$

where we use different arbitrary constants, since there is no reason to suppose that they would necessarily be the same. The general solution is therefore

$$\begin{aligned} x &= A\,e^{(-b/2 + i\omega)t} + B\,e^{(-b/2 - i\omega)t} \\ &= e^{-bt/2}[A\,e^{i\omega t} + B\,e^{-i\omega t}] \end{aligned}$$

which can also be written, using the expansion $e^{i\omega t} = \cos(\omega t) + i\sin(\omega t)$

$$x = e^{-bt/2}[E\cos(\omega t) + F\sin(\omega t)]$$

where E and F are new arbitrary constants. Now replace E by $H\sin\alpha$ and F by $H\cos\alpha$. H and α can be determined using the relations $E^2 + F^2 = H^2(\cos^2\alpha + \sin^2\alpha) = H^2$ and $\tan\alpha = E/F$. Then, using the standard trigonometrical identity

$$\sin(\omega t + \alpha) = \sin\alpha\cos(\omega t) + \cos\alpha\sin(\omega t)$$

we can rewrite the above expression as

$$x = H\,e^{-bt/2}\sin(\omega t + \alpha).$$

This solution consists of two functions multiplied together: an exponential decline and an oscillatory function of constant amplitude. Therefore the solution is oscillatory and the oscillations have exponentially declining amplitude. The motion is called *damped harmonic motion*. Thus we see that the introduction of the friction term in the differential equation has had the expected effect—it has resulted in converting simple harmonic motion to damped harmonic motion, and the extent of the damping $e^{-bt/2}$ is related to the magnitude of the coefficient, b, of the friction term $b\,dx/dt$.

This example illustrates the strength of the second approach based on complex variables. It is able to provide solutions which are oscillatory of constant amplitude, oscillatory of exponentially declining or increasing amplitude, and, as we shall see in the next section, exponentially declining solutions with no oscillatory aspect.

8.3.6 Solution of the homogeneous, second order, linear, differential equation with constant coefficients

The box below summarises the steps involved in the solution of any homogeneous second order linear differential equation, following the methods outlined in the previous sections. The equation is said to be *homogeneous* because there is zero rather than a function of t on the right hand side, linear because only first powers of y appear in the equations, and with constant coefficients because the coefficients a, b, c are constants rather than functions of t.

General solution of the homogeneous, second order, linear, differential equation with constant coefficients

$$a\frac{\mathrm{d}^2y}{\mathrm{d}t^2} + b\frac{\mathrm{d}y}{\mathrm{d}t} + cy = 0$$

with a, b, c constants.

Principles of general solution

Substitute in the solution $y = A\exp(\lambda t)$ where A is an arbitrary constant:

$$A(a\lambda^2 + b\lambda + c)\,\mathrm{e}^{\lambda t} = 0.$$

It follows that, for this to be a solution for all t and A, λ must satisfy the *auxiliary equation*

$$a\lambda^2 + b\lambda + c = 0.$$

Dividing through by a, this is more conveniently written for our later purposes in terms of the sum, $T = -b/a$, and the product, $D = c/a$, of the roots of the equation:

$$\lambda^2 - T\lambda + D = 0.$$

The two roots are given by

$$\lambda_1, \lambda_2 = \frac{T \pm \sqrt{T^2 - 4D}}{2}.$$

Then we obtain the forms of solution given below.

Forms of solution

(1) $T^2 > 4D$, i.e. *roots of auxiliary equation are real and different*, $\lambda_1 \neq \lambda_2$, then solution is *a sum of exponentials* i.e.

$$y = A_1\,\mathrm{e}^{\lambda_1 t} + A_2\,\mathrm{e}^{\lambda_2 t}$$

where A_1, A_2 are arbitrary constants, to be determined by the initial conditions.

Cases

$D > 0$, λ_1, λ_2 both of the same sign
$T > 0$, both roots positive, so y grows without limit as t increases ($y \to \pm\infty$ as $t \to \infty$).

$T < 0$, both roots negative, so y declines to zero as t increases ($y \to 0$ as $t \to \infty$).
$D < 0$, λ_1, λ_2 of different sign, so there is one positive root, and hence y grows without limit as t increases ($y \to \pm\infty$ as $t \to \infty$).

(2) $T^2 = 4D$, i.e. *roots of auxiliary equation are real and identical,* $\lambda_1 = \lambda_2 = \lambda$, then solution is of the form

$$y = A_1 \, e^{\lambda t} + A_2 t \, e^{\lambda t}.$$

Why this form of solution? First of all, if the two roots are equal, the solution

$$y = A_1 \, e^{\lambda t} + A_2 t \, e^{\lambda t}$$

can be shown by substitution to satisfy the differential equation. Secondly, it is a way of including two arbitrary constants in the general solution. Two constants are necessary so that any two initial conditions, e.g. on the initial position and initial velocity, can be satisfied.

Cases

$T > 0$, common root positive, so y grows without limit as t increases ($y \to \pm\infty$ as $t \to \infty$)
$T < 0$, common root negative, so y declines to zero as t increases ($y \to 0$ as $t \to \infty$).

(3) $T^2 < 4D$, i.e. *roots of auxiliary equation are complex,* $\lambda_1 = \alpha + i\beta$, $\lambda_2 = \alpha - i\beta$ then solution is *an exponential* $\times$ *an oscillatory term*
i.e.

$$y = e^{\alpha t}(A_1 \cos \beta t + A_2 \sin \beta t)$$

where A_1, A_2 are again arbitrary constants. An alternative form of solution which demonstrates how the two arbitrary constants determine the amplitude scaling and phase of the solution is

$$y = A_1 \, e^{\alpha t}[\sin \beta(t - A_2)].$$

Cases

$T > 0$, $\alpha > 0$, so y oscillates with increasing amplitude as t increases.
$T < 0$, $\alpha < 0$, so y oscillates with decreasing amplitude as t increases; ($y \to 0$ as $t \to \infty$).
$T = 0$, simple harmonic motion as in the undamped pendulum.

Summary diagram of qualitative behaviour of solutions

The qualitative behaviour of the solution can be determined for any specific equation by evaluating and determining which region of the (T, D) plane the case falls in. This diagram (Figure 8.2) can be obtained by considering the various cases individually as above.

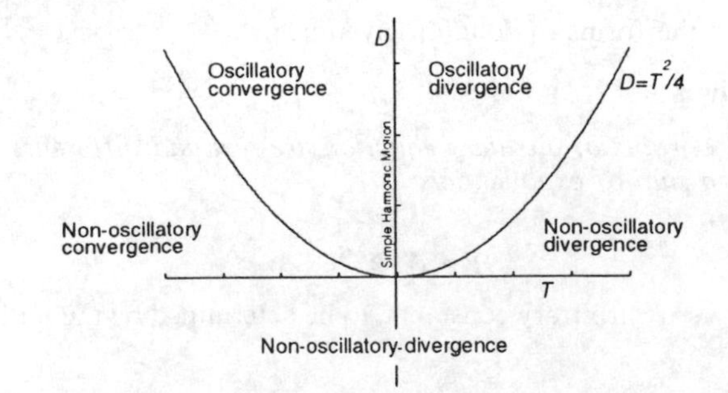

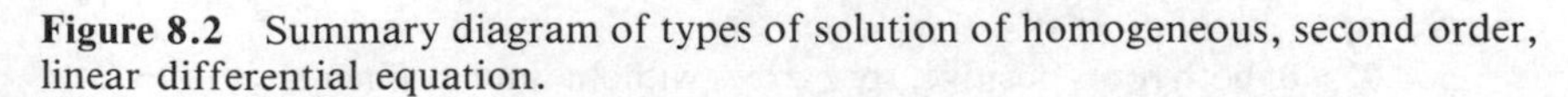

Figure 8.2 Summary diagram of types of solution of homogeneous, second order, linear differential equation.

8.4 AN EXAMPLE OF A LINEAR SYSTEM: A COMPARTMENT MODEL OF DRUG TURNOVER

We now consider a physiological example of modelling biological interaction. In this case the interaction consists of flows of material through several 'pools'. The pools are interacting in the sense that material is continually interchanging between them. A pool is a general term which we will discuss later, but for the moment a pool can be considered as a reservoir of a fixed volume containing fluid in which the material of interest is suspended or dissolved.

The example from pharmacokinetics is of drug turnover after a single injection into the bloodstream. If a fixed quantity of a drug is injected into the bloodstream of an animal, called pool X, in a single quick injection, the first thing which happens is that the drug is thoroughly mixed so as to reach a homogeneous concentration, x, within the whole of the blood volume, and this would occur (for an animal such as a sheep or a human) in a matter of a minute or so. The concentration of the drug in the bloodstream then begins to decay as it is absorbed by other tissues, assumed for the moment to be a single homogeneous unit called pool Y, in which the concentration is y. In many cases, the rate of loss of drug is proportional to the concentration in the bloodstream, x, but as soon as the concentration in pool Y builds up to an appreciable level, there is also a backward flow, proportional to the concentration there, y, into the bloodstream. We assume that the drug is irreversibly converted into a different form from pool Y, and that this is the only site where drug is lost during the course of the experiments being modelled. We also assume that the rate of loss of drug from pool Y is proportional to the concentration of the drug in pool Y. This interaction can be modelled by two linear differential equations, as follows.

Compartment model for two interacting compartments X and Y, with eventual loss from compartment Y

$$\frac{\mathrm{d}x}{\mathrm{d}t} = -k_{yx}x + k_{xy}y$$

$$\begin{pmatrix}\text{Rate of}\\ \text{change of}\\ \text{concn. in X}\end{pmatrix} = -\begin{pmatrix}\text{transfer}\\ \text{to Y}\\ \text{from X}\end{pmatrix} + \begin{pmatrix}\text{transfer}\\ \text{to X}\\ \text{from Y}\end{pmatrix}$$

$$\frac{\mathrm{d}y}{\mathrm{d}t} = -k_{xy}y + k_{yx}x - k_{0y}y$$

$$\begin{pmatrix}\text{Rate of}\\ \text{change of}\\ \text{concn. in Y}\end{pmatrix} = -\begin{pmatrix}\text{transfer}\\ \text{to X}\\ \text{from Y}\end{pmatrix} + \begin{pmatrix}\text{transfer}\\ \text{to Y}\\ \text{from X}\end{pmatrix} - \begin{pmatrix}\text{loss}\\ \text{from Y}\end{pmatrix}$$

where x, y is the concentration in compartments X, Y respectively, and k_{xy} is the rate constant to X from Y, etc.

The k's are known as rate constants of the interchanges between the compartments. Thus the flow to Y from X is given by $k_{yx}x$ and that to X from Y by $k_{xy}y$, and this explains these terms appearing on the right-hand sides of the equations above. The rate at which the drug flows from pool Y to the outside of the system (where the drug is used up, or irreversibly bound) is $k_{0y}y$. Other information which is required to determine a solution of these equations is the dose of the drug, D, and the blood volume, V. Note that if there were no return flow from compartment Y to X, then we would have the simpler single exponential case, whose solution is given in Chapter 1.

8.5 ANALYTICAL SOLUTION OF LINEAR SYSTEMS

Example 8.1 illustrates the method of solution of a specific numerical case of the model for drug turnover.

Example 8.1 General solution of a linear system from first principles

Obtain a solution from first principles of the system of differential equations

$$\frac{\mathrm{d}x}{\mathrm{d}t} = -0.3x + 0.05y \tag{1}$$

$$\frac{\mathrm{d}y}{\mathrm{d}t} = 0.3x - 0.20y \tag{2}$$

with initial conditions: $x(0) = x_0$ and $y(0) = 0$.

(a) Differentiate equation (1)

$$\frac{\mathrm{d}^2x}{\mathrm{d}t^2} = -0.3\frac{\mathrm{d}x}{\mathrm{d}t} + 0.05\frac{\mathrm{d}y}{\mathrm{d}t}. \tag{3}$$

(b) Obtain an expression for $\mathrm{d}y/\mathrm{d}t$ *in terms of* x *and* $\mathrm{d}x/\mathrm{d}t$, *by elimination of* y *from equations (1) and (2).*
From (1)

$$y = 20\frac{\mathrm{d}x}{\mathrm{d}t} + 6x \tag{4}$$

Substituting in (2)

$$\frac{\mathrm{d}y}{\mathrm{d}t} = 0.3x - 0.2\left(20\frac{\mathrm{d}x}{\mathrm{d}t} + 6x\right)$$

$$= -0.9x - 4\frac{\mathrm{d}x}{\mathrm{d}t}.$$

Then substituting in (3)

$$\frac{\mathrm{d}^2x}{\mathrm{d}t^2} + 0.3\frac{\mathrm{d}x}{\mathrm{d}t} - 0.05\left(-0.9x - 4\frac{\mathrm{d}x}{\mathrm{d}t}\right) = 0$$

i.e.

$$\frac{\mathrm{d}^2x}{\mathrm{d}t^2} + 0.5\frac{\mathrm{d}x}{\mathrm{d}t} + 0.045x = 0$$

(c) *Find the general solution of the resulting 2nd order linear equation in x.*
The auxiliary equation is

$$\lambda^2 + 0.5\lambda + 0.045 = 0$$

and the roots are

$$\lambda_1, \lambda_2 = \frac{-0.5 \pm \sqrt{0.5^2 - (4)(0.045)}}{2}$$
$$= -0.25 \pm \sqrt{0.0175}.$$

Therefore the general solution is

$$x = A_1 \, e^{\lambda_1 t} + A_2 \, e^{\lambda_2 t}$$

where A_1, A_2 are arbitrary constants.

(d) *Find the general solution for y by using equation (4).*

$$y = 20 \frac{dx}{dt} + 6x = 20(\lambda_1 A_1 \, e^{\lambda_1 t} + \lambda_2 A_2 \, e^{\lambda_2 t}) + 6(A_1 \, e^{\lambda_1 t} + A_2 \, e^{\lambda_2 t})$$
$$= (20\lambda_1 + 6) A_1 \, e^{\lambda_1 t} + (20\lambda_2 + 6) A_2 \, e^{\lambda_2 t}$$

(e) *Use the initial conditions to determine the arbitrary constants.*
The initial conditions are that $x = x_0$ and $y = 0$. Hence

$$x_0 = A_1 + A_2$$
$$0 = (20\lambda_1 + 6)A_1 + (20\lambda_2 + 6)A_2$$

and the arbitrary constants satisfying these conditions are

$$A_1 = \frac{(\lambda_2 + 6/20)x_0}{\lambda_2 - \lambda_1} \qquad A_2 = \frac{-(\lambda_1 + 6/20)x_0}{\lambda_2 - \lambda_1}.$$

The solution in the general 2-variable homogeneous case, given below, is obtained in the same way as the above example.

General formula for the solution of the 2-variable homogeneous linear system

	Vector and matrix notation
The equations are $\frac{dy_1}{dt} = a_{11}y_1 + a_{12}y_2$ $\frac{dy_2}{dt} = a_{21}y_1 + a_{22}y_2$	$\frac{d\mathbf{y}}{dt} = \mathbf{A}\mathbf{y}$ where $\mathbf{y} = \begin{pmatrix} y_1 \\ y_2 \end{pmatrix}, \quad \mathbf{A} = \begin{pmatrix} a_{11} & a_{12} \\ a_{21} & a_{22} \end{pmatrix}.$ **A** is often referred to as the *community matrix*.

The general solution is $$y_1 = A_1 e^{\lambda_1 t} + A_2 e^{\lambda_2 t}$$ $$y_2 = \frac{A_1(\lambda_1 - a_{11}) e^{\lambda_1 t} + A_2(\lambda_2 - a_{11}) e^{\lambda_2 t}}{a_{12}}$$ where A_1, A_2 are arbitrary constants; λ_1 and λ_2 are the two roots or solutions of the quadratic equation (the *auxiliary equation*) $$\lambda^2 - T\lambda + D = 0$$ where $$T = \text{Trace}(\mathbf{A}) = a_{11} + a_{22}$$ and $$D = \text{Determinant}(\mathbf{A}) = a_{11}a_{22} - a_{21}a_{12}$$ i.e. $$\lambda_1, \lambda_2 = \frac{T \pm \sqrt{T^2 - 4D}}{2}.$$	The general solution is $$\mathbf{y} = A_1 e^{\lambda_1 t}\mathbf{v}_1 + A_2 e^{\lambda_2 t}\mathbf{v}_2$$ i.e. $$\mathbf{y} = \begin{pmatrix} y_1 \\ y_2 \end{pmatrix} = A_1 e^{\lambda_1 t}\begin{pmatrix} v_{11} \\ v_{12} \end{pmatrix} + A_2 e^{\lambda_2 t}\begin{pmatrix} v_{21} \\ v_{22} \end{pmatrix}$$ where A_1, A_2 are arbitrary constants; and $\mathbf{A}$ has two linearly independent eigenvectors $\mathbf{v}_1$, $\mathbf{v}_2$, with corresponding eigenvalues λ_1, λ_2. This formulation easily generalises to the case of three or more differential equations.

8.5.1 Stable and unstable equilibria

An *equilibrium* is a point defined in the variable space (y_1, y_2) at which the system is at rest, i.e. one at which both $\mathrm{d}y_1/\mathrm{d}t = 0$ and $\mathrm{d}y_2/\mathrm{d}t = 0$. These points are sometimes referred to as *singular* or *stationary points* of the system. They are obviously of great interest as these are points which the system might ultimately tend to and remain at. For example, all homogeneous linear systems (such as that considered in the previous section) have equilibria at the origin (0,0), as at that point $\mathrm{d}y_1/\mathrm{d}t = 0$ and $\mathrm{d}y_2/\mathrm{d}t = 0$. An important aspect of such equilibria which we shall consider later is their stability. What we mean by stable and unstable here is best illustrated by considering the equilibrium of a ball on a hilly terrain, which we will suppose for our present purposes to be fairly smooth. In theory, the ball could stay at rest at any point of zero slope, i.e. any *horizontal* point on the terrain, in particular at the bottoms of hollows and on the tops of hills. An equilibrium at the top of a hill would be *unstable* since, if the ball were slightly disturbed, it would roll away, into an adjacent valley. An equilibrium in a hollow would be *stable* since if disturbed, the ball would return to equilibrium. Any equilibrium on a surface that is horizontal for some distance in all directions would be *neutral* since, if disturbed, the ball would neither move further away from, nor return to, the original equilibrium. As well as roughly spherical or ellipsoidal hollows and hilltops, which correspond to stable and unstable equilibria respectively, more complex forms such as *saddle-point instabilities* are possible. The next few sections contain depictions of the various types of solution of the linear system, which are models for the different types of *local* stability or instability of linear and non-linear systems.

8.5.2 Graphical display of solutions—phase plane plots

One way of plotting out solutions is simply as plots of each response variable against time. Another simple way to plot two variables which are varying in time is to plot one against the other, and to trace out the path as time passes, just as journeys are drawn out on maps where the x and y coordinates represent the distances with respect to a fixed origin along an East–West and a North–South axis respectively. These are known as phase plane plots. Some examples follow which illustrate the types of behaviour possible in linear systems plotted in these ways. Further properties and uses of phase plane plots are discussed in Section 8.9.2.

Example 8.2 Solution of a linear system with specified starting values

Obtain the solutions (x, y) of the linear system

$$\frac{dx}{dt} = -1.5x + 0.5y, \qquad \frac{dy}{dt} = -0.5x - 0.5y$$

which start at $(x, y) = (1, 1)$, $(0, 1)$, $(1, 0)$. Plot these out as separate plots of x against t and y against t. Infer the phase plane plot from the separate time course plots.

The trace of the community matrix $\mathbf{A}$ is $T = -1.5 - 0.5 = -2$ and the determinant is $D = (-1.5)(-0.5) - (0.5)(-0.5) = 0.75 + 0.25 = 1$, and so the auxiliary equation is

$$\lambda^2 + 2\lambda + 1 = 0$$
$$(\lambda + 1)^2 = 0$$

i.e.

$$\lambda_1 = \lambda_2 = -1.$$

Using the results of Section 8.3.6, the solution for x is

$$x = A\,e^{-t} + Bt\,e^{-t}.$$

The solution for y is given from the 1st differential equation as

$$y = 2\frac{dx}{dt} + 3x = 2(-A + B - Bt)\,e^{-t} + 3(A + Bt)\,e^{-t} = (A + 2B)\,e^{-t} + Bt\,e^{-t}$$

The solution starting at $(x, y) = (1, 1)$ is obtained by substituting in the general solution

$$1 = [A + B \times 0]e^0 = A$$
$$1 = (A + 2B)e^0 + B \times 0 \times e^0 = A + 2B$$

i.e.

$$B = 0$$

Hence the solution starting at $(1, 1)$, plotted in Figure 8.3, is

$$x = e^{-t}, \qquad y = e^{-t}.$$

By similar analysis, the solution starting from $(0, 1)$ is

$$x = t\,e^{-t}/2, \qquad y = (1 + t/2)\,e^{-t}$$

and that from $(1, 0)$ is

$$x = (1 - t/2)\,e^{-t}, \qquad y = -t\,e^{-t}/2.$$

The sketches of the plots of the solutions against t, and the phase plane plots, are as given in Figure 8.3.

The above is an example of what is termed a *stable node*: the solutions move into the equilibrium point in a smooth non-oscillatory fashion. Two other examples are given in Figure 8.4 which are derived in a similar way, and a complete list of all the possible types of behaviour then follows.

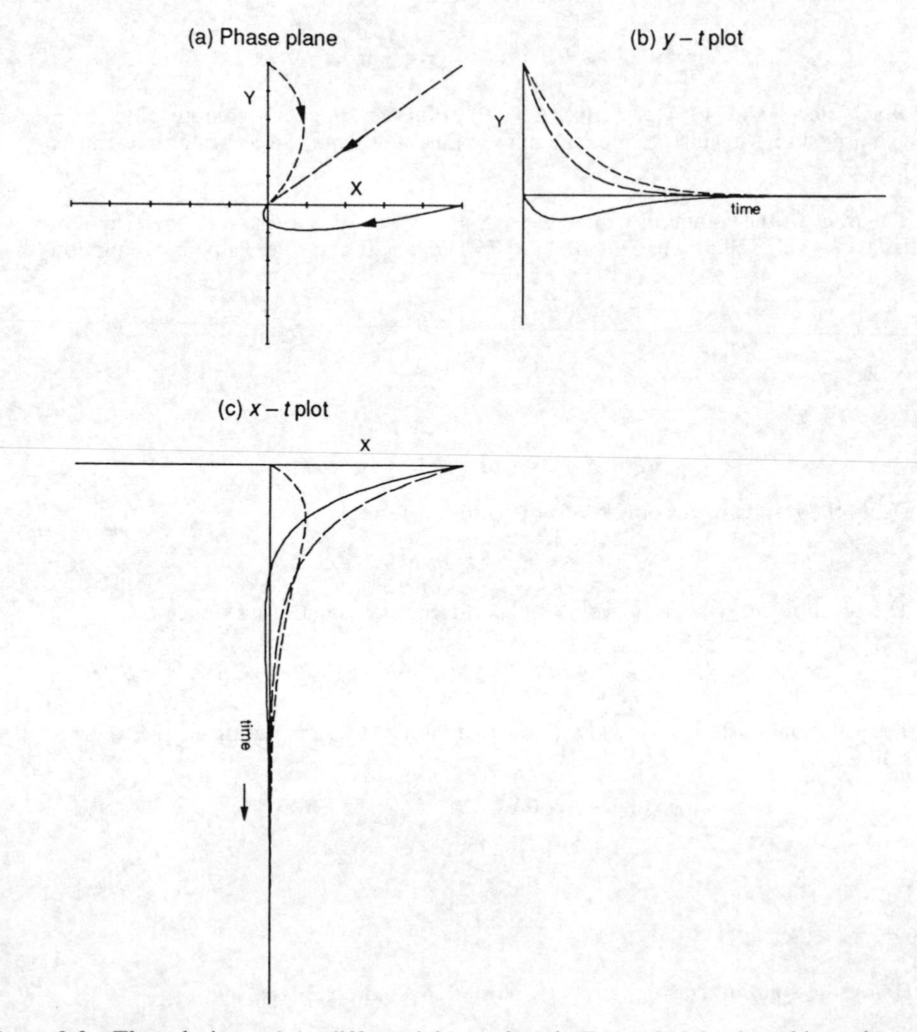

Figure 8.3 The solutions of the differential equations in Example 8.2, plotted in various ways.

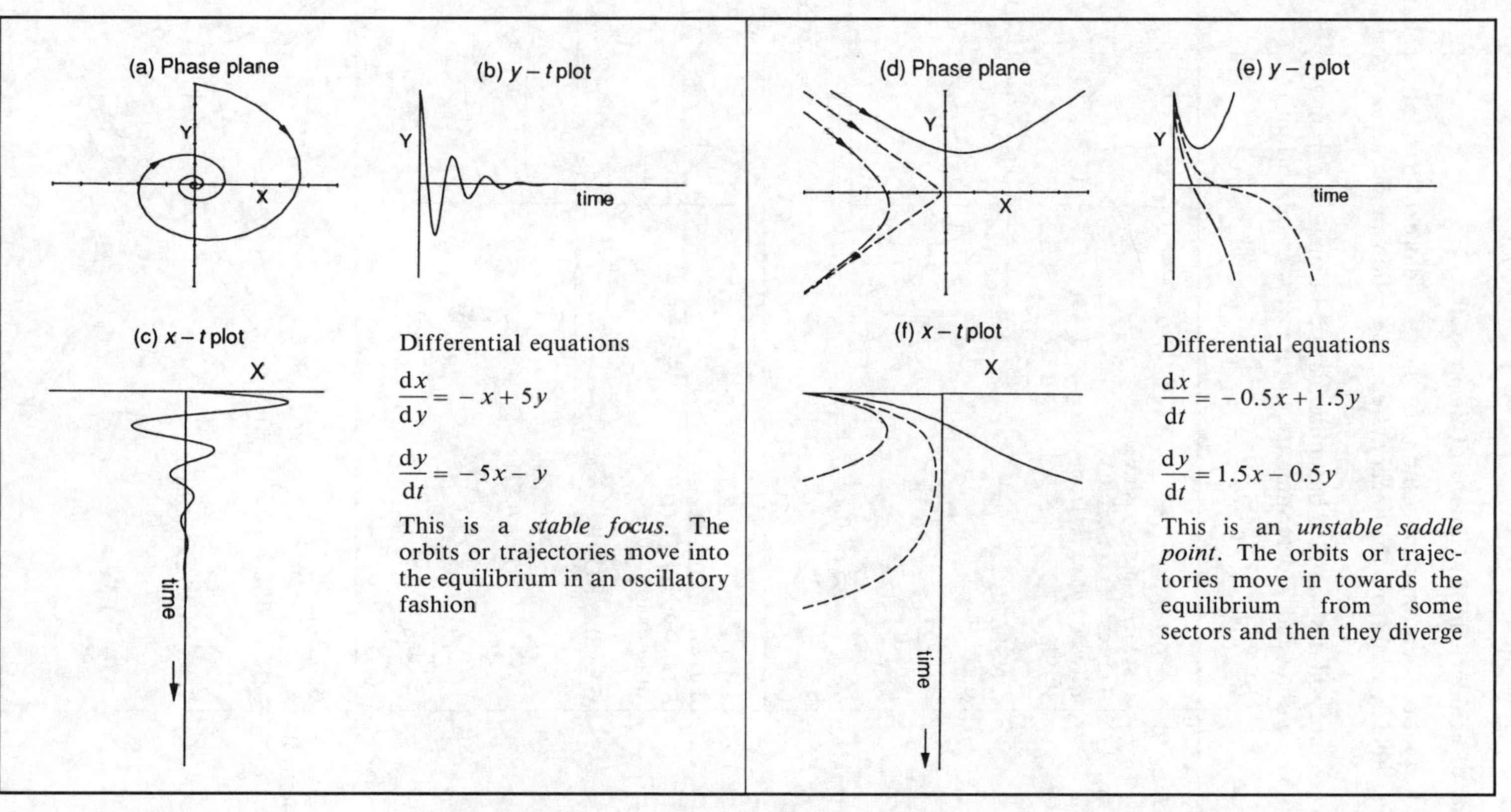

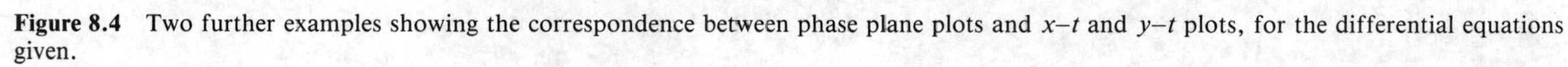

Figure 8.4 Two further examples showing the correspondence between phase plane plots and x–t and y–t plots, for the differential equations given.

8.5.3 Kinds of equilibrium of solutions of linear systems

By relating the type of solution to the solution of the second order differential equation in a single variable derived from the linear system, the following broad categories can be obtained.

These diverse possibilities can also be summarised in a single diagram (Figure 8.5) in which the value of D and T determine the nature of the equilibrium.

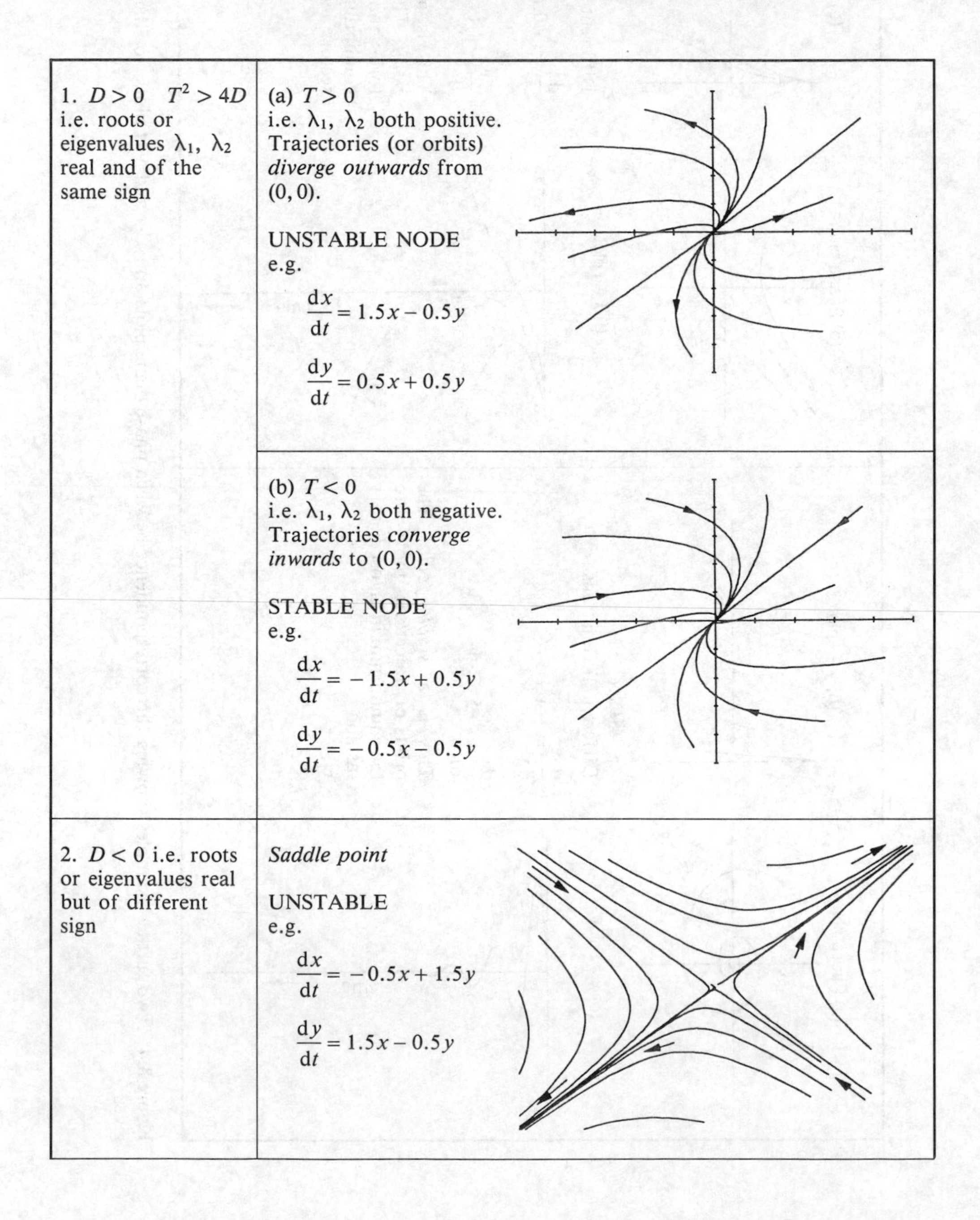

1. $D > 0 \quad T^2 > 4D$ i.e. roots or eigenvalues λ_1, λ_2 real and of the same sign	(a) $T > 0$ i.e. λ_1, λ_2 both positive. Trajectories (or orbits) *diverge outwards* from (0, 0). UNSTABLE NODE e.g. $\frac{dx}{dt} = 1.5x - 0.5y$ $\frac{dy}{dt} = 0.5x + 0.5y$
	(b) $T < 0$ i.e. λ_1, λ_2 both negative. Trajectories *converge inwards* to (0, 0). STABLE NODE e.g. $\frac{dx}{dt} = -1.5x + 0.5y$ $\frac{dy}{dt} = -0.5x - 0.5y$
2. $D < 0$ i.e. roots or eigenvalues real but of different sign	*Saddle point* UNSTABLE e.g. $\frac{dx}{dt} = -0.5x + 1.5y$ $\frac{dy}{dt} = 1.5x - 0.5y$

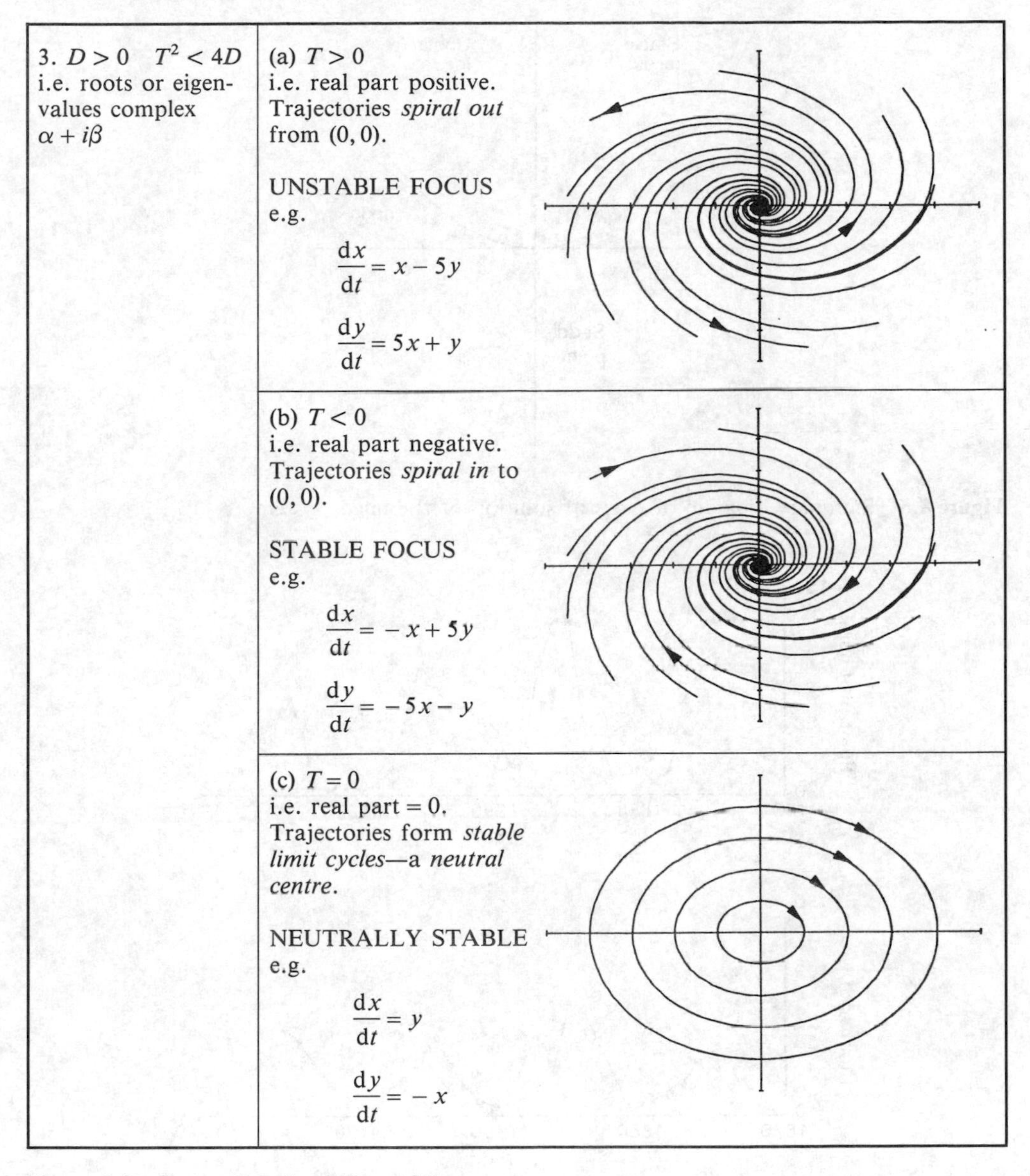

Figure 8.5 Categories of solution of linear systems

8.6 A SYSTEM OF DIFFERENTIAL EQUATIONS APPLIED TO PREDATOR–PREY INTERACTION

We now move on to consider a further model in which there are two populations of response variables which vary with time, and which interact, an ecological example which has been extensively modelled: the interaction between a predator and its prey. Celebrated data are those of the Canadian lynx and the snowshoe hare which, if the trapping records are anything to go by, have undergone remarkably regular oscillations over a long period (see Figure 8.6). The cycles seem to be approximately synchronised,

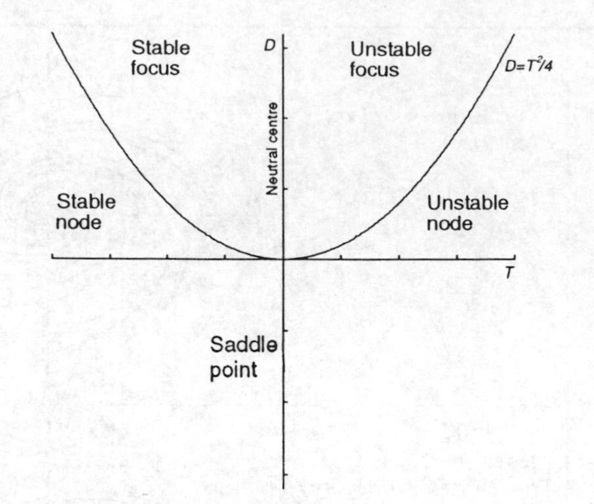

Figure 8.5 Summary diagram of types of solution of the linear system.

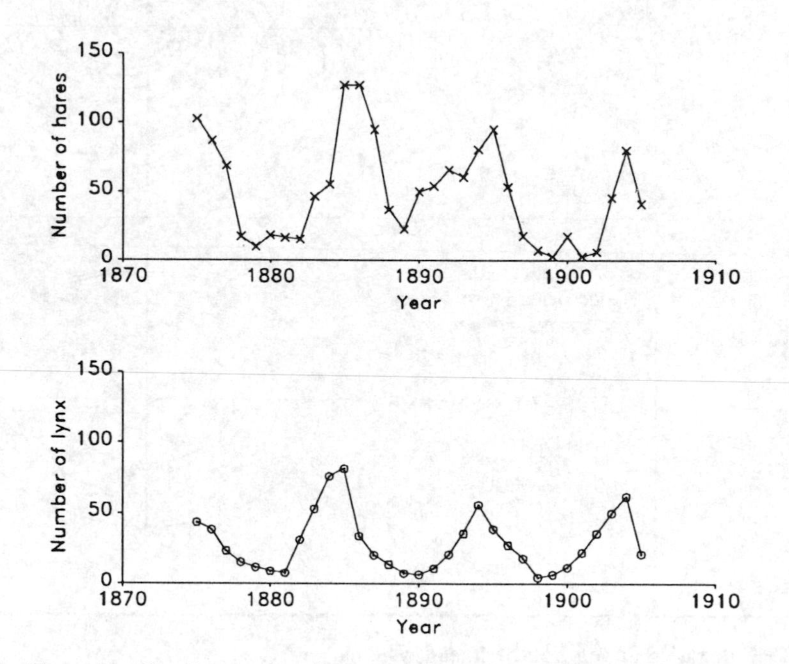

Figure 8.6 Trap records of snowshoe hares and lynx (kept by the Hudson Bay Company in Canada) plotted against year (data from Elton & Nicholson (1942)).

and early in the century Volterra (1926) and Lotka (1925) independently devised differential equations representing the simplest possible case of predator–prey interaction, in which cycles also occur.

The basic Lotka–Volterra model is a simpler model than the models of predation proposed in Chapter 10. Although unrealistic and in a sense pathological, it is used here to illustrate some of the properties of systems of differential equations. Let P and H be the density of predators and prey respectively at time t. It is possible to model these two populations separately ignoring any interaction between them. The predator population would decline in the absence of any predation and the simplest model we could use for this is exponential decline $\mathrm{d}P/\mathrm{d}t = -r_P P$ where r_P is the relative mortality rate, which can be solved to give $P = P(0)\,\mathrm{e}^{-r_P t}$, where $P(0)$ is the predator density at time $t = 0$. We now consider the prey population. In the absence of predation, this population would be likely to increase exponentially, with a differential equation $\mathrm{d}H/\mathrm{d}t = r_H H$, where r_H is the relative growth rate of the prey population. The solution is $H = H(0)\,\mathrm{e}^{r_H t}$, where $H(0)$ is the prey density at time $t = 0$.

How are these differential equations and their solutions modified when we introduce predation? We first consider how we would model the effect of predation on the prey population. It seems plausible that each predator would take prey at a rate proportional to prey density, i.e. predation depresses the growth rate of the prey population by a term which is proportional to the product of the density of predators and the density of prey. So $r_H H$ becomes $r_H H - aPH$ and the equation for the prey population becomes

$$\frac{\mathrm{d}H}{\mathrm{d}t} = (r_H - aP)H.$$

We would also expect that there would be a similar effect, but with a different constant of proportionality, on the growth rate of the density of predators. So $r_P P$ becomes $r_P P - bHP$. Thus we obtain the simple Lotka–Volterra model.

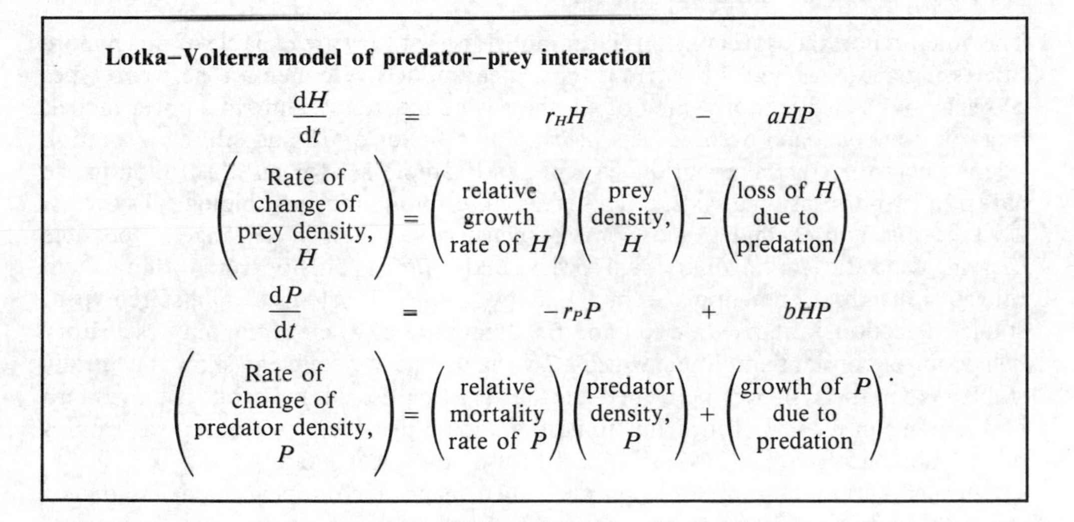

This model for predation is the simplest possible. It seems plausible that the number of encounters between predator and prey should be proportional to the product of the prey and the predator densities, HP, at least at moderate densities. There would be departures from this at low and high densities of prey, as predators would probably search more thoroughly at low prey densities in order to ensure a sufficient food intake, and similarly would reduce their search effort at high prey densities as they became satiated. These and other complications are discussed in Chapter 10.

8.7 LINEAR AND NON-LINEAR SYSTEMS

We now come to an important practical distinction. The techniques we can use for assessing the properties of a model consisting of a system of differential equations depend on whether the system is linear or non-linear. What do we mean by a linear and a non-linear system?

8.7.1 Discriminating non-linear and linear models

A simple way to assess whether a differential equation model is linear or not is to see whether the variables occur only as a sum of terms each of which is a constant times a variable, e.g. $2x + 3y$ rather than $2xy + 3$. By this yardstick, the Lotka–Volterra equations are non-linear because the variables H and P do not occur linearly. The terms reflecting the effects of predation in the differential equations are the products HP multiplied by constants a and b in the two equations. The constants do not matter, but it is the fact that the variables H and P do not occur linearly just as H and P but as the product HP that causes the problems. If H^2 or $\sin(H)$ had been in the equations, there would be similar difficulty. As the Fitzhugh–Nagumo model (described in Section 8.11) contains a cubic expression in v in the differential equation for dv/dt, it also is non-linear. On the other hand, the model used for drug turnover is linear, as the variables, x and y, only occur in linear combinations.

8.7.2 Properties of non-linear models

The other important aspect of non-linear models is that they are capable of much more interesting and meaningful behaviour than linear models. We mention here two types of behaviour as illustrations. First of all, there is no inherent scaling in a linear model; any solution for x and y could be scaled up by a factor of 10 and still be a solution of the linear differential equation system. See Example 8.3 for an illustration in the case of a two-dimensional system. This is not appropriate for most biological systems. To take just two examples, most environments have a restricted range of possible carrying capacities for any individual species, and neurones only operate within certain ranges of membrane potential. Secondly, many biological systems exhibit stereotyped, stable oscillations* of fixed period and fixed amplitude, whereas the only oscillatory behaviour of constant amplitude which a 2-variable linear system can show is neutrally stable. When the system is perturbed from its existing oscillatory orbit, it moves into and remains in its new orbit. Thus to accommodate just these two observed features of real biological systems, we are forced to use non-linear models.

Another important aspect to consider when using a non-linear model is that if the populations or components of the system being modelled are themselves heterogeneous, then the behaviour of the average need not be at all like the average of the behaviours of the separate strata if they were modelled separately (see Section 8.14.2).

Example 8.3 Linear systems and scaling effects

In the case of the general linear system with an equilibrium at (0, 0), show that the

---

* Stable oscillations are oscillations which the system returns to when slightly perturbed from the oscillatory pattern of fluctuation.

slope of any trajectories along a line of constant slope through the equilibrium point is constant. What are the implications of this?

The differential equations of the system are

$$\frac{dy_1}{dt} = a_{11}y_1 + a_{12}y_2 \qquad \frac{dy_2}{dt} = a_{21}y_1 + a_{22}y_2.$$

The slope of the trajectory at any point (y_1, y_2) is given by

$$\frac{dy_2}{dy_1} = \frac{a_{21}y_1 + a_{22}y_2}{a_{11}y_1 + a_{12}y_2}$$

Now consider the point which is c times further out from the equilibrium than (y_1, y_2) i.e. (cy_1, cy_2). The slope there is

$$\frac{dy_2}{dy_1} = \frac{a_{21}cy_1 + a_{22}cy_2}{a_{11}cy_1 + a_{12}cy_2} = \frac{a_{21}y_1 + a_{22}y_2}{a_{11}y_1 + a_{12}y_2}$$

which is the same as previously. The consequence of this is that the trajectory starting at (cy_1, cy_2), for $c \neq 1$ is identical with that starting at (y_1, y_2) except that it is scaled up by a factor c. Thus there is no possibility of inherent scaling in a linear model; any scale of the behaviour, whatever the behaviour is, about the equilibrium is possible.

8.8 TECHNIQUES FOR NON-LINEAR MODELS

The unfortunate thing about *non-linear* differential equations like the Lotka–Volterra equations is that they cannot be solved analytically as we did the differential equation for the motion of the pendulum, or the differential equation of growth with constant relative growth rate. For example, the solution of the differential equation we considered above for the growth of the prey in the absence of predation, $dH/dt = r_H H$, is $H = H(0)\exp(r_H t)$. For some special classes of differential equations and systems of differential equations, we can obtain explicit solutions such as this which involve the parameters such as r_H. This is a very powerful approach in that it enables us to see quickly and easily what the solution is for any numerical value of the starting point, e.g. $H(0) = 3$, and for any particular numerical value of r_H, e.g. $r_H = 0.10$, which is $H = 3\exp(0.10t)$. This is an ideal way to proceed, where we harness the power of mathematics to obtain the general form of the solution as a first step. Linear *systems* of differential equations can also be solved in this way (see Section 8.5).

For non-linear systems like the Lotka–Volterra model on the other hand, an analytical solution cannot be obtained. There are two complementary approaches. The first is a simple numerical solution for particular values of the parameters and for particular starting values of H and P. For details about methods of numerical solution see Section 8.13. An example for the Lotka–Volterra model for parameter values $r_P = 0.7$, $b = 0.025$, $r_H = 0.7$, $a = 0.055$, and starting values $H(0) = 5$, $P(0) = 5$ is given in Figure 8.7. This is interesting and useful, and shows that the solution of the differential equations starting from these values and with this selection of parameter values is similar in overall behaviour to the data. However, there are many possible values of the starting points of the numerical solutions, and also many possible values of the parameters. How does the solution change when the parameter values are varied? How does it change as we change the starting points? These and related matters are considered in Section 8.8.1 onwards.

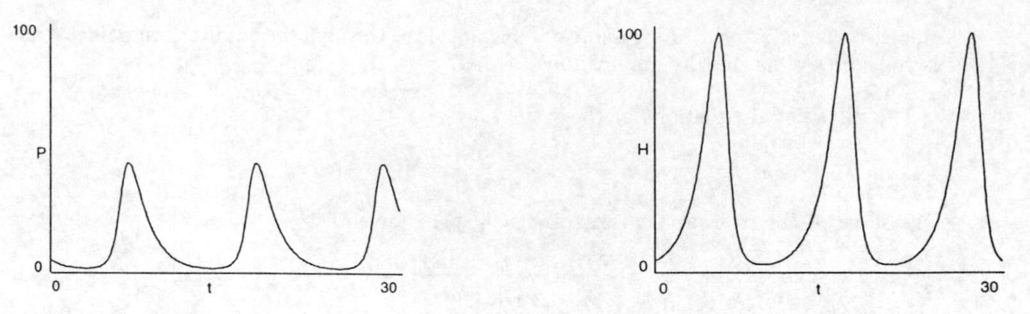

Figure 8.7 (H, t) and (P, t) plots of that solution of the Lotka–Volterra predation model with parameter values given in the text which starts at the point $(H, P) = (5, 5)$.

8.8.1 Qualitative behaviour of solutions

For non-linear systems, another approach consists of applying a battery of techniques designed to elucidate the *qualitative* rather than the detailed quantitative behaviour of the solutions. In the early stages of a modelling exercise, we are often not interested in the exact solution of the equations but only in the general pattern of behaviour of the solution. We usually do not know the exact values of the parameters of the model anyway, or cannot estimate them very well. For example, in the case of the lynx/hare data, the parameters r_H, r_P, a, b are not known at all precisely. In these circumstances, what we wish to know is (i) that the model can reproduce behaviour that is qualitatively similar to that we observe in the real data, (ii) that the behaviour does not change dramatically as the parameters are changed slightly, and (iii) that there are no implausible features of the behaviour of the model either for some sets of parameter values, or for certain values of H and P.

This approach involves the display of numerical solutions in various ways, both as plots against time as above, referred to in this case as $H - t$ and $P - t$ plots, and as *phase plane* plots, i.e. plots of P against H. The approach is also of use for determining whether there are equilibrium solutions of the equations and for examining the nature of the equilibria.

8.9 GRAPHICAL DISPLAYS OF MODELS

The first question then is how can we display the numerical solutions of the model, so that we can consider its behaviour, and also so that we might compare the observed data with the proposed model.

8.9.1 Variable *vs* t plots

The most obvious plots we can make are of each variable against time. Figure 8.6 gives a variable–t or v–t plot of the lynx/hare data for both variables, and it is immediately obvious that the two species are cycling and are very approximately synchronised. This is also true of the v–t plots for the model (Figure 8.7).

8.9.2 Phase plane plots

Phase plane plots have already been introduced and numerous examples given in the context of linear systems (Sections 8.5.2 onwards). A further illustration, this time for the Lotka–Volterra model, is given in Figure 8.8. On this figure, the same time course as for the variable-t plots of the model given in Figure 8.7 is plotted. Such maps in the phase plane or phase plane plots enable us to see directly many aspects of the behaviour of the solutions. For example, they show whether the components of the system vary in a positively or negatively correlated fashion, or whether they vary independently; whether they settle into cyclical orbits and if so, how they cycle—are they in or out of phase—or alternatively whether they approach equilibrium. In the case of the lynx/hare data, a cycle is indicated by the plotted solution being a closed trajectory.

Elements and properties of phase plane plots

Multiple trajectories
One big advantage of phase plane plots is that they enable us to plot trajectories starting at a number of points so that we can see how the the behaviour of the system varies as we change the starting point. A number of examples are plotted for the Lotka–Volterra model in Figure 8.9. Here the trajectories form a series of concentric anticlockwise cycles on an (H, P) plot. The bunching up of the trajectories in the regions closest to the axes and considerable spacing out on the other side of the equilibrium suggest that slight changes in the number of predators and prey at small numbers could induce substantial changes in the peak numbers of both species. The concentric nature of the trajectories suggests—if it is maintained into the close vicinity of the equilibrium (it is, although not plotted here)—that the equilibrium will be neutrally stable, and in fact the model shows neutrally stable limit cycles. If the model populations are perturbed from one limit cycle due to chance influences, they will remain in the new limit cycle until a further influence not contained in the model shifts them out of it.

Nullclines and singular points
Often a useful first step is to draw in on the phase plane plot curves called

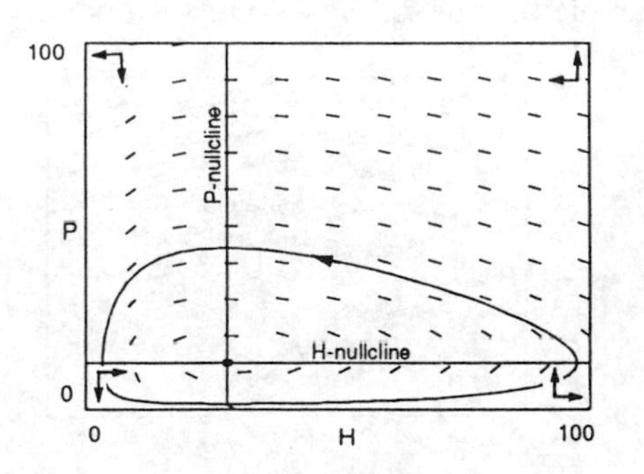

Figure 8.8 A phase plane plot of the solution of the Lotka–Volterra model corresponding to the time series plots in Figure 8.7. Also plotted are the non-zero P- and H-nullclines, and direction fields positioned at the intersections of a 10 by 10 grid over the area of the plot.

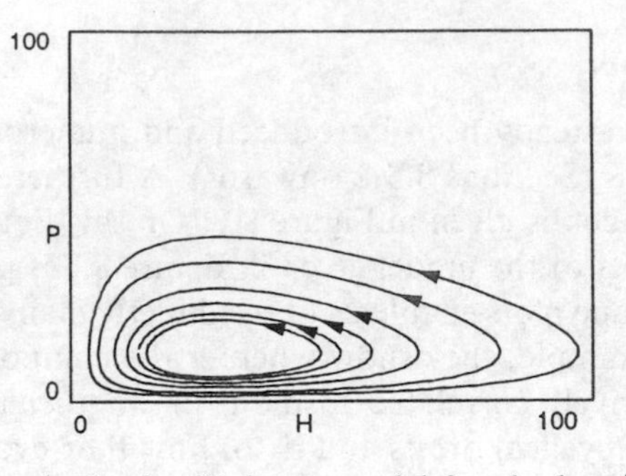

Figure 8.9 Trajectories for the Lotka–Volterra model for the lynx/hare data.

nullclines, connecting all the points in the phase plane for which the rate of change of each of the variables is zero. For example, in the case of the Lotka–Volterra model, the H-nullcline is the line connecting up all the values for which $dH/dt = 0$, i.e. $H(r_H - aP) = 0$, i.e. $H = 0$ and $P = r_H/a$. The equations of the P-nullcline are $P = 0$ and $H = r_P/b$. In this case, there are two nullclines for each variable. The non-zero nullclines are plotted on Figure 8.8.

There are two useful properties of the nullclines. The first is that they intersect at *equilibrium values*, as at the intersection both derivatives are zero. Equilibrium values are of obvious importance, as they tell us that it is possible within a range of values to obtain static solutions, and where they are. Other terms for equilibrium values are *stationary* or *singular points*. So it can be a quick graphical method of finding equilibrium values. The second is that in each region of the phase plane of which the nullclines form boundaries, the direction of flow is determined approximately. There are four possibilities ($H\downarrow P\uparrow$), ($H\uparrow P\uparrow$), ($H\uparrow P\downarrow$) and ($H\downarrow P\downarrow$) or graphically presented (↵↑), (↑↳), (↓↱) and (↰↓). So if the phase plane plot is being drawn manually, rather than using some computer program, sketching in the nullclines and noting these directions (Figure 8.8) can be very helpful for understanding the broad nature of the flow.

Direction flows

Instead of plotting trajectories starting at an arbitrary set of points, or a set determined by consideration of a particular model, we can plot them starting at a grid of points regularly spaced over the whole phase plane. These are called *direction flows*. See Figure 8.10 for an example for the Lotka–Volterra model.

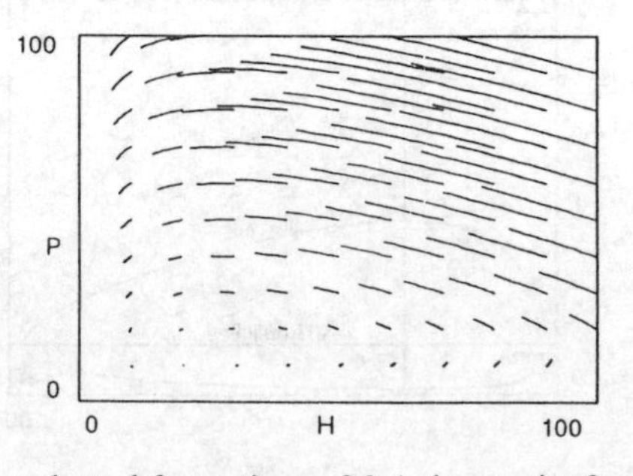

Figure 8.10 Direction flows plotted for a time of 0.1 time units for the Lotka–Volterra model for the lynx/hare data, starting at the points of a 10×10 grid over the area of the phase plane. The direction of flow can be determined since the initial points lie on the grid.

Typically each trajectory is plotted for a relatively short time, and such plots then enable us to see two features of the behaviour of a model. They enable us to trace approximate trajectories over whichever regions of the whole plane we wish, by mentally interpolating between the flows starting at the grid points. Secondly, unlike single trajectories plotted out for a long time, they give us some information about the rates at which the variables are changing. The regions of the plane in which the flows are longest are those in which the rates of change are the greatest. The direction flows for the lynx/hare model show rather slow movement for low values of H and P but much faster movement when both are high, as might be expected from an examination of the model, and they reveal that the trajectories correspond to anticlockwise cyclical movement, confirmed by the plots of individual trajectories in Figure 8.9.

Direction fields

A related concept to direction flow is that of a *direction field*. This consists of line segments all of the same arbitrary length plotted out at each point of a grid indicating only the direction of the trajectory through each point (see Figure 8.8). These are much easier to calculate than direction flows, requiring only one evaluation of the gradient for every grid point. However, as both are often evaluated using a computer program, this difference is of no practical significance. In this case, they confirm the generally cyclical appearance of the model behaviour. Note also in Figure 8.8 that the direction fields at the nullclines are parallel to the axes, as they should be by definition.

8.10 LOCAL STABILITY ANALYSIS

We have seen above how to obtain graphically the equilibrium or singular points of a system of two differential equations. The coordinates in the phase plane of the singular points can also often be found algebraically, or failing that numerically. For example, for the Lotka–Volterra equation we need to solve the two equations

$$(r_H - aP)H = 0 \qquad (-r_P + bH)P = 0.$$

The first equation is satisfied when either $P = r_H/a$ or $H = 0$; the second when either $H = r_P/b$ or $P = 0$. There are only two ways simultaneously to satisfy both of the equations. So we have two possible singular points $(H, P) = (r_P/b, r_H/a)$, $(0, 0)$.

It is not sufficient just to find the equilibrium values as it is possible that the equilibrium values are unstable, and so in any practical situation could not be held. How then do we apply the notions of stability and instability discussed in the context of a linear system (in Section 8.5.1) to the two equilibria of the non-linear Lotka–Volterra model? We set out the procedure in Example 8.4 below. The rationale of this method is that we convert the non-linear differential equations into approximate linear differential equations which hold only in the vicinity of the equilibrium point, and then consider the stability of the resulting linearised system.

Example 8.4 Determining the stability of equilibria of the Lotka–Volterra predation model

Stage 1. Find the equilibrium values by solving the equilibrium conditions for the two differential equations

$$\frac{dH}{dt} = (r_H - aP)H = 0$$

$$\frac{dP}{dt} = (-r_P - bH)P = 0.$$

The first condition is satisfied when either $P = r_H/a$ or $H = 0$; the second condition is satisfied when either $H = r_P/b$ or $P = 0$. If we take the first equilibrium of the first equation $P_{E1} = r_H/a$, the only way of simultaneously satisfying the equilibrium condition of the second equation is $H_{E1} = r_P/b$. Taking the second equilibrium of the first equation, $H_{E2} = 0$, the only way of simultaneously satisfying the equilibrium condition of the second equation is $P_{E2} = 0$. Thus the two equilibria are $(H, P) = (r_P/b,\ r_H/a)$ and $(0, 0)$.

Stage 2. Rewrite the differential equations in terms of the small *departures from the equilibrium values* (and form a linear approximation), i.e. rewrite in terms of a linear expression in h and p where $h = H - H_E$ and $p = P - P_E$, where (H_E, P_E) are the coordinates of the equilibrium.

Stage 2 Method (a). Direct substitution illustrated on the first equation which is transformed from

$$\frac{dH}{dt} = H(r_H - aP) \text{ to}$$

$$\frac{d}{dt}(H_E + h) = \frac{dh}{dt} = (H_E + h)[r_H - a(P_E + p)]$$

$$= [H_E(r_H - aP_E)] + [h(r_H - aP_E - ap) + H_E(-ap)]$$

The contents of the first square bracket are zero since (H_E, P_E) is an equilibrium point, and so the equation reduces to

$$\frac{dh}{dt} = h(r_H - aP_E) + p(-aH_E).$$

if we ignore terms of higher order than the first in h and p.

Stage 2 Method (b). Taylor series expansion illustrated on the second equation. The Taylor series expansion of any function, $f(H, P)$ about the point (H_E, P_E) is given by

$$f(H,P) \cong f(H_E, P_E) + h\frac{\partial f(H_E, P_E)}{\partial H} + p\frac{\partial f(H_E, P_E)}{\partial P} + \begin{pmatrix}\text{terms of second order}\\ \text{in } h \text{ and } p\end{pmatrix}.$$

For the second differential equation,

$$f(H, P) = P(-r_P + bH) \text{ so}$$

$$\frac{\partial f(H_E, P_E)}{\partial H} = bP_E, \qquad \frac{\partial f(H_E, P_E)}{\partial P} = -r_P + bH_E.$$

Hence the second equation reduces to

$$\frac{dp}{dt} = h(bP_E) + p(-r_P + bH_E).$$

since $f(H_E, P_E) = 0$, and we are ignoring higher order terms.

Stage 3. Determine the qualitative behaviour of the resulting linear system. We saw (in Section 8.5) that the behaviour depends on the value of the trace and the determinant of the matrix, **A**, given for this model by

$$\mathbf{A} = \begin{pmatrix} r_H - aP_E & -aH_E \\ bP_E & -r_P + bH_E \end{pmatrix}$$

known as the *community matrix.*

The solution is

$$h = A_1 e^{\lambda_1 t} + A_2 e^{\lambda_2 t}$$
$$p = B_1 e^{\lambda_1 t} + B_2 e^{\lambda_2 t}$$

where λ_1, λ_2 are eigenvalues of the matrix, **A**, i.e. the roots of the quadratic equation

$$\lambda^2 - T\lambda + D = 0$$

where $T = \text{trace}(\mathbf{A})$ and $D = \det(\mathbf{A})$
i.e.

$$T = r_H - aP_E - r_P + bH_E, \qquad D = (r_H - aP_E)(-r_P + bH_E) - (-aH_E)(bP_E).$$

Consider the two equilibria separately.

Equilibrium 1. $(H_E, P_E) = (r_P/b, r_H/a)$

$$T = r_H - r_H - r_P + r_P = 0$$
$$D = (r_H - r_H)(-r_P + r_P) + a(r_H/a)b(r_P/b) = r_P r_H.$$

By reference to Figure 8.5, the solution of the linearised differential equations for small perturbations from this equilibrium is thus a neutral centre, i.e. both variables exhibit simple harmonic motion—periodic fluctuations of constant amplitude. The solution is of the same form for both H and P, except that they are out of phase, and thus we obtain elliptical orbits of constant amplitude. Thus the equilibrium is said to be neutral. Once the solution is perturbed from the equilibrium, it neither returns towards nor moves further away from the equilibrium. It stays in its new orbit.

Equilibrium 2. $(H, P) = (0, 0)$. Thus $T = r_H - r_P$, $D = (r_H)(-r_P) = -r_H r_P$

By reference to Figure 8.5, as $D < 0$, the solution of the linearised equation is a saddle point, and therefore this equilibrium is unstable. Thus if there were a small simultaneous introduction of predator and prey, this analysis would indicate that the populations would grow. However, as we saw in Chapter 4 when discussing demographic stochasticity, it is at such small population densities that the deterministic model is most unreliable. So this result is of no value in interpreting any real fluctuations, except to show that it is *possible* for substantial populations of predator and prey to grow from very small numbers.

Another important aspect of the above analysis is that, for the non-linear system, we are only making inferences about *local behaviour*. Once the system has moved away from the immediate vicinity of the equilibrium, the linear approximate model ceases to hold and the behaviour of the system cannot be predicted from it. An example of this is seen in the analysis which follows of the Fitzhugh–Nagumo model of nerve-cell firing.

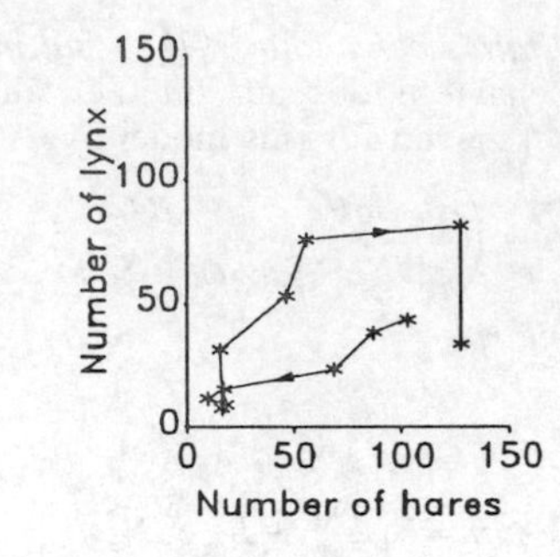

Figure 8.11 A phase plane plot of a section of the lynx/hare data.

8.10.1 Do hares eat lynx?

One interesting thing that the lynx–hare phase plane plots (of P on the y-axis against H on the x-axis as in Figure 8.11) reveal is an important discrepancy between the data and an otherwise reasonably well-fitting model which is not very obvious from the $H - t$ and $P - t$ plots. The *model* shows periodic anti-clockwise cycles; first the prey population increases, then the predator, causing the prey to decline after which the predator population follows suit. This seems quite plausible. However, for at least one of the cycles the *data* follow a clockwise cycle, which could be obtained in this model only if hares were the predators and lynx the prey. The article entitled 'Do hares eat lynx?' by M. E. Gilpin, 1973, discusses the problem and provides a possible resolution: the numbers perhaps reflect the fluctuations in the habits of the human hunters whose catch records constitute the data more than the dynamics of the lynx and the hare.

8.11 EXCITABILITY AND CYCLES: THE FITZHUGH–NAGUMO MODEL OF NEURONE BEHAVIOUR

8.11.1 Introduction

Generally speaking neurones are continually receiving inputs from other neurones, and often remain passive until the total stimulation from the other neurones reaches a threshold. Then the neurone itself fires, i.e. it undergoes a rapid reversal of its membrane potential and relays copies of this along its axon to its terminals for onwards transmission to other neurones (see Figure 8.12 for a schematic representation of a neurone). Figure 8.13 shows the kind of change which occurs in membrane potential during such a discharge. In the 1950s great strides were made towards understanding the events which are involved, in the case of the giant squid axon, in a series of ingenious experiments and mathematical models of Hodgkin and Huxley (1952). The details of their models are too advanced to discuss here, but for our present purposes a simpler model which captures most of the essential behaviour of the Hodgkin–Huxley model will suffice, devised independently by Fitzhugh and Nagumo (Fitzhugh, 1961; Nagumo *et al.*, 1962). We ignore spatial aspects of the transmission of the discharge along the axon. The resulting model is only applicable to the space-clamped axon, where the axon is specially prepared so that all points on it are at the same potential. This model involving two non-linear differential equations is as follows. The

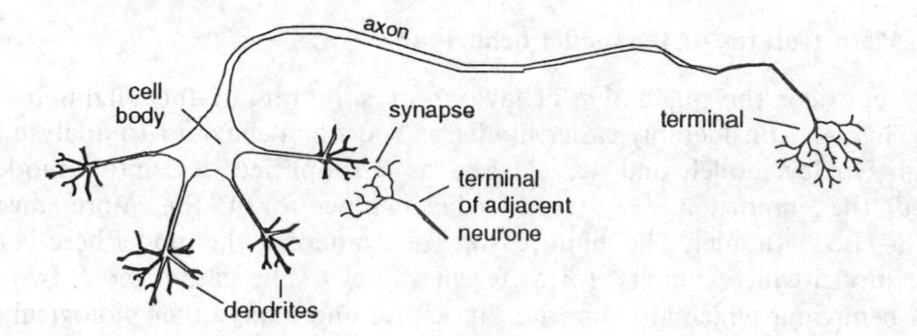

Figure 8.12 The main components of a neurone.

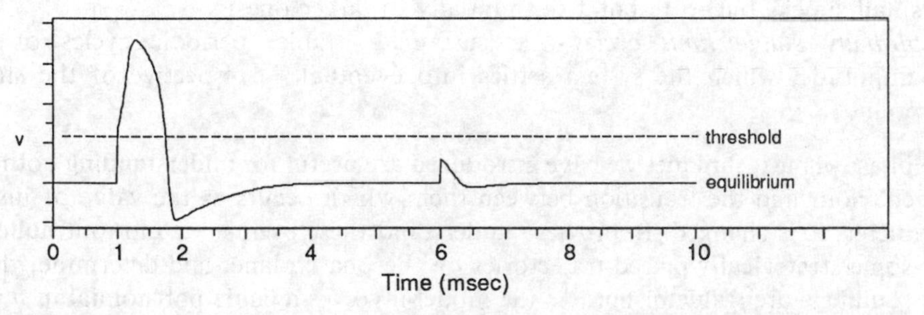

Figure 8.13 Membrane potential changes during the firing of a neurone, as a result of a supra- and sub-threshold stimulation, at $t = 1$ and $t = 6$ respectively.

first variable, v, is the departure of the membrane potential from its equilibrium; the second is a general recovery variable reflecting voltage-dependent changes in ion conductances. The model is linear apart from the cubic term in v in the first equation. This gives the model its interesting behaviour. If there were no w term in the first equation, then the equation would describe bistable dynamics. There are three equilibria at $v = 0$, $v = v_1$, and $v = v_2$, and only the outer two are stable (if the w-term is absent and $I = 0$). Suppose $0 < v_1 < v_2$. Then if v is moved from its equilibrium at zero up to a level below v_1, it will return to zero; if perturbed from equilibrium beyond v_1, it will move towards and remain at the stable equilibrium at v_2. This is the feature of the dynamics of v which is responsible for the phenomenon of excitability in the full equations which is discussed below.

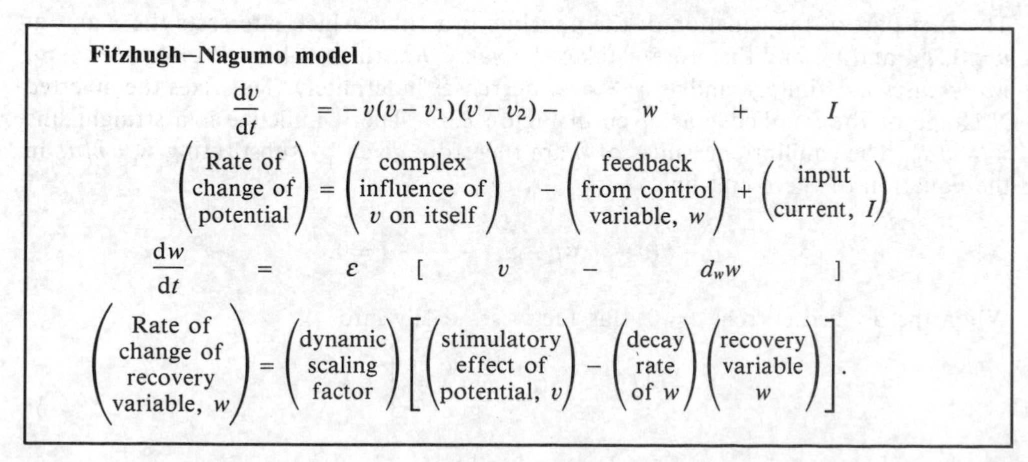

8.11.2 Main features of the model behaviour

We now consider the qualitative behaviour of solutions of the Fitzhugh–Nagumo model. This is a considerably easier model of neurone behaviour to analyse than the Hodgkin–Huxley model, and was devised as a simplified descriptive model which exhibited the important features of the behaviour of the more mechanistic Hodgkin–Huxley model. The main reason for examining the model here is that it is a simple model which exhibits, for different values of the parameter I, two separate types of behaviour which are important in neural and many other biological systems:

(1) *excitability*—the propensity of a system to return very quickly to equilibrium after small shocks but to respond substantially to larger ones;
(2) *globally stable limit cycles*—i.e. sustained, stable, periodic cycles of fixed amplitude, which the system settles into eventually irrespective of the starting point.

The phase plane techniques we have introduced are useful for understanding both types of behaviour and the transition between them which occurs as the value of just one parameter, I, is changed. In order to understand these events, we plot out nullclines, and some strategically placed trajectories on the phase plane, and determine whether any equilibria are stable or not. As the model involves a cubic polynomial in v which is not readily amenable to analytical solution, we use mainly numerical techniques, carried out in a numerical and graphical package for solution of differential equations (such as *Phaseplane*).

8.11.3 Nullclines and equilibria

The v-nullcline is given by

$$\frac{dv}{dt} = -v(v - v_1)(v - v_2) - w + I = 0$$

i.e.

$$w = -v(v - v_1)(v - v_2) + I.$$

The first part of the equation of the nullcline is a cubic which intersects the v-axis at $v = 0$, v_1, and v_2, and therefore w takes the value I at these values of v. As $v \to -\infty$, w increases indefinitely, and as $v \to \infty$, w decreases indefinitely. This fixes the inverted N shape of the v-nullcline as given in Figure 8.14. The w-nullcline is a straight line $w = v/d_w$. The equilibrium values of v are therefore given by substituting $w = v/d_w$ in the equation of the v-nullcline,

$$-v(v - v_1)(v - v_2) - \frac{v}{d_w} + I = 0.$$

When the applied current $I = 0$, this factorises easily into

$$-v\left[(v - v_1)(v - v_2) + \frac{1}{d_w}\right] = 0$$

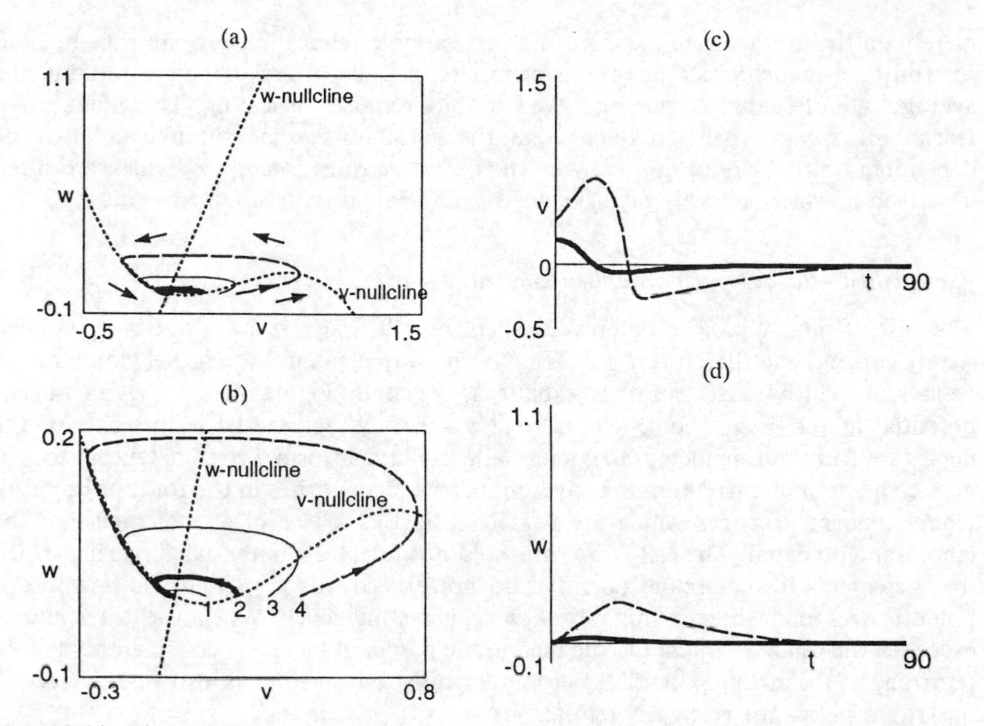

Figure 8.14 Fitzhugh–Nagumo model when the applied current $I = 0.0$. (a) a phase plane plot on which are plotted the v- and w-nullclines, and, given in more detail in (b), trajectories starting at $(0.1, 0)$, $(0.2, 0)$, $(0.3, 0)$ and $(0.4, 0)$, indicated by 1, 2, 3, 4. (c) and (d) give (v, t) and (w, t) plots respectively for the two trajectories, marked in bold or dashed lines in (a) and (b) starting at $(0.2, 0)$ and $(0.4, 0)$. The arrows in (a) indicate the general direction of flow in each region of the phase plane.

which has a solution $v = 0$. The quadratic factor remaining

$$v^2 - v(v_1 + v_2) + v_1 v_2 + \frac{1}{d_w} = 0$$

has solutions

$$v = \frac{v_1 + v_2 \pm \sqrt{(v_1 + v_2)^2 - 4v_1 v_2 - 4/d_w}}{2}.$$

For most biologically meaningful values of the parameters, these other roots are complex, and so are of no interest, and similarly there is only one real root of the cubic which results when $I \neq 0$. Thus there is a single equilibrium of the model for most values of I, which is at $v = 0$ when $I = 0$, and is at positive values of v when $I > 0$ and at negative values when $I < 0$.

In the remainder of this section, unless otherwise stated, we consider the specific instance of the model when the parameters take the following values: $\varepsilon = 0.02$, $v_1 = 0.2$, $v_2 = 1$, $d_w = 0.5$. The very low value for ε ensures that the dynamics of the recovery variable, w, are rather slow. ε and similar rate parameters are convenient parameters to include in a model as changes in them have no effect on the positions of the nullclines, and sometimes no effect on the qualitative pattern of behaviour, but merely on the average rate at which the two variables change. They can thus be used for tuning a model which has the appropriate overall pattern of behaviour, but the average rate of change of one or more variables requires re-scaling. This illustrates a feature of *some* dynamic models: that the variables can be subdivided into sets depending on the relationship between their average rates of change, which are often described as variables with *fast dynamics*, or variables with *slow dynamics*.

Excitability—the case of no applied current, I = 0

The case of most biological interest is when the cell is at rest and there is no applied steady current, and therefore $I = 0$. What we need to explain is the rather standardised behaviour which most neurones exhibit as given in Figure 8.13. The membrane potential at rest is at about -70 mV (at $v = 0$ since our model is in terms of the departure form resting membrane potential), and under normal circumstances receives no steady input of current, but receives inputs via its dendrites in the form of impulses from adjacent neurones which are perceived by the cell as brief increments of the membrane potential. The cell has a threshold of membrane potential for firing: when the increments in membrane potential do not exceed this threshold, the membrane potential responds by declining back to the resting level. When the threshold is exceeded the cell itself fires, i.e. the membrane potential increases considerably before returning to the resting potential. Before returning to rest, there is also a characteristic overshoot below the resting potential.

In this case ($I = 0$) the nullclines intersect on the left hand falling arm of the v-nullcline near the minimum (see Figures 8.14(a) and 8.14(b)), and therefore at values of w which are also less than the maximum value achieved on the middle portion of the v-nullcline. The single equilibrium of the system is at $(v, w) = (0, 0)$ and this equilibrium is stable. The eigenvalues of the linearised system are $(-0.105 + 0.105i, -0.105 - 0.105i)$, and thus the linearised system behaves locally like a stable focus. We consider the effects of perturbing the system from this equilibrium by incrementing v by varying amounts: $\Delta v = 0.1, 0.2, 0.3, 0.4$. Figures 8.14(a) and 8.14(b) demonstrate the results on the phase plane, and, in Figure 8.14(c) and 8.14(d), as $v - t$ and $w - t$ plots for two cases $\Delta v = 0.2$ and 0.4. For $\Delta v = 0.1$, the perturbed starting point is $(0.1, 0)$. In the sector containing this point, $\mathrm{d}v/\mathrm{d}t < 0$ and $\mathrm{d}w/\mathrm{d}t > 0$, and so there is a move upwards and to the left until the w-nullcline is crossed, and then w starts to decrease. The next event is when the v-nullcline is crossed, this time at negative values of v, and then the system starts to home in on the equilibrium $(0, 0)$. For a starting point at $(0, 0.2)$, the pattern is almost the same. This time the starting point is on the v-nullcline, and therefore the trajectory is initially parallel to the w-axis. As soon as it moves away from the v-nullcline, it is in the region of falling v and increasing w, and therefore the solution is then like that starting at $(0.1, 0)$, although the trajectory is radially further out from the origin until the two trajectories fuse close to the origin.

The trajectory starting at $(0.3, 0)$ is qualitatively different: this trajectory starts in the region of increasing v and increasing w, and therefore the trajectory initially moves up and to the right. In this case it is only for a brief period, as the v-nullcline is fairly quickly crossed and then the solution behaves as in the two previous cases. However, as the starting point moves further to the right, a new pattern begins to emerge. The trajectory starting at $(0.4, 0)$ moves upwards and to the right until it crosses the

v-nullcline on its rightmost falling portion, and only then does v begin to decline, and in this case, there is a substantial overshoot of $v = 0$, since the only way to return to the origin is over the top of the hump of the v-nullcline, and at these values of w, the speed of decline of v is very great, resulting in an overshoot, and v cannot increase anyway until the v-nullcline is crossed. Figure 8.14 illustrates the plots of v vs t and w vs t for the two trajectories starting at (0.2, 0) and (0.4, 0). The plot for v vs t is quite like that of the neurone whose behaviour is sketched in Figure 8.13.

As stated above, this sudden change in behaviour is dependent on the shapes of the nullclines and on the different scales of the dynamics of the two variables, w having very much slower dynamics than v. If the relative speed of w is decreased further by setting ε to 0.002, then the split between just sub- and just supra-threshold stimulations becomes more marked (see Figure 8.15).

Sustained periodic oscillations—the case of constant positive applied current, $I = 0.7$

We now increase I to 0.7, i.e. we apply a substantial steady current to the cell (and revert to $\varepsilon = 0.02$). The effect of this change in current is to move the v-nullcline upwards in the phase plane so that it intersects the line $v = 0$ (not plotted) at (0.0, 0.7), as in Figure 8.16, rather than (0.0, 0.0), as before. The intersection is now on the middle arm of the v-nullcline, and the resulting equilibrium at (0.370, 0.740) is unstable. The eigenvalues of the linearised system are 0.159 and 0.108 which are characteristic of an unstable node. A number of trajectories demonstrate that whatever the starting point, the system develops periodic oscillations of constant amplitude. If the system is perturbed from the cycle, it returns, and therefore this is a stable limit cycle. Figure 8.16 gives $v - t$ and $w - t$ plots for two of the trajectories indicated by bold and dashed lines. These indicate that the cycles are indeed strictly periodic. Thus as a result of a change in just one parameter the solution has completely changed its character. This behaviour is also demonstrated by real neurones. If given a sustained input current, many neuronal systems do go into sustained regular firing of this kind.

The further point should be made that in this case the local instability in the vicinity of the only equilibrium is no guide to the global stability. Indeed it quite frequently

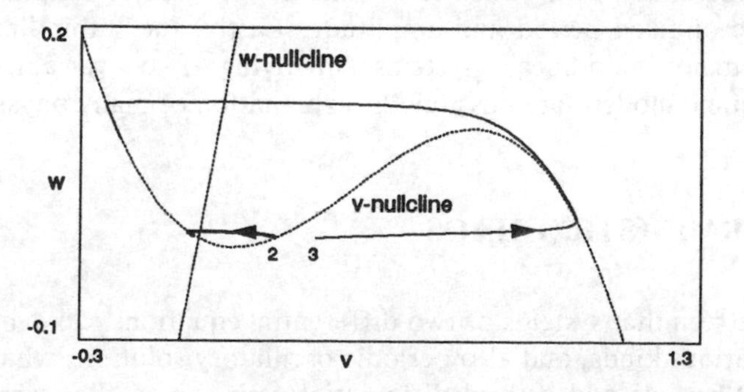

Figure 8.15 A phase plane plot of the Fitzhugh–Nagumo model with applied current $I = 0.0$, with 10-fold slower dynamics for w than in Figure 8.14, indicated by ε which takes the value 0.002. Note the much sharper bifurcation between responses to sub- and supra-threshold stimulations.

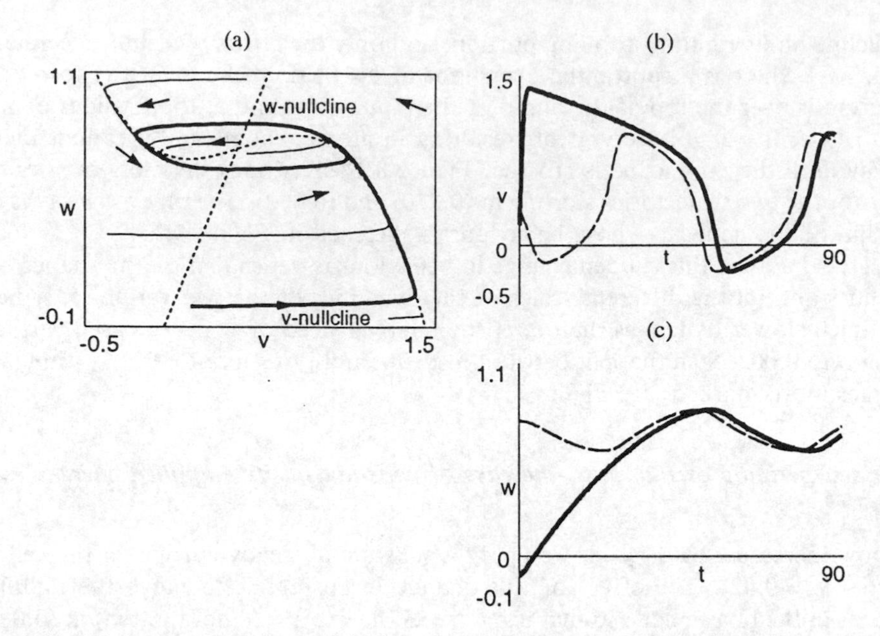

Figure 8.16 Fitzhugh–Nagumo model when the applied current $I = 0.7$. (a) a phase plane plot on which are plotted the v- and w-nullclines, and trajectories starting at various points. The arrows indicate the general direction of the flow in each region of the phase plane. (b) and (c) give (v, t) and (w, t) plots respectively for the two trajectories, marked in bold and dashed lines in (a) starting at a point near the origin and near the unstable equilibrium.

happens that a stable limit cycle surrounds a locally unstable equilibrium. The cyclical behaviour in the present model is also different from that of the undamped pendulum and of the simple Lotka–Volterra model. There the oscillations were neutrally stable, and took on an amplitude dependent on the starting point: if the pendulum were pulled out further before releasing, the amplitude of the resulting simple harmonic motion would be greater. Similarly for the Lotka–Volterra model, if the predator population were suddenly increased at a point in the cycle corresponding to high predator density, then the fluctuations which followed would be of greater amplitude. Here, the oscillations are of fixed period and amplitude. Lastly, the stable limit cycles which occur with many non-linear systems involving two variables—besides the Fitzhugh–Nagumo model—are possibly the explanation of many physiological clocks.

8.12 DETERMINISTIC CHAOS

So far we have seen that systems of two differential equations can exhibit equilibrium solutions of various kinds, and also periodic oscillatory solutions which can be stable or unstable. When we add a third differential equation to the system, a new kind of behaviour is possible, in which the component variables fluctuate erratically apparently without any pattern. That this can occur with very simple, straightforward-looking equations is rather a surprise. It was first discovered by a meteorologist,

Edward Lorenz, in 1963 but did not make much of a mark in the mathematical community until ten years later. He produced the equations

$$\frac{\mathrm{d}x}{\mathrm{d}t} = -10x + 10y$$

$$\frac{\mathrm{d}y}{\mathrm{d}t} = 28x - y - xz$$

$$\frac{\mathrm{d}z}{\mathrm{d}t} = \frac{8}{3}z + xy$$

as a stripped-down version of the differential equations of convection currents in the atmosphere. The constants are chosen so as to represent convection just after it has passed through the transition from steady to unsteady convection. Let us look at the solutions in Figure 8.17. If we start the equations at the point (2, 0, 0), we see that the system fairly quickly settles into periodic solutions with steadily increasing amplitude for all three variables. Until $t = 16$, nothing untoward seems to be happening: all three variables are oscillating with slowly increasing amplitude of oscillations. Then x and y jump up to oscillate just once in a different range of values, followed by further one-cycle switches up and down between the two ranges of values of x and y and then a return to higher values for a few cycles. Then apparently random switching between the two ranges occurs. If we look at phaseplane plot of y (ordinate) against x and z (ordinate) against x, we get another view of what is happening. In each plot, there are two series of rings joined at the centre. The solution cycles round one ring for a while, then jumps across to the other for one or more oscillations before returning. There appear to be some points on the trajectories which are very sensitive and with just a slight change the solution could readily move over to the other cycle. The two rings occupy the same range of z, but different ranges of x and y. So all we observe in z is a change in amplitude of the oscillation, whereas with x and y we see jumps between two different ranges of values.

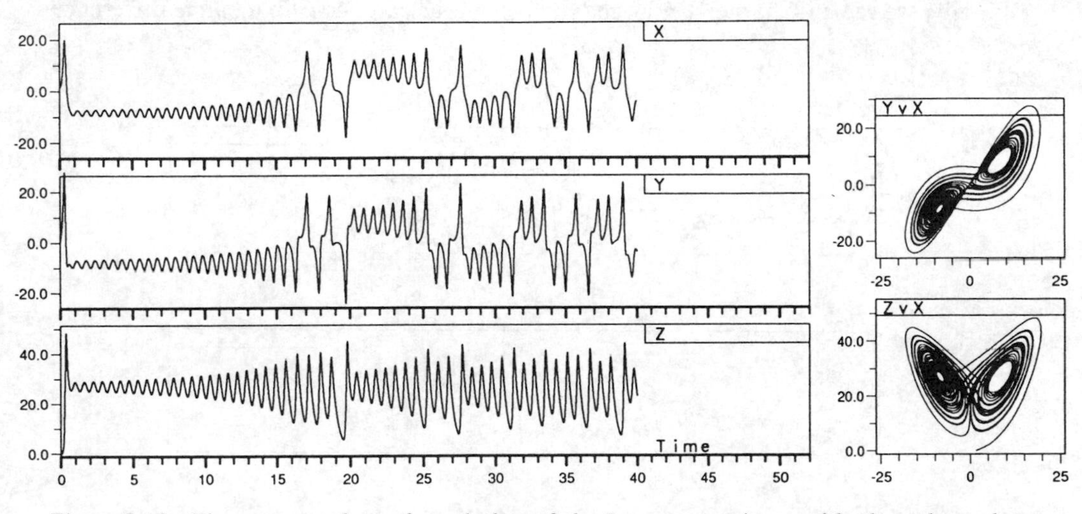

Figure 8.17 Time course plots of a solution of the Lorenz equations, with phaseplane plots.

A second important feature of this type of dynamics called *deterministic chaos* is that if the starting point is perturbed very slightly, the solution remains close to the unperturbed solution for some time, but then diverges from it and diverges substantially. Figure 8.18 illustrates the solution for x starting at $(2, 0, 0)$ together with that starting at $(2.001, 0, 0)$. Until about $t = 25$, the two solutions stay close together, but then diverge markedly. There are major consequences of the behaviour of such models for anyone attempting to make predictions using them. Starting with approximate knowledge of the state variables—as is almost always the case for any system outside a laboratory—short-term predictions can be made quite accurately, but beyond a certain time, no useful predictions can be made. Lorenz termed this the 'butterfly effect'. His original aim was to predict the weather, but he felt that with dynamics of this kind, it was virtually impossible to predict the weather very far into the future. As a delightful illustration of this, he described how the effects on the weather of a butterfly flapping its wings would be inconsequential for a short period of time, but that after a certain time, this initially tiny influence could make a very substantial unpredictable difference to the outcome. Thus although his model weather system precisely followed an extremely simple deterministic model, the outcome after a period of time was so sensitive to the exact initial values of the variables that no useful predictions could be made.

The solution to which a system tends after a period of time is termed an *attractor*. Sometimes there is an initial transient element in a system's behaviour, but the most interesting part of the behaviour is frequently that to which the system tends after a period of time, and this is what is meant by an attractor. The above discussion adds to our existing catalogue of attractors. We have met so far *limit point attractors*, or simple equilibrium points, and *limit cycle attractors*, such as the stable limit cycle which eventually occurs in the Fitzhugh–Nagumo model when subjected to a sufficiently large continuous current. The chaotic behaviour of the Lorenz equations is an example of a *strange attractor*. Some models have a number of attractors and the system will tend to one or the other depending on where it starts from. The catchment area of each attractor is known as the *basin of the attractor*.

Finally, we saw in Chapter 1 how chaotic behaviour can occur in a *single* difference

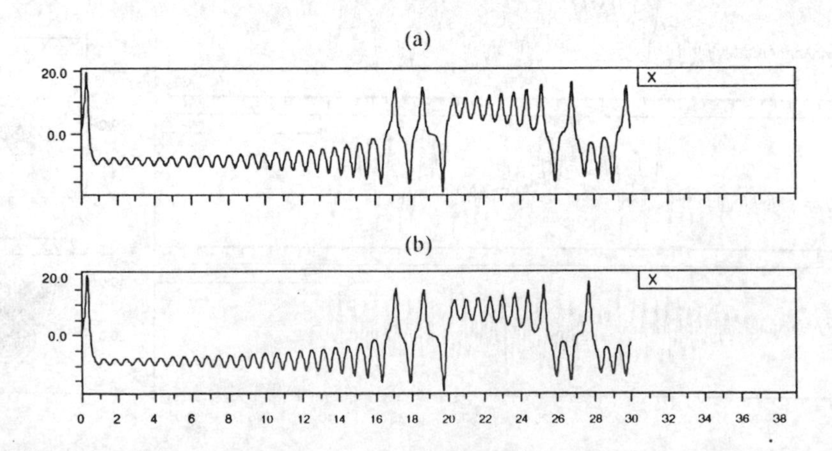

Figure 8.18 Two solutions of the Lorenz equations starting at (a) $(2, 0, 0)$ and (b) $(2.001, 0, 0)$.

equation, when we considered the logistic map. It is interesting that almost the simplest difference equation which incorporates negative feedback should show chaotic behaviour, whereas *three* differential equations are necessary. There have been many more interesting discoveries over the last few years about subtle patterns in the way in which chaos develops in many model systems. The interested reader is referred to the by now extensive literature on this subject for information about these discoveries, and for a more precise definition of the phenomenon of chaos itself (Stewart (1990) is a very readable introduction).

8.13 APPROXIMATE, NUMERICAL SOLUTIONS OF SIMULTANEOUS DIFFERENTIAL EQUATIONS

In Chapter 1, in the context of models of population growth and decline, we introduced simple methods of numerical solution of first order differential equations in which we essentially approximated the differential equation by a difference equation and then simulated this difference equation approximation. Such methods are available to us here also. We outline the method for two non-linear differential equations.

Euler's method

Suppose the two differential equations are:

$$\frac{dy_1}{dt} = f(y_1, y_2), \qquad \frac{dy_2}{dt} = g(y_1, y_2).$$

By replacing the differential term by the ratio of two small increments,

$$\frac{dy_j}{dt} = \frac{\Delta y_j}{\Delta t}$$

rearranging and adding $y_{j,i}$ to both sides of the two equations ($j = 1, 2$), we obtain the standard form of Euler's rule.

Euler's method of numerical integration

$$y_{1,i+1} = y_{1,i} + \Delta y_{1,i} = y_{1,i} + f(y_{1,i}, y_{2,i})\Delta t$$
$$y_{2,i+1} = y_{2,i} + \Delta y_{2,i} = y_{2,i} + g(y_{1,i}, y_{2,i})\Delta t$$

where $(y_{1,1}, y_{1,2}, y_{1,3} \ldots), (y_{2,1}, y_{2,2}, y_{2,3} \ldots)$ are predictions of y_1 and y_2 at equally spaced values of time, Δt apart.

These equations can be used to step through a simulated solution of the equations. Table 8.1 and Figure 8.19(a) show the application of Euler's method to the numerical solution of the Fitzhugh–Nagumo equations given in Section 8.11

$$\frac{dv}{dt} = v(v - 0.1)(1 - v) - w + 0.1, \qquad \frac{dw}{dt} = 0.001(v - 2.5w).$$

Table 8.1 Applications of Euler's method to the numerical solution of the Fitzhugh–Nagumo equations (see text)

i	t	v	$f(v, w)$	Δv	w	$g(v, w)$	Δw
1	0	0.000	0.100	0.200	0.0000	0.0000	0.0000
2	2	0.200	0.116	0.232	0.0000	0.0002	0.0004
3	4	0.432	0.181	0.362	0.0004	0.0005	0.0009
4	6	0.794	0.212	0.424	0.0013	0.0008	0.0016
5	8	1.219	−0.201	−0.402	0.0028	0.0012	0.0024
6	10	0.817	0.202	0.404	0.0053	0.0008	0.0016
7	12	1.221	−0.209	−0.418	0.0069	0.0012	0.0024
8	14	0.803	0.202	0.404	0.0093	0.0008	0.0016
9	16	1.207	−0.187	−0.374	0.0108	0.0012	0.0024
10	18	0.833	0.189	0.378	0.0132	0.0008	0.0016
11	20	1.211			0.0148		

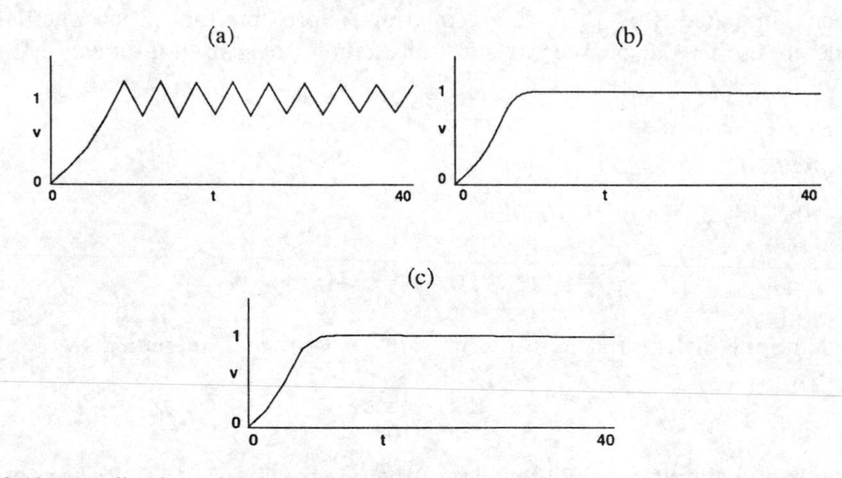

Figure 8.19 Application of Euler's method with (a) $\Delta t = 2$ (b) $\Delta t = 0.1$ and (c) Gear's method with maximum $\Delta t = 2$.

We start at $t = 0$, with $(v, w) = (0, 0)$. The step length is much larger than would normally be used in practice, so as to illustrate some problems with Euler's method with these equations. The values of the right hand sides of the differential equations (from Section 8.11) are listed in the columns headed $f(v, w)$, $g(v, w)$, and the increments in v and w over the next time step are given by these multiplied by 2. The new value is given by adding the increment and the current value. In this example, the step length is too long, and the same method with a step length of 0.1 gives a much better approximation to the behaviour of the differential equation.

8.13.1 Accuracy, computation time and choice of algorithm and step length

In general it is not necessary to know the details of the numerical methods used for the solution of a system of differential equations. However, it is important to be important to be aware of some of the problems which might arise. The accuracy of the solution by simple techniques such as *Euler's method* is related to the step length, Δt, and generally we can improve the accuracy sufficiently for almost any problem except certain difficult cases by making Δt smaller and smaller. However, the cost of this could be greater computer time, and there are other techniques which perform very well without such a large overhead in computing time. Some are extensions of the *Runge–Kutta technique* which is an efficient recursive method of approximating a Taylor series expansion (Press *et al.*, 1989).

8.13.2 Stiff systems of differential equations

The type of differential equation system which is particularly difficult to deal with is one which has extremes of dynamic behaviour, i.e. there are periods of time during which the response is changing slowly, and other periods, possibly very brief, during which the response is changing extremely quickly. This is what is meant by a *stiff system* of differential equations. The problem with most numerical solution methods is that the very small step length Δt which is appropriate during the bursts of speed would take an inordinately long computation time for the periods of slow change. If the step length is increased to improve the computation time, the fast dynamics will be very poorly reproduced. The answer is to use an adaptive method, i.e. one which moves with a small step up and down steep slopes, lengthening its pace when it encounters plains and plateaux. An adaptive method which usually gives satisfactory results is *Gear's method*. For further details of this and other methods, the reader is advised to consult Press *et al.* (1989).

Figures 8.19 (a), (b) and (c) illustrate the solution of the Fitzhugh–Nagumo equations

$$\frac{dv}{dt} = v(v - 0.1)(1 - v) - w + 0.1, \qquad \frac{dw}{dt} = 0.001(v - 2.5w)$$

A simple non-adaptive method (Euler's method) was used with two step lengths (a) $\Delta t = 2$, (b) $\Delta t = 0.1$; also (c) Gear's method, which is particularly good for stiff differential equations, with maximum step length $\Delta t = 2$. It is clear that Euler's method with the larger step length performs very poorly at the point where the slope changes suddenly at about $t = 7$. Gear's method performs almost as well as Euler's method with a much smaller step length.

8.14 SYSTEMS OF NON-LINEAR DIFFERENCE EQUATIONS

Systems of difference equations are used for describing changes in the sizes of interacting populations or processes measured at *discrete*, usually equally spaced times, e.g. the interaction between a parasitoid and its host where both species have relatively short and synchronised breeding seasons (Chapter 10). In realistic models the interactions involve non-linear functions so that changes are best described by systems of non-linear difference equations. In this section we deal briefly with the analysis of a system of two such equations. These methods can be extended to more than two equations but this is beyond the scope of the book. For two response variables X and Y the general form of a system of two non-linear difference equations is given by

$$X_{n+1} = f(X_n, Y_n)$$
$$Y_{n+1} = g(X_n, Y_n),$$

where f and g are non-linear functions. See Section 10.4 for the use of difference equations to describe host–parasitoid interaction.

Stability analysis of a system of two non-linear difference equations

Numerical methods can be used to examine the dynamical properties of a model for particular parameter values but it is also useful to find the general conditions for which the model has a stable equilibrium. To do this we use local stability analysis and consider a small perturbation of the population values about the equilibrium. For the general formulation the method is as follows.

Stage 1. The equilibrium values, obtained by setting $X_{n+1} = X_n = X_E$ and $Y_{n+1} = Y_n = Y_E$, therefore satisfy the equations

$$X_E = f(X_E, Y_E)$$
$$Y_E = g(X_E, Y_E).$$

Stage 2. Consider small displacements about the equilibrium values given by $X_n = X_E + x_n$ and $Y_n = Y_E + y_n$, and substitute in the equations so that

$$X_E + x_{n+1} = f(X_E + x_n, Y_E + y_n)$$
$$Y_E + y_{n+1} = g(X_E + x_n, Y_E + y_n).$$

Stage 3. Expand the right-hand sides about the equilibrium values as a Taylor series to give

$$X_E + x_{n+1} \approx f(X_E, Y_E) + x_n \frac{\partial f}{\partial X} + y_n \frac{\partial f}{\partial Y}$$

$$Y_E + y_{n+1} \approx g(X_E, Y_E) + x_n \frac{\partial g}{\partial X} + y_n \frac{\partial g}{\partial Y},$$

where the partial derivatives are evaluated at the equilibrium values.

Stage 4. Using the equations for the equilibrium values from Stage 1 gives the two *linear* difference equations

$$x_{n+1} = a_{11}x_n + a_{12}y_n$$
$$y_{n+1} = a_{21}x_n + a_{22}y_n,$$

where $a_{11} = \partial f/\partial X$, $a_{12} = \partial f/\partial Y$, $a_{21} = \partial g/\partial X$, $a_{22} = \partial g/\partial Y$.

Stage 5. The solution of the equations is

$$x_n = C_{11}\lambda_1^n + C_{12}\lambda_2^n$$
$$y_n = C_{21}\lambda_1^n + C_{22}\lambda_2^n,$$

where the C_{ij} are constants determined by the initial conditions. λ_1 and λ_2 are the roots of the auxiliary equation

$$\lambda^2 - T\lambda + D = 0.$$

where $T = a_{11} + a_{22}$ and $D = a_{11}a_{22} - a_{12}a_{21}$ (the trace and the determinant of the community matrix).

Condition for a stable equilibrium is then that the magnitude of the roots of the auxiliary equation are less than one. It can be shown that for this to be so

$$2 > 1 + D > |T|.$$

Example 8.5 Stability analysis of the Nicholson–Bailey host–parasitoid model

In Chapter 10 we present the Nicholson–Bailey model for the dynamics of a host–parasitoid interaction in which the density of the host (H_n) and the parasitoid (P_n) at time n are described by the pair of non-linear difference equations

$$H_{n+1} = RH_n \, e^{-aP_n}$$

$$P_{n+1} = cH_n(1 - e^{-aP_n}),$$

where R is the intrinsic net reproductive rate of the host, c is the average number of parasitoids emerging from a parasitised host and a measures the searching efficiency of the parasitoid—the function $1 - e^{-aP_n}$ relates the fraction of hosts parasitised to the density of parasitoids. Applying the stability analysis to this model leads to the following.

Step 1. Find equilibrium values

$$H_E = RH_E \, e^{-aP_E}; \qquad P_E = cH_E(1 - e^{-aP_E})$$

i.e.

$$P_E = \frac{\ln R}{a}; \qquad H_E = \frac{R \ln R}{ca(R-1)}$$

Step 2. Evaluate derivatives at equilibrium and find the elements of the community matrix

$$\frac{\partial f}{\partial H} = R \, e^{-aP_E}, \qquad \frac{\partial f}{\partial P} = -RaH_E \, e^{-aP_E}$$

$$\frac{\partial g}{\partial H} = c(1 - e^{-aP_E}), \qquad \frac{\partial g}{\partial P} = acH_E \, e^{-aP_E}$$

so that

$$a_{11} = 1, \quad a_{12} = -aH_E, \quad a_{21} = c(R-1)/R, \quad a_{22} = acH_E/R.$$

Step 3. Find the trace and determinant of the community matrix

$$T = a_{11} + a_{22} = 1 + acH_E/R = 1 + \ln R/(R-1)$$

$$D = a_{11}a_{22} - a_{12}a_{21} = acH_E/R + acH_E(R-1)/R = R \ln R/(R-1).$$

Step 4. Check for stability: $2 > 1 + D > |T|$.

For a viable host population we must have $R > 1$. The system is *unstable* because $D = R \ln R/(R-1)$ exceeds one so that the first inequality is not satisfied. Some numerical values are: $R = 1.01$, $D = 1.005$; $R = 1.10$, $D = 1.048$; $R = 2$, $D = 1.386$; $R = 5$, $D = 2.011$; $R = 10$, $D = 2.558$ but more generally we can show that $R \ln R/(R-1)$ is an increasing function of R. In other cases we might also need to check the second inequality, but in this case it is not necessary since violation of the first is sufficient to demonstrate instability.

Global dynamical behaviour

In general, for an unstable system we can iterate the equations to examine the global dynamic behaviour of the model. Since a *single* non-linear difference equation can display a remarkable range of dynamic behaviour including limit cycles and chaos (Chapter 1) it is not surprising that a system of non-linear difference equations can do the same. Chapter 10 gives some illustrations.

8.15 COMPLICATIONS

8.15.1 Stochastic effects

We have discussed in Chapter 4 how random variation can impact on the course of development of a dynamic model. There we discussed some different types of stochastic element in population models: demographic stochasticity which has greater consequences at low densities, environmental stochasticity which can substantially affect populations at low or high densities, and measurement error which affects our measurements but is not part of the process itself. Because of their greater complexity, random effects can influence the elements of non-linear models of biological interaction in more subtle ways. We have discussed above how the Fitzhugh–Nagumo model has regions of the phase plane in which its behaviour is very sensitive to any perturbations—and this feature was used in the model of the phenomenon of excitability; also we saw that at other times—during the refractory period—the system is quite resistant to any perturbations. The stable limit cycle behaviour which occurs when a steady current is applied also appears to be quite resistant to perturbation, apart from possible disturbance of the phase. The Lorenz equation has two areas of oscillation and random perturbation could shift the system from one oscillatory range to the other. This system as we have seen is easily perturbed. Thus a diversity of responses to random perturbations can occur, from rugged indifference to acute sensitivity. More complex models have basins of attraction of different attractors—whether limit points, limit cycles or strange attractors—and a perturbation could easily move the system from one basin of attraction to another. It is possible that a system locked onto a strange attractor—thus exhibiting chaotic dynamics—could be shifted by a perturbation to a stable limit cycle, or a stable limit point! In this way, random perturbations could abolish fluctuations rather than, as might be expected, induce them. Thus all kinds of reversible and irreversible changes to the prevailing pattern of behaviour can occur as a result of random perturbations. Randomly disturbed non-linear systems have a very rich repertoire of behaviour.

8.15.2 Heterogeneity

As well as being subject to temporal variation, most populations or components of systems are heterogeneous to some degree, resulting in what we shall term here strata. These are subsets of each population or component which it is reasonable to regard as homogeneous. If we know sufficient about the nature of the heterogeneity and the interaction between the strata of each population or element of the system, we can model the strata separately. With linear dynamic models, the behaviour of the average over strata is often a reasonably good approximation to the average of the behaviours of the separately modelled strata. But with non-linear models this is not at all the case, and the only way to obtain an understanding is to model each stratum separately. Some work has recently been done using cellular automata (see Chapter 12), or more generally, parallel processors. In the simplest version of such models, each stratum is effectively run on a separate processor, with appropriate communication reflecting the natural links between strata being set up between processors. There is little possibility of progress without such computer simulation, except in very special models for which analytical solutions have been obtained.

EXERCISES

8.1 Find the general solution of the following differential equations:
(a) $2y'' + 5y' + 2y = 0$;
(b) $y'' - 6y' + 13y = 0$;
(c) $y'' + 3y' + 2y = 0$, where $y' = \mathrm{d}y/\mathrm{d}t$, $y'' = \mathrm{d}^2y/\mathrm{d}t^2$.

8.2 Find the solutions of the following equations which satisfy the conditions given
(a) $2y'' + 5y' + 2y = 0$, $y(0) = 1$, $y'(0) = 0$;
(b) $y'' - 6y' + 13y = 0$, $y(0) = 0$, $y(1) = 1$;
(c) $y'' + 3y' + 2y = 0$, $y(0) = 0$, $y'(0) = 0$.

8.3 *Solution of the linear, non-homogeneous, 2nd order differential equation*
(a) Find the general solution of the differential equation

$$\frac{\mathrm{d}^2y}{\mathrm{d}t^2} + 3\frac{\mathrm{d}y}{\mathrm{d}t} + 2y = 0 \tag{1}$$

(b) Find the values of C and D such that

$$y = C\cos t + D\sin t$$

is a solution of the differential equation

$$\frac{\mathrm{d}^2y}{\mathrm{d}t^2} + 3\frac{\mathrm{d}y}{\mathrm{d}t} + 2y = \cos t. \tag{2}$$

(c) Show that if $y_1(t)$ is a general solution of equation (1) and if $y_2(t)$ is a specific solution (any solution not containing any arbitrary constants—known in general as a *particular integral*) of equation (2), then $y_1(t) + y_2(t)$ is a general solution of equation (2).
(d) Hence find the general solution to equation (2).

8.4 (a) Using the method outlined in Exercise 8.3, obtain the general solution to the differential equation

$$\frac{d^2y}{dt^2} - 2\frac{dy}{dt} + 2y = e^t.$$

(b) Find the solution which satisfies the initial conditions $y(0) = y'(0) = 0$.

8.5 Small departures (x and y) from equilibrium for a biological system, comprising two species, satisfy the following differential equations

$$\frac{dx}{dt} = -x + 3y, \qquad \frac{dy}{dt} = 3x - y.$$

(a) Sketch the nullclines for both variables on a phase plane plot.
(b) Find the second order differential equation satisfied by x and obtain its general solution. Obtain the general solution for y from the solution for x.
(c) Determine whether the equilibrium is stable or not.
(d) Find the equations of the trajectories (i.e. separate expressions for x and y in terms of t) which start from the points $(-1, 2)$ and $(-1, 0)$ at $t = 0$.
(e) Use these and any other specific trajectories which you find necessary to describe the behaviour of the solution near the equilibrium.
(Cambridge, NST, Part IA, Quantitative Biology, 1991.)

8.6 (a) Explain the assumptions underlying the Lotka–Volterra predator–prey equations:

$$\frac{dN}{dt} = aN - \alpha NP, \qquad \frac{dP}{dt} = -bP + \beta NP$$

where a, b, α, and β are positive constants. In what circumstances can these equations plausibly be applied to animals with discrete breeding periods?
(b) Show that there are two possible states of equilibrium, one unstable and one neutrally stable.
(c) Describe the variation of P and N with time at the neutrally stable equilibrium point and explain the result with a biological example.
(Cambridge, NST, Part IA, Biological Mathematics, 1980.)

8.7 In a laboratory experiment, the time scale for birth and death of predators is much longer than that for prey so that the number (P) of predators remains constant throughout the experiment. It is found that the number of prey (N) varies exponentially in time t according to the formula

$$N = N_0 e^{\alpha t},$$

where α is a parameter which is found to vary linearly with P according to $\alpha = a - bP$, where a and b are constants.
(a) Find the first order ordinary differential equation governing $N(t)$ which does not include the initial value of N.
(b) Interpret the meanings of the various terms in the equation.
(c) Show that a stationary population is possible if P takes a value P_E to be specified.
(d) If at time $t = t_0$ the number of predators is suddenly doubled to $2P_E$, calculate the time at which the number of prey drops to half the initial value, N_0.
(Cambridge, NST, Part IA, Biological Mathematics, 1984.)

8.8 Two populations of prey N and predators P are described by the Lotka–Volterra equations

$$\frac{\mathrm{d}N}{\mathrm{d}t} = aN - \alpha NP$$

$$\frac{\mathrm{d}P}{\mathrm{d}t} = -bP + \beta NP.$$

(a) In terms of these symbols, state how long it would take
(i) for N to increase by a factor e in the absence of P, and
(ii) for P to decrease by a factor e in the absence of N.
(b) Show that these two populations naturally oscillate about their equilibrium magnitudes with a period $2\pi/\sqrt{ab}$.
(c) Explain in graphical, analytical and biological terms why you might expect a phase difference in the oscillations of the populations.
(Cambridge, NST, Part IA, Biological Mathematics, 1981.)

8.9 A swimming pool is infested with algae whose population is $N(t)$. The owner attempts to control the infestation with an algicidal chemical, poured into the pool at a constant rate. In the absence of algae, the chemical decays naturally; when algae are present it is metabolised by them and kills them. The equations of the rates of change of $N(t)$ and the concentration of the chemical in the pool, $C(t)$, are

$$\frac{\mathrm{d}N}{\mathrm{d}t} = aN - bNC$$

$$\frac{\mathrm{d}C}{\mathrm{d}t} = Q - \alpha C - \beta NC,$$

where a, b, Q, α, β are positive constants. Discuss the meaning of each term in these equations.
Show that this system has two positive equilibria if $Q > \alpha a/b$, and in that case one of the equilibria is stable. What happens if $Q < \alpha a/b$?
(Cambridge, NST, Part IA, Biological Mathematics, 1982.)

8.10 The Gompertz model of predator–prey interaction is one in which the effect of one species on the growth rate of another is logarithmic, i.e. the differential equation for the prey population, $N(t)$, is

$$\frac{\mathrm{d}N}{\mathrm{d}t} = aN - \alpha N \log P$$

where $P(t)$ is the predator population and a, α are positive constants. Write down the corresponding equation for $P(t)$, introducing whatever constants are necessary and explaining (briefly) the meaning of each term in the equation.
By making the substitutions $x = \log N$, $y = \log P$, show that the equation can be solved exactly to give

$$y = \frac{a}{\alpha} + A \cos \omega t + B \sin \omega t$$

where ω is a given constant (to be found) and A, B are constants which will be determined by the initial conditions. Write down the corresponding expression for x.
Suppose that, at $t = 0$, the prey population is at its equilibrium value, but that there is a great influx of predators. Show that if the initial predator population exceeds a certain value (to be found) the prey will become extinct.
(Cambridge, NST, Part IA, Biological Mathematics, 1979.)

8.11 A specific case of the Fitzhugh–Nagumo model of neural firing is written as a system of two differential equations

$$\frac{dv}{dt} = -v(v-0.2)(v-1) - w + I$$

$$\frac{dw}{dt} = \varepsilon(v - 2.5w)$$

in which v represents the departure of membrane potential from equilibrium, w a recovery variable, I is the magnitude of any applied current and ε is a constant, usually small.

(a) By trial and error using a numerical simulation package, or otherwise, determine the range of values of I for which the equilibrium is unstable and hence the range within which the solution is sustained periodic oscillations.

(b) For a value of I in the centre of the range determined in (a), show how the amplitude of the cycle in v is dependent on the value of ε, the velocity-scaling factor for w.

8.12 The van der Pol system of two non-linear differential equations is given by

$$C\frac{dV}{dt} = I - \varepsilon\left(\frac{V^3}{3} - V\right), \qquad L\frac{dI}{dt} = -V$$

where L, C and ε are constants. The two variables V and I are the voltage and current respectively. ε is a parameter which usually takes small values.

(a) Show that when $\varepsilon = 0$, the equations reduce to those for simple harmonic motion,

$$\frac{d^2V}{dt^2} + \frac{V}{LC} = 0$$

and that therefore the van der Pol oscillator with $\varepsilon > 0$ could be said to be a form of damped harmonic motion, in which the damping is a function of V. What is the form of the damping function?

(b) Determine the equilibrium solutions, and assess whether they are stable or not.

(c) Sketch the nullclines on a phase plane diagram.

(d) Using Phaseplane, or a similar computer package, describe the transition from simple harmonic motion when $\varepsilon = 0$ to the solution when ε takes the values 0.1 and 1.

(e) What are the major differences between the solutions for the three values of ε, 0.0, 0.1 and 1?

8.13 Two species compete for the same food and are preyed on by the same predator. Discuss the assumptions and approximations involved in a model in which their population sizes (N_1, N_2, P respectively) are given by the following equations:

$$\frac{dN_1}{dt} = (a_1 - b_1N_2 - c_1P)N_1$$

$$\frac{dN_2}{dt} = (a_2 - b_2N_1 - c_2P)N_2$$

$$\frac{dP}{dt} = (-\alpha + \beta N_1 + \gamma N_2)P.$$

Show that an equilibrium with no population zero is possible if

$$\alpha b_1 b_2 - \beta b_1 a_2 - \gamma a_1 b_2 < 0$$

is satisfied, along with two other inequalities which should be specified.

Explain why this equilibrium will be stable only if none of the eigenvalues of the following matrix has positive real part:

$$\begin{pmatrix} 0 & -N_{10}b_1 & -N_{10}c_1 \\ -N_{20}b_2 & 0 & -N_{20}c_2 \\ P_0\beta & P_0\gamma & 0 \end{pmatrix}$$

where N_{10}, N_{20}, P_0 are the equilibrium values of N_1, N_2, P. By considering the sum and product of the eigenvalues, deduce that the equilibrium is unstable. What ecological implications does this have?

9 Biochemistry and Physiology

9.1 INTRODUCTION

A multitude of interactions occur among the components of an organism and between them and the organism's environment—from molecules and cells upwards. At the most elementary level, during the construction of the basic building materials of cells or of metabolism of stored energy supplies, they take the form of chemical or similar bonding interactions between molecules. At a slightly higher level, there are interactions between cells and molecules: such as those between cells and extraneous molecules, e.g. between antibody and antigen in immune reactions; or constructive interactions between cells and intrinsic materials such as hormones prior to some homeostatic response. Then there are the many processes which operate almost at the level of the whole animal; such as the circulation of the blood transporting information and materials from one place to another, and the neural control of the organism via the central nervous system.

This chapter gives some simple models for the dynamics, and the eventual equilibria reached, of some physiological and biochemical processes involving more than one response variable. This is in contrast to Part I of this book which was concerned essentially with the analysis of the behaviour of single populations of organisms, cells or molecules. We start by outlining some simple mechanisms of chemical reaction and applying them to determine the initial velocity of reaction in enzyme catalysed reactions in *enzyme kinetics*. For some biological processes, the interest is in what static equilibrium is eventually reached, rather than the rate at which the process occurs or the course by which the equilibrium is reached. The dynamics are sometimes so fast that there is little to be gained by analysing them. We only consider the equilibrium eventually attained in the section on *ligand–receptor interaction*: the binding between cell receptors and external molecules. We discuss assays for quantifying the binding ability of receptors of both animals and plants, and for measuring the concentrations of hormones and similar molecules by the method of *radioimmunoassay*. *Compartmental models* are general models of the rate of interchange of material between different forms or locations within the body of an animal or plant, and are useful in studies in which a tracer is used to assess rates of flow, interchange and loss for organisms in a steady state. Finally, we briefly consider some models of oscillators and clocks.

9.2 CHEMICAL KINETICS

Molecular interactions are the basic building and metabolic mechanisms of all organisms, but we only have space to describe a few elementary principles of biochemistry. We start with models of chemical and enzyme kinetics.

9.2.1 First order, unimolecular reactions

The simplest possible chemical reaction in which substance A is converted into B, and there is no back-reaction from B to A, is:

$$A \xrightarrow{k} B.$$

The Law of Mass Action states that the rate of a chemical reaction is proportional to the product of the concentrations of the reacting substances. In this case there is only reactant, and the forward velocity of the reaction (i.e. the rate of loss of A and the rate of production of B) at any time is proportional to the remaining concentration of A. There is a convention that names of substances are usually written in upper case, e.g. A and B, and their molar concentrations are usually written in lower case, e.g. a and b respectively. The rate of production of B is equal to the rate of utilisation of A, and so the sum total amount of A and B is constant (and so $a + b = \text{constant} = a_0$) and the differential equations which describe the reaction are as follows.

Example of first order reaction

$$A \xrightarrow{k} B$$

molar concentrations a b

$$\text{Rate or velocity} = \frac{da}{dt} = -ka.$$

As $a + b = a_0$, the equation could just as easily be written

$$\frac{db}{dt} = k(a_0 - b).$$

k = first order rate constant of the reaction (dimensions t^{-1}).

The first differential equation is solved (as in Chapter 1) to give

$$a = a_0 \, e^{-kt},$$

the familiar case of exponential decay. Using the fact that the total concentration of A and B is constant, the solution for b is therefore

$$b = a_0(1 - e^{-kt})$$

which is a curve known as the *monomolecular curve* which rises with ever decreasing slope to an asymptote a_0 (see Figure 1.8). k is known as the first order rate constant of the reaction.

Chemical reactions can be classified by their order or their molecularity. The *order* of a reaction is given by the number of molar concentrations which are multiplied together to obtain the rate of the reaction. Thus the above reaction is *first order*. The *molecularity* of a reaction is the number of molecules which are altered by the reaction. For single-step reactions the order is usually the same as the molecularity. However, many reactions consist of a number of stages and it is not usually possible to define an order of the whole reaction.

9.2.2 Second order reactions

The most usual type of bimolecular second order reaction is of the form

$$\mathrm{A} + \mathrm{B} \xrightarrow{k} \mathrm{C} + \mathrm{D}$$

with molar concentrations $a \quad b \qquad c \quad d$

where the rate of production of C is given by

$$\frac{\mathrm{d}c}{\mathrm{d}t} = kab$$

and k now is a second order rate constant. For every molecule of A and B which react together one molecule of C and D is produced. Thus $\mathrm{d}a/\mathrm{d}t = \mathrm{d}b/\mathrm{d}t = -\mathrm{d}c/\mathrm{d}t = -\mathrm{d}d/\mathrm{d}t$. These equalities enable us to obtain a differential equation in just one of the molar concentrations. Some reactions occur in both directions, i.e. as soon as the products C and D are formed they start to react together to produce A and B, and are denoted

$$\mathrm{A} + \mathrm{B} \underset{k_{-1}}{\overset{k_{+1}}{\rightleftharpoons}} \mathrm{C} + \mathrm{D}$$

and in this case the rate of production of C is affected by the forward and back reactions:

$$\frac{\mathrm{d}c}{\mathrm{d}t} = k_{+1}ab - k_{-1}cd.$$

The rate of production of A is similarly written:

$$\frac{\mathrm{d}a}{\mathrm{d}t} = -k_{+1}ab + k_{-1}cd.$$

These differential equations are non-linear because of the product terms and require individually tailored, usually approximate solution. Note that the units of the second order rate constants, k_{+1}, k_{-1}, of the reaction are different from those of first order rate constants: in order to balance up the units on both sides of the equation, the units are $\mathrm{M}^{-1}t^{-1}$, where M denotes molar concentration.

9.2.3 Enzyme catalysed reactions

Extensive studies were carried out at the end of the nineteenth century on the rates of enzyme–substrate reactions, principally concerned with the mechanism of fermentation. Some progress was made, but it was not until the experiments of Michaelis and Menten (1913) in which acetate buffers were used to control the pH of the reaction, and in which the *initial* velocity of reaction was measured at different substrate concentrations that a fully satisfactory model was obtained. The mechanism, which had already been proposed by earlier workers but had not been supported by such definitive experiments (Henri, 1902), is

$$\mathrm{E} + \mathrm{S} \underset{k_{-1}}{\overset{k_{+1}}{\rightleftharpoons}} \mathrm{ES} \xrightarrow{k_2} \mathrm{E} + \mathrm{P}$$

molar concentrations $\quad e \quad s \qquad x \qquad e \quad p$

Applying the Law of Mass Action to this reaction sequence, equimolar concentrations of E and S react together to produce the compound ES, with a forward rate constant k_{+1}, i.e. at a rate $k_{+1}es$. As soon as any of this is formed, it begins to dissociate back to E and S at a rate $k_{-1}x$, but also is involved in a further reaction to produce E and P at a rate k_2x. Thus we can write down differential equations in the four molar concentrations

$$\frac{dx}{dt} = k_{+1}es - k_{-1}x - k_2x \tag{9.1a}$$

$$\frac{de}{dt} = -k_{+1}es + k_{-1}x + k_2x \tag{9.1b}$$

$$\frac{ds}{dt} = -k_{+1}es + k_{-1}x \tag{9.1c}$$

$$\frac{dp}{dt} = k_2x \tag{9.1d}$$

This seems a formidable set of differential equations to solve, but we can simplify them fairly readily. First of all, equation (9.1d) is in dp/dt, and the variable p does not appear in any of the other three equations, and so the first three equations can be solved separately and then this one used to obtain dp/dt. Secondly we note from (9.1a) and (9.1b) that $de/dt = -dx/dt$, and so integrating we get $e = c - x$ where c is the arbitrary constant of integration. At $t = 0$, $e = e_0$ and $x = 0$ therefore $c = e_0$, i.e. $e = e_0 - x$. Substituting for e into equations (9.1a) and (9.1c) we get

$$\frac{dx}{dt} = k_{+1}(e_0 - x)s - (k_{-1} + k_2)x$$

$$\frac{ds}{dt} = -k_{+1}(e_0 - x)s + k_{-1}x$$

So we have reduced the original set of four differential equations to two. These equations are non-linear—since they involve the product xs on the right hand sides—and cannot be solved exactly.

Various approximations have been used to solve them, and we give here the Briggs and Haldane (1925) treatment. Lin and Segel (1973) give a more rigorous derivation based on singular perturbation methods. ES is formed as a result of the first step of the reaction, and lost as a result of the second, and Briggs and Haldane argued that a steady state would be quickly achieved in which the concentration of ES is constant. This is not so unreasonable because (a) the rate constants of the first reaction are usually much greater than that of the second and (b) E occurs on the left and right hand sides of the equation of the two-stage reaction, i.e. as the second reaction proceeds and ES is lost, more E is available to fuel its further production (see also Example 9.1). Thus, by setting $\mathrm{d}x/\mathrm{d}t = 0$, we convert the differential equation into an ordinary equation:

$$k_{+1}(e_0 - x)s - (k_{-1} + k_2)x = 0.$$

Solving for x, we get

$$x = \frac{k_{+1}e_0 s}{k_{-1} + k_2 + k_{+1}s}$$

and so from equation (9.1d) we obtain an expression for the rate of production of p,

$$\frac{\mathrm{d}p}{\mathrm{d}t} = k_2 x = \frac{k_2 k_{+1} e_0 s}{k_{-1} + k_2 + k_{+1}s}$$

$$= \frac{k_2 e_0 s}{\dfrac{k_{-1} + k_2}{k_{+1}} + s}.$$

This is usually referred to as the *initial rate of production* of P. According to the above derivation, it is the rate of production of P immediately after the steady state in which the concentration of ES is constant has been established. The function of the rate constants of the reaction in the denominator of this expression

$$K_m = \frac{k_{-1} + k_2}{k_{+1}}$$

is known as the *Michaelis constant* of the reaction. The more usual way in which the expression for the initial rate of production is written is given in the box below. The value of this equation is that it applies to a much wider range of more complex enzyme catalysed reactions than the simple two stage one presented here, and then $V_{\max}$ and K_m are more complex functions of the rate constants of the component reactions.

Standard form of the Michaelis–Menten equation

$$v = \frac{V_{\max} s}{K_m + s}$$

v = initial rate of production of product P (i.e. initial velocity of reaction)
s = substrate molar concentration
$V_{\max}$ or V = maximum velocity
K_m = *Michaelis constant* = substrate concentration at which initial velocity equals $V_{\max}/2$.

The curve (see Figure 9.1) has the following properties.

(1) It forms a monotonically increasing curve of decreasing slope as s increases, starting from the origin.
(2) It eventually reaches an asymptote, given by $V_{\max}$. This is shown by allowing s to take a very high value much greater than K_m, in which case

$$v \cong V_{\max} s/s = V_{\max}.$$

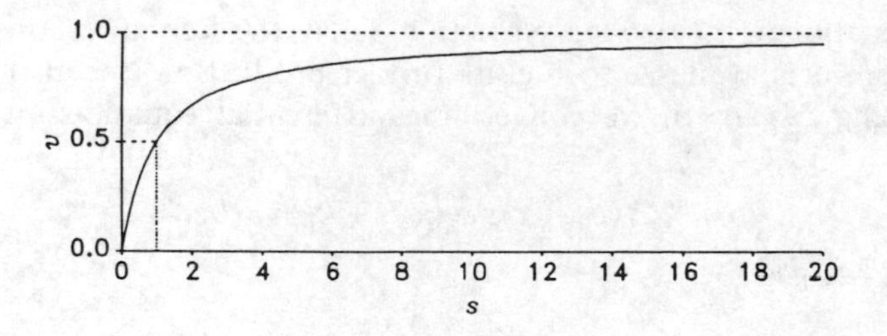

Figure 9.1 The relationship between v the initial rate of an enzyme catalysed reaction and s the substrate concentration.

(3) By substituting $s = K_m$ into the equation we identify the parameter K_m as the value of s at which $v = V_{max}/2$.

(4) One aspect of the curve which is not immediately apparent is that the asymptote is only reached at very high values of s. Even at $s = 10K_m$, the value of the response v is still almost 10% below its maximum, V_{max}. The curve is identical in algebraic form to the one used for ligand–receptor assay considered in Section 9.3.

Example 9.1 Numerical integration of the Michaelis–Menten model

We demonstrate by numerical integration of the differential equations that the Briggs–Haldane assumption that the concentration of ES is constant during the course of the reaction is reasonable in the case when $k_{+1} = 10$, $k_{-1} = 0.1$, $k_2 = 0.001$, $e_0 = 0.1$, with starting values $x = 0$, $s = 5$.

Figure 9.2 shows the phase plane plots and plots of x and s against t, which were obtained by Gear's method for integration of differential equations, which adjusts

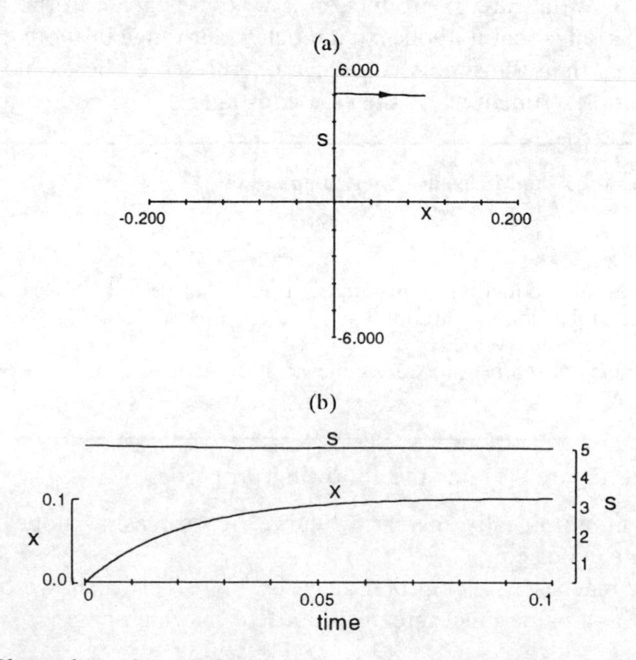

Figure 9.2 (a) Phase plane plot and (b) x–t and s–t plots of the solution starting at $x = 0$, $s = 5$.

the step size locally in time to accommodate the speed of events (see Chapter 8). The expected short-term equilibrium is

$$x = \frac{k_{+1}e_0 s}{k_{+1}s + k_{-1} + k_2} = \frac{10 \times 0.1 \times 5}{10 \times 5 + 0.1 + 0.001} = \frac{5}{50.101} = 0.1$$

which agrees with the x–t plot.

Example 9.2 Fitting the Michaelis–Menten equation

Kuhn (1923) obtained the following data for the velocity, v, of an enzyme-catalysed reaction at various substrate concentrations, s.

v	21.5	21.0	19.0	16.5	14.5	11.0	8.5	7.0
s	0.1970	0.1385	0.0678	0.0417	0.0272	0.0145	0.0098	0.0082

The model fitted by least squares (Chapter 11) is

$$v = \frac{23.6s}{0.0175 + s}.$$

9.3 LIGAND–RECEPTOR INTERACTION

Cells often communicate with each other and with important elements of their environment as a result of the binding of a specific part of the surface of the external object called a *ligand* by a complementary site on the surface of the cell called a *receptor*. The characteristic feature of ligand–receptor interaction is that the ligand fits into and binds to the receptor in a specific and exclusive manner. Often the binding is a precursor to some other action on the part of the cell depending on the nature of the external material. Here we are concerned only with modelling the ligand–receptor interaction, the first stage in the process. In animals, hormones are one of the commonest ligands released into the bloodstream after some stimulus; they often initiate action at various points so that the body can mount an appropriate response to the stimulus. The first step at the site of action of the hormone is for the hormone molecules to be bound by the cells there. In the case of the immune system foreign material (antigens) are bound by antibody (see Example 9.3). Drugs often exert their action first of all by binding to cells. Ligand–receptor interaction has been less widely studied in plants mainly because plant cells have a tough outer wall forming a barrier which makes study of the receptors which are usually found inside rather more difficult.

9.3.1 General theory

We now consider the simple case of one type of ligand and receptor. The binding is characterised by a relationship between the amount of the ligand which is bound at various concentrations of free ligand. This is usually measured by the use of radioactively labelled ligand, administered according to an experimental protocol that we shall deal with later. The binding usually involves a number of forces which are individually weak compared with covalent bonds (hydrogen bonds, electrostatic, van der Waals and hydrophobic forces) but as there are so many of them they can lead to a considerable binding energy. The non-covalent bonds are critically dependent on distance, d, between the interacting groups, being proportional to $1/d^2$ for electrostatic forces and to $1/d^7$ for van der Waals forces. Thus they require a close fit between the ligand and receptor and it is this which gives the bond its narrow specificity. Fortunately for us, the binding still follows the Law of Mass Action, and so we can use the theory we have outlined in Section 9.2 on chemical kinetics.

Suppose we have a ligand G and receptor R, the reaction can be written

$$\mathrm{G} + \mathrm{R} \underset{k_{-1}}{\overset{k_{+1}}{\rightleftharpoons}} \mathrm{GR}$$

where k_{+1} and k_{-1} are the rate constants of the forward and backward reactions, and so the Law of Mass Action gives us the differential equation of the progress of the reaction

$$\frac{\mathrm{d}x}{\mathrm{d}t} = k_{+1}gr - k_{-1}x$$

where r, g and x are the molar concentrations of R, G and the product GR. We are not interested in this case in the dynamics of the binding, but in determining the concentration of bound ligand x at equilibrium as a function of the free ligand, g. By setting the derivative equal to zero, we obtain the solution $x = grk_{+1}/k_{-1} = gr/K_D$, where K_D is known as the *dissociation constant* of the reaction. The total amount of the receptor is given by $r_T = r + x$ and hence by substituting for r in the above we get

$$x = \frac{1}{K_D}\, g(r_T - x) \quad \text{or} \quad x = \frac{r_T g}{K_D + g}$$

which is more usually written as follows.

Ligand–receptor equilibrium model

$$B = \frac{B_{\max} F}{K_D + F} = \frac{nF}{K_D + F}$$

B = molar concentration of *bound ligand*
F = molar concentration of unbound or *free ligand*
$B_{\max} = n$ = the maximum value of B attainable which occurs when all the receptors are bound to ligand, or density of sites
K_D = the *dissociation constant* of the binding reaction
$K_A = 1/K_D$ = the *affinity* or association constant of the binding reaction.

Note that $B_{\max} = r_T$. The affinity or association constant, K_A, which is the reciprocal of the dissociation constant, K_D, is widely used in immunology.

This is the same hyperbolic relationship as that between the initial velocity of an enzyme catalysed reaction and the substrate concentration and it has similar properties (see Figure 9.3). It forms a monotonically increasing curve of decreasing slope as F increases, eventually reaches an asymptote, given by $B_{\max}$, K_D = the value of F at which

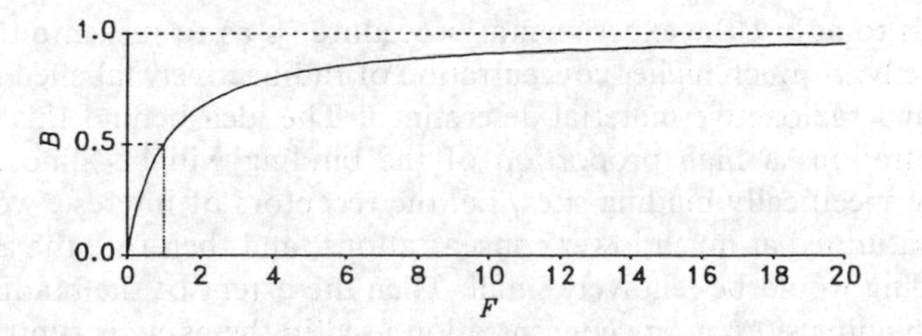

Figure 9.3 The relationship between bound (B) and free (F) ligand in the ligand–receptor assay model discussed in Section 9.3.1.

$B = B_{max}/2$; and the asymptote is only reached at very high values of F. Even at $B = 10K_D$, the value of the response F is still almost 10% below its maximum, B_{max}.

Saturation

The saturation of the binding is defined as the proportion of sites which are bound, i.e. B/B_{max}. In some assays it happens that the concentrations of free ligand which are used are very low compared with K_D. The model can then be simplified as in that case the denominator can be written approximately as $K_D + F \approx K_D$ and so the relationship is a straight line through the origin $B \approx B_{max}F/K_D$. When this happens, the particular receptors are said to be *unsaturated* or *unsaturable* (within the range of receptor and ligand concentrations used in the study). On the contrary, when concentrations near to K_D are used, the full model is required and then the receptors are said to be *saturable.*

Departures from the simple deterministic model

Binding is usually measured by adding a known concentration of a mixture of the normal ligand and a radioactive isotope of the ligand in a liquid vehicle to a mixture containing a known concentration of receptors. After a suitable time in which an equilibrium has been reached, the supernatant is washed away, and the radioactivity in the residue used as a measurement of the amount of radioactively labelled ligand bound to the receptors, from which the total ligand bound can be calculated.

Non-specific binding

A commonly occurring complication is that binding occurs to the tissue apart from at the sites of specific interest, and sometimes occurs to the glass of the tube in which the sample is contained. Any binding to the glass can be determined simply by using samples which contain only radioactively labelled ligand and measuring the resulting binding of radioactive material. The non-specific binding to the tissue cannot be separated off in this way. However in any well designed assay, this is usually small relative to the specific binding within the range of concentrations of ligand used in the assay, and also has a K_D value which is high compared with the K_D of the receptors of interest, i.e. non-specific binding is unsaturable within that range of concentrations. This can be dealt with in two ways.

The first is to adjust the experimental procedure so as to measure the non-specific binding directly. A much higher concentration of radioactively labelled ligand is added and the bound radioactive material determined. The idea behind this is that at such high concentrations a high proportion of the binding will be almost entirely non-specific. The specifically binding sites, i.e. the receptors of interest, would have been completely saturated at much lower concentrations, and therefore the amount of such specific binding would be relatively small. Then the extent of the binding by the non-specifically binding sites at any concentration used in the assay is subtracted from the total amount bound before fitting the model given in the above box. This method suffers from the drawback that a sufficiently high concentration of ligand is required to ensure that most of the binding is non-specific, and that at these concentrations it might happen that the non-specifically binding sites are becoming saturated. This simple correction is not then available.

A second procedure is not to attempt to measure non-specific binding separately, but to incorporate it into the model, and determine its magnitude by fitting the enlarged model

$$B = \frac{nF}{K_D + F} + C_{ns}F,$$

where $C_{ns} = n_{ns}/K_{ns}$ and n_{ns} and K_{ns} are the density and the dissociation constant of the non-specific binding. This usually requires data at a greater number of concentrations than would fitting the simple model.

A plot which has come to be widely used for visual presentation of the results of ligand–receptor assays is known as the *Scatchard plot*. This plot of B/F against B—which is approximately a straight line if the simple model without any complications holds, see Figure 9.4(a)—suffers from the disadvantage that if there is random variability affecting the measurements of the bound fraction, then this source of variation would affect both axes of the plot. Thus standard methods for fitting straight lines would possibly give misleading results if based on these transformed variates. However, for some assay protocols, F is affected by measurement error anyway (see Section 9.3.2), so, if the main source of variation is measurement error, special estimation methods would be needed for methods utilising just F as the explanatory variable.

A major use of the Scatchard plot is as a diagnostic of the type of departure from the simple model (other diagnostic plots which are useful in this context are discussed in Chapter 11). In that case, the straight line is often modified into an upwardly concave curve, as in Figure 9.4(b). However, such a curve can be obtained for reasons other than the presence of unsaturable non-specific binding (Model D2 in Table 9.1): there might be heterogeneity in the K_D values of the specifically binding sites. A number of simple models exist for such heterogeneity. The two most useful ones are (a) that two classes of sites exist each with its own K_D (Model D5), and (b) that a continuous distribution of K_D values occurs (Model D3, as discussed in Nisonoff & Pressman (1958)). A further possible reason is negatively cooperative binding in which binding at one site makes it less likely that binding will occur at some other sites (Model D4(a)). On the other hand, a Scatchard plot which exhibits downward concavity could mean the presence of positively cooperative binding (Figure 9.4(c), Model D4(b) and Section 9.3.4).

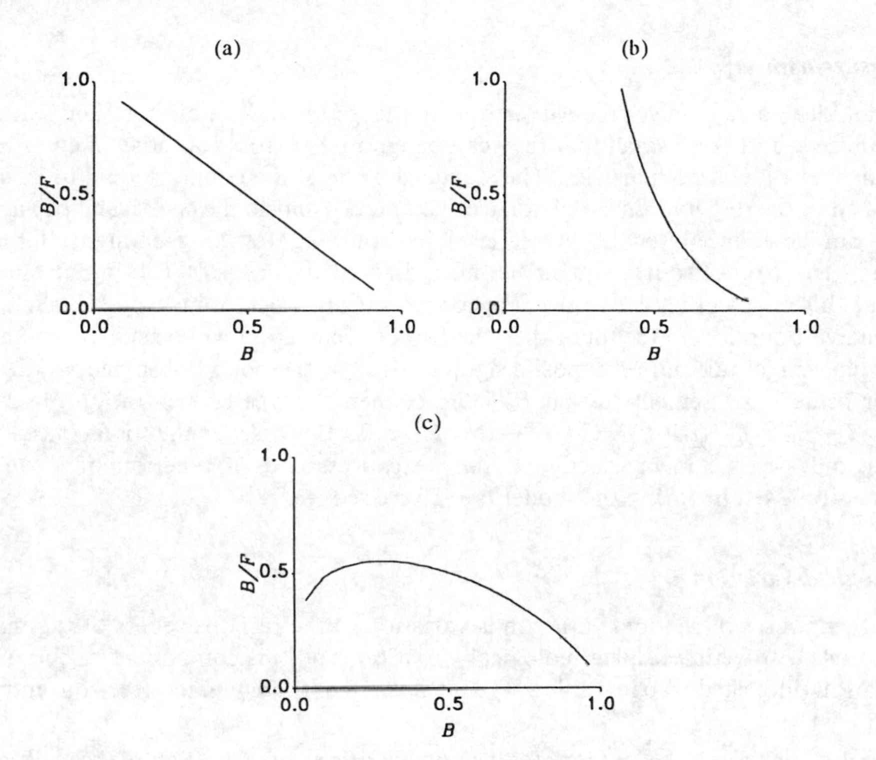

Figure 9.4 Scatchard (i.e. B/F vs B) plots which occur for models involving: (a) one class of sites; (b) heterogeneity of forms D3 or D5 or non-specific binding (D2), or negatively cooperative binding; (c) positively cooperative binding; occasionally concave-downwards Scatchard plots occur because the non-specific binding has been overestimated and subtracted from the bound values before plotting.

9.3.2 Sources of variation in ligand–receptor assays

A second complication common to many assay systems is random variation, and here it is important to distinguish between measurement error and biological variability in response. Because of the slightly unusual protocol for carrying out binding studies, if measurement error occurs at all, it usually affects two variables in the model, and hence complicates the fitting procedure. If on the other hand there is random variability in response of the material in the assay and little or no measurement error, the assay is affected by random components in a simpler way.

One protocol which is often used is that a fixed amount of ligand (a constant proportion of which is radioactively labelled) is successively diluted by a fixed dilution factor so that the concentration of label in the samples constitutes a dilution series, i.e. the concentrations are $D, D\alpha, D\alpha^2, D\alpha^3 \ldots$ where D is the initial concentration and α the dilution factor. A constant amount of each of these labelled samples is added to a fixed concentration of a sample of the receptor material. After the binding has occurred, the sample is centrifuged, the supernatant liquid is washed off and the amount of radioactive label in the fraction left is determined, from which the bound concentration can be estimated.

Measurement error

In an ideal assay, measurement errors in the estimated amount bound would be minimised, and be so small that they can be ignored. Errors can arise in counting the number of radioactive particles. The standard error of the count is equal to its square root, since the distribution is Poisson (see Chapter 3) and so the coefficient of variation $\sqrt{r}/r$ can be reduced to any preset level by counting for long enough. But errors arising from other sources might be more difficult to control. The important point about this protocol is that as the free concentration is determined by subtracting the estimated bound from the total, then the free concentration will be subject to an error of equal magnitude but of opposite sign, i.e. if B_o = the bound observed, and ε is the error in its measurement so that $B_o = B + \varepsilon$, then the free concentration observed is $F_o = T - B_o = T - (B + \varepsilon) = (T - B) - \varepsilon = F - \varepsilon$. In this case, as the total concentration of ligand, T, is known exactly, B can be expressed as a function of T and this expression used in fitting the model (see Exercise 9.8).

Biological variation

Another source of random variation is variation in the responsiveness of the material being used, sometimes called biological variation, and this component of the random variation (indicated by δ in Table 9.1) does not affect the estimated free concentration.

Table 9.1 Summary table of forms for the deterministic and random part of the mathematical models of ligand–receptor interaction

Deterministic components		
D1 One homogeneous class of sites $B = \dfrac{nF}{K_D + F}$	D2 Non-specific binding $B = \dfrac{nF}{K_D + F} + C_{ns}F$	
D3 Continuous variation in K_D $B = \dfrac{nF^a}{K_{Do}^a + F^a}$	D4 Co-operativity $B = \dfrac{nF^a}{K_D^a + F^a}$ D4(a) Negative $a < 1$ D4(b) Positive $a > 1$	
D5 Two homogeneous classes of sites, dissociation constants K_{D1} and K_{D2} $B = \dfrac{n_1F}{K_{D1} + F} + \dfrac{n_2F}{K_{D2} + F}$		
Random components		
R1 Measurement error	R2 Variability in response	R3 Both sources of random variation
$B_0 = f(F_0 + \varepsilon) + \varepsilon$ $\varepsilon \sim N(0, \sigma_e^2)$	$B = f(F) + \delta$ $\delta \sim N(0, \sigma_b^2)$	$B_0 = f(F_0 + \varepsilon) + \varepsilon + \delta$

concentration. The reason can be seen by considering the case when there is no measurement error. Then the true concentration bound in each individual sample can be determined exactly and therefore the true concentration of free ligand is also determined exactly. It is necessary though to have an additive random term in the model for B to cater for the random variation. If for example the density of sites varies from sample to sample so that $E[N] = \mu_n$, $\text{var}[N] = \sigma_n^2$ where we write capital N rather than lower case to emphasise that now N is a random variable, then the variance of the *true* bound concentration from sample to sample would be

$$\text{var}[\delta] = \left(\frac{F}{K_D + F}\right)^2 \text{var}[N] = \left(\frac{F}{K_D + F}\right)^2 \sigma_n^2.$$

Example 9.3 Analysis of an assay of antibody–antigen interaction

The following data come from an assay of the binding of antibody to antigens on cells, in which the interest is in estimating the number of sites on the antigen (the receptor) and the affinity with which they bind an antibody (the ligand) of the immunoglobulin Igm family. Figure 9.5 displays the data and fitted model.

Data

B (μg)	0.0098	0.0088	0.0218	0.0211	0.0363	0.0364	0.0439	0.0496
F (μg)	0.0053	0.0054	0.0187	0.0194	0.0702	0.0822	0.2778	0.3469

The curve fitted by non-linear least squares, as discussed in Chapter 11, assuming that virtually all the variation is biological variation, and that therefore there is relatively little measurement error, is

$$B = \frac{0.050F}{0.026 + F}.$$

The B_{max} can be converted to the number of sites per cell by multiplying by (no of moles)/μg $= 3.33 \times 10^{12}$ and dividing by the number of cells used in the assay $= 1.08 \times 10^7$ giving about 15 400 sites per cell.

9.3.3 Summary of models

A summary of some simple models is given in Table 9.1.

9.3.4 Cooperative binding

Curved Scatchard plots can arise from positive and negative cooperativity. A well-known and important example of positively cooperative binding in which the binding at one site increases the chances that binding will occur at another site is that between haemoglobin and oxygen. Haemoglobin circulating in the bloodstream binds oxygen in the lungs and then discharges it where it is needed in various tissues. It achieves high efficiency in this task by utilising cooperative binding, and also by the binding being partially regulated by carbon dioxide concentration. Another case occurs when binding at one site interferes with binding at an adjacent site. Then negative cooperativity is said to operate.

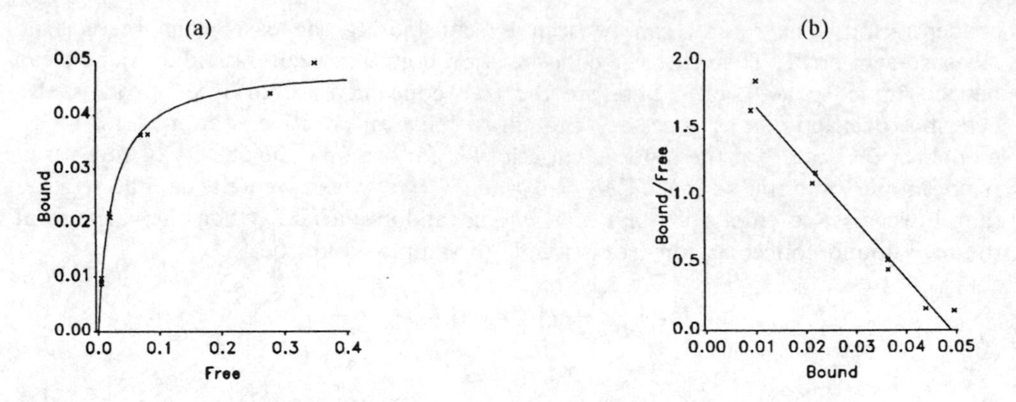

Figure 9.5 Plots of (a) B against F and B/F against B (a Scatchard plot) for the data of Example 9.3.

As oxygen is a gas, unlike the ligands we have already considered, we use partial pressure of oxygen, denoted by pO_2, rather than concentration in the binding equation. It is also convenient to discuss this phenomenon using the term *saturation* which is defined as the fractional occupancy of the oxygen binding sites on the haemoglobin or in our terms, $P_B = B/B_{max} = B/n$. In contrast to haemoglobin, the quite closely related myoglobin does not exhibit cooperative binding, and therefore the binding equation is

$$P_B = \frac{pO_2}{K_D + pO_2}.$$

This follows directly from

$$B = \frac{nF}{K_D + F}$$

after substituting $P_B = B/n$ and $pO_2 = F$, which in turn results from the simple reaction equation

$$\mathrm{Mb} + \mathrm{O_2} \rightleftharpoons \mathrm{MbO_2}.$$

The reaction equation for haemoglobin binding is

$$\mathrm{Hb} + m\mathrm{O_2} \rightleftharpoons \mathrm{Hb(O_2)}_m$$

which leads to the expression for the saturation

$$P_B = \frac{(pO_2)^m}{(P_{50})^m + (pO_2)^m}. \tag{9.2}$$

This is a relationship with an asymmetrical sigmoid appearance, displaying a definite point of inflexion, as compared with the hyperbolic appearance of the ordinary saturation curve, such as that for myoglobin. P_{50} is the partial pressure of O_2 at which 50% of the sites are occupied. The derivation of the equation for P_B is similar to the case when $m = 1$, and is left as an exercise for the reader (Exercise 9.9).

Equation (9.2) can be transformed into a straight line relationship by first writing

$$\frac{P_B}{1 - P_B} = \left(\frac{pO_2}{P_{50}}\right)^m$$

and then taking logs of both sides

$$\log\left(\frac{P_B}{1 - P_B}\right) = m \log pO_2 - m \log P_{50}.$$

A plot of $\log(P_B/(1 - P_B))$ against $\log pO_2$, called a *Hill plot*, is a straight line with slope m. The slope (calculated at the value $P_B = 0.50$ if the relationship is slightly curved) is called the *Hill coefficient*. This increases with the extent of the cooperativity, up to a maximum possible value of the total number of sites. The Hill coefficient for haemoglobin is 2.8. Thus the binding of the oxygen at one site of the haemoglobin molecule in the lungs once started tends to proceed more quickly because of this facilitation, and similarly once the conditions are such that the oxygen molecules start to be unloaded in the tissues, further unloading is rendered more likely.

We can calculate by how much the efficiency of blood for transporting oxygen is increased by cooperativity. Let the partial pressure of oxygen in the lungs be p_L and in muscle be p_M, then the saturation in the lungs and in the muscles are given by

$$P_{BL} = \frac{p_L^m}{P_{50}^m + p_L^m}, \qquad P_{BM} = \frac{p_M^m}{P_{50}^m + p_M^m}.$$

Plausible values for the partial pressure of oxygen are 100 torrs in the lungs and 20 torrs in muscle, and $P_{50} = 26$ torrs. The differences between the proportion of sites bound in the two locations gives a measure of the efficiency of transportation, and if this is calculated for a series of values of the Hill coefficient, m, using the above formulae we obtain the values of proportion of carrier sites bound in the two locations shown in Table 9.2.

Thus haemoglobin with its Hill coefficient of 2.8 has 97.8% of its sites bound in the lung and 32% in muscle, and so discharges the difference in the muscle (assuming that none is lost on the way to the muscle), whereas for myoglobin with a Hill coefficient of 1, the same figures would be 79% and 43%, thus transporting a smaller amount of oxygen per site. Various other mechanisms, such as a facilitation of discharge in the presence of high concentrations of carbon dioxide, reinforce this transport mechanism. Recent studies of the structure of haemoglobin have uncovered a fascinating story of how structural changes in the haemoglobin molecule consequent upon the first oxygen

Table 9.2 Percentage of carrier sites bound in lungs and muscle using the values for partial pressure of oxygen and P_{50} given in the text

	Hill coefficient, m										
	1.0	1.3	1.6	1.9	2.2	2.5	2.8	3.1	3.4	3.7	4.0
Lungs	79	85	90	93	95	96.7	97.8	98.5	99.0	99.3	99.5
Muscle	43	42	40	38	36	34	32	31	29	27	26
Difference	36	43	50	55	59	63	66	68	70	72	74

molecule being bound produce this effect, resulting in the fourth molecule being bound about 300 times as tightly as the first.

As discussed above a value of m less than one could indicate negative cooperativity. Another explanation is random variability from site to site in K_D values. The greater the variability the smaller the value of m (see Nisonoff & Pressman, 1958).

9.4 RADIOIMMUNOASSAY

The discussion in the previous section has been about the binding of ligand and receptors with the object of characterising the ability of receptors within an animal or plant tissue to bind the ligand. The focus of interest has been on the receptors. But ligand–receptor binding has also been extremely fruitfully exploited for assessing concentrations of ligands such as hormones in samples of blood, plasma, cerebrospinal fluid and other body fluids. For about thirty years, radioimmunoassay, RIA for short, has been the standard technique for estimating such concentrations. Now a technique called ELISA (enzyme-linked immunosorbent assay) which measures the binding using a fluorescent label, rather than a radioactive one, is also widely used. Both these assay systems have been developed to the point where they are extremely sensitive and highly specific, and are thus invaluable tools in medicine and physiology. Although the technology for quantifying the binding is different, the basic principles of the two methods are similar, and we only consider RIA in the remainder of this section.

The assay of a substance Q utilises an antibody which binds quite specifically to Q, the degree of such binding being related as we have seen above to the concentration of Q present in the sample. In any assay, Q is present in two forms, a radioactively labelled and an unlabelled form. A known concentration, Q_L, of labelled Q is added to a sample containing an unknown concentration, Q_U, of unlabelled Q—the concentration which we are trying to determine. The bound and free fractions can usually be separated by some physical means; if, for example, the free is soluble or in suspension, and the bound can be precipitated, then the sample can be spun and the supernatant liquid can be removed. The fraction of the amount in the sample which is bound—which is assumed to be the same for the labelled and unlabelled forms, since they behave identically in the chemical reactions involved—is a function of the total concentration of Q present. If we know the relationship between the fraction bound and the total concentration present, it is possible in any specific case to determine the total concentration of Q from the fraction bound. Then the concentration of the initial unknown sample can be determined by the difference between the determined total concentration and the concentration of labelled Q which has been added.

9.4.1 The RIA protocol

There are usually two stages in any radioimmunoassay. The first stage is the determination of what is called a *standard curve*. For this purpose some *standard samples* are employed in which the concentrations of hormone, Q_U are known (the subcript U stands for unlabelled). The standard curve relating amount of label bound, B_L, to the known concentration of added hormone Q_U in these standard samples can be determined in any manner, so long as a reliable predictive curve is obtained. The second stage is for the samples with unknown concentration of the hormone to be put through an identical experimental procedure, and from the amount bound in these samples and the standard curve obtained in the first stage, the hormone concentrations in these samples determined. The standard curve needs to be determined for each individual assay as the curve will vary from occasion to occasion depending on the exact conditions involved. It is much easier to determine the curve appropriate in the conditions prevailing and use this on the unknowns, than tightly to control the conditions so as to obtain some ideal standard conditions which would be necessary if some predetermined standard curve were used.

9.4.2 The standard curve for radioimmunoassay

How do we determine the standard curve and then use it to estimate hormone concentrations in unknown samples? The most general expression in use is the log-logistic proposed by Rodbard *et al.* (1968)

$$B_L = a + b\,\frac{e^Z}{1 + e^Z},$$

where $Z = c - d \ln Q_U$ and Q_U is the unlabelled hormone concentration in the sample. This model has four unknown parameters a, b, c, and d, i.e. constants within any one assay which need to be determined for each assay in some way. This model is related to the logistic growth curve considered in Chapters 1 and 4. Figure 9.6 gives an example of such a curve: an S-shaped curve with lower and upper asymptotes of a and $a + b$ respectively. Curves of this form turn out to be quite satisfactory in a wide variety of situations. They are empirically justified, and no other justification is needed. However, we feel more confident when using a model when we have a mechanistic explanation, and the next few paragraphs discuss this.

We have seen above that the relationship between bound, B, and free, F, is

$$B = \frac{B_{\max} F}{K_D + F}.$$

The usual approach to RIA is to simplify this relationship by ensuring that the F is

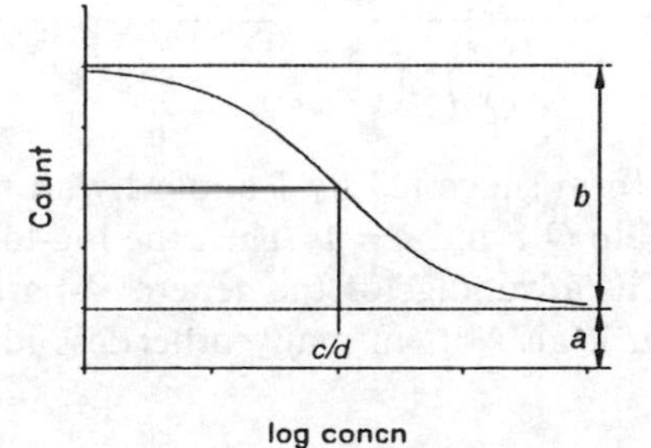

Figure 9.6 An example of a standard curve for radioimmunoassay (see Section 9.4.2). The lower asymptote is a, the difference between lower and upper asymptotes is b and the point of inflexion of the curve occurs at ln(concn) = c/d.

so large that we are operating effectively on the upper asymptote of this curve, so that $B \cong B_{max}$.

The following argument depends on saturation being achieved, and also on very simple expressions about the extent to which unlabelled hormone is diluted by labelled hormone. This type of assay is therefore both a *saturation* and a *dilution assay*. So if we add subscripts L and U to denote labelled and unlabelled respectively,

$$B_L + B_U = B_{max} \tag{9.3}$$

and also

$$\frac{B_L}{B_L + B_U} = \frac{Q_L}{Q_U + Q_L} \tag{9.4}$$

where Q_L and Q_U are the labelled and unlabelled total concentrations (the sum of bound and free). This is just saying that the proportion of the ligand bound which is labelled is the same as the proportion of the total amount of ligand present which is labelled. In equations (9.3) and (9.4), B_L is known because it is directly measured and Q_L is also known because a known amount of labelled Q of known specific activity (radioactivity per unit weight) is added. B_{max} would in general not be known but would be a constant. Using equation (9.3), and substituting for $B_L + B_U$ in equation (9.4) gives

$$B_L = \frac{B_{max} Q_L}{Q_U + Q_L}.$$

Recapitulating for a moment, B_L is the response which is measured, the radioactive count of the bound fraction, and Q_U is what we want to know, the unknown concentration of Q. B_{max} is an unknown constant, and Q_L is the concentration of label added and therefore known. So we now almost have the relationship which we required.

We used the log of the unknown concentration of (unlabelled) hormone, i.e. $\ln Q_U$ in the above model, so it seems sensible to insert this here. $\exp(\ln Q) = Q$ for any Q, and substituting in the above expression we obtain

$$B_L = \frac{B_{max}\left(\dfrac{Q_L}{Q_U}\right)}{1 + \left(\dfrac{Q_L}{Q_U}\right)} = \frac{B_{max}\, e^{[-\ln Q_U + \ln Q_L]}}{1 + e^{[-\ln Q_U + \ln Q_L]}}.$$

This is of the log-logistic form suggested by Rodbard with the parameters taking the values $a = 0$, $b = B_{max}$, $c = \ln Q_L$, and $d = 1$. Thus the log-logistic model includes this model as a special case. The advantage of the general 4-parameter log-logistic model is that it is so flexible that it also fits in many other less ideal conditions.

Example 9.4 An insulin assay

The following data due to R.S. Yalow (from Meinert & McHugh, 1968) are the standards data for an early insulin radioimmunoassay.

Proportion counts bound	Concentration of unlabelled insulin ($m\mu g\ ml^{-1}$)
0.673	0.05
0.657	0.10
0.641	0.15
0.629	0.20
0.608	0.30
0.583	0.40
0.579	0.50
0.574	0.75
0.545	1.00
0.495	1.25
0.462	1.50
0.430	2.00
0.370	2.50
0.364	3.00
0.326	4.00
0.240	5.00

The model (fitted by non-linear least squares) to these data, using $\log_{10}(\text{concn})$, is

$$\text{count} = -1.41 + 2.10\,\frac{\exp[-1.59(\log_{10}(\text{concn}) - 1.53)]}{1 + \exp[-1.59(\log_{10}(\text{concn}) - 1.53)]}.$$

The figure illustrates the conversion of an observed count of 0.430 for an 'unknown' sample to give an estimated insulin concentration of 2.019. The lower asymptote, -1.41, is determined extremely poorly in this case because there is virtually no data beyond the point of inflection of the logistic curve. The sole purpose here, however, is to obtain a descriptive curve over the range of the data, and so this is of no consequence.

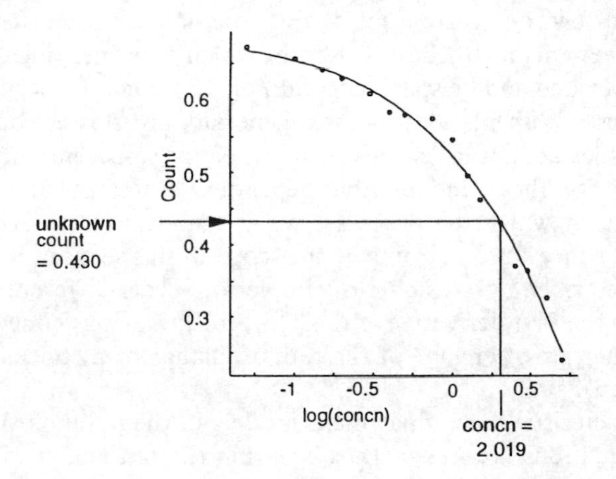

Figure 9.7 The standards data due to R.S. Yalow for an early insulin assay, together with the fitted standards curve. The conversion of an unknown count to an insulin concentration is also illustrated.

Complications and quality control. Considerable efforts have been dedicated to ensuring the reliability of immunoassays, and many safeguards both in the experimental technique and in the statistical analysis are now adopted routinely. For a discussion of these matters the reader is referred to Rodbard (1974).

9.5 COMPARTMENTAL MODELS

9.5.1 Introduction

The compartments referred to in this general class of model can take many forms. They are often physical spaces within which the substance of interest resides, the model being then concerned with the transfer of the substance between the different spaces. Alternatively the two compartments could be two different chemical or physical forms of a substance (e.g. a drug and its metabolites), in which case the model is concerned with interconversion between the different forms. The main characteristics of compartment models are as follows.

(1) The first important aspect of a compartment, which applies to the above two interpretations equally, is that it can be considered as *homogeneous*, and well mixed.
(2) A compartment exchanges with other compartments and the environment in such a way that the rate of change of concentration within the compartment is a function of the concentration in that compartment and in compartments to which it is connected. Thus it could be said that the forces driving the dynamics are *local*.
(3) Thirdly we are concerned with the *steady state dynamics* of these transfer processes, in which the rates at which things happen do not change with time. Inanimate objects when in a steady state are static, but for living organisms there are always some processes which are ticking over steadily—the processes of metabolism, absorption of food, clearance of waste products by the bloodstream to name just a few. For quite a lot of the time animals are not in a steady state, for example they might just have eaten, in which case their metabolic rate would probably be elevated and be starting to fall, or they might be becoming active after a period of rest. With plants processes generally are slower, but they also have periods when they are not in a steady state: when the sun eventually reaches a plant, its rate of photosynthesis and all other dependent processes suddenly rise. In these cases, the rates at which the processes we are modelling are occurring would be changing, and hence would be outside the scope of this section. Systems in a steady state can be described by a number of simple first order differential equations, i.e. involving only the first derivative, e.g. $\mathrm{d}y/\mathrm{d}t$, of the various concentrations (unlike those where the rate of change of the rate of change of a concentration, $\mathrm{d}^2y/\mathrm{d}t^2$, is also involved).
(4) Another characteristic of compartment models in their standard form is that the differential equations are *linear*, thus ensuring that an analytical solution can be obtained. This is a real restriction on their applicability as most biological processes are non-linear. However, a number of useful physiological models describing transfer of material from the bloodstream, including drugs, fall into this category.

We introduce the ideas in the context of a model of a single compartment, which can be handled by the mathematical methods of Chapter 1 and then extend the methods to cope with the more interesting case of two intercommunicating compartments.

A physical analogy

A physical analogy will perhaps clarify the features of this class of models. We could construct a working physical model of any compartmental model with linear kinetics (i.e. one which can be expressed as a set of first order linear differential equations) by representing each compartment by a fixed volume of fluid in a closed container with a perfectly efficient stirrer so that at any time the concentration of the material of interest, M, is completely uniform within the container. The containers are connected by pipes with pumps which maintain a constant fixed flow rate of fluid (and therefore also of our material M) between each pair of containers which are connected in the model. The concentrations of the material M in every compartment do not change with time, although they would in general be different between compartments; similarly for the flow rates—constant in time but different between pairs of compartments.

The use of tracers

In order to obtain estimates of the flow rates and volumes, a method called the *single injection method* could be used. In this a small quantity of a tracer is injected into one compartment at a particular time, and then the time course of the change in concentration of that tracer is followed at various points in the system, quite often just at one point. The tracer behaves exactly as the material in the pool or compartment, but its concentration can easily be determined. Tracers are often radioactive isotopes of the tracee, or a vehicle carrying a dye; in both cases, the amount present in a sample can easily be determined by counting the number of radioactive particles emitted per second, or by optical means in the case of a dye. A small quantity is used so as to have very little or no effect on the kinetics. The differential equations and the solutions can be used to model the changes in concentration of injected material in each compartment. The reason why the rate of loss of marker is proportional to the concentration of the marker in the compartment is that fluid is being extracted from the compartment at a fixed rate and therefore the quantity of marker being extracted is proportional to its concentration in the compartment. Another way of delivering tracer is the *continuous infusion method* in which the tracer is infused at a constant rate at one point in the system until an equilibrium concentration is reached in each compartment. From the time course of the changes in concentration, or possibly just from the equilibrium concentrations in each compartment, estimates of the flows through various parts of the system can be obtained.

9.5.2 A single compartment

Rate constants and flows

First consider a compartment or pool of fluid of fixed volume, V (ml), into which fluid flows at a fixed inflow rate, F (ml min^{-1}) (see Figure 9.8). As the volume of the pool

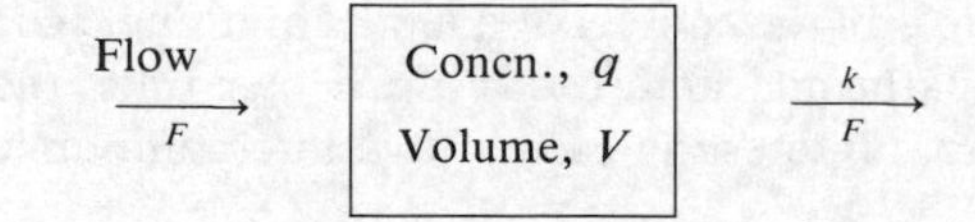

Figure 9.8 Flow through a single pool.

is fixed, then there must be a constant outflow rate of F. What fraction of the pool flows out per minute? In one minute this is clearly F/V. Consider a specific example in which $V = 100$ ml, and $F = 10$ ml min^{-1}, then the fractional rate of outflow is $10/100 = 0.10$ min^{-1} or 10% min^{-1}. This quantity expressed as a proportion, the fractional rate of outflow, is known as the *rate constant*, and is usually denoted by k.

Estimation of flow rate, F

Usually it is not possible to measure F directly, and the method then employed is to inject a *small* quantity of a tracer. Suppose that we inject a quantity, D, of tracer more or less instantaneously, and then repeatedly sample the pool subsequently, and determine the *concentration* of tracer present in each sample, q at time t. This gives rise to exponential decline of tracer, $q = q_0 \, e^{-kt}$, where q_0 is the initial concentration of tracer in the pool. Hence we see k can be estimated as the negative of the *slope* of the plot of the natural logarithm of the amount of tracer against time. Then to estimate F, we merely multiply V by our estimated value of k. Note that the only time-varying quantity in this case is the concentration of tracer, which decays with time. The system being modelled is in a *steady-state*. This is an important requirement for almost all tracer work. If the system being modelled is not stable, then the rate of decay of the tracer could change because the system itself is changing, rather than just because the concentration of tracer in the pool is falling.

Estimation of pool size, V

It might be thought that if the pool size is not known then all is lost, but one advantage of this method is that the pool size can be estimated. If we were to take a sample of fluid from the pool at time zero, immediately after injection of the dose of tracer D, the concentration of tracer in that sample would be given by $q_0 = D/V$ and so an estimate of V is given by $V = D/q_0$. However, it might take a minute or so in mammal studies, when the site of injection is the blood, for the injected dose (assumed to be injected at a particular *instant* in time) to mix thoroughly in the pool, and so q_0 might not be directly measurable. However, it can be determined by extrapolating the fitted line on a log-scale back to $t = 0$, or as an estimated parameter in the fitted model.

Many unconnected inputs and outputs

The same model as in this section holds in the case when there are many routes of input and output, provided that the inputs and outputs are unconnected. In order to determine the total k-value of the output, we merely add the individual k-values of the individual outputs. The total inflow is similarly the sum of all the inflows. Note that the total rate of inflow in this case is dictated by the total rate of outflow, as the pool size is assumed not to change. It is as if the material is instantaneously pulled in by the vacuum which would have occurred if material had flowed out without any replacement.

9.5.3 Two interchanging compartments

We now consider a more complex two-compartment model, derive the equations for the concentration of tracer in both compartments after a single injection into the first compartment, and consider how we can use data on the decay of tracer in compartment A to estimate various parameters of the model, such as rate constants and compartment volumes. Figure 9.9 schematically depicts the model in which the compartments, A and B, interchange material, input from the outside occurs in compartment A and loss from the system occurs from compartment B. Another common two-compartment model has loss to the outside of the system from compartment A, but there is no essential difference in methods for handling the two models.

Suppose we inject a dose, D, of tracer in compartment A at time $t = 0$, and that the concentrations of tracer in the two compartments at time t are q_a and q_b. Then the tracer concentrations in A and B satisfy the following equations.

$$\begin{pmatrix}\text{Rate of}\\ \text{increase of}\\ \text{tracer concn.}\\ \text{in A}\end{pmatrix} = \begin{pmatrix}\text{Loss from}\\ \text{A}\end{pmatrix} + \begin{pmatrix}\text{Gain from}\\ \text{B}\end{pmatrix}$$

$$= -\begin{pmatrix}\text{Rate}\\ \text{constant}\\ \text{to B}\\ \text{from A}\end{pmatrix}\begin{pmatrix}\text{Concn. of}\\ \text{tracer}\\ \text{in A}\end{pmatrix} + \begin{pmatrix}\text{Rate}\\ \text{constant}\\ \text{to A}\\ \text{from B}\end{pmatrix}\begin{pmatrix}\text{Concn. of}\\ \text{tracer}\\ \text{in B}\end{pmatrix}.$$

Then the differential equations for the concentrations in the two compartments are as follows.

Differential equations for general 2 compartment system (linear kinetics)

Differential equations for tracer concentration

$$\frac{dq_a}{dt} = -k_{ba}q_a + k_{ab}q_b$$

$$\frac{dq_b}{dt} = k_{ba}q_a - k_{ab}q_b - k_{0b}q_b$$

k_{xy} = rate constant for flow to X from Y
q_x = concentration of tracer in compartment X.

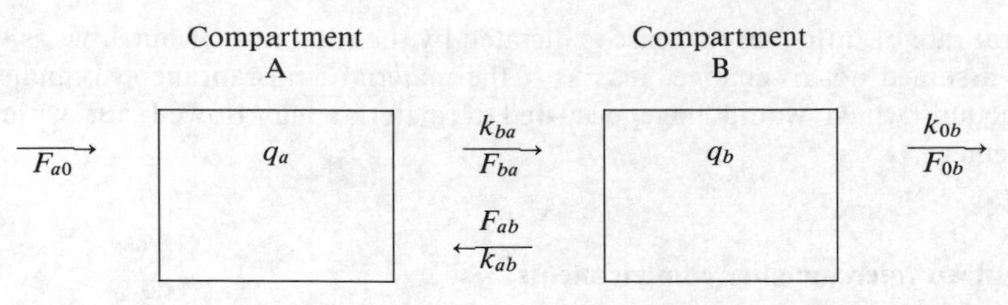

Figure 9.9 The compartment diagram for a two-compartment system, with outflow from compartment B.

Note that the recommended convention which we follow for the rate constants is that k_{ab} denotes the rate constant of the flow *to* A *from* B. Some texts reverse the subscripts. Using the result from Section 8.5, we obtain a solution

$$q_a = A_1 e^{\lambda_1 t} + A_2 e^{\lambda_2 t}$$

$$q_b = \frac{A_1(\lambda_1 + k_{ba})e^{\lambda_1 t}}{k_{ab}} + \frac{A_2(\lambda_2 + k_{ba})e^{\lambda_2 t}}{k_{ab}}$$

where λ_1 and λ_2 satisfy the auxiliary (or characteristic) equation

$$\lambda^2 + (k_{ba} + k_{ab} + k_{0b})\lambda + k_{ba}k_{0b} = 0$$

i.e.

$$\lambda_1, \lambda_2 = \frac{-(k_{ba} + k_{ab} + k_{0b}) \pm \sqrt{(k_{ba} + k_{ab} + k_{0b})^2 - 4k_{ba}k_{0b}}}{2}. \tag{9.5}$$

Now using the initial conditions we need to determine the values of the arbitrary constants. At time $t = 0$, $q_a = q_{a0} = \text{Dose/volume}$, $q_b = 0$, and therefore, substituting these values in the solution, we obtain

$$q_{a0} = A_1 + A_2 \qquad 0 = \frac{A_1(\lambda_1 + k_{ba}) + A_2(\lambda_2 + k_{ba})}{k_{ab}}$$

from which A_1 and A_2 can be determined:

$$A_2 = -A_1\left(\frac{\lambda_1 + k_{ba}}{\lambda_2 + k_{ba}}\right)$$

hence

$$q_{a0} = A_1\left(1 - \frac{\lambda_1 + k_{ba}}{\lambda_2 + k_{ba}}\right),$$

and therefore

$$A_1 = \frac{(\lambda_2 + k_{ba})q_{a0}}{\lambda_2 - \lambda_1}; \qquad A_2 = \frac{-(\lambda_1 + k_{ba})q_{a0}}{\lambda_2 - \lambda_1}.$$

So the solution is as given in the box below.

Solution of general two-compartment system

$$q_a = \frac{q_{a0}[(\lambda_2 + k_{ba})e^{\lambda_1 t} - (\lambda_1 + k_{ba})e^{\lambda_2 t}]}{\lambda_2 - \lambda_1}$$

$$q_b = \frac{q_{a0}(\lambda_1 + k_{ba})(\lambda_2 + k_{ba})[e^{\lambda_1 t} - e^{\lambda_2 t}]}{k_{ab}(\lambda_2 - \lambda_1)}$$

where q_a, q_b = concentrations in compartments A and B respectively
Initial concentration in compartment A = q_{a0}
Initial concentration in compartment B = 0
and λ_1, λ_2 = the roots, given in (9.5), of the auxiliary equation.

9.5.4 Estimating the parameters from measurements on one compartment

For some models it is possible from measurements on just one compartment to estimate some of the rate constants and compartment sizes. An example of the analysis of a two-compartment model assuming that a satisfactory model can be fitted to the tracer concentration in one of the compartments is given in Example 9.5 and illustrates the kind of inferences which are possible.

Example 9.5 Decay of BSP in calves

Harrison *et al.* (1986) used the dye sulphobromopthalein (BSP) in experiments to assess the liver blood flow. They assumed a two-compartment model with compartment A being plasma, compartment B hepatocytes, and loss being entirely from hepatocytes into bile. Figure 9.10 illustrates measurements on the decay of BSP in plasma from one calf. The concentrations are in units of mg/ml × 10^2.

t (min)	Concentration
1	10.2
2	7.67
3	5.76
4	4.50
5	3.56
6	2.77
7	2.30
8	1.84
9	1.46
10	1.26
13	0.77
16	0.52
19	0.39
22	0.28
25	0.22
28	0.17
31	0.14

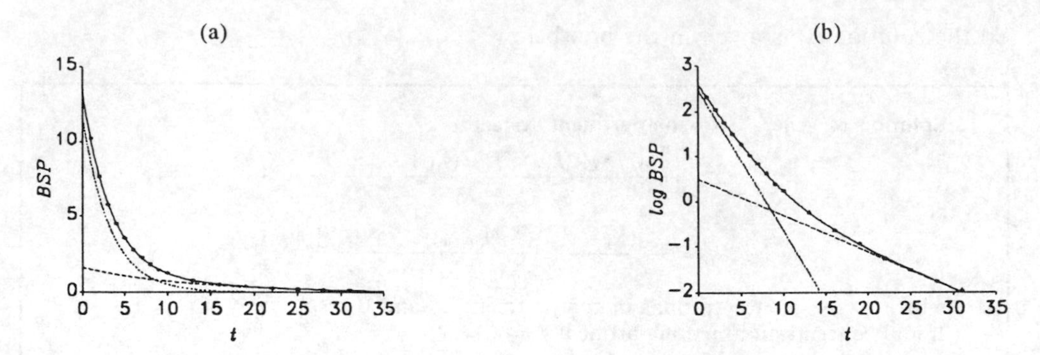

Figure 9.10 (a) The BSP decay data, and the fitted curve (solid line), together with the two component negative exponentials, component 1 with the faster decay, component 2 with the slower. (b) The same data and fitted curves after the y-scale has been logarithmically transformed.

The model fitted to these data by log least squares (with adjusting constant, $c = 0.5$, see Section 11.5.3) is

$$q_a = 13.34(0.878e^{-0.312t} + 0.122e^{-0.0808t}) \tag{9.6}$$

Calculation of plasma volume

The calf weighed 72.2 kg and the dose was 396.8 mg, and so the volume of the plasma is given by $V_{\text{plasma}} = 396.8/13.34 \times 100 = 2975$ ml or 41.2 ml/kg.

Estimation of rate constants

(1) *Estimation of k_{ba}*. The initial relative rate of change is given by the differential of the expression inside the brackets in equation (9.6) at $t = 0$, i.e. $(0.878)(0.312) + (0.122)(0.0808) = 0.2838$. This equals k_{ba}, since the only route from A is that to B.
(2) *Estimation of k_{0b}*. From the general solution of the two-compartment system, the product of the roots of the auxiliary equation and hence of the exponents in the solution $= D$ (in the notation of Section 8.5) $= k_{ba}k_{0b} = (0.312)(0.0808) = 0.02521$. Hence $k_{0b} = 0.0252/0.284 = 0.0887$.
(3) *Estimation of k_{ab}*. Similarly the sum of the two exponents, $0.312 + 0.0808 = 0.3928 = T = k_{ba} + k_{0b} + k_{ab}$, and therefore

$$k_{ab} = 0.393 - k_{ba} - k_{0b} = 0.393 - 0.284 - 0.0887 = 0.020.$$

By means such as these for this particular model, the main aspects of the dynamics are derived from measurements on compartment A alone.

It seems surprising on first acquaintance what information can be gleaned from measurements on just one of the compartments, and the technique has proved very useful in many physiological investigations. There are, however, four major problems with the technique which limit its usefulness.

(1) Quite often the model is only linear in rather restricted circumstances, and as soon as the non-linearity becomes substantial, analytical solutions are no longer available nor are approximately valid linear solutions useful. A typical non-linearity in physiological modelling arises from Michaelis–Menten type elimination kinetics, which are the simplest description of capacity-limited rates, as in the

following example. We specify the modification to the equations of the standard model, as a result of the elimination from compartment A to B being capacity-limited:

$$\frac{dq_a}{dt} = -\frac{V_m q_a}{K_m + q_a} + k_{ab} q_b$$

$$\frac{dq_b}{dt} = \frac{V_m q_a}{K_m + q_a} - k_{ab} q_b - k_{0b} q_b$$

This modification means that solution of the differential equations must be numerical and if there is a requirement to fit the model to data, then special techniques which are usually extremely computationally intensive must be used.

(2) If the model is linear, it is sometimes not known, nor can it be determined from the available data, how many interchanging compartments are involved. Using the wrong model would often result in incorrect inferences.

(3) If the model structure is correctly determined, it might be that important details such as rate constants and compartment sizes cannot be estimated from measurements on one compartment. Either some constants must be known *a priori*, or measurements are required on other compartments. This is known as the *deterministic or structural identifiability* problem.

(4) There are quite often statistical problems in estimating the parameters of the multiple exponential curves which arise in the analysis. This is often referred to as the *numerical identifiability* problem. Data which are little affected by random variability, and in the appropriate ranges of time, are required to obtain estimates with sufficiently narrow standard errors. The implications of the random variability for the fitting procedure are briefly discussed in the next section.

Godfrey (1983) and Shipley and Clark (1972) discuss some of these problems and provide further references.

Random elements in the two-exponential model

A brief discussion of methods of fitting the two-exponential model is given in Chapter 11, but it is appropriate here to discuss possible forms for the random residual component in the model. There are two unusual aspects involved in statistical analysis of the two-exponential curve which need to be considered carefully.

(1) *Disparity in residual variation at different times.* There is sometimes a very great range in the responses measured: early in the experiment the responses might be moderate or high, whereas at longer times, the responses are often very small in comparison. Usually the residual variation—whether it be measuring error or biological variation—is related to the average level, and often the coefficient of variation is approximately constant, except that there is a minimum value of residual variance at high values of t. So the residual variation is usually much greater for the earlier higher points. This is not academic, as the assumptions made about the residual variation, if an appropriate method of fitting the curve is used, impact substantially on the estimates obtained.

(b) *Earlier points are more important for the larger k-value.* The other unusual aspect of the model is that data in different regions of the curve have marked influences on estimates of different parameters in the model. The early values are usually very important in fixing the magnitude of the larger k-value, whereas the later ones have more influence on the smaller k-value. This can be seen by examining the plot of the two components of the fitted curve in Example 9.5. The data for $t > 20$ do not have any impact at all on the fit of component 1, since component 1 is effectively zero in that time range, whereas those before $t = 5$ have the greatest effect on component 1. We shall see in Chapter 11 that the variance of the random component in each part of the curve is important in determining an appropriate weighting in any weighted fitting procedure. Thus if, as a result of an incorrect assumption about the magnitude of the variance, an inappropriate weighting is used, this could have a big effect on the estimated model.

9.6 OSCILLATORS AND CLOCKS

9.6.1 Types of oscillators

There are many types of oscillators in physiological and biochemical systems, most models for which are too advanced to discuss here, often involving more than two components. We briefly mention some examples.

Neural excitability and membrane oscillators

We have already considered the Fitzhugh–Nagumo model of nerve cell firing in the previous chapter, and demonstrated that it is capable of two different kinds of behaviour depending on the nature of the input. Intermittent impulses result in either a swift return to equilibrium, or a substantial reversal of the membrane potential depending on the size of the impulses and the state of the system, demonstrating in its simplest form the phenomenon of excitability. A steady positive current input to the Fitzhugh–Nagumo model will induce continued oscillations of fixed amplitude and period. Membrane oscillators occur in many cells, as well as chemical oscillations within the cell. In some cases, as the two are linked it is not known with any certainty whether the membrane or cytoplasmic chemical oscillations are the most important.

Chemical oscillators

The Lotka–Volterra model for predation was also initially used by Lotka for a three-stage chemical reaction, which reduced to the two well-known differential equations. More recently, some quite complex models of biochemical systems have been devised and tested. One of the simplest is that of glycolysis, observed in yeast cells and extracts, which exhibits both oscillatory behaviour and excitability (Goldbeter, 1980). Berridge and Rapp (1979) give a survey of cellular oscillators, and Edelstein-Keshet (1988) discusses conditions for the existence of oscillations in chemical systems.

Delayed-feedback models

Glass and Mackey (1978) have shown how oscillatory behaviour in physiological control systems associated with various diseases can be induced in single-variable models by introducing a time lag. Delayed feedback has also been demonstrated in other systems to result in cycles (see also Chapters 1 and 13).

9.6.2 Linked oscillators

Oscillators do not normally function in isolation; they are linked to some extent to outside influences. These outside influences can be substantial shocks, or smaller random fluctuations in conditions, or they can be other oscillators which are of different phase or period. It is interesting to see how oscillators behave when linked in various ways; phase adjustment is a common response. An example is the case of two Fitzhugh–Nagumo oscillators of different periods, which—when linked by even weak positive electrical connections—tend to come into phase (see Figure 9.11). A related matter is the sensitivity of pulsatile systems to incoming random stimulations. In Chapter 7, we discuss the relationship between the mean frequency of a Fitzhugh–Nagumo neurone and the frequency of Poissonian stimulations.

Invertebrate neural systems have been shown to exhibit a range of involved cyclical behaviours (see Selverston & Moulins, 1985), some of which are due to the bursting properties of individual neurones, and others to properties of the network. These types of cyclical behaviour appear to characterise particular circuits involving named cells and are associated with specific functions in some well-studied ganglia. An example is the nine-celled cardiac ganglion of the lobster, which produces bursts of impulses that drive the heart muscle at frequencies between 20 and 50 per minute.

9.6.3 Disturbed oscillators and phase resetting

Winfree (1980, 1987) gives many examples of the effect of sudden disturbances on biological rhythms. One is the resetting, by a brief pulse of light in a dark environment, of the phase of the circadian clock involved in the timing of a fruit fly's emergence from its pupal case. When a large collection of pupae are kept in total darkness, the pupae largely emerge in bursts approximately every 24 hours. In experiments reported by Winfree, pulses of light of different durations were given at various points in the normal 24 h cycle, and the resetting of the time of emergence noted. For short stimuli given at any time of the cycle, little change in emergence time from the normal occurred. However, as the stimulus lengthened, this picture changed, and a more complex pattern emerged: for a range of stimulus lengths, when the stimulus was applied before a particular point in the cycle, the pupae tended to emerge at a constant time after the application of the light pulse. When applied after this point in the cycle, the pupae tended to emerge at a fixed but different interval after the application of the pulse. What was very surprising was that for a particular stimulus length given at one particular time of the cycle, the extent of the phase resetting could not be predicted at all. The flies so treated emerged subsequently at all times of the day. It is also clear from the experiments that their internal clock had not been destroyed, but merely put into neutral as it were, as a subsequent flash of light could restore the clock's rhythm

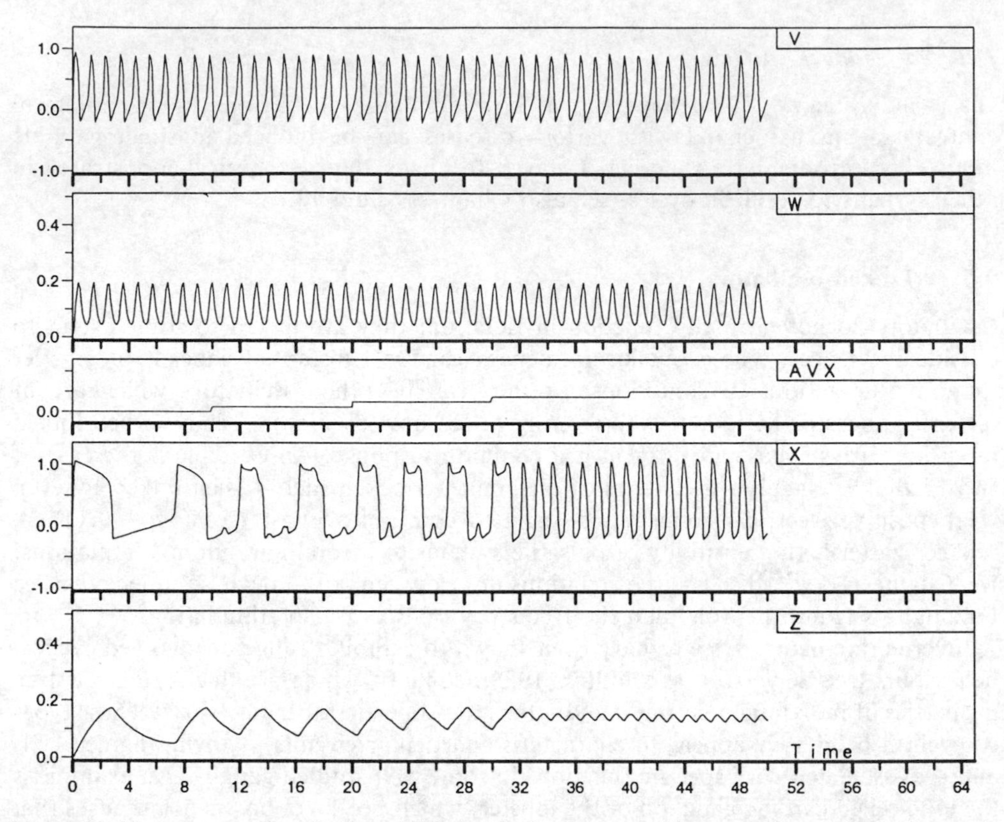

Figure 9.11 An example of linked Fitzhugh–Nagumo oscillators. Plots of oscillator 1, consisting of variables V and W, and oscillator 2 (X and Z) with a different period are given. Until $t = 10$, the two are unlinked, and then a link, AVX, of gradually increasing strength between V and X is made. X eventually has the same period and phase as V.

by bringing the flies back into an approximately synchronised bursting pattern of emergence.

This can be partially understood by considering the model of the frictionless pendulum with which we began Chapter 8. Assume that there is an associated clock which ticks every time the pendulum passes through its lowest point from left to right. If the pendulum is given a small impulse at any point of its swing, it will make little difference to the subsequent timing of the tick, and this is similar to the effect of a brief flash of light on the fruit fly. However, if an impulse is given to the pendulum weight when the pendulum is in a vertical position, and of a magnitude which is just sufficient to stop the motion of the pendulum, the associated clock will stop. There is only one position at which this can be done—the bottom of the swing of the pendulum—and the impulse has to be of exactly the right magnitude. Otherwise the phase of the clock (indicated by the timing of the tick) is reset, but it goes on swinging, and as the period is determined by the length of the pendulum, it goes on ticking with the same frequency. If the impulse applied at the bottom of the swing had not been quite sufficient to stop the swinging of the pendulum, the pendulum would then tend to swing with a very small amplitude, and the phase would be dependent on the exact point to which the impulse pushed the pendulum before it came to rest; hence the indeterminacy in the phase (see Murray (1988), Chapter 8).

Phase resetting can occur with almost any oscillating system which also has an equilibrium. For example, the Fitzhugh–Nagumo (FN) model discussed in Chapter 8, when subject to a positive applied current. Such a system goes into continuous cycling of fixed period and amplitude as is demonstrated in Figure 8.16. If a pulse consisting of a positive shift in v (i.e. to the right in the phase plane diagram, Figure 8.16(a)) occurs, it disturbs the periodic motion, but the system eventually returns to the periodic trajectory, and the time at which v passes through zero is shifted along the time axis, i.e. the phase is reset. However, if the system is perturbed when it is at a point on the trajectory which is exactly to the left of the equilibrium, and it is perturbed by an amount which is just sufficient to move it to the equilibrium, the periodic firing would in theory be abolished. However, in practice as this equilibrium is unstable, the system would eventually move away from the equilibrium and resume the periodic firing. But the phase (the time at which v passes through zero) would then be totally unpredictable as it is critically dependent on the exact point in the vicinity of the equilibrium to which the impulse moves the trajectory.

A more interesting example from the point of view of modelling of real biological systems was examined by Best (1979). He considered the more complex Hodgkin–Huxley (HH) model of neurone firing. This model involves four differential equations, and its behaviour is very similar to that of the Fitzhugh–Nagumo equations. When the neurone is subjected to a positive applied current, the system goes into continuous fixed period oscillations in a very similar fashion to the FN model. But for our present purposes, the main difference is that if the leakage current involved in the HH model is made sufficiently large, the equilibrium at the centre of the limit cycle in the phase plane is no longer unstable, but becomes stable, and has a small area of the phase plane around it called the basin of attraction of the stable equilibrium. If the system finds itself in this basin, then it moves towards the stable equilibrium. If it is outside, it would move on to the stable limit cycle, oscillatory solution. For models of this type, one possible outcome of a phase-resetting experiment is therefore complete and permanent abolition of the oscillation. This does occur with some real biological systems, and models of this type are consequently of use in interpreting such behaviour.

EXERCISES

9.1 Two substances react chemically according to the formula:

$$\mathrm{A}+\mathrm{B} \underset{k_2}{\overset{k_1}{\rightleftharpoons}} \mathrm{C}$$

where k_1 and k_2 are the rate constants. Derive equations for the molar concentrations a, b, c of substances A, B, C and show that if $a = a_0, b = b_0, c = 0$ at $t = 0$ then

$$\frac{\mathrm{d}a}{\mathrm{d}t} = -k_1 a^2 + a[k_1(a_0 - b_0) - k_2] + k_2 a_0.$$

In the particular case where $b_0 = (1-\gamma)a_0$ and $k_2/k_1 = \gamma a_0$ show that a satisfies

$$\frac{a/a_0 - \sqrt{\gamma}}{a/a_0 + \sqrt{\gamma}} = \frac{1-\sqrt{\gamma}}{1+\sqrt{\gamma}}\,e^{-2a_0k_1t\sqrt{\gamma}}.$$

Sketch the graph of a, b and c against time.
(Cambridge, NST, Part IA, Biological Mathematics, 1980.)

9.2 State the Law of Mass Action, and explain what is meant by the *molecularity* of a single-step reaction.

The reaction

$$H_2 + I_2 \xrightarrow{k} 2HI$$

may be assumed to be simple and irreversible. Let the concentrations of hydrogen, iodine and hydrogen iodide at time t be $a(t)$, $b(t)$ and $p(t)$ respectively, and let their concentrations at $t=0$ be a_0, b_0 and p_0. Write down equations for the rate of change of $a(t)$, $b(t)$ and $p(t)$ and calculate the concentrations $a(t)$, $b(t)$ as functions of time.

Without making detailed calculations, explain why a similar analysis aimed at finding the concentration of atomic bromine as a function of time in the reaction

$$2Br \rightarrow Br_2$$

would lead to a qualitatively different answer.
(Cambridge, NST, Part IA, Biological Mathematics, 1986.)

9.3 An enzyme E catalyses a reaction in which a substrate is converted to a product P via an intermediate complex C:

$$S + E \underset{k_{-1}}{\overset{k_{+1}}{\rightleftharpoons}} C \xrightarrow{k_2} E + P.$$

What assumptions are required in order to apply Michaelis–Menten theory to this reaction? What is the key deduction from these assumptions? Explain briefly how this deduction may be justified.

Using this deduction, derive the formula

$$c = \frac{e_0 s}{s + K_m}$$

for the concentration c of the complex in terms of the concentrations of the enzyme, e_0, and substrate, s, where K_m is a constant. Show that the rate of the reaction, v, is given by

$$v = \frac{e_0 k_2 s}{s + K_m}.$$

Draw a sketch of v as a function of s, and describe two different plots where this graph may be represented as a straight line.
(Cambridge, NST, Part IA, Biological Mathematics, 1987.)

9.4 A drug is injected intramuscularly at time $t = 0$. The half-life for uptake by the blood is T_1; the half-life for subsequent biological decay of the drug in the bloodstream is T_2. Write down differential equations satisfied by the concentrations of the drug in the muscle and blood compartments (C_m and C_b respectively), and deduce that C_b varies with time according to the equation

$$C_b = \frac{aC_0}{a-b}\,(e^{-bt} - e^{-at}),$$

where a and b should be identified in terms of T_1 and T_2. What does C_0 represent?

In a particular case, $T_1 = 30$ min, $T_2 = 4$ h, and the blood concentration must be kept below a value of 10 mg/ml. What is the maximum permitted value of C_0? In this case at what time should a further injection be made to stop the blood concentration falling below 4 mg/ml? (Cambridge, NST, Part IA, Biological Mathematics, 1988.)

9.5 From the equation for the interaction between a ligand, G, and a set of receptors, R,

$$G + R \underset{k_2}{\overset{k_1}{\rightleftharpoons}} GR$$

obtain the usual relationship between the concentration of bound ligand, B, and free ligand, F

$$B = \frac{B_{\max} F}{K_D + F}.$$

Sketch the curve, and discuss its main properties. Given data on concentrations of B and F, describe a method for estimating the parameters K_D and $B_{\max}$.

9.6 The following data are the concentration of free T3 (thyroxin) and those of T3 bound to receptors in pig skeletal muscle, obtained in two separate assays. The data were obtained in assays of two samples of tissue from animals which received different treatments. Assuming that these differences would be reflected in further animals which have received the two treatments, what are your conclusions about the effect of the treatment on the binding characteristics of the tissue? In particular

(a) Plot out B against F.

(b) Construct two plots of transformations of the x- and y-scales which would produce straight lines if the standard model for ligand receptor binding holds.

(c) Use these plots to estimate the maximum binding ($B_{\max}$) and dissociation constants (K_D) for the binding of T3 in these tissues (but see also Exercise 11.2).

Treatment A

Free	2437	1218	611	315	169	96	24.6
Bound	1.270	0.910	0.667	0.402	0.248	0.153	0.037

Treatment B

Free	4939	2469	1238	627	325	175	174	99	25.6
Bound	0.982	0.790	0.589	0.409	0.252	0.142	0.149	0.099	0.023

9.7 Outline the main steps in the radiommunoassay of hormone concentration in a sample of blood. State the usual mathematical model used in such assays. Radioimmunoassay is a *saturation, dilution assay*. Explain the statement and, assuming its truth, derive a mathematical expression for the relationship between the concentration of bound hormone and the concentration of hormone in the assayed sample.

The following is data from an assay of thyrotrophin given in Healey (1972). Analyse the data, estimating a standard curve from the standards data and then use the standard curve to estimate the thyrotrophin concentration for the samples of unknown concentration. Give some attention to the quality control of the c.p.m. figures.

Standards

Conc.	0	0.1	0.5	1	2	5	10	16	20	50	100	blank
obs. (c.p.m.)	37	4789	4979	4769	4270	4462	3374	2793	2482	2111	1676	686
	5076	4928	4961	5131	4571	3939	3500	2966	3423	2278	1904	700

Unknowns

(c.p.m.)						
	4779	5599	3799	70	6305	4657
	3913	4944	3802	8375	5013	4602

9.8 Rewrite the standard model for ligand–receptor binding,

$$B = \frac{B_{\max} F}{K_D + F}$$

in terms of the total concentration of ligand, $T = B + F$, and solve the resulting equation for B in terms of T. If the usual protocol is used, and measurement errors in B (and hence equal and opposite errors in F) are the main source of variation, how would you use this result to estimate $B_{\max}$ and K_D?

9.9 Derive the expression for the saturation P_B for the cooperative binding between haemoglobin and oxygen at a partial pressure of pO_2

$$P_B = \frac{(pO_2)^m}{(P_{50})^m + (pO_2)^m}$$

given in Section 9.3.4, assuming that m molecules of oxygen bind to each molecule of haemoglobin. In your answer, you should (a) write down the differential equation for the molar concentration of the product, y; (b) find an expression for the equilibrium; (c) substitute for the free molar concentration of haemoglobin in the resulting expression to obtain a formula involving the total molar concentration of haemoglobin.

9.10 In many physiological and/or pharmacological studies of mammals, the following equations are used to model the concentration (C_i) of injected material in a number of 'compartments'.

For each i $(= 1, 2, \ldots, n)$

$$\frac{\mathrm{d}C_i}{\mathrm{d}t} = \sum_{j=1}^{n} L_{ij} C_j$$

where the coefficients L_{ij} are fixed.

The solutions are assumed to have the form

$$C_i = \sum_{j=1}^{n} a_{ij} \, \mathrm{e}^{\lambda_j t}$$

where a_{ij} are constants for $i = 1, \ldots, n$.

(a) For the case $n = 3$ show that, if the constants a_{ij} are not zero, then λ_j for $j = 1, 2, 3$ must be the roots of the determinant

$$\begin{vmatrix} L_{11} - \lambda & L_{12} & L_{13} \\ L_{21} & L_{22} - \lambda & L_{23} \\ L_{31} & L_{32} & L_{33} - \lambda \end{vmatrix} = 0.$$

(b) Calculate λ_j in the case where $L_{11} = -4$, $L_{12} = -3$, $L_{13} = L_{31} = L_{23} = L_{32} = 0$, $L_{21} = -2$, $L_{22} = -6$, $L_{33} = -4$.

(c) To find the ultimate rate of decay of any injected material, calculate the form of C_i as $t \to \infty$. Is this the same in all the compartments?

9.11 Consider an enzyme E on which there are two sites to which the substrate S can bind, so it can exist in three states: pure E, a complex C_1 with one site occupied by S, and complex C_2 with both sites occupied. The chemical reactions can be represented as

$$S + E \underset{k_{-1}}{\overset{k_{+1}}{\rightleftharpoons}} C_1 \xrightarrow{k_{+2}} E + P$$

$$S + C \underset{k_{-3}}{\overset{k_{+3}}{\rightleftharpoons}} C_2 \xrightarrow{k_{+4}} C_1 + P,$$

where P is the eventual product.

Write down the equations which govern these reactions. Making the same approximations as in Michaelis–Menten theory, show that if the initial enzyme concentration is e_0, then the concentration of enzyme when the substrate concentration is s_0 is

$$e = e_0\left[1 + \frac{s_0}{K_m} + \frac{s_0^2}{K_m K_m'}\right]^{-1},$$

where

$$K_m = \frac{k_{-1} + k_{+2}}{k_{+1}}, \qquad K_m' = \frac{k_{-3} + k_{+4}}{k_{+3}}.$$

Show too that the velocity of the reaction, $v(= \mathrm{d}p/\mathrm{d}t)$, is given by

$$v = \frac{es_0}{K_m}\left(k_{+2} + k_{+4}\frac{s_0}{K_m'}\right).$$

(Cambridge, NST, Part IA, Biological Mathematics, 1979.)

10 Ecology and Epidemiology

10.1 INTRODUCTION

This chapter is about models for describing how interactions between individual organisms of different categories or species affect the dynamics of whole populations. For example, the effects of predator–prey interaction on fluctuations in population size of both species, or of contacts between susceptible and infected individuals on the growth of an epidemic or the spread of a disease. These models involve more than one response variable in contrast to the models described in Part I for populations of one species or type of individual.

In Sections 10.2–10.4 we present models for the ecological interactions of competition, predation and parasitism. The study of interaction is basic to ecology and is of fundamental and practical significance. For example, in Darwin's *Theory of Natural Selection* competition amongst genotypes is seen as the potential driving force of evolution. More practically, certain predatory insects can be of value for biological control of pest species; yields of crops, such as wheat, may be affected when growing in competition with weeds. Ecological communities are usually complex, containing many species interacting in a temporally and spatially varying environment. However, realistic descriptions of ecosystems involving dynamic stochastic models with many response variables are very complex mathematically and beyond the scope of this book. We adopt a much simpler approach and deal with pairwise interactions using deterministic models and assuming a spatially homogeneous environment. Nevertheless, even these relatively simple models for two species can produce interesting and sometimes unexpected behaviour so that there is a need to consider simple cases in their own right and as a first step towards more complex models. We describe dynamic models using differential equations for competition and predation and difference equations for parasitism, where often the insects involved have discrete non-overlapping generations. We also present equilibrium models, with no explicit time element, for describing the effects of plant competition.

Section 10.5 is about epidemic models and, in particular, modelling the spread of an epidemic in a population of fixed size where the disease is transmitted through contacts between individuals. We present a model which uses differential equations to describe the course of an epidemic in a large homogeneous population. Then, we consider a model with demographic stochasticity to examine the effect of chance variations in a small population, or at the beginning of an epidemic in a large population when the *initial* number of infected individuals could be small.

At the end of each section, more complex models including stochasticity, heterogeneity and spatial effects are discussed.

10.2 COMPETITION MODELS

Although different species usually exploit different resources, there are many situations where there is some overlap which can lead to competition *between* species, sometimes called *interspecific competition*. The resources might be light, nutrients and water in the case of plants, or food and breeding sites for animals. Competition between species may have a similar effect to that between individuals of the same species (*intraspecific competition*) by causing reduced growth of one or both populations. An important difference, however, is that interspecific competition can lead to extinction of one species by another. In this section we present some simple models for competition between two species. These are descriptive models for the effects of competition rather than mechanistic models of the underlying processes.

10.2.1 The Lotka–Volterra model for two-species competition

A basic model for the interaction of two competing species was proposed by Lotka (1925) and, independently, by Volterra (1926). The approach extends the logistic model for limited population growth of a single species in which the growth rate is given by

$$\frac{\mathrm{d}N}{\mathrm{d}t} = rN\left(1 - \frac{N}{K}\right)$$

where N denotes population density at time t, r is the intrinsic relative growth rate and K is the carrying capacity. The logistic model does not incorporate a specific mechanism for limited population growth but merely describes its effect as a relative growth rate which decreases linearly with population density, i.e.

$$\frac{1}{N}\frac{\mathrm{d}N}{\mathrm{d}t} = r - \frac{rN}{K}$$

or

Relative growth rate
= Intrinsic relative growth rate
– Reduction in relative growth rate due to effect of species on itself.

This reduction in relative growth rate with increased density is a simple model for the effect of competition between individuals of the same species. The approach can also be used to measure the effects of competition between species by adding a further reduction in the relative growth rate of one species which is proportional to the density of the other species. For the interaction of species 1 with species 2 we write

$$\frac{1}{N_1}\frac{\mathrm{d}N_1}{\mathrm{d}t} = r_1 - \frac{r_1 N_1}{K_1} - \frac{r_1 \alpha_{12} N_2}{K_1}$$

or

Relative growth rate of species
= Intrinsic relative growth rate
– Reduction in relative growth rate due to effect of species 1 on itself
– Reduction in relative growth rate due to effect of species 2 on species 1.

The parameter α_{12} is called a *competition coefficient* and measures the competitive effect of species 2 on species 1. When α_{12} exceeds one, species 1 suffers a greater competitive effect from species 2 than it does from itself. A similar coefficient α_{21} measures the competitive effect of species 1 on species 2. This is the basis of the Lotka–Volterra model for two species competition.

Lotka–Volterra model for two species competition

$$\frac{dN_1}{dt} = r_1 N_1 \left[1 - \frac{(N_1 + \alpha_{12} N_2)}{K_1}\right], \qquad \frac{dN_2}{dt} = r_2 N_2 \left[1 - \frac{(N_2 + \alpha_{21} N_1)}{K_2}\right]$$

where N_1 and N_2 are the population densities of species 1 and 2, r_1 and r_2 are the intrinsic relative growth rates, K_1 and K_2 are the carrying capacities in single population growth, and α_{12} and α_{21} are the competition coefficients.

In the absence of competition from species 2 (i.e. $\alpha_{12} = 0$ or $N_2 = 0$), species 1 exhibits logistic population growth increasing to its carrying capacity K_1. Similarly, with no competition from species 1, the density of species 2 will increase and stabilise at its carrying capacity K_2.

For specified values of the parameters and initial population densities the behaviour of the model can be studied by solving the differential equations of population change using numerical methods. Figure 10.1 shows examples for a range of values of the competition coefficients. This illustrates that the eventual outcome can either be coexistence or extinction of one species by the other. A more systematic study of the behaviour of the model and the conditions for the various outcomes is given below using phase plane analysis.

Phase plane analysis

The qualitative dynamic behaviour of the Lotka–Volterra model can be examined by constructing a phase plane diagram showing the nullclines, direction flows and the equilibrium values or steady states (see Chapter 8).

The nullclines for species 1 are obtained by setting dN_1/dt to zero, i.e.

$$\frac{dN_1}{dt} = r_1 N_1 \left[1 - \frac{(N_1 + \alpha_{12} N_2)}{K_1}\right] = 0,$$

so that $N_1 = 0$ or $N_1 = K_1 - \alpha_{12} N_2$. The first case corresponds to absence of species 1 and growth in a single population of species 2. For the two species model the most interesting nullcline is the straight line which intercepts the N_1-axis at $N_1 = K_1$ and the N_2-axis at $N_2 = K_1/\alpha_{12}$. Similarly, the nullclines for species 2 are obtained from

$$\frac{dN_2}{dt} = r_2 N_2 \left[1 - \frac{(N_2 + \alpha_{21} N_1)}{K_2}\right] = 0,$$

so that $N_2 = 0$ or $N_2 = K_2 - \alpha_{21} N_1$. The nullcline for species 2 involving the two species densities is then a straight line which intercepts the N_1-axis at $N_1 = K_2/\alpha_{21}$ and the N_2-axis at $N_2 = K_2$.

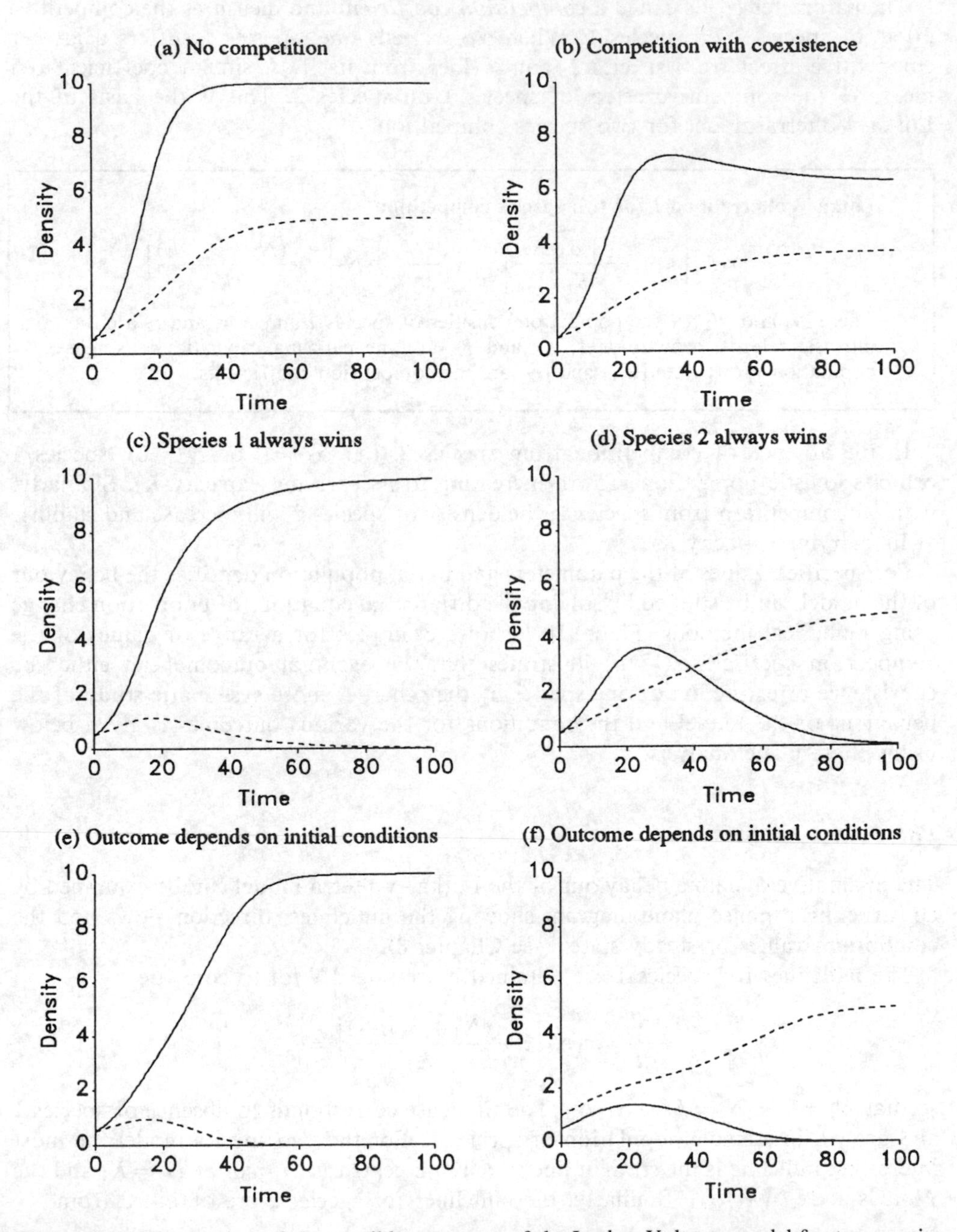

Figure 10.1 Illustration of the possible outcomes of the Lotka–Volterra model for two species competition. In each case: species 1 (solid line), $r_1 = 0.20, K_1 = 10$; species 2 (broken line), $r_2 = 0.10, K_2 = 5$. For (a) to (e), $N_1(0) = 0.5$, $N_2(0) = 0.5$. (a) $\alpha_{12} = \alpha_{21} = 0$; (b) $\alpha_{12} = 1, \alpha_{21} = 0.2$; (c) $\alpha_{12} = 1.5, \alpha_{21} = 0.8$, (d) $\alpha_{12} = 3.0, \alpha_{21} = 0.4$; (e) $\alpha_{12} = 4.0, \alpha_{21} = 1.5$; (f) competition coefficients as in (e) only initial conditions differ, $N_1(0) = 0.5$, $N_2(0) = 1.0$.

Figure 10.2 illustrates the four possibilities for the relative positions of the two nullclines: (a) nullcline for species 1 lies wholly above that for species 2; (b) nullcline for species 1 lies wholly below that for species 2; (c) nullcline for species 1 crosses that for species 2 with shallower slope; (d) nullcline for species 1 crosses that for species 2 with steeper slope. The nullclines divide the phase plane diagram into regions within which species 1, or species 2, is either increasing or decreasing, so that direction flows of the system can be sketched in. Figure 10.2 gives direction flows over a regular grid of points generated by computer. The flows indicate the eventual state of the system, or outcome of competition, for the different relative positions of the nullclines. Table 10.1 gives a summary of these outcomes which are discussed below. The stability properties can be confirmed by local stability analysis (Chapter 8).

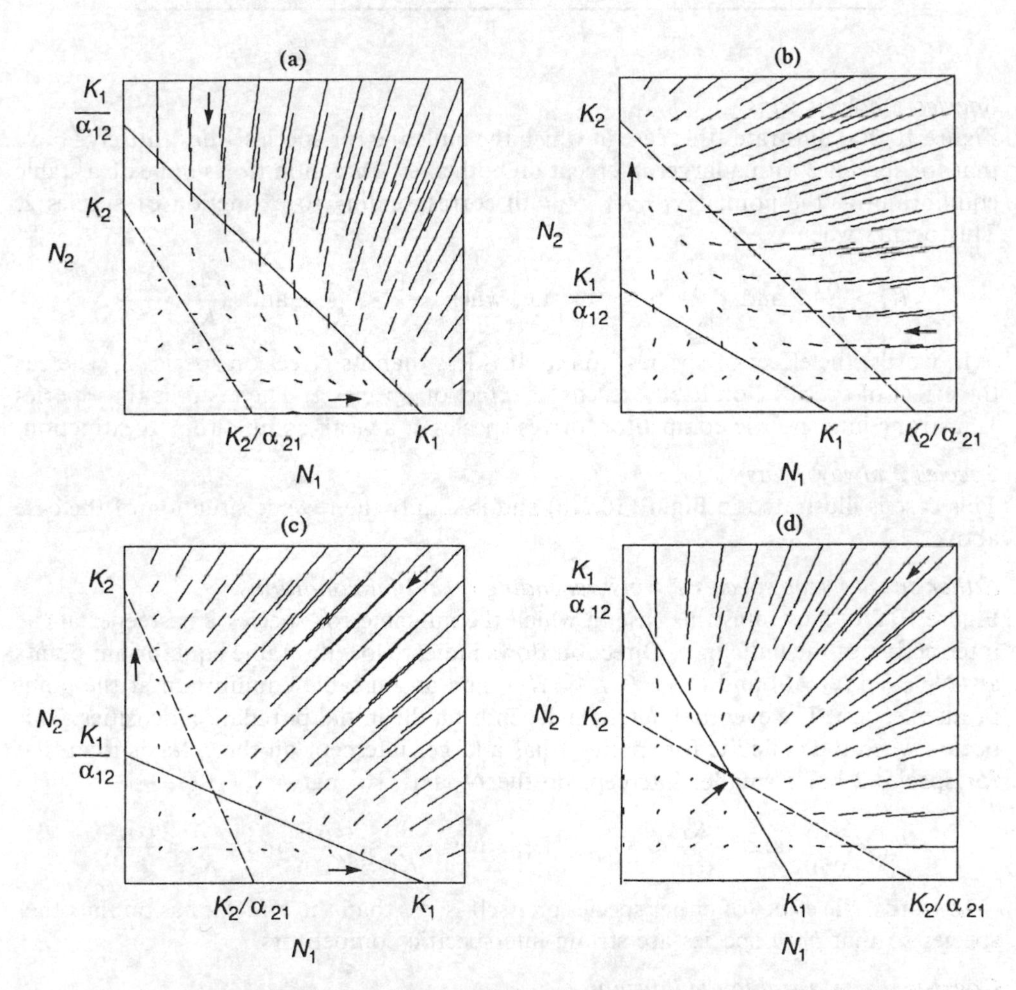

Figure 10.2 Phase plane plots, nullclines and direction flows to illustrate the conditions for the possible outcomes of the two species Lotka–Volterra model. Nullclines: species 1 (solid line); species 2 (broken line). (a) Species 1 always wins; (b) species 2 always wins; (c) outcome depends on initial conditions; (d) coexistence as a stable equilibrium. See text.

Table 10.1 Summary of the outcomes of the Lotka–Volterra model for two species competition

Competitive effect of species 1 on species 2	Competitive effect of species 2 on species 1: Weak $\frac{\alpha_{12}}{K_1} < \frac{1}{K_2}$	Competitive effect of species 2 on species 1: Strong $\frac{\alpha_{12}}{K_1} > \frac{1}{K_2}$
Weak: $\frac{\alpha_{21}}{K_2} < \frac{1}{K_1}$	Coexistence as a stable equilibrium	Species 2 always wins
Strong: $\frac{\alpha_{21}}{K_2} > \frac{1}{K_1}$	Species 1 always wins	Winner depends on initial conditions

Species 1 always wins

Figure 10.2(a) illustrates this case in which the nullcline for species 1 lies entirely above that for species 2 with a larger intercept on both axes. Direction flows indicate a stable equilibrium at the point $(N_1 = K_1, N_2 = 0)$ corresponding to extinction of species 2. This occurs when

$$K_1 > \frac{K_2}{\alpha_{21}} \quad \text{and} \quad \frac{K_1}{\alpha_{12}} > K_2, \qquad \text{i.e. when} \quad \frac{1}{K_1} < \frac{\alpha_{21}}{K_2} \quad \text{and} \quad \frac{1}{K_2} > \frac{\alpha_{12}}{K_1}.$$

In words, the effect of species 1 on itself is less than its effect on species 2, whereas the effect of species 2 on itself exceeds its effect on species 1. The result is that species 1, a strong interspecific competitor, drives species 2, a weak competitor, to extinction.

Species 2 always wins

This case is illustrated in Figure 10.2(b) and is simply the reverse situation of the case above.

Either species 1 or species 2 wins depending on initial conditions

Figure 10.2(c) illustrates this case in which the nullcline for species 2 is steeper at the intersection of the nullclines. Direction flows indicate locally stable equilibrium points at $(N_1 = K_1, N_2 = 0)$ and $(N_1 = 0, N_2 = K_2)$, and an unstable equilibrium at the point of intersection. The eventual outcome depends on the initial population densities. This occurs when the nullcline for species 1 has a larger intercept on the N_1-axis than that for species 2 but a smaller intercept on the N_2-axis, so that

$$K_1 > \frac{K_2}{\alpha_{21}} \quad \text{and} \quad \frac{K_1}{\alpha_{12}} < K_2, \qquad \text{i.e. when} \quad \frac{1}{K_1} < \frac{\alpha_{21}}{K_2} \quad \text{and} \quad \frac{1}{K_2} < \frac{\alpha_{12}}{K_1}.$$

In words, the effect of either species on itself is less than the effect it has on the other species so that *both* species are strong interspecific competitors.

Coexistence as a stable equilibrium

Figure 10.2(d) shows this case for which the nullclines cross with that for species 1 at a steeper angle. Direction flows indicate that the point of intersection is a stable equilibrium. This occurs when the nullcline for species 1 has a smaller intercept on the

N_1-axis than that for species 2 and a larger intercept on the N_2-axis so that

$$K_1 < \frac{K_2}{\alpha_{21}} \quad \text{and} \quad \frac{K_1}{\alpha_{12}} > K_2, \qquad \text{i.e. when} \quad \frac{1}{K_1} > \frac{\alpha_{21}}{K_2} \quad \text{and} \quad \frac{1}{K_2} > \frac{\alpha_{12}}{K_1}.$$

Here, the effect of either species on itself exceeds the effect it has on the other species. Interspecific effects are relatively small and a stable coexistence is possible.

Example 10.1 Two species competition model fitted to data from Gause's experiments on yeasts

Gause (1934) carried out a series of experiments on the competition between the two species of yeast *Saccharomyces cerevisiae* and *Schizosaccharomyces kephir*. Yeasts were grown separately in pure cultures and then in mixture. For pure cultures the growth of both species was well described by the logistic model (Figure 10.3(a)). In mixture, the effects of competition were apparent as the amount of each species increased to a level well below the carrying capacity observed in pure culture (Figure 10.3(b)). Gause used data on growth in the pure cultures to estimate the intrinsic relative growth rates and the carrying capacities, and growth in mixture to estimate the competition coefficients. The estimates were:

Saccharomyces cerevisiae	$r_1 = 0.218\ \text{h}^{-1}$	$K_1 = 13.0$	$\alpha_{12} = 3.15$
Schizosaccharomyces kephir	$r_2 = 0.061\ \text{h}^{-1}$	$K_2 = 5.8$	$\alpha_{21} = 0.439$

Schizosaccharomyces has a low intrinsic growth rate and a low carrying capacity in pure culture but is a relatively strong competitor in mixture, exerting more competitive influence on *Saccharomyces* than *Saccharomyces* does on itself. By contrast, *Saccharomyces* grows relatively quickly to a higher carrying capacity in pure culture but is a relatively weak competitor in mixture. Figure 10.3(b) shows the population growth in mixture for the Lotka–Volterra model over the 60 hours of the experiment using the above parameter values and initial densities of 0.45 units of yeast. There is reasonable agreement between the observed densities and those calculated from the model although the model consistently overestimates the density of *Saccharomyces* in the early stages of growth. For *Saccharomyces*, the fitted densities show a slight decline towards the end of the series and if the fitted model is extended beyond the duration of the experiment the decline in *Saccharomyces* continues and the density of *Schizosaccharomyces* increases until the curves cross after about 190 hours. This is, however, not to say that this outcome would have been observed in an experiment of longer duration. Finally, we can use the parameter estimates and the results of the phase plane analysis to predict the outcome of competition. This gives the following:

$$\frac{\alpha_{12}}{K_1} = \frac{3.15}{13.0} = 0.242; \qquad \frac{1}{K_2} = \frac{1}{5.8} = 0.172$$

and

$$\frac{\alpha_{21}}{K_2} = \frac{0.439}{5.8} = 0.0757; \qquad \frac{1}{K_1} = \frac{1}{13.0} = 0.0769.$$

For these values, Table 10.1 confirms the numerical finding that the *eventual* outcome of the model is extinction of *Saccharomyces* (species 1) by *Schizosaccharomyces* (species 2). The qualitative prediction based on the phase plane analysis is, however, a very poor guide to the observations over the duration of the experiment. A further point is that the second inequality concerning the competitive effect of species 1 on species 2 falls very close to the boundary between extinction of species 1 and an eventual winner which depends on the initial conditions. Two implications are: (a) different outcomes may have been observed

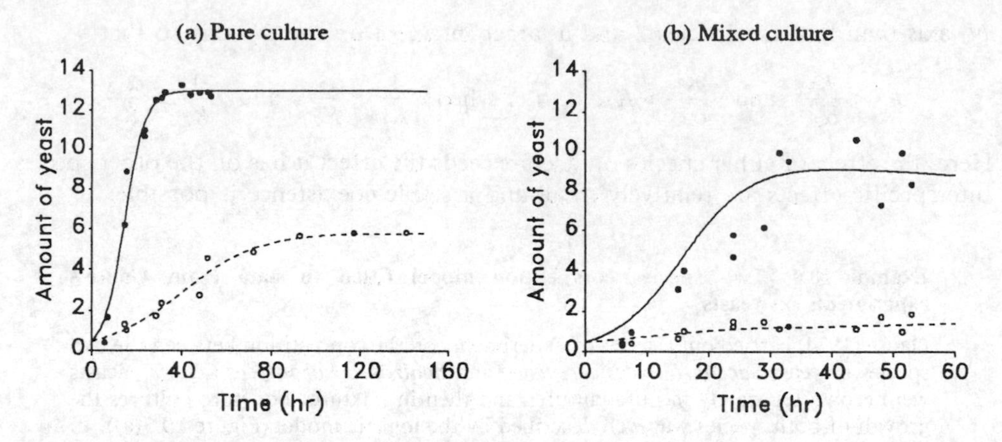

Figure 10.3 Gause yeast data with logistic curves fitted to growth in pure cultures and the Lotka–Volterra two species competition model fitted to growth in mixed culture. *Saccharomcyes cerevisiae* (solid line); *Schizosaccharomyces kephir* (broken line).

using different initial densities; (b) stochastic variation could be important; (c) predictions could be sensitive to estimation errors in the competition coefficients (see Chapter 11 for a discussion of fitting the model).

10.2.3 Discussion and more complex models

Stochastic effects

A prediction of the Lotka–Volterra model is that when there is strong competition between species the outcome depends on the initial conditions. As a consequence the outcome could be affected by chance variations in species densities. Park (1954), who worked on the flour beetles *Tribolium confusum* and *T. castaneum*, found that the outcome of competition could not always be predicted and there appeared to be a chance element: sometimes *confusum* won, sometimes *castaneum*. He identified the mechanism of competition as a form of mutual predation, i.e. both species had a greater competitive effect on each other than on themselves suggesting that the outcome would depend on the initial conditions. Barnett (1962) describes a stochastic model which shows that chance effects from demographic stochasticity can lead to extinction of the species with an apparent initial advantage (see Renshaw, 1991). Also, because the outcome of the competition depends on the values of the competition coefficients and the carrying capacities, any environmental stochasticity inducing random variation in the values of these parameters would introduce a further chance element into the eventual outcome.

Models with more than two species

Gilpin (1975) discusses an extension of the Lotka–Volterra model to systems of three or more competing species and shows that in some situations stable limit cycles can occur.

10.2.4 Yield–density models for inter-plant competition

Plants and trees are usually rooted in the ground at fixed positions so that in competition for resources, such as light and soil nutrients, spacing and size of neighbouring individuals are important. This is not to say that the situation is a static one because individual growth and thinning-out through mortality are part of a complex dynamic process. Two somewhat different approaches to modelling this process can be identified. In the first, average responses in the population, such as mean plant weight or mortality rate, are related to population characteristics such as overall planting density or the amount of a nutrient applied to all the plants in the plot. The second approach attempts to relate the size and growth of individual plants to the size and position of their neighbours. The models may describe the population at a given time or they could be dynamic models based on rules which determine growth of individuals from a given set of prevailing conditions (Hegyi, 1974; Ford & Diggle, 1985). In this section we shall concentrate on the first approach for describing the effects of competition at the population level. In particular, we present a model for competition between two species growing in mixture.

Many studies of growth in plant monocultures have found a decreasing relationship between the mean yield per plant and the planting density, an effect which is attributed to competition. To describe this relationship Watkinson (1980) proposed the following model

$$\overline{W} = \frac{W_{\max}}{(1 + aN)^b}$$

where $\overline{W}$ is the mean plant weight at harvest, N is the plant density at harvest and $W_{\max}$ is the mean weight in absence of competition. The parameter a is referred to as the ecological neighbourhood area and b is called the resource utilisation constant (Figure 10.4(a)). At high densities the logarithm of mean plant weight decreases linearly as the logarithm of density with slope $-b$, i.e.

$$\log_e \overline{W} \approx \log_e W_{\max} - b \log_e a - b \log_e N.$$

When the line is projected backwards it crosses the horizontal $W = W_{\max}$ at density of $N = 1/a$. In this sense the ecological neighbourhood area a measures the area required by a plant to achieve maximum size, or the reciprocal of the density at which individual plant size begins to be depressed (Figure 10.4(b)). In reality, the mean yield will be subject to random variation but we deal with a deterministic model appropriate for a very large population in which individual variation is effectively ironed out.

Firbank & Watkinson (1985) have extended the above model for the effect of competition in a monoculture to describe the effect of competition between two species growing in a mixture. Their model involves two equations relating mean plant weight of each species to their densities in mixture given by

$$\overline{W}_1 = \frac{W_{1\max}}{[1 + a_1(N_1 + \alpha_{12}N_2)]^{b_1}}; \quad \overline{W}_2 = \frac{W_{2\max}}{[1 + a_2(N_2 + \alpha_{21}N_1)]^{b_2}}$$

where $\overline{W}_1$, $W_{1\max}$, N_1, b_1 and a_1 have the same interpretation as in monoculture. The new parameters α_{12} and α_{21} are called the *competition coefficients*. α_{12} measures the competitive effect of species 2 on species 1 with a larger value than one indicating that

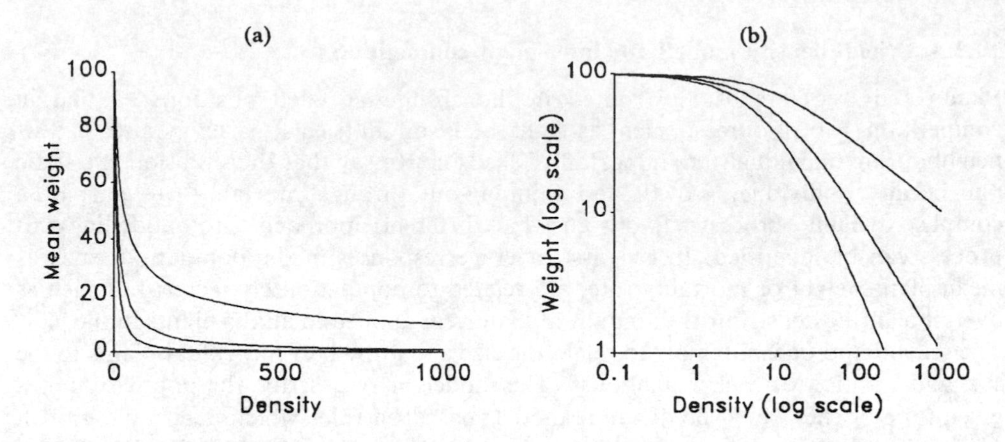

Figure 10.4 Illustration of a yield–density model relating mean plant weight to harvest density for $W_{max} = 100$, ecological neighbourhood area $a = 0.1$ and values of the resource utilisation constant $b = 0.5$ (top), 1.0 (middle), 1.5 (bottom).

for a given density the competitive effect of species 2 on species 1 is greater than the competitive effect of species 1 on itself. Similarly, α_{21} measures the competitive effect of species 1 on species 2.

Example 10.2 Two species yield–density model fitted to data on winter wheat and corncockle

Firbank & Watkinson (1985) fitted the two species yield–density model to data on winter wheat, *Triticum aestivum* and corncockle *Agrostemma githago*. Species were grown both in monoculture and in 1:1 mixtures at densities of 16, 40, 120, 400 and 1200 seeds m^{-2} and shoot dry weights of the plants were measured at harvest. Yield–density curves were fitted to the data from monocultures to estimate the mean weights in absence of competition (W_{1max} and W_{2max}), the ecological area constants (a_1 and a_2), and the resource utilisation constants (b_1 and b_2). Data from the mixtures relating mean weight to combinations of plant densities were then used to estimate the competition coefficients. The estimated parameters were as follows:

Winter wheat	$W_{1max} = 46.8$ g	$a_1 = 0.24\ m^2$	$b_1 = 0.66$	$\alpha_{12} = 1.63$
Corncockle	$W_{2max} = 31.7$ g	$\alpha_2 = 0.0063\ m^2$	$b_2 = 0.72$	$\alpha_{21} = 0.41$

The estimated competition coefficients show that the competitive effect of corncockle on winter wheat (measured by α_{12}) was greater than the effect of wheat on itself and that competitive effect of wheat on corncockle was less than the effect of other corncockles. Figure 10.5 illustrates the fitted model as a pair of response surfaces which show how mean plant weight of each species varies with combinations of densities in mixture. The model provides a quantitative description which embodies effects of both inter- and intraspecific competition and which may be used to examine the effect of changes in one or both densities. Furthermore, estimated model parameters could be used to compare the performance of the same species in different situations or with other species. This example forms only a part of the analysis carried out by Firbank and Watkinson who also applied similar models to describe the effects of competition on mortality and total stand yield.

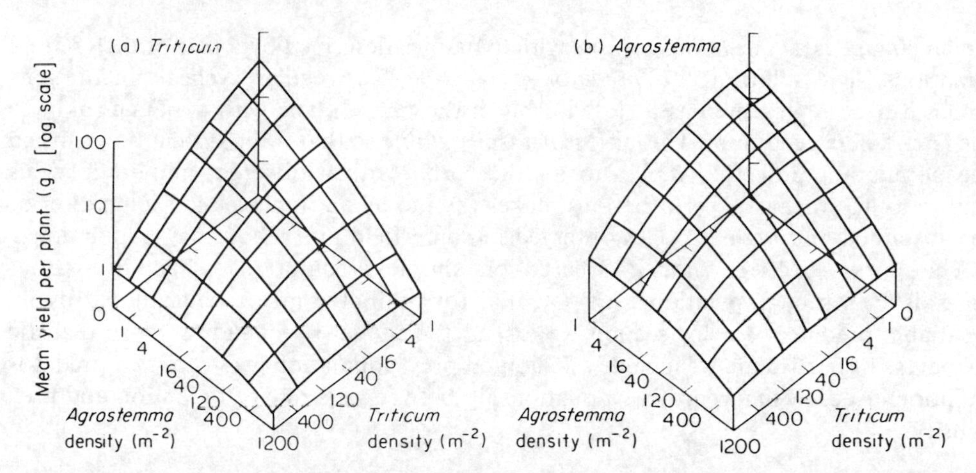

Figure 10.5 Illustration of the two species yield–density model showing yield per plant for different combinations of densities of winter wheat (*Triticum aestivum*) and corncockle (*Agrostemma githago*). Reprinted with permission from the *Journal of Applied Ecology*, 1985, 22.

10.3 PREDATOR–PREY MODELS

Predators may be carnivores such as lions and ladybirds, which kill and consume their prey, or they may be herbivores, such as zebras and greenfly, which eat green plants or fruits, often damaging but not killing the plants. A different form of predation is insect parasitism in which an insect parasitoid lays its eggs in, on, or near a host which is subsequently killed by the emergent larvae. In this section we present some simple mathematical models for predator–prey interaction which apply to situations where prey are actually consumed by the predator. Models for the host–parasitoid interaction are described in Section 10.4.

Basic equations for changes in density of the predator and its prey are as follows:

Change in the density of prey
= Increase of prey in absence of predators
− Consumption of prey by predators

Change in the density of predators
= Increase in the number of predators from ingesting prey
− Deaths of predators

This says nothing about the details of the interaction but it is a useful general framework for developing models because it identifies the four key components in the interaction. In Chapter 8 we presented the Lotka–Volterra model as a particular case of this general formulation in which the four components are modelled as follows: (a) the prey population increases exponentially in the absence of the predator; (b) each predator consumes prey at a rate directly proportional to prey density; (c) without prey, the predator population declines exponentially; (d) the reproduction rate per predator is directly proportional to prey density. The model applies to populations in which change is a continuous process with dynamics described by a pair of differential equations. The solution is a *predator–prey cycle*, i.e. regular cyclical fluctuations which can be represented as a closed orbit in the phase plane with dynamics akin to the frictionless pendulum. The cycle is *neutrally stable* so that if the system is perturbed the population trajectory moves to another orbit with a different amplitude. This sensitivity to stochastic perturbations makes the model an implausible explanation of the sustained and regular cycles observed in some field and laboratory populations.

The Lotka–Volterra model is based on simple assumptions which are rather unrealistic: a prey population with potential for unlimited growth but limited by an insatiable predator. In this section we extend the model by including more realistic elements. In particular, we include self-limited prey population growth and a predator satiation effect by incorporating a relationship between the rate of predation and prey density.

10.3.1 Lotka–Volterra model with limited prey population growth

In the basic Lotka–Volterra model for the predator–prey interaction the growth of the prey population is limited entirely by predation, i.e. in the absence of the predator the prey population would increase exponentially. More realistically, prey population growth is likely to be limited in the absence of the predator. A simple model is logistic prey population growth with a linear decrease in relative growth rate with increased prey density. The equations for the rate of change of prey density (H) and predator density (P) are then

$$\frac{\mathrm{d}H}{\mathrm{d}t} = r_H H\left(1 - \frac{H}{K}\right) - aHP$$

$$\frac{\mathrm{d}P}{\mathrm{d}t} = -r_P P + bHP$$

The model includes the extra parameter K, the carrying capacity of the prey population in the absence of the predator. The remaining parameters are as before: r_H is the intrinsic relative growth rate of the prey; r_P is the mortality rate of the predator without prey; a measures the rate of consumption of prey by the predator; b measures the conversion of prey consumed into the predator reproduction rate. The effect of incorporating limited prey growth is studied below by using phase plane analysis. The important result is that the change leads to a stable equilibrium. This contrasts with the behaviour of the basic model with unlimited prey population growth in the absence of the predator which produces a neutrally stable cycle with periodic fluctuations in both prey and predator densities.

Phase plane analysis

To construct the phase plane diagram for the model we find the nullclines, direction flows and the equilibrium, or stationary, values. The prey nullclines correspond to those points in the phase plane diagram for which the rate of change of prey density is zero, i.e.

$$\frac{dH}{dt} = r_H H\left(1 - \frac{H}{K}\right) - aHP = 0.$$

This occurs when

$$H = 0 \quad \text{or} \quad r_H K - r_H H - aPK = 0.$$

The first case implies absence of prey and therefore eventual extinction of the predator. The second case gives the prey nullcline as

$$P = \frac{r_H}{a} - \frac{r_H H}{aK}.$$

This is the equation of a straight line which intercepts the P-axis at $P = r_H/a$ and the H-axis at $H = K$.

The predator nullcline is obtained in a similar way as follows. First,

$$\frac{dP}{dt} = -r_P P + bHP = 0$$

so that

$$P = 0 \quad \text{or} \quad -r_P + bH = 0$$

with predator nullcline for both predator and prey present given by

$$H = \frac{r_P}{b}.$$

This is a straight line which runs parallel to the P-axis.

Figure 10.6(a) shows the nullclines on a phase plane diagram with prey density along the horizontal axis. The vertical line of the predator nullcline is drawn to the left of the intercept of the prey nullcline at ($P = 0, H = K$) so that $r_P/b < K$. This is necessary for the predator to maintain or increase its numbers since if the predator nullcline falls further to the right, i.e. $r_P > bK$, then the maximum rate at which prey can be consumed and converted into reproduction is not sufficient to offset the losses through mortality. The nullclines divide the phase plane diagram into regions of increasing, or decreasing, densities for predator and prey. Figure 10.6(a) summarises the pattern as direction flows from points on a grid and shows an anticlockwise spiral which converges on the point of intersection of the two nullclines. This indicates a stable equilibrium with the predator keeping the prey at a constant level below its potential carrying capacity. The graphical illustration can be confirmed by mathematical stability analysis. Figure 10.6(b) and (c) shows a particular trajectory in phase space with the corresponding time plot of H and P for some values for the model parameters. The series shows a rapidly damped oscillation to the equilibrium level.

10.3.2 Functional response of predators to prey density

A key component of models for the interaction of a predator and its prey is the relationship between the predation rate and prey density. In the Lotka–Volterra model

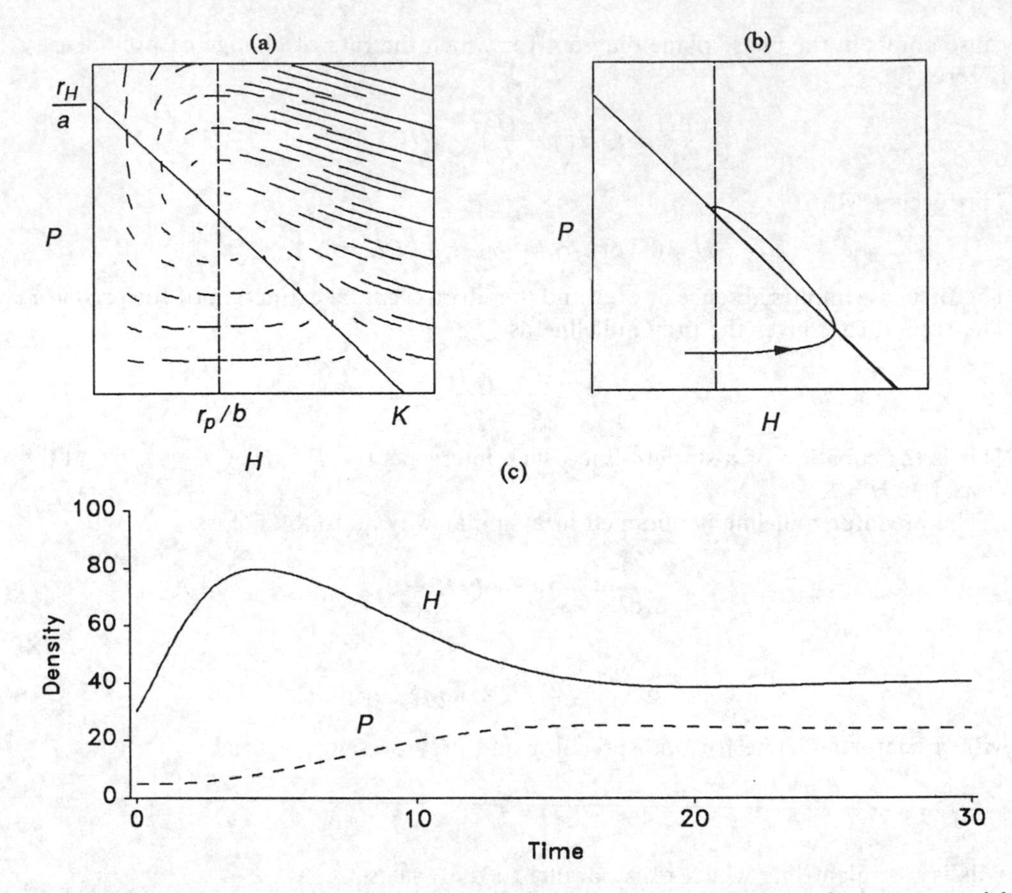

Figure 10.6 Illustration of the dynamic behaviour of the Lotka–Volterra predator–prey model with limited prey population growth. (a) Phase plane diagram with direction flows—prey nullcline (solid line), predator nullcline (broken line); (b) particular trajectory in the phase plane; (c) time plot corresponding to (b). Model parameters: $r_H = 1.0$, $K = 100$, $a = 0.025$, $r_P = 0.20$, $b = 0.005$. Starting values $H(0) = 30$, $P(0) = 5$.

the number of prey consumed per unit time by each predator increases in direct proportion to prey density. More realistically, there must be some upper limit on the predation rate and many experimental studies have confirmed this (Example 10.3). The relationship between the predation rate and the prey density is called the *functional response*, usually denoted by $f(H)$. Functional responses can be incorporated into the basic Lotka–Volterra model to produce more realistic models of the predator–prey interaction. Section 10.3.3 presents the Holling–Tanner model as a particular case.

In general, the detailed shape of the functional response curve depends on the predator and prey species involved. Holling (1959) identified three broad types which show different patterns of steady increase to an upper limit (Figure 10.7).

Type I shows a linear increase to a plateau. In the type II response the predation rate increases at a decreasing rate. For this case, Holling proposed the so-called 'disc

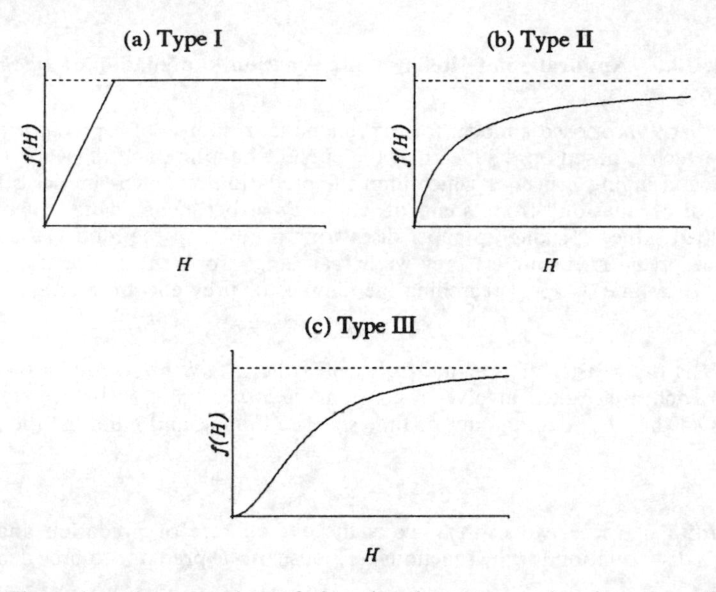

Figure 10.7 Three broad categories of functional response showing the relationship of predation rate to prey density.

equation' in which the increase follows a rectangular hyperbola, i.e. a functional response of the form

$$f(H)=\frac{wH}{D+H}.$$

The parameter w is the maximum rate of predation and w/D measures the predation rate at low prey density. Holling developed a simple model in which the parameters can be related to the rate at which the predator encounters prey while searching and the handling time for each prey item (Example 10.3). Holling's equation has been found to provide a good fit to experimental data on a wide range of invertebrate predator species. An alternative model for the type II response is the asymptotic exponential function

$$f(H)=w(1-e^{-H/D})$$

which has a faster approach to the upper limit than the rectangular hyperbola. For the type III functional response the increase of predation rate with prey density follows a sigmoidal curve—for example,

$$f(H)=\frac{wH^2}{D^2+H^2}.$$

More realistic models for the predator–prey interaction can be developed by modifying the basic Lotka–Volterra model to include a functional response. Section 10.3.3 describes one of the many possibilities.

Example 10.3 Application of Holling's disc equation to predation of *Daphnia* by damselflies

Holling (1959) proposed a model for the functional response of a predator to prey density which is based on the idea that the pursuit, handling and digestion of prey are time-consuming processes which limit the predation rate. The model is known as the 'disc equation' from some of Holling's experiments which involved a blindfolded subject picking up paper discs from a table. In the model, the rate at which the predator encounters prey while searching is proportional to the density of prey. In a time T_s spent searching the number of prey encountered is given by

$$N_e = aHT_s,$$

where H is the density of prey and a is called the *attack rate* in units of area^{-1} time^{-1}. Each prey taken involves a constant *handling time*, h, before searching resumes so that the total amount of time spent searching and handling the prey is then

$$T = T_s + ahHT_s.$$

The ratio of these two expressions is the overall rate of predation and gives Holling's disc equation for the functional response of the predator to prey density as

$$f(H) = \frac{aH}{1 + ahH},$$

where H is prey density, a is the attack rate and h is the handling time. This is the equation of a rectangular hyperbola which increases with decreasing slope to an asymptote $1/h$, i.e. the upper limit set on the predation rate by the handling time per prey item.

Many experimental studies have found functional responses similar to Holling's disc equation. Figure 10.8 shows an example for data on predation of *Daphnia* by the damselfly *Ischnura elegans* over a 24 h period. The fitted curves are Holling's disc equation with the following parameters:

Daphnia mean length (mm)	Attack rate (a)	Handling time (h)
1.1	0.940	0.070
1.7	0.421	0.153

In this case attack rates for the larger prey are about half those for the smaller prey while handling times are approximately double.

10.3.3 Holling–Tanner model

The Holling–Tanner model extends the basic Lotka–Volterra predator–prey models by incorporating more realistic components for the interaction of predator and prey. It involves the following modifications.

(1) Holling's disc equation as a functional response relating predation rate to prey density.
(2) A component for the effect that the density of the predator has on its own growth. This is a complex situation involving many factors such as the effect of food intake on survival and reproduction, and competition amongst the predators for food and other resources. In the model, these effects are described by a predator relative

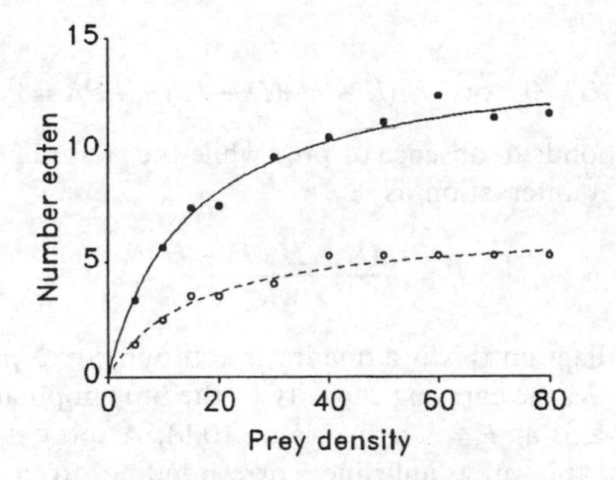

Figure 10.8 Functional response of the damselfly *Ischnura elegans* (9th instar) feeding on *Daphnia* over a 24 h period (after Thompson, 1975). Closed symbols-solid line: mean prey length = 1.1 mm; open symbols-broken line: mean prey length = 1.7 mm. Fitted curves are Holling's disc equation.

growth rate which decreases as the ratio of the density of predator to prey increases so that the growth of the predator population follows the logistic model with a carrying capacity which is proportional to the current density of prey.

The equations of the Holling–Tanner model describing the rate of change in density of predator and prey are

$$\frac{\mathrm{d}H}{\mathrm{d}t} = r_H H\left(1 - \frac{H}{K}\right) - \frac{wHP}{D+H}$$

$$\frac{\mathrm{d}P}{\mathrm{d}t} = s_P P\left(1 - \frac{bP}{H}\right),$$

where H and P are densities of prey and predator, r_H is the intrinsic relative growth rate of the prey population, K is the prey carrying capacity in absence of the predator and s_P is the intrinsic relative growth rate of the predator, i.e. with unlimited food and no effect of its own numbers. The parameter b is the number of prey required to support each predator at equilibrium $P = H/b$. Note that the parameter s_P has a different interpretation from the mortality rate r_P in the Lotka–Volterra model. To analyse the Holling–Tanner model we begin with a phase plane diagram to explore the qualitative behaviour then we present some trajectories for particular values of the model parameters.

Phase plane analysis

To construct the phase plane diagram of the model we first find the nullclines corresponding to the points for which the rate of change in density is zero. For the prey population

$$\frac{\mathrm{d}H}{\mathrm{d}t} = r_H H\left(1 - \frac{H}{K}\right) - \frac{wHP}{D+H} = 0.$$

This gives

$$H = 0 \quad \text{or} \quad r_H(K - H)(D + H) - wPK = 0.$$

The first case corresponds to absence of prey while the second gives the prey nullcline for the predator–prey interaction as

$$P = \frac{r_H(K - H)(D + H)}{wK}.$$

On the phase plane diagram this is a quadratic contour (i.e. a parabola) which meets the H-axis when $H = K$, the carrying capacity of the prey population in absence of the predator, and the P-axis at $P = r_H D/w$ (Figure 10.9). At prey density $H = (K - D)/2$, there is a peak in the prey nullcline corresponding to a maximum value of $r_H(K + D)^2/4wK$.

The nullclines for the predator are obtained from

$$\frac{\mathrm{d}P}{\mathrm{d}t} = s_P P\left(1 - \frac{bP}{H}\right) = 0$$

so that

$$P = 0 \quad \text{or} \quad H - bP = 0$$

with the non-zero predator nullcline given by

$$P = H/b.$$

On the phase plane diagram this is a straight line through the origin with slope $1/b$ (Figure 10.9). There is an equilibrium point where the line crosses the prey nullcline. The nature of the equilibrium and the dynamic behaviour of the model is summarised below.

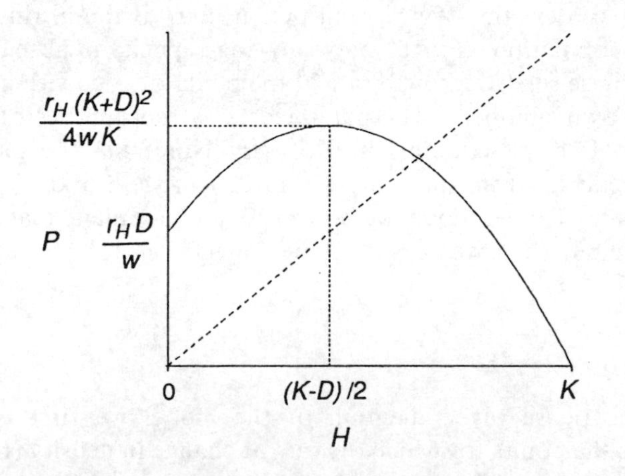

Figure 10.9 Illustration of the nullclines for the Holling–Tanner predator–prey model. Prey nullcline (solid line); predator nullcline (broken line). In this case the intersection of the nullclines is to the right of the peak in the prey nullcline but in other cases it can fall to the left, depending on the position of the peak and the slope of the predator nullcline (see Figure 10.10).

Summary of dynamic behaviour of Holling–Tanner model

The stability properties of this model were examined by Tanner (1975) using mathematical stability analysis. We simply summarise the results and illustrate them using simulated data for particular values of the parameters.

(1) If the intersection of the nullclines is to the right of the peak in the prey nullcline there is a stable equilibrium point. This is illustrated by the direction flows in Figure 10.10(a) and the anticlockwise spiral trajectory towards the equilibrium point in Figure 10.10(b). This case occurs when the slope of the predator nullcline is *less* than the slope of the line from the origin to the point at the peak of the prey nullcline, i.e.

$$\frac{1}{b} < \frac{r_H(K+D)^2}{2wK(K-D)}$$

or

$$\frac{r_H b}{w} > \frac{2(K/D)(K/D-1)}{(K/D+1)^2}.$$

This condition for stability is therefore determined by the two quantities $r_H b/w$ and K/D.

(2) If the intersection of the nullclines is to the left of the peak of the prey nullcline there are two possibilities:

(i) there is a stable equilibrium point if the ratio of the intrinsic rate of increase of the predator to that of the prey, s_P/r_H, exceeds some critical value;
(ii) when s_P/r_H is smaller than the critical value the system settles into a stable limit cycle, but as the prey carrying capacity, K, increases the amplitude of the limit cycles increases and extinction of the system occurs.

The critical value is a complicated function of the quantities $r_H b/w$ and K/D but as the carrying capacity of the prey population, K, increases the critical value approaches an upper limit of $r_H b/w$ and the condition for a stable equilibrium becomes

$$s_P/r_H > r_H b/w.$$

These cases are illustrated in Figure 10.10(a)–(f). In Figure 10.10(a) and (b) the intersection of the nullclines falls to the right of the peak in the prey nullcline. Direction flows indicate a stable equilibrium (Case 1). In Figure 10.10(c)–(f) the peak of the prey nullcline has been pushed to the right by increasing the carrying capacity, K, of the prey population. In Figure 10.10(c) and (d) the intrinsic growth rate of the predator has been chosen to be relatively large and the direction flows in 10.10(c) indicate a damped oscillation to a stable equilibrium value (Case 2(i)). In Figure 10.10(e) and (f) the intrinsic rate of increase of the predator is half of that in Figure 10.10(c) and (d). Direction flows indicate a stable limit cycle which is illustrated by the particular trajectory in Figure 10.10(f) (Case 2(ii)). Figure 10.11 shows the cycle as a time plot.

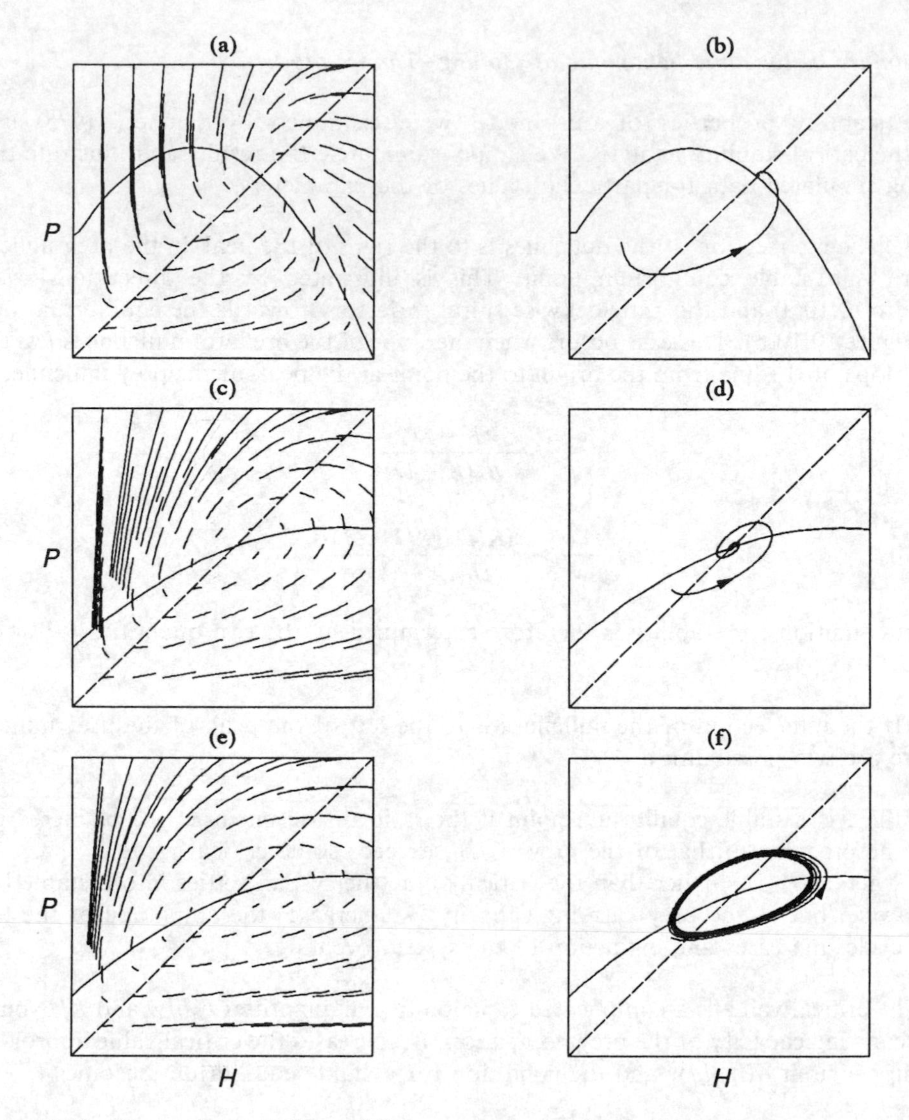

Figure 10.10 Illustration of the stability properties and the dynamic behaviour of the Holling–Tanner model using phase plane diagrams with direction flows and individual trajectories. Prey nullcline (solid line); predator nullcline (broken line). (a) and (b): $r_H = 1.0$, $K = 5, D = 1, w = 0.5, s_P = 0.50, b = 1, H(0) = 0.9, P(0) = 2.0$; (c) and (d): $r_H = 1.0, K = 10, D = 1$, $w = 1.0, s_P = 0.50, b = 1, H(0) = 1.7, P(0) = 2.0$; (e) and (f): $r_H = 1.0, K = 10.0, D = 1, w = 1.0$, $s_P = 0.25, b = 1, H(0) = 1.7, P(0) = 2.0$.

10.3.4 Discussion and more complex models

Population cycles

This section has dealt with two models of predator–prey interaction which add more realistic elements to the basic Lotka–Volterra model. The first model imposes a limit on prey population growth and leads to a stable equilibrium, which contrasts with the

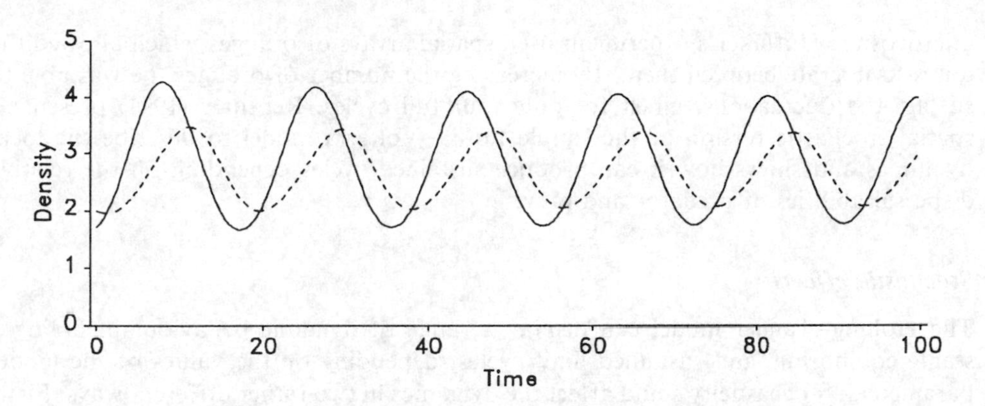

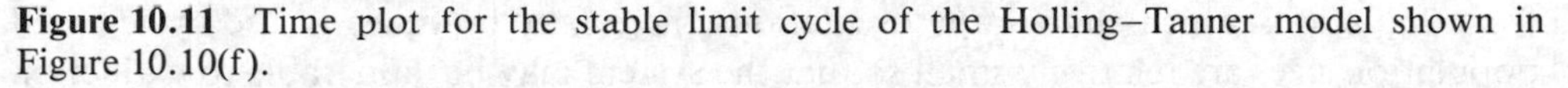

Figure 10.11 Time plot for the stable limit cycle of the Holling–Tanner model shown in Figure 10.10(f).

neutral cycle of the basic Lotka–Volterra model. The Holling–Tanner model includes a functional response and an effect of self-limited predator population growth and can generate a range of dynamic behaviour including stable equilibrium, stable limit cycles and divergent cycles leading to eventual extinction.

Other deterministic predator–prey models

The Holling–Tanner model is just one of many models which have been proposed for predator–prey interaction. May (1976) gives a catalogue of models of the form

$$\frac{\mathrm{d}H}{\mathrm{d}t} = rH\left(1 - \frac{H}{K}\right) - PF(H, P); \qquad \frac{\mathrm{d}P}{\mathrm{d}t} = PG(H, P).$$

where F and G are functions of predator and prey density. A particular type of model, which has been used for the plant–herbivore interaction, is the *laissez-faire* system in which the herbivores (i.e. the predators) do not interfere with each other. In this case G does not depend on P. *Laissez-faire* models which include a functional response of the predator to prey density can also produce a range of dynamic behaviour including stable equilibrium, stable limit cycles and divergent oscillations (Caughley, 1976).

For a graphical method of analysing the predator–prey interaction for more general models, see Rosenzweig & MacArthur (1963), Rosenzweig (1969) and Maynard Smith (1974).

Spatial effects

In the predator–prey models given here there is no explicit spatial element, i.e. the dynamic equations are in terms of the overall densities, essentially assuming that individuals are mixing homogeneously. Often, this will be unrealistic as interactions occur locally in different parts of the population. Huffaker (1958) carried out a series of fascinating experiments to examine the effect of the spatial distribution on the interaction of the predatory mite, *Typhlodromus occidentalis*, with the six-spotted mite *Eotetranychus sexmaculatus*. He set out to test whether a complex habitat could help to sustain predator–prey cycles which rapidly went extinct in simple experimental microcosms. Huffaker's experiment used spatial arrays of oranges which allowed the mites to migrate between them. By increasing the number of oranges, he was able to sustain the fluctuations, albeit for only four full cycles. Renshaw (1991) presents a spatial stochastic version of the simple Lotka–Volterra model to describe the local dynamics and shows how it can produce sustained cycles depending on the relative dispersal abilities of predator and prey.

Stochastic effects

The Holling–Tanner model can display a range of dynamic behaviour including a stable equilibrium and sustained limit cycles, depending on the values of the model parameters. Stochasticity could affect the dynamics in two rather different ways. First, for populations which show large oscillations there may be times in the cycle when population sizes are relatively small so that the system may be vulnerable to extinction due to the effects of demographic stochasticity (Chapter 4). Secondly, because of environmental stochasticity the parameters of the process may vary with time so that the conditions for, say, a stable equilibrium may be satisfied in some years and not others. In general, it is difficult to predict what the effect might be but particular cases could be examined by simulating data from models which include plausible amounts of stochastic variation in the parameter values. Note that a random element can sustain fluctuations in an otherwise damped deterministic model (Chapter 13, Example 13.3).

10.4 HOST–PARASITOID MODELS

Parasitoids are insects which are free-living in their adult stage but which lay their eggs in the larvae or pupae of other insects, or sometimes spiders or woodlice. The larval parasitoids develop on, or within, their *host*, consuming and eventually killing it before pupating. Often only one parasitoid develops from each host but in some species there may be several. In either case the number of parasitised hosts in one generation is closely related to the number of parasitoids in the next generation. The direct link between the dynamics of the parasitoid and its host makes the host–parasitoid interaction especially suitable for both experimental work and the development of mathematical models. Furthermore, an important area of application is the biological control of insect pests. In this section we briefly describe some of the basic models for host–parasitoid interaction. For a more detailed discussion see Hassell (1978).

We begin with some data from an experimental study of the parasitoid braconid wasp *Heterospilus prosopidis* and its host the azuki bean weevil *Callosobruchus chinensis* (Utida, 1957). Populations were maintained under constantly controlled conditions of temperature, humidity, light, food and space. Figure 10.12 shows the densities in host and parasitoid populations in an experiment which spanned 112 generations. These fluctuate in a rather regular manner but with a varying amplitude. In two other experiments, lasting fewer than 25 generations, the host became extinct followed by eventual extinction of the parasitoid. In a third series the parasitoid became extinct after 15 generations after which the host maintained a steady level. Thus, even in a relatively controlled and constant environment the host–parasitoid interaction displays a range of interesting population dynamics.

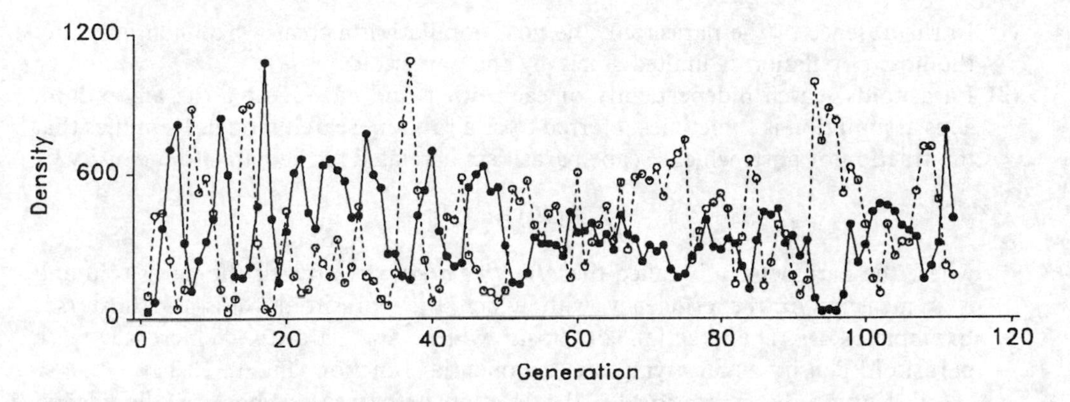

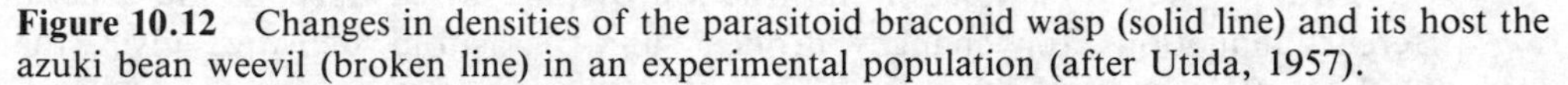

Figure 10.12 Changes in densities of the parasitoid braconid wasp (solid line) and its host the azuki bean weevil (broken line) in an experimental population (after Utida, 1957).

Many host–parasitoid interactions involve non-overlapping generations reproducing at regular intervals. Changes in population size can therefore be considered at discrete time intervals and described by using difference equations. A model for population change in female densities is then

$$H_{t+1} = RH_t f(H_t, P_t)$$

i.e.

Density of hosts at time $t+1$ = Host net fecundity
× Density of hosts at time t × Fraction of hosts *not* parasitised

and

$$P_{t+1} = cH_t[1 - f(H_t, P_t)]$$

i.e.

Density of parasitoids at time $t+1$ = Parasitoid net fecundity
× Density of hosts at time t × Fraction of hosts parasitised

where H_t and P_t are the densities of female hosts and parasitoids at time t, R is the host net reproductive rate, c is the parasitoid net fecundity, i.e. the mean number of female parasitoids emerging from a single host which survive to the next generation, $f(H_t, P_t)$ is the fraction of hosts *not* parasitised. The fraction of hosts which *are* parasitised depends on the encounter rate between parasitoid and host and is written as a function of both host and parasitoid densities. It may also depend on other factors which affect the efficiency of search of the parasitoid. To use this general formulation we need to specify the form of the function which determines the fraction of hosts parasitised.

10.4.1 The Nicholson–Bailey model

A basic model for the host–parasitoid interaction was developed by the biologist A.J. Nicholson and the physicist V.A. Bailey in 1935. The model is deterministic so that it applies to large populations of parasitoids and hosts. It involves the following assumptions.

(1) In the absence of the parasitoid, the host population increases exponentially, i.e. the host population is limited solely by the parasitoid.
(2) Parasitoids search independently of each other and encounter hosts at random. This assumption is sometimes referred to as a random search model. It implies that the fraction of hosts which escape parasitism is related to the parasitoid density by

$$f(P_t) = e^{-aP_t}$$

where the parameter a is called the *effective area of search* of the parasitoid and is a measure of the efficiency with which the parasitoid finds the host (see Example 10.4). The fraction of hosts which are parasitised increases with parasitoid density as an asymptotic exponential function (Figure 10.13).

Substituting the expression for the fraction of parasitised hosts in the general difference equations for changes in host and parasitoid densities gives the Nicholson–Bailey model.

Nicholson–Bailey host–parasitoid model

$$H_{t+1} = RH_t\,e^{-aP_t}$$
$$P_{t+1} = cH_t(1 - e^{-aP_t})$$

where H_t and P_t are the host and parasitoid densities, R is the intrinsic net reproductive rate of the host, c is the average number of parasitoids emerging from a parasitised host which join the next generation and a is the *effective area of search*.

Dynamic behaviour of the Nicholson–Bailey model

Figure 10.14 shows some simulated data for the Nicholson–Bailey model with a host net reproductive rate $R = 2$, a parasitoid fecundity $c = 1$ and an effective area of search $a = 0.07$, starting from host and parasitoid densities of 22 and 11 per unit area. Both

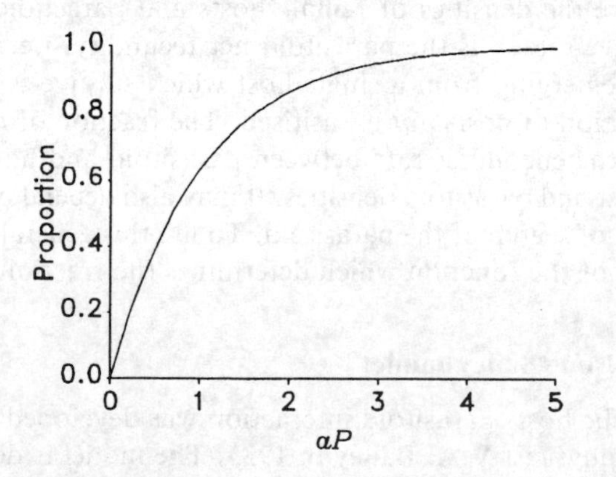

Figure 10.13 Relationship between the fraction of hosts parasitised and the product of parasitoid density (P) and effective search area (a) for the Nicholson–Bailey model.

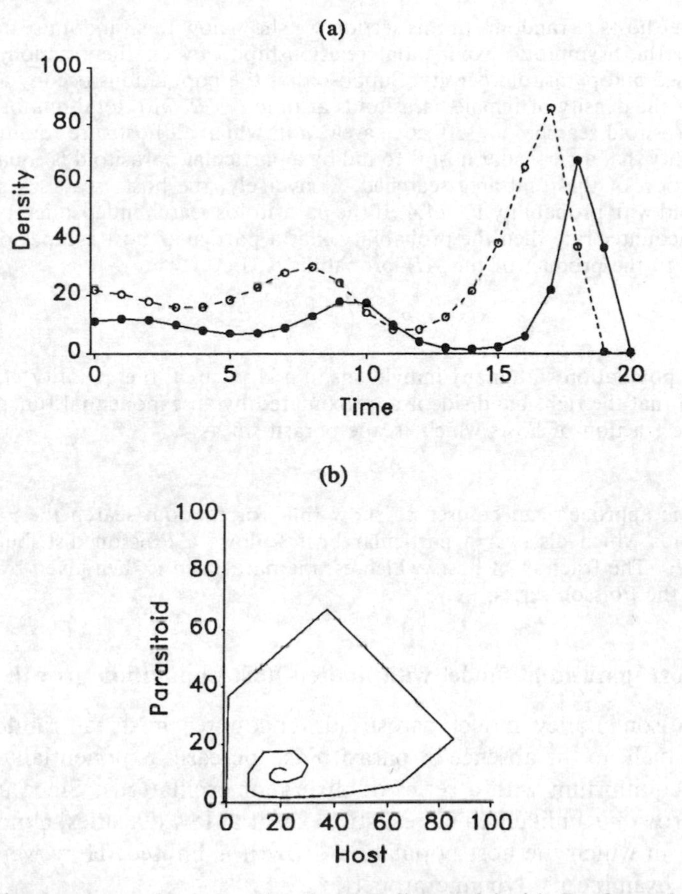

Figure 10.14 Simulated data from the Nicholson–Bailey model with $R = 2$, $c = 1$ and $a = 0.07$, showing time plots for parasitoids (solid line) and hosts (broken line) with corresponding trajectory in a phase plane diagram.

populations show a series of divergent oscillations, i.e. an unstable equilibrium. The Nicholson–Bailey model has a single equilibrium at

$$H_E = \frac{R \log_e R}{ac(R-1)}; \qquad P_E = \frac{\log_e R}{a}$$

and local stability analysis shows that the equilibrium is always *unstable* (Chapter 8). Also, simulations show that the solution is globally unstable so that the model cannot account for the sustained and contained fluctuations in the interaction between the braconid wasp and the azuki bean weevil (Figure 10.12). In Section 10.4.2 we consider extensions to the model and examine the dynamic behaviour.

Example 10.4 Derivation of the fraction of hosts parasitised for the random search model

In the Nicholson–Bailey model, parasitoids search independently of each other and encounter hosts at random. In this section we show how this random search model leads to the asymptotic exponential relationship between the fraction of hosts parasitised and parasitoid density. Suppose that the populations occupy an area A and that the density of female parasitoids at time t is P_t with total number AP_t. If each parasitoid searches an 'effective area' a in which all hosts are found then the probability that a particular host is found by a particular parasitoid is equal to a/A, the fraction of the total area searched. Conversely, the host escapes a particular parasitoid with probability $1 - a/A$. If the parasitoids search independently with the same encounter rate then the probability that a particular host escapes parasitism is equal to the product of the AP_t probabilities, i.e.

$$f(P_t) = \left(1 - \frac{a}{A}\right)^{AP_t}.$$

For a population with many individuals in a large area, the quantity a/A will be small so that the right-hand side is approximated by an exponential function. This gives the fraction of hosts which escape parasitism as

$$f(P_t) = e^{-aP_t}.$$

The same approach can be used to show that for random search the number of parasitoids which discover a particular host follows a Poisson distribution with mean aP_t. The fraction of hosts which escape parasitism is then given by the zero term in the Poisson series, i.e. e^{-aP_t}.

10.4.2 A host–parasitoid model with limited host population growth

In the Nicholson–Bailey model parasitoids encounter hosts at random in a host population which, in the absence of parasitoids, increases exponentially. The result is an unstable equilibrium with a series of divergent oscillations. Since unlimited host population growth is unlikely to be realistic, except at low densities, a more reasonable model is one in which the host population growth is limited. Here we consider such a model and examine its dynamic properties.

Beddington *et al.* (1975) proposed an extension of the Nicholson–Bailey model in which the relative growth rate of the host population, in the absence of the parasitoid, decreases with density. The model includes the discrete logistic equation for limited population growth in which the relative growth rate decreases exponentially with density. The difference equations describing the change in host and parasitoid densities from one generation to the next are

$$H_{t+1} = H_t\, e^{r(1-H_t/K)-aP_t}$$
$$P_{t+1} = cH_t(1 - e^{-aP_t}).$$

The parameter K is the equilibrium population density of the hosts in the absence of the parasitoid and r is related to the intrinsic net reproductive rate of the host by $r = \log_e R$.

We recall from Chapter 1 that non-linear discrete difference equations for single populations can display a wide range of dynamic behaviour including stable equilibria, 2-point, 4-point, and higher order limit cycles, and chaos. From this we might expect the above model to have complicated dynamics and this turns out to be the case. Before characterising the behaviour of the model we present three series of simulated data. These correspond to a host population with an intrinsic rate of increase $r = 1$ (i.e. $R = 2.72$), equilibrium density $K = 100$ in the absence of the parasitoid, a

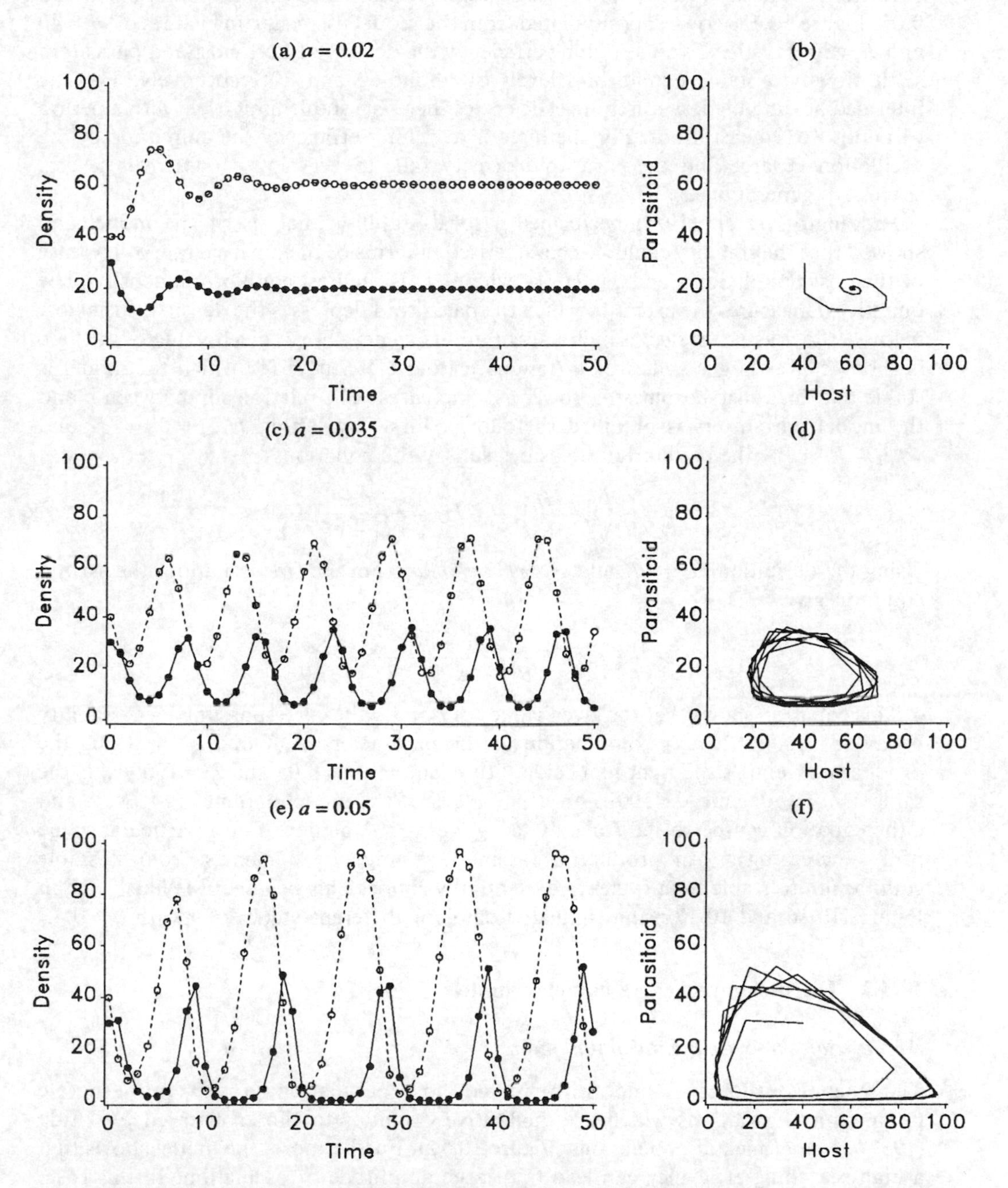

Figure 10.15 Simulated data from the Nicholson–Bailey model with discrete logistic growth in the host population showing time plots and phase plane trajectories for different values of the area of search a. In each case $r = 1, K = 100, c = 1, H_0 = 40, P_0 = 30$. Hosts (open symbols–broken line); parasitoids (closed symbols–solid line).

parasitoid fecundity of $c=1$ and three searching efficiencies with $a=0.02$, 0.035 and 0.05. Figure 10.15 shows data simulated from the model using starting values of $H=40$ and $P=30$. For the lowest searching efficiency the density of both host and parasitoid settle down to stable equilibrium levels of about 60 and 20 respectively. For the intermediate and highest searching efficiencies there is a stable limit cycle with a period of about 8–10 generations. For the highest searching efficiency the amplitude of the oscillation is large and the parasitoid density falls to very low values over several successive generations.

Beddington *et al.* (1975) performed a local stability analysis of the model and showed that the stability could be characterised in terms of the intrinsic rate of increase of the host population r and $q=H_E/K$, where H_E is the host equilibrium density. The quantity q measures the extent to which the parasitoid depresses the density of the host below K; low values reflect a high parasitoid efficiency. For a given value of r, there is a range over larger values of q (low parasitoid efficiency) for which the model is stable but this range becomes narrower as r increases. The relationship between q and the model parameters is obtained as follows. First, by setting $H_{t+1}=H_t=H_E$ and $P_{t+1}=P_t=P_E$, the equilibrium densities satisfy the equations

$$P_E=\frac{r}{a}\left(1-\frac{H_E}{K}\right);\qquad H_E=\frac{P_E}{c(1-\mathrm{e}^{-aP_E})}.$$

Using the definition $H_E=qK$ and $P_E=r(1-q)/a$ from the first equation, the second equation gives

$$acK=\frac{r(1-q)}{q[1-\mathrm{e}^{-r(1-q)}]}.$$

This equation shows that for given values of r and q the corresponding local stability behaviour applies to any combination of the parameters a, K and c for which the product acK equals the right hand side. For example, $a=0.05$ and $K=100$ yields the same as $a=0.01$ and $K=500$. For given values of the model parameters r, K, a and c the equation can be solved for q by using Newton's method. For a particular value of q, increasing r can produce a fascinating range of dynamics from a stable equilibrium to stable limit cycles and eventually chaos. This behaviour is illustrated in Figures 10.16 and 10.17 using simulated data for different values of r with $q=0.4$.

10.4.3 Discussion and more complex models

Application to experimental populations

The Nicholson–Bailey model with limited host population growth can generate population fluctuations which are qualitatively similar to those observed by Utida (1957) in experimental populations (Figure 10.12). Furthermore, the model shows that a high searching efficiency can lead to a large amplitude of oscillation. In this case, numbers may reach very *low* values making the system vulnerable to extinction from the effects of demographic stochasticity. We recall that in one of Utida's experiments, the parasitoid became extinct after which the host maintained a relatively steady level, whereas in two other experiments the host became extinct followed by extinction of the parasitoid.

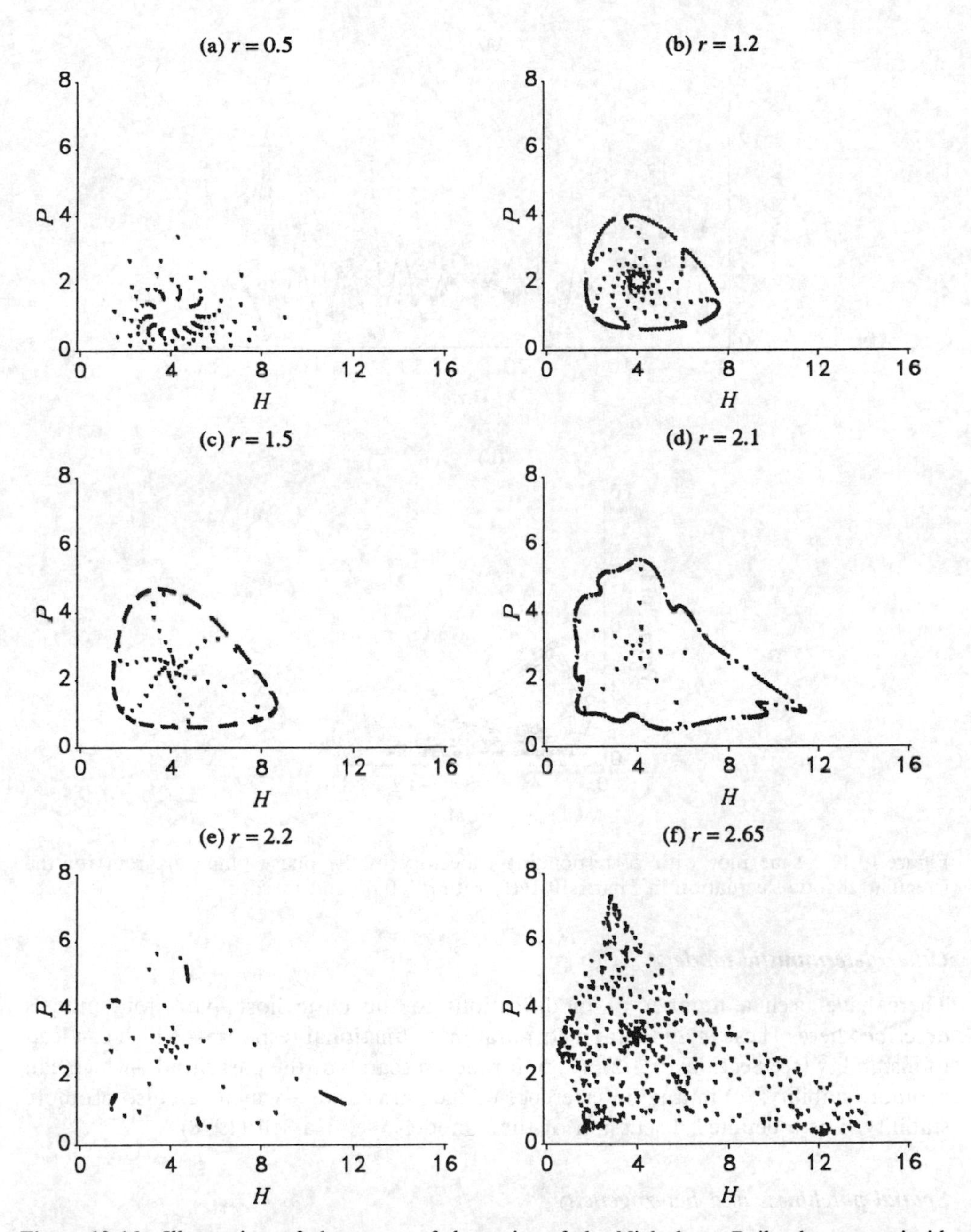

Figure 10.16 Illustration of the range of dynamics of the Nicholson–Bailey host parasitoid model with limited host population growth. The figure shows phase plane diagrams for simulated data using $q = 0.4$ and different values of r. In each case, $K = 10$, the parasitoid fecundity $c = 1$ and the search area a is chosen so that $q = 0.4$. Starting values: (a) $H_0 = 9$ and $P_0 = 1$; (b)–(f) a displacement of 0.1 away from the equilibrium. (a) A stable equilibrium with values falling on an anticlockwise spiral giving the impression of a series of arms; (b)–(d) stable limit cycles with spokes showing movement away from the equilibrium; (e) a 5-point limit cycle; (f) an irregular chaotic fluctuation. After Beddington *et al.* (1975).

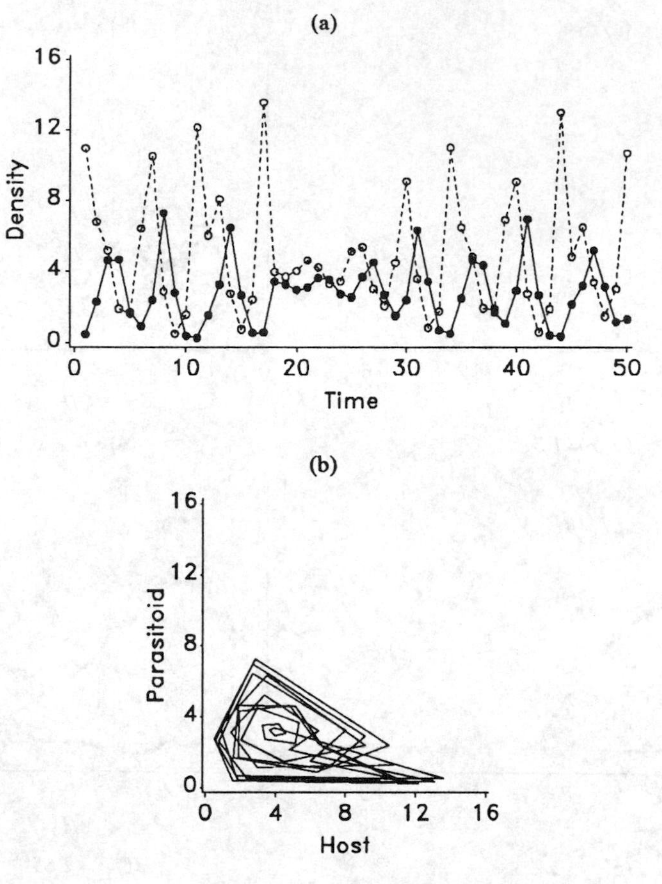

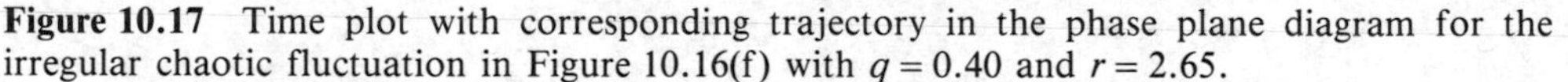

Figure 10.17 Time plot with corresponding trajectory in the phase plane diagram for the irregular chaotic fluctuation in Figure 10.16(f) with $q = 0.40$ and $r = 2.65$.

Other deterministic models

There have been a number of modifications to the basic host–parasitoid models described here. These include: (a) incorporating a functional response—which can lead to instability (see Section 10.4.2); (b) non-random search of the parasitoid—which can promote stability; (c) mutual interference of the parasitoids—which can also promote stability. For a detailed discussion of these models, see Hassell (1978).

Spatial patchiness and heterogeneity

Pacala *et al*. (1990) consider a model with spatial patchiness in which parasitoids search at random within patches so that at the patch level the dynamics are described by the Nicholson–Bailey model. They show that if the coefficient of variation of the density of searching parasitoids in the vicinity of each host exceeds one, then the system will be stable. In other words, the Nicholson–Bailey model, which is unstable in a homogeneous population, can be stable when the distribution of parasitoids is patchy. This illustrates the important general idea that heterogeneity can promote stability.

Spatial structure

Hassell *et al.* (1991) also considered the Nicholson–Bailey model in a spatial context but allowing local host and parasitoid populations to move between neighbouring patches. They simulated data on an $n \times n$ array of patches and found that populations persisted in larger arrays (15×15) when a relatively small fraction of hosts dispersed to neighbouring patches (see Hassell *et al.*, 1991 for details). Thus, by introducing local migration they were able to promote stability. Also, the persistent deterministic systems settled down to one of three broad types of *spatial* pattern: (a) spiral waves; (b) crystal lattice; (c) spatial chaos, with an erratic pattern which changed from generation to generation.

10.5 EPIDEMIC MODELS

The spread of epidemic diseases is one of the major problems of medical science. For example, AIDS (acquired immune deficiency syndrome) is now increasing in many countries of the world and many other diseases such as malaria and influenza take a substantial toll on human life. Animals too are susceptible to epidemic disease—for example, the British rabbit population was almost wiped out by the introduction of myxomatosis from South America. Epidemiology is the scientific study of the occurrence, transmission and control of disease. In epidemiology, mathematical models are important for quantifying patterns of spread of the disease and in understanding the process of transmission and spread. In this section we present some simple models.

We deal with epidemic models for a homogeneous population of fixed size where the disease is spread by contacts between individuals—for example, a measles epidemic in a boarding school. We begin with a deterministic model which uses differential equations to describe the spread in a large population. Then we consider a model with demographic stochasticity because in the early stages of an epidemic there may be a small number of infected individuals so that chance events may be important.

10.5.1 Terminology and general concepts

First, it is useful to give some terminology for describing the transmission and course of an infectious disease in a susceptible individual who has been exposed to infection. Following exposure there is a *latent period* until the individual becomes infectious—i.e. capable of infecting susceptibles. This is followed by an *infectious period*, during which the individual can pass on the disease to susceptibles—an actively infectious individual is called an *infective*. The period from infection to the appearance of symptoms is the *incubation period*. At this stage the infectious individual may be isolated to prevent further spread of the disease.

10.5.2 A simple deterministic epidemic model

A basic model for the spread of an epidemic within a population was developed by Kermack and McKendrick in 1927 while working on measles epidemics. The model deals with a disease, such as measles, which is transmitted through contact with infected individuals. At a given time, individuals in the population are classified into one of three classes: *susceptibles* who can catch the disease; *infectives* who can pass on the disease to others; *removals* who have had the disease but are no longer exposed to infection because of immunity, death or isolation. This scheme is sometimes called a SIR model because of the possible progression of individuals from susceptible to infective to removal.

The model is deterministic, assuming a large population of fixed size N in which the number of susceptibles, S, infectives, I, and removals, R, at some time t are changing smoothly as continuous variables with $N = S + I + R$. The rate of change of the number of individuals in the three classes is determined by the following assumptions.

(1) Susceptibles join the infective class at a rate which is proportional to the product of the numbers of susceptibles and infectives, i.e. βSI where β is referred to as the *infection rate*. This would occur if susceptibles and infectives were homogeneously mixed so that each infective encounters susceptibles at an average rate proportional to their numbers. Infected susceptibles immediately become infective.
(2) Infectives are removed at a rate proportional to their numbers, i.e. γI, where γ is the *removal rate*. This rate is inversely proportional to the time which elapses before removal, e.g. before the appearance of symptoms.

Combining the above assumptions gives the net rate of increase of infectives as $\beta SI - \gamma I$. Changes in the number of susceptibles (S), infectives (I) and removals (R) are then described by the differential equations as follows.

Kermack and McKendrick deterministic SIR epidemic model

$$\frac{dS}{dt} = -\beta SI$$

$$\frac{dI}{dt} = \beta SI - \gamma I$$

$$\frac{dR}{dt} = \gamma I$$

where S, I and R are the number of susceptibles, infectives and removals, β is the *infection rate* and γ is the *removal rate*.

In studying the course of an epidemic some important questions are as follows.

(1) What are the conditions for an epidemic to ensue?
(2) How many people will catch the disease and, in particular, will there be any susceptibles left when the epidemic has died out?
(3) What is the pattern in time of reported cases?

Conditions for an epidemic: threshold effect

An important property of the model is the so-called *threshold theorem* which says that the epidemic will not get started unless the initial number of susceptibles exceeds a certain threshold value. This follows from the equation for the rate of change of the number of infectives which is positive when $\beta SI - \gamma I > 0$ or $S > \gamma/\beta$. The critical parameter is the ratio of the removal rate to the infection rate, called the *relative removal rate* ($\rho = \gamma/\beta$). The threshold theorem states that an epidemic occurs when the initial number of susceptibles exceeds the relative removal rate, i.e.

$$S_0 > \rho.$$

It follows that for a fixed relative removal rate a large population will be more vulnerable to an epidemic outbreak than a small one. The threshold theorem is sometimes stated in terms of the so-called *reproductive rate*, $R_0 = \beta S_0/\gamma$. This is the number of infections produced by a single infective in a population of S_0 susceptibles before being removed. This follows because the length of the infectious period is $1/\gamma$, the reciprocal of the removal rate, during which time an infected individual will encounter a number $R_0 = \beta S_0/\gamma$ susceptibles. Essentially, the population of infectives grows when each of its members reproduces itself by infecting at least one susceptible before being removed.

Threshold theorem for an epidemic

An epidemic occurs when the reproductive rate exceeds unity, i.e.

$$R_0 = \frac{\beta S_0}{\gamma} > 1$$

where β is the infection rate, γ is the removal rate and S_0 is the initial number of susceptibles.

Severity of the epidemic

If an epidemic does become established how large will it be, i.e. how many susceptibles will eventually catch the disease? To answer this question for the Kermack and McKendrick model we derive an expression for the number of susceptibles left when the epidemic has died out. The steps are as follows.

- Divide the equation for the rate of change of susceptibles by that for the rate of change of removals to eliminate the number of infectives and obtain

$$dS/dR = -S/\rho$$

  where $\rho = \gamma/\beta$ is the ratio of the removal rate to the infection rate.
- Separate the variables and integrate to give

$$S = S_0 \, e^{-R/\rho}$$

  where S_0 is the number of susceptibles at time zero.
- Use the fact that at the end of the epidemic the number of infectives is zero, i.e. $I_\infty = 0$, so that

$$N = S_\infty + R_\infty.$$

- Substitute for R_∞ in the above expression for S to give the number of susceptibles at the end of the epidemic as

$$S_\infty = S_0 \, e^{-[N - S_\infty]/\rho}.$$

Often, the initial number of infectives is small so that S_0 is approximately equal to N and the proportion of susceptibles which contract the disease is close to $\pi = 1 - S_\infty/S_0$. Substituting in the above expression shows that the proportion of susceptibles ultimately infected satisfies the equation

$$\pi = 1 - e^{-S(0)\pi/\rho} = 1 - e^{-R_0\pi}$$

where R_0 is the reproductive rate. For a given value of R_0, the equation can be solved using Newton's method. Table 10.2 and Figure 10.18 show how the proportion increases with the reproductive rate. When the number of susceptibles is relatively high, or the removal rate relatively low, epidemics are more likely to start and result in a large proportion of infected individuals. An interesting point is that not everybody is necessarily infected—for a relatively small reproductive rate the epidemic 'dies out' from a lack of infectives.

The spread of the epidemic and the epidemic curve

In practice, the course of an epidemic is usually recorded in terms of the number of new cases per day, week or perhaps month, the so-called *epidemic curve*. This is the rate at which infectives are removed and is proportional to the number of infectives, i.e.

$$\frac{dR}{dt} = \gamma I.$$

Table 10.2 The proportion of susceptibles ultimately infected in the simple deterministic SIR epidemic model for given values of the reproductive rate

Reproductive rate $R_0 = S_0/\rho$	1.00	1.05	1.25	1.50	2.00	3.00	4.00
Proportion of susceptibles ultimately infected	0.00	0.09	0.37	0.58	0.80	0.94	0.98

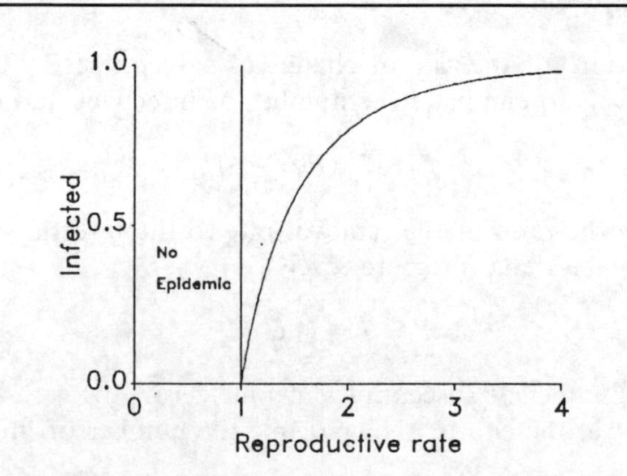

Figure 10.18 Relationship between the proportion of susceptibles ultimately infected and the reproductive rate for the simple deterministic SIR epidemic model.

Using the fact that: (a) $I = N - S - R$; and (b) $S = S_0 \mathrm{e}^{-R/\rho}$, the epidemic curve is given by

$$\frac{\mathrm{d}R}{\mathrm{d}t} = \gamma(N - R - S_0 \mathrm{e}^{-R/\rho}).$$

For specified values of population size, infection rate (β), removal rate (γ), and hence $\rho = \gamma/\beta$, and the initial number of susceptibles (S_0), the equation can be solved using numerical methods. This in turn gives values for S and I. Figure 10.19(a)–(c) shows the course of a model epidemic in a population of size 500, starting with a single infected individual with infection rate $\beta = 0.001$, removal rate $\gamma = 0.1$, and hence relative removal rate $\rho = 100$. The threshold condition for an epidemic is satisfied because the initial number of susceptibles is larger than the relative removal rate or, equivalently, the reproductive rate $R_0 = \beta S_0/\gamma = 499/100 = 4.99$ exceeds one. The large reproductive rate results in a severe epidemic with only a small proportion of susceptibles avoiding infection (Table 10.2). The epidemic curve has a single mode and a skewed appearance with a rapid rise to the peak followed by a more prolonged decline. Figure 10.19(d)–(f) shows the spread in a smaller population of 150 individuals starting with a single infected individual using the same rates of infection and removal as in the larger population. In this case the reproductive rate of 1.49 is closer to the threshold value but is still large enough for an epidemic to ensue. The result is an epidemic with about 58% of the susceptibles contracting the disease. In this case, the epidemic curve has a more symmetrical shape which is typical for reproductive rates close to one. Figure 10.19 also illustrates the following two further properties of the epidemic curve.

(1) The peak in the epidemic curve occurs when the number of susceptibles is equal to the relative removal rate. To see this, use the equations for $\mathrm{d}R/\mathrm{d}t$ and $\mathrm{d}I/\mathrm{d}t$ to obtain

$$\mathrm{d}^2R/\mathrm{d}t^2 = 0 = \gamma\ \mathrm{d}I/\mathrm{d}t = \gamma(\beta SI - \gamma I),$$

so that at the peak $S = \gamma/\beta$.

(2) Initially, the epidemic curve increases approximately exponentially. This follows from the fact that in the early stages of the epidemic the number of susceptibles changes rather slowly so that

$$\frac{\mathrm{d}I}{\mathrm{d}t} \sim [\beta S_0 - \gamma]\, I$$

or

$$I(t) \sim I_0\ \mathrm{e}^{[\beta S_0 - \gamma]t}.$$

Figure 10.19(c) and (f) shows the approximately linear increase in the logarithm of the epidemic curve.

Example 10.5 Deterministic SIR epidemic model fitted to Eyam plague data

Eyam, a village in England, suffered an outbreak of the Great Plague in the late seventeenth century. It is thought that the plague came from London in tailors cloth containing plague-carrying rat fleas. Out of 350 villagers only 83 survived. The course of the epidemic was recorded by the young rector of the village, the Reverend

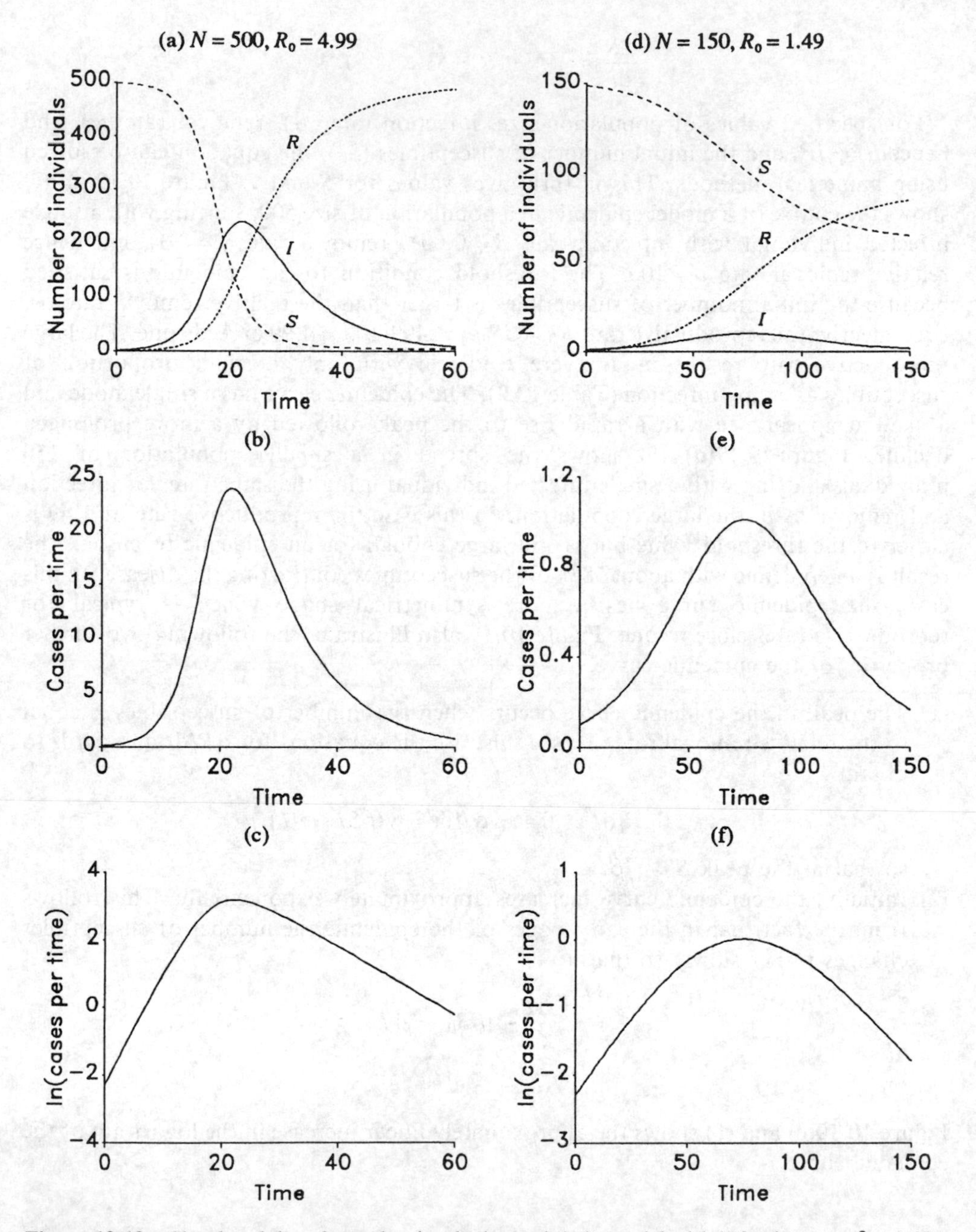

Figure 10.19 Simulated data from the simple deterministic model with infection rate $\beta = 0.001$, removal rate $\gamma = 0.1$ starting with a single infected individual for two populations with different reproductive rates.

William Mompesson who isolated the community and thereby prevented spread to other villages. Raggett (1982) discusses data and models for the Eyam plague, and the following analysis is based on his paper. After a period of fairly limited infection the major part of the epidemic ran from June to October 1666. Figure 10.20 shows the number of susceptibles and infectives recorded at approximately 15 day intervals. At the start of the period there were 7 infected individuals in a total of 261 and at the end of the epidemic there were 83 remaining susceptibles. To estimate the relative removal rate Raggett used the equation for the number of susceptibles, S_∞, remaining at the end of the epidemic, i.e.

$$S_\infty = S_0\, e^{-[N - S_\infty]/\rho}.$$

Using $S_\infty = 83$, $S_0 = 254$ and $N = 261$ gives an estimated relative removal rate of $\rho = \gamma/\beta = 159$. To estimate the removal rate, γ, Raggett simulated data from the model using a relative removal rate of 159 for a range of values of γ and selected the value which minimised the sum of the squared deviations of the observations from the simulated values. This gave an estimated removal rate of $\gamma = 0.090$ per day which agrees closely with the estimated infection period of 11 days and corresponding removal rate $1/11 = 0.091$ per day. Figure 10.20 shows that there is good agreement between the observed numbers of susceptibles and infectives and those calculated from the fitted model.

10.5.3 Simple epidemic models with demographic stochasticity

In the deterministic epidemic model the course of the epidemic changes smoothly in time. The model applies when the number of susceptibles, infectives and removals are relatively large so that they can be regarded as continuous variables. At a detailed level, however, the epidemic is a process consisting of discrete individuals subject to the chance events of infection or removal, i.e. there is inherent demographic stochasticity (Chapter 4). In large homogeneous populations the effects of demographic stochasticity are reduced by the aggregation of many events, but in small populations this is not the case and the effects may be more important. This consideration applies to the early stages of an epidemic where the number of infectives may be small and the justification for a deterministic model is less clear. In this section we explore this

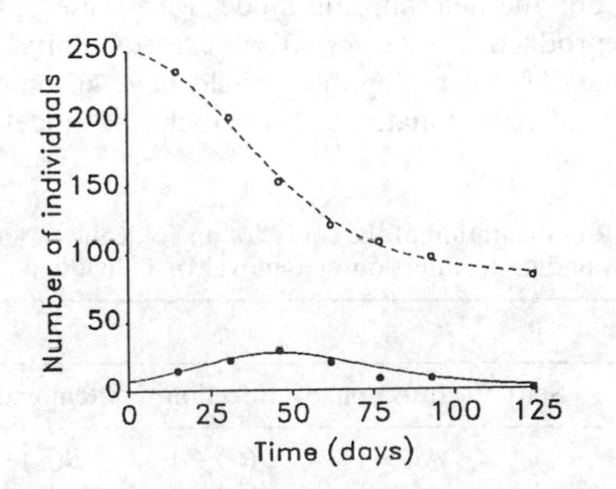

Figure 10.20 Simple deterministic SIR epidemic model fitted to Eyam plague data. Susceptibles (broken line); infectives (solid line).

by using a simple epidemic model with demographic stochasticity to allow for chance effects when numbers of individuals are small.

We assume the SIR structure in which the population contains susceptibles which may become infectives and eventually removals. We regard the changes of the number of individuals in each class as a series of discrete events (Table 10.3).

To develop a stochastic model to describe the changes in numbers of susceptibles, infectives and removals we recall the two key components of the deterministic model. These are: (a) susceptibles join the infectives at a rate βSI where β is the *infection rate*, i.e. the number of susceptibles becoming infected in small time interval $(t, t+\delta t)$ is equal $\beta SI\delta t$; (b) infectives are removed at rate γI where γ is the *removal rate*, i.e. the number of infectives removed in time interval $(t, t+\delta t)$ is equal to $\gamma I\delta t$. In the stochastic model, the changes in a small interval of time $(t, t+\delta t)$ are discrete events which occur as follows.

(1) The probability of an infection is $\beta SI\delta t$.
(2) The probability of a removal is $\gamma I\delta t$.
(3) The probability of more than one event is zero, i.e. the situation stays the same with probability $1-\beta SI\delta t-\gamma I\delta t$.
(4) The events of infection and removal occur *independently* from one time interval to the next.

In the stochastic model events occur randomly in time with rates determined by the current number of susceptibles, infectives and removals. At a particular time the overall rate at which events occur is equal to $\beta SI+\gamma I$. The time to the next event then follows an exponential distribution with mean $1/(\beta SI+\gamma I)$. When an event occurs the chance of an infection depends on the relative rates of infection and removal, i.e. an infection occurs with probability $\beta SI/(\beta SI+\gamma I)$. This gives the basis for simulating the epidemic. We start with an initial number of susceptibles and infectives, sample the time to the first event at random from the exponential distribution and determine an infection or removal at random. The number of susceptibles and infectives is then updated and the process is repeated.

To illustrate the properties of the model we simulate an epidemic in a population of size $N=150$, infection rate $\beta=0.001$, removal rate $\gamma=0.1$ and starting with a single infected individual. For the deterministic model with these parameters an epidemic ensues since the reproductive rate $R_0=1.49$ exceeds unity. At the end of the deterministic epidemic 58% of susceptibles would have become infected. Table 10.4 summarises the results of 100 simulations of the stochastic model and demonstrates the

Table 10.3 Representation of the course of an epidemic as series of discrete events corresponding to infection or removal of individuals

Time	0	t_1	t_2	t_3	t_4
Event	Start	Infection	Infection	Removal	Infection
Susceptibles	S_0	S_0-1	$S(t_1)-1$	$S(t_2)$	$S(t_3)-1$
Infectives	I_0	I_0+1	$I(t_1)+1$	$I(t_2)-1$	$I(t_3)+1$
Removals	0	0	0	1	1

Table 10.4 Summary of the outcomes in 100 realisations of a stochastic epidemic model starting with a single infected individual, 149 susceptibles, infection rate $\beta = 0.001$ and removal rate $\gamma = 0.1$

Number of individuals ultimately infected	Frequency	Percentage of susceptibles ultimately infected	Mean time to end of epidemic
1	36	0.6	3.4
2–10	24	2.9	14.7
11–20	2	8.4	33.5
21–30	3	17.9	63.4
31–40	2	24.8	91.0
41–50	0	0.0	—
51–60	1	34.9	76.6
61–70	1	44.3	121.4
71–80	4	51.7	95.6
81–90	4	55.4	106.9
91–100	13	63.7	136.0
101–110	8	71.3	152.3
111–120	2	75.8	157.0
121–130	0	0.0	—
131–140	0	0.0	—
141–150	0	0.0	—

important property of the model that the number of susceptibles which become infected form two broad groups of values, one with very few infected individuals the other with many. Broadly speaking, the epidemic either takes off and is severe, or it fizzles out in the early stages. This behaviour contrasts with that of the corresponding deterministic model for which a severe epidemic always ensues.

Threshold effects and extinction probabilities

The results of the simulations (Table 10.4) show that for the stochastic model the outcome of the epidemic is subject to chance and an epidemic may not occur even when the reproductive rate exceeds unity. The conditions can be stated more precisely as follows.

(1) When the reproductive rate R_0 exceeds one the epidemic may infect only a few individuals with *approximate* probability given by

$$P(\text{no epidemic}) = \frac{1}{R_0^{I_0}},$$

where I_0 is the initial number of infectives. For the example reported in Table 10.4, $R_0 = 1.49$, $I_0 = 1$ which gives probability of $1/1.49 = 0.67$. This compares with the 60% of epidemics with fewer than 10 (7%) of individuals ultimately infected.

(2) If the epidemic survives the initial stages, the proportion who become infected is given approximately by that for the deterministic model. In Table 10.4 an average of 62% of susceptibles ultimately became infected in epidemics with more than 30% of susceptibles ultimately infected which compares with the 58% infected in the deterministic model.

Figure 10.21 shows the course of a stochastic epidemic which survived the early stages and resulted in a higher percentage of infectives (62%) than in the corresponding deterministic model (58%). The irregular traces reflect the changes in the number of susceptibles, infectives and removals as a series of discrete steps. Figure 10.21(b)–(d) shows the changes in each category with the corresponding curve for the deterministic model. Other stochastic simulations would in general produce different patterns. The selected example is intended to illustrate that when the stochastic epidemic survives its early stages the corresponding deterministic model captures the broad trends. For the chosen example the number of removals lies wholly above the deterministic curve—an above average number of early chance removals is maintained throughout. In other cases the stochastic curve could fall below that for the deterministic model.

Variation in the incubation period

In the simple deterministic epidemic model infectives are removed at a constant per capita rate, i.e. when there are I infectives the rate of removal is equal to γI. In the

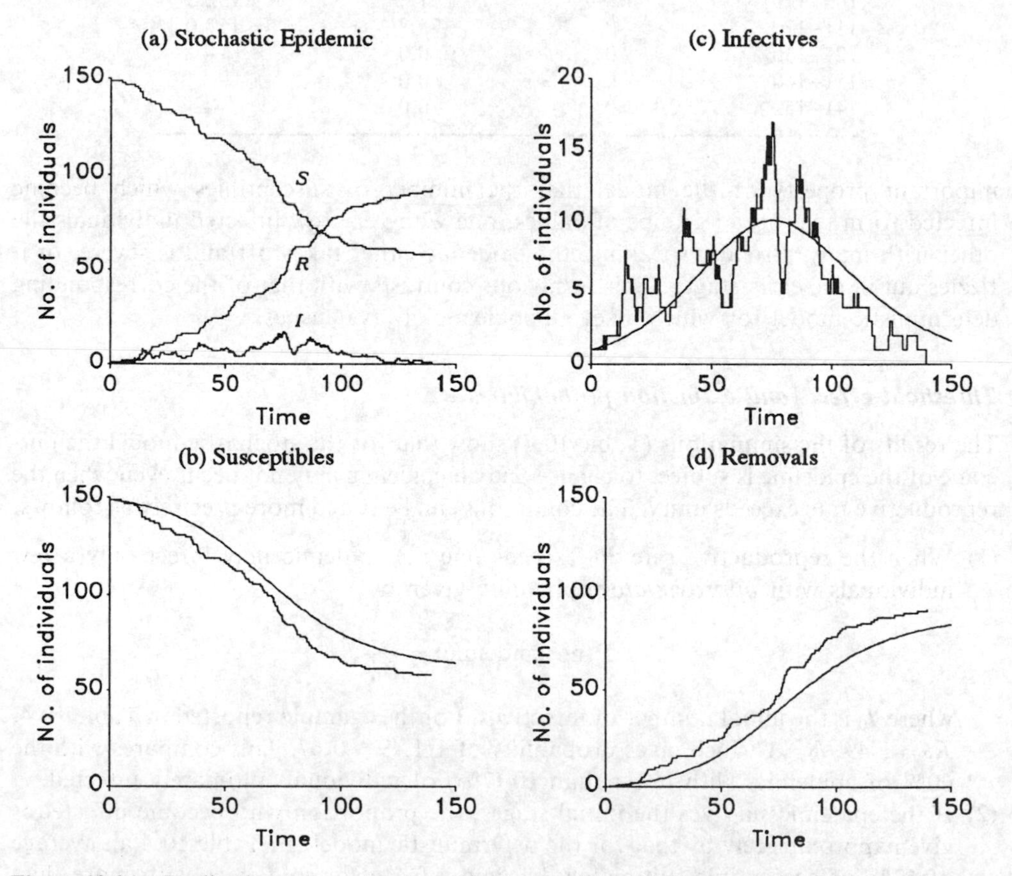

Figure 10.21 Realisation of a simple epidemic model with demographic stochasticity. (a) Changes in the number of susceptibles, infectives and removals; (b)–(d) separate components with corresponding deterministic curves.

model with demographic stochasticity the corresponding assumption is that removals occur independently and at random at the same per capita removal rate γ. As a consequence, the incubation period of each infective, i.e. the time from infection to the appearance of symptoms and removal, follows an exponential distribution with mean $1/\gamma$. This assumption implies that the rate at which symptoms appear does not depend on how long the individual has been infected which is rather unrealistic and lacks empirical justification.

The point is illustrated using information on the incubation period of individuals who became infected with AIDS via blood transfusion. Table 10.5 shows data on the time from transfusion during April 1978 and February 1986 to diagnosis during January 1982 and June 1986 (Medley *et al.*, 1987). A problem with interpretation is that individuals transfused later have been at risk of developing AIDS for a much shorter length of time so that the observed frequency distribution of incubation periods is misleading. However, it is apparent by examination of the data in each row of the table that an exponential distribution is a poor description of the distribution. For example, in those patients transfused in 1982 the observed frequency indicates a peak at 3–4 years.

A more realistic model is one in which the rate at which symptoms appear in an infective increases with the time since infection. To illustrate this idea consider the fate of a *large* number of individuals infected at time $t = 0$ and suppose that symptoms appear at rate directly proportional to the time since infection. Then, at time t, the number of infected individuals still without symptoms, x, satisfies the differential equation

$$\frac{\mathrm{d}x}{\mathrm{d}t} = -atx$$

where a is a rate constant. To solve the equation, separate the variables and integrate both sides to give the number of infected individuals without symptoms by time t as

$$x = x_0 \, \mathrm{e}^{-at^2/2}.$$

Dividing this by x_0 gives the proportion of individuals whose time to the appearance

Table 10.5 Time to diagnosis of AIDS in patients infected via blood transfusion (Modified from Medley *et al.*, 1987)

Year of transfusion	Time to diagnosis (years)							
	0–1	1–2	2–3	3–4	4–5	5–6	6–7	7–8
1978	0	0	0	0	0	1	0	1
1979	0	0	0	2	2	5	5	3
1980	0	0	2	10	7	8	3	—
1981	0	5	7	15	16	16	—	—
1982	1	7	18	25	13	—	—	—
1983	3	20	23	23	—	—	—	—
1984	2	26	14	—	—	—	—	—
1985	5	7	—	—	—	—	—	—
1986	1	—	—	—	—	—	—	—

of symptoms is greater than t, i.e. one minus the cumulative distribution function of the time to appearance of symptoms. So, the probability density function of the time to appearance of symptoms is given by

$$f(t) = at\, e^{-at^2/2}.$$

This is a unimodal distribution with a mode at $t = 1/\sqrt{a}$ and which is skewed to the right with mean equal to $\mu = \sqrt{\pi/2a}$. Figure 10.22 shows both the p.d.f. and the c.d.f. for the case $a = 0.10$ with mode equal to 3.16 and mean equal 3.96 together with the exponential distribution having the same mean. There are marked differences between models which could be relevant to the spread of an epidemic. For example, in the exponential model about 20% of individuals show symptoms by time 1, whereas with a non-constant rate of appearance only about 5% show symptoms, leaving more individuals to spread the infection. For further discussion of the effects of the form of the incubation distribution on the spread of AIDS, see Isham (1988). In the above analysis the figures are purely fictional and chosen for illustration. For a detailed discussion of the blood transfusion data, the reader should refer to the paper by Medley *et al.* (1987) where a range of models was fitted which suggested that the incubation period varied with age.

10.5.4 Discussion and more complex models

The simple epidemic models of this section apply to the spread of a disease in a population of fixed size in which individuals are classified as either a susceptible, infective or removal and where the infection rate and the removal rate are constant. Most situations are much more complex than this. Some aspects are briefly discussed below.

Heterogeneity

The population may be heterogeneous with some individuals being more prone to infection than others. For example, in the spread of AIDS it is known that individuals

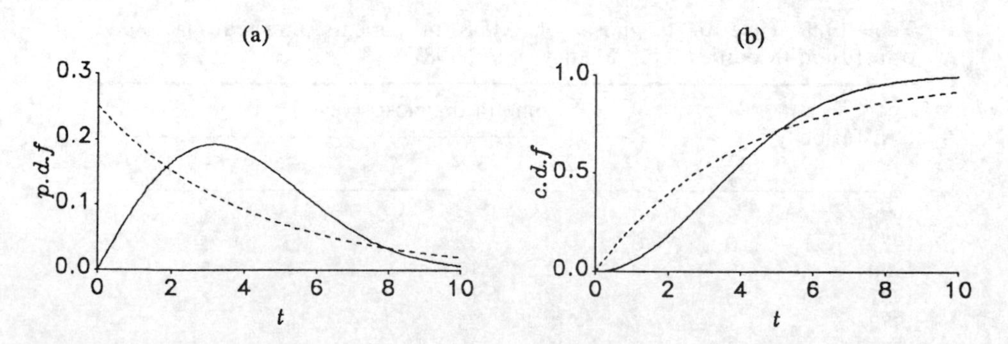

Figure 10.22 Probability density function (a) and cumulative distribution function (b) of incubation times for two models. (i) Rate of appearance of symptoms increases linearly with time since infection (solid line); (ii) constant rate of appearance of symptoms with exponential distribution of times (broken line). Both distributions have mean equal to 3.96.

vary widely in their levels of sexual activity (May & Anderson, 1987; Peto, 1986). Heterogeneity can lead to an initial period of rapid spread, compared with a homogeneous population, followed by a slower rate. For a discussion of the effect of heterogeneity in modelling the transmission of AIDS, see Isham (1988).

Vectors of transmission

Many diseases are carried and transmitted by an insect or some other vector, e.g. malaria by the mosquito (Ross, 1910). In these cases the life-cycle of the carrier as well as the epidemiology of the disease needs to be taken into account. For examples of models and their applications, see Anderson (1982). For a range of models for the spread of rabies by the fox, see Bacon (1985).

Spatial models and stochastic effects

There is often a spatial element in the transmission of the disease—for example, rabies is transmitted by the fox which is a territorial animal. In this case, realistic models of spread include spatial effects and contact between neighbouring individuals (Mollison & Kuulasmaa, 1985). Mollison (1981) stresses that demographic stochasticity cannot easily be neglected for spatial processes in which individuals interact with a relatively small number of neighbours and what goes on at the front of the process, where numbers are small, is of particular importance. This contrasts with the relatively small effect of demographic stochasticity in a large, homogeneous population.

EXERCISES

10.1 *Amensalism* is a form of competition between two species in which species 1 has a competitive effect on species 2, whereas species 2 does not affect species 1.

(a) Write down the equations for the Lotka–Volterra two species competition model to describe the amensal relationship between two species.

(b) Construct phase plane diagrams to analyse the possible outcomes of competition in relation to the values of the parameters of the model.

10.2 In the Lotka–Volterra predator–prey model with self-limited prey population growth, prey (H) and predator (P) densities are given by

$$\frac{\mathrm{d}H}{\mathrm{d}t} = rH\left(1 - \frac{H}{K}\right) - aPH; \qquad \frac{\mathrm{d}P}{\mathrm{d}t} = -sP + bPH,$$

(a) Show that the trace (T) and determinant (Δ) of the community matrix are given by

$$\mathrm{T} = -rH_e/K; \qquad \Delta = abP_eH_e$$

where H_e and P_e are equilibrium densities of prey and predator, respectively.

(b) Hence show that there is a stable equilibrium with non-zero prey and predator densities when $s < bK$.

10.3 Consider the predator–prey model

$$\frac{\mathrm{d}H}{\mathrm{d}t} = rH\left(1 - \frac{H}{K}\right) - \frac{wPH}{(D+H)}; \qquad \frac{\mathrm{d}P}{\mathrm{d}t} = -sP + \frac{bPH}{(D+H)},$$

where H and P are the densities of prey and predator, respectively.

(a) Briefly give the biological interpretation of the components and the parameters of the model.

(b) Derive expressions for the H and P nullclines and sketch them on a phase plane diagram with P along the y-axis. Show that when $b/s > 1 + D/Kn$, then both prey and predator have non-zero equilibrium densities given by

$$H_e = \frac{sD}{(b-s)}; \qquad P_e = \frac{r(K-H_e)(D+H_e)}{Kw}.$$

(c) Show that the trace (T) and the determinant (Δ) of the community matrix are given by

$$\mathrm{T} = \frac{rH_e(K-D-2H_e)}{K(D+H_e)}; \qquad \Delta = \frac{bH_e(K-H_e)}{K(D+H_e)}.$$

Deduce that when the intersection of the nullclines is to the right of the peak in the prey nullcline then the equilibrium is stable. Hence show that the condition for a stable equilibrium is that

$$b/s - 1 > D/K > (b/s-1)/(b/s+1).$$

Sketch the region of stability in the $(b/s, D/K)$ plane.

10.4 Using Phaseplane or some similar package, carry out the following simulation study to illustrate the stability properties and the dynamic behaviour of the predator–prey model in Exercise 10.3. Fix the values of the parameters: $r = 1, K = 10, s = 1, w = 3$, and simulate time plots of data for each of the 8 combinations of the parameter values $b = 2, 3$ and $D = 1.5, 3.0, 4.5$ and 6.0, over the range $t = 0$–100. In each case, start the series with values at half the equilibrium densities (see Exercise 10.3). Then, classify the dynamic behaviour for different combinations of $(b/s, D/K)$ as: (i) damped oscillations to a stable equilibrium; (ii) stable limit cycles; (iii) unstable divergent oscillations leading to low prey densities and eventual extinction. Compare your findings with the conditions for a stable non-zero equilibrium (Exercise 10.3).

Comment on the effect of increased predator efficiency on the dynamics of the system.

10.5 In a simple host–parasitoid model with non-random search, the number of encounters per host follows a negative binomial distribution with mean aP_t and index of dispersion k, where a is the effective area of search and P_t is the density of the parasitoid at time t. For this model, the fraction of hosts not parasitised is related to parasitoid density by $f(P_t) = (1 + aP_t/k)^{-k}$. The difference equations relating the densities of hosts and parasitoids in successive generations are

$$H_{t+1} = RH_t(1 + aP_t/k)^{-k}; \qquad P_{t+1} = cH_t[1 - (1 + aP_t/k)^{-k}],$$

where R is the intrinsic growth rate of the host and c is the net fecundity of the parasitoid.

(a) Show that there is an equilibrium with non-zero parasitoid and host densities given by

$$P_e = k(R^{1/k} - 1)/a; \qquad H_e = RP_e/(c(R-1)).$$

Comment on the model when k takes very large values.

In the following, consider the particular case: $R = 2, c = 1, a = 0.025$.

(b) Illustrate the general property that, as k decreases, the host and parasitoid densities become exceedingly large.

(c) Simulate data from the model with $k = 0.8, 0.9, 1.0, 1.1$ and 1.2 and starting values equal to half the equilibrium values to illustrate: (i) for $k < 1$ there is a stable equilibrium; (ii) for $k > 1$ the equilibrium is unstable. For the stable cases, show that low starting values can lead to initial oscillations of large amplitude.

(d) For $k = 1$, simulate data using starting values of: $(H_0, P_0) = (20, 10), (40, 20), (60, 30), (80, 40)$ and $(100, 50)$. Comment on the dynamics.

10.6 Discuss ways in which demographic stochasticity could be incorporated into the Nicholson–Bailey host–parasitoid model with self-limited host population growth.

10.7 Show that the Lotka–Volterra predator–prey models with unlimited, and limited, prey population growth are both equivalent to epidemic models for a population in which susceptibles are regarded as prey and infectives as predators.

For the predator–prey model with limited population growth, find the condition in the equivalent epidemic model for persistence of the disease at a stable level.

10.8 Consider a modification of the SIR epidemic model for a population of fixed size in which removals return to become susceptibles at a constant rate, α, so that the equations for the number of susceptibles (S), infectives (I) and removals (R) are given by

$$\frac{dS}{dt} = -\beta SI + \alpha R; \qquad \frac{dI}{dt} = \beta SI - \gamma I; \qquad \frac{dR}{dt} = \gamma I - \alpha R$$

where $S + I + R = N$.

(a) Construct a phase plane diagram for I (y-axis) and S (x-axis) showing nullclines and equilibrium states. What is the condition for a non-zero equilibrium value for I?

(b) Show that the trace (T) and the determinant (Δ) of the community matrix are given by

$$\mathrm{T} = -(\beta I_e + \alpha); \qquad \Delta = (\gamma + \alpha)\beta I_e$$

where I_e is the equilibrium value. Deduce that the non-zero equilibrium is always stable and sketch direction flows on the phase plane diagram to illustrate that the approach to equilibrium can be oscillatory.

10.9 Suppose that in the SIR epidemic model a proportion p of the population has been immunised against infection. Show that an epidemic amongst the remaining susceptibles will be prevented when

$$p > 1 - \frac{1}{R_0}$$

where R_0 is the reproductive rate of the disease in the population with no immunisation.

Calculate p for a population of size $N = 200$, infection rate $\beta = 0.001$ per day and removal rate $\gamma = 0.10$ per day. Repeat the calculation for a population of size $N = 1000$ and comment on the result.

10.10 Consider the following simplified model for the spread of AIDS through the sexually active population. Let w be the number of people who have developed AIDS, x the number known to contain the virus but who have not yet developed AIDS, y the number infected with the virus unknown to themselves, and z the number of non-infected people. It is assumed that a negligible number of those known to be infected continue to infect others. In a given time period, a proportion of those with the virus are tested and correctly diagnosed, a proportion of those with the virus develop AIDS and a proportion of those with the disease die. Obtain the equations

$$\frac{dw}{dt} = \mu(x + y) - pw; \qquad \frac{dx}{dt} = ky - \mu x; \qquad \frac{dy}{dt} = -(k + \mu)y + \lambda yz; \qquad \frac{dz}{dt} = -\lambda yz$$

interpreting the positive constants μ, p, k, and λ and stating any further assumptions you make.

Initially, $w = w_0$, $x = x_0$, $y = y_0$ and $z = z_0$ and these values are such that w, x and y are initially increasing, while z is decreasing. Describe qualitatively what will happen to the population and give rough sketches of plausible behaviour of y and z against t.

Show that the only possible steady states of the system are $(w, x, y, z) = (0, 0, 0, z_1)$, for any z_1, but that these equilibria are unstable if $z_1 > \varepsilon$ where $\varepsilon = (k + \mu)/\lambda$.

Obtain an equation relating y and z and hence show that

$$y = y_0 + z_0 - z + \varepsilon \ln \frac{z}{z_0}.$$

Noting that z will continue to decrease until $y = 0$, show that if ε is small then the final size of the population will be given approximately by

$$z = z_0 \exp[-(y_0 + z_0)/\varepsilon].$$

What attributes of the disease render it dangerous to humanity as a whole? (Cambridge NST, Part IA, Biological Mathematics, 1989.)

11 Advanced Model Fitting

11.1 INTRODUCTION

We first discussed methods for fitting models to data in Chapter 5. We fitted models with deterministic parts consisting of just a mean, a straight line, a quadratic curve or a non-linear function. The simplest formal method introduced was least squares. By this procedure parameter values are selected so as to minimise the sum of the squared deviations of the observations about the fitted values of the deterministic part of the model. The assumption is that each observation is made up of an *underlying* value which lies on a curve—the deterministic part of the model—and a *random* departure. A further assumption in ordinary least squares is that the variance of the random departure is constant over the whole range of the data. In this chapter we extend these methods to fit some of the more complex models discussed in various more recent chapters. These are: (a) the logistic-binomial model for bioassay and with ecological and medical applications from Chapter 4; (b) the rectangular hyperbola used in enzyme kinetics and ligand–receptor assay, and (c) the sum of two exponentials used in compartment modelling from Chapter 9; and finally, (d) the dynamic models of Chapters 8, 9 and 10. We shall see that there is more than one way of fitting some of these models, e.g. the rectangular hyperbola, and that some belong to one or more general classes of models. For example, there is the class of models which can be reduced by transformation to linear models, called *generalised linear models*—including (a) and (b) above. Some models cannot be transformed in this way and belong to the class of *non-linear models*, e.g. the sum of two exponentials. These are also discussed in some detail. Finally, we consider methods for fitting *dynamic* ecological and physiological *models*, which have their own particular problems such as serial dependence amongst the random components.

11.2 LINEARISATION

For some models it is possible to transform the response variable or the explanatory variable or both so that the relationship between the transformed variables is linear. An example is exponential growth of a population in which size at time t is given by

$$N(t) = N(0)e^{rt}$$

so that the logarithm of population size has a straight line relationship to time with slope r, the intrinsic relative growth rate, i.e.

$$\ln N(t) = \ln N(0) + rt.$$

Taking logarithms in this context is useful: (a) as a check on whether growth is exponential—a straight line after transformation indicating exponential growth; (b) for estimating the relative growth rate by fitting a straight line to the transformed data. This is an example of the Linearisation technique.

Linearisation is a technique for transforming the response, and in some cases the explanatory variable, to achieve a linear relationship:

(a) between response and explanatory variable, or more generally
(b) between response and model parameters.

Uses

(a) a diagnostic check of the adequacy of the original model
(b) estimating the parameters in the model by fitting the resulting linear model.

In practice, the observations will be subject to random variation so that statistical methods are needed to fit the linear model. In this section we discuss some properties of linearisation using for illustration the rectangular hyperbola. Then, in the next section, we show how to deal with the problems this creates when fitting the resulting straight line or more generally the resulting linear model.

11.2.1 Linearisation of the rectangular hyperbola

There are many examples in biology of response curves which have the shape and mathematical form of the rectangular hyperbola. Some are: (a) the Michaelis–Menten equation relating the velocity of an enzyme catalysed reaction to the substrate concentration (Chapter 9); (b) the relation between the concentration of bound and free ligand in ligand–receptor interaction (Chapter 9); (c) the response of the rate of photosynthesis to the incident light level; (d) the functional response of predation rate to prey density (Chapter 10). The interpretation of the parameters of the model depends on the particular application but the form of the relationship is the same in all four examples

$$y = \frac{\alpha x}{\beta + x}$$

where y is the response variable and x is the explanatory variable. The function increases at a decreasing rate to approach an upper asymptote α and attains the value $\alpha/2$ when $x = \beta$. Surprisingly, there are several different ways to linearise the rectangular hyperbola. In this section we present two, illustrated using some data from enzyme kinetics (Figure 11.1). The two transformations are given in Table 11.1, and discussed below.

Lineweaver–Burk plot

The Lineweaver–Burk plot (Lineweaver & Burk, 1934) is based on the oldest and most widely used linearising transformation of the rectangular hyperbola. It involves a plot of $1/y$ against $1/x$ obtained by simply inverting the above relationship. Comparing the expression in Table 11.1 with the equation of a straight line $y = mx + c$ shows that the slope is β/α and the intercept $1/\alpha$. The parameters of the hyperbolic relationship can therefore be estimated from the slope and intercept: $\hat{\alpha} = 1/\text{intercept}$ and $\hat{\beta} = \text{slope}/\text{intercept}$. Apart from some scatter this plot for the enzyme kinetic data

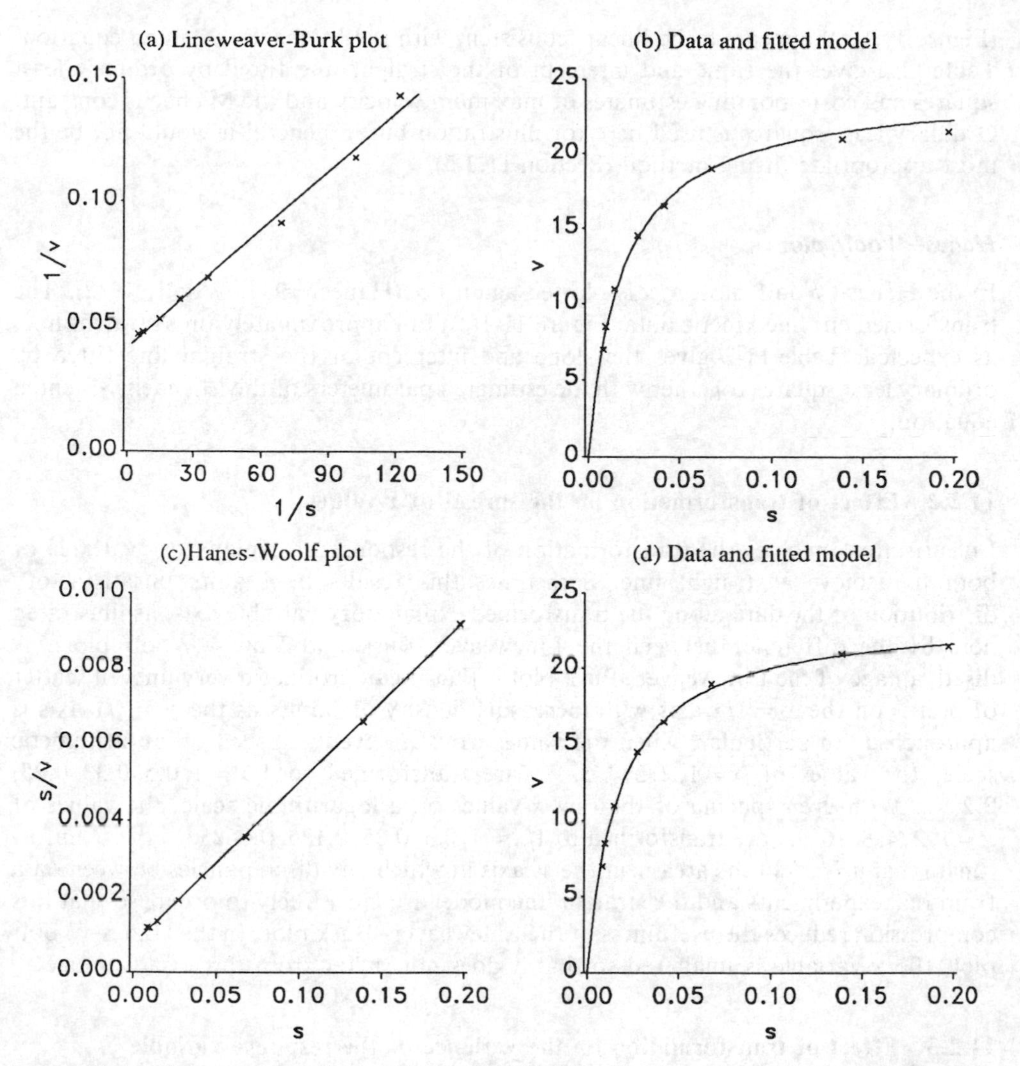

Figure 11.1 Lineweaver–Burk and Hanes–Woolf plots for the enzyme kinetic data (Example 9.2) with straight lines fitted by ordinary least squares. Corresponding Michaelis–Menten curves shown with observed data.

Table 11.1 Two ways to linearise the hyperbolic relationship $y = \alpha x/(\beta + x)$

Method	Linearisation $y = c + mx$	Intercept on x-axis $(-c/m)$	Intercept on y-axis (c)	Slope (m)
Lineweaver–Burk	$\left(\frac{1}{y}\right) = \frac{1}{\alpha} + \frac{\beta}{\alpha}\left(\frac{1}{x}\right)$	$-\frac{1}{\beta}$	$\frac{1}{\alpha}$	$\frac{\beta}{\alpha}$
Hanes–Woolf	$\left(\frac{x}{y}\right) = \frac{\beta}{\alpha} + \frac{1}{\alpha}(x)$	$-\beta$	$\frac{\beta}{\alpha}$	$\frac{1}{\alpha}$

(Figure 11.1(a)) appears to be linear, consistent with the Michaelis–Menten equation. Table 11.2 gives the slope and intercept of the straight line fitted by ordinary least squares and corresponding estimates of maximum velocity and the Michaelis constant. Ordinary least squares is used here for illustration but in general it would not be the most appropriate fitting method (Section 11.3.3).

Hanes–Woolf plot

In the Hanes–Woolf plot, x/y is plotted against x (Hanes, 1932; Woolf, 1932). The transformed enzyme kinetic data (Figure 11.1(c)) fall approximately on a straight line, as expected. Table 11.2 gives the slope and intercept of the straight line fitted by ordinary least squares together with the estimated parameters of the Michaelis–Menten equation.

11.2.2 Effect of transformation on the spread of *x*-values

Linearisation involves the transformation of the response or explanatory variable or both to achieve a straight line. Sometimes this results in a quite unsatisfactory distribution of the data along the transformed explanatory variable axis, as illustrated here by the difference between the Lineweaver–Burk and Hanes–Woolf plots. A disadvantage of the Lineweaver–Burk plot is that it can produce a very uneven scatter of points on the $x_t = 1/x$ axis with increasing density of points as the $y_t = 1/y$ axis is approached. In particular, when the values of x are evenly spaced on an arithmetic scale, the values of $x = 1, 2, 3, 4, 5, \ldots$ are transformed to $1/x = 1, 0.5, 0.33, 0.25, 0.2, \ldots$ With even spacing of the raw x-values on a logarithmic scale, the values of $x = 1, 2, 4, 8, 16, \ldots$ are transformed to $1/x = 1, 0.5, 0.25, 0.125, 0.0625, \ldots$. It is unfortunate that it is often the area near the y_t axis in which any discrepancies between data from real experiments and the straight line model are most likely to occur, so that this compression reduces the usefulness of the Lineweaver–Burk plot. In the Hanes–Woolf plot, the x variable is unaltered so that it does not suffer from this disadvantage.

11.2.3 Effect of transformation on the variance of the response variable

Another consequence of transforming a response variable is that the variance of the response variable is affected. More importantly the variance is usually different at

Table 11.2 Results of fitting the Michaelis–Menten equation $v = Vs/(K_m + s)$ to the enzyme–substrate data (Figure 11.1) using two linearisation methods

Method	Linearisation $y = c + mx$	Estimated parameters			
		Slope	Intercept	$\hat{V}$	$\hat{K}_m$
Lineweaver–Burk plot	$\left(\frac{1}{v}\right) = \frac{1}{V} + \frac{K_m}{V}\left(\frac{1}{s}\right)$	0.000790	0.0409	24.45	0.0193
Hanes–Woolf plot	$\left(\frac{s}{v}\right) = \frac{K_m}{V} + \frac{1}{V}(s)$	0.0426	0.000736	23.47	0.0173

different values of x, and so a model in which the variance is the same for all values of x can be converted to one with a marked trend in variance as x_t increases. This is important for any linearisation technique in which a straight line is subsequently fitted using ordinary least squares, a technique which essentially assumes that the variance of the response is constant. Here we show how transformation can affect the variance of the response variable and consider the implications for fitting a rectangular hyperbola using the Lineweaver–Burk plot.

The effect of the Lineweaver–Burk transformation on the variation in the response variable is illustrated by showing how an error band of constant width—which could be indicating, for example, one standard deviation of the random component in the untransformed model—is distorted into a divergent error band (Figure 11.2). The

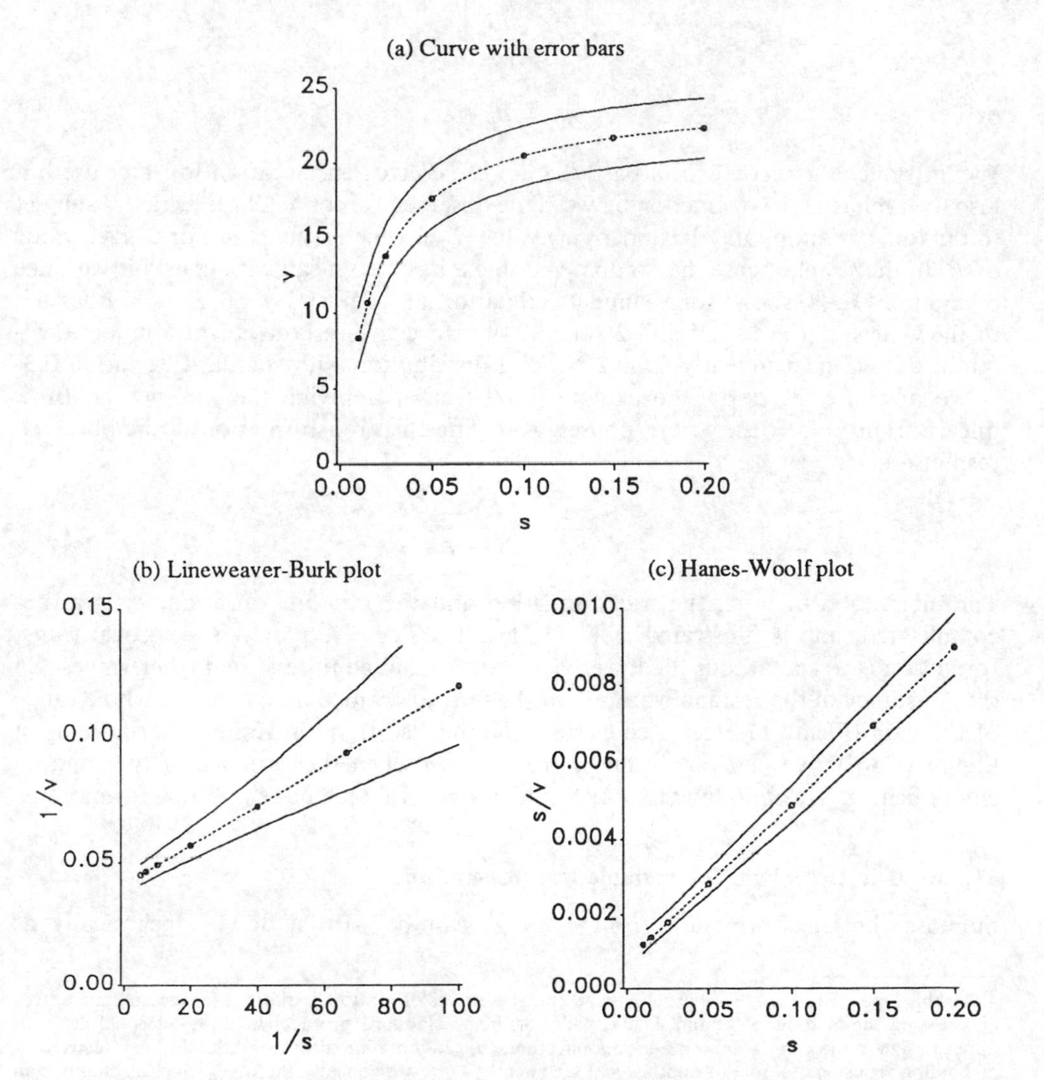

Figure 11.2 Effect of the Lineweaver–Burk plot and the Hanes–Woolf plot applied to the rectangular hyperbola with an error band of constant width.

effect is more pronounced in the Lineweaver–Burk plot than it is in the Hanes–Woolf plot. One way to allow for this in fitting the straight line is to weight the points differentially in the fitting method, giving more weight to the points which are more reliable, i.e. those with smaller variance (as discussed further in Section 11.3.1). First we need to examine in more detail the effect of the transformation on the variance of the response variable.

We consider the application of the Lineweaver–Burk plot to fitting the simple ligand–receptor binding model where we assume that there is just one random component.* We think of the *observed* concentration of bound ligand (B) as being made up of an *underlying* systematic component determined by the rectangular hyperbola plus a random disturbance about the line reflecting the biological variation, i.e.

$$B_{\text{obs}} = \frac{nF}{K_D + F} + \varepsilon$$

or

$$B_{\text{obs}} = B_u + \varepsilon.$$

We introduce the second form partly as a more convenient notation for later use but also to emphasise the distinction between the observed response (B_{obs}), which is subject to random variation, and the underlying value (B_u), which is constant for a given value of F. In the simplest case the random variable ε has zero mean with constant variance σ^2. Figure 11.3(a) shows some simulated data for a model with $n = 5$, $K_D = 3$. For each of the values of $F = 1$, 2, 5, 10, 20 and 30 there are five responses with random deviations drawn independently from a Normal distribution with standard deviation 0.3.

We now use this model to explore the effect of applying the Lineweaver–Burk linearisation to the observed responses. Applying this transformation to the observed response gives

$$\frac{1}{B_{\text{obs}}} = \frac{1}{B_u + \varepsilon}.$$

The interaction between the transformation and the random variation, ε, is rather complicated, and is illustrated in Figure 11.3(b). The effect of the reciprocal transformation is to spread out the lower values of Y, and compress the higher values. So the constancy of the residual variance of the untransformed data over the whole range of the data (Figure 11.3(c)) is converted to a marked trend in residual variance as in Figure 11.3(d). We can obtain an *approximation* when the variance of the random component, ε, is *small*, but first we consider a general technique for transformations.

11.2.4 The Delta rule for variable transformations

Suppose the transformation from Y to Y_t can be written down algebraically as

---

* For this type of data there are usually two different sources of random variation: (a) measurement error in assessing the amount of bound ligand, which with the standard protocol involves an equal error of opposite sign in the x-variable, the free concentration; (b) random biological variability in the effectiveness of binding from one sample to another (see Chapter 9). Here we consider the case, which is common in physiological experiments, where the major source of random variation is biological and the measurement error is so small that it can be ignored.

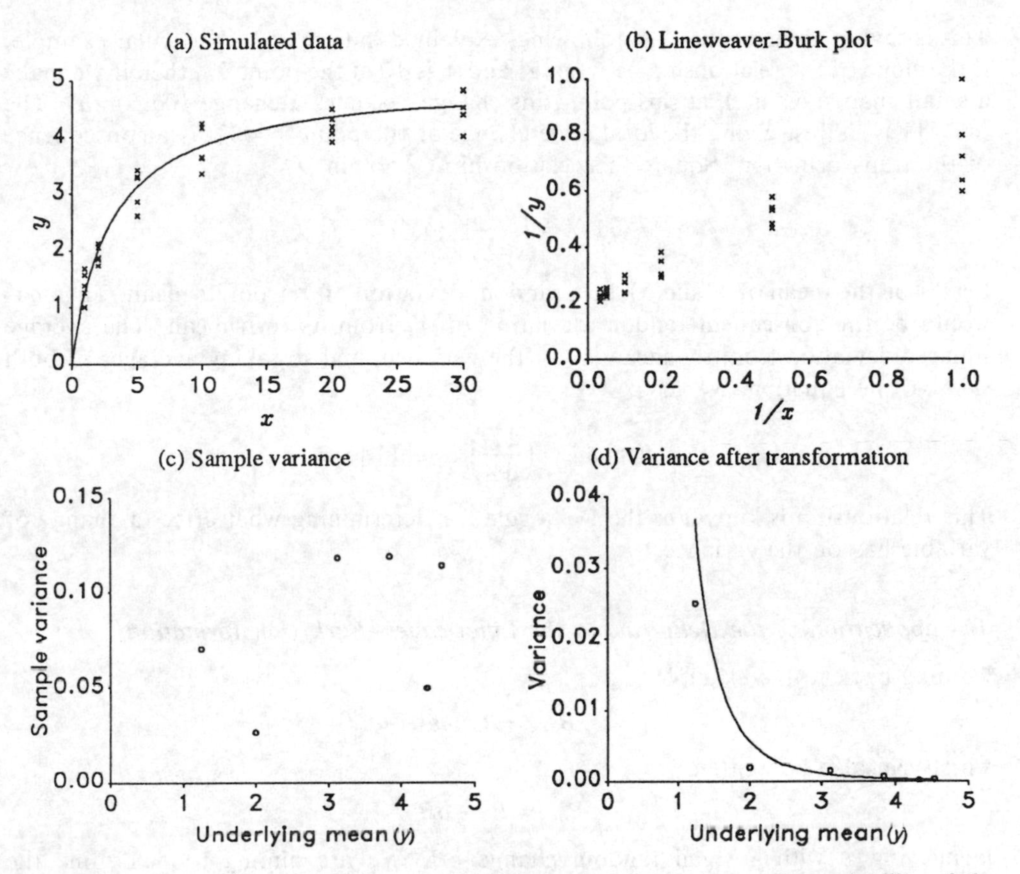

Figure 11.3 Illustration of the effect of the Lineweaver–Burk plot on the variability of the responses about a rectangular hyperbola. (a) Simulated data of five independent responses for each level of x sampled from a Normal distribution with variance equal to 0.09; (b) Lineweaver–Burk plot of simulated data—note the increased scatter of points as $1/x$ increases; (c) *sample* variances of the simulated observations for each level of x against mean response—note how they vary about the true value of 0.09; (d) *sample* variances of the transformed responses together with a curve showing the theoretical approximation derived from the Delta rule.

$$Y_t = f(Y), \qquad \text{e.g.} \qquad Y_t = 1/Y, \qquad Y_t = \log_e Y$$

then

$$\delta Y_t = \left(\frac{dY_t}{dY}\right)_{Y_0} \delta Y$$

$$\text{i.e.} \quad \begin{pmatrix}\text{The change} \\ \text{in } Y_t \\ \text{resulting from} \\ \text{a small change} \\ \delta Y \text{ in } Y\end{pmatrix} = \begin{pmatrix}\text{slope of} \\ Y_t \text{ as a} \\ \text{function of } Y \\ \text{evaluated at } Y_0\end{pmatrix} \times \begin{pmatrix}\text{the small change} \\ \delta Y \text{ in } Y\end{pmatrix}.$$

This is saying no more than the following, explained in terms of a particular example: if the slope of the relationship between Y_t and Y is 10 at the point Y_0, then if we make a small change δY in Y at this point, this change becomes a change $10\delta Y$ in Y_t. The $(\mathrm{d}Y_t/\mathrm{d}Y)_{Y_0}$ is just giving the local scale change at the point $Y = Y_0$ as a consequence of the transformation. Square this relationship to obtain

$$(\delta Y_t)^2 = \left(\frac{\mathrm{d}Y_t}{\mathrm{d}Y}\right)^2_{Y_0} (\delta Y)^2.$$

Let Y_0 be the mean of Y and δY be a random deviation of Y from its mean. Then δY_t would be the consequent random deviation of Y_t from its own mean. The average squared deviation from the mean is just the variance, and so taking averages of both sides of the equation, we get

$$\mathrm{var}\,[Y_t] = \left(\frac{\mathrm{d}Y_t}{\mathrm{d}Y}\right)^2_{Y_0} \mathrm{var}\,[Y].$$

This relationship is known as the Delta rule for determining what effect a change of variable has on the variance.

The application of the Delta rule to the Lineweaver–Burk transformation

Coming back to the example,

$$B_{\mathrm{obs}} = B_u + \varepsilon$$

which can also be written

$$B_{\mathrm{obs}} = B_u + \delta B$$

identifying ε with a small random change δB. We are aiming to determine the underlying and random components of a similar relationship in the transformed random variable $B_t = 1/B$:

$$B_{t,\mathrm{obs}} = B_{t,u} + \delta B_t.$$

The underlying component transforms straight across $B_{t,u} \approx 1/B_u$ provided ε is small. The random component in the transformed scale is given by the delta rule as

$$\delta B_t = \frac{\mathrm{d}}{\mathrm{d}B}\left(\frac{1}{B}\right)_{B=B_u} \delta B$$

$$= \left(-\frac{1}{B_u^2}\right)\delta B$$

$$\mathrm{var}\,[B_{t,\mathrm{obs}}] \approx \frac{\mathrm{var}\,[B_{\mathrm{obs}}]}{B_u^4} = \frac{\sigma^2}{B_u^4}.$$

Figure 11.3(b) and (d) illustrate this result using simulated data. This is only true when $\mathrm{var}\,[\varepsilon] = \sigma^2$ is small. This analysis shows that when the variance of the responses about the rectangular hyperbola is constant then linearisation leads to a non-constant variance about the resulting straight line. We need to make a special allowance for this when fitting the straight line (using the method of *weighted least squares* as described in the next section).

11.3 WEIGHTED LEAST SQUARES

11.3.1 Fitting a straight line by weighted least squares

Ordinary least squares involves finding the line which minimises the sum of the squared deviations of the observations about the fitted line (see box below). The fitted line passes through the centroid of the data and is rotated about the centroid so as to minimise the sum of the squared deviations. For any given sample, the contribution of each observation to the sum of squared deviations will in general be different. The important point, however, is that the squared deviations are given *equal* weight in calculating the sum of squares—in the expression for S on the left of the box below, the sum is formed of terms like $(y_i - \hat{\alpha} - \hat{\beta}x_i)^2$ with no w_i multiplier which is found in the equivalent expression on the right. This approach is reasonable when the variance of the observations about the line is constant, in which case there is no compelling reason to give more weight to one squared deviation than to another.

Now consider the case of non-constant variance in which the variation about the line in some observations is greater than in others (for example, see Figure 11.3(b)). Observations with larger variance are less reliable for estimating the position of the line since they are subject to larger random fluctuations about the line. Fitting the line by *weighted least squares* allows for this effect. The principle of fitting a straight line by weighted least squares is to minimise a *weighted* sum of squared deviations of the observations about the fitted line in which the weight of each squared deviation is inversely proportional to the variance of the corresponding observation (see box). Thus the more variable, and in a sense less reliable, points are given less weight in the fitting. In practice the difficulty of applying the principle of weighted least squares is that the variance of the observations is generally unknown. This problem is discussed in Section 11.3.2.

Comparison of ordinary and weighted least squares for fitting a straight line

The sample of n data is given by $(x_1, y_1), \ldots, (x_n, y_n)$.

[*Ordinary least squares*]
Minimise the sum of squares

$$S = \sum_{i=1}^{n} (y_i - \hat{\alpha} - \hat{\beta}x_i)^2.$$

Estimated slope and intercept

$$\hat{\beta} = \frac{\sum_{i=1}^{n} y_i(x_i - \bar{x})}{\sum_{i=1}^{n} (x_i - \bar{x})^2}; \qquad \hat{\alpha} = \bar{y} - \hat{\beta}\bar{x}$$

where $\bar{x}$ and $\bar{y}$ are the averages of x and y respectively.

[*Weighted least squares*]
Minimise the *weighted* sum of squares

$$S_w = \sum_{i=1}^{n} w_i(y_i - \alpha - \beta x_i)^2$$

with *weights* inversely proportional to the variance of each observation, i.e.

$$w_i = \frac{1}{\text{var}[Y_i]}.$$

Estimated slope and intercept

$$\hat{\beta} = \frac{\sum_{i=1}^{n} w_i y_i(x_i - \bar{x}_w)}{\sum_{i=1}^{n} w_i(x_i - \bar{x}_w)^2}; \qquad \hat{\alpha} = \bar{y}_w - \hat{\beta}\bar{x}_w$$

where $\bar{x}_w = \Sigma w_i x_i / \Sigma w_i$ and $\bar{y}_w = \Sigma w_i y_i / \Sigma w_i$ are *weighted* averages of x and y respectively.

11.3.2 Appropriate choice of weights in weighted least squares

In weighted least squares, we normally weight the squared deviations about the fitted line using weights which are inversely proportional to the variance of the corresponding observation, if we can determine what these variances are. Other weights can be used, but these are not optimal. An illustration of the appropriateness of weighting inversely proportional to variance is given in Section 11.7.1. The example is intended to illustrate the choice of weights in weighted least squares by using a situation in which the weights can be calculated from the known relationship between the variances of the observations. In practice, however, the form of the variance and hence the weights are not usually known, so how do we then apply weighted least squares?

Estimating a variance function (or modelling the variance as a function of x)

In some situations there may be a pattern or structure in the variability of the observations—for example, there might be a smooth relationship between var $[\varepsilon]$ and the explanatory variable, x, or between var $[\varepsilon]$ and the mean of Y, for a given x. In these cases it may be possible to postulate a variance function to describe the form of the variation and then from this to determine appropriate weights. In Section 11.2.4 we illustrated how a variance function could be obtained for a Lineweaver–Burk plot. In other cases the form of the data may suggest a plausible working model for the variance function—for example, in counts which follow a Poisson distribution the variance is equal to the mean. In the next three sections, we develop this approach and show how it can be applied to a wide range of situations.

11.3.3 Iterative weighted least squares

In Section 11.3.1, we outlined the principle of fitting a straight line by weighted least squares, which involved minimising a weighted sum of the squared deviations of the observations about the line, using weights which are *inversely* proportional to the variance of the responses. We have already shown how we can linearise the rectangular hyperbola using the Lineweaver–Burk transformation (taking reciprocals). We showed that when the variance of the response (B) about the hyperbola is constant then the variance of the reciprocal ($1/B$) about the resulting straight line decreases approximately as $1/B_u^4$, where B_u is the *underlying* response. In this case the appropriate weights for fitting the straight line by weighted least squares are given by

$$w = B_u^4 = \left(\frac{nF}{K_D + F}\right)^4.$$

An immediate problem is that we do not know the weights since they depend on the unknown parameters n and K_D. The way around this difficulty is to use the following iterative scheme.

Iterative weighted least squares

Step 1. Fit the line using equal weights $w = 1$, to obtain initial estimates of the slope and intercept and hence initial estimates of n and K_D.
Step 2. Use the current estimates of n and K_D to calculate a new set of weights and then fit the line using the new weights to obtain new estimates.
Step 3. Repeat step 2 as many times as necessary.

Usually the estimates eventually converge to values which do not change from one iteration to the next (to within a prespecified tolerance), and these are the appropriate estimates.

Table 11.3 shows an application of the method to some ligand–receptor assay data (Figure 11.4). After five iterations the parameter estimates have settled down to the required values.* Iterative weighted least squares forms the basis of a method for fitting a class of models referred to as *generalised linear models*. These models are discussed in detail in Section 11.4.

11.3.4 Iterative weighted least squares for fitting models with a variance–mean relationship: the logistic-binomial model

In the previous section, weighting was used to allow for a non-constant variance about the line caused by transforming the original responses which were assumed to have constant variance. In some models for describing random variation, the variance of the

Table 11.3 Fitting the simple ligand–receptor model to the data in Example 9.3 by applying iterative weighted least squares to the Lineweaver–Burk plot. After five iterations the estimates have converged to the required values

Transformed data		Initial weights	Weights at each iteration number ($\times 10^8$)				
$x = 1/F$	$y = 1/B$		1	2	3	4	5
188.68	102.04	1	0.7	0.5	0.5	0.5	0.5
185.19	113.64	1	0.8	0.6	0.6	0.6	0.6
53.48	45.87	1	22.4	19.8	19.4	19.3	19.3
51.55	47.55	1	24.2	21.5	21.1	21.0	21.0
14.25	27.47	1	169.3	178.2	177.7	177.6	177.6
12.17	27.47	1	195.1	208.8	208.5	208.5	208.5
3.60	22.78	1	372.0	432.6	435.9	436.2	436.2
2.88	20.16	1	394.6	462.7	466.7	467.1	467.1
Estimated slope		0.4653	0.5111	0.5170	0.5177	0.5178	0.5178
Estimated intercept		21.095	20.087	20.025	20.018	20.017	20.017
Estimated n		0.0474	0.0498	0.0499	0.0500	0.0500	0.0500
Estimated K_D		0.0221	0.0254	0.0258	0.0259	0.0259	0.0259

* A conspicuous feature of the analysis is the very large range (approximately 500-fold) in the values of the weights, with observations for larger values of bound ligand receiving most weight. This property of the method is unsatisfactory in some situations since the fitted model could be rather sensitive to particular observations, for example a 'rogue' data point coming from a contaminated sample.

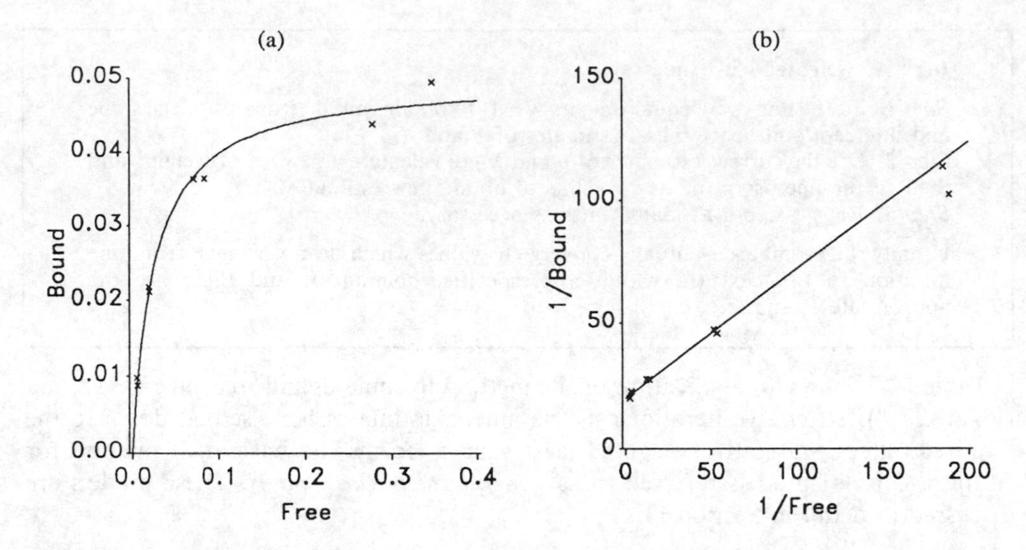

Figure 11.4 (a) Ligand–receptor data with fitted model; (b) Lineweaver–Burk plot.

original response is not constant but is related to the mean. Examples include the Poisson distribution for count data for which the variance is equal to the mean; and the binomial distribution for proportions derived from binary responses in which the variance of the *observed* proportion, P, is given in terms of the *underlying* proportion or probability of a response, p, by $\text{var}[P] = p(1-p)/n$. A particular model we have already discussed and shown to be very useful is the logistic-binomial model in which

$$p = \frac{\exp(\alpha + \beta x)}{1 + \exp(\alpha + \beta x)}$$

and where the distribution of the observed proportion is obtained from a binomial distribution with underlying proportion p.

For such situations where there is a clearly defined variance–mean relationship in the original, *untransformed* responses, how can the method of iterative weighted least squares as applied to a linearised response/explanatory variable relationship be used to fit the model? We now have two features to contend with. We first of all have to linearise the model, and work out how the linearisation has affected the variance to mean relationship, as we did for the hyperbola above. Secondly, we also have a variance to mean relationship in the original, untransformed response variable, which seems to complicate matters.

However, it causes no new problems, and can be handled by methods similar to the rectangular hyperbola. The derivation of the variance of the linearised response variable is slightly more complex, but apart from that the method is the same. The model can be fitted by iterative weighted least squares using the inverse of the variance of the linearised response variable. Nowadays, the logistic-binomial model is usually fitted as a special case of a wider class of models called generalised linear models, using widely available software. The algorithm employed is a form of iterative weighted least squares similar to the method described above. The more refined algorithm uses estimates of the transformed responses which cope with proportions of zero and one for which the logit transformation of the observed proportions can not be applied (Section 11.4.2). Example 11.1 shows the results of an application of the method to some insulin bioassay data using the Genstat package.

11.4 GENERALISED LINEAR MODELS

The logistic-binomial model, the straight line-Normal model and the rectangular hyperbola-Normal model all belong to the wider class of so-called *generalised linear models*. In this section we briefly describe the key components of the models. For a more detailed discussion of this important class of models see McCullagh & Nelder (1989).

Recall that a *linear model* is defined for a statistician as one for which the systematic part of the model is a linear combination of the model *parameters*. Thus in the straight line-Normal model, the value of a response variable (Y) is related to the corresponding value of an explanatory variable (x) by

$$Y = \alpha + \beta x + \varepsilon$$

where ε is a random variable which follows a Normal distribution with mean zero and constant variance. The straight line-Normal model is *linear* because the systematic part of the model is a linear function of the model *parameters*, i.e. $L = \alpha + \beta x$. Another linear model is the quadratic-Normal model given by

$$Y = \alpha + \beta_1 x + \beta_2 x^2 + \varepsilon$$

where the underlying relationship with x is curvilinear but the model is specified as linear because the parameters α, β_1 and β_2 occur as a linear combination.

11.4.1 Components of generalised linear models

Generalised linear models are extensions of linear models in which the underlying systematic component, μ, is a specified function, f, of a linear combination of the parameters, L, also referred to as a *linear predictor*

$$\mu = f(L), \text{ so } Y = \mu + \varepsilon = f(L) + \varepsilon.$$

The random effect ε may follow one of a specified number of distributions, called error distributions. So we obtain the following definitions of the components $f(L)$ and error distribution for two important examples, the straight line-Normal model and the logistic-binomial, given above the middle line in Table 11.4.

Link function

The *link function* of a generalised linear model transforms the underlying part of the model into a linear function of the parameters. If the mean response, or expected value of Y, is given by $\mu = f(L)$ then the link function is defined to be the function g such that $g(\mu) = L$, i.e. the inverse function of f. In the straight line-Normal model the mean response *is* a linear function of the parameters, i.e. $\mu = L$, and therefore the link

Table 11.4 Components of a generalised linear model for two models $Y = f(L) + \varepsilon$

	Straight line-Normal	Logistic-binomial
$f(L)$	L	$\frac{e^L}{1 + e^L}$
Distribution of ε	Normal	Binomial
Link function	$g(\mu) = \mu$	$g(\mu) = \log_e[\mu/(1-\mu)]$
Error distribution	Normal	Binomial
Variance function	σ^2, constant	$\mu(1-\mu)/n$

function is simply the identity function $g(\mu) = \mu$. For the logistic-binomial the link function is the logit transformation (see Section 4.3.5) given by

$$g(\mu) = \log_e\left(\frac{\mu}{1-\mu}\right) = \alpha + \beta x = L.$$

Error distribution and variance function

The *error distribution* of the generalised linear model specifies the random component of the model. The terminology *error* is rather misleading since the random component need not be plain error; it can be random variation from any source. Some error distributions, or distributions of the random component, induce a specific relationship between the variance of the response and the mean response which is characteristic of that distribution. This function is called a *variance function*. For the logistic-binomial model the error distribution is the binomial, and the variance function is therefore $V(p) = p(1-p)/n$ or in our present terminology $V(\mu) = \mu(1-\mu)/n$, where n is the number of independent responses from which the proportion is derived.*

11.4.2 Fitting generalised linear models

Generalised linear models are fitted using an iterative weighted least squares algorithm. The algorithm is similar to, but not exactly the same as, the empirical approach of applying iterative weighted least squares to the linearised responses which we used in fitting the rectangular hyperbola (Section 11.3.3). Using a statistical package such as Genstat or GLIM all that is required is to specify the appropriate link function and error distribution. Once this is done, the fitting is a little more complicated from the point of view of the user than fitting an ordinary linear model. Example 11.1 shows an application using Genstat to fit the logistic-binomial to some insulin bioassay data.

---

* Here the $V(\mu)$ is specifying a variance function; i.e. that the variance is a function of μ. $V(\mu)$ does not signify 'variance of the random variable μ', since μ is not a random variable, but a parameter of the model. We follow the convention in this book of writing the variance of the random variable X as var$[X]$, the square brackets here denoting an operator on the random variable rather than a function of the random variable.

Example 11.1 Logistic-binomial model fitted to insulin bioassay data—using Genstat to fit a generalised linear model

Genstat code for fitting the logistic-binomial model:

```
MODEL [LINK = logit; DISTRIBUTION = binomial] Y = r; NBINOMIAL = n
FIT x
```

where r, n and x are variates storing the number of individuals responding (r) out of a total number (n) for a given value of the explanatory variable (x).

Genstat output for the insulin bioassay data (see Chapter 9)

```
***** Regression Analysis ******
   Response variate:   r
     Binomial totals   n
        Distribution   Binomial
       Link Function   Logit
        Fitted terms   Constant, x
*** Summary of Analysis ***
   Dispersion parameter is 1

                                     mean
              d.f.    deviance   deviance
Regression       1     70.6464    70.6464
Residual         6      0.3656    0.06093
Total            7     71.0120   10.14457
*** Estimates of regression coefficients ***
              estimate     s.e.        t
Constant        -5.271    0.736    -7.16
x                2.197    0.298     7.38
```

Discussion of the results of Example 11.1

Model specification

The response is R, the number of animals out of n at each dose responding to the treatment. We use capital R to emphasise here that we are referring to the random variable, rather than the data, or observed values. The distribution of R at a specified dose is binomial, total n, and proportion, p, given by the logistic function

$$p = \frac{e^{\alpha + \beta x}}{1 + e^{\alpha + \beta x}}.$$

Thus the error distribution is specified as binomial, and the link function as described above is the logit function.

Fitted model

The log odds form of the fitted logistic model is given by

$$\log_e \frac{p}{1 - p} = -5.271 + 2.197x,$$

where x is the log dose, with corresponding logistic curve for the probability of a reaction

$$p = \frac{e^{-5.271 + 2.197x}}{1 + e^{-5.271 + 2.197x}}.$$

Uncertainty in parameter estimates and adequacy of fit
The estimated slope against the log dose (2.197) is high compared with its standard error (0.298), as shown by the *t*-value ($7.38 = 2.197/0.298$). So there is strong evidence for a real effect of log dose. It is almost always a good idea to plot out the fitted model and the observed data, especially as this can be done automatically in many packages. Figure 11.5 illustrates this and shows that the logistic response curve provides a good fit.

Analysis of deviance table
An analysis of deviance table is given in the lines headed 'd.f. deviance mean deviance'. This is a generalisation of the analysis of variance we have considered in Chapters 6 and 7. The analysis of variance is applicable to models with Normally distributed random components, whereas the analysis of deviance is a generalisation applicable to generalised linear models with other random component distributions.

One line in the table gives the residual deviance (residual deviance = 0.3656 on 6 degrees of freedom). The residual deviance of a generalised linear model is a measure of the discrepancy between the observations and the fitted values. Residual deviance is analogous to the residual sum of squares

$$\text{Residual s.s.} = \Sigma\,(y_{\text{observed}} - y_{\text{fitted}})^2$$

which is used for models in which the distribution of the random component, ε, of the response is Normal and its variance is constant, but residual deviance is calculated in a different way in the general (i.e. non-Normal) case. The reason why residual s.s. is not an appropriate measure of discrepancy for generalised linear models is that var $[\varepsilon]$ is not always constant but usually depends on the underlying mean value. Residual deviance is similar to a weighted sum of squares of deviations with weights determined by var $[\varepsilon]$. The precise calculation of the deviance uses the statistical concept of *likelihood* (Section 11.7.2) but we do not here go into any technical detail. In this case, the residual deviance is quite small, and hence we might infer that the model is quite a good fit.

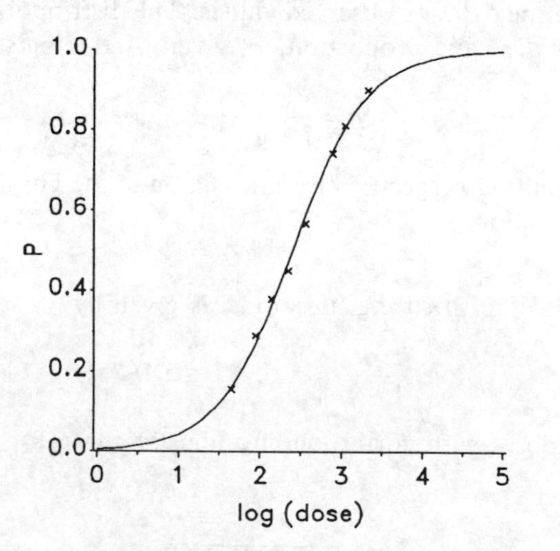

Figure 11.5 The logistic response curve fitted to the insulin bioassay data.

The table also shows how the total deviance in the data is split up into a portion due to the fitted deterministic component of the model (termed 'regression') and a portion representing the residual random component. The total deviance is the deviance after fitting just a constant for the deterministic part of the model, i.e. with no slope against x. This is quite large (71.012 on 7 degrees of freedom) indicating that some further term is necessary in the model. Most of the deviance is accounted for by the regression; in other words, by the fitted deterministic component of the model.

Tests of the relationship using the analysis of deviance and chi-square or F-distributions

One other result about deviances is that the null hypothesis—that the coefficient of the explanatory variate, x, is zero—can be tested by comparing the regression deviance with the percentage points of a chi-squared distribution (except in the case of the Normal error distribution when the ratio of the regression mean square to the residual mean square should be compared with the percentage points of the F-distribution). For example, in this case, the line in the table labelled 'regression', which gives the change in deviance due to including $\beta \times \log(\text{dose})$ in the model as 70.65 on 1 degree of freedom should be compared with the 5, 1 and 0.1% points of chi-square on 1 degree of freedom which are 3.84, 6.64 and 10.83 respectively. 70.65 is higher than all of these tabulated percentage points, and so there is strong evidence for a real relationship with $x = \log(\text{dose})$, as we would expect in a well designed bioassay.

The analysis of deviance table in cases of more complex models can be used to assess the contribution to the goodness of fit of a number of different terms in the model. We illustrate such use of analysis of deviance in Chapter 14 for analysing descriptive models of complex relationships in which there are a potentially large number of explanatory variables.

Iterative fitting procedure

You will also notice that the details of the iterations are usually hidden from the user. Usually a few iterations are sufficient for generalised linear models. This is in contrast with fitting non-linear models by non-linear least squares. Non-linear least squares involves a different kind of iterative procedure, discussed in the next section: a search through the parameter space to find the values of the parameters which correspond to the least squares solution. For a moderate number of parameters these searches can take very many iterations, and convergence is generally much more difficult to achieve than in the iterative weighted least squares procedure for generalised linear models. Thus, if it is possible to fit the model as a generalised linear model it is usually more efficient computationally.

11.5 FITTING NON-LINEAR MODELS USING LEAST SQUARES

So far in this chapter we have concentrated on fitting linear models or models which can be linearised by transformation of the response variable. In this section we take up again methods for the fitting of non-linear models introduced in Chapter 5. We recall that, in the context of fitting models, a non-linear model means *a model which is not linear in its parameters*. An example is the asymptotic exponential, or monomolecular, curve given by

$$y = A(1 - e^{-kx})$$

which is used for describing the kinetics of a first order chemical reaction (Chapter 9) or the growth of an organism (Chapter 1). In this section we apply the method of least squares—which when applied to a non-linear model is sometimes referred to as non-linear least squares—to other non-linear models and discuss more general aspects of the method. Our treatment is brief and illustrative, with little technical detail. For a more comprehensive and detailed account, see Ross (1990).

11.5.1 Fitting the rectangular hyperbola by non-linear least squares

We begin with a relatively straightforward application of non-linear least squares to fit a rectangular hyperbola to some ligand–receptor assay data (from Chapter 9). This is a non-linear model although it can be linearised and fitted as a generalised linear model. Remarkably the two approaches give the same results as we shall see. Our aim here, however, is to illustrate the more general method of non-linear least squares, the main use of which is fitting models which cannot be linearised.

The model is

$$B_i = \frac{nF_i}{K_D + F_i} + \varepsilon_i, \qquad i = 1, \ldots, m,$$

where each data point is identified by the subscript i. We assume that the random variation has zero mean and constant variance and therefore fit the model by unweighted or ordinary least squares. This involves finding the values of the parameters which minimise the sum of squares of deviations

$$S = \sum_{i=1}^{m} \left(B_i - \frac{\hat{n}F_i}{\hat{K}_D + F_i} \right)^2.$$

We use an iterative search technique in the parameter space to locate the minimum of the least squares surface because finding the minimum of S by partial differentiation, as we did previously in Chapter 5, does not result in least squares equations for the parameter estimates which can be solved explicitly. Initial estimates to use as starting values for the search can be guessed from a graph of the data or obtained by using various *ad hoc* techniques. Figure 11.6 shows a contour diagram of the least squares surface as a function of the two parameters in the neighbourhood of the minimum. This is an elongated dished-shaped surface with a minimum located in the bottom of the dish.

Example 11.2 gives the output from fitting the model using the statistical package Genstat. The output is typical of other packages and includes: (a) an analysis of variance table which apportions the total variation into a regression sum of squares due to fitting the model and a residual sum of squares formed from the sum of the squared deviations from the fitted model—this partitions the variance rather than the deviance as the distribution of ε is Normal, and its variance is constant; (b) parameter estimates with *approximate* standard errors; (c) estimated correlations between the parameter estimates. These aspects are discussed further in the next section. Note that the

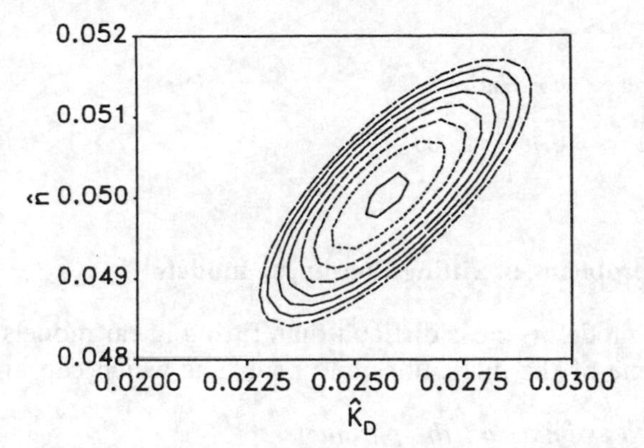

Figure 11.6 Contour plot of the least squares surface for the ligand binding assay data. Contour levels increasing in steps of 0.0000005 from the innermost level of 0.0000185. Minimum sum of squares of 0.0000181 occurs at parameter estimates $\hat{K}_D = 0.0259$, $\hat{n} = 0.050$.

parameter estimates obtained by non-linear least squares are very similar to those obtained by applying iterative weighted least squares to the Lineweaver–Burk transformed variables (Section 11.3.3). In fact, fitting the rectangular hyperbola by non-linear least squares is equivalent to fitting a generalised linear model with reciprocal link function and Normal error distribution.

Example 11.2 Fitting the simple ligand-receptor model by non-linear least squares using Genstat

Genstat code for fitting the model

```
EXPRESSION mm; VALUE = !E(fitted = n * Free/ (Kd + Free))
MODEL [DISTRIBUTION = normal] Bound; FITTEDVALUES = fitted
RCYCLE n, Kd; INITIAL = 0.05, 0.026
FITNONLINEAR [PRINT = model, summary, estimates, correlation; CALCULATION = mm]
```

where Bound and Free are variates storing the measurements of bound and free ligand.

Genstat output

```
***** Nonlinear Regression Analysis ******

     Response variate:    Bound

*** Summary of Analysis ***

                 d.f.        s.s.         m.s.        v.r.
Regression         2    0.00810594    0.405E-02    1350.40
Residual           6    0.00001801    0.300E-05
Total              7    0.00812395    0.102E-02

Percentage variance accounted for 98.7

*** Estimates of parameters ***

        estimate       s.e.
n       0.05008     0.00148
Kd      0.02590     0.00279
```

*** Correlations ***

estimate	ref	correlations	
n	1	1.00	
Kd	2	0.757	1.000
		1	2

11.5.2 General problems of fitting non-linear models

Fitting non-linear models is more difficult than fitting linear models. In this section we briefly discuss some of the difficulties and problems which can arise in practice.

Inadequate data to estimate all the parameters

A crucial consideration when fitting a non-linear model is whether the data are sufficient to estimate all the parameters in the model. If this is not the case then formal fitting procedures may fail or give misleading results. For example, the rectangular hyperbola

$$y = \frac{\alpha x}{\beta + x}$$

has two broad phases: (a) an initial increase with slope α/β at $x = 0$; (b) an asymptotic phase where y increases rather slowly towards the asymptote α. To fit the model satisfactorily the data must cover both phases of growth. For example, data which are restricted to a narrow range in the early increase phase can be used to estimate the initial slope α/β, but will contain little information on the value of the asymptote. Similarly, observations confined to the large values of x estimate the asymptote but contain less information on the initial slope. With a restricted range of observations fitting by non-linear least squares could result in the following problems.

(1) The iterative search technique does not converge.
(2) Estimates of α or β or both have large standard errors.
(3) A high correlation occurs between the parameter estimates. This could arise, for example, if only the single quantity α/β can be estimated from data over a limited range (as discussed above) so the data are consistent with many possible combinations of values of α and β. If $\hat{\alpha}$ by chance happens to take a high level, then $\hat{\beta}$ will need to increase so that their ratio is consistent with the tightly estimated ratio, α/β.
(4) The scope for checking the form of the assumed model is reduced since very different models may look quite similar over restricted ranges.
(5) The estimates may be highly sensitive to the choice of model. For example, asymptotes estimated by fitting the monomolecular curve and the rectangular hyperbola may be substantially different when the data just cover the early initial growth phase.

The problem of having too many parameters and too few data to estimate them is best considered at the *design stage* of the investigation. Data should be collected so as to span as far as possible the essential features of the response, if these are known or can be guessed in advance.

Choice of starting values for parameter estimates

Iterative search techniques for fitting non-linear models need initial values of the parameters to start the search. Such starting values can often be guessed from a plot of the data, or might be known from previous results or obtained by other means. Difficulties in ascertaining plausible values may mean that the model cannot be satisfactorily fitted. Some computer packages have facilities for automatic determination of starting values for a range of standard models.

Lack of convergence

In some cases the iterative search technique may fail to converge. Possible causes are as follows.

(1) Poor choice of starting values very far away from the minimum of the least squares surface. This may be remedied by restarting the search at a more appropriate point. Care in choosing good starting values usually results in much better performance of the iterative search routine, particularly with models which have more than one or two parameters and a moderately high level of random variation.
(2) A least squares surface which is irregular or with extensive quite flat areas for which the minimum is difficult to locate. There may, for example, be a ridge on the surface so that the minimum value is consistent with many combinations of the parameter values. A contour diagram of the surface (if in two dimensions, i.e. for two estimated parameters) will show this up. This is likely to indicate that there are insufficient data to estimate all the parameters in the model. In some cases convergence may be achieved by writing the model in an alternative but equivalent form by using different parameters. For example, a rectangular hyperbola expressed in terms of α and β can also be written using α and α/β. Some computer programs fit models by using so-called *stable parameters* (see Ross, 1990) to help facilitate convergence. The results are then back-converted at the end of fitting into the values of the parameters which have a biological interpretation.
(3) A small data set with a large amount of residual variation. Quite often for such data sets the least squares surface is quite irregular. If the residual variation is large, then non-linear models with more than two parameters often require very large samples of data covering areas which are appropriate for fixing all the parameters.

Sampling properties of estimates

Estimated parameters of non-linear models are subject to random errors but their sampling distributions are usually too complicated to derive using analytical methods. There is no exact theory analogous to that for the linear-Normal models (as in Chapter 5). We recall that, for example, in the straight line-Normal model the sampling distribution of the estimated slope of the line fitted by least squares is Normal with mean equal to the true slope (i.e. the estimate is unbiased) and standard error which is a known function of the variance about the line. By using the residual mean square to estimate the variance, we can calculate a 95% confidence interval for the true slope as *estimated slope* $\pm\, t \times$ *standard error*.

For non-linear models parameter estimates are not in general unbiased, standard errors are *approximate*, and sampling distributions are not Normal, sometimes being highly skew. The output of statistical packages for fitting non-linear models provide standard errors for the parameter estimates. These standard errors are based on approximations which hold when there are a large number of observations throughout a range sufficient to fix the parameters. Confidence intervals calculated as *parameter estimate* $\pm$ $t \times$ *approximate standard error* can, for these reasons, therefore be very approximate.

In any particular case, the properties of the sampling distribution of the parameter estimates can be examined by using *computer simulation*. The approach is to pretend that the true model we are interested in making inferences about, and which is of course unknown, is the same as the fitted model; then to investigate by simulation in that approximately correct context the properties of the estimation procedure we are using. We repeatedly generate samples of data from the fitted model, obtain the parameter estimates for each sample of simulated data and build up a histogram of the sampling distribution of the estimate.

11.5.3 Using a logarithmic transformation to stabilise the random variation without linearising the model

Section 11.2.3 shows that transformation usually scales the variance differentially at different points of the curve so that after transformation the variance is not constant. However, it can happen in some cases that a transformation can convert a heterogeneous variance about the untransformed curve to a homogeneous one about the straight line resulting from the transformation. At the beginning of Section 11.2 we showed that an exponential increase or decrease can be linearised by taking logarithms. If, additionally, the standard deviation of the untransformed observations is proportional to the mean, i.e. the coefficient of variation is constant ($=k$, say) and small, the log transformation will also make the variance constant. To see this we apply the Delta Rule (Section 11.2.4) so that

$$\text{var}[\ln Y] \approx \text{var}[Y]\left(\frac{\text{d} \ln Y}{\text{d}Y}\right)^2_{Y=Yu}$$

and

$$\text{var}[\ln Y] \approx \frac{\text{var}[Y]}{(\text{mean}[Y])^2} = k^2.$$

The other situation which can occur is that the log transformation can stabilise the variance of the random component, but not linearise the deterministic part. When this occurs the model can be fitted by first taking the logarithms of the responses, and then fitting by ordinary (i.e. unweighted) least squares to the resulting logged responses. The fit still has to be done by a searching technique as in Section 11.5, but the transformation has taken care of any weighting problems. This fitting technique is called *log least squares*.

Example 11.3 shows an application of log least squares to fitting a sum of two exponentials to data on the decay of BSP in calves (from Chapter 9). The model is fitted by minimizing the sum of squares

$$S = \sum_{i=1}^{n} \{\ln(qa_i) - \ln(\widehat{qa0}) - \ln[\hat{p}\ e^{-\hat{r}_1 t} + (1-\hat{p})\ e^{-\hat{r}_2 t}]\}^2.$$

Sometimes a quantity c is added to the values before taking logarithms. This avoids giving too much weight to values of the responses near to zero which could otherwise lead to large negative values of the logarithm. The results of adding c first in this example are reported in Chapter 9, but in this case the modification results in very similar parameter estimates.

Example 11.3 Fitting a sum of exponentials to the decay of BSP in calves by log least squares using Genstat

Note that the plot of the logarithm of BSP against time clearly shows that the data span a sufficient range to warrant fitting the sum of two exponentials (Chapter 9).

Store the quantity of BSP and time t in variates qa and t.

Genstat code for fitting the model

```
CALC lqa = LOG(qa)
SCAL qa0, r1, r2, p
EXPRESSION sume; VALUE = !E(fitted = LOG(qa0) + LOG(p*EXP(-r1*t)   \
                                  + (1 - p)*EXP(-r2*t)))
MODEL [DISTRIBUTION = normal] lqa; FITTEDVALUES = fitted
RCYCLE qa0, r1, r2, p; INITIAL = 13, 0.3, 0.1, 0.9
FITNONLINEAR [PRINT = model, summary, estimates, correlation;   \
              CALCULATION = sume]
```

Genstat output

```
***** Non-linear Regression Analysis ******

   Response variate:     lqa

***  Summary of Analysis  ***

                d.f.        s.s.          m.s.         v.r.
Regression        4     31.080870     7.7702174    20883.96
Residual         13      0.004837     0.0003721
Total            17     31.085707     1.8285710

Percentage variance accounted for 100.0

***  Estimates of parameters  ***

        estimate       s.e.
qa0      13.222       0.223
r1       0.30588     0.00607
r2       0.07753     0.00235
p        0.88648     0.00641
```

*** Correlations ***

estimate	ref	correlations		
qa0	1	1.000		
r1	2	0.804	1.000	
r2	3	0.469	0.821	1.000
p	4	−0.267	−0.723	−0.964
		1	2	3

Interpretation of the results

The true value of p is clearly neither one nor zero (one and zero are about 18 and 130 standard errors away from the estimate), and therefore the two exponentials are necessary. $\hat{p}$ and $\hat{r}_2$ are highly negatively correlated, suggesting that the data might not be fully adequate to resolve these two parameters. Thus if the fitted model had involved a slightly higher proportion of the first exponential, there would have been a compensating reduction in the estimated parameter $\hat{r}_2$. However, the standard errors of both parameter estimates are quite low, and so this is not a problem. The parameters are in the main estimated quite precisely and the residual standard deviation is quite low. The residual standard deviation equals $\sqrt{0.0003721} = 0.019$ or about 2%.

11.6 FITTING DYNAMIC MODELS

The various procedures for fitting models discussed so far have been illustrated using data from essentially static situations. Typically, we have dealt with models involving a single response variable which is made up of an underlying systematic component plus a random disturbance, i.e.

$$Y_i = f(x_i, \theta) + \varepsilon_i$$

where $f(x_i, \theta)$ is a known function of an explanatory variable (x_i). Fitting the models involves estimating one or more parameters denoted by θ. The basic methods of ordinary least squares and weighted least squares assume that the random disturbances vary independently of each other. Observations on dynamic processes by their nature exhibit a sequential dependence. It is, however, important to distinguish the *deterministic dependence*, described either by differential equations or by difference equations, from the *statistical dependence* which refers to the correlations between the random departures of the observations about some underlying value.

Statistical independence of the observations implies that the random departures vary independently of each other. When the major source of random variation is measurement error the assumption of statistical independence may be reasonable. However, random variation can affect dynamic processes in many different ways so that often the random components of the observations are not statistically independent. For example, a random departure affecting the true value of the response variable at one particular time—and not just the measurement of it—thereby becomes part of the process and so affects subsequent observations, as discussed in Chapter 4. An example is the effect of a bad winter on the course of the dynamics of a population over a number of years following. If a severe winter depressed the population in year 1970 say, then for slowly reproducing species, the population might take some time to recover, so that by 1974 it is still suffering the results of the winter of 1970. For fitting a dynamic model, an important consideration is the relative magnitude of the independent random variation, arising for example from measurement error, and variation which is either part of the underlying process, such as demographic variation, or becomes incorporated into the underlying process and which then leads to correlated random variation.

In this section we deal briefly with methods for fitting dynamic models. We describe two different approaches and illustrate them for single population models. Then we deal briefly with some of the additional problems of fitting models for interacting populations.

11.6.1 Longitudinal data and cross-sectional data

For fitting dynamic models it is useful to distinguish between two different types of data: *longitudinal data*—a series of measurements in time made on the *same* individual, or sector of a population or total population; *cross-sectional data*—a series of measurements in time made on *different* individuals, or samples of individuals, sampled from the same population. These types of data have already been discussed in Chapter 7, to which the reader should refer for further details. The essential difference when fitting models is that longitudinal data are serially dependent on each other, whereas cross-sectional data are statistically independent.

11.6.2 Fitting dynamic models to cross-sectional data

For cross-sectional data the observations are statistically independent so that many of the methods of the earlier sections apply. The basic model is that each observation is made up of an *underlying* value plus a *random independent* disturbance, i.e.

$$Y_{\text{obs}} = Y_u + \varepsilon.$$

The underlying part of the model describes the time course of trends in average values of Y written

$$Y_u = f(t, \theta_1, \theta_2, \ldots, \theta_p)$$

where $\theta_1, \theta_2, \ldots, \theta_p$ denote the p model parameters, and f is a function which is either known, or can be calculated by methods discussed next, involving both t and the p parameters. For simplicity in the initial presentation, we collapse the list of parameters, $\theta_1, \theta_2, \ldots, \theta_p$, to one, θ. Thus in this shorthand, θ can stand for one or for many parameters. The dynamics of the model are often expressed in terms of a differential equation, or a difference equation involving model parameters. For specified parameter values, or for the current estimates of the parameters, $\hat{\theta}$, the equation can be solved, either analytically or numerically to find the underlying response $f(t, \hat{\theta})$ as a function of time.

To fit the model we then use weighted (or unweighted as appropriate) least squares and thus find the values of the parameters which minimise

$$S_w = \sum_{i=1}^{n} w_i \{y_i - f(t_i, \theta)\}^2.$$

The choice of weights will depend on the variation about the underlying part of the model. If its variance is related to the mean response then iterative weighted least squares is used to fit the model.

Example 11.3 shows an application of this approach to fitting the decay in time of BSP in calves. The underlying part of the model is a sum of two exponentials—which is very conveniently in this case the solution which can be written down as an explicit algebraic expression derived from a pair of linear differential equations for describing the flow of BSP. The analysis uses log least squares to allow for a standard deviation of the random component which increases in proportion to the mean. In this application we assume that the major source of random variation is measurement error in BSP and treat observations at different times as independent, i.e. as cross-sectional data.

11.6.3 Fitting dynamic models to longitudinal data

For longitudinal data the observations are in general not statistically independent so that the method used for cross-sectional data of fitting an underlying trend in time assuming independent random departures is often not appropriate. For example, over certain time periods the observed trend may be consistently lower or higher than the postulated underlying value due to the continuing effects of some past random disturbance. Also, if there is an underlying periodic change with time the observations may become grossly out of phase with the fitted values according to the model, as a result of a disturbance which does not of course affect the fitted model.

Stepwise prediction method

One approach to fitting models using longitudinal data is to take advantage of the fact that successive observations are dependent and to use the value of *the immediately preceding observation* to calculate the fitted value at a given time. In other words, we treat the observation at the preceding time in the data as a *known* value of the population response variable and use it as a starting value for calculating the fitted value corresponding to the current observation. A potential problem is that measurement error attached to each observation will obscure the true values of the previous observations but for the moment we ignore this complication and describe the method assuming that there is no measurement error.

To illustrate the method, consider a dynamic stochastic model in which the response at a given time t depends on the value of the response at time $t-1$ plus a random disturbance, i.e.

$$Y_t = f(Y_{t-1}, \theta) + \varepsilon_t$$

where θ denotes, as above, one or more model parameters. We assume that the function f is either known explicitly or can be calculated numerically for known values of the parameters by integrating a differential or difference equation using Y_{t-1} as a starting value. In this model, the *random variation is incorporated into the dynamics* since ε_t affects not only Y_t but also future observations. Thus observations on the model will not be statistically independent.

Suppose now that we have data $y_1, y_2, \ldots, y_n$ to fit the model. The basic approach is then to use the model to calculate fitted values for each value of y_t and then to find the parameter values which minimise a weighted sum of squared deviations of the observations about the fitted values. In other words, fit the model by minimising

$$S_w = \sum_{t=2}^{n} w_t\{y_t - f(y_{t-1}, \theta)\}^2$$

with respect to θ. The choice of weights will be determined by the form of the random variation in y_t given the previous value y_{t-1}. This approach contrasts with the method for cross-sectional data (Section 11.6.2) in which the fitted values are obtained from the model as a function of time.

Example 11.4 Fitting a logistic model of yeast growth

The experiment of Carlson (Example 1.6) on the growth of yeast is an example of a longitudinal study. A measured amount of wort was seeded with a few yeast cells and the amount of yeast in the same colony was measured every hour for 18 hours, and the growth approximately followed a logistic pattern. Thus it would be appropriate to fit a logistic model using the stepwise prediction method. The amount of random variation in that dataset is rather small, and so, to illustrate the method, we simulate further data using the model which Carlson found, but with greater random variation. We use the model

$$\frac{dN}{dt} = 0.54N\left(1 - \frac{N}{665}\right)$$

but impose on the simulated values at each stage multiplicative random variation with a coefficient of variation of 0.10. The random variation becomes part of the process, so that these randomly perturbed simulated values are used as starting values to obtain the simulated value at the next time step, which is then itself randomly perturbed. The data obtained from one such simulation are given in Table 11.5.

Table 11.5 Simulated data from Carlson's fitted model of the logistic growth of yeast

t (h)	N	t (h)	N
0	9.6	10	469
1	15	11	435
2	29	12	504
3	50	13	560
4	80	14	607
5	137	15	662
6	197	16	670
7	271	17	708
8	353	18	777
9	420		

In fitting this non-linear model by the stepwise prediction method, we determine the fitted values of N_i at t_i, from the value N_{i-1} and the current values of the parameters K, r in the iterative process, using the formula

$$N_i = \frac{K}{1 + (K/N_{i-1} - 1)\, e^{-r(t_i - t_{i-1})}}.$$

The same formula but starting at $t_0 = 0$,

$$N_i = \frac{K}{1 + (K/N_0 - 1)\, e^{-r(t_i - t_0)}}$$

could also be used to determine the fitted values, in an iterative fit. As the random component was multiplicative, the models were fitted in each case to $\log(N_i)$. The model fitted in this way, also shown in Figure 11.7, appears to be a poorer fit from about 12 h onwards, demonstrating the superiority of the stepwise prediction method in this case. The parameters fitted by both procedures are as follows. The residual mean square is smaller for the stepwise prediction method, giving a numerical indication of the method's superiority.

	Stepwise prediction method		Simulation from $t = 0$	
Parameter	Estimate	s.e.	Estimate	s.e.
K	662	47.0	629.7	24.6
r	0.5422	0.0368	0.5546	0.0118
Residual m.s.	0.00673	on 16 d.f.	0.01135	on 16 d.f.

Measurement error in longitudinal data

In presenting the stepwise prediction method of fitting models to longitudinal data we ignored the possibility of measurement error in the observations. This allows us to treat

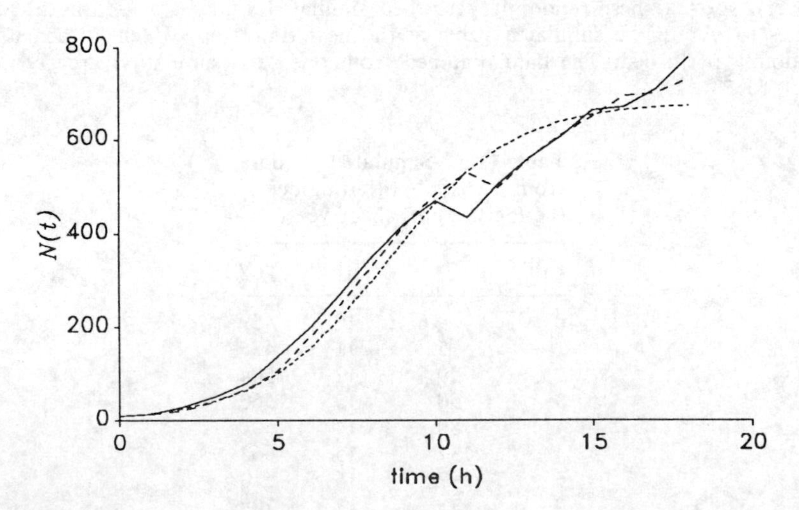

Figure 11.7 Simulated data using model fitted to Carlson's data (solid line), with the model fitted by the stepwise prediction method (dashed line), and fitting using simulation from $t = 0$ (dotted line).

the observation at a particular time as a *known* value of the population response variable and to use it as a starting value for calculating the fitted value corresponding to the next observation. These calculations would be affected by any measurement error because the population responses would then not be known exactly. If the magnitude of the measurement error is small relative to the value of the population response then it might be reasonable to ignore the effect. The problem thus requires a consideration of the *relative* magnitude of the measurement error, and the random variation in the underlying process.

Methods for fitting dynamic models when there is measurement error are beyond the scope of this book. It should be emphasised that unless you have replicate measurements at each time point—which enable you separately to assess the variance of the measurement error assuming that it follows a simple model—there is no possibility of separating out the measurement error and the random component in the underlying process. Even when such information is available, estimation is likely to be difficult in other than the very simplest situations, especially if the variance of the random element in the underlying process is large and the model for the underlying response is non-linear.

Time series models

Methods for fitting dynamic models in which the observations are not statistically independent have been developed under the general heading of *Time Series Analysis.* Chapter 13 presents an introduction to time series models and gives an application to a series of grouse bag records. For further discussion of time series the reader is referred to Chatfield (1984) and Diggle (1990).

11.6.4 Fitting dynamic models for biological interactions

The methods for fitting dynamic models described in the previous sections and illustrated for single population models can also be applied to interacting populations, or other cases of biological interaction. We discuss the special aspects in the context of models of interacting populations but they are equally applicable to other interactions. Some aspects which may be particularly important are as follows.

Weighted least squares with several response variables

Fitting dynamic models of interacting populations involves several response variables. Therefore we need to consider how to combine criteria for fitting models, such as least squares, when there is more than one response variable. The problem is similar to that of deciding how much weight to give the different observations when fitting a single population model. This would be an important consideration if one response variable was an order of magnitude larger than the other. Fitting to the logarithms or square roots of the raw data are two ways to deal with differential scaling and to give relatively less weight to larger observations. In general, however, the choice of weights requires consideration of the random variation in the data. The point is illustrated in Example 11.5 using Gause's yeast data.

Fitting by the stepwise prediction method

This method using the preceding observation in the series as a starting value of simulations for the purpose of calculating fitted values extends to situations when there are several response variables. Thus, the fitted value of a given response depends not only on its own preceding value but also on the preceding values for the other populations.

Fitting to one response only, conditioning on the values of the other responses

This method might also apply even when there is no available model for one of the populations provided that a series of responses is available, matching the responses of the variable we wish to model. In this case we use the responses of population 1 simply as a predictive explanatory variable to calculate fitted values for the other series, without trying to model population 1's dynamics. This might well be a good method in general, as it reduces the calculations which need to be performed and the number of parameters which need to be estimated in one estimation run.

Lack of analytical solutions and the need for numerical methods

Dynamic models are usually formulated in terms of differential equations in continuous time, or difference equations in discrete time. Usually the equations cannot be solved analytically, except in simple cases such as exponential or logistic growth. This is also the case for models of interacting populations which generally require numerical solutions of systems of equations. This is straightforward in principle but often requires extensive computations. Nowadays, there are computer programs and packages for fitting systems of ordinary differential equations,* and this might be the best way for an inexperienced user to start fitting these models.

Many parameters to be estimated and design considerations

Often even a relatively simple system of equations will contain a large number of parameters. For example, the Lotka–Volterra model for two-species competition involves six parameters to describe the dynamics of a system with only two response variables corresponding to the densities of the two species measured at selected times. This raises questions about what variables should be measured so that the model can be fitted satisfactorily. In the model for two-species competition, four of the parameters relate to the dynamics of the growth of each species in single populations. These parameters can be estimated from separate data on growth of single species and this was what Gause did in his experiments. The remaining two parameters of the model relate to the competitive effects of the species on each other. These parameters can in principle be estimated from growth in mixtures assuming that the other four parameters are known.

The important point is that fitting a complex model involves much wider considerations than simply choosing a statistical method. The problem involves more

* For example, the BMDP program AR is a general-purpose program for fitting a non-linear model by least squares in which models can be specified indirectly as a system of ordinary differential equations.

basic questions about the design of the study and what data should be collected in the first place.

Example 11.5 Fitting the two-species competition model to Gause's yeast data

To illustrate some of the aspects of fitting dynamic models discussed in the previous section we consider the Lotka–Volterra model for two-species competition and the application by Gause to the growth in mixture of the two yeasts *Saccharomyces cerevisiae* and *Schizosaccharomyces kephir* (see Chapter 10 for details).

The model involves the pair of differential equations

$$\frac{\mathrm{d}N_1}{\mathrm{d}t} = r_1 N_1 \left\{ 1 - \frac{(N_1 + \alpha_{12} N_2)}{K_1} \right\}; \qquad \frac{\mathrm{d}N_2}{\mathrm{d}t} = r_2 N_2 \left\{ 1 - \frac{(N_2 + \alpha_{21} N_1)}{K_2} \right\}$$

where N_1 and N_2 are the two response variables, the densities of the two species. The model has six parameters: r_1 and r_2 are the intrinsic relative growth rates of the two species; K_1 and K_2 are the corresponding carrying capacities; α_{12} and α_{21} are the competition coefficients—α_{12} measures the competitive effect of species 2 on species 1. Each species grown in the absence of the other as a single population follows a logistic population growth curve.

To examine the model Gause grew the two species of yeasts both in pure culture and mixed culture. Because of the difficulty of taking observations upon a particular culture the experiments involved a large number of test-tubes each containing either a pure or mixed culture inoculated at the start of the experiment. Growth was monitored by taking measurements on *different* tubes at selected times, i.e. the data were cross-sectional. With this approach observations at successive times are independent of each other. For each species, growth in pure culture was used to estimate the intrinsic relative growth rate (r) and the carrying capacity (K) by fitting logistic curves. Growth in mixed cultures was then used to estimate the competition coefficients assuming values r and K from the growth in pure culture. To do this Gause used the model to obtain equations for the competition coefficients given by

$$\alpha_{12} = \frac{K_1}{r_1 N_1 N_2} \left\{ r_1 N_1 \left(1 - \frac{N_1}{K_1} \right) - \frac{\mathrm{d}N_1}{\mathrm{d}t} \right\}; \qquad \alpha_{21} = \frac{K_2}{r_2 N_1 N_2} \left\{ r_2 N_2 \left(1 - \frac{N_2}{K_2} \right) - \frac{\mathrm{d}N_1}{\mathrm{d}t} \right\}.$$

Then the right hand side of each expression was calculated from the data by estimating growth rates at different points along the curve. For α_{21} this gave a mean of 0.439 with a corresponding value for α_{12} of 3.15. These were the values used to illustrate the fit of the model (Chapter 10). Gause's approach to estimating the competition coefficients makes use of a particular property of the model, namely that simple expressions can be found relating parameter values to quantities which can be measured.

A more general approach to estimating the parameters is to apply weighted least squares and find the values of the competition coefficients for which a weighted sum of squares of deviations of the observations from the fitted values is least. This requires: (a) a suitable choice of weights; (b) numerical methods to find the fitted values for different combinations of parameter values. For illustration we have applied the method using two weighting schemes. The first scheme uses weights inversely proportional to the fitted value which is equivalent to assuming that the variance of the response is proportional to the mean. This is plausible since population size was estimated from a cell count which in a well-mixed sample would vary as a Poisson distribution, i.e. with variance equal to the mean. The second scheme uses log least squares, i.e. ordinary unweighted least squares applied to the log transformed data and fitted values. This is almost equivalent to using weights inversely proportional to the square of the mean response and is appropriate when the coefficient of variation of the response is constant.

Table 11.6 Sum of squared deviations of observations about the fitted values for the Lotka–Volterra two-species competition model fitted to Gause's yeast data by weighted least squares over a range of values for the competition coefficient

Method 1. Weighted least squares with weights inversely proportional to the mean response

α_{21}	α_{12}						
	2.0	2.5	3.0	3.5	4.0	4.5	5.0
0.35	9.01	6.93	6.03	7.01	10.99	19.82	36.68
0.40	9.84	7.37	5.83	5.76	8.06	14.70	26.74
0.45	11.13	8.32	6.24[G]	5.31*	6.24	10.21	19.26
0.50	12.93	9.79	7.23	5.58	5.40	7.62	13.84
0.55	15.29	11.81	8.81	6.53	5.42	6.20	10.15
0.60	18.31	14.46	10.99	8.13	6.21	5.81	7.90

Method 2. Log least squares

α_{21}	α_{12}						
	2.0	2.5	3.0	3.5	4.0	4.5	5.0
0.35	4.91	4.48	4.23	4.26	4.67	5.66	7.49
0.40	5.07	4.50	4.09	3.92	4.09	4.75	6.14
0.45	5.53	4.80	4.21[G]	3.83	3.75	4.10	5.08
0.50	6.31	5.39	4.61	4.01	3.68	3.73	4.32
0.55	7.44	6.32	5.32	4.49	3.90	3.64*	3.85
0.60	8.93	7.58	6.36	5.28	4.42	3.85	3.71

[G]—cell close to the values used by Gause; *—cell with smallest sum of squares.

For both schemes the weighted sum of the squared deviations of the observations from the fitted values was calculated over a grid of values for the competition coefficients. Fitted values were obtained by using the Runge–Kutta numerical method to solve the system of equations with starting values of 0.45 units as used in the experiment. Table 11.6 gives the results. Using weights inversely proportional to the fitted values leads to estimates of the competition coefficients close to 3.5 and 0.45, very similar to those obtained by Gause. For log least squares the minimum of the least squares surface occurs with estimates of about 4.5 and 0.55. The surface is, however, rather flat in this region indicating relatively large uncertainty in the estimates. The analysis illustrates the general point that estimates will depend on the chosen method of fitting the model.

11.7 APPENDICES

11.7.1 The choice of weights in weighted least squares

In weighted least squares, we normally weight the squared deviations about the fitted line using weights which are inversely proportional to the variance of the corresponding observation, if we can determine what these variances are. Other weights can be used, but these are not optimal in a sense which we now explain. To motivate this choice of weights we give a numerical example involving two sets of simulated data from the straight line-Normal model. Both sets involve the same values of the explanatory variable x but in one set there is equal replication with three values of y for each value of x while in the other there is unequal replication ranging from 1 to 6 values depending on x (Table 11.7 and Figure 11.8).

Let us consider two scientists one of whom has the full sets of data in these two cases, and the other who only has the means—the raw data having been lost. Suppose also that they know from previous experience that the variance about the line of the *raw* data is constant for all values of x. This would not be true for the means in the second example as the replication is different for some values of x. How should they each fit the line?

Scientist 1, working with the raw data, fits the lines by ordinary least squares, i.e. giving all points equal weight, since the variance of each observation is the same.

Scientist 2, working with the means at each value of x, tries two strategies: (a) fitting the lines to the means giving equal weight to each mean, i.e. fitting to the means using ordinary least squares; (b) fitting the lines to the means, weighting each mean by the number of observations of which it is composed.

The results both scientists obtained are given in Table 11.8.

Scientist 1 has adopted a reasonable strategy as there is no reason to give any point a different weight to any other, and his fitted lines are correctly estimated. Approach (b) of scientist 2 produces the same results in both cases as scientist 1, and therefore approach (b) is the right one. It does not matter for Example 1, as all the means were based on equal levels of replication; but for Example 2, giving equal weight produces incorrect results. Strategy (b) of scientist 2 is to use weights equal to the number of observations which went into each mean.

Table 11.7 Two sets of simulated data from the straight line-Normal model used to illustrate the principle of weighted least squares in which observations are weighted inversely in proportion to their variance. See text and Figure 11.8 for details

Equal replication at each x-value				Unequal replication at each x-value			
x	n	Raw data (y)	Mean (y)	x	n	Raw data (y)	Mean (y)
0	3	9.9, 10.5, 9.9	10.10	0	1	10.0	10.00
0.125	3	10.3, 10.1, 10.7	10.37	0.125	1	9.8	9.80
0.250	3	10.6, 11.1, 10.2	10.63	0.250	1	10.0	10.00
0.375	3	10.5, 10.7, 10.9	10.70	0.375	1	10.1	10.10
0.500	3	11.2, 10.9, 10.3	10.80	0.500	3	10.8, 10.4, 11.4	10.87
1.0	3	11.1, 11.7, 11.3	11.37	1.0	1	11.3	11.30
1.5	3	12.5, 12.2, 11.9	12.20	1.5	3	12.9, 12.1, 12.3	12.43
2.0	3	13.3, 12.6, 12.4	12.77	2.0	6	12.8, 13.3, 12.0 13.5, 12.9, 12.5	12.83

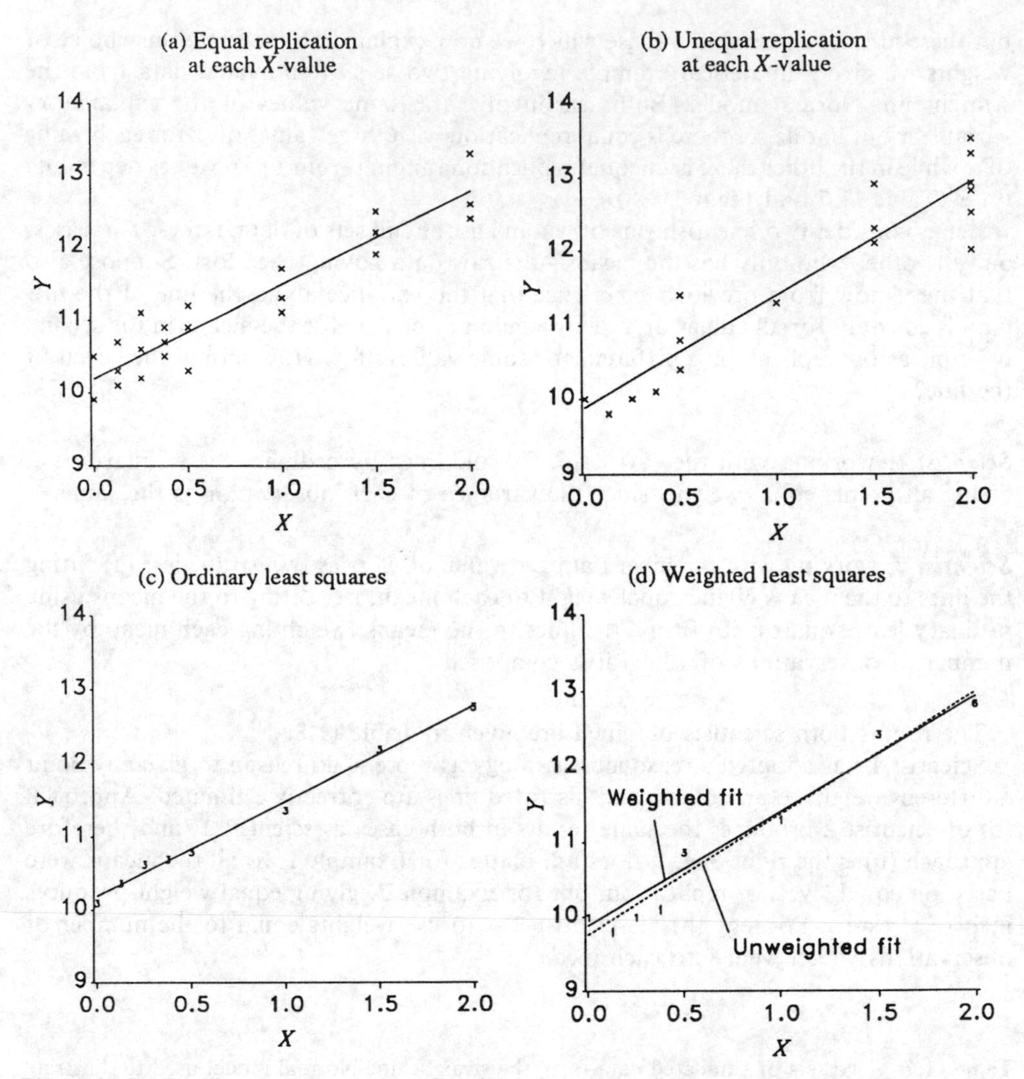

Figure 11.8 To illustrate the choice of weights in weighted least squares. Top figures show simulated data from the straight line-Normal model with: (a) equal replication; (b) unequal replication at each value of x, and with lines fitted by ordinary least squares, as scientist 1 would have done. Bottom figures show mean values of y for each value of x with lines fitted by the two approaches of scientist 2: (a) ordinary least squares; (b) weighted least squares with weights equal to the number of observations at the corresponding value of x. With unequal replication the weighted fit to the means produces the same line as that obtained by applying ordinary least squares to the original observations.

This strategy can also be described as weighting inversely proportionally to the variance. The variance is obtained from the formula for the variance of the mean of a random sample as follows

$$\text{var}\,[\overline{Y}_i] = \frac{\sigma^2}{n_i}$$

where σ^2 is the variance of a *single* observation about the line, $\overline{Y}_i$ is the mean of the

Table 11.8 Equations of lines fitted by different methods to the two sets of data in Table 11.7

Fitting method	Response variable	Equal replication at each x-value	Unequal replication at each x-value
Scientist 1			
Ordinary least squares	Raw data	$y = 10.19 + 1.30x$	$y = 9.88 + 1.53x$
Scientist 2			
(a) ordinary least squares	Mean	$y = 10.19 + 1.30x$	$y = 9.75 + 1.61x$
(b) Weighted least squares	Mean	$y = 10.19 + 1.30x$	$y = 9.88 + 1.53x$

observations at x_i, n_i is the number of observations. Fitting proportionally to the inverse of the variance weights by n_i/σ^2. An apparent difficulty is that the σ^2 is not known but this is not a problem since multiplying the weights by an arbitrary constant does not affect the fitted line. Each mean is therefore weighted by the number of replicates on which it is based, i.e. $w_i = n_i$.

11.7.2 Fitting models by maximum likelihood

In the text, we have dealt with fitting models by the methods of ordinary or weighted least squares. In this section we briefly discuss another widely used method: *maximum likelihood*. The basic idea of fitting a model by maximum likelihood is to find those values of the parameters for which the data are *most likely* to have occurred.

Before defining the concept of likelihood and the principle of maximum likelihood estimation we illustrate the idea with a simple example. Suppose that in 6 independent trials, each with the same probability of success, we observe 2 successes. If p is the true probability of success at each trial then the probability of 2 successes out of 6 trials is given by the appropriate term of the binomial distribution as

$$\begin{aligned} P(R = 2) &= {}^6C_2 p^2(1-p)^4 \\ &= 15p^2(1-p)^4. \end{aligned}$$

This expression for the probability of the observed data given a particular value of the true probability, p, can also be thought of the other way round: it can also be viewed as a way of telling us how the probability of observing what we have observed could be dependent on p. Having observed 2 successes out of 6—let us call this event E—it gives us the probability of having observed E for each possible value of the parameter, p. Figure 11.9 shows a plot of this function. It has a maximum at $p = 0.333$, and falls off on both sides of this. As the probability of E is rather low for $p = 0.95$, we might say that the parameter value 0.95 is rather unlikely to be the true parameter value; similarly, $p = 0.333$ is the most likely true parameter value. More generally, we could say that the curve gives us a plot of the likelihood of each possible parameter value, and thus is referred to as the *likelihood* of p given the data. Values of p which have a higher value of this function, a higher likelihood, are thought to be more likely to be the true value of p.

The *maximum likelihood estimate* of p is the value which maximises the likelihood. Figure 11.9 shows the likelihood of p for 2 successes out of 6 trials as a curve with a single peak at $p = 0.333$ (the position of the maximum can in general be obtained by

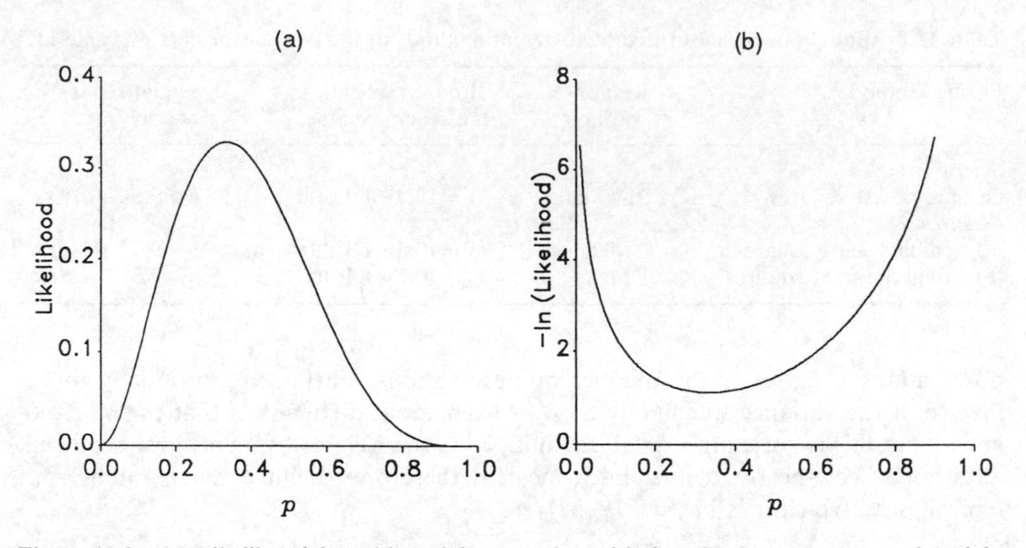

Figure 11.9 (a) Likelihood for a binomial proportion with data for 2 successes out of 6 trials; (b) − log likelihood.

differentiating the likelihood, rather than by graphical methods). In this case the maximum likelihood estimate for p is the common-sense estimate, i.e. the observed proportion of successes. In practice, it is more convenient to maximise the *log* likelihood or equivalently to minimise *minus* the *log likelihood* (Figure 11.9). Furthermore, the log likelihood is important for studying the properties of maximum likelihood estimates.

Likelihood and maximum likelihood estimation

Let $\mathbf{y} = (y_1, y_2, \ldots, y_n)$ denote a set of observations with

Either	*or*
for *discrete data*, probability distribution $P(\mathbf{y} \mid \theta)$ (which means probability of y given a value of the parameter θ).	for *continuous data*, probability density function $f(\mathbf{y} \mid \theta)$ (probability density of y given a value of the parameter θ).

(1) The function which expresses the probability of the data as a function of the parameters. i.e.

$$L(\theta) = P(\mathbf{y} \mid \theta) \qquad\qquad L(\theta) = f(\mathbf{y} \mid \theta)$$

is called the *likelihood*.

(2) The *maximum likelihood estimate* of θ is that value of $\hat{\theta}$ for which the likelihood is maximised.

Maximum likelihood, least squares and iterative weighted least squares

For some models fitting by maximum likelihood is equivalent to least squares or iterative weighted least squares. Two important cases are as follows.

(1) For linear and non-linear models with Normally distributed responses with constant variance fitting by maximum likelihood is equivalent to ordinary least squares.
(2) For the logistic-binomial model and some other standard generalised linear models, fitting by maximum likelihood is equivalent to iterative weighted least squares.

To illustrate the equivalence of maximum likelihood and least squares for fitting linear-Normal models we use the straight line-Normal model

$$Y_i = \alpha + \beta x_i + \varepsilon_i, \qquad i = 1, \ldots, n$$

where the random deviations ε_i are independent values from a Normal distribution with mean zero and variance σ^2. In other words, for a given value of x_i, the response follows a Normal distribution with mean $\alpha + \beta x_i$ and variance σ^2, independently of the other responses.

The probability distribution for the set of observations $y_1, \ldots, y_n$ formed from the product of the corresponding Normal probability density functions (see Chapter 4) so that the likelihood of the parameters is given by

$$L(\alpha, \beta, \sigma) = \prod_{i=1}^{n} \frac{1}{\sqrt{2\pi}\,\sigma} \exp[-(y_i - \alpha - \beta x_i)^2/2\sigma^2]$$

To fit by maximum likelihood we find the parameter values which maximise the likelihood or, equivalently, which minimise minus the logarithm of the likelihood given by

$$-\ln L = \frac{1}{2\sigma^2} \sum_{i=1}^{n} (y_i - \alpha - \beta x_i)^2 + n \ln(\sqrt{2\pi}\,\sigma).$$

So, fitting by maximum likelihood is equivalent to fitting the line which minimises the sum of the squared deviations of the observations about the fitted line, i.e. fitting by ordinary least squares.

The above analysis refers to estimating the parameters in the systematic part of the model. We can also apply the principle of fitting by maximum likelihood to estimate the variance of the random component σ^2.

EXERCISES

11.1 Fit the Michaelis–Menten equation to the enzyme kinetics data in Example 9.2 using ordinary least squares.

11.2 Suggest why the approximate methods of estimating B_{max} and K_D in Exercise 9.6 might be misleading, and outline the principles of a better method. Using a statistical package, carry out this analysis and compare your conclusions with those using the graphical approach in Exercise 9.6.

11.3 The table below contains data on the height of microvilli in relation to the distance along the villus (Example 4.3, Figure 4.6). Fit by ordinary least squares the logistic-Normal model

$$H = \alpha + \beta \frac{e^{\gamma(d-m)}}{1 + e^{\gamma(d-m)}} + \varepsilon$$

where $\varepsilon \sim N(0, \sigma^2)$. Check your estimates against those in Example 4.3. Comment on the relatively high negative correlation between the estimates of α and β.

Distance	−270	−226	−189	−149	−100	−88	−55	−47	−26	0	0	0
Height	0.65	0.56	0.70	0.92	0.87	0.95	0.76	1.12	1.32	0.87	0.88	0.90
Distance	43	49	72	104	120	158	177	228	264	300	385	422
Height	1.24	1.19	1.17	1.49	1.45	1.34	1.72	2.28	2.10	2.05	2.23	1.78
Distance	485	527	587	638	705	783	880	982	1263	1513		
Height	2.08	1.77	2.13	2.22	2.25	1.91	2.14	2.04	2.12	1.98		

11.4 The table below shows data on the number of kills of the beetle *Tribolium confusum* after 5 hours' exposure to gaseous carbon disulphide (CS_2) at various concentrations in a bioassay (Bliss, 1935). Fit a logistic response curve relating the proportion of beetles dying to the log dose of carbon disulphide by fitting the logistic-binomial model as a generalised linear model. Use the fitted model to estimate the LD_{50}, i.e. the median lethal log dose.

Log_{10} CS_2 mg l^{-1}	1.6907	1.7242	1.7552	1.7842	1.8113	1.8369	1.8610	1.8839
Number of beetles	29	30	28	27	30	31	30	29
Number dead	2	7	9	14	23	29	29	29

Two other models for the dose–response curve which can be fitted as generalised linear models are: (i) the cumulative Normal with link function equal to the inverse of the cumulative Normal distribution, i.e. $g(p) = \Phi^{-1}(p) = a + bx$; (ii) the double exponential model with link function equal to the complementary log-log transformation, i.e. $g(p) = \ln[-\ln(1-p)]$. Fit these two models to the above data and compare the fitted models with the fitted logistic.

11.5 For the data in Exercise 4.8 on survival of sparrowhawks in relation to age, fit a logistic-binomial model in which

$$\text{logit}(p) = \alpha + \beta_1 + \beta_2 x^2$$

where p is the probability of survival and x is the age of the bird.

11.6 Some species of nematodes have a remarkable capacity to survive desiccation by entering a state of anhydrobiosis. The table below gives some data on the survival of the species *Ditylenchus* sp. *B* showing the number of animals surviving in five groups of 50 animals dehydrated for different periods of time.

Time (days)	20	40	80	120	160
Number surviving	48	45	36	20	3

A plausible model to relate survival to the length of time of exposure is the Weibull survivorship curve in which the proportion surviving after time t is given by

$$p(t) = \exp[-(t/c)^b].$$

(a) Show that the model can be fitted as a generalised linear model and apply the method to the above data. Use the fitted model to estimate the median survival exposure time.
(b) How would you fit an alternative model in which the proportion surviving is given by

$$p(t) = \frac{1}{(1 + at^k)}.$$

11.7 The table below contains data on the progress of annual moult in a population of common seals as the number of seals which have completed their moult in a series of samples of animals sighted at times during the moulting period.

Date of sample	25/7	2/8	8/8	14/8	22/8	29/8	4/9	10/9	16/9	24/9
Number of seals seen	17	17	19	33	23	26	15	22	6	6
Number moulted	0	0	5	16	21	26	13	18	6	6

A simple model to describe the seasonal pattern of moult assumes that the date of moult completion varies between animals according to a Normal distribution with mean μ and standard deviation σ. In other words, the proportion of animals which have completed their moult by time t is given by the cumulative Normal distribution function

$$p(t) = \Phi\left(\frac{t-\mu}{\sigma}\right).$$

Show how the field observations can be used to estimate the mean and standard deviation of the moult completion date, stating the assumptions involved in the analysis?

11.8 Apply the delta rule to show that fitting the logistic-binomial by weighted least squares applied to the linearised observed proportions requires weights given by

$$w_i = n_i p_i (1 - p_i).$$

Show that the corresponding weights for a generalised linear model with complementary log-log link function, $g(p) = \ln[-\ln(1-p)]$, are given by $w_i = n_i(1-p_i)[\ln(1-p_i)]^2/p_i$.

11.9 *Log-linear models for counts*. Log-linear models for count data are generalised linear models with a log link function and a Poisson error distribution, i.e.

$$Y = f(L) + \varepsilon,$$

where $E[Y] = \mu = f(L) = e^L$, $g(\mu) = \ln L$, $Y \sim Poisson(\mu)$ and $V(\mu) = \mu$.

Log-linear models are widely used for analysing proportions based on counts in contingency tables. To illustrate the analysis, consider the problem of comparing the proportion of multiple births in flocks of sheep from three areas (Section 6.4.6, Example 6.6) using the following data

	Area A	Area B	Area C
Single births	20	31	15
Multiple births	10	6	6
Total	30	37	21

First, we have a model in which L is a linear function of area and type-of-birth effects, i.e.

$$\text{Model 1: } L_{ij} = m + b_i + a_j,$$

where m, b_i and a_j are the overall mean, the effect of birth type i and the effect of area j respectively. A second model is one containing an interaction term, i.e.

$$\text{Model 2: } L_{ij} = m + b_i + a_j + (ab)_{ij}.$$

Equality of proportions of multiple births corresponds to Model 1 holding. Model 2 would always hold exactly for any data set so that a test of equality of proportions can be performed by comparing the difference in residual deviance of Model 1 and that of Model 2 (which is zero) with chi-squared, in this case on 2 degrees of freedom.

(a) Show that for Model 1 the underlying proportion of multiple births in each area is the same and given by $p = e^{b_2}/(e^{b_1} + e^{b_2})$.

(b) Use a statistical package, such as GENSTAT or SAS, to fit the log-linear model and confirm that the residual deviance is equal to the deviance chi-squared, or G-statistic, calculated in Example 6.6. (For an introduction to fitting log-linear models to contingency table data using GENSTAT, see Digby *et al.* (1989).)

11.10 Using Carlson's yeast data (Example 1.6, Table 1.2), apply the stepwise fitting procedure to fit the logistic model assuming the form

$$\ln N_t = \ln\left\{\frac{K}{\left[1+\left(\frac{K}{N_{t-1}}-1\right)e^{-r}\right]}\right\} + \varepsilon_t,$$

where N_t is the amount of yeast observed at time t (h) and $\varepsilon_t \sim N(0, \sigma^2)$. You should obtain estimates of the intrinsic rate of increase, $\hat{r}$, of 0.54 per hour and of the carrying capacity $\hat{K}$, of 664. Interpret the parameter σ^2.

In what situations is this stepwise procedure preferable to fitting the logistic curve by ordinary least squares to the raw observations?

11.11 Fitting a straight line by weighted least squares involves finding the line for which a weighted sum of the squared deviations of the observations from the fitted line is least, i.e. minimising

$$S_w = \sum_{i=1}^{n} w_i(y_i - \hat{\alpha} - \hat{\beta}x_i)^2,$$

where the w_i are specified weights.

Show that the estimated slope and intercept are given by

$$\hat{\beta} = \frac{\sum_{i=1}^{n} w_i y_i(x_i - \bar{x}_w)}{\sum_{i=1}^{n} w_i(x_i - \bar{x}_w)^2}; \qquad \hat{\alpha} = \bar{y}_w - \hat{\beta}\bar{x}_w,$$

where $\bar{x}_w = \Sigma\, w_i x_i / \Sigma\, w_i$ and $\bar{y}_w = \Sigma\, w_i y_i / \Sigma\, w_i$ are weighted averages of the x and y values.

11.12 In some experiments the variance of the response in a Michaelis–Menten relationship is proportional to the square of the mean response. This assumption is sometimes expressed as a model with a multiplicative error term in which the velocity of the reaction and the substrate concentration are related by

$$v_i = \frac{Vs_i}{K_m + s_i}(1 + \varepsilon_i), \qquad i = 1, \ldots, n,$$

where the ε_i are random disturbances with zero mean and constant variance. Show that fitting the model by weighted least squares involves minimising

$$S_w = \sum_{i=1}^{n}\left(\frac{K_m v_i}{Vs_i} + \frac{v_i}{V} - 1\right)^2$$

with respect to V and K_m, and that the estimates are given by

$$\hat{V} = \frac{\Sigma(v_i^2/s_i^2)\Sigma\, v_i^2 - (\Sigma\, v_i^2/s_i)^2}{\Sigma(v_i^2/s_i^2)\Sigma\, v_i - \Sigma(v_i^2/s_i)\Sigma(v_i/s_i)}; \qquad \hat{K} = \frac{\Sigma\, v_i^2\, \Sigma(v_i/s_i) - \Sigma(v_i^2/s_i)\Sigma\, v_i}{\Sigma(v_i^2/s_i^2)\Sigma\, v_i - \Sigma(v_i^2/s_i)\Sigma(v_i/s_i)}$$

(Hint: Let $a = K_m/V$ and $b = 1/V$ and minimise with respect to a and b.)

Apply the method to the following simulated data from a model with $V = 20$, $K_m = 0.02$ and $\text{var}[\varepsilon_i] = 0.01$.

s_i	0.01	0.01	0.02	0.02	0.04	0.04	0.08	0.08	0.16	0.16	0.32	0.32
v_i	5.6	6.5	10.2	9.9	13.4	15.0	13.4	16.5	16.7	17.5	21.1	17.9

11.13 Let $(x_1, y_1), (x_2, y_2), \ldots, (x_n, y_n)$ denote a sample of observations from the straight line-Normal model.

(a) Show that the maximum likelihood estimator of the variance of the random component is given by

$$\widehat{\sigma^2} = \frac{1}{n} \sum_{i=1}^{n} (y_i - \hat{\alpha} - \hat{\beta} x_i)^2,$$

where $\hat{\beta}$ and $\hat{\alpha}$ are the estimated slope and intercept of the line fitted by ordinary least squares.

(b) Comment on the difference between the maximum likelihood estimator of the variance and the residual mean square of the analysis of variance.

Part IV

ADVANCED TOPICS

12 Transport and Diffusion

12.1 INTRODUCTION

Transport is one of the fundamental processes of biology. Animals move from regions of high to low density because of shortage of resources, or in a trial-and-error search for better conditions. They also need to search for food and for a mate. Plant seeds disperse to avoid competing with the parent plant. On a smaller scale, materials need to be transported from the site of synthesis within the organism to the utilisation site, for metabolism or growth or some other need. This chapter is about how to model mechanisms of transport mathematically, and determine the spatial patterns of abundance or concentration which result.

Diffusion, the central process we discuss in this chapter, is a form of transport due to the continual random motion of particles. This ranges from the Brownian motion of very small particles (named after the biologist R. Brown who first wrote about the random motion of pollen grains in a dish of water in the early nineteenth century) to the random walks of invertebrates and small animals, such as those of the larvae of *Trichostrongylus retortaeformis* (see Example 12.1). In these cases, the random motion usually results in a rate of spread from one region to another which is proportional to the concentration difference between the two regions. Typically any movement is over a short distance, unless helped by some external agent. Skellam (1951) discussed the spread of the oak in Britain after the final recession of the Ice Age in about 18 000 BC until it was abundant at the time of the Roman invasion, and concluded that help from rooks, squirrels or other animals was necessary. Diffusion of substances due to continual, random atomic movement is also usually only effective over very short distances (see Section 12.5.1). In Section 12.2, we derive the distribution of abundance resulting from a random walk in one dimension, then obtain a partial differential equation describing deterministic diffusion, which itself is solved to give the same Normal distribution which resulted from the random walk. Diffusion in two dimensions is discussed in Section 12.4, and applied to the movement of *Trichostrongylus retortaeformis*.

If material has the capability of reproducing itself from small concentrations, then the rate of spread will be increased. In Section 12.5.2, we discuss how the muskrat spread from a few releases in 1905 in Central Europe to cover a large part of Europe by 1927. The exponential growth of population density at the leading edge of the spreading population—where competition is likely to be virtually zero—is an important facilitator of the spread. Diffusion also has the capability of destabilising an otherwise spatially homogeneous interaction between two chemicals or animal populations, resulting in waves of abundance which could provide a simple explanation of animal coat markings or clumping of plankton in the sea. This phenomenon known as reaction-diffusion is briefly discussed in Section 12.5.4.

Many other transport processes—such as epidemic spread and cancer growth—require stochastic models of a different kind, and these are briefly discussed in Section

12.6. A further distinction is whether the movement is directional or not. Partly because of the limited power of movement in undirected diffusion, a small consistent directional component, even if apparently outscaled by random motion over a short time, can sometimes be the major ultimate determinant of the speed and direction of travel. We discuss in Section 12.6.2 how birds such as the Manx shearwater might achieve such remarkable navigational success on their transatlantic journey from America to their British nesting area by using rather limited, error prone, directional abilities.

Finally, in Section 12.7, we discuss computer simulation of spatio-temporal processes using cellular automata. This technique has provided a means of efficiently simulating deterministic diffusion and more complex spatial processes using parallel computers. Study of cellular automata has also cast light on how, from extremely simple local rules of interaction, emergent behaviour of considerable complexity and on varying scales can occur.

12.2 RANDOM WALKS, BROWNIAN MOTION AND DIFFUSION OF ORGANISMS

The irregular and continual motion of pollen particles observed under a microscope and the wanderings of *Trichostrongylus* (see Example 12.1) have much in common, despite the considerable difference of scale: both involve random, irregular movements of effectively point objects, the mathematical term for which is *random walk*. We first consider a simple example of a random walk in one dimension, i.e. along a straight line.

12.2.1 The one-dimensional random walk

Suppose a bird perches on a telegraph wire and starts out from the pole which we regard as a kind of origin. Suppose also that the bird is rather active and jumps at random to the left or to the right once every second. The jumps are of a fixed length, s, either to the left with probability 0.5, or to the right with probability 0.5. What course would the bird be likely to take? What would be the distribution of its position after n seconds? Figure 12.1(a) illustrates a number of instances of such walks along the wire. It is a fairly straightforward calculation in probability to work out the possible positions and the probability of each position and these are given in Table 12.1.

In order to study more closely the properties of such a random walk, let us first consider the distribution of the number of right moves, R, in the n seconds. We consider the more general case in which the probability of a jump to the right is p (now not necessarily equal to 0.5). This is exactly the same as the distribution of the number of successes in a fixed number of trials, n, where the probability of a success is p. We know from Chapter 3 that this follows the binomial distribution, parameters n and p. The position of the bird relative to the telegraph pole is of course given by the difference between the number of rightward and leftward jumps, times the size of the jump, s,

$$X = [R - (n - R)]s = (2R - n)s.$$

So the distribution of position after n jumps is just a scaled (i.e. multiplied by $2s$)

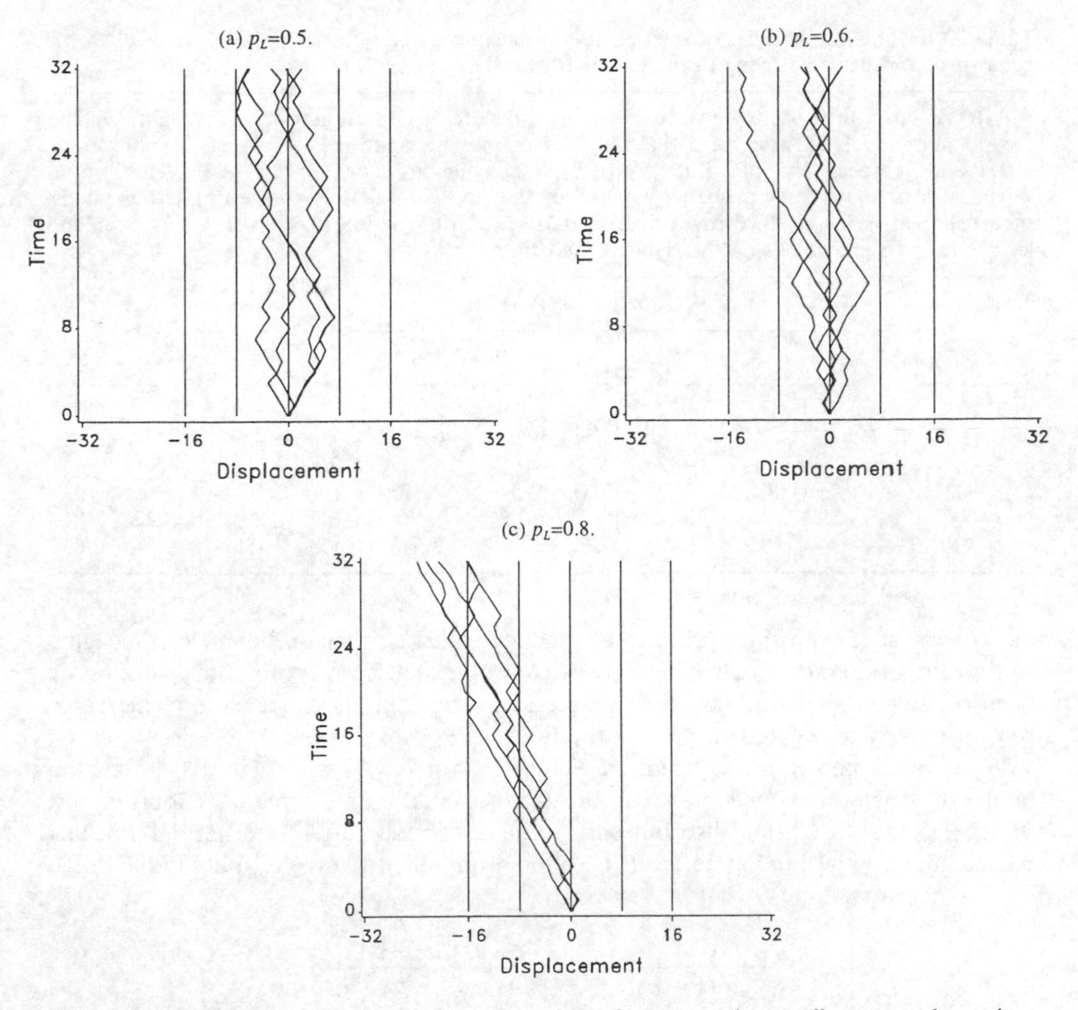

Figure 12.1 This figure illustrates the progression of some random walks over a long time period for a number of cases, including the symmetrical case (a) above, and two asymmetrical ones, (b) and (c).

binomial, with its mean displaced. From this, we are able to calculate the mean position and the variance of position as time increases. We know that for a binomial distribution, parameters n and p,

$$\mathrm{E}[R] = np$$
$$\mathrm{var}[R] = np(1-p),$$

and so we obtain

$$\mathrm{E}[X] = (2\mathrm{E}[R] - n)s = (2np - n)s = ns(2p-1) = ts(2p-1),$$
$$\mathrm{var}[X] = 4s^2 \ \mathrm{var}[R] = 4s^2 np(1-p) = 4s^2 tp(1-p)$$

substituting t for n to get the last expressions. So, if $p = 0.5$ the mean position (over a large number of random realisations of the model) will remain at the starting point,

Table 12.1 The probabilities of each position in a simple one-dimensional random walk, with equal probabilities of moving to the left or to the right, at each jump

After 1 jump there is a 0.5 chance of having moved either to the left or to the right. On the second jump, each of these has a chance of 0.5 of moving right or left, giving probabilities at -2, 0 and 2 respectively of 0.25, $(0.25+0.25)$, 0.25 which are the 0.25, 0.5, 0.25 given. At time three, each of these has a probability of 0.5 of moving to the left or to the right, thus giving the probabilities in the third row of $(0.25)(0.5)$, $(0.25)(0.5)+(0.5)(0.5)=0.125$, 0.375 on the left, with the same values on the right of the origin.

Time	Position										
	$-5s$	$-4s$	$-3s$	$-2s$	$-s$	0	s	$2s$	$3s$	$4s$	$5s$
$t=0$						1					
$t=1$					0.5		0.5				
$t=2$				0.25		0.5		0.25			
$t=3$			0.125		0.375		0.375		0.125		
$t=4$		0.0625		0.25		0.375		0.25		0.0625	
$t=5$	0.031		0.156		0.313		0.313		0.156		0.031

but otherwise, the particle will have a tendency to drift either to the left or the right, so that the mean moves with a velocity of $(2p-1)s$. The spread of the distribution of position also increases as time increases; the standard deviation increases proportionally to the square root of the time.

We know from Chapter 3 (compare Figure 3.3 for $n=12$ with Figure 3.7) that the binomial distribution for large n can be approximated by a Normal distribution and the approximating Normal distributions for the distribution after a substantial time has passed are also given in Figure 12.2(b). So the probability that the bird lies in the range $(x, x+\delta x)$ after time t would be given by

$$\frac{1}{2s\sqrt{2\pi p(1-p)t}}\exp\left[\frac{-(x-ts(2p-1))^2}{8ts^2p(1-p)}\right]\delta x.$$

Now suppose that N individuals are initially released at time zero, and they each carry out independent random walks of this nature then the *density* of birds at a distance x away from the starting point is given by

$$\frac{N}{2s\sqrt{2\pi p(1-p)t}}\exp\left[\frac{-(x-ts(2p-1))^2}{8ts^2p(1-p)}\right].$$

In the special case when $p=\frac{1}{2}$, this simplifies to

$$\frac{N}{s\sqrt{2\pi t}}\exp\left(\frac{-x^2}{2ts^2}\right)$$

Thus we see that the variance, ts^2, increases linearly with time, and the mean position stays the same. Using the property of the Normal distribution that 95% of cases lie within 1.96 standard deviations of the mean, we would expect to find 95% of the birds within $1.96\ s\sqrt{t}$ at any time, so at times $t=10, 20\ldots$ the range containing 95% of birds is as follows, in the case when $s=1$.

t	10	20	30	40	50	100	1000	10 000	100 000
Range	6.2	8.8	10.7	12.4	13.9	19.6	62.0	196	620

Thus the edge of the distribution moves outwards increasingly slowly with time. We shall see later that this places a real limit on the possibility of moving great distances by diffusion alone.

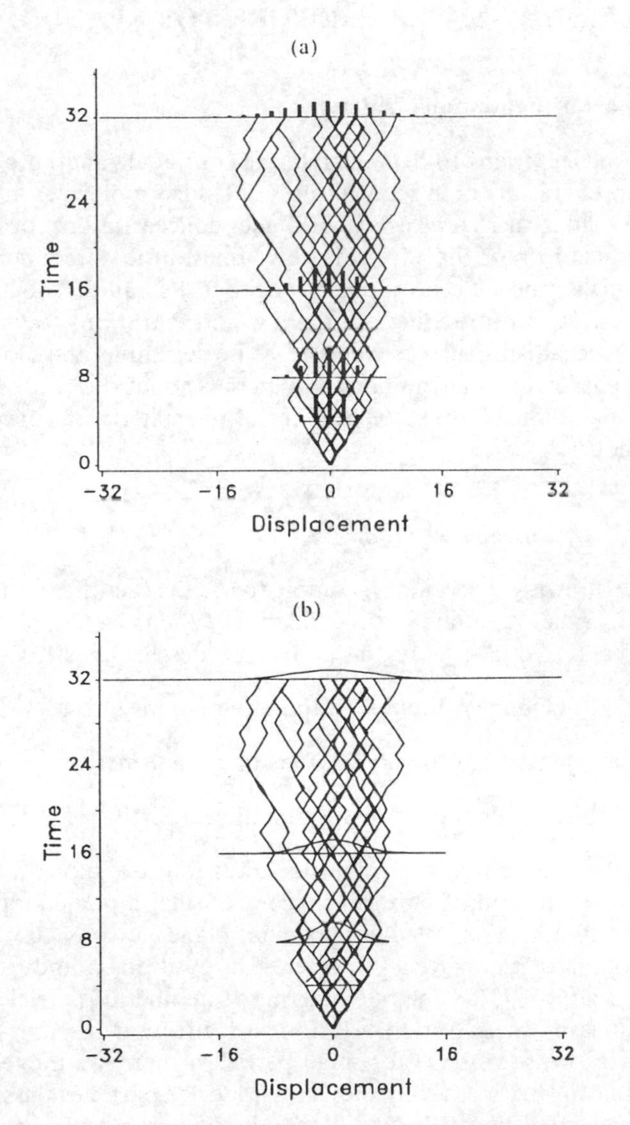

Figure 12.2 Twenty random walks with P(left) = $p_L = 0.5$ at each time interval. Superimposed are (a) the binomial distributions and (b) Normal approximations to these binomials at time $t = 4$, 8, 16, 32.

It is useful for subsequent comparisons to rewrite this equation more generally for the case when the steps are of length Δx which take place every Δt seconds:

$$\frac{N}{\Delta x}\sqrt{\frac{\Delta t}{2\pi t}}\exp\left[\frac{-x^2\Delta t}{2(\Delta x)^2 t}\right].$$

This transformation is a result of replacing the spatial step size s by Δx and t by $t/\Delta t$, since the increment in time used in the above derivation is 1.

12.3 A DETERMINISTIC FRAMEWORK FOR SPATIAL FLOW

12.3.1 Transport of continuous material

R. Brown used pollen grains to demonstrate the continual random buffetings that all molecules get in liquids at room temperatures. All such molecules are moving to and fro all the time, and if there are regions of higher concentrations locally, this random movement will tend to blur the edges of the volumes of different concentrations and eventually to equalise the concentrations throughout the liquid. The dissolved material is then said to diffuse from the high to the low concentrations. Some substances are transported by mechanisms such as convection. For example, the bloodstream is used as a means of transporting many important materials about the body. How might these processes be modelled on a large scale, thinking of the material as continuous and using deterministic models?

12.3.2 The equation of conservation

The first stage in drawing up a model is simply to use the fact that accumulation within an infinitesimally small region is just the balance of the flow in and the flow out plus any birth or creation within the region minus any death or loss:

$$\text{Change} = \text{Input} - \text{Output} + \text{Creation} - \text{Loss},$$

equivalent to the relationship for populations of organisms:

$$\text{Growth} = \text{Immigration} - \text{Emigration} + \text{Births} - \text{Deaths}.$$

We first restrict our attention to flow in one direction, i.e. along a tube of uniform cross-sectional area, A, and of uniform shape. Consider a plane perpendicular to the x-direction at position x, and an adjacent similar plane δx away at $x + \delta x$. Let $u(x, t)$ be the concentration of the particles of interest at position x, and $J(x, t)$ be the flux of these particles at time t, i.e. the rate in terms of number of particles which pass the point per unit area of cross-section in a forward direction per unit time (see Figure 12.3). In reality as we saw above for small particles, there are movements backward and forward, and so flux is actually the *net* number of particles passing in a forward direction per unit time. Normally flux is in units of number of particles (or mass of material) per unit area (of cross-section) per unit time, and the convention is that it is usually measured in a forward direction, i.e. in this case as x increases.

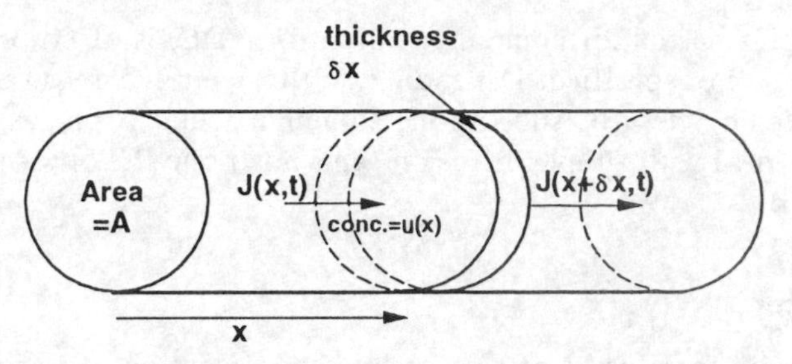

Figure 12.3 Flow along a tube.

Applying the above input/output relationship to this infinitesimally thin slice, we obtain

$$\text{Change in } u = \text{Inflow} - \text{Outflow} + \text{Net creation}$$

i.e.

$$A\ \delta x\ \delta u = J(x,t)A\ \delta t - J(x+\delta x, t)A\ \delta t + A\ \delta x\ R(x,t)\ \delta t$$

where $R(x,t)$ is the net rate of creation of particles per unit volume. Thus dividing throughout by δt,

$$A\ \frac{\delta u(x,t)}{\delta t}\ \delta x = AJ(x,t) - AJ(x+\delta x, t) + AR(x,t)\ \delta x$$

and taking the limit as $\delta t \to 0$, using the fact that $\lim_{\delta t \to 0} [\delta u(x,t)/\delta t] = \partial u/\partial t$, we get

$$A\ \frac{\partial u(x,t)}{\partial t}\ \delta x = AJ(x,t) - AJ(x+\delta x, t) + AR(x,t)\ \delta x.$$

Further dividing through by δx

$$A\ \frac{\partial u(x,t)}{\partial t} = \frac{AJ(x,t) - AJ(x+\delta x, t)}{\delta x} + AR(x,t)$$

then letting $\delta x \to 0$, we get

$$A\ \frac{\partial u(x,t)}{\partial t} = -A\ \frac{\partial J(x,t)}{\partial x} + AR(x,t).$$

If A varies along the length of the tube or even with time, we could easily accommodate this by inserting $A = A(x,t)$ inside the partial differentiation signs, and we obtain *the one-dimensional balance or conservation equation.*

One-dimensional balance or conservation equation

$$\frac{\partial}{\partial t}\ [A(x,t)u(x,t)] = -\frac{\partial}{\partial x}\ [A(x,t)J(x,t)] + A(x,t)R(x,t)$$

where $J(x,t)$ = flux per unit area (of cross-section)
= number of particles which cross the plane in a positive direction per unit time per unit area of cross-section
$u(x,t)$ = concentration of particles per unit volume of fluid
$A(x,t)$ = cross-sectional area of the container
$R(x,t)$ = net creation of particles per unit volume at position x and time t.

As it stands, this partial differential equation (or PDE) is of no predictive value as $J(x, t)$, the flux, is not specified. Depending on the agent which is responsible for the motion, we can complete it by substituting some particular form for $J(x, t)$. We next consider how to deal with simple diffusion, and postpone the other forms for $J(x, t)$ to Section 12.5.4.

12.3.3 Diffusion

As diffusion occurs from areas of high to low concentration, it is not surprising that flux is proportional to the concentration gradient. This is known as Fick's law, and diffusion which satisfies it is known as Fickian diffusion.

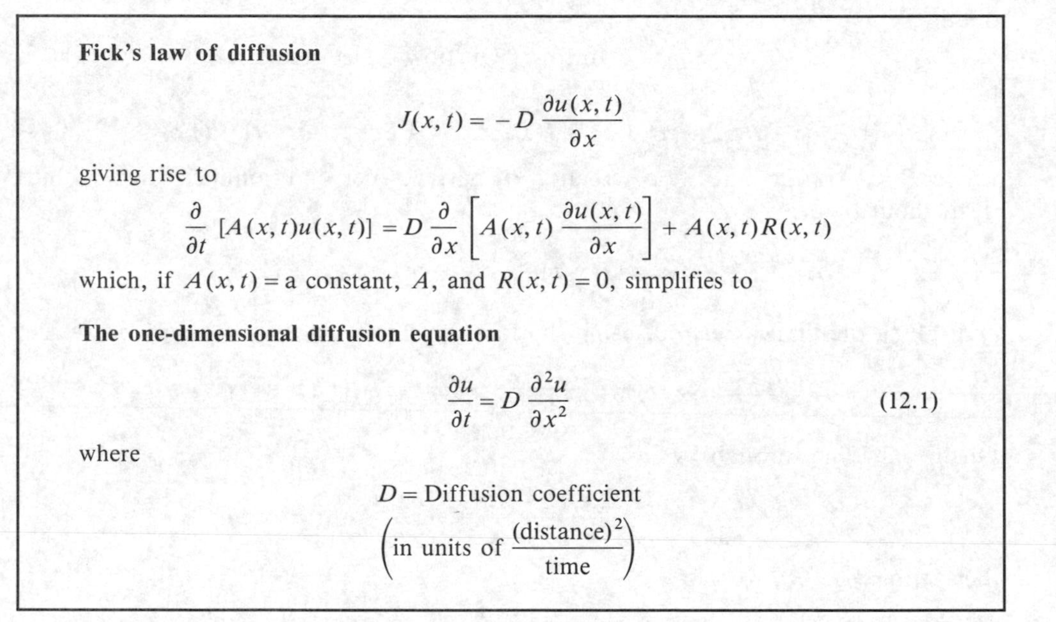

Fick's law of diffusion

$$J(x,t) = -D\frac{\partial u(x,t)}{\partial x}$$

giving rise to

$$\frac{\partial}{\partial t}[A(x,t)u(x,t)] = D\frac{\partial}{\partial x}\left[A(x,t)\frac{\partial u(x,t)}{\partial x}\right] + A(x,t)R(x,t)$$

which, if $A(x, t) =$ a constant, A, and $R(x, t) = 0$, simplifies to

The one-dimensional diffusion equation

$$\frac{\partial u}{\partial t} = D\frac{\partial^2 u}{\partial x^2} \qquad (12.1)$$

where

$$D = \text{Diffusion coefficient}$$
$$\left(\text{in units of } \frac{(\text{distance})^2}{\text{time}}\right)$$

It is worth thinking what this form of the diffusion PDE in one dimension means. The rate of change of concentration at a point is proportional to the curvature in the relationship of concentration to distance. Thus if the curvature, $\partial^2 u/\partial x^2$, is zero, i.e. $u(x, t)$ is a horizontal function of x or a straight line relationship in x, $\partial u/\partial t$ would be zero, and so there would be no change (see Figure 12.4). If curvature is high and positive (the rate of change of $\partial u/\partial x$ with distance is positive), such as you might get at the base of a narrow steep-sided valley in u, then $u(x, t)$ would increase at that point, at a rate proportional to the sharpness of the V-shape at the valley bottom; at points on the shoulders of the adjacent hills u would fall again proportionally to the rate of change of gradient on the shoulder. The situation is reversed for a steep-sided hill. What does seem counter-intuitive is that when the fall in u is linear, then $\partial u/\partial t$ is zero; the reason is that diffusion is taking place down the gradient but the flux due to diffusion is constant along the gradient, so that whatever a point receives from the left, it loses at the same rate to the right. The diffusion constant tells us how sensitive the system is to the spatial curvature of the concentration. This fits in with it also being a measure of the speed of the diffusion.

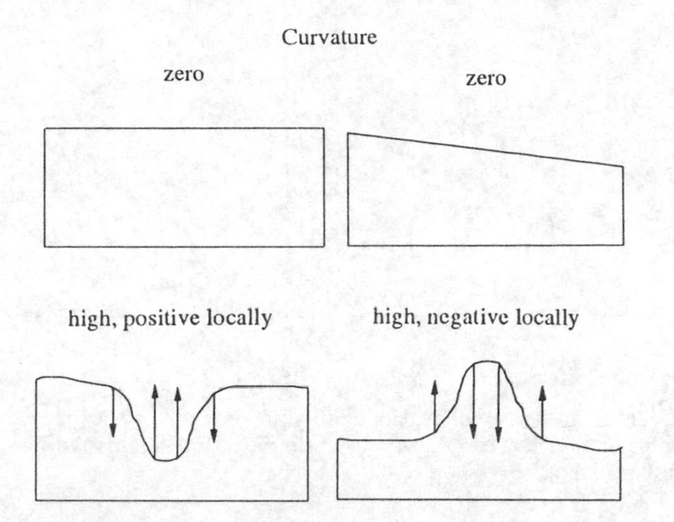

Figure 12.4 The relationship of rate of change of u with time (indicated by the length of the arrow) and the curvature of u.

12.3.4 Solution of the deterministic diffusion equation

Methods for analytical and numerical solution of PDEs are introduced in the Appendix, and some examples of solution of the one-dimensional diffusion equation

$$\frac{\partial u}{\partial t} = D \frac{\partial^2 u}{\partial x^2}$$

are given there. The ease with which a solution can be found is dependent on the initial conditions as well as any boundary conditions and the form of the differential equation itself. In this section we discuss the solution in one of the classic applications—that of spread from a point source over an effectively infinite region in both directions; this solution can then be compared with the solution derived in Section 12.2.1 for the one-dimensional random walk. The solution is

$$u(x, t) = k \frac{\exp(-x^2/4Dt)}{\sqrt{t}},$$

where k is an arbitrary constant. It can be verified that the solution satisfies the PDE for one-dimensional diffusion by calculating the partial differentials and substituting in the equation.

$$\frac{\partial u}{\partial x} = \frac{-2kx}{4Dt} \frac{\exp(-x^2/4Dt)}{\sqrt{t}} = -\frac{kx \exp(-x^2/4Dt)}{2Dt^{3/2}}$$

and

$$\frac{\partial^2 u}{\partial x^2} = \left[\left(-\frac{k}{2Dt^{3/2}}\right) + \left(-\frac{kx}{2Dt^{3/2}}\right)\left(-\frac{2x}{4Dt}\right)\right]\exp(-x^2/4Dt)$$

$$= -\frac{k \exp(-x^2/4Dt)}{2Dt^{3/2}}\left(1 - \frac{x^2}{2Dt}\right)$$

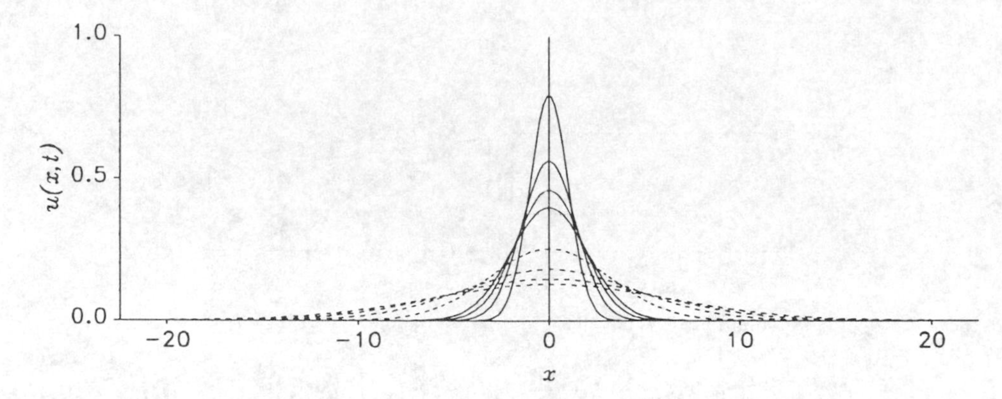

Figure 12.5 Solution of the one-dimensional diffusion equation: distribution of concentration starting from a point source at $x = 0$ at $t = 0$, at times $t = 1, 2, 3, 4$ (solid lines) and $t = 10, 20, 30, 40$ (broken lines).

whereas

$$\frac{\partial u}{\partial t} = \left[\left(-\frac{k}{2t^{3/2}}\right) + \left(\frac{k}{t^{1/2}}\right)\left(\frac{x^2}{4Dt^2}\right)\right]\exp(-x^2/4Dt)$$

$$= -\frac{k\exp(-x^2/4Dt)}{2t^{3/2}}\left(1 - \frac{x^2}{2Dt}\right) = D\frac{\partial^2 u}{\partial x^2}.$$

Hence the solution satisfies the PDE. The solution is a Normal distribution (see Chapter 3) with zero mean, and standard deviation $\sqrt{2Dt}$. As t takes very small values, it can be seen that the spread would tend to zero, and thus at $t = 0$, the distribution is effectively concentrated at a point. Thus the solution also satisfies the initial conditions. The solution is given graphically for various values of t in Figure 12.5.

Comparison of the solution with that for the one-dimensional random walk shows the same equation resulting from the two quite different approaches if you set $D = (\Delta x)^2/(2\Delta t)$. This gives us the relationship between the size of the diffusion coefficient and the step length and interval between steps. To some extent, this result justifies the application of the deterministic differential equation to the more plausibly stochastic situations of animal dispersal and of diffusion resulting from Brownian motion. The deterministic model appears to capture quite well the features of dispersal when a large number of individuals diffuse as a result of individual, independent random motions. However, there are many situations where the simple deterministic diffusion model is inappropriate. From Section 12.4.3 onwards we discuss some of the limitations of the basic diffusion model, and suggest how they might be overcome.

12.4 DIFFUSION IN TWO DIMENSIONS

12.4.1 The two-dimensional random walk

The simple random walk, of the type outlined above along a line, when carried out in a plane results in the same form of differential equation and solution as the deterministic equation of diffusion given in Section 12.2.1. In the simplest generalisation the only difference is that in the plane we allow movements in four directions at right angles with specified probabilities, rather than just right and left movements. Figure 12.6 gives some examples of this type of random walk. During the first time period ($t = 1 \ldots 32$), the paths are concentrated in quite a small area, with much tracing and retracing of paths. During the second period, some individuals begin to spread out. At longer times, some become detached from the central group, and the occasional one (not illustrated here) moves far away. Section 12.4.2 shows how to determine the pattern of abundance which results using the deterministic diffusion equation approximation, and shows that the solution is a Normal distribution as in the one-dimensional case.

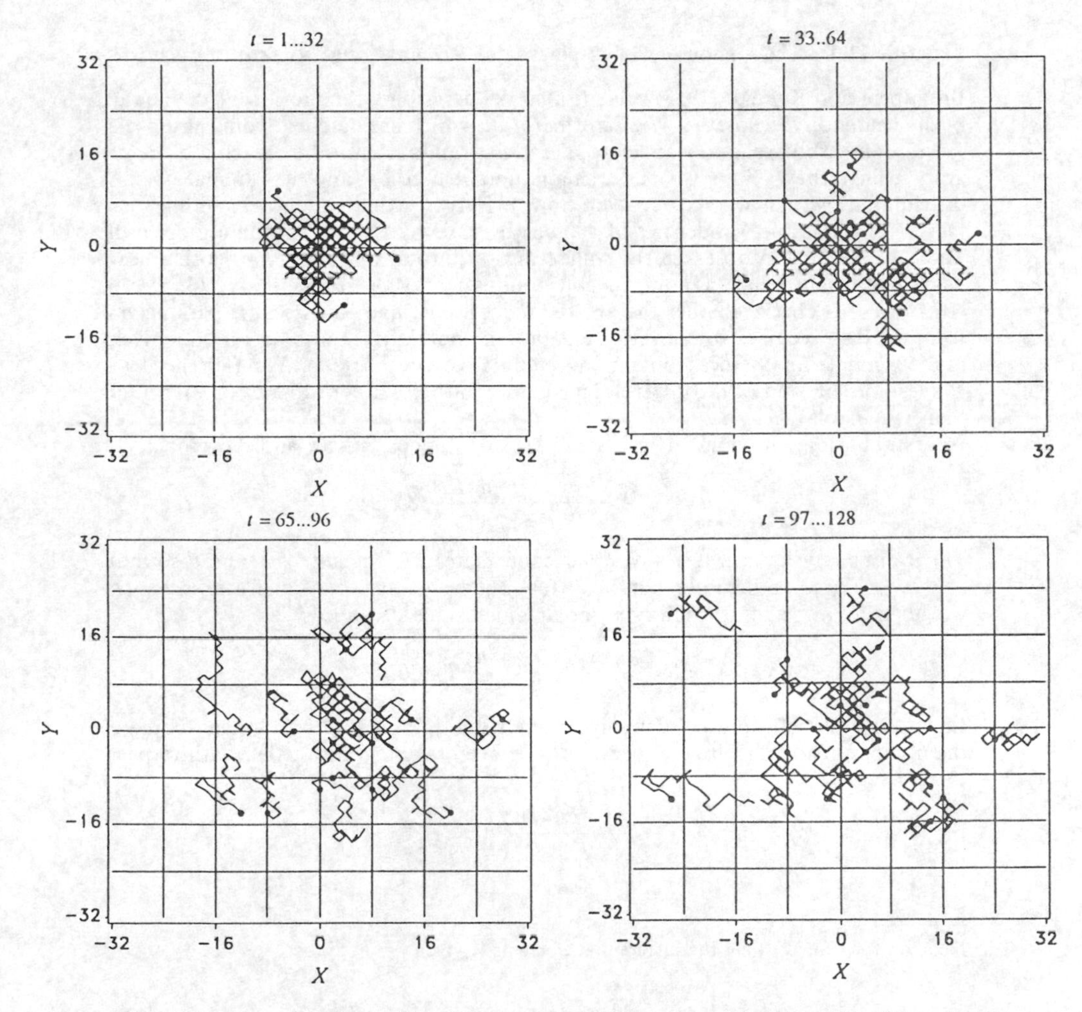

Figure 12.6 Twenty random walks with equally likely jumps to the NE, SE, SW and NW at each time, followed over four time periods from $t = 0$ to $t = 128$. The random walks are perturbed by a further very slight random movement to enable individual walks to be seen more easily. The position of each walk at the end of each time period is indicated by a circle.

12.4.2 Differential equations of diffusion in a plane

The simple deterministic diffusion equation generalises easily to higher dimensions. Equation (12.1) becomes

$$\frac{\partial u}{\partial t} = D\left(\frac{\partial^2 u}{\partial x^2} + \frac{\partial^2 u}{\partial y^2}\right)$$

in two dimensions, assuming isotropy, and the solution for diffusion in an infinite plane from a release at the origin (demonstrated by substitution in the PDE) is

$$u = \frac{k}{t} \exp\left[\frac{-(x^2 + y^2)}{4Dt}\right].$$

Example 12.1 The wandering of the larvae of *Trichostrongylus retortaeformis*

Broadbent and Kendall (1953) present a model describing the motion of the larvae of the helminth *Trichostrongylus retortaeformis* which are hatched from eggs in the faeces of a herbivore, and wander apparently at random until they reach a blade of grass which they climb on and remain until eaten by another animal. What distribution would we expect and how effective would such wandering be? Observations on the wanderings of 400 larvae made by H. D. Crofton are given in Table 12.2. Crofton observed the wanderings of a total of 400 larvae, noting how many had reached each annulus of outer radius 1 . . . 12 at $t = 5, 10, 15, 20, 25, 30$ s. In order to check whether the above model holds, and to assess its parameters from the data we need to convert the data into cumulative numbers within a given radius, and from these to obtain the cumulative proportions. We first need to convert the solution of the differential equation into polar coordinates (r, θ), rather than rectangular (x, y).

First, consider the number in a small area of dimensions $\mathrm{d}x$ and $\mathrm{d}y$ which is

$$\mathrm{d}u = \frac{k \exp[-(x^2 + y^2)/4Dt]}{t} \,\mathrm{d}x\, \mathrm{d}y.$$

This can be rewritten in terms of polar coordinates $x = r\cos\theta$, $y = r\sin\theta$ so that $r^2 = x^2 + y^2$ and taking note that the differential elements need transforming also,* we get the number in a small area, polar dimensions $\mathrm{d}r$ and $\mathrm{d}\theta$, as

$$\mathrm{d}u = \frac{kr \exp(-r^2/4Dt)}{t} \,\mathrm{d}r\, \mathrm{d}\theta.$$

Integrating over all values of θ involves merely multiplying by 2π and then to obtain the number inside the circle of radius R, we need to integrate further with respect to r

$$u_{\mathrm{cum}}(R) = 2\pi k \int_0^R \frac{r \exp(-r^2/4Dt)}{t} \,\mathrm{d}r$$

* The Jacobian for transforming the differential elements is given by

$$\mathrm{d}x\,\mathrm{d}y = \begin{vmatrix} \dfrac{\partial x}{\partial r} & \dfrac{\partial x}{\partial \theta} \\ \dfrac{\partial y}{\partial r} & \dfrac{\partial y}{\partial \theta} \end{vmatrix} \mathrm{d}r\,\mathrm{d}\theta = \begin{vmatrix} \cos\theta & -r\sin\theta \\ \sin\theta & r\cos\theta \end{vmatrix} \mathrm{d}r\,\mathrm{d}\theta$$

$$= (r\cos^2\theta + r\sin^2\theta)\,\mathrm{d}r\,\mathrm{d}\theta = r\,\mathrm{d}r\,\mathrm{d}\theta$$

Table 12.2 Crofton's data on the wanderings of *Trichostrongylus retortaeformis* (the unit of length is 0.7 mm)

Time (s)	Number of larvae in annulus with outer radius											
	1	2	3	4	5	6	7	8	9	10	11	12
5	53	124	118	81	21	3	—	—	—	—	—	—
10	32	60	142	88	41	20	11	6	—	—	—	—
15	20	31	69	124	78	45	31	—	2	—	—	—
20	11	40	65	94	96	47	19	6	10	7	5	—
25	11	33	54	72	87	69	37	20	8	8	—	1
30	13	22	43	40	77	90	53	34	17	3	7	—

$$= 2\pi k\left(-\frac{4Dt}{2}\right)\int_0^R \frac{1}{t}\exp\left(\frac{-r^2}{4Dt}\right)\mathrm{d}\left(-\frac{r^2}{4Dt}\right)$$

$$= -4\pi kD\exp(-r^2/4Dt)\,\big|_0^R = 4\pi kD(1-\exp(-R^2/4Dt))$$

The total over the whole plane is given using the above equation by letting $R \to \infty$, and this is seen to be $4\pi kD$. Thus if $P(R,t)$ is the cumulative proportion within a radius R of the origin at time t, then

$$P(R,t) = 1 - \exp(-R^2/4Dt)$$

and so

$$\ln[1 - P(R,t)] = \frac{-R^2}{4Dt}.$$

Hence plotting the sample values of $\ln[1 - P(R,t)]$ against R^2/t we should obtain a straight line if the model is correct. Figure 12.7 is such a plot, and it appears that although there are possibly some departures from the model, that it might act as good first approximation.

12.4.3 How robust is the deterministic diffusion approximation to the random walk?

A critical reader might be rather uncertain about the value of the two-dimensional

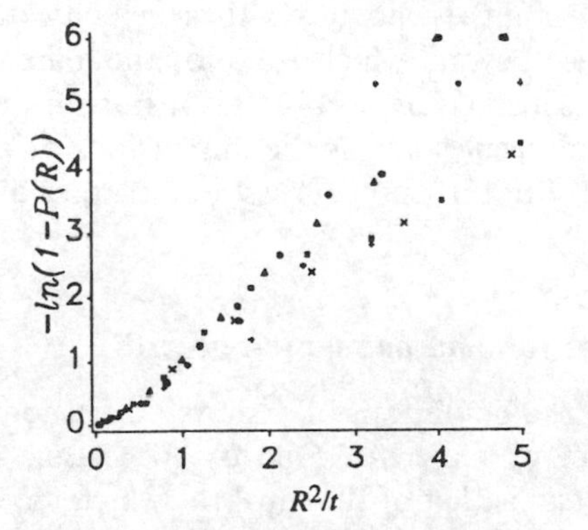

Figure 12.7 Plot of $-\ln[1 - P(R,t)]$ against R^2/t for the *Trichostrongylus* data for various times.

random walk model outlined above, armed with objections of the type: (a) animals generally do not follow the compass directions in their movements; (b) they usually do not move in a homogeneous medium, but in one in which obstacles cause diversions to be made; (c) their movements—corresponding to the steps in the above model—are usually of variable sizes. Skellam (1973) discusses these objections coming to the conclusion that they do not affect the *overall*, *large-scale* pattern of the outcome materially. He also discusses a model in which successive steps are not independent; the idea being that an organism is perhaps more likely to continue *roughly* in the same direction than choose a direction completely at random. That is, it is likely that the *changes* of direction would be independent and at random, but the changes might be restricted rather than spanning the four points of the compass. He also considers other fine details of the process coming to the conclusion 'If we are primarily concerned with animal behaviour studies or the fine-scale pattern of density, it is clearly much better to construct the diffusion model in relation to the realities of the grass-roots situation than to borrow some simple model, such as $\partial f/\partial t \propto \partial^2 f/\partial x^2$, from physical science or mathematical textbooks and trust in its applicability.' We consider some of these models in Section 12.6 and also briefly discuss computer simulation in Section 12.7.

12.5 EXTENSIONS OF DIFFUSION MODELS

12.5.1 The limitations of diffusion

Pure molecular diffusion can only achieve very small-scale movements. Typical diffusion constants of some biological molecules of different sizes are given in Table 12.3.

Using the solution to the one-dimensional equation above, the equation is the same, apart from a constant, as that of the probability density function of a Normally distributed random variable with mean zero, and variance $2Dt$. Thus if any of the substances in Table 12.3 were released at a point, the distribution after t seconds would have a mean equal to its starting point, but with a variance equal to $2Dt$. About 5% (2.5% in each tail) of a Normal distribution lie beyond two standard deviations. So this position (the 2.5% point) is moving with a velocity of $2\sqrt{2Dt}/t = 2\sqrt{2D/t}$. Using this formula, the distance beyond which 2.5% of the material released is found after various periods of time is also given in Table 12.3. It is seen that unaided diffusion can disperse most of these molecules across the dimensions of a typical cell (up to about 10 μ) in under a second. But diffusion across much larger distances would take a very long time.

12.5.2 Combined diffusion and population growth

We remarked above on the limitations of a simple diffusion process for long-distance spread of any organism or molecule. One of the classic examples of diffusion as applied to animal dispersal is that of the muskrat which spread across entire countries within a decade or two. The spread was so fast because the new centres of population acted as breeding colonies in their own right, and this very much accelerated the

Table 12.3 Typical diffusion constants for some biological molecules diffusing through water at 20 °C, and distances moved by the edge of the distribution (as determined by two standard deviations) in various times. (Reproduced with permission from *Molecular Biophysics*, Addison-Wesley, Reading, USA (1962) by R.B. Setlow & E.C. Pollard, Table 4–2, p. 83)

Molecule	Molecular weight	D, $m^2\ s^{-1} \times 10^{11}$	Distance moved (μ) in		
			1 ms	1 s	1 h
Glycine	75	95	2.8	87	5200
Arginine	174	58	2.2	68	4100
Cytochrome C	13 000	10.1	0.90	28	1700
CO-haemoglobin	68 000	6.2	0.70	22	1300
Urease	480 000	3.5	0.53	17	1000
TMV	40 000 000	0.53	0.21	7	390

spread. If the population locally is reproducing at a rate, ρ, and the density is denoted by N, then the growth rate for unaided diffusion is incremented by ρN, and so we obtain the equation

$$\frac{\partial N}{\partial t} = D\frac{\partial^2 N}{\partial x^2} + \rho N$$

in one dimension, and

$$\frac{\partial N}{\partial t} = D\left[\frac{\partial^2 N}{\partial x^2} + \frac{\partial^2 N}{\partial y^2}\right] + \rho N$$

in two dimensions.

Example 12.2 The spread of the muskrat *Ondata zibethica L.*

A few muskrat were accidentally released in 1905 in central Germany, and spread out to cover a large part of Germany and Austria by 1927. Ulbrich (1930) gives data, from which Skellam (1951) extracted the boundaries given in Figure 12.8(a).

Skellam (1973) used the two-dimensional version of the PDE which combines diffusion and exponential growth as a model for this. For simplicity we consider the one-dimensional version of the equation when a number N_0 are released at time zero. The solution is

$$N(x, t) = \frac{N_0}{2\sqrt{\pi D t}} \exp\left[\rho t - (x^2/4Dt)\right],$$

Skellam was particularly interested in the speed at which the muskrat enlarged its range. We can obtain a velocity from the last expression, by inserting a specific value for the density, let us say N_1, and then solving the equation for x in terms of t. This then gives us an expression for the speed of movement of the point at which the density has just reached N_1. This could be thought of as the *velocity of spread*, although it is tied to the particular value we choose for N_1. Solving the equation we obtain

$$2\sqrt{\pi D}\frac{N_1}{N_0} = \exp\left(\rho t - \frac{1}{2}\ln t - \frac{x^2}{4Dt}\right)$$

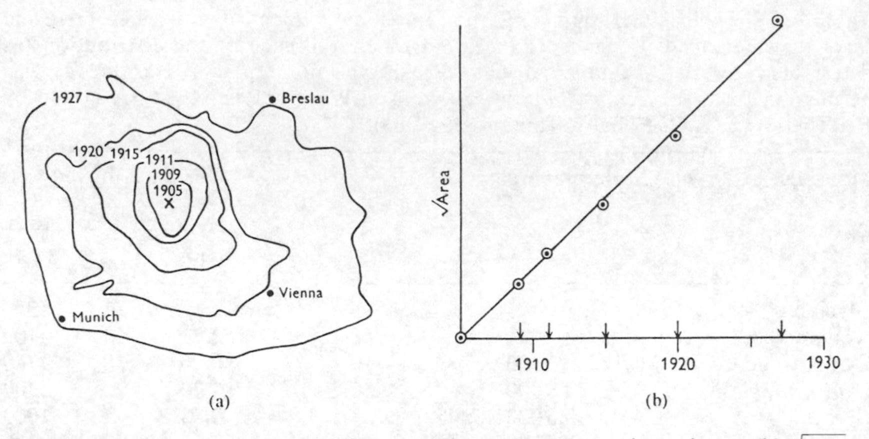

Figure 12.8 (a) The outer contours of the muskrat range at various times; (b) $\sqrt{\text{area}}$ plotted against date. Reprinted, with permission, from J.G. Skellam (1951) Random dispersal in theoretical populations. *Biometrika*, **38**, 196–218.

which can be rearranged to give

$$\frac{x^2}{4D} = \rho t^2 - \frac{t}{2} \ln t - t \ln\left(2\sqrt{\pi D}\,\frac{N_1}{N_0}\right).$$

For very large t, this simplifies to

$$x^2 \cong 4D\rho t^2$$

or

$$x \cong 2\sqrt{D\rho}\,t$$

and hence the asymptotic velocity is $2\sqrt{D\rho}$, now independent of N_1. By a similar argument for the more complex two-dimensional case, we can show that the square root of the area covered grows at a constant rate, and Skellam confirmed that this was the case for the muskrat (see Figure 12.8(b)).

12.5.3 Reaction-diffusion: morphogenesis and animal coat-patterns

So far we have considered simple diffusion, and the effect on the course of the diffusion process of facilitation by population growth. Now we come to consider the problem the other way round. What is the effect of diffusion on the population process? Generally speaking it has been thought that diffusion would even out the spatial distribution resulting from processes operating at different rates within different regions. However, this is not always so: diffusion can sometimes destabilise otherwise homogeneous processes.

In this section we introduce one simple mechanism which has been suggested as a possible cause of some of the more obvious patterns of nature such as the stripes on a zebra and the pattern of bristles on a *Drosophila*. Alan Turing's seminal paper in 1952 first outlined the mechanism whereby the reactions between chemicals, called morphogens, when destabilised by diffusion can account for the periodicity which is an essential feature of such patterns. In this section, we do no more than illustrate the phenomenon. To obtain more detail, consult Meinhardt (1982) or Murray (1989).

In the absence of diffusion, the differential equations of the kinetics of two chemicals might be written

$$\frac{\mathrm{d}u_1}{\mathrm{d}t} = f_1(u_1, u_2)$$

$$\frac{\mathrm{d}u_2}{\mathrm{d}t} = f_2(u_1, u_2),$$

where u_1 and u_2 are the concentrations of the chemicals, and f_1 and f_2 are functions of u_1 and u_2 and the various rate constants determined by the nature of the chemical reactions. Under certain conditions, these equations will have equilibrium solutions in (u_1, u_2). If we now assume that the chemicals are constrained to a thin tube—we consider the one-dimensional case for simplicity—but can diffuse along the tube in both directions, the equations then would incorporate diffusion terms

$$\frac{\partial u_1}{\partial t} = f_1(u_1, u_2) + D_1 \frac{\partial^2 u_1}{\partial x^2}$$

$$\frac{\partial u_2}{\partial t} = f_2(u_1, u_2) + D_2 \frac{\partial^2 u_2}{\partial x^2}.$$

Turing showed that under certain conditions, such a uniform state would break down, and result in waves of concentration of the chemicals along the length of the tube, as a result of the diffusion.

In order to demonstrate this, following the presentation of Maynard Smith (1965), consider the linearised versions of the above equations, expressed in terms of small departures from the equilibrium $w = u_1 - u_{1E}$ and $v = u_2 - u_{2E}$

$$\frac{\partial w}{\partial t} = aw + bv + D_1 \frac{\partial^2 w}{\partial x^2}$$

$$\frac{\partial v}{\partial t} = cw + dv + D_2 \frac{\partial^2 v}{\partial x^2}.$$

In the absence of diffusion (i.e. $D_1 = D_2 = 0$) the equations have an equilibrium at $w = 0$, $v = 0$. Consider the case when $a = 1$, $b = -1$, $c = 1$, $d = 0$, i.e.

$$\frac{\partial w}{\partial t} = w - v + D_1 \frac{\partial^2 w}{\partial x^2}$$

$$\frac{\partial v}{\partial t} = w + D_2 \frac{\partial^2 v}{\partial x^2}.$$

Then if we let $D_1 < D_2$, i.e. v diffuses faster than w, we could obtain patterns as given in Figure 12.9: starting from $w = 0$, $v = 0$ for all x, a stationary wave of growing amplitude develops as a result of a small local perturbation. A perturbation in w at time 0 grows with time as the coefficient a is positive. As $c = 1$, v grows also, but as the rate of diffusion of v is greater than that of w, at $t = 1$ v is more spread out. For some values of x at this time, $w = 0$ and v is positive. Since $b < 0$, the result is a depression of w in this region, so we obtain the situation at $t = 2$. Diffusion continues, and so the pattern has spread out a little also. At $t = 2$, w is now negative at the edges,

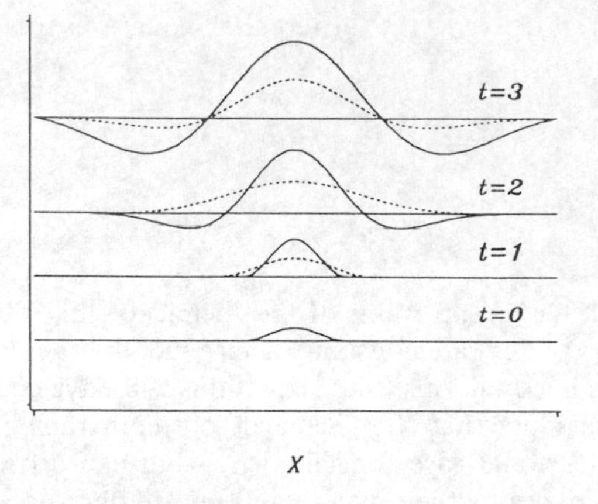

Figure 12.9 The development of periodic pattern in the departures from the equilibrium $(w, v) = (0, 0)$ for all values of x, with increasing time, according to the model outlined in Section 12.5.3. w is indicated by a solid line in the figure, v by a broken line.

and this exerts a downward effect on v as $c = 1 > 0$, especially as $d = 0$, and so w is the only influence on v. Thus v becomes negative, and this causes a further decrease in w, which in turn causes a further depression in v. If the linear equation held over the whole range and not just for small values of w and v this would continue until the diffusion forces and these forces tending to increase the amplitude of the fluctuations are in balance. Our analysis only covers small departures from equilibrium, but similar waves can be obtained from the full equations. Such waveform chemical gradients could be the starting points for patterns of differentiation of cells, resulting in the familiar coat patterns of many organisms.

12.5.4 Convection and force fields

Diffusion often operates in conjunction with other forces or agents. The conservation equation can also be applied to other means of transport such as convection without any appreciable diffusion being present. In this section, we briefly consider some of these cases.

Convection

In this case, as the particles of the substance move with their containing fluid, the flux is proportional to the velocity of the moving fluid

$$J(x, t) = \beta u(x, t) v(x, t)$$

and so we obtain, in the case where $A(x, t) =$ a constant A and $R(x, t) = 0$,

$$\frac{\partial u}{\partial t} = -\beta \frac{\partial (uv)}{\partial x}.$$

Combined diffusion and convection

In many cases of convection, diffusion occurs also. Convection by the bloodstream is the dominant component of the transport of many hormones from the site of release to the site of action. But it is interesting to know how a bolus released over a very small interval of time might spread out by diffusion within the bloodstream before it reaches its target. The equation for combined convection and diffusion can be obtained by including fluxes due to both types of transport and hence we get an equation of the form

$$\frac{\partial u}{\partial t} = D\frac{\partial^2 u}{\partial x^2} - \beta\frac{\partial (uv)}{\partial x}.$$

Force fields

Here, the agent responsible for the motion is not a non-zero gradient in concentration of the transported substance as in diffusion, nor the movement of the medium in which it as it were hitches a lift as in convection, but some other force which could be gravity, or an electrostatic or magnetic field, or a gradient in the concentration of a chemical other than the transported substance as in chemotaxis. Some force field, $\psi(x,t)$, must be invoked, and the flux is a product of the local concentration, the gradient of ψ, and some constant of proportionality, β, so we obtain

$$J = \beta u(x,t)\frac{\partial \psi(x,t)}{\partial x}.$$

This leads to the equation of motion as follows

$$\frac{\partial u}{\partial t} = -\frac{\partial}{\partial x}\left[\beta u\frac{\partial \psi}{\partial x}\right]$$

in the case when $A(x,t)$ is constant, and $R(x,t)=0$.

12.6 STOCHASTIC MODELS

Several authors (apart from Skellam, see Section 12.4.3) have shown how limited the potential of the deterministic diffusion equation is, whether aided by supplementary agents or not. It cannot adequately model the spread of many important larger scale biological phenomena, for example colonisation, spread of epidemics and tumour growth. We have seen above how random walks of the Brownian motion type can be modelled by deterministic diffusion equations. This is related to the fact that diffusion is essentially a small-scale phenomenon that can mimic processes which are the result of many small random steps. Many mechanisms of dispersal do not conform to the Brownian motion type nor do they result in a Normal distribution—the outcome of spatially unlimited diffusion when the initial source is a point.

Mollison (1977) discusses models for ecological and epidemic spread. He argues that models which are both *stochastic and non-linear* are necessary in order to capture many of the essential features of these phenomena. We have discussed elsewhere the need for non-linear models. Without them, we do not have access to a simple growth formula like the logistic equation, being restricted to exponential growth. It is also especially necessary to incorporate demographic stochasticity—the departures from the deterministic predictions which occur at low densities because organisms usually only come in whole numbers—into models of spatial spread because the leading edge of the process frequently consists of very small numbers of individuals. Other detailed surveys of more complex spatial stochastic models are given by Liggett (1985) and Renshaw (1986, 1991).

One of the simplest stochastic models which has the potential for describing a wide range of phenomena is the *contact birth process*. In this, individuals have fixed locations in some space. These individuals give birth to others according to what is called a *simple birth process*, and each of the newly born individuals is assigned a location relative to its parent at birth by drawing at random from a specified probability distribution called a *contact distribution*, which can be of *any* form. The term contact distribution probably originates in epidemiology. Diffusion birth processes are a restricted version in which each offspring performs independent Brownian motion relative to its parent before settling. Although it seems more restrictive to allow only one move of the offspring from the parent, the contact birth process will accommodate a much wider range of behaviour, and is therefore more useful for many models of ecological and epidemic spread. For example, if distributions with long tails were allowed, we could model processes in which occasionally much greater distances were covered, and this would allow faster spread than a diffusion process. The simplest variants of contact birth processes restrict locations to the points of a regular grid, commonly square or hexagonal. We consider such a model for tumour growth in the next section. In the subsequent section, we consider a simple stochastic model involving a more substantial directional component, which has been used to model bird navigation.

12.6.1 A stochastic contact process on a grid as a model for tumour growth

Bjerknes & Iversen (1968) suggested a possible mechanism for tumour induction in the basal layer of an epithelium according to which abnormal cells escape from the control of cell division normally exercised by the secreted chemical chalone. It is in the basal layer that cells divide, and in the higher layers that they differentiate. The differentiated cells in the higher layers secrete chalone, which diffuses back to the basal layer and acts as an effective controlling agent for multiplication of normal cells there.

Cells tend to pack into a honeycomb pattern in layers, and when a cell divides in the basal layer, it and its daughter remain in the basal layer, displacing a neighbouring cell to a higher layer. In a very simplified mathematical model of the basal layer and of tumour development proposed by Williams and Bjerknes (1972), the centres of the cells are assumed to be situated on a regular hexagonal grid. Abnormal cells divide κ times as fast as normal cells; κ, a quantity greater than 1, was termed by them the 'carcinogenic advantage'. Cells are assumed to divide independently according to a Poisson process with rate 1, the abnormal cells having a rate κ, i.e. the times between cell division follow negative exponential distributions with means 1 and $1/\kappa$ for normal and abnormal cells respectively. The questions which can be asked of such a model are: (a) what are the chances that an abnormal cell line will die out; (b) what is the growth rate of the abnormal cell line; (c) what spatial configuration does the abnormal cell line take up?

The probability that the entire abnormal clone will disappear is $1/\kappa$, and simulated configurations for normal and abnormal cells are given in Figure 12.10, for κ taking the values infinity, 2 and 1.1. The only cells which matter in the development of the

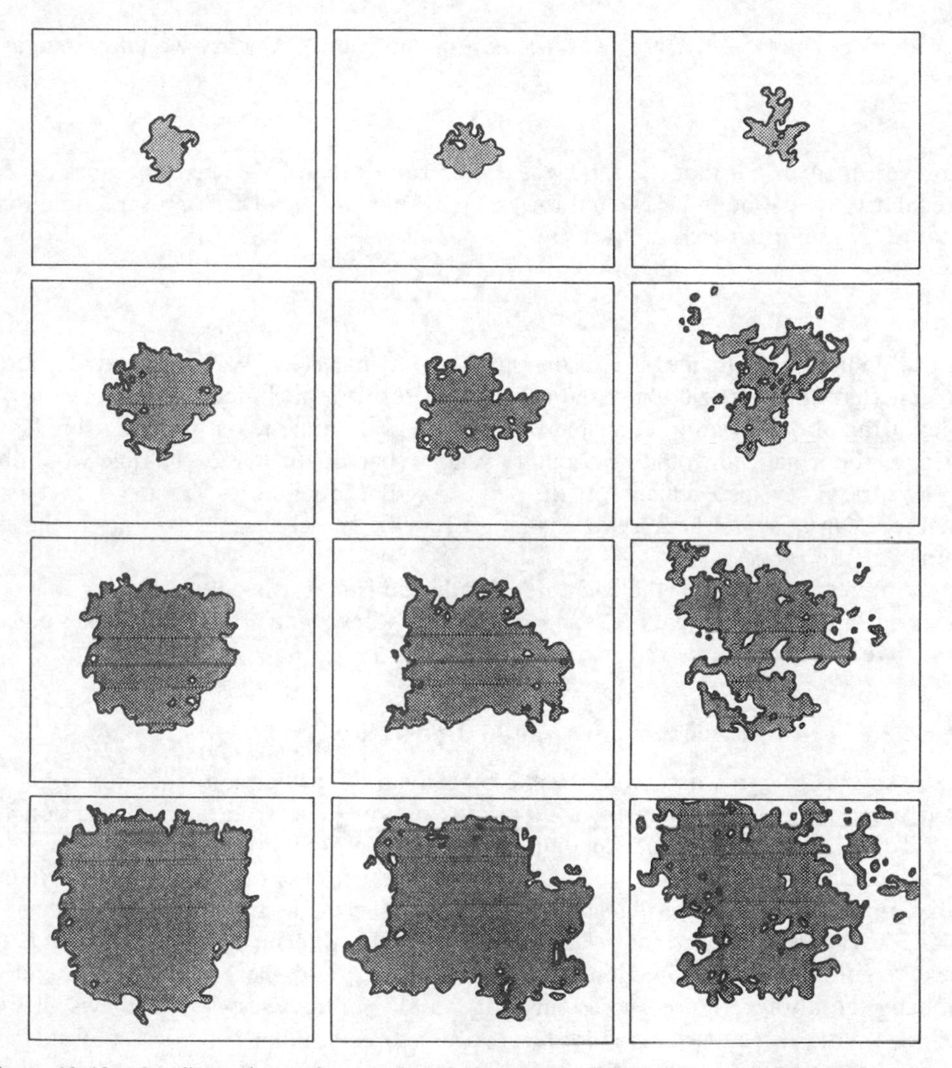

Figure 12.10 Configurations of normal and abnormal cells in the basal layer for simulations of the model in Section 12.6.1. The boundaries of the abnormal cells are drawn in, abnormal cells being indicated by shaded, and normal cells by white areas. The three columns correspond to different values of κ, on the left $\kappa = \infty$, in the middle $\kappa = 2$, and on the right, $\kappa = 1.1$. The four rows show the configurations when the number of abnormal cells first reach 100, 400, 900 and 1600. Reprinted by permission from *Nature*, vol. **236**, pp. 19–21; Copyright © 1972 Macmillan Magazines Limited.

clone are those on the boundary of the abnormal cells, because the interior and exterior cells merely replicate themselves; only when cells on either side of the boundary divide can the result be a change of cell type. Then, calling the total number of abnormal cells N, and the number of peripheral abnormal cells n, we would expect that

$$\frac{\mathrm{d}N}{\mathrm{d}t} = (\kappa - 1)n$$

since the extra advantage to the abnormal cells for each boundary position which they already occupy is $\kappa - 1$. One might also expect, since the boundary would on average be circular, that

$$n \propto N^{0.5}$$

however, it turns out that the exponent is 0.55 rather than 0.5. This is because of the irregularity of the boundaries, and the result is that the mean first-passage time to a total of N abnormal cells is given by

$$\bar{t} = \frac{1.25 N^{0.45}}{\kappa - 1}$$

in time units equal to mean division time for normal cells. Williams and Bjerknes suggest that about 50 000 abnormal cells would result in a clinically manifest growth (measuring about 2 mm in diameter). For $\kappa = 1.1$, the time taken to reach this from a single abnormal cell would be about 13 years, whereas for $\kappa = 2$, the time would be 1.33 years. They also remark on the very irregular boundaries for $\kappa = 1.1$. These therefore can arise not because of any heterogeneity, merely because of the stochastic nature of the process.

An interesting aspect of the computer simulation is that it is only necessary to store the positions of the boundary cells, updating these whenever a replacement takes place. This saves storage space, and execution time.

12.6.2 Directed Brownian motion and bird migration

Many species of birds accomplish remarkable feats of navigation when returning to their breeding grounds often from thousands of miles away. In an investigation by G.V.T. Matthews and D.G. Kendall reported in 1974, mathematical models were constructed in order to assess whether it is possible, with the rather limited cues likely to be available, to achieve the same navigational success as many bird species.

The key assumption is that at any point on the Earth's surface the bird is able to make a guess as to the direction in which home lies; and the guesses are generally, though not always, more successful than totally blind guesses. Matthews (1955) collected data, reproduced in Figure 12.11, of the direction in which released birds flew relative to the target. Examination of this figure reveals that quite frequently the birds

Figure 12.11 Matthews' data on release angle of Manx shearwaters, relative to the direction of home which is indicated by the vertical line in the central circle. Reproduced, with permission, from *Bird Navigation*, by G.V.T. Matthews, published by Cambridge University Press, 1955.

travel in almost the right direction, but occasionally are out by up to 180°. Kendall used a Wrapped Normal distribution to describe the distribution centred on the direction of home. A Wrapped Normal distribution is a Normal distribution with mean zero and specified standard deviation (we use λ for this) wrapped round a circle of unit perimeter, denoted $WN(\lambda)$ until the tails of the distribution meet. The value of λ used in the simulations, estimated using Matthews' data, is 0.202.

The model used incorporates the following features.

(1) The bird sets out in what it considers to be the right direction and flies in a straight line at a uniform speed, V, on the first leg of the journey, of length L.
(2) At the end of this leg, the bird reassesses the direction of home, makes a correction and sets off again on another leg of the same length.
(3) The bird makes errors in assessing the direction which follow a distribution of a particular type. The error distribution is the same throughout the journey, and the errors are independent between legs.
(4) The cycle is repeated until the bird finds itself within sighting distance, a, of its target, and immediately homes in on it.

The birds' motion was simulated using this model (termed Manx Motion by Kendall), and 500 artificial flights were made using the Titan computer of the University of Cambridge. The parameters used in the simulations were:

V = flying speed = 40 m.p.h.
L = length of leg between re-checks = 20 miles
D = release-point to target distance = 200 miles
b = distance at which the bird is lost to sight = 2 miles
a = distance at which the bird sights the target = 10 miles.

The aspects of the individual flights which were recorded were as follows.

θ_0 = the angle which the radius to the bird when lost to sight makes with the true direction from release-point to target-centre.
T = the time taken to reach the target.
Y = the maximum lateral displacement of the flight path joining the release-point to the target-centre.
ϕ = the angle between the radius from the target-centre to the point at which the target perimeter is first crossed, and the line joining the target-centre and the release-point. These are illustrated in Figure 12.12.

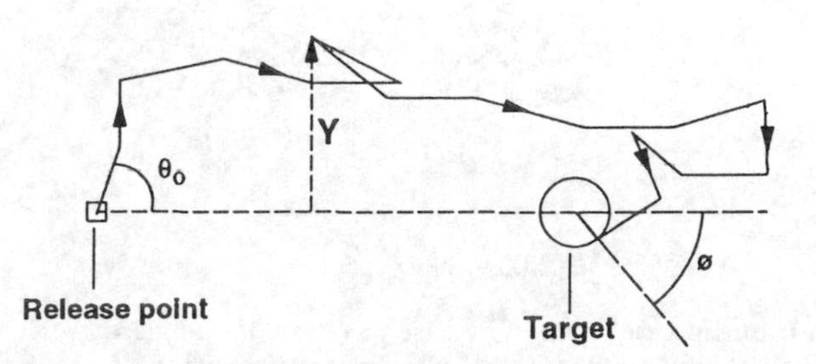

Figure 12.12 Parameters of individual flight which were measured.

Figure 12.13 displays the results of these simulations for 100 such flights. The other 400 flights produced similar results. The important results are that:

(1) All 500 birds made it to their target;
(2) The ϕ's are widely scattered (indicating that a large proportion of the birds overshot the target region to some extent)—more so than the θ_0's;
(3) The distribution of T is highly skew, with a long tail towards high values—some flights were much less successful than others;
(4) The birds seem to have diverged most (measured in terms of lateral displacement from a straight path) in the middle to later parts of their flights, only really starting to home in on the target when they were almost overshooting it.

Kendall carried out much more analysis, including simulations using a slightly different model. He also developed an approximate diffusion model of Brownian motion supplemented by an attractive polar drift. Assuming this model, he demonstrated conclusively that the bird will always reach its target provided that the drift is strictly positive, even if it is very small.

He also demonstrated the existence of a *circle of confusion*. While the bird is some way from the target, the polar drift still tends to move it on balance in the right direction. When the bird gets close to the target—within Kendall's circle of confusion—the Brownian motion becomes much more important and tends to dominate. If the radius of the circle of confusion is smaller than the target-radius a, the bird will cross the target-radius first, and then immediately home in to the target-centre. If on the other hand, the circle of confusion lies outside the target, after first entering the circle of confusion the bird will tend to fly outside then back again, possibly many times, before reaching the target and then homing in on the centre. In this case we would expect the distribution of the angle ϕ to be uniform on the complete circle, so that there is no preferred direction. The radius of the circle of confusion for the simulated model is about 8.7 miles, which is quite close to the target-radius of 10 miles, and this is consistent with the almost uniform appearance of the distribution of ϕ obtained in the simulations.

A general formula for expected length of time before hitting the target was obtained, assuming the diffusion model holds. The interesting thing about the simulations is that,

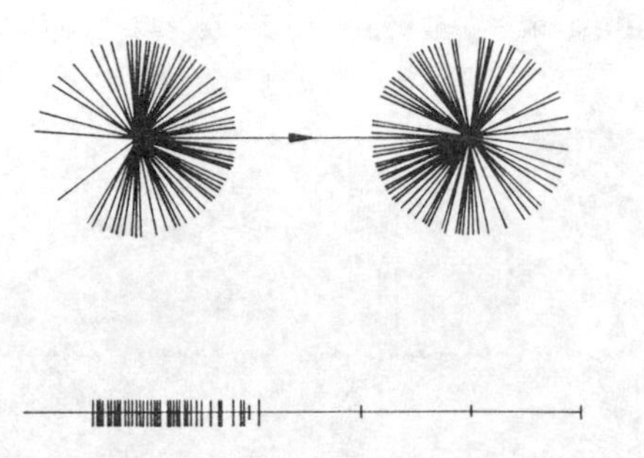

Figure 12.13 Distributions of angle of release θ_0 (top left), angle of arrival ϕ (top right) and T for 100 simulated flights. Reproduced, with permission, from D.G. Kendall, (1974) Pole-seeking Brownian motion and bird navigation. *J. Roy. Statist. Soc.* B, **36**, 365–402.

despite in some ways departing substantially from the theoretical diffusion model (the length of the leg before changing direction was 20 miles compared with the 200 miles distance between release-point and target-centre), the average hitting-time was only 1.14 times the expected value for the Brownian motion model.

12.7 COMPUTER SIMULATION OF SPATIO-TEMPORAL PROCESSES

Just as there are many models, so there are also many approaches to the simulation of spatial processes. Some examples we have seen already. The precise simulation of a spatial/temporal model in all its fine detail requires both substantial computing time and a large amount of memory, whether real or virtual. Considerable efforts have been invested in finding short cuts that attempt to scale down these requirements to a level that can be satisfied with current computers. One technique which has been used quite widely for modelling and simulating some classes of spatially distributed systems is the cellular automaton.

12.7.1 Cellular automata

Cellular automata were devised by Stanislaw Ulam and John von Neumann in the late 1940s so as to provide a model for spatially extended complex systems in which the conventional distinction in computing between hardware and software was blurred. Von Neumann's interest was in constructing self-reproducing structures which could be used for understanding biological problems. Today cellular automata are used as general computational devices for simulating spatial processes, and are of practical value for solving diffusion problems in physical systems. They consist of a lattice of sites, at each of which a variable resides which can take a finite set of possible values. The variables are updated—synchronously at regularly spaced times—using the same update rule at each site. The rules are local in time and space: the value at a site at time t is dependent only on the values at neighbouring sites at time $t-1$. They can be one-dimensional, or in higher dimensions although the cases which have been most intensively studied have been in one and two dimensions.

Cellular automata

Cellular automata are models of *processes defined at the points of a regular grid*. The simplest, in a way the standard, version has the following properties.

- Associated with each grid point is a variable which takes one of a *finite number of discrete values*. So it might be a simple binary variable denoting presence or absence, or could be a count, or an approximate representation of concentration of a chemical.
- The values at each point are *updated simultaneously at discrete, usually regularly spaced, times*.
- The updating rule is the *same for every point*.
- The rule involves the current *value at the point*, and the *values at* a specified set of *near neighbours*.
- Usually the updating rule involves only the *current values*, or the *last few values*.
- The updating rule is *deterministic*.

Example 12.3 'Life'

One of the earliest widely used cellular automata was that devised by John Conway in the 1960s, when the associated game called 'Life' had a considerable following. This is a very simple two-dimensional automaton which can be operated on a fairly low power personal computer (see Gardner, 1971). The rules are as follows:

(1) The automaton operates on a square grid; thus the points situated at the centre of every square each have eight neighbours, four with a side in common, and four with only a corner in common.
(2) Each point can take one of two values; it can be either alive or dead.
(3) The update rule determines whether a cell is alive or dead in the next time period according to the number of living neighbours at present:

Number of neighbours alive	2	3	0, 1, 4, 5, 6, 7, 8
State next time	Unchanged	Alive	Dead

All cells are updated simultaneously at the tick of the clock, at a time interval depending on the speed of the computer, or the preference of the user. Despite its extreme simplicity, this automaton has quite a wide repertoire of behaviour, which has been extensively studied. The following types of behaviour occur: *stable isolated configurations*, including some very simple ones such as squares and hexagons; *oscillators* with various periods; and *gliders* which drift across the grid at a constant rate as well as other more complex forms of structured behaviour. These patterns of behaviour are not of particular interest in themselves. The important thing is that this extremely simple model has a surprisingly wide range of behaviour patterns.

Cellular automata can thus be thought of as deterministic, spatially extended, dynamic systems, discrete in both time and space, having a discrete set of possible values at each point, and involving only local interaction. The reasons for some of these restrictions—and others listed above—are partly concerned with the ease of simulation of the automaton on existing hardware, but also because they make theoretical study of the properties more manageable. The updating rules are usually converted into a *look-up table*. This is a table which the automaton uses to obtain the new value at any grid point depending on the current or past few values at the point or at its neighbours. Thus it is a very quick matter to update the automaton for every point of the grid. The rules are usually simple in the sense that they depend on few bits of information about each of a few neighbours for each of a few recent times. The look-up tables would become enormous if any of these restrictions were relaxed too much. This use of a look-up table for updating means that the number of possible values allowed at each point has to be limited. High speed in updating every point of the grid is important if the simulations are not to take an inordinately long time.

12.7.2 Classes of emergent behaviour and dynamical systems

Cellular automata are defined by local rules of interaction. The models for two-dimensional cellular automata result in look-up tables which only involve the immediate neighbours—either four or eight—of the cell in question. The definition is entirely local and yet, even for simple cellular automata such as 'Life' discussed above, behaviour on a larger scale often occurs. Such behaviour is said to be *emergent behaviour*, since it emerges out of the local behaviour. The gliders which drift across the screen at a uniform velocity are examples of behaviour on the scale of the whole automaton, which arises solely as a consequence of a particular initial configuration and the local rules—which do not contain a specific directional component!

One reason for the importance of cellular automata over more flexible parallel models is that they are more amenable to analysis of their emergent behaviour. Stephen Wolfram (1984, 1985) has identified—as a result of extensive simulations—four qualitatively different classes of behaviour displayed by one-dimensional cellular automata. He found that there is a link with the behaviour of homogeneous dynamical systems (as demonstrated by systems of differential and difference equations, see Chapter 8).

Cellular automata classes of emergent behaviour

Class 1. Evolution to a spatially homogeneous steady state, in which all cells are the same. The comparable behaviour of a dynamical system is a limit point.

Class 2. Simple, separated, stable or periodic structures, typically with low period. The comparable behaviour of a dynamical system is a limit cycle.

Class 3. Chaotic, i.e. aperiodic, patterns, some of which are spatially regular and others spatially irregular. Some of the regular patterns which occur are self-similar fractals, i.e. their form is the same when viewed at several magnifications. The comparable behaviour of a dynamical system is chaos.

Class 4. Localised structures, including propagating structures, which take on more complex patterns than the above. There is no equivalent comparable behaviour in dynamical systems.

The reader is referred to Wolfram (1984, 1985, 1986), Langton (1986) and Toffoli & Margolus (1987) for other results about cellular automata. Many of the restrictions of the standard cellular automaton—when seen as a biological simulation tool—have been relaxed as more powerful hardware appears. Features such as asynchronous updating at random times using stochastic rules, a much larger number of states at each point and some degree of more remote intercommunication are available now. Cellular automata have not been widely used in biological modelling so far, but with these generalisations, and as theoretical understanding improves, their use is likely to increase.

Langton (1986) discusses the potential of these systems as models of the synthesis of some of the basic molecules of living organisms, and as models of the sophisticated emergent behaviour seen in social insect colonies, even though the behaviour repertoire of individual insects is rather limited. Inghe (1989) has shown how a computer simulation of the growth of laterally spreading clonal organisms could be implemented using a cellular automata model. Pandey (1989) has modelled the development of the immune response in a retrovirus system using cellular automata. Hassell, Comins & May (1991) have used cellular automata involving qualitative rules of interaction between patches for assessing the behaviour of spatially distributed host–parasitoid systems. By this means, they were able to provide a check on more specific models involving Nicholson–Bailey dynamics.

12.8 APPENDICES

12.8.1 Analytical solution of partial differential equations

Partial differential equations (PDEs) are equations containing partial derivatives.

Notation

The *subscript notation* saves a great deal of use of the partial ∂ notation, replacing $\partial u/\partial t$ simply by u_t, and $\partial^2 u/\partial x^2$ by u_{xx}. As an example of a PDE consider the simple diffusion equation $\partial c(x,t)/\partial t = D\partial^2 c(x,t)/\partial x^2$ which can also be written as $c_t = Dc_{xx}$.

Differences between ordinary and partial differential equation

Arbitrary constants and arbitrary functions
The ordinary differential equation (ODE) $\mathrm{d}u/\mathrm{d}t = 3t$ has a general solution $u = 3t^2/2 + C$, where C is an arbitrary constant. The arbitrary constant is necessary because specifying the rate at which u changes does not completely specify u, but it does specify the solution down to an arbitrary *additive constant*. This constant is determined by the value of u we start from, specified by the initial condition. If the initial condition is that $u = 0$ when $t = 0$, then substituting in we get $0 = 0 + C$ and therefore $C = 0$.

PDEs can be integrated in similar ways, except that the 'constants' become more interesting.
If

(i) $u = 3x + t$ then $\partial u/\partial x = 3$;
(ii) $u = 3x + t^2$ then $\partial u/\partial x = 3$;
(iii) $u = 3x + 4t^2 + 5t^3 + 6t^4$ then $\partial u/\partial x = 3$.

From these examples, we conclude that the general solution of the PDE $\partial u/\partial x = 3$ is $u = 3x +$ (any function of t) written as $u = 3x + k(t)$ where $k(t)$ is known as an arbitrary function of t. Thus as with ODEs, there are general solutions containing arbitrary elements and, as we see above, these are of a more general form than arbitrary constants called *arbitrary functions* which are additive in the scale of the response variable.

Initial and boundary conditions
With ODEs there are solutions to specific problems which are obtained by fixing the arbitrary constants so as to satisfy the initial conditions. In PDEs, speaking informally, we have conditions at the edges for each of the variables, as it were. Those for t, we call initial conditions, and those for spatial variables are called boundary conditions.

Initial conditions specify the state of the system at zero time, and therefore need to specify the state of the system for all values of x at $t = 0$. They might be as simple as $u = 0$ for all x, at $t = 0$. Or might specify some functional form: $u = \sin x$ for $-2\pi \leqslant x < 2\pi$, and $u = 0$ otherwise at $t = 0$.

Boundary conditions are different from initial conditions, in that they *can* specify conditions at every spatial boundary (whereas the initial conditions usually only specify the state of the system at zero time). Also these boundary conditions can be fixed, e.g. $u = 0$ at $x = 0$ and $u = 3$ at $x = 5$, for all values of t or variable in time, e.g. $u = \sin(t)$ at $x = 0$ and $u = 0$ at $x = 5$ for all values of t. They might also be in terms of the flux across the boundary, e.g. if the boundary is a perfect insulator, there would be zero flux across it. Alternatively no boundaries might be set, and therefore no boundary conditions. Sometimes solutions are required in a medium which is effectively infinite in extent, for example an infinitely long tube in one dimension. We examine (in Section 12.4.2) solutions in which diffusion from a point source spreads over a plane of infinite extent.

Analytical techniques for solution: separation of variables

There is a whole battery of techniques for solving PDEs of which we only consider one: separation of variables. The subject is generally beyond the scope of this book, but we outline one example so that interested readers can appreciate the remarkable mathematical theory which is used and the interesting solutions which can occur.

Obtaining the general solution

Rather than consider a general problem we apply the method to one of the most important PDEs which we study in this book: the simple one-dimensional diffusion equation, $u_t = Du_{xx}$. The technique attempts to express the solution as a product $u(x, t) = X(x)T(t)$. We use the notation $X'(x) = \mathrm{d}X/\mathrm{d}x$ $(= \partial X/\partial x)$ but the partial ∂ is unnecessary because X is a function of x only.

Substituting $u_t = X(x)T'(t)$, and $u_{xx} = X''(x)T(t)$ into the equation we obtain

$$X(x)T'(t) = DX''(x)T(t).$$

Hence, dividing both sides by $X(x)T(t)$,

$$\frac{T'(t)}{T(t)} = D\frac{X''(x)}{X(x)}.$$

Now the left hand side only depends on t and the right hand side depends only on x, and as the equation is true for all x and t, they must both be equal to a fixed constant, say k. So we get

$$\frac{T'(t)}{T(t)} = k$$

$$D\frac{X''(x)}{X(x)} = k$$

and so

$$T'(t) - kT(t) = 0$$

$$DX''(x) - kX(x) = 0.$$

We have thus converted the PDE into two ODEs, and we now proceed to solve these. The first in T can be solved by separation of variables.

$$\frac{\mathrm{d}T}{\mathrm{d}t} = kT$$

with solution

$$T = T(0)\ \mathrm{e}^{kt}.$$

In order for this solution to be bounded as $t \to \infty$, k must be negative, and so set $k = -Dm^2$. Then solving the second equation

$$X'' + m^2 X = 0$$

by the methods of Chapter 8, we get

$$X = A\ \sin(mx) + B\ \cos(mx),$$

and hence the overall solution is

$$u = T(0)\ \mathrm{e}^{-Dm^2 t}[A\ \sin(mx) + B\ \cos(mx)]\,.$$

Imposing the boundary and initial conditions

Suppose the problem is such that we are only interested in the solution for values of x between $x = 0$ and $x = 1$, and at these boundaries, the concentration is zero at both ends throughout the study. Also suppose we start the problem with u spread over this axis according to a function which we do not wish to specify at the moment, call it $f(x)$. So the boundary and initial conditions are as follows.

Boundary conditions: $u(0, t) = 0$, $u(1, t) = 0$ for all t

Intitial conditions: $u(x, 0) = f(x)$, $\quad 0 \leqslant x \leqslant 1$

Satisfying the boundary conditions

The solution we found above was

$$u = T(0)\ \mathrm{e}^{-Dm^2 t}[A\ \sin(mx) + B\ \cos(mx)]\,.$$

We impose the boundary conditions that $u(0, t) = 0$, $u(1, t) = 0$ for all t, by substituting $x = 0$ and $x = 1$ into the general solution and equating the resulting expressions to the values given which are zero in each case. Hence we get

$$u(0, t) = T(0)\ \mathrm{e}^{-Dm^2 t}[A\ \sin(0) + B\ \cos(0)] = T(0)\ \mathrm{e}^{-Dm^2 t}(0 + B) = 0.$$

Hence $B = 0$.

$$u(1, t) = T(0)\ \mathrm{e}^{-Dm^2 t}[A\ \sin(m) + B\ \cos(m)] = T(0)\ \mathrm{e}^{-Dm^2 t} A\ \sin(m) = 0.$$

Hence either $\sin(m) = 0$ or $A = 0$ but we do not choose the alternative $A = 0$, since this would mean that the solution takes the value zero for all time. Hence, $\sin(m) = 0$, i.e. $m = \pm\pi,\ \pm 2\pi,\ \pm 3\pi, \ldots$. Note that we set $k = -Dm^2$, and that nothing so far in the problem has determined the value of k and hence m, and so we are free to choose the values of m which satisfy this boundary condition, namely, $m = \pm\pi$, $\pm 2\pi$, $\pm 3\pi, \ldots$. So if we write the general form of this as $m = n\pi$, where n can be a positive or negative integer, we get the following form for the general solution which also satisfies the boundary conditions

$$u(x, t) = u_n(x, t) = A_n \, e^{-D(n\pi)^2 t} \sin(n\pi x).$$

Note that we have put a subscript n as an indicator of which of the many solutions we are referring to. Any solution of this form will satisfy the differential equation and the boundary conditions. Also a sum of solutions of this form will still satisfy the differential equation and the boundary conditions. This can be demonstrated by substituting into the differential equation and the boundary conditions the solution $u = u_1 + u_2$ and showing that the differential equation and the boundary conditions are still satisfied, by using the fact that u_1 and u_2 satisfy them. We see below that the solution which also satisfies the initial conditions can be built up from a sum of solutions of this form.

Satisfying the initial conditions

We now need to find a sum of selected members of the set of u_n which also satisfies the initial conditions. Suppose that, by a lucky coincidence, the initial condition was

$$u(x, 0) = 3 \sin(\pi x) + 2 \sin(3\pi x),$$

then the choice would be straightforward. We need a sum of two components as the solution

$$u(x, t) = A \, e^{-D\pi^2 t} \sin(\pi x) + B \, e^{-D(3\pi)^2 t} \sin(3\pi x).$$

Substituting into the boundary condition, we get

$$u(x, 0) = A \, e^0 \sin(\pi x) + B \, e^0 \sin(3\pi x),$$

from which we can infer the appropriate values of A and B: $A = 3$, $B = 2$.

So the solution would be

$$u(x, t) = 3 \, e^{-D\pi^2 t} \sin(\pi x) + 2 \, e^{-D(3\pi)^2 t} \sin(3\pi x).$$

What do we do if the initial condition is that u is some other function than a simple cosine or sine function $u(x, 0) = f(x)$? Then we can make use of a Fourier Series expansion of $f(x)$, i.e. an expansion as a sum of sine and cosine terms, using Fourier's Theorem.

Fourier's Theorem

If $f(x)$ is a continuous function on the interval $[-L, L]$, the Fourier Series of f is the trigonometric series

$$f(x) = \frac{a_0}{2} + \sum_{n=1}^{\infty} \left(a_n \cos \frac{n\pi x}{L} + b_n \sin \frac{n\pi x}{L} \right)$$

where the a_n's and b_n's are given by the formulae

$$a_n = \frac{1}{L} \int_{-L}^{L} f(x) \cos \frac{n\pi x}{L} \, dx, \qquad b_n = \frac{1}{L} \int_{-L}^{L} f(x) \sin \frac{n\pi x}{L} \, dx, \qquad n = 0, 1, 2 \ldots.$$

By this means we can determine the coefficients a_n and b_n, and hence obtain an expansion of $f(x)$ in terms of cos and sin terms, and from that, by replacing $A\ \sin(mx)+B\ \cos(mx)$ by $e^{-Dm^2t}[A\ \sin(mx)+B\ \cos(mx)]$, get the solution of the PDE

$$f(x)=\frac{a_0}{2}+\sum_{n=1}^{\infty} e^{-(n\pi)^2Dt}\left(a_n \cos\frac{n\pi x}{L}+b_n \sin\frac{n\pi x}{L}\right).$$

So from any well defined function which serves as initial conditions at $t=0$, we can obtain the solution of the PDE by this means.

Elementary properties of the solution

Let us consider the simpler example above, rather than the general form, as the important properties can be more easily gauged from the solution to the problem with initial condition

$$u(x,0)=3\ \sin(\pi x)+2\ \sin(3\pi x)$$

namely,

$$u(x,t)=3\ e^{-D(\pi)^2t}\ \sin(\pi x)+2\ e^{-D(3\pi)^2t}\ \sin(3\pi x).$$

Note that the effect of the diffusion in this case is to multiply each sine term by a negative exponential function of t. The effect of these in turn is to cause each component to decay towards zero, so that at infinite time, the solution is zero. Also the term in $\sin(3\pi x)$ has an exponential declining multiplier $e^{-D(3\pi)^2t}$ attached to it as opposed to the multiplier $e^{-D(\pi)^2t}$ attached to the $\sin(\pi x)$ term, so that the former decays much faster, and after a time, the solution will be dominated by the sine term with the lower coefficient. This is illustrated in Figure 12.14.

This makes sense. The effect of diffusion in this simple situation is to blur any sharp

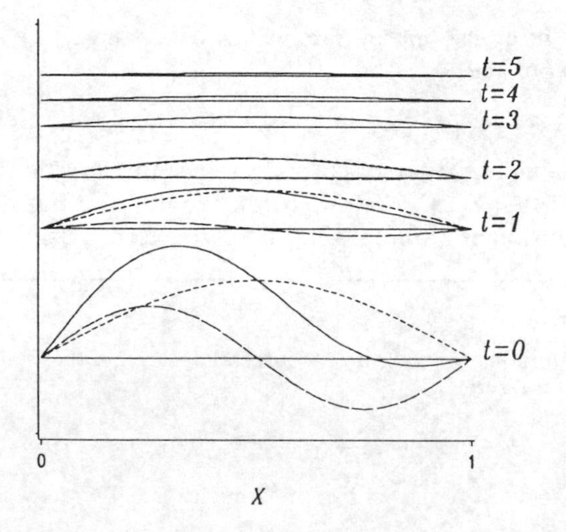

Figure 12.14 Demonstration of solution of the PDE in the case where D = 0.7, for times $t=0$, 1, 2, 3, 4, 5. The two components of the solution are indicated by broken lines, and the sum of these by a solid line.

edges. The higher order sine terms in the specification of the initial condition are representing those elements which are more spatially variable and so we might expect that they would disappear more quickly than the lower order components.

12.8.2 Numerical solution of diffusion equations

Many PDEs are difficult or impossible to solve analytically, and so we are forced then to fall back on numerical solution or simulation. There are many different types of techniques for the numerical solution of PDEs, and we briefly mention just one here.

The solution we obtained in the previous section was for a PDE with both initial and boundary conditions, as indicated in Figure 12.15. The diagram indicates that the solution is fixed along the lines $t = 0$ and $x = 0$ and $x = 1$. The method used for numerical solution of the ODE

$$\frac{dy}{dt} = \frac{e^{2t} + 3}{\cos(t)}$$

given an initial condition $y = 0$ at $t = 0$ was to step along the t-axis from $t = 0$ in small steps, and use the differential equation to obtain the value of the solution y at $t + \Delta t$ from the solution at t. In the case of this PDE, the method has to be more complex: we need to step sequentially through time from the initial condition, but also need to ensure that the boundary conditions at $x = 0$ and $x = 1$ are satisfied. The method is to impose a grid on the (x, t) space as in Figure 12.15, and to calculate the value in column 1 from those in column 0, by a finite difference formula relating the element in the ith position in column 1 to the elements in positions $(i - 1)$, i, and $(i + 1)$ in column 0. Then

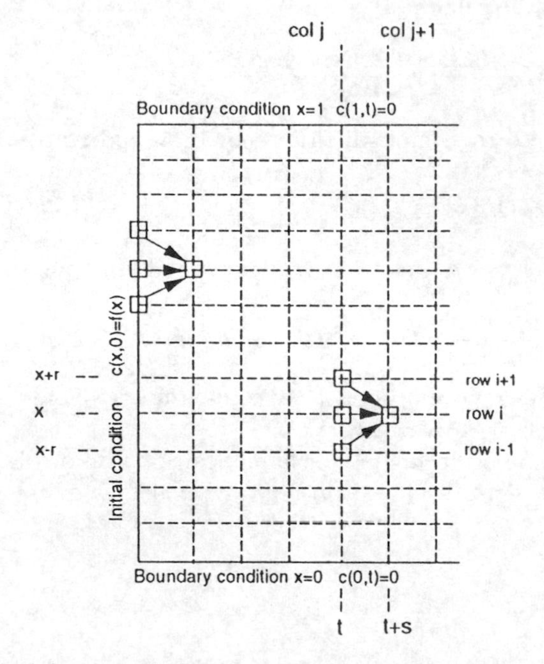

Figure 12.15 The principle of methods of numerical solution of PDEs, with particular initial and boundary conditions.

this is repeated by calculating the values in column 2 from those in column 1, and so on until the whole grid has been evaluated. The values in the 1st row and the $(n-1)$th row at any time (i.e. in any column) are related to the values in the end rows in the previous column, and this ensures that the boundary conditions are met approximately.

The finite difference formulae can easily be derived from the definition of the partial derivatives. Suppose that the spacing of the grid in the x-direction is s and that in the t-direction is r. Then we replace the partial derivative with respect to t by

$$c_t = \frac{c(x, t+r) - c(x, t)}{r}.$$

This reflects the definition of partial derivative

$$c_t = \lim_{r \to 0} \left[\frac{c(x, t+r) - c(x, t)}{r} \right].$$

Similarly we replace c_{xx} by

$$\begin{aligned} c_{xx} &= \frac{c_x(x+s, t) - c_x(x, t)}{s} \\ &= \frac{1}{s} \left[\frac{c(x+2s, t) - c(x+s, t)}{s} - \frac{c(x+s, t) - c(x, t)}{s} \right] \\ &= \frac{c(x+2s, t) - 2c(x+s, t) + c(x, t)}{s^2} \end{aligned}$$

which is the same in the limit as

$$= \frac{c(x+s, t) - 2c(x, t) + c(x-s, t)}{s^2}.$$

Substituting these into the simple diffusion equation and rearranging, we get

$$c(x, t+r) = c(x, t) + D \frac{r}{s^2} [c(x+s, t) - 2c(x, t) + c(x-s, t)]$$

Expressing this in terms of values in the ith row and jth column, we get

$$c_{i,j+1} = c_{i,j} + D \frac{r}{s^2} [c_{i+1,j} - 2c_{i,j} + c_{i-1,j}]$$

by which means we can move progressively across the grid from $t = 0$ (column 0) until the required time has been reached.

This is only the simplest technique for numerical solution of PDEs of this type. The reader is referred to Press *et al.* (1989) for further details of this and more complex techniques.

EXERCISES

12.1 This simulation exercise demonstrates the relationship between the distribution of total distance moved in various random transport processes along a line in time t seconds, and the features of the random process. Take a random sample of nt steps of lengths sampled from a distribution with p.d.f. $f(x)$, and calculate the total distance moved by summing over the nt step lengths. Consider the following cases, at times $t = 1, 5, 10$:

(i) $n = 12$ steps per second, steplength, $X \sim$ uniform on $[-5/12, +7/12]$;
(ii) $n = 1$ step per second, steplength, X is exponential, so that $f(x) = \lambda\, e^{-\lambda x}$, for $\lambda = 1$;
(iii) $n = 12$ steps per second, steplength $X \sim$ uniform on $[-0.5, +0.5]$;
(iv) $n = 1$ step per second, steplength as in (ii), i.e. $f(x) = \lambda\, e^{-\lambda x}$, except that $\lambda = 2^{1/2} = 1.4142$, and the steps can be positive as well as negative, with equal probability, $p = 1/2$.

(a) Verify that the mean and variance of the distance moved in time t is the same in cases (i) and (ii); and also in cases (iii) and (iv).
(b) Obtain 200 realisations of each, and calculate the mean and variance of the distance reached.
(c) Plot out the distributions of distance which result and, using Normal probability plots, assess their closeness to Normality.
(d) Discuss your findings, particularly comparing (i) and (ii) and separately (iii) and (iv).

12.2 Discuss the following topics.
(a) 'Diffusion is essentially a small-scale phenomenon.'
(b) 'Diffusion always has the effect of making the density more uniform throughout the available space.'

12.3 *A more realistic model of animal movement.* The usual diffusion model for spread of animal populations in two dimensions is thought by some biologists to be rather unrealistic since animals do not move in totally unrelated directions each time they reassess their position and direction of movement. A more likely mechanism is that animals tend to continue moving in the same direction, but random variation of some magnitude is imposed on this. Suppose that the species under consideration moves in steps of constant length d, and that the *change in direction*, θ_j, at the jth step follows a distribution which is symmetric about zero, i.e. $E[\theta_j] = 0$. Various distributions are possible, but we use here a Wrapped Normal, $WN(\lambda)$, for simplicity, which is a Normal distribution with standard deviation λ radians wrapped round a circle of unit radius. For λ quite small there is a strong directional component, and as λ increases, the strength of the directional component is reduced. Let R_n be the distance moved from the starting point after n steps. For $d = 1$ and $\lambda = 0.1$, 0.2, 0.4, investigate by simulation the dependence of the distribution of R_n^2, in particular, the average squared distance moved, $E[R_n^2]$, on the value of λ. In particular,
(a) Show that at the first step the x, y coordinates of the point reached are ($d \cos \theta_1$, $d \sin \theta_1$) where $\theta_1, \theta_2, \theta_3, \ldots$ are the random angles from the wrapped Normal distribution in order of generation;
(b) Show that the second point reached has x, y coordinates ($d \cos \theta_1 + d \cos[\theta_1 + \theta_2]$, $d \sin \theta_1 + d \sin[\theta_1 + \theta_2]$);
(c) Show that $\rho = E[\cos \theta_j]$, which is the equivalent of a variance for random variables measured on a circle, is given in terms of λ approximately by

$$\rho \cong 1 - \frac{\lambda^2}{2} + \frac{\lambda^4}{8}$$

for small λ;
(d) Hence compare your result with that obtained by Skellam (1973)

$$E[R_n^2] = s^2\left(\frac{1+\rho}{1-\rho}\right)\left[n - \frac{2\rho(1-\rho^n)}{(1+\rho)(1-\rho)}\right].$$

12.4 A pharmaceutical company produces a drug P by flowing a solution of substrate, S along a set of tubes with enzymes that convert S to P bound to the tube walls. Let the average speed of flow be u and let the rate at which S is converted to P, per unit volume per unit time, at any position down the tubes, be

$$\frac{Vs}{s+K_m},$$

where s is the local concentration of S, and V, K_m are constants. Assuming that conditions are steady, that the solution is well mixed across the tube (so that s is a function only of x, the distance along the tube), and that diffusion along the tube is negligible, show that s satisfies the differential equation

$$u\frac{\mathrm{d}s}{\mathrm{d}x} = -\frac{Vs}{s+K_m}.$$

If the substrate concentration at the inlet to the system is s_0, show that the length of tube required to convert 99% of S into P is approximately

$$\frac{u}{V}(s_0 + K_m \log_e 100).$$

(Cambridge, NST, Part IA, Biological Mathematics, 1982.)

12.5 *Flux across a membrane.* In a counter-current exchanger, the transfer of a substance from one stream to the other can approach 100%. This is not true for a co-current exchanger drawn below.

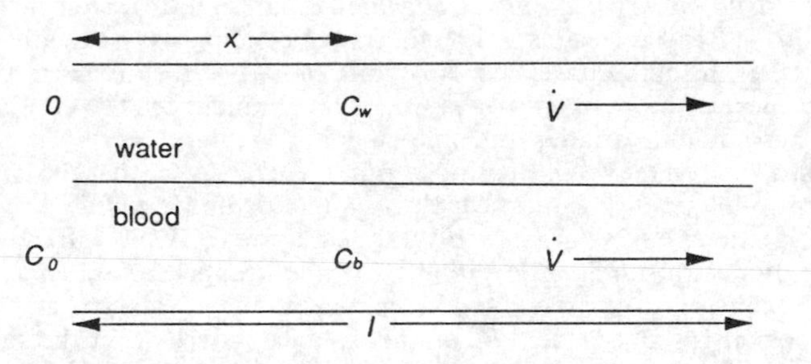

Blood and water flow in the same direction along parallel streams of length l, separated by a membrane across which a substance can diffuse. The volume flow rate is $\dot V$ in each stream. The concentration of the diffusing substance is C_0 in the blood and zero in the water as each enters the exchanger. The diffusive flux across the membrane per unit length of the exchanger is $\alpha(C_b - C_w)$, where $C_b(x)$ and $C_w(x)$ are the concentrations of the substance, at distance x along the exchanger, in the blood and water, respectively.

(a) Assuming steady-state conditions, show that C_b and C_w satisfy the equations

$$-\dot V\frac{\mathrm{d}C_b}{\mathrm{d}x} = \alpha(C_b - C_w) = \dot V\frac{\mathrm{d}C_w}{\mathrm{d}x}.$$

(b) Show that

$$\frac{\mathrm{d}C_b}{\mathrm{d}x} = -\frac{\alpha C_0}{\dot V}\exp\left(-\frac{2\alpha}{\dot V}x\right).$$

(c) Use this relationship to integrate the diffusive flux along the length of the exchanger and show that the total rate of transfer, R, of the substance from blood to water is

$$R = \frac{\dot V C_0}{2}\left[1 - \exp\left(-\frac{2\alpha}{\dot V}l\right)\right].$$

(d) What is the maximum percentage transfer from blood to water, and under what conditions will this occur?

(Cambridge, NST, Part IA, Quantitative Biology, 1991.)

12.6 Show that a solution of the equation

$$\frac{\partial z}{\partial x} = \frac{1}{a}\frac{\partial z}{\partial t},$$

where a is a constant, is $z = f(x + at)$, where the function f is arbitrary. If $z = \exp(-x^2)$ at time $t = 0$, find $z(x, t)$ for all x, t. If x is distance, and t time, how would you describe this solution?

12.7 *The wave equation*. The partial differential equation

$$\frac{\partial^2 u}{\partial t^2} = c^2 \frac{\partial^2 u}{\partial x^2}$$

occurs widely in models involving moving waves.

(a) Using the Chain Rule, express $\partial^2 u/\partial t^2$ and $\partial^2 u/\partial x^2$ in terms of partial derivatives with respect to z and w, where $z = x + ct$, $w = x - ct$.

(b) Hence apply this transformation to the PDE and show that the general solution of the equation is $u = f(z) + g(w) = f(x + ct) + g(x - ct)$, where f and g are arbitrary functions.

(c) Interpret the terms of this solution.

(d) If the system starts with u and u_t satisfying the following initial conditions

$$u(x, 0) = \sin x, \qquad u_t(x, 0) = 0$$

what is the solution at any time, t?

12.8 *Examples of Life Structures*.

(a) Follow the rules of the game of *Life* given in Example 12.3 to determine the fate of the following configurations in six generations, assuming they remain isolated:

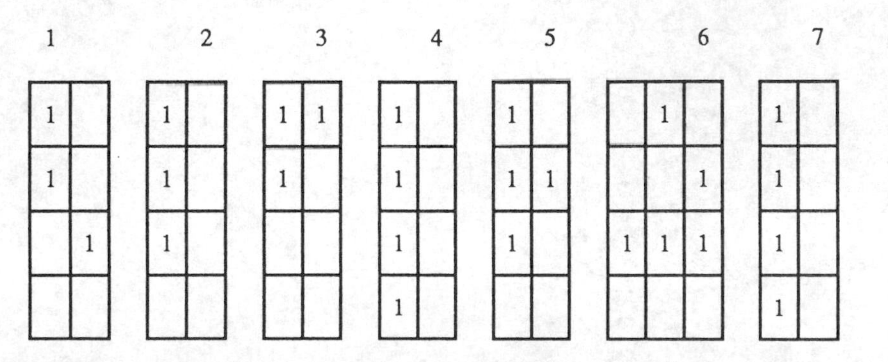

(b) Which die out in six generations, which are stable or periodic, and which are still evolving?

(c) What is the relevance of these and similar examples for modelling spatially distributed biological phenomena?

(See Gardner (1983), for discussion of these and many other configurations.)

12.9 *Fractals and cellular automata*. This exercise involves a cellular automaton on a line, with just two states, 1 or 0, starting from a single isolated 1 surrounded by zeros, using the following rule, known as a *Parity Rule*: if there are an odd number of *neighbouring* 1's, the position becomes or remains 1, otherwise it remains or becomes 0. By neighbouring, we mean adjacent in this context of a line. This rule can also be written in terms of all the possible transitions as follows:

111	110	101	100	011	010	001	000
↓	↓	↓	↓	↓	↓	↓	↓
0	1	0	1	1	0	1	0

This rule can thus be characterised among all two-valued 3-neighbourhood cellular automata rules by the binary number 01011010: the outcome from the possible 3-neighbourhoods ordered as above from 111 to 000.

The transitions for the first few times are therefore:

0	0	0	1	0	0	0	$t = 0$
0	0	1	0	1	0	0	$t = 1$
0	1	0	0	0	1	0	$t = 2$
1	0	1	0	1	0	1	$t = 3$

(a) Continue the pattern for a further 20 generations.

(b) Examine the resulting pattern of the automaton's growth in time. Two properties are of interest.

Property 1. If continued indefinitely, it exhibits the self-similarity characteristic of fractal patterns, i.e. similarity of pattern on each scale. In particular this automaton generates one of the most famous fractals, known as Sierpinski's gasket.

Property 2. Emergent behaviour which can occur with cellular automata: despite the local nature of the rule, large scale pattern is emerging. This also emerges even when starting from a random initial pattern, rather than a single cell set to 1 and all the remainder set to zero, as here.

(c) Perform the same analysis for the rules characterised, as described above, by the binary numbers: (i) 00110010, (ii) 01111110, (iii) 10110010.

(d) Discuss the value of these automata as models of biological phenomena. (For further discussion of fractals, see Peitgen *et al.* (1992).)

13 Statistical Analysis of Pattern and Sequence: Temporal and Spatial Series and DNA Sequences

13.1 INTRODUCTION

A major theme of this book is that observations on biological populations and processes often involve an element of random variation. This variability tends to obscure real differences and in Chapter 6 we presented some statistical methods for comparing population parameters such as the mean and variance and for detecting differences using random samples. In other situations we are more interested in the pattern or structure of a population or process, either in the spatial distribution or in the temporal changes. Typically, such a pattern is also partially obscured by random variation so that there is a need for methods of statistical analysis to tease out any pattern which is present. An example in this chapter involves a study of the spatial pattern of the nests of certain species of ground-nesting ducks where we are interested in whether the nests are spaced out with respect to birds of the same species or birds of other species. Visual inspection of a distribution map suggests some regularity but also shows some randomness which renders our interpretation less certain, so we need objective methods to help us decide whether or not the regularity could have arisen entirely by chance.

If a pattern is detected using statistical methods or is perhaps fairly obvious—the eye is very good at detecting pattern but it can be deceived and find patterns where none exists—a second question is whether its features can be described by a mathematical model. Ideally, we would build a mechanistic model based on postulates about the underlying processes but often these processes are poorly understood so that we must use more descriptive models. Such descriptive models can, however, be useful for: (a) summarising data; (b) comparing different but related sets of data; (c) providing a framework for analysing different features of the same pattern—for example, a spatial pattern of objects which are spaced out with respect to their neighbours on a local scale, but which are unevenly distributed on a larger scale. Furthermore, even relatively simple models for spatial patterns, are often too complicated to study using analytical techniques so that computer simulation is an important tool in the analysis.

This chapter deals with models and methods for detecting and describing pattern from observations on spatially or temporally distributed biological populations and processes in which there is a stochastic element. There are three main sections dealing with three broadly different types of data. Section 13.2 discusses the analysis of *time series* in which a series of observations is made at equally spaced times—for example, grouse bags on a moor over several successive years or the metabolic rate of a rat measured at 10-minute intervals. In particular, we describe models for serial dependence and ways of detecting and modelling *cycles* in time series data. Section 13.3 deals with data which form a series of point *events* in space, such as the location of ducks' nests in a breeding area, although many of the methods also apply to point events in time, such as the times of firing of a nerve cell, or in one-dimensional space such as locations in a linear habitat. We outline the use of spatial randomness as a standard against which pattern in distribution maps of points is detected and of simulation models for describing non-random point patterns. In Section 13.4 we discuss some methods for the analysis of DNA sequences where the observations are strings of the bases, e.g. ATTGCCGT, and where a basic problem is to compare different sequences, to measure the similarity between them and to detect pattern in the similarity of two sequences.

13.2 TIME SERIES ANALYSIS

A *time series* is a set of measurements made sequentially in time. Such series are familiar from everyday experience. For example, in his book *The Visual Display of Quantitative Information*, Tufte (1983) reports that in a random sample of 4000 graphics drawn from 15 of the world's newspapers and magazines published during 1974–80, more than 75% were plots of time series. Many of the time series which occur in the media arise from economics and show, for example, share prices on successive days, monthly export figures or annual company profits. Time series also occur widely in biology with measurements of different kinds taken at both the population and individual level over a range of time scales. For example, a demographic study of an insect population may involve a time series of population sizes recorded over many years, together with supplementary data from series of weather variables such as temperature and rainfall. In a physiological investigation the levels of luteinising hormone may be recorded in blood samples taken from women at 10-minute intervals over several hours.

In this section we present some basic methods for the analysis of time series data. First, we briefly discuss the types of data for which the methods are suitable and some of the objectives of the analysis. Then we describe the different kinds of variation which are typically present in time series data. We introduce the important idea of *autocorrelation* for measuring the serial dependence in so-called *stationary* time series with no systematic elements such as trends or seasonal effects. We develop *autoregressive schemes* which are basic stochastic models for the analysis of stationary time series and apply the ideas to a 73-year long series of data on grouse shooting bags to investigate the phenomenon of grouse population cycles. Finally, we outline the consequences of autocorrelation for sampling biological populations and for the application of standard statistical methods which assume a random sample of independent observations.

Time series is a large subject area and we shall deal only with a few of the basic ideas. For a more detailed introduction see Kendall (1976), Chatfield (1984) and Diggle (1990).

13.2.1 Time series data, examples and objectives

Types of data

We deal with methods of analysis for time series data in which the measurements are made at equally spaced times.

Discrete time series
Figure 13.1 gives an example involving the grouse bags from moors of some Scottish estates. The data are displayed as a *time plot* showing the observations plotted against time. In this example, the time series is inherently discrete since the shooting season is an annual event.

Continuous time series
In other cases time series data may be obtained by sampling an underlying continuous process at regular intervals. Figure 13.2 shows mean monthly sea-surface temperatures observed at a site in the Galapagos Islands during 1965–90. The sea-surface temperature will vary in a fairly smooth and continuous way but for purposes of recording, temperatures are taken at discrete times. In other situations the series may consist of a continuous recording such as an electrocardiograph trace but for analysis it is required to sample at times along the trace. The sampling interval must be short enough to avoid missing important features of the series such as cycles or periodic pulses but not too short so that storage of data, computation and presentation becomes a problem.

Types of observation
The observations in the time series may be effectively measured at a point in time or they may represent an accumulation of some quantity over an interval of time, e.g. daily rainfall data. A somewhat different type of time series data are the times at which some event, such as the firing of a nerve cell, occurs. In this case the variation between times at which events occur is an integral part of the process and different methods of analysis are required (see Section 13.3).

Objectives

There are several possible objectives in the analysis of a time series. In some situations we may wish to determine whether there is any pattern or structure in the series, as

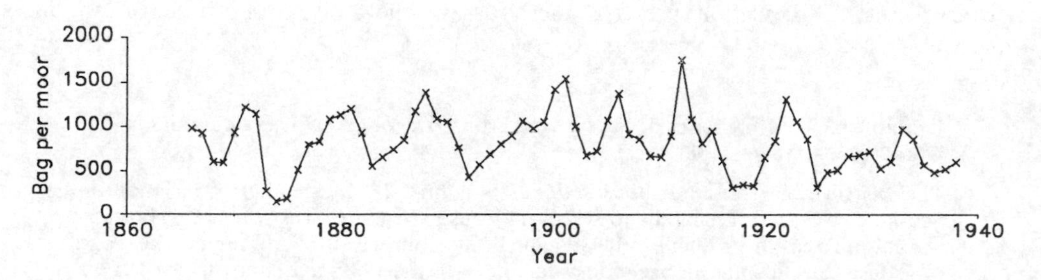

Figure 13.1 Time series of grouse bags on moors at the Atholl estate, 1866–1938.

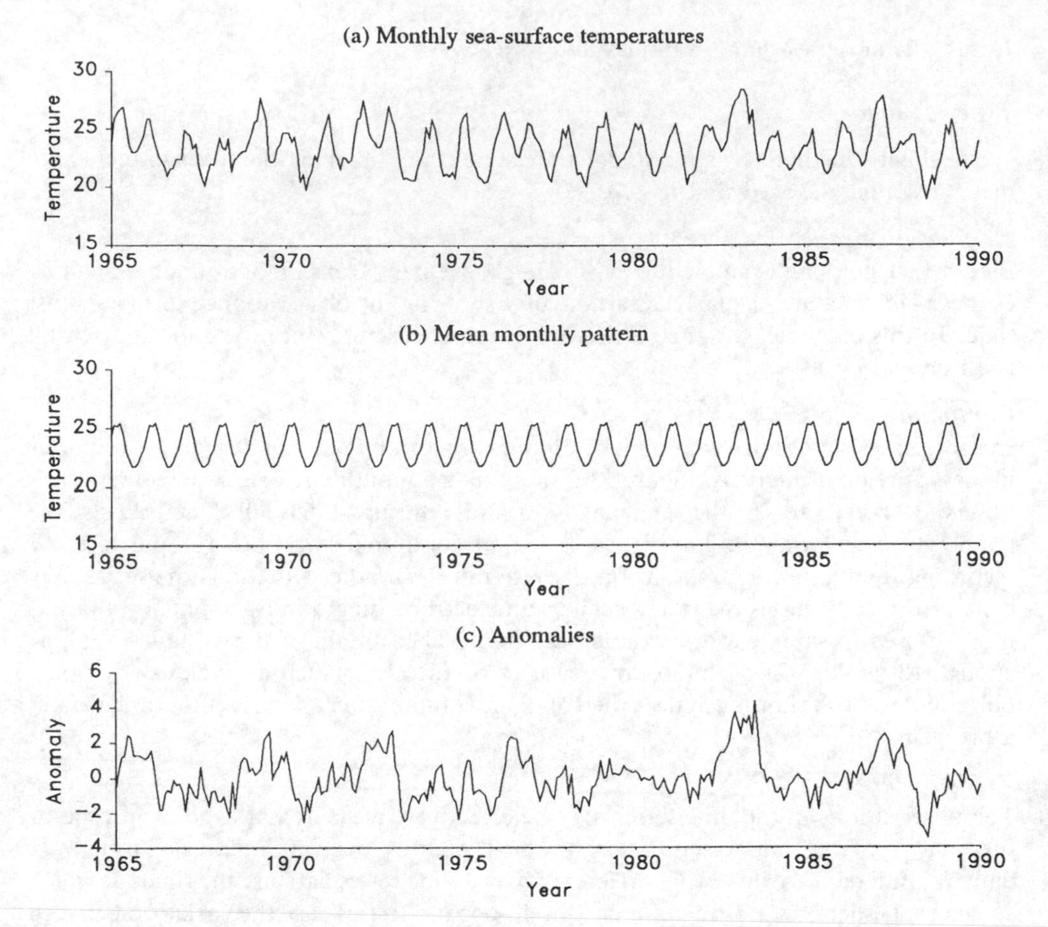

Figure 13.2 Time series of sea surface temperatures observed on the shore at Academy Bay, Santa Cruz, Galapagos, 1965–1990. (a) Original series; (b) mean monthly temperatures; (c) anomalies. Reprinted by permission of the Charles Darwin Research Station.

opposed to say a completely random distribution of observations. More often, however, some sort of pattern can be detected which raises the question of whether it can be described by a mathematical model. Such a description could be useful for: (a) comparison of time series with other sets of data; (b) helping to understand the underlying mechanisms which generated the pattern; (c) predicting future observations. This section presents methods which apply to both detection and description.

Example 13.1 Time series of bags of red grouse *Lagopus scoticus* on moors of the Atholl estate, 1866–1938

The British red grouse *Lagopus scoticus* is a bird of the heather moorlands of Scotland, Northern England and Ireland. Grouse are game birds which are shot for 'sport' in a season starting on the so-called 'Glorious Twelfth' of August. Records of grouse bags at some grouse moors date back to the 1830s and many to the 1850s. Figure 13.1 shows a series from the moors on the Atholl estate, Scotland from

1866–1938 (Mackenzie, 1952). The bag is expressed as the total number of birds shot per moor to allow for slight variations in the number of moors involved. Although grouse bags are likely to be affected by factors such as the bag limit set on individuals and weather, the grouse moors are carefully managed and the total bag is regarded as a fairly reliable index of population size. A conspicuous feature of the series is the cyclical appearance of the fluctuations with peaks and troughs at fairly regular intervals. These are sometimes called 'quasi population cycles'—the name reflects the irregularity in the series and distinguishes them from regular cycles with fixed amplitude and period. Quasi population cycles occur in other species and explaining their cause is an interesting ecological problem. We shall use the grouse bag data to illustrate some of the methods of time series analysis. In particular, we illustrate the idea of *autocorrelation* for measuring the serial dependence in the series and for suggesting models to describe the data. We then consider *autoregressive schemes*, which are stochastic models for describing the serial dependence in time series, and fit a simple model to the grouse bag data. Then we discuss the limitations of the model for grouse population cycles and describe a more mechanistic model.

13.2.2 Types of systematic and random variation in time series data

Time series can display different types of systematic and random variation. Before presenting models for describing these patterns of variation we give a broad classification of the various types. This illustrates a general approach to time series analysis in which the observations are *decomposed* into different contributions from systematic trends, seasonal effects and a random component. This leads to the important idea of a *stationary time* series in which there are no seasonal effects or systematic trends.

Seasonal effects

Seasonal effects are familiar from everyday experience—summer months are warmer than winter months and show consistent patterns from one year to the next. Figure 13.2(a) shows a time series of monthly sea-surface temperatures which show seasonal variation. In Figure 13.2(b) the seasonal pattern is summarised as a mean monthly temperature, averaged over the whole run of years, which varies in a regular manner by about 4 °C. In some situations seasonal effects are not of primary interest and we wish to remove them either to make comparisons between different parts of the series or to analyse another component of variation. In the sea-surface temperature series the variation between years is of biological importance because of the years of 'El Nino' when sea-surface temperatures rise and the marine iguanas and other marine species suffer heavy mortality and almost no individual growth. To analyse the annual variation it is useful to remove the seasonal pattern by expressing the original data as a deviation, or anomaly, from the average sea-surface temperatures for each month. Thus, from the original time series we obtain a derived series which is more suitable for analysis (Figure 13.2(c)). In this example seasonal effects refer to months which have a period of one year but more generally a seasonal effect comprises a regular cyclical fluctuation with a fixed period which may be 24 hours, 12 months, etc.

Cycles and quasi-cycles

Some series display cyclical variation which appears to be the result of an underlying biological mechanism rather than simply reflecting a known periodicity such as a seasonal effect. Often both the period and amplitude are rather irregular and fluctuations tend to drift out of phase compared with the clockwork of a regular cycle with fixed period and amplitude. Such series are sometimes said to show quasi-cycles. The effect appears to be present in the grouse bag series data but the irregularity in the series raises the question of whether the apparent cycle is real or not, i.e. can it be discriminated from pure random variation? If the cycle is real then what is the process which drives it? These aspects of the grouse data are dealt with later in this section.

Trends

Trends in time series are usually thought of as a smooth underlying long-term change. Depending on the length of the series, however, an apparent trend may be part of a cyclical fluctuation with a relatively long period so that on some time scales trends and seasonal effects may be indistinguishable. In some series the trend may be of prime interest. For example, if the aim is to use the time series to predict future observations finding the form of the trend may be crucial for obtaining accurate forecasts. In other situations the trend may be of secondary importance to the pattern of the variation about it.

Residual or random variation

Residual or random variation in a time series refers to irregular fluctuations over and above any systematic trends or seasonal variation. In this context the word 'random' needs to be qualified because it does not mean that the series lacks structure or pattern. To illustrate this idea consider the series in Figure 13.3. This is simulated data but the pattern is fairly typical of, say, observations on the metabolic rate of an inactive rat made at 1-minute intervals, or luteinising hormone levels in blood samples taken at 10-minute intervals from healthy women in the early follicular phase of the menstrual cycle. This series shows neither systematic trends nor seasonal effects, but rather a

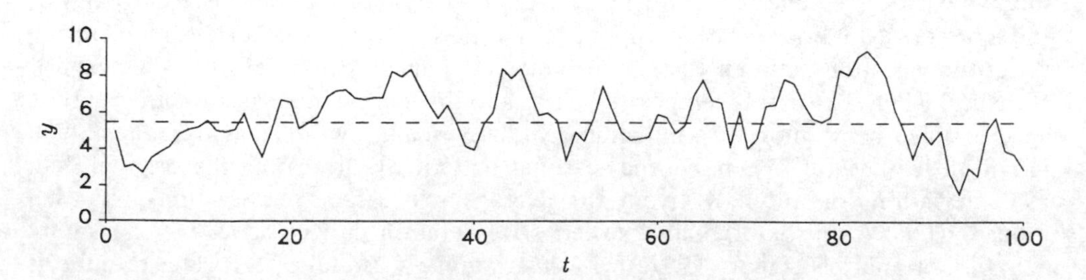

Figure 13.3 Time series of simulated data to illustrate a series with no trend or systematic variation but which shows serial dependence in the random component with successive observations positively correlated (one observation per time unit).

slowly varying irregular fluctuation about a constant mean level. Nevertheless, there is marked serial dependence in that successive observations show a clear tendency to lie on the same side of the mean. This is the effect of positive autocorrelation which we describe in Section 13.2.4. For now the main point of the example is to illustrate the idea that the random component of the observations in a time series refers to random variation which possibly has serial dependence.

Stationary time series

In a stationary random time series there is no systematic trend or seasonal variation. This is illustrated by the series in Figure 13.3 which appears to be stationary and, in fact, was produced by a model with no systematic component except for a single mean. In general, a stationary series is a term used for a series with random variation with or without serial dependence. A further requirement for stationarity is that the variability of the observations and the pattern of serial dependence does not vary in time. In Section 13.2.4 we describe the fundamental concept of *autocorrelation* for measuring the pattern of serial dependence in stationary time series and in Section 13.2.5 we present *autoregressive schemes* which are descriptive stochastic models for stationary series.

A stationary time series can arise in two rather different ways. First, the original series may be stationary—for example, that shown in Figure 13.3. Second, a series may be stationary after a trend or seasonal effect has been removed—for example the series of sea-surface temperature anomalies appears to be stationary (Figure 13.2(c)). Section 13.2.3 contains some techniques which can be useful for rendering a series stationary or approximately so.

White noise

The simplest example of a stationary random series is when the observations are distributed *independently* with the same distribution, sometimes referred to as *white noise*. This is written as

$$Y_t = \mu + Z_t$$

where Z_t is a random term with mean zero and variance σ^2, often assumed to be from a Normal distribution, i.e. $Z_t \sim N(0, \sigma^2)$. Independence of the observations implies that there is no serial dependence so that in most cases the model is unrealistic for time series data. It is, however, often used as a null hypothesis in tests of non-randomness in time series (see Section 13.2.4).

Decomposing the variation: trend + seasonality + random

The above discussion itemises and illustrates the major components of systematic trends, seasonal effects and stationary random variation in time series data. Not all of these components need be present in a particular case but for some applications it is useful to think of a general model which includes each term. One approach is to write the observations as the sum of two systematic components, for trend and seasonal effects, and a random component, in a model

$$Y_t = m_t + s_t + R_t$$

where m_t is the underlying mean at time t, which may be a trend, s_t is a deviation from the mean representing the seasonal effect at time t and R_t is random term which, in general, follows a stationary random series.

13.2.3 Descriptive methods and models for systematic trends and seasonal effects

Broadly speaking a systematic trend or seasonal effect is either central to the analysis, e.g. for predicting future observations, or it is a nuisance, e.g. when analysing annual variation about a pronounced seasonal pattern (Figure 13.2). There is therefore a need for techniques for emphasising, describing and removing trends and seasonal effects. In this section we give a brief summary; for more discussion and illustrations, see Diggle (1990).

Smoothing: moving averages and curve fitting

Sometimes the random variation in a time series makes it difficult to see clearly underlying trends so that a smoothing technique may be useful either for emphasising the trend or for describing it.

Moving averages
One way of smoothing a time series is to calculate a running mean or moving average. For example, a 3-point running mean is obtained as the average of three consecutive observations to give a smoothed series with terms

$$\overline{Y}_t = \frac{(Y_{t-1} + Y_t + Y_{t+1})}{3}.$$

The basic idea is that by calculating a mean, the local random fluctuations about any underlying trend are ironed out (recall that the variance of the mean decreases with increasing sample size). It is easy to see that this works by applying it to simulated data from a series with some trend plus white noise. In situations where deviations about the trend are of most interest we may calculate a deviation, or residual, of each observation from the smoothed series and think of each observation as having a 'smooth' component and a 'rough' component. Note that smoothing a series induces serial dependence because successive smoothed values have some of the original observations in common. More generally, a moving average is a weighted linear combination of the observations.

Curve fitting
Another way to smooth a series is to fit a smooth curve to the observations. For this exercise the models presented in Chapters 1 and 4 may be useful but in general it may be difficult to find an appropriate form of curve especially for a long series. Also, if

new observations are added to the series the curve will need to be updated. One way round this problem is to use *splines* which are polynomials fitted to short sections of the time series and joined together to form a smooth curve. For many biological examples, an appropriate model for the trend would involve one or more non-linear difference or differential equations (Section 8.7) but this can make the analysis rather complex (Section 11.6).

Differencing

Differencing is a way of *removing* a trend in a time series. The simplest case is the *first difference* obtained as the difference between successive observations, i.e.

$$D_t = Y_t - Y_{t-1}.$$

The first difference is effective for removing a *linear* trend in time when the observations are equally spaced. To see this write the 'smooth' part as a linear trend so that

$$Y_t = \alpha + \beta t + R_t$$

to give $D_t = R_t - R_{t-1}$, which is all 'rough' with no trend term. Note that successive differences are, in general, serially dependent because they have one rough term in common. To see the effect, plot first differences for some simulated data from a series of white noise.

13.2.4 Autocorrelation: describing serial dependence in stationary series

Autocorrelation refers to the serial dependence of the observations in a time series. In this section we describe the autocorrelation function (ACF) and the partial autocorrelation function (PACF) for measuring autocorrelation. These functions are used for summarising patterns of serial dependence and for suggesting models to describe the data.

Autocorrelation coefficients and the autocorrelation function (ACF)

The pattern of autocorrelation in a time series is often described by a set of quantities called the *autocorrelation coefficients* or *serial correlation coefficients*. These measure the correlation between observations at different times apart or *lags*. To illustrate the calculation of the autocorrelation coefficients we consider the series of simulated data shown in Figure 13.3. This series shows no long-term trends or seasonal pattern of variation but we can see from the time plot that there is marked serial dependence as observations close together in time tend to fall on the same side of the series mean. This is demonstrated by a scatter diagram in which the observation at time t is plotted against the observation at time $t + 1$ (Figure 13.4(a)). Similar diagrams for observations at times apart, or lags, of 2, 3 and 4 show the same effect but with an increased scatter as the lag increases (Figure 13.4(b)–(d)). This is the effect of autocorrelation. In this case the autocorrelation for observations at lags 1, 2, 3 and 4 is positive but in other cases there could be negative autocorrelation and we shall meet examples of this.

Figure 13.4 illustrates the autocorrelation in the series but how can we measure it? To do this we make use of the ordinary correlation coefficient r between two

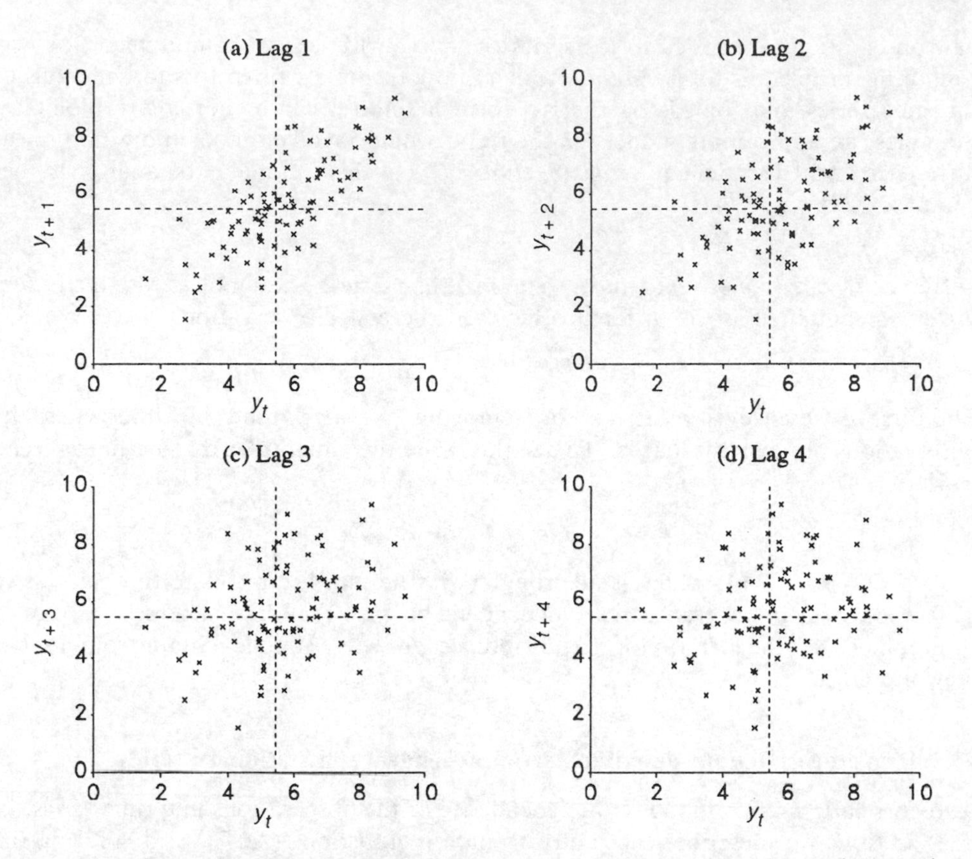

Figure 13.4 Scatter diagrams of observations at time t plotted against observations at times $t+1$, $t+2$, $t+3$ and $t+4$ for the data in Figure 13.3.

variables x and y, calculated from a sample of n pairs of values $(x_1, y_1), (x_2, y_2), \ldots, (x_n, y_n)$ as

$$r = \frac{\sum_{i=1}^{n} (x_i - \bar{x})(y_i - \bar{y})}{\sqrt{\sum_{i=1}^{n} (x_i - \bar{x})^2 \sum_{i=1}^{n} (y_i - \bar{y})^2}},$$

where $\bar{x}$ and $\bar{y}$ are the mean values of x and y respectively (Chapter 4). The numerator of the coefficient is the sum of the product of the deviations of the x and y values from their respective means so that for deviations of like sign the contribution is positive and for those of opposite sign it is negative. The denominator is proportional to the product of the sample standard deviations and scales the coefficient to lie between the extremes of -1 and 1 which occur when the points fall on a straight line.

To measure the autocorrelation in a time series with n observations $(y_1, y_2, \ldots, y_n)$ we first form $(n-1)$ pairs of values from consecutive observations as $(y_1, y_2), (y_2, y_3), \ldots, (y_{n-1}, y_n)$. Then, we calculate a correlation coefficient regarding the first observation as x and the second observation as y to give the autocorrelation between y_t and y_{t+1} as

$$r_1 = \frac{\sum_{t=1}^{n-1} (y_t - \bar{y}_1)(y_{t+1} - \bar{y}_2)}{\sqrt{\sum_{t=1}^{n-1} (y_t - \bar{y}_1)^2 \sum_{t=1}^{n-1} (y_{t+1} - \bar{y}_2)^2}},$$

where $\bar{y}_1$ is the mean of the first $(n - 1)$ observations and $\bar{y}_2$ is the mean of the last $(n - 1)$ observations. This coefficient which measures the correlation between successive observations is called the *autocorrelation coefficient* or *serial correlation coefficient* of lag 1. In practice, the number of observations in the time series is usually quite large so that the two means $\bar{y}_1$ and $\bar{y}_2$ are approximately equal to the overall mean of the series ($\bar{y}$) and the autocorrelation coefficient is then calculated from the simpler looking formula

$$r_1 = \frac{\sum_{t=1}^{n-1} (y_t - \bar{y})(y_{t+1} - \bar{y})}{\sum_{t=1}^{n} (y_t - \bar{y})^2}.$$

The calculation of the autocorrelation coefficient of lag 1 between successive observations in the series extends to observations at different times apart.

Autocorrelation and the autocorrelation function (ACF)

Autocorrelation coefficient of lag k

$$r_k = \frac{\sum_{t=1}^{n-k} (y_t - \bar{y})(y_{t+k} - \bar{y})}{\sum_{t=1}^{n} (y_t - \bar{y})^2}$$

where $\bar{y}$ is the mean value of the series $y_1, y_2, \ldots, y_n$.
The autocorrelation function (ACF), or correlogram, is a graph showing the autocorrelation coefficient r_k plotted against the lag k.

Nowadays, statistical packages can be used to calculate the autocorrelation function up to a specified lag. Figure 13.5(a) shows the ACF for the time series in Figure 13.3 with coefficients up to lag 20. For this series there is curvilinear decline in the autocorrelation as the time lag increases with relatively small values from about lag 5 onwards. We sometimes refer to this as a pattern of *short-term correlation*. Other series can give quite different patterns of autocorrelation. Some examples are as follows.

Alternating series. If successive observations tend to lie on different sides of the mean then the ACF will alternate between negative (lag 1) and positive values (see Section 13.2.5).

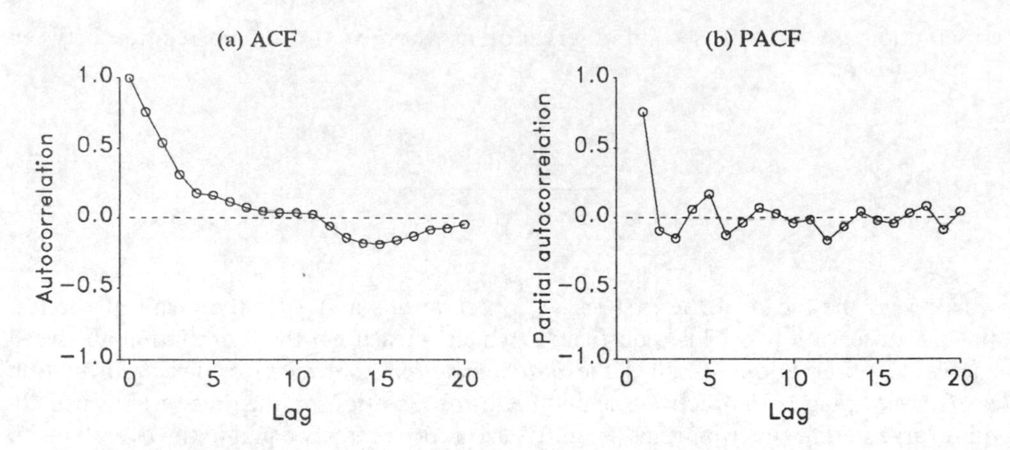

Figure 13.5 (a) Autocorrelation function (ACF); (b) partial autocorrelation function (PACF) for the time series in Figure 13.3.

Trends. Series with trends generally produce autocorrelation coefficients which decrease rather slowly to zero. Usually the ACF is calculated for a stationary series with no systematic trend or seasonal variation.
Cyclical series. For series with seasonal fluctuations or quasi-cycles the autocorrelation function will also show oscillations with similar frequency. Example 13.2 illustrates the effect for the grouse bag data.
White noise. In a hypothetically infinitely long series of white noise the autocorrelation coefficients are zero. Coefficients calculated from shorter series are subject to random variation (see comments below on estimating autocorrelation coefficients).

Partial autocorrelation and the partial autocorrelation function (PACF)

An allied concept to autocorrelation is partial autocorrelation. This is similar to the ordinary notion of partial correlation which measures the correlation between two variables x_1 and x_2 when the effect of a common correlation with a third variable x_3 has been excluded (Sokal & Rohlf, 1981). In a time series, partial autocorrelation refers to the autocorrelation at different lags while allowing for the effects of autocorrelation at intermediate lags. In other words, the partial autocorrelation between observations y_{t+2} and y_t is the autocorrelation between y_{t+2} and y_t after allowing for the autocorrelation between (y_{t+2} and y_{t+1}) and (y_{t+1} and y_t).

To measure partial autocorrelation we use the ideas of ordinary partial correlation and the result that in a sample of observations on three variables x_1, x_2 and x_3 with pairwise correlations r_{12}, r_{13} and r_{23} the partial correlation coefficient between variable 1 and 3, allowing for the effects of variable 2, is defined to be

$$r_{13.2} = \frac{r_{13} - r_{12}r_{23}}{\sqrt{(1 - r_{12}^2)(1 - r_{23}^2)}}\,.$$

To calculate the partial autocorrelation coefficient of lag 2 we represent the series of observations as triplets $(y_1, y_2, y_3), (y_2, y_3, y_4), \ldots, (y_{n-2}, y_{n-1}, y_n)$ and apply the formula for partial correlation using the autocorrelation coefficients in place of the pairwise correlations. This gives

Partial autocorrelation coefficient of lag 2

$$r_{2.1} = \frac{r_2 - r_1^2}{1 - r_1^2}.$$

where r_1 and r_2 are the autocorrelation coefficients of lag 1 and 2.

Partial autocorrelation extends to longer time lags with coefficients which form the partial autocorrelation function. We omit the details partly because the formula is rather complicated and also because the function can be calculated using statistical packages. The general idea, however, is that the *partial autocorrelation* of lag k measures the autocorrelation of lag k allowing for the effects of autocorrelation at shorter lags. Some packages include in the partial autocorrelation function a coefficient for lag 1, which is by definition the autocorrelation coefficient of lag 1.

Figure 13.5(b) shows the partial autocorrelation function for the series in Figure 13.3. Note the difference between the patterns of autocorrelation and partial autocorrelation. In this case the partial autocorrelation coefficients of lag 2 or more are all relatively small, indicating that the pattern of the autocorrelation is due to the indirect effects of the autocorrelation between successive values. This is an important special case and suggests a model in which the observation at a particular time is determined, apart from random variation, by the value of the observation at the previous time (see Section 13.2.5). Example 13.2 illustrates another case where the pattern of partial autocorrelation suggests that the observations at a given time may be determined by the values of the previous *two* observations (see Section 13.2.5).

Sampling errors in estimating autocorrelations

In practice, autocorrelation coefficients are calculated from series of finite length—in the above example a series of 100 observations. Sometimes we refer to these as *sample autocorrelation coefficients* to emphasise that the coefficients are subject to sampling variation. For example, a coefficient calculated using the observations in the first half of the series will, in general, be different from that based on data in the second half. In the analysis of time series data we often think of the sample autocorrelation coefficient as an *estimate* of the autocorrelation coefficient for some model of the observations, i.e. the coefficient calculated from a hypothetically infinitely long series of data generated by the model. In this case the sample values will be subject to estimation errors. The sampling theory for autocorrelation coefficients is rather complicated and beyond the scope of this book so we simply summarise some results of practical importance.

(1) For a series of white noise, i.e. independently distributed observations with the same distribution, the sample autocorrelations are distributed *approximately* with mean zero and variance $1/n$, where n is the length of the series. So for a long series a coefficient of given lag will fall in the range $\pm 2/\sqrt{n}$ in about 95% of cases. Sometimes these limits are superimposed on a plotted ACF to help interpretation.

The 95% confidence interval refers only to a single coefficient and not to the set of values in the ACF so the chances of at least one value falling outside the range will be larger than 5%.

(2) The sample autocorrelations are biased. For example, in a series of white noise with zero autocorrelation, the mean value of r_k is approximately equal to $-1/(n-1)$, so that the true value is slightly underestimated.

(3) The sample autocorrelations are not independent, i.e. they are correlated. This is illustrated in Figure 13.5(a) where successive values tend to fall on the same side of some underlying trend.

Example 13.2 Analysis of the Atholl grouse bag series: autocorrelation and partial autocorrelation functions

Figure 13.6 shows the autocorrelation function (ACF) and the partial autocorrelation function (PACF) for the grouse bag data. The most striking feature is that the cyclical appearance of the time plot is reflected in the pattern of autocorrelation which shows a damped periodic fluctuation. This effect is typical of time series data with periodic variation—a sustained regular cycle in a series is reflected by a corresponding regular cycle in the autocorrelation function. Damping arises when the series is not strictly periodic with variation in the times between peaks and troughs. The partial autocorrelation function shows a relatively large negative value of -0.41 at lag 2 years compared with the corresponding autocorrelation of 0.12. So after allowing for the effect of a positive autocorrelation of lag 1 year, the autocorrelation between observations two years apart is negative. After lag 2 years the pattern of partial autocorrelation is irregular with small values. One interpretation of this is that the grouse bag in a particular year is essentially related to those in the previous two years with rather little direct dependence on bags from three or more years ago. Section 13.2.5 formulates this idea as a descriptive statistical model and Example 13.3 discusses the application of the model for describing the variation in grouse bag numbers.

One aspect of the analysis is whether the autocorrelation coefficients are really different from zero, or whether the values could have arisen as a result of sampling fluctuations in a series of white noise with no serial dependence. In this case the rather marked pattern in the time plot as well as other information on grouse

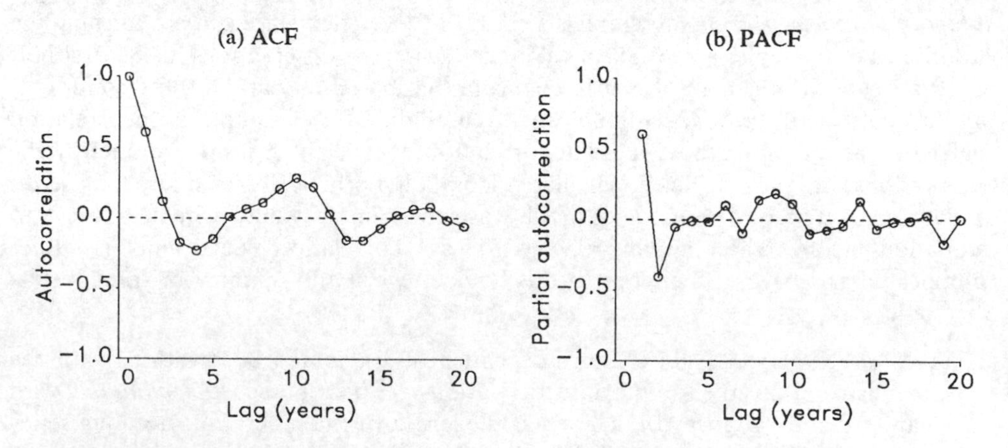

Figure 13.6 Autocorrelation function (ACF) and partial autocorrelation function (PACF) for the Atholl grouse bag time series data.

population dynamics make the hypothesis of zero autocorrelation untenable. In other situations when this is not so we might apply a test of zero autocorrelation to safeguard against spurious effects. To do this we use the result that for a long series of white noise with n observations the autocorrelation coefficient is approximately Normally distributed with mean zero and variance $1/n$, so that in approximately 95% of cases the values will lie within $\pm 2/\sqrt{n}$. For the grouse series $n = 73$, which gives a range of ± 0.23. It seems clear that the relatively large value of 0.61 for the autocorrelation correlation coefficient of lag 1 is sufficient to demonstrate that there is some autocorrelation: the corresponding standardised Normal deviate is given by $z = (0.61 - 0)/0.12 = 5.08$ ($p < 0.001$). Note that if a non-zero autocorrelation of lag 1 is detected then it is inappropriate to apply the test to other autocorrelation coefficients because the null model of a series of white noise no longer holds. A more appropriate null model must include some autocorrelation of lag 1 (see Section 13.2.5).

13.2.5 Autoregressive schemes: descriptive stochastic models for stationary time series

Autoregressive schemes are models for describing the serial dependences in stationary time series, i.e. series in which there are no systematic trends or seasonal effects. For series with trends or seasonal variation autoregressive schemes can be applied to the residuals after removing either trends or seasonal effects. The general approach is to relate the value of an observation at time t to the values of previous observations at times $t-1, t-2, \ldots, t-p$, where p is called the *order* of the scheme: AR(p) stands for an autoregressive scheme of order p. We start with the simplest case, the first-order scheme, in which the observation at a given time depends on the previous value in the series.

First-order autoregressive scheme, AR(1)

We begin by reiterating that autoregressive schemes are models for the serial dependence in stationary time series with no systematic trend or seasonal variation. To develop the model it helps to imagine a very long time plot with observations fluctuating about some mean value. In the first-order autoregressive scheme the deviation of an observation from the series mean is the sum of: (a) a quantity directly proportional to the deviation of the previous observation; (b) a random value which is independent of the previous values in the series. The scheme thus embodies some systematic dependence on past history which is blurred by the effects of random disturbances. If we denote the observation at time t by Y_t, then the first-order autoregressive scheme is written as

$$Y_t - \mu = \alpha(Y_{t-1} - \mu) + Z_t$$

where α is called the *autoregression coefficient*, μ is the underlying mean of the series and Z_t is a random variable which is *independent* of the previous values in the series and follows a Normal distribution with mean zero and constant variance, i.e. $Z_t \sim N(0, \sigma^2)$. Often the autoregressive scheme is presented for the particular case where the mean is zero (i.e. $\mu = 0$) when Y_t is usually interpreted as a deviation from the series mean rather than as an absolute quantity.

The form of the autoregressive model is rather like a simple linear regression model with slope equal to α. A fundamental difference, however, is that Y_t is not being related to some other explanatory variable (x) but on past values in the series—hence the term *autoregressive* or regression on itself. It must be stressed that results from standard regression theory which assume that the observations are independent do not, in general, apply to the analysis of time series where the observations are autocorrelated. This serial dependence can be seen by iterating the model a few times as follows

$$Y_1 - \mu = \alpha(Y_0 - \mu) + Z_1$$
$$Y_2 - \mu = \alpha(Y_1 - \mu) + Z_2 = \alpha^2(Y_0 - \mu) + \alpha Z_1 + Z_2$$
$$Y_3 - \mu = \alpha(Y_2 - \mu) + Z_3 = \alpha^3(Y_0 - \mu) + \alpha^2 Z_1 + \alpha Z_2 + Z_3$$

A random 'kick', Z_1, at time 1 not only affects the value of Y_1 but also subsequent values Y_2 and Y_3, so that the observations are serially dependent. Similarly, the other random kicks are incorporated into the process as illustrated in Table 13.1.

The contribution of a random kick to subsequent observations is successively multiplied by the autoregression coefficient, α. To keep the contribution finite the absolute value of α must be less than one, and this is, in fact, the condition for the series to be stationary. In this case the contribution of a random kick to subsequent observations decreases exponentially so that observations which are far apart have only a small amount of random variation in common and are therefore almost independent of each other. More specifically, we can use analytical methods to show that for the first-order autoregressive scheme the autocorrelation coefficient of lag k is given by

$$\rho_k = \alpha^k.$$

The Greek symbol rho is used for the autocorrelation coefficient of the scheme, i.e. the autocorrelation coefficient in a hypothetical series of infinite length. This is distinct from the sample autocorrelation coefficient r_k which is calculated from a sample series and is subject to sampling variation.

Figure 13.3 shows the series used to illustrate the idea of autocorrelation (Section 13.2.4). These data were simulated from a first-order autoregressive scheme with $\alpha = 0.8$ and $\sigma = 0.9$. The series shows a marked pattern of positive autocorrelation which decreases with increasing lag. Note that the partial autocorrelation coefficients

Table 13.1 Illustration of how random variation is incorporated into the AR(1) model leading to serial dependence

T	Random kick at time t and its contribution at subsequent times T						
	$t=1$	$t=2$	$t=3$	$t=4$	$t=5$	.	$t=n$
1	Z_1	—	—	—	—	.	—
2	αZ_1	Z_2	—	—	—	.	—
3	$\alpha^2 Z_1$	αZ_2	Z_3	—	—	.	—
4	$\alpha^3 Z_1$	$\alpha^2 Z_2$	αZ_3	Z_4	—	.	—
5	$\alpha^4 Z_1$	$\alpha^3 Z_2$	$\alpha^2 Z_3$	αZ_4	Z_5	.	—
.	.	.	.	.	.	.	.
n	$\alpha^{n-1} Z_1$	$\alpha^{n-2} Z_2$	$\alpha^{n-3} Z_3$	$\alpha^{n-3} Z_4$	$\alpha^{n-4} Z_5$	.	Z_n

for lag 2 and more are close to zero with no apparent pattern. This reflects the fact that the model has a 'memory' of only one time step and once the serial dependence between successive observations is allowed for the autocorrelation at longer lags is zero. Figure 13.7 shows simulated data from a series with $\alpha = -0.50, \sigma = 0.90$. In this case the autocorrelations tend to alternate between positive and negative values as the series tends to fluctuate above and below the mean from one time to the next. As in the above example the partial autocorrelation coefficients for lags 2 and more are close to zero reflecting the single time-step memory of the model.

Figure 13.8 shows the sample autocorrelation functions for the above two series compared with the function for the scheme, i.e. $\rho_k = \alpha^k$. For the series with $\alpha = 0.8$, the sample ACF shows a steady decline but with values below the theoretical curve. This arises partly because the sample autocorrelation coefficients are not independent of each other, i.e. they are themselves autocorrelated, so they are not scattered randomly about the theoretical curve. In this case successive departures tend to fall on the same side of the line. In shorter series there may also be an appreciable bias in the estimates. For the series with $\alpha = -0.50$ the sample ACF at longer lags shows marked sampling fluctuations compared with the theoretical curve which is rapidly damped to near zero. This emphasises that even with a time series which appears to be quite long ($n = 100$), the sampling errors in the ACF may be appreciable.

Finally, we note that for the first-order scheme the variance of the observations about the mean of the series depends on the variance of the random component and

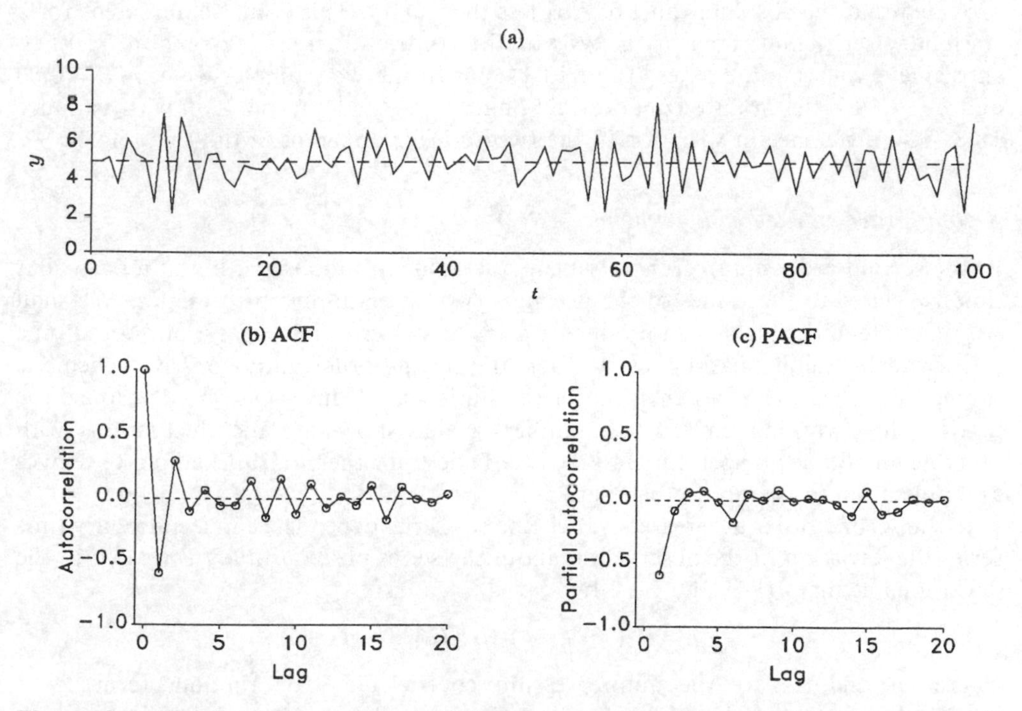

Figure 13.7 Simulated data from an AR(1) scheme with $\alpha = -0.50, \sigma = 0.90$. (a) Time plot; (b) autocorrelation function; (c) partial autocorrelation function.

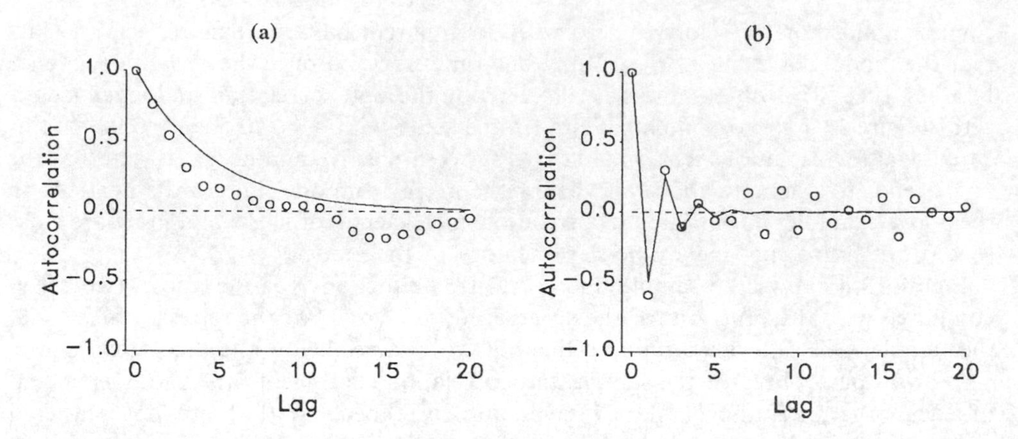

Figure 13.8 Sample autocorrelation functions calculated from simulated data for a series of length $n = 100$ together with theoretical functions for two AR(1) schemes. (a) $\alpha = 0.8$; (b) $\alpha = -0.5$.

the value of the autoregressive coefficient. Using analytical methods the variance is given by

$$\text{var}[Y] = \frac{\sigma^2}{(1 - \alpha^2)}$$

provided that the absolute value of α is less than unity. This is the requirement for a stationary series illustrated by analysis of the contributions of the random kicks to subsequent values in the series (Table 13.1). For the above examples we have $\alpha = 0.80$ and $\sigma = 0.90$ with variance equal to 2.25, and $\alpha = -0.50$ and $\sigma = 0.90$ with variance 1.08. The difference in variation in the two series is apparent in the time plots.

Second-order autoregressive scheme, AR(2)

In the second-order autoregressive scheme the value of an observation at a particular time is related to the values of the previous *two* observations in the series. We shall see that extending the lag can produce a periodic pattern of autocorrelation and time series which exhibit quasi-cycles. In fact, the second-order autoregressive scheme is sometimes called the Yule scheme after the statistician Udny Yule who developed the model when working on sunspot numbers which show periodic fluctuations with variable amplitude and period. Here we briefly describe the model. Example 13.3 gives an application to the grouse bag data.

In the second-order autoregressive scheme for serial dependence in a stationary time series the deviation of the observation about the series mean at time t is related to the deviations at times $t - 1$ and $t - 2$ by

$$Y_t - \mu = \alpha_1(Y_{t-1} - \mu) + \alpha_2(Y_{t-2} - \mu) + Z_t,$$

where α_1 and α_2 are the autoregression coefficients. The random term Z_t is independent of the previous values in the series and follows a Normal distribution with mean zero, i.e. $Z_t \sim N(0, \sigma^2)$. As for the first-order scheme we emphasise the difference between the model and an ordinary regression model of Y on two explanatory variables x_1 and x_2. In the regression model the observations are statistically independent whereas in the autoregressive scheme there is serial dependence because a random kick, Z_t, at time t affects subsequent observations.

The properties of the time series produced by the second-order autoregressive scheme depend on the values of the autoregression coefficients. A detailed discussion is beyond the scope of this book and we simply quote the main results (Box & Jenkins, 1970).

(i) $-2 < \alpha_1 < 2$ and $\alpha_2 < 1 - \lvert\alpha_1\rvert$	Stationary series;
(ii) $0 < \alpha_1 < 2$ and $\alpha_1^2 + 4\alpha_2 > 0$	ACF positive decreasing;
(iii) $-2 < \alpha_1 < 0$ and $\alpha_1^2 + 4\alpha_2 > 0$	ACF alternately positive and negative decreasing;
(iv) $-2 < \alpha_1 < 2$ and $\alpha_1^2 + 4\alpha_2 < 0$	ACF damped sine wave with period > 1.

Case (iv) is particularly interesting because it leads to quasi-cycles in the time series. Figure 13.9(a) illustrates this with simulated data from the model

$$Y_t - 6 = 1.5(Y_{t-1} - 6) - 0.5(Y_{t-2} - 6) + Z_t,$$

where $Z_t \sim N(0, 1)$. This series shows quasi-cycles with mean distance between peaks

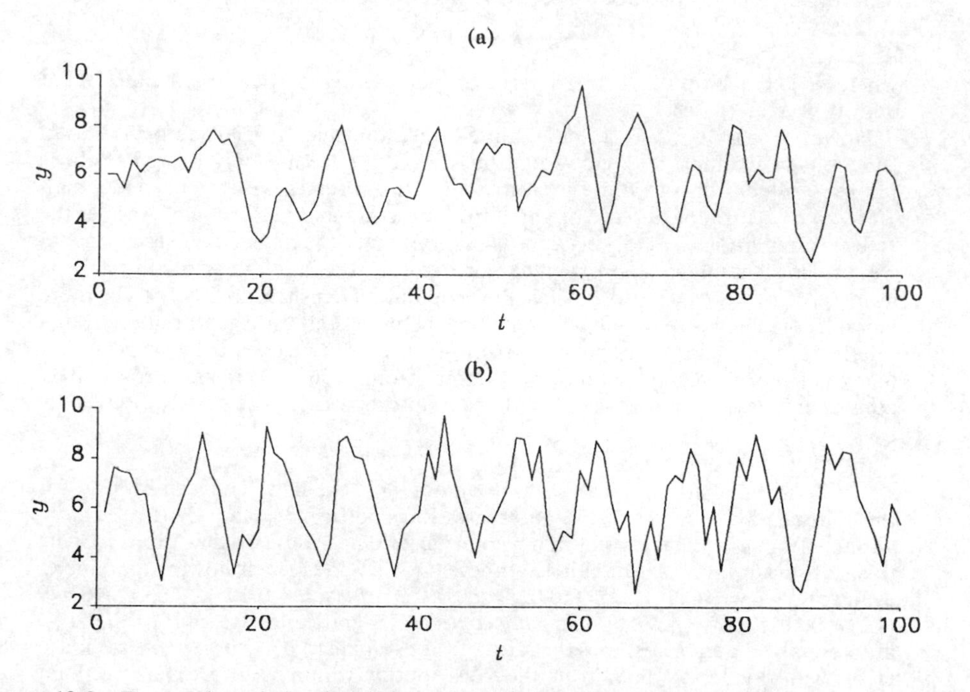

Figure 13.9 Two series of simulated data showing irregular periodic fluctuations: (a) AR(2) model; (b) harmonic series with added random variation. See text.

of about 10 time units. Figure 13.9(b) shows for comparison a harmonic series with added random variation simulated from the model

$$Y_t = 6 \sin \frac{\pi t}{5} + Z_t,$$

where $Z_t \sim N(0, 1)$. The harmonic series shows peaks at more or less regular intervals whereas in the quasi-cycles of the autoregressive scheme the amplitude, the period and the phase are variable.

Fitting autoregressive schemes

Nowadays, there are statistical packages with algorithms for fitting autoregressive schemes—for example Minitab and Genstat. They are usually fitted as a special case of the so-called autoregressive integrated moving average scheme (ARIMA).

Example 13.3 Analysis of Atholl grouse bag series: fitting a second-order autoregressive scheme

To illustrate the fitting of the AR(2) model we use the Atholl grouse bag data. First note that if the shooting pressure remained fairly constant over the period then the grouse bag in a particular year may be assumed to be roughly proportional to the population size in that year. If this is so we are essentially trying to build a demographic model for change in population size. In Chapter 1 we discussed some simple models for species with a short annual breeding season, such as the grouse, in which successive population sizes are related by difference equations of the form

$$N_t = RN_{t-1}F(N_{t-1}),$$

where R is the intrinsic geometric rate of increase and $F(N_{t-1})$ is a function of density in year $(t-1)$.

However, the analysis of the autocorrelation and partial autocorrelation functions in Example 13.2 shows that bags in year t are not only related to those in year $t-1$ but also to those in year $t-2$. This suggests a different model with $F(N_{t-1}, N_{t-2})$. In the absence of any information about $F(.)$ we could consider the AR(2) formulation as a working approximation. Since the model involves a product we use logarithms to approximate more closely the additive structure of the autoregressive scheme. Also, since higher population densities are likely to be more variable than lower ones, logarithms may help to stabilise the variance over a relatively large range of population sizes—in the grouse series the bag per moor displays approximately 10-fold variation from 156 to 1756. From these considerations we consider the second-order autoregressive scheme AR(2) given by

$$Y_t - \mu = \alpha_1(Y_{t-1} - \mu) + \alpha_2(Y_{t-2} - \mu) + Z_t,$$

where Y_t is the logarithm of the bag per moor, μ is the underlying mean level for the scheme, α_1 and α_2 are the autoregression coefficients, and Z_t is a random variable from a Normal distribution with mean zero and standard deviation σ. Using the statistical package Minitab to fit the model gave the following parameter estimates: $\hat{\mu} = 6.62$ (s.e. $= 0.072$); $\hat{\alpha}_1 = 0.979$ (s.e. $= 0.105$); $\hat{\alpha}_2 = -0.479$ (s.e. $= 0.105$); $\hat{\sigma}^2 = 0.0950$. An *approximate* significance test for the two autoregression parameters uses the ratio of the estimate to its *approximate* standard error, with p-value obtained from the t-distribution with $n-3$ degrees of freedom. For the lag 2 coefficient, $t = -4.56$ (d.f. $= 70$, $p < 0.01$). The method tests for a non-zero value of the coefficient of order 2 after fitting the AR(1) scheme and shows that a first-order scheme is not sufficient to describe the pattern of variation in the grouse bag series. The fitted model is

$$Y_t - 6.62 = 0.979(Y_{t-1} - 6.62) - 0.479(Y_{t-2} - 6.62) + Z_t$$

or
$$Y_t = 3.31 + 0.979Y_{t-1} - 0.479Y_{t-2} + Z_t$$

where the random term Z_t has an estimated variance of 0.095.

Figure 13.10(b) shows a series of simulated data using the fitted model and illustrates that the model captures the quasi-cyclical appearance of the field data Figure 13.10(a)). Figure 13.10(c) shows the autocorrelation function and Figure 13.10(d) the partial autocorrelation for the simulated data: these are similar to those of the observed series (Figure 13.6).

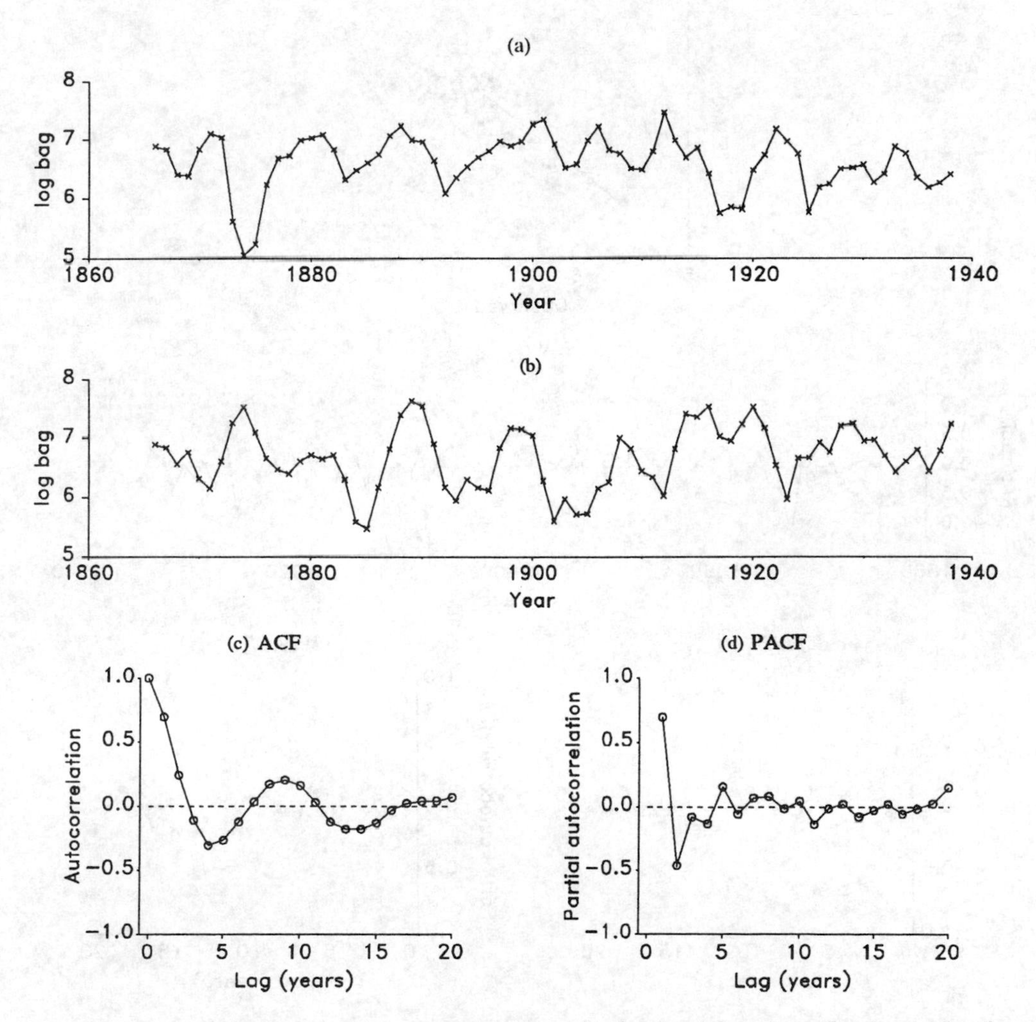

Figure 13.10 (a) Time series of log grouse bag data; (b) simulated data from fitted second-order autoregressive model; (c) autocorrelation function and (d) partial autocorrelation function for simulated data.

To examine the fit of the model in more detail we compare the observed values with those predicted from the fitted model, i.e.

$$\hat{y}_t = 3.31 + 0.979y_{t-1} - 0.479y_{t-2}.$$

Figure 13.11(a) shows fitted values plotted against the observed. The plot suggests that the model tends to overestimate the bag at low values and underestimate the bag at high values. The mean absolute difference between predicted and observed is 0.23. An absolute error of this size in the logarithm corresponds to an average absolute percentage error of 25% for the actual bag. A further check on the model is to examine the residuals, calculated as the difference between observed and fitted values, i.e. $y_t - \hat{y}_t$. When the model holds we expect the residuals to behave

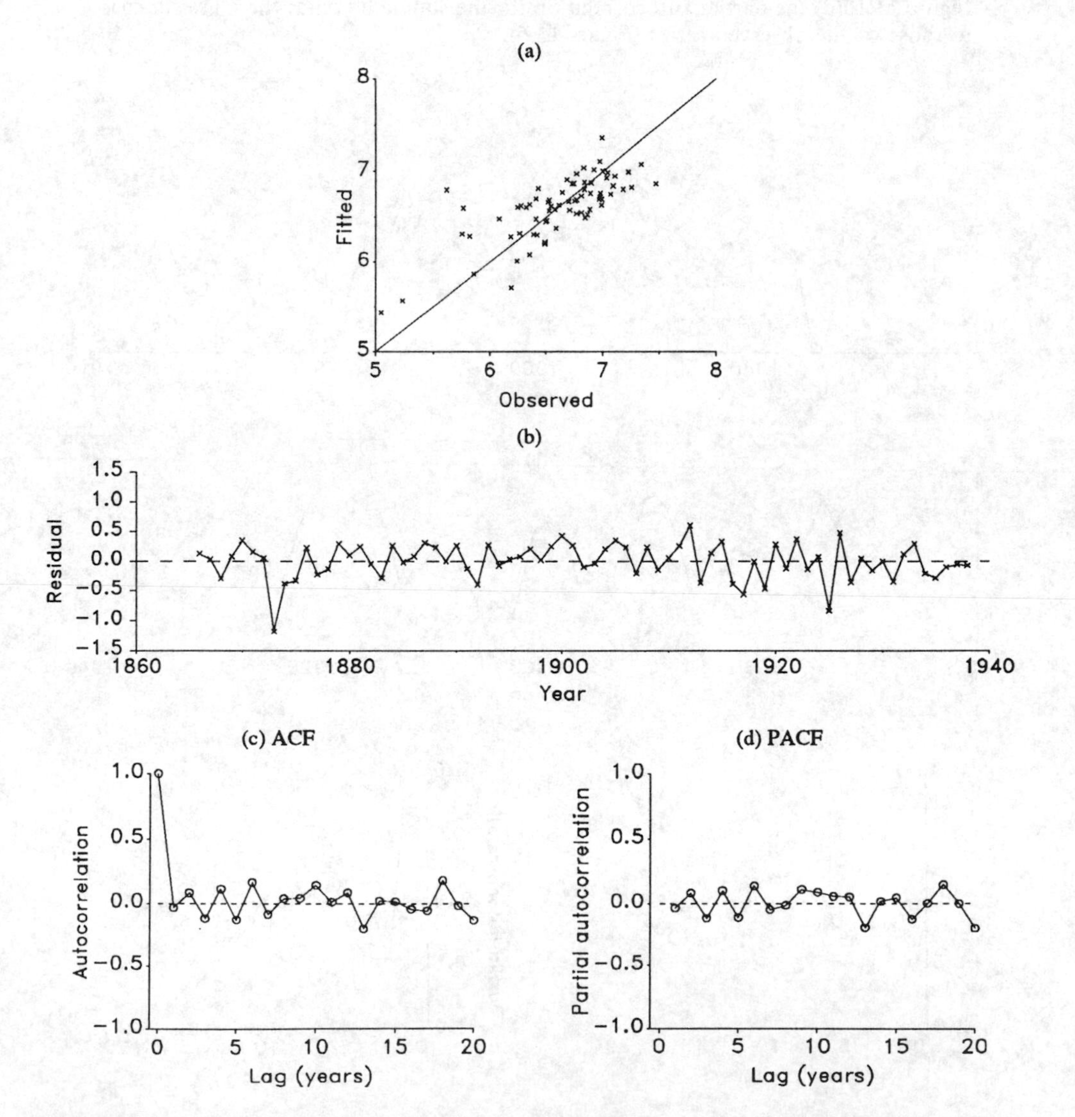

Figure 13.11 Illustration of the second-order autoregressive model fitted to the red grouse bag data. (a) Fitted values plotted against observed values; (b) time plot of residuals; (c)–(d) autocorrelation function and partial autocorrelation function for the residuals.

approximately as a series of white noise. This can be checked with a time plot and by calculating the autocorrelation function. Figure 13.11(b)–(c) shows that the residuals are consistent with a series of white noise.

Limitations of the AR(2) model for grouse population cycles

The analysis in Example 13.3 suggests that the second-order autoregressive scheme is a plausible description of the grouse bag data. However, as a general model for grouse population cycles the model is unsatisfactory for two reasons. First, it is purely empirical and lacks a mechanism for causing, or driving, the cycle. Second, without the random component the cycles are damped and the model population settles down to a stable value (Figure 13.12(a)). This behaviour stems from the fact that the model is linear with dynamics which depend on the solution of a linear second order difference equation. Either the deterministic model converges or it diverges—a

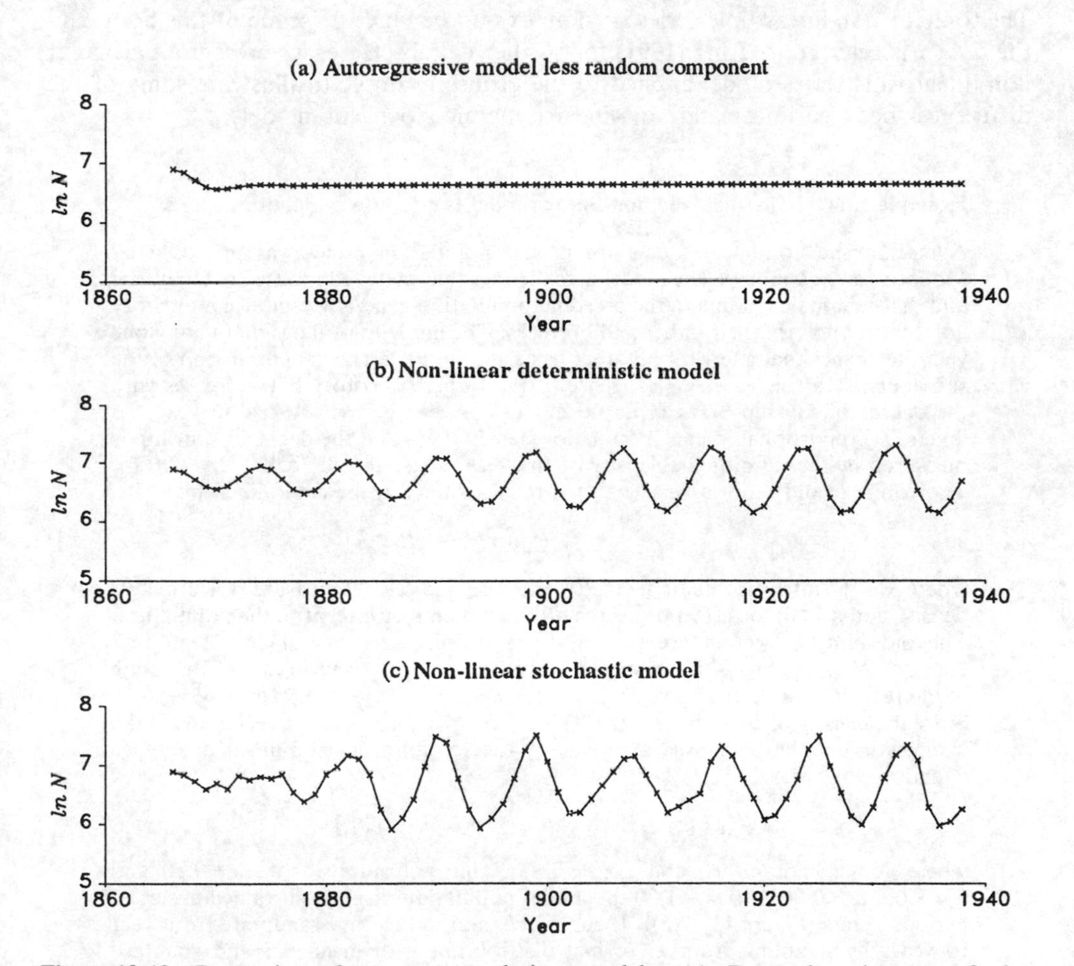

Figure 13.12 Properties of grouse population models. (a) Damped trajectory of the deterministic component of the AR(2) scheme; (b) sustained cycles of a mechanistic non-linear model; (c) irregular quasi-cycles of a stochastic non-linear model.

sustained cycle requires a root of the characteristic equation *exactly* equal to one which is possible mathematically but highly unlikely to occur. A more realistic model which is non-linear and mechanistic is described in Example 13.4

13.2.6 Non-linear time series

In the previous section we saw how the second-order autoregressive scheme is unsatisfactory for describing grouse population cycles because it is a linear model. More generally, linear models are unrealistic for most biological processes. This has been illustrated in this book where we have met many examples of non-linear models both for single populations (e.g. the logistic model for limited population growth, Chapter 1) and interacting populations and processes (e.g. the Fitzhugh–Nagumo model, Chapter 8). When such non-linear dynamic models also incorporate substantial stochastic components, they become what are known as non-linear time series models. The topic of non-linear time series is complex and beyond the scope of this book and the reader is referred to Tong (1991) for further details. Here, we present a particular non-linear time series model applied to the grouse example to illustrate some of the differences between linear and non-linear dynamic stochastic models.

Example 13.4 A mechanistic non-linear model for grouse population cycles

A mechanistic non-linear model for grouse population cycles was proposed by Mountford *et al.* (1989). The model uses the fact that male red grouse are territorial and that new male recruits to the breeding population generally establish a territory adjacent to those of their father and brothers. The model postulates that in any one year the chance that a territorial father has a son in the territorial population of the subsequent year increases with the size of the father's territory but decreases with the degree of kinship between its neighbours. Average territory size in year t is inversely proportional to cock population density (N_t) and the degree of kinship is measured by the ratio of the density of first-year cocks to old cocks ($R_t = Y_t/O_t$). The young to old ratio in year $(t+1)$ is then assumed to be determined by

$$R_{t+1} = \lambda \exp[-\exp(\alpha N_t + \beta R_t)],$$

where λ is the intrinsic rate of geometric increase and $\alpha(>0)$ measures the effect of density and $\beta(<0)$ the effect of degree of kinship on recruitment to the population. The old birds of year t are the survivors of birds from year $(t-1)$ so that $O_t = s_{t-1}N_{t-1}$ where s_{t-1} is the survival rate from year $(t-1)$ to year t. The young to old ratio in year t is then given by $R_t = Y_t/O_t = (N_t - O_t)/O_t = N_t/(s_{t-1}N_{t-1}) - 1$. Since the density of cocks in year $(t+1)$ is $N_{t+1} = Y_{t+1} + O_{t+1} = O_{t+1}(1 + R_{t+1})$ the changes in population density are given by the second order non-linear difference equation

$$N_{t+1} = s_t N_t\{1 + \lambda \exp[-\exp(\alpha N_t + \beta R_t)]\},$$

where R_t depends on N_t and N_{t-1}. For a constant survival rate $s = 0.50$ and $\lambda = 3.00$, $\alpha = 0.0013$, $\beta = -1.00$ the model population displays sustained cycles of period 8 years (Figure 13.12(b)). If survival is allowed to vary randomly from year to year, by sampling from a Normal distribution with mean 0.5 and standard deviation 0.1, the cycle persists with varying amplitude (Figure 13.12(c)). The model can produce cycles either in both the deterministic or the stochastic version and contrasts with the dynamics of the linear autoregressive scheme.

13.2.7 Autocorrelation: consequences for estimating means and variances

In Sections 13.2.4 and 13.2.5 we used the idea of autocorrelation and the sample autocorrelation function to analyse the serial dependence in a stationary time series. Autocorrelation also has important implications for the rather different problem of estimating the mean and variance of a population from a serial sample. This is because when there is autocorrelation the observations are not statistically independent so that the theory for simple random sampling which assumes that the observations are independent does not apply. In this section we briefly summarise the effect of autocorrelation when estimating the population mean and the variance.

First, recall that for a simple random sample: (a) the variance of the sample mean ($\bar{Y}$) is equal to the population variance divided by the sample size, i.e. σ^2/n; (b) the sample variance s^2 is an unbiased estimator of the population variance (Chapter 7). Consider now a stationary time series in which the observations vary randomly about a mean but with positive autocorrelation, i.e. successive observations tend to lie on the same side of the mean. The effect of this is that the sample mean of n consecutive observations will be more variable than a sample of n independent observations in which deviations from the mean tend to cancel each other out and thereby reduce the variance. Also, because successive observations tend to be similar, the sample variance of n consecutive observations will be less than that for n independent observations. So, with *positive* autocorrelation the variance of the sample mean is inflated and the sample variance *underestimates* the population variance. The effect is that the calculated confidence intervals on the population mean are on average too narrow. When the autocorrelation is negative, the converse is true and the confidence intervals tend to be too wide. This informal argument is confirmed by using analytical methods to calculate the variance of the sample mean and the expectation of the sample variance in a series of autocorrelated observations. The results are

$$\mathrm{var}[\bar{Y}] = \frac{\sigma^2}{n}[1 + (n-1)\bar{\rho}]; \qquad \mathrm{E}[s^2] = \sigma^2(1 - \bar{\rho}),$$

where the $\bar{\rho}$ is the average of the $n(n-1)/2$ correlations between all pairs of observations. For a particular pattern of autocorrelation we can use these formulae to calculate the amount by which the variance is inflated, the underestimation of the population variance and the combined effect on the standard error of the sample mean. Table 13.2 shows some calculations for a first-order autoregressive scheme with positive autocorrelation, decreasing exponentially with increasing lag. Note that, as the series length increases, the bias in the estimated population variance is reduced, but that the standard error of the mean may still be grossly misleading.

Example 13.5 Effect of spatial autocorrelation on point quadrat sampling

Point quadrat sampling is a method used by ecologists for measuring vegetation cover. The technique is to lower a pin onto the sward at randomly selected locations and record the contacts made with different species. For a particular species the proportion of contacts is used to estimate the cover. In some schemes a frame about 1 m long, holding 10 equally spaced pins, is used at each location. The motivation for this is that once a location has been reached it is worth taking several observations rather than just one. Due to patchiness in the vegetation cover,

Table 13.2 Effect of autocorrelation on the variance of the sample mean, the expectation of the sample variance and the standard error of the sample mean for observations on a first-order autoregressive scheme. ρ_1 is the autocorrelation between successive observations. (Variance and expectation when compared with a simple random sample of the same size)

	Variance of the sample mean (ratio actual:random)			Expectation of sample variance (ratio actual:random)			Standard error of sample mean (ratio actual:apparent)		
	sample size			sample size			sample size		
ρ_1	10	30	50	10	30	50	10	30	50
0.0	1.00	1.00	1.00	1.00	1.00	1.00	1.00	1.00	1.00
0.1	1.20	1.21	1.22	0.98	0.99	1.00	1.11	1.11	1.06
0.3	1.73	1.82	1.83	0.92	0.97	0.98	1.37	1.37	1.37
0.5	2.60	2.87	2.92	0.82	0.94	0.96	1.78	1.75	1.74
0.7	4.16	5.15	5.36	0.65	0.86	0.91	2.53	2.45	2.42
0.9	7.28	13.25	15.42	0.30	0.58	0.71	4.90	4.79	4.67

observations on pins in a frame at one location are likely to be more similar than those on pins at different locations. This gives rise to autocorrelation between scores on pins in the same frame. Figure 13.13 illustrates the effect for two species (Rothery, 1974). Estimated autocorrelation coefficients between scores on pins in a frame decrease with distance between pins.

The pattern of autocorrelation has a bearing on the efficiency of the choice of the number of pins to use in a frame. For example, might it be better to use 2 pins or perhaps 1 pin per frame? The question can be examined by comparing the sampling variances of the estimated cover for different schemes with a standard scheme such as 100 frames of 10 pins. If the data are scored as 1 for a contact and 0 if not, then the score on a randomly selected pin follows a binomial distribution with mean p and variance $p(1-p)$, where p is the probability of a contact (i.e. the cover). Note that the distribution of the total score in a frame of 10 pins is *not* Binomial with variance $10p(1-p)$ because scores on pins in the same frame are not independent of each other due to the spatial autocorrelation. However, for a scheme containing

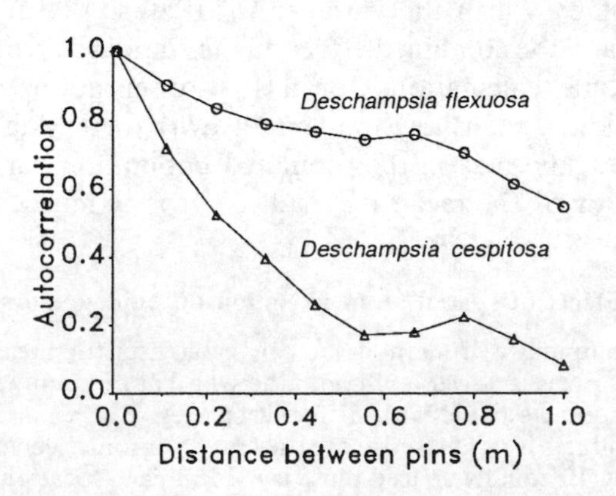

Figure 13.13 Pattern of autocorrelation between scores on pins in frames of point quadrats in relation to distance apart along the frame.

Table 13.3 Effect of changing the number of pins in a point quadrat frame

Species	Cover	Average autocorrelation		Number of frames to give same precision as 100 10-pin frames	
		Pins per frame		Pins per frame	
		10	2	2	1
Deschampsia cespitosa	0.14	0.40	0.09	120	219
Deschampsia flexuosa	0.26	0.79	0.54	95	124
Festuca rubra	0.32	0.33	0.37	172	251
Agrostis tenuis	0.34	0.15	0.08	226	418
Galium saxatile	0.39	0.26	0.19	180	302
Festuca ovina	0.69	0.46	0.34	131	195

N frames of n pins we can use the theory given in Section 13.2.7 to show that the sampling variance of the estimated cover is given by

$$\text{var}\,[\hat{p}] = \frac{p(1-p)}{Nn}\,[1 + (n-1)\bar{\rho}_n]\,,$$

where $\bar{\rho}_n$ is the average of the pairwise autocorrelation coefficients in a frame of n pins. Table 13.3 shows average correlations in 2-pin and 10-pin schemes and the estimated number of 1-pin and 2-pin frames required to give the same precision as 100 frames with 10 pins per frame for a range of species. Note that the average correlation is usually less in the 2-pin frame because the pins are more spaced out. Generally, reducing the number of pins per frame requires an increase in the number of frames. However, in many cases the increase is probably offset by having far fewer pins to record. A remarkable result is that for one species there is a *reduction* in the number of required 2-pin frames. This paradox is explained by the fact that for a series of autocorrelated observations the sample mean is not the most efficient estimator of the population mean. Observations at the end of the frame should, in theory, be given more weight than internal ones depending on the autocorrelation. However, the gain in precision is slight and probably not of any practical interest since the autocorrelation will generally be unknown.

13.3 ANALYSIS OF SERIES OF POINT EVENTS IN SPACE OR TIME

13.3.1 Introduction: examples and objectives

Many biological investigations involve observations of a series of *events* occurring in space or time. For example, the events may be the nests of species of ground-nesting ducks distributed over an area, spikes on a polygraph record of a neurone firing during some interval, or saplings distributed within a sample plot in a forest (Figure 13.14). We deal with situations in which the size, shape and orientation of the events can be ignored so that the events can be represented as points in space or time. This could, for example, apply when the times or distances between the events are large compared with the size of the events.

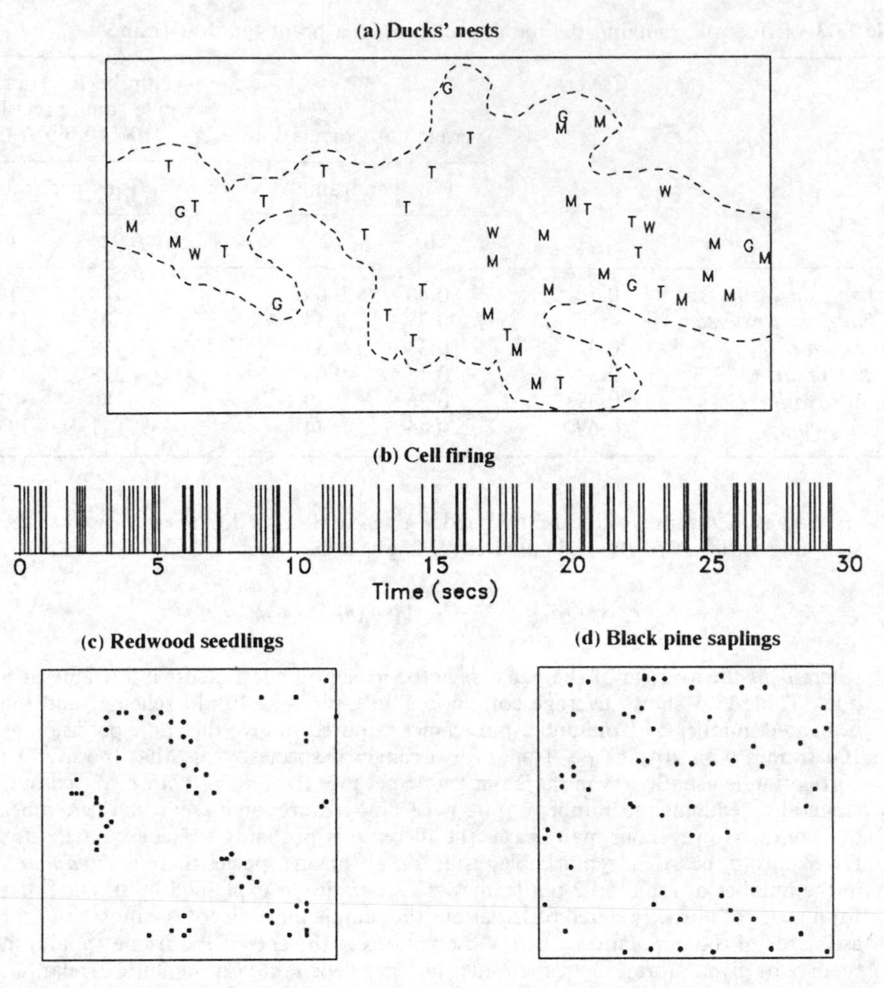

Figure 13.14 Examples of series of events in space or time. (a) Locations of the nests of some ground-nesting species of ducks: T, tufted duck; M, mallard; G, gadwall; W, widgeon; (b) firing pattern of an oxytocin magnocellular neurosecretory cell; (c) locations of 62 redwood seedlings in a square of side 23 m (Strauss, 1975; Ripley, 1977); (d) locations of 65 Japanese black pine saplings in a square of side 5.7 m (Numata, 1961; Diggle, 1983).

Detecting pattern

Often the main interest is trying to detect some special feature in the pattern of events. In the example involving duck's nests (Figure 13.14(a)) a pertinent question is whether the nests of a species are spaced out with respect to those of the same species or those of other species. Such spacing could indicate that birds repel each other through some behavioural response which may have evolved to reduce predation or parasitism or for some other reason. In any situation a vital first step is to plot the data and look at the pattern since the human eye is very good at detecting features. This approach works well for the redwood seedlings which show marked clusters (Figure 13.14(c)). Other cases may be less clear-cut especially as the eye often imposes pattern which is not present. For example, how would you pronounce on the pattern of Japanese black pine saplings displayed in Figure 13.14(d)?

In many situations the processes which underlie the series of events will contain elements, possibly random, which tend to obscure features of interest. For example, the distribution of duck's nests is likely to be affected by variations in the suitability of nesting habitat due to unknown factors which cannot be measured. So, there is a need for methods of analysis to tease out the features of interest from observed patterns which include an element of random variation. Also, even in relatively clear-cut cases such methods serve as a check on subjective assessments based on visual displays. One basic approach is to start with the null hypothesis that there is no pattern at all, i.e. that the events have been generated by a completely random process, and then to construct a significance test with high power for rejecting the null hypothesis when the special features are present (Section 13.3.5).

Describing point patterns

Often a test for non-randomness is a preliminary to the development of a model to describe the data. Such a model may help to understand the underlying biological mechanisms or it could be an empirical description which may be useful for characterising and comparing patterns from different but related situations. The development of *point processes* as models for *point patterns* is a difficult topic and we shall only present a few basic ideas. Computer simulation plays an important role and makes the models more accessible since situations which are too complicated for theoretical analysis can sometimes be analysed relatively easily by computer simulation.

In this section we present some methods for analysing series of events in space and time. We shall use the duck nest data to illustrate these methods so that there is an emphasis on two-dimensional spatial data. However, the concepts and techniques also apply to one dimension either in space or time, or even three-dimensional space. For a more detailed discussion of the statistical analysis of spatial point patterns see Diggle (1983) and Ripley (1981).

Example 13.6 Ducks' nests on St Serf's Island, Loch Leven, Scotland

Newton and Campbell (1975) report a study of the breeding biology of several species of ground-nesting ducks on St Serf's Island, Loch Leven. In one year, measurements of the nest positions in a fairly uniform area of *Deschampsia* tussocks were recorded to find out if the ducks of a particular species were spaced out with respect to birds of the same species or birds of other species. Spacing may be an adaptation to avoid predation since clusters of nests may be conspicuous and more vulnerable to a predator which concentrates its search around a nest already found. Also, spacing may have important consequences for limiting the size of the breeding population.

Figure 13.14(a) shows the locations of nests for all species on part of the study area. We shall consider the question of whether nests of ducks of a given species are spaced out with respect to each other and, in particular, look at the tufted duck (Figure 13.15). From visual inspection of the map, the birds appear to be rather regularly spaced but not uniformly over the area. In other words, the pattern appears to be locally regular but patchy on a larger scale. In Section 13.3.5 we develop a procedure to detect this regularity in an objective way.

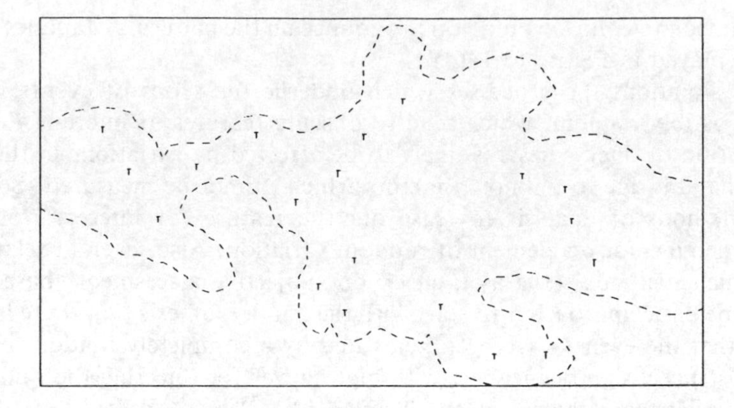

Figure 13.15 Distribution map showing the locations of tufted ducks' nests.

13.3.2 Complete spatial randomness and the Poisson process in two dimensions

In Chapter 3 we developed the Poisson process for point events occurring completely at random and with uniform rate in time. Here we present the two-dimensional version of the Poisson process in which point events occur at random and with uniform rate in space—sometimes referred to as *complete spatial randomness*. Complete spatial randomness is a key idea for analysing spatial point patterns because it defines a hypothesis of absence of pattern and is used as a yardstick against which the spatial distribution of events can be tested.

The two-dimensional Poisson process has two properties:

(1) The probability of an event occurring in a small area δA is a constant $\lambda \delta A$, where λ is the *rate* or *intensity* of the process;
(2) The occurrence of an event in any small area is independent of whether any events have occurred in any other area.

Clearly, the independence of events is an approximation for objects such as birds nests but it may be acceptable when the size of the events is small compared with the average distance between events.

For the Poisson process in time, the number of events occurring per unit time follows a Poisson distribution with mean λ and a similar result applies to the two-dimensional case for the number of events occurring per unit area. Also, for the Poisson process in time, the times between events follow an exponential distribution with mean $1/\lambda$. For the two-dimensional case the corresponding result is that the square of the distance from an event of the process to its nearest event, i.e. the squared nearest neighbour distance, follows an exponential distribution with mean $1/\pi\lambda$. To prove this result we derive the cumulative distribution function of the square of the nearest neighbour distance, i.e. the probability that the squared nearest neighbour distance is less than or equal to some specified value. For an event E, the square of the distance to the nearest event will be greater than w when the circle of radius $\sqrt{w}$ centred on E contains no other events. This is the probability that the area πw contains no events—the area occupied by the point event E at the centre of the circle is zero. For the two-dimensional Poisson process the number of events per unit area follows a Poisson distribution with mean λ so the probability of no events in an πw area is $e^{-\lambda \pi w}$. This is the probability that the square of the nearest neighbour distance is greater than w, so the probability that it is less than or equal to w is $1 - e^{-\lambda \pi w}$. Therefore, the cumulative distribution function of the squared nearest nearest neighbour distance for an event in a Poisson process is

$$P(W \leqslant w) = F(w) = 1 - e^{-\lambda \pi w}, \qquad w \geqslant 0.$$

i.e. an exponential distribution with mean $1/\lambda\pi$ (Chapter 3). A remarkable corollary is that for the two-dimensional Poisson process the square of the distance from any *point* in the plane to the nearest *event* of the process also follows a negative exponential distribution with mean $1/\lambda\pi$.

In deriving the exponential distribution for the nearest neighbour distances in the spatial Poisson process we are essentially assuming that the events are distributed in a plane of infinite extent since there is no upper limit on the range of the distribution. This assumption breaks down for a mapped point pattern in a bounded region such as the ducks' nests. In this case the nearest neighbour distance from an event will be affected by the presence of a boundary—for those events near the edge the distance will tend to be inflated. These problems can, however, be surmounted by using the computer to simulate complete spatial randomness by the process of locating events at random in the study area.

13.3.3 Complete spatial randomness in mapped point patterns

In the previous section we discussed complete spatial randomness and the two-dimensional Poisson process where events occur at random and independently with constant rate in an infinite plane. This leads to variation both in the number of events per unit area and their locations. In a mapped point pattern the *number* of events is usually regarded as fixed and complete spatial randomness describes how the given number of events are located, i.e. over the area at random, uniformly and independently. This is the hypothesis of no pattern at all which provides the starting point for detecting features of interest. Here we deal briefly with the procedure of simulating complete spatial randomness in a region within a specified and possibly irregular boundary. The technique is important because it forms the basis of *Monte Carlo tests* for detecting special features of the pattern.

We start by simulating complete spatial randomness of points in a square of unit side. For each point the x and y coordinates are chosen independently from a uniform distribution on the interval $(0, 1)$ by using a pseudo-random number generator. Figure 13.16(a) shows a realisation of the process with 100 points. It is important to note that complete spatial randomness does not imply an observed even distribution of points although, on average, any point is equally likely to fall anywhere in the area. In fact, the occurrence of points close together and some apparent open space is entirely expected simply by chance and the eye can easily be fooled.

The method for locating points at random in a square is easily extended to other areas. For example, in a rectangle of length L and breadth B, we can select x and y coordinates from uniformly distributed variates on $(0, L)$ and $(0, B)$ or multiply values from $(0, 1)$ by L and B respectively. In practice, we often have to deal with areas with irregular boundaries such as the duck nesting habitat (Example 13.6). One approach

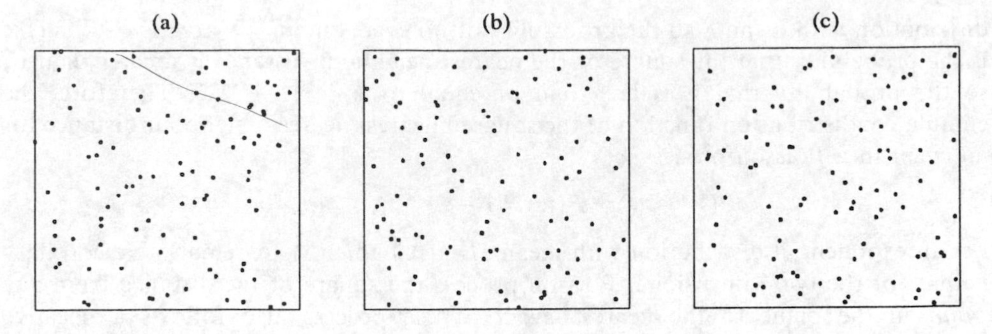

Figure 13.16 Illustration that a completely random distribution remains random when the area is uniformly stretched or compressed. Each picture is derived from realisations of complete spatial randomness using 100 points in areas of equal size but different shapes. (a) Original shape; (b) originally twice as long and half as high; (c) originally twice as high and half as long. Areas in (b) and (c) have been uniformly deformed to make them square. The resulting spatial distribution of points appears random in both cases. After Skellam (1972).

is randomly to place points in a rectangle surrounding the area and reject those falling outside of the required zone.

Unlike familiar patterns the completely random distribution of points remains random when uniformly stretched or compressed in any direction (Figure 13.16). Skellam (1972) notes that this property suggests that it may be more profitable to think of pattern as being absent in the random case than to think that we are dealing with a particular kind of pattern.

13.3.4 A catalogue of spatial point patterns

We have developed the two-dimensional Poisson process and the idea of complete spatial randomness as a way of defining *absence* of pattern in a spatial distribution. In this section we consider situations where pattern is present and give a catalogue of some particular model types. We begin by discussing the features of the various point patterns in Figure 13.17. These are derived from simulated data and we shall reveal the methods used in the simulation to illustrate in an informal way two important ideas: (a) of regarding a spatial point pattern as a *realisation* of some underlying process by which the points were located; (b) the problem of inferring properties of these processes from features in the observed pattern. Models for simulating the data are discussed later in Section 13.3.7.

In Figure 13.17(a)–(c) each square contains 100 points but the distributions are markedly different.

In Figure 13.17(b) the points were located uniformly at random and independently in the square to produce a realisation of complete spatial randomness. Any semblance of pattern is illusory and it is important to appreciate that although the underlying process involves a uniform rate for locating the points, 'close' pairs of points and 'empty space' can occur by chance.

Figure 13.17(a) has a uniform and regular appearance with points spaced out over the square. Comparison with the realisation of complete spatial randomness (Figure 13.17(b)) emphasises the absence of pairs of points falling close together.

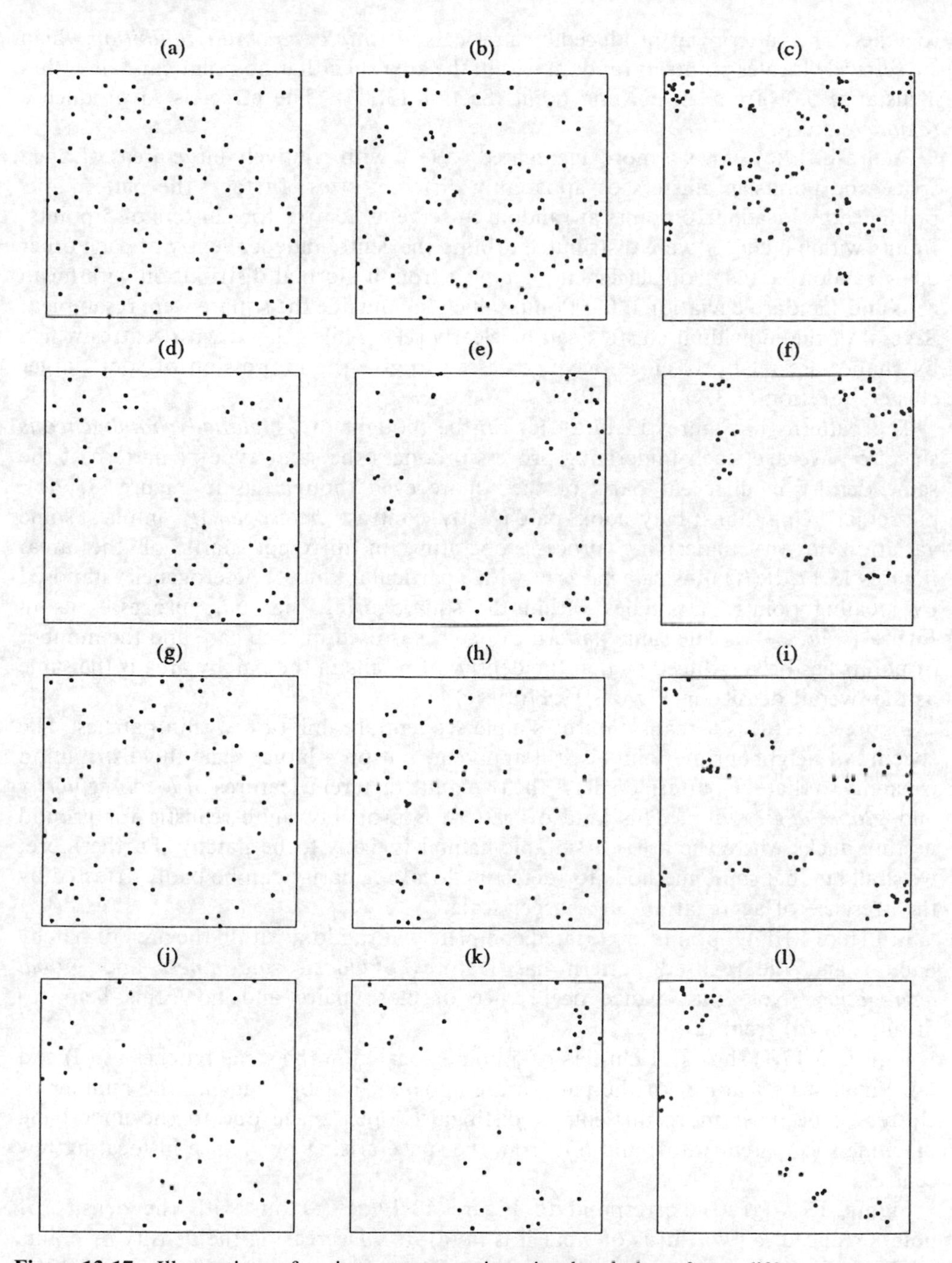

Figure 13.17 Illustration of point patterns using simulated data from different processes. *Simple sequential inhibition*: (a) homogeneous habitat; (d) patchy habitat; (g) homogeneous habitat, reduced density; (j) patchy habitat, reduced density. *Complete spatial randomness*: (b) homogeneous habitat; (e) patchy habitat; (h) homogeneous habitat, reduced density; (k) patchy habitat, reduced density. *Cluster process*: (c) homogeneous habitat; (f) patchy habitat; (i) homogeneous habitat, reduced density; (l) patchy habitat, reduced density. See text for details and discussion.

together. The pattern was produced by a process of *simple sequential inhibition* which successively locates points at random within the area such that no point can fall within a distance 0.06 from an existing point (Section 13.3.7). The effect is to produce a *regular* pattern.

Figure 13.17(c) shows a more aggregated pattern with relatively large areas of open space and points in clusters of apparently different sizes. In fact, the pattern was produced by locating 20 points at random to serve as centres for clusters of 5 points. Points within a cluster were distributed around the centre independently of each other using random x and y displacements sampled from a Normal distribution with mean zero and standard deviation 0.03. Points which fell outside the square were resampled. Several of the individual clusters can be clearly seen while others, with centres which by chance are relatively close, have coalesced to give the impression of some larger clusters (Section 13.3.7).

The patterns in Figures 13.17(a)–(c) can be thought of as *globally homogeneous* since, on average, each underlying process produces the same type of pattern at the same density in different parts of the square even though, as in Figure 13.17(c), particular realisations may look patchy. By contrast *heterogeneity* implies some variation in the underlying process operating in different parts of the area. Figures 13.17(d)–(f) illustrate patterns with a particular kind of heterogeneity imposed by locating points in patches within the square using the same processes as in Figure 13.17(a)–(c). The same pattern of patches is used in each case and the number of points has been reduced so that the density of points in the patchy area is the same as the overall density in Figures 13.17(a)–(c).

Figure 13.17(d) is a realisation of simple sequential inhibition within patches. The spacing of neighbouring points is still apparent, but on a larger scale the distribution is patchy so that the pattern displays the two quite different features of *local regularity* and *global aggregation*. This kind of pattern is probably quite realistic for ground nesting ducks where the area of suitable habitat is likely to be patchy. Furthermore, we shall see that some methods for detecting local regularity can be badly affected by the presence of aggregation on a larger scale.

In Figure 13.17(e) points are located completely at random within the area of patchy habitat and the realised pattern has features of *local randomness* and *global aggregation*. Note the chance occurrence of close pairs and how this helps to distinguish (d) from (e).

Figure 13.17(f) shows 11 clusters of 5 points located in the same patches as (d) and (e). Small scale features of the pattern are apparent but by reducing the number of clusters it becomes more difficult to distinguish open space due to the underlying patchiness (apparent in (d) and (e)) from the space arising by chance in locating few centres.

Figures 13.17(g)–(l) correspond to Figures 13.17(a)–(f) but with the density of points reduced to two thirds of what it is in (a)–(f). Decreasing the density of points tends to dilute the appearance of regularity. The difference between Figure 13.17(g) and (h) is less marked than between (a) and (b), although some effect is still apparent. The same comment applies to detecting local regularity with global aggregation. These situations emphasise the need for objective methods of analysis.

Note also the rather similar appearance of Figure 13.17(f), which has relatively low density with clustering in patches, and Figure 13.17(i) which is globally homogeneous with a lower density of clusters. Again, the underlying difference is blurred by the low density of clusters. This illustrates the great difficulty of detecting heterogeneity when clustering is present and vice versa.

Figures 13.17(d)–(f) and 13.17(j)–(l) illustrate the situation where there is no obvious structure in the large scale heterogeneity—merely a collection of suitable patches. In other situations global heterogeneity may take the form of trends reflecting, for example, a fertility gradient. In Figure 13.18, points were located randomly but with a trend in density.

13.3.5 Monte Carlo tests for spatial pattern

In the study of the pattern of ducks' nests we are interested in whether the nests are spaced out, i.e. we wish to detect regularity. In other situations we may wish to detect clustering, heterogeneity or some other feature of interest. One approach is to compare the observed pattern of points with that produced simply by locating the same number of points at random within the area, i.e. use complete spatial randomness as a yardstick for gauging pattern. This begs two questions. First, how can we measure the departure of the observed pattern from the random arrangement? Second, since the patterns produced by placing points at random are subject to stochastic variation, how can we assess the importance of any measured departure? To do this we construct a significance test (Chapter 6) as follows.

(1) Set up the null hypothesis of complete spatial randomness, i.e. that the observed pattern is one realisation of a process of locating points at random, uniformly and independently in the area.
(2) Select a sensible test-statistic for detecting the feature of interest. For example, to detect regularity of spacing we could use the average of the distances from each point to its nearest neighbour—large values indicating a more regular than random pattern. There are many other possibilities some of which are illustrated in Example 13.7.
(3) Obtain the p-value, i.e. the probability of a test-statistic as or more extreme than that observed when the null hypothesis is true. Often in tests for spatial pattern it is not possible to calculate p-values using analytical methods. We can, however, *estimate* the p-value by repeatedly locating sets of points at random and thereby

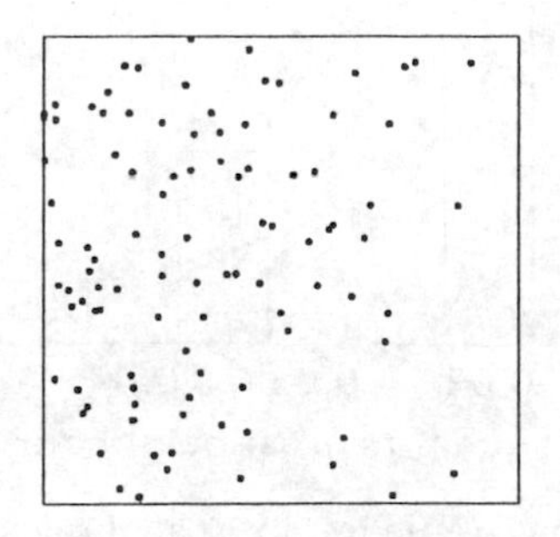

Figure 13.18 Observed pattern of points placed at random but with a trend in density decreasing from left to right.

building up a picture of the sampling distribution of the test statistic on the null hypothesis. This is referred to as Monte Carlo testing.

To illustrate the principle and mechanics of the Monte Carlo test procedure consider a test for regularity in the pattern of points in Figure 13.19(a). The points were positioned by adding random disturbances to the points on a 4×4 square grid to produce a rather marked regular pattern for illustration. To apply the Monte Carlo test we first need to choose a test-statistic which will be relatively sensitive to detecting regularity. One such test-statistic which has often been used is the mean of all the squared distances from each point to its nearest neighbour, or mean squared nearest neighbour distance, with large values detecting regularity. This usage stems partly from the fact that for the Poisson process the distribution of the mean squared nearest neighbour distance can be derived using analytical methods. We could equally well have used the mean nearest neighbour distance or some other plausible statistic.

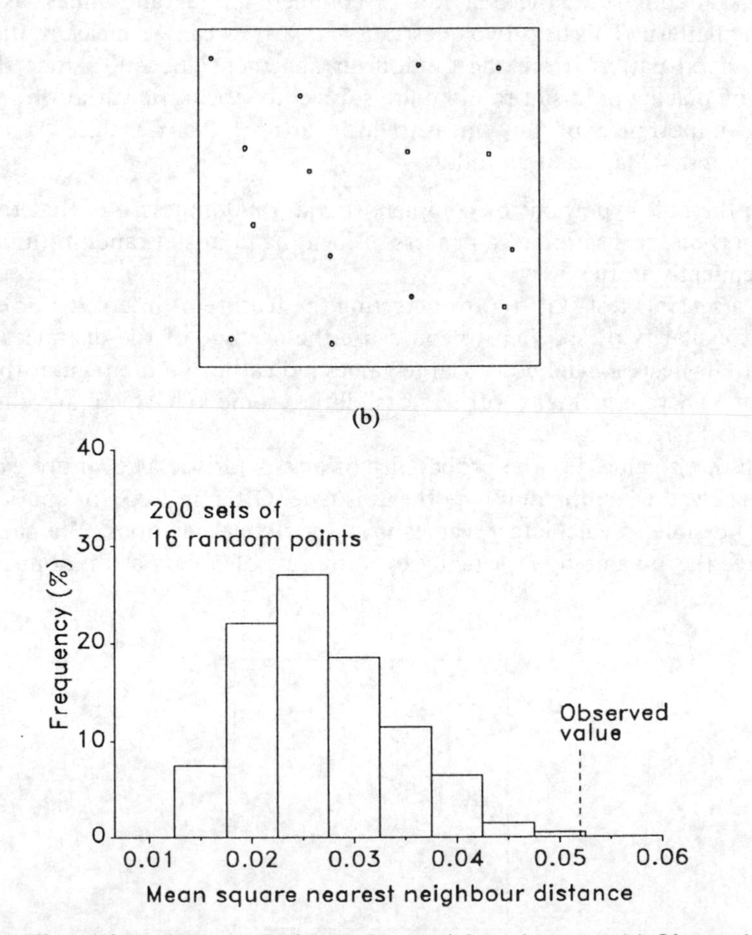

Figure 13.19 Illustration of Monte Carlo test for spatial randomness. (a) Observed pattern of 16 points; (b) histogram of the mean squared nearest neighbour distance for 200 sets of 16 points selected at random within the area.

The steps in the Monte Carlo test are then as follows.

Step 1. Calculate the value of the test-statistic for the observed configuration, M_{obs}.
Step 2. Locate $n = 16$ points at random in the area and calculate the corresponding value of the test-statistic, M_1.
Step 3. Repeat step 2 many times (say 200, but see below) to produce a reference set of statistics $M_1, \ldots, M_{200}$.
Step 4. Estimate the p-value for the test as the proportion of the values in the reference set greater than or equal to the observed value. Reject the null hypothesis at the 5% level if the estimated p-value is less than or equal to 0.05.

For the data in Figure 13.19(a) the calculated value of the mean square distance is $M_{obs} = 0.0520$. Figure 13.19(b) shows a histogram of the reference set of 200 values for random patterns. Only 1 value exceeded the observed value of 0.0520 giving an estimated p-value of $1/200 = 0.005$. We conclude that the observed pattern is unlikely to have occurred by chance and reject the null hypothesis of complete spatial randomness.

In the Monte Carlo test the p-value is estimated by simulation and is therefore subject to error. We make this error relatively small by taking a sufficiently large number of simulations. For example, if the true p-value is p, then the estimated p-value based on N simulations is a binomial proportion with variance $p(1-p)/N$. So, for example, with $N = 200$ the standard error of the estimate is equal to $\sqrt{0.05 \times (1-0.05)/200} = 0.015$ which gives the p-value to within about 3%. If required the error can be reduced by performing more simulations.

Example 13.7 Analysis of the tufted duck nest data: a test of randomness of nest spacing

This section describes an application of the Monte Carlo test to the tufted duck nest data. The null hypothesis is that of a completely random distribution of nests within the area of suitable habitat. The alternative considered is that of some process which tends to space out the nests. Such a process could involve interactions between neighbouring birds which suggests a test based on nearest neighbour distances. We consider three test-statistics (Brown, 1975): (a) M, the arithmetic mean of the squared nearest neighbour distances—large values indicate a pattern which is more regular than random; (b) S, the square of the coefficient of variation of the squared nearest neighbour distances—regularity is indicated by relatively little variation in the nearest neighbour distances; (c) G, the ratio of the geometric mean of the squared nearest neighbour distances to their arithmetic mean—this statistic takes its maximum value of unity when the nearest neighbour distances are all equal so that *large* values indicate a regular pattern (this result can be checked with a few numerical examples). If we suppose that the nearest nest to the ith nest is at a distance d_i, then the test-statistics are as follows:

$$M = \frac{\sum_{i=1}^{n} d_i^2}{n}; \qquad S = \frac{\sum_{i=1}^{n} (d_i^2 - M)^2}{(n-1)M^2}; \qquad G = \frac{\left(\prod_{i=1}^{n} d_i^2\right)^{1/n}}{M},$$

where n is the number of nests.

The Monte Carlo test involves: (a) calculating the values of each test-statistic for the observed pattern of 19 points; (b) taking 200 sets of 19 points located at random in the area and calculating the corresponding values of the three test-statistics;

(c) for each statistic finding the number of simulated values as, or more extreme, than the observed value (i.e. *larger* values of M and G, *smaller* values of S). Table 13.4 summarises the results.

Both the statistics S and G which measure the relative variability in the nearest neighbour distances provide strong evidence of nest spacing. The observed value of S is much less than its mean value for a random distribution of points and out of 200 simulations none of the values were less than the observed. The situation is similar for G. By contrast the observed value of the mean squared nearest neighbour is actually much *less* than its mean value for a random distribution and is exceeded by 142 values in the simulations. The poor performance of this statistic for detecting the regularity is due to the fact that although the nests are regularly spaced on a small scale they are aggregated on a larger scale. The effect of clustering on a larger scale is to *reduce* the size of the mean squared distance relative to a random distribution and to cancel the increase due to the regularity. In this case neither type of pattern is detected.

Although S and G have successfully demonstrated the non-randomness in the distribution there are some circumstances where the statistics may not work well. In particular, the test based on S is more sensitive to occasional large distances, for example arising from isolated nests, because these tend to *increase* the value of the statistic. On the other hand the test based on G is more sensitive to small distances which tend markedly to *reduce* the value of test-statistic. In general, it is unrealistic to expect a single test-statistic to perform well in all cases and a more pragmatic approach is to apply different approaches to help gain a fuller appreciation of the nature of the situation and the potential problems of interpretation of the data.

Finally, as a word of caution, we note that tests based on properties of nearest neighbour distances in the Poisson process can be misleading if applied to the analysis of a mapped pattern. One reason is that the presence of the boundary inflates the value of nearest neighbour distances relative to those of the Poisson process with the same density—points near the boundary may have had a close neighbour when set in the context of an unbounded region. This effect is illustrated by the above example: the overall density of points is $19/4572.5 = 0.00416\ \mathrm{m}^{-2}$ for which a Poisson process at the same density has mean squared nearest neighbour distance equal to $1/0.00416\pi = 76.5\ \mathrm{m}^2$ compared with a value of $119.8\ \mathrm{m}^2$ estimated from 200 simulations of 19 points located at random. A further reason why the results of the Poisson process are not applicable to the analysis of mapped patterns is that the observed nearest neighbour distances in a map of points are not *independent* of each other.

13.3.6 Empirical cumulative distribution function of distances in the analysis of point patterns

In the analysis of a mapped point pattern it is often useful to calculate the empirical cumulative distribution function (ECDF) of distances between points, i.e. the function

Table 13.4 Results of a Monte Carlo test of randomness of nest spacing for tufted ducks

Test statistic	Observed value	200 sets of 19 points located at random: Mean value	Number more extreme than the observed value
M (m^2)	98.4	119.8	142 ($\geqslant$observed)
S	0.178	1.54	0 ($\leqslant$observed)
G	0.926	0.50	0 ($\geqslant$observed)

derived from the data showing the proportion of values less than or equal to a given distance. This could be calculated using: (a) the nearest neighbour distances; (b) *all* the distances between individuals (i.e. the inter-event distances); (c) distances of the nearest individuals to extra points located in the area (i.e. point to nearest event distances).

An important use of the ECDF is to provide a visual summary of the pattern which can be compared with ECDFs for random arrangements, or with data simulated by some model of non-randomness. Since for any stochastic model the ECDF will vary from one simulation to the next we construct a reference set of ECDFs for comparison. A simple way to proceed is then to plot the ECDFs on the same diagram. A practical problem, however, is that this quickly produces a rather confused picture so that some summary of the reference set of ECDFs is required. One approach is to use: (a) the mean ECDF; (b) *upper* and *lower simulation envelopes*, obtained as the maximum and minimum values of the ECDFs at a particular distance. Departures from the model are indicated by the observed ECDF straying outside the simulation envelopes.

Analyses based on different distances will generally reveal different aspects of the spatial pattern. For example, point to nearest event distances contain information about the 'empty spaces' in the region compared with event to event distances which reflect local pattern within groups of individuals. Ideally, we would examine ECDFs based on all three types of distance and Diggle (1983) illustrates this approach.

Example 13.8 Analysis of the tufted duck nest data: application of the empirical cumulative distribution of nearest neighbour distances to assess non-randomness

For the tufted duck data the observed nearest neighbour distances (m) in ascending order are: (7.14, 7.14), (7.29, 7.29), 8.23, 8.56, (9.41, 9.41), 9.73, 9.82, (10.26, 10.26), 10.41, 11.58, 12.41, 12.82, 14.35, where numbers in parentheses correspond to pairs of nests which are nearest neighbours of each other. Table 13.5 gives the ECDF.

The ECDF is illustrated in Figure 13.20(a) together with a corresponding ECDF calculated from the nearest neighbour distances in a set of 19 points located at random in the area. The observed ECDF is markedly different from that for a random distribution with a smaller proportion of relatively small and large values. In other words the spread of the observed distribution is smaller than for the random pattern. To assess the effect of variation in the ECDF for the random pattern we take repeated sets of points at random and calculate the ECDF for each one. For plotting out we compute the average ECDF, and its upper and lower envelopes. Figure 13.20(b) shows the plot based on 20 simulations of complete spatial randomness. It is worth noting that although the observed ECDF appears quite different from the average curve for a random distribution the simulation envelopes are quite wide.

13.3.7 Models for point patterns and processes

When a point pattern is markedly non-random it is sometimes useful to see if it can

Table 13.5 ECDF of nearest neighbour distances for tufted duck data

d	7.14	7.29	8.23	8.56	9.23	9.41	9.73
ECDF(d)	2/19	4/19	5/19	6/19	8/19	10/19	11/19
d	9.82	10.26	11.41	11.58	12.41	12.82	14.35
ECDF(d)	12/19	14/19	15/19	16/19	17/19	18/19	1

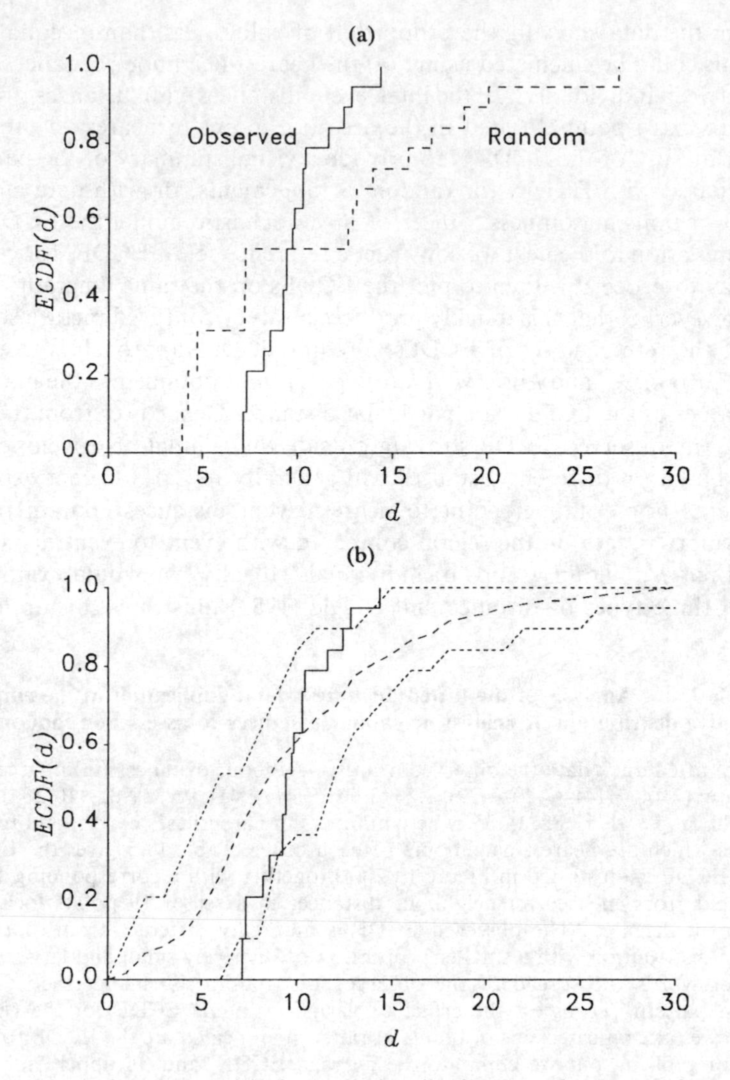

Figure 13.20 Comparison of empirical cumulative distribution of nearest neighbour distances for the tufted duck nest data with that based on 20 simulations of locating points at random. (a) ECDF for a typical set of random points; (b) average ECDF and simulation envelopes for 20 sets of points located at random.

be described by a mathematical model. More precisely, we mean a model of the *underlying processes* and regard the observed pattern as data for fitting the model. Such models can aid an understanding of the underlying biological processes or they could provide a quantitative description for comparing different but related sets of data. In some cases the analysis of a particular aspect of a pattern may depend on an appropriate model for some other aspect. For example, in the duck nest data the question of whether nests from ducks of different species are spaced out requires some consideration of the regular spacing of ducks of the same species—it would be naive to assume a random distribution of tufted duck nests to see if they were spaced out with respect to mallards. The problem is a general one for patterns involving several types of points.

The mathematical theory of point processes is difficult and often intractable but this need not deter us. Many models can be formulated as a set of rules for the *process* of locating points so that they can be implemented and analysed by computer simulation. There are, however, some problems to consider when using simulation. First, since we are dealing with stochastic models different runs of the same model will produce different observed patterns. This means that we need to be able to summarise the variation in our simulated data for comparison with the observed pattern. We have already illustrated one approach which uses the empirical cumulative distribution function (ECDF) of distances (Section 13.3.6). A second problem arises in fitting the models, i.e. estimating the values of the parameters in the model. In many cases formal statistical methods have not been developed so we adopt *ad hoc* empirical methods for matching the model with data.

Regular patterns and simple sequential inhibition

Although the regimented regularity of the forest plantation is rare in nature, individuals are often spaced out as a result of processes such as competition and territorial behaviour. When a regular pattern is the result of interactions between individuals, mechanistic models may be difficult to develop because the nature of the interactions is usually complex and poorly understood. In these cases progress may be possible by constructing simple descriptive models for the outcome of the interactions. As an example, we present the simple sequential inhibition model in which individuals are successively located at random in a region but subject to a minimum distance apart. First, we give a skeleton outline of the procedure for simulating the process of simple sequential inhibition then we describe an application of the model to the tufted duck nest data.

The simple sequential inhibition model has a single parameter which is the minimum allowable distance ($d_{\min}$) between a pair of events. Around each point there is a circular zone within which no other event may fall and for this reason the model is sometimes called a *hard core model*. To simulate the process of simple sequential inhibition for n events we proceed as follows:

Step 1. Locate an event at random in the area.
Step 2. Successively locate further events at random, rejecting any which fall within a distance $d_{\min}$ of an existing event, until there are n events in the area.

By computing distances from a new event to those already located and rejecting those which fall too close to any established event we avoid the awkward problem of demarcating the area available to each new event. A potential difficulty is that the process may take much CPU time to complete as the area fills up and more and more events are rejected especially if they are densely packed. In practice, it would be wise to monitor the progress of the computer program. The simple sequential inhibition model assumes that the circle of inhibition is a constant. A possible modification would be to allow the radius to vary between individuals, say by sampling the value from some distribution with specified mean and variance.

Example 13.9 Analysis of the tufted duck nest data: fitting the simple sequential inhibition model

The Monte Carlo test demonstrates that the tufted ducks' nests are spaced out compared with complete spatial randomness. This raises the question of how well the observed pattern can be described by a model such as simple sequential inhibition. We can address this question by simulating data from the process and comparing the results with the observed pattern by using the empirical cumulative distribution function. To do this we need an estimate of the spacing parameter $d_{\min}$. We proceed by trial and error starting with the observed minimum spacing between a pair of nests, i.e. 7.14 m, and updating the estimate by comparing the observed empirical cumulative distribution function with that based on simulations of the model.

Figure 13.21(a) shows the observed empirical cumulative distribution function together with an average value and simulation envelopes calculated from 20 simulations of the process using a minimum nest spacing equal to that observed. For the small nearest neighbour distances there is reasonable agreement between the observed picture and the model—this is partly to be expected because the inhibition radius used in the simulations was set equal to the observed minimum spacing. The observed ECDF is, however, displaced to the left of the fitted value—this may be due to a bias in the estimate of the minimum nest spacing since the value obtained from the data must necessarily exceed a fixed minimum value, if one exists. To allow for this effect the analysis was repeated using a smaller minimum distance of 6.5 m and this gives a close match for smaller distances (Figure 13.21(b)). However, comparison of observed and fitted ECDFs shows fewer larger distances than would have been expected from the model. This suggests a pattern of local regularity, due to neighbouring birds spacing out their nests, with some aggregation on a larger scale so that regularly spaced nests tend to occur in groups. This could be reflecting variation in suitable habitat or perhaps a form of social behaviour, or even both. These possible interpretations illustrate the general point that a particular observed spatial pattern may be the result of several quite different biological processes which may be indistinguishable from the distribution map alone. The situation requires data of a different kind to help interpretation.

Aggregated patterns, heterogeneity and clustering

An aggregated point pattern appears as a patchy distribution of points. More precisely, aggregation implies that the variation in the density of points is larger than expected in a random distribution of the same density. One possible cause of aggregation is environmental heterogeneity—for example, variation in soil fertility leading to a patchy distribution of plants. A different source of aggregation is clustering caused by some biological mechanism—for example, local dispersal of seeds so that seedlings grow in the vicinity of their parents. Despite the fundamental difference between environmental heterogeneity and clustering mechanisms, both can give rise to similar patterns so that their effects may be indistinguishable from a distribution map—essentially clusters of points can mimic variation in density and *vice versa* (see Figures 13.17(f) and (i)).

For building models of environmental heterogeneity much will depend on the particular situation. There may be both systematic effects in the form of trends and a random component with spatial autocorrelation. Sometimes a simple empirical

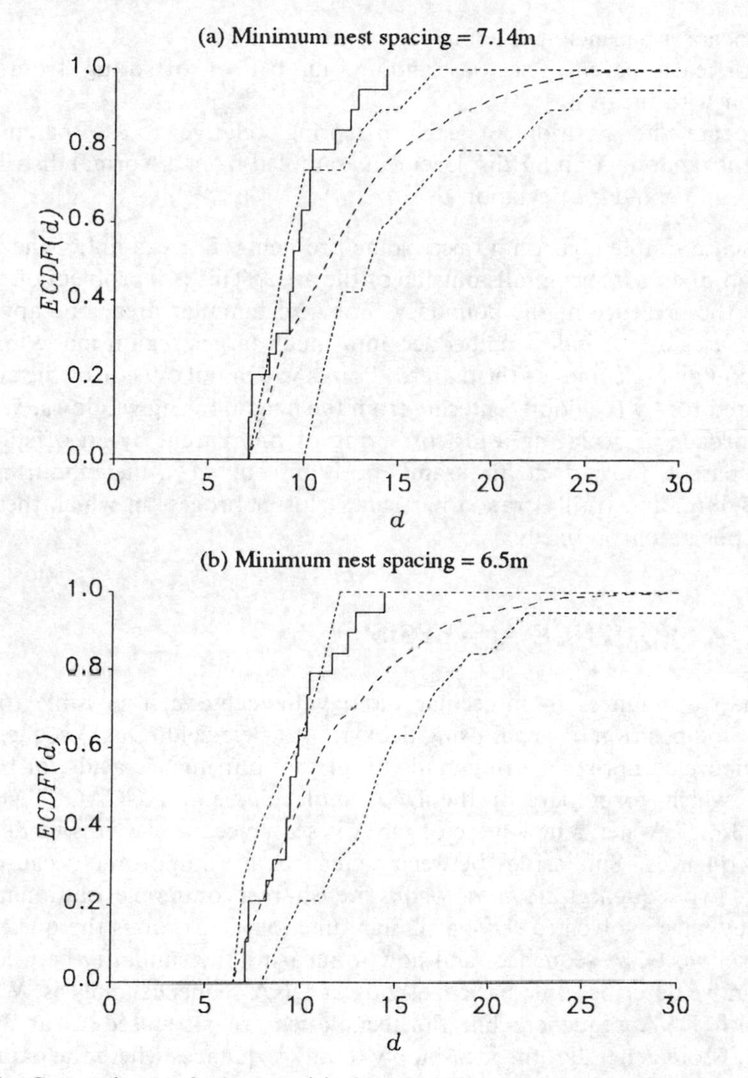

Figure 13.21 Comparison of the empirical cumulative distribution function of nearest neighbour distances for the tufted duck nest data with that based on simulations of the simple sequential inhibition model for different values of the minimum nest spacing: (a) 7.14 m; (b) 6.5 m. Solid line shows observed ECDF; broken lines are mean and simulation envelopes for 20 realisations of the model.

description will suffice but in other situations it may be necessary to consider the effect of external processes. Such detailed considerations are beyond the scope of this section. Instead, we discuss a simple mechanistic model for clustering, the *Poisson cluster process*. In this model a number of parents are located at random in an area and each parent gives rise to a number of offspring which are located around the position of the parent according to some distribution. Both the number of offspring per parent and their location may be fixed, or the values may be sampled from some probability distribution. A particular case of the Poisson cluster process can be simulated using the following steps.

Step 1. Locate n parents at random.
Step 2. For each parent select a random number of offspring from a Poisson distribution with mean μ.
Step 3. Locate the position of each offspring relative to its parent by using independent random X and Y displacements sampled from a Normal distribution with mean zero and standard deviation σ.

Even such a simple algorithm raises some problems. For example, what happens if the location of an offspring falls outside of the area? This is a problem of edge-effects caused by the presence of the boundary. For a rectangular area, one approach is to regard the area as a window embedded in a much larger region made up of similar windows containing copies of the pattern. Points which fall outside of the area are then compensated for by the points entering from the neighbouring windows. We can think of the approach as replacing a lost offspring of one parent by an offspring from a different parent located at the same position but in a neighbouring window. Figures 13.17(c) and (i) illustrate a particular cluster process in which the number of offspring per parent is fixed.

13.4 DNA SEQUENCE ANALYSIS

Recent major advances in molecular biology have made it possible to study the molecular composition of genes using deoxyribonucleic acid (DNA) sequence data. A DNA sequence comprises a string made up of the four nucleic acids, or bases, A, C, G and T which form part of the DNA molecule, e.g. AGCACCTACCAGT . . . (Section 13.4.1). When a new piece of DNA is sequenced a search is made for similar existing sequences. Similarities between sequences are important because they may suggest the two sequences are *homologous*, i.e. share a common evolutionary ancestor, or that they have evolved to serve a similar function. This raises the question of how to compare two DNA sequences and how to measure the similarity between them. In this section we describe some basic methods of DNA sequence analysis. We shall deal mainly with DNA sequences but the techniques are also used to analyse protein sequences. More generally, the ideas apply to any sequences where the data are strings of elements or symbols, e.g. ABCDEFGH . . ., from some alphabet. Applications also arise in speech processing (Hunt *et al.*, 1983) and in the analysis of bird song (Bradley & Bradley, 1983)

13.4.1 Elements of molecular sequences

Figure 13.22 illustrates the structure of the DNA molecule (Alberts *et al.*, 1983). This comprises two repetitive sugar-phosphate chains joined together by the nucleotide bases adenine (A) always paired with thymine (T), and cytosine (C) always paired with guanine (G). The two strands of complementary base pairs are linked by hydrogen bonds and twisted into the form of a double helix in the cell's nucleus. Each strand has a direction labelled as '5′ to 3′' with complementary strands running in opposite directions. DNA replicates by breaking the hydrogen bonds to separate the two complementary strands. Each strand then acts as a template for the formation of a new

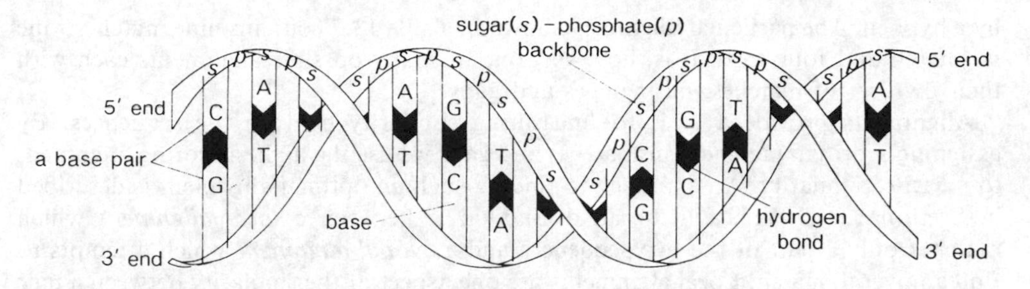

Figure 13.22 Representation of the structure of a DNA molecule. Scale: 1 helical turn = 3.4 nm. Reproduced with permission from *Biometric Bulletin*, **6**(4), 1989.

DNA molecule by the sequential addition of deoxyribonucleoside triphosphates. DNA replication is extremely accurate with fewer than one mistake in 10^9 nucleotides added, but very rarely bases may be skipped or added, a T placed instead of an C, or an A instead of a G. Such *mutations* are then faithfully copied in future cell generations and have important consequences for sequence analysis.

13.4.2 Alignments for comparing two sequences

Some basic methods for comparing DNA sequences and for measuring the similarity between them can be illustrated by considering the comparison of two sequences. The problem arises because any natural correspondence between the sequences is not known in advance and changes may have occurred to either sequence through mutation and loss or gain of material. An additional problem is that the sequences may be of different lengths. In general, a method for comparing two sequences must allow for the fact that:

(1) Regions of similarity may begin at different positions along the sequence;
(2) Elements of either sequence may have been changed by mutation (*substitutions*);
(3) For both sequences elements may have been lost (*deletions*) or gained (*insertions*);
(4) The lengths of the sequences may be different.

Global and local sequence alignments

One way of comparing two sequences is to construct an alignment. This matches the two sequences by sliding one relative to the other and allowing for gaps in either sequence. Table 13.7 illustrates the idea for the sequences in Table 13.6. Matching elements in the alignment are indicated by a colon, mismatches by a space and gaps

Table 13.6 Two short artificial DNA sequences used to illustrate sequence alignment and dot-matrix plots

Sequence *x*:	A	G	A	A	C	C	C	T	A	G	T	A		
Sequence *y*:	G	A	C	C	A	T	A	G	C	G	T	A	G	T	C

by a hyphen. The particular alignment shown in Table 13.7 contains nine matches, one mismatch and four gaps. It is, however, one of many possible alignments each with their own set of matches, mismatches and gaps.

Alignments provide a basis for measuring similarity between the sequences. By assigning scores to matches, mismatches and gaps we use the highest scoring alignment to measure similarity. An approach to finding such an optimal alignment is described in Section 13.4.3. A useful broad distinction is between a *local alignment* which matches only a part of the two sequences and a *global alignment* which attempts to find an overall match. Local alignments are one aspect of the similarity between a pair of sequences which may be of significance in themselves or they may be used as a starting point for constructing a global alignment and measuring overall similarity. An optimal global alignment of two sequences indicates the number of mutations needed to convert one sequence into the other so that it could be thought of as a measure of evolutionary distance between the two sequences.

Table 13.7 Illustration of an alignment for the two sequences in Table 13.6

Sequence x	A	G	A	A	C	C	C	T	A	G	—	—	T	A			
		:	:		:	:		:	:	:			:	:			
Sequence y:	—	G	A	—	C	C	A	T	A	G	C	G	T	A	G	T	C

Colon (:), base match; space ( ), mismatch; hyphen (—), gap

13.4.3 Dot-matrix plots

The dot-matrix plot is a method for displaying sequences which is useful for making preliminary comparisons and for finding alignments. Table 13.8 shows a dot-matrix plot for the sequences in Table 13.6. It is constructed as follows:

(1) One sequence runs across the top of the matrix and the other sequence runs down the left hand side;
(2) Elements in the horizontal sequence label the columns of the matrix whereas elements of the vertical sequence label the rows;
(3) Dots in the matrix indicate that the corresponding sequence elements are the same, blanks show that they are different.

The dot-matrix is useful for depicting subsequences common to both sequences since these appear as a diagonal series of dots. When comparing long sequences the dot-matrix may appear rather congested so that it can be sometimes useful to apply a filter and only to plot dots which are part of diagonals of a specified minimum length.

A key property of the dot-matrix is that it can be used to represent an alignment of the two sequences as a *path* traced through the matrix with the following properties:

(1) The path moves either down a rightward diagonal, horizontally to the right or vertically downwards;
(2) Diagonal segments correspond to a match or mismatch depending on the entry in the dot-matrix;
(3) *Horizontal* steps in the path correspond to a gap in the *vertical* sequence whereas *vertical* segments denote gaps in the *horizontal* sequence.

Table 13.8 Dot-matrix plot for two short artificial DNA sequences. Underscore indicates a path through the matrix for the alignment in Table 13.7

	A	G	A	A	C	C	C	T	A	G	T	A
G		$\underline{\cdot}$								$\cdot$		
A	$\cdot$		$\underline{\cdot}$	$\underline{\cdot}$					$\cdot$			$\cdot$
C					$\underline{\cdot}$	$\cdot$	$\cdot$					
C					$\cdot$	$\underline{\cdot}$	$\cdot$					
A	$\cdot$		$\cdot$	$\cdot$			$\underline{\ }$		$\cdot$			$\cdot$
T								$\underline{\cdot}$			$\cdot$	
A	$\cdot$		$\cdot$	$\cdot$					$\underline{\cdot}$			$\cdot$
G		$\cdot$								$\underline{\cdot}$		
C					$\cdot$	$\cdot$	$\cdot$			$\underline{\ }$		
G		$\cdot$								$\underline{\cdot}$		
T								$\cdot$			$\underline{\cdot}$	
A	$\cdot$		$\cdot$	$\cdot$					$\cdot$			$\underline{\cdot}$
G		$\cdot$								$\cdot$		
T								$\cdot$			$\cdot$	
C					$\cdot$	$\cdot$	$\cdot$					

Table 13.8 shows the path for the alignment in Table 13.7. For a global alignment, i.e. an overall match of the two sequences, the path starts either along the top of the matrix or down the left hand side and finishes along the bottom or far right hand side. In the next section we show how these paths can be used to construct algorithms to find optimal alignments.

13.4.4 Dynamic programming algorithms for finding optimal alignments

For a pair of sequences we can construct many possible alignments so which do we choose? Ideally, the alignment should have as many matches and as few mismatches or gaps as possible. One approach is to assign scores to matches, mismatches and gaps and to find the alignment, or alignments as there may be more than one, with the largest score. In this section we describe an algorithm for doing this and show how it can be applied to finding optimal global and local alignments.

Needleman and Wunsch algorithm for optimal global alignments

A basic method for finding an optimal global alignment was described by Needleman and Wunsch (1970) and Sellers (1974). This works by assigning scores to matches, mismatches and gaps and finding the alignment, or alignments, with maximum score. The key to searching for the maximum score is that an alignment can be represented as a path traced through the dot-matrix in which diagonal segments of the path correspond to matches or mismatches depending on the element in the dot-matrix, and horizontal and vertical lines denote gaps in the two sequences (Section 13.4.3). As the path is traced out, a score is accumulated by adding the scores from its matches, mismatches and gaps of various lengths. Finding an optimal alignment involves finding a path through the dot-matrix with largest accumulated score. Sometimes there will be more than one optimal alignment with the same maximum score.

To apply the Needleman and Wunsch algorithm we need numerical values for the scores but to illustrate the principle of the method we use symbols for scores as follows.

(1) s_{ij} denotes the score for a match or mismatch between the ith element of one sequence and the jth element of the other.
(2) W_k denotes the penalty for a gap of length k. The most frequently used gap-weighting function is of the form $W_k = a + bk$ with a weight for each occurrence of a gap (a) and a weight for each element in the gap (b).

Now consider possible paths through the matrix and the problem of finding the one with largest accumulated score. The ingenious solution is to take each cell in turn and to find the approach to it for which the accumulated score is largest. This leads to a recursive algorithm for calculating a maximum cumulative score and corresponding route for each cell of the dot-matrix. To derive the algorithm we start by noting that any path which passes through a particular cell of the dot-matrix must approach it diagonally from above left, horizontally from a cell to the left in the same row, or vertically downwards from a cell in the same column. Each approach carries an accumulated score—a path down the diagonal involves either a match or a mismatch whereas those coming from the horizontal and vertical contain a gap penalty. Table 13.9 shows the different approaches of paths to cell (i, j) together with their cumulative score S_{ij}.

The Needleman and Wunsch algorithm selects the route which maximises the cumulative score for each cell. The end of an optimal path occurs at a cell in the final row or column of the matrix with largest score. The optimal alignment is obtained by retracing the route to the cell. Since the algorithm computes cumulative scores recursively from previous scores we need initial scores to start it off. Usually, the scheme is initialised with $S_{00} = 0$ and $S_{0k} = S_{k0} = -W_k$, i.e. a misalignment at the end is counted as a gap.

Numerical example of the Needleman and Wunsch algorithm

For illustration we apply the method to the sequences in Table 13.6 using the following scores: $s_{ij} = 3$ for a match; $s_{ij} = -1$ for a mismatch; penalty for a gap length

Table 13.9 Different ways in which a path through the dot-matrix can approach a particular cell (i, j) with corresponding cumulative score S_{ij}

Approach to cell (i, j) of path	Cumulative score S_{ij}	Needleman & Wunsch
Diagonally from cell $(i-1, j-1)$	$S_{i-1,j-1} + s_{ij}$	Maximise
Horizontally from cell $(i, j-k)$	$S_{i,j-k} - W_k$	S_{ij}
Vertically from cell $(i-k, j)$	$S_{i-k,j} - W_k$	

s_{ij}, score for match/mismatch in cell (i, j); W_k, penalty for a gap of length k.

$k = W_k = 3 + k$. Figure 13.23 shows how the calculations proceed starting from the top left hand corner of the dot-matrix. For example, to obtain the cumulative score S_{11} we can move: (a) diagonally from S_{00} to give $S_{11} = S_{00} + s_{11} = 0 - 1 = -1$; (b) vertically downwards from S_{01}, with gap penalty $W_1 = 4$, to give $S_{11} = S_{01} - W_1 = -4 - 4 = -8$; (c) horizontally across from S_{10}, with gap penalty $W_1 = 4$, to give $S_{11} = S_{10} - W_1 = -4 - 4 = -8$. In this case the move which maximises the cumulative score is from S_{00}. Using the value $S_{11} = -1$ we can now compute in turn: $S_{12} = -1$ (diagonal move from S_{01}); $S_{21} = -1$ (diagonal move from S_{10}) and then $S_{22} = -2$ (diagonal move from S_{11}). In each case the direction from where the best path is coming is indicated by the term t_{ij} of a *traceback matrix* which is: (a) zero for a diagonal move; (b) $+k$ for a vertical move with k steps; (b) $-k$ for a horizontal move with k steps. These trace terms are used to retrace the steps of any particular path. Note that there may be more than one path which maximises the cumulative score. For example, to calculate S_{24} we first obtain $S_{14} = -5$. Then, $S_{24} = -5 + 3 = -2$ (diagonally from S_{13}) or $S_{24} = 2 - 4 = -2$ (horizontally from S_{23}).

Table 13.10 shows the completed array of cumulative scores. The maximum cumulative score is 13 and occurs in the final column of the matrix. A path through the dot-matrix with this score is indicated. Table 13.11 shows the traceback matrix for this example. Note that there are some cells with more than one direction for the same cumulative score, illustrating that there is more than one optimal alignment. In this case the different possibilities correspond to small deviations from a particular path through the matrix but in other cases the routes could be very different.

Qualifications on the 'optimal alignment'

The Needleman and Wunsch algorithm finds an optimal alignment, or perhaps more than one, but it is important to realise that the alignment is optimal only for the choice of similarity scores for matches and mismatches and for the gap-weighting function used. A different set of scores and weights will, in general, produce a different optimal alignment. In practice, the calculations will usually be performed using a computer

	x	A	G	A	.
y	$S_{00} = 0$	$S_{01} = -4$	$S_{02} = -5$	$S_{03} = -6$	.
G	$S_{10} = -4$	$s_{11} = -1$ $t_{11} = 0$ $S_{11} = -1$	$s_{12} = 3$ $t_{12} = 0$ $S_{12} = -1$	$s_{13} = -1$ $t_{13} = -1$ $S_{13} = -5$	. .
A	$S_{20} = -5$	$s_{21} = 3$ $t_{21} = 0$ $S_{21} = -1$	$s_{22} = -1$ $t_{22} = 0$ $S_{22} = -2$	$s_{23} = 3$ $t_{23} = 0$ $S_{23} = 2$	. .
C	$S_{30} = -6$	$s_{31} = -1$ $t_{31} = +1$ $S_{31} = -5$	$s_{32} = -1$ $t_{32} = 0$ $S_{32} = -2$	$s_{33} = -1$ $t_{33} = +1$ $S_{33} = -2$	. .
.	.	.	.	.	.

Figure 13.23 Illustration of calculations in the Needleman and Wunsch algorithm (see text for details).

Table 13.10 Cumulative similarity scores for the Needleman and Wunsch algorithm to compare two short sequences. Underlined bold numerals show a path of an optimal global alignment with maximum score of 13 given by

Sequence *x*:	A	G	A	A	C	C	C	T	A	G	—	—	T	A			
		:	:		:	:		:	:	:			:	:			
Sequence *y*:	—	G	A	—	C	C	A	T	A	G	C	G	T	A	G	T	C

	x	A	G	A	A	C	C	C	T	A	G	T	A
y	0	**−4**	−5	−6	−7	−8	−9	−10	−11	−12	−13	−14	−15
G	−4	−1	**−1**	−5	−6	−7	−8	−9	−10	−11	−9	−13	−14
A	−5	−1	−2	**2**	**−2**	−3	−4	−5	−6	−8	−7	−9	−10
C	−6	−5	−2	−2	1	**1**	0	−1	−5	−6	−7	−8	−9
C	−7	−6	−6	−3	−3	4	**4**	3	−1	−2	−3	−4	−5
A	−8	−4	−7	−3	0	0	3	**3**	2	2	−2	−3	−1
T	−9	−8	−5	−5	−4	−1	−1	2	**6**	2	1	1	−1
A	−10	−6	−9	−2	−2	−2	−2	−2	2	**9**	5	4	4
G	−11	−10	−3	−6	−3	−3	−3	−3	1	5	**12**	8	7
C	−12	−11	−7	−4	−7	0	0	0	0	4	**8**	11	7
G	−13	−12	−8	−8	−5	−4	−1	−1	−1	3	**7**	7	10
T	−14	−13	−9	−9	−9	−5	−5	−2	2	2	6	**10**	6
A	−15	−11	−10	−6	−6	−6	−6	−6	−2	5	5	6	**13**
G	−16	−15	−8	−10	−7	−7	−7	−7	−3	1	8	5	9
T	−17	−16	−12	−9	−11	−8	−8	−8	−4	0	4	11	8
C	−18	−17	−13	−12	−10	−8	−5	−5	−5	−1	3	7	10

Table 13.11 Traceback matrix for application of the Needleman and Wunsch algorithm (Table 13.10). Diagonal move (0); k horizontal moves $(-k)$; k vertical moves (k). Two numbers indicates two possible moves with the same cumulative score. Underlined bold numerals show route of path in Table 13.10

	A	G	A	A	C	C	C	T	A	G	T	A
G	0	**0**	−1	−1	−3	−4	−5	−6	−7	0	−1	−2
A	0	0	**0**	−**1**, 0	−2	−3	−4	−5	−6, 0	−7	−8	0
C	1	0	1	0	**0**	0	0	−1	−2	−3	−4	−5
C	2	0, 1	0, 2	0, 1	0	**0**	0	−1	−2	−3	−4	−5
A	0	0, 2	0	0	1	0	**0**	0	0	−1	−2	0
T	1	0	4	0, 1	0, 2	0, 1	0	**0**	−1	−2, 0	0	−4
A	0	0, 1	0	0	0, 3	0, 2	0, 1	1	**0**	−1	−2	0
G	1	0	1	0	0	0, 3	0, 2	2	1	**0**	−1	−2
C	2	1	0	0	0	0	0	3	2	**1**	0	−1, 0
G	3	0, 2	0	0, 1	1	0	0	0	3	0, **2**	0	0
T	4	3	0, 2	0, 1	2	0, 1	0	0	4	3	**0**	0
A	0	4	0	0	3	0, 2	0, 1	1	0	4	1	**0**
G	1	0	1	0	0	0	2	2	1	0	2	1
T	2	1	0	0, 1	0	0	0	0	2	1	0	2
C	3	2	3	0	0	0	4	3	2	2	1	1

program which allows the user to specify scores and weights so the effects of changing the parameter values can be explored. Fitch & Smith (1983) have shown how sensitive an alignment can be to small changes in the parameters and the reader should refer to their paper for some practical guidelines. In practice, sequence similarity searching requires experience of the methods being used and a good knowledge of the biology of the situation.

Smith and Waterman algorithm for finding local alignments

The basic Needleman and Wunsch algorithm for finding an optimal global alignment was modified by Smith and Waterman (1981) for optimal local alignments. Cumulative similarity scores are constructed using the recursive formulae of the Needleman and Wunsch algorithm but with the following modifications.

(1) Cumulative similarity scores are initialised so that $S_{0k} = S_{k0} = 0$, i.e. there is no penalty for a gap at the beginning of the alignment.
(2) If a cumulative similarity score becomes negative it is reset to zero indicating zero cumulative similarity. This emphasises local similarity since the effects of more remote dissimilarities are removed.
(3) The end of an optimal local alignment occurs at the largest score in the cumulative similarity matrix. This can be anywhere in the matrix and is not necessarily in the final row or column as it is in the Needleman and Wunsch search for an optimal global alignment. There may be more than one maximum score and the alignment will, in general, depend on the choice of scores for matches and mismatches and the gap-weighting function.

Table 13.12 shows an example from the original paper by Smith and Waterman comparing two short RNA sequences. The matrix was calculated using scores of 3 for a match, -1 for a mismatch and a penalty for a gap of length k equal to $W_k = 3 + k$. The largest similarity score is 10 and this marks the end of the optimal local alignment. We then use the traceback matrix to find the optimal path through the matrix and to construct the corresponding alignment.

13.4.5 Estimating the statistical significance of sequence similarities

A question in comparing two sequences is whether the similarity is statistically significant, i.e. is it greater than we would expect by chance in *random* sequences. To tackle this question we need to specify what we mean by randomness in DNA sequences. The difficulty arises because DNA sequences have a complex composition in which the base frequencies can vary and neighbouring nucleotides are not independent—for example, some pairs of bases are more likely to be nearest neighbours than others.

Random shuffling method

The simplest approach which has been widely used is to ignore the complexity of the DNA structure and to compare the observed similarity with the values obtained by randomly permuting the elements of the sequence, i.e. shuffling the elements at

Table 13.12 Application of the Smith and Waterman algorithm to two short sequences of RNA. Bold underlined numerals show path of optimal local alignment

Sequence x	G	C	C	—	U	C	G
	:	:	:		:		:
Sequence y	G	C	C	A	U	U	G

	x	C	A	G	C	C	U	C	G	C	U	U	A	G
y	0	0	0	0	0	0	0	0	0	0	0	0	0	0
A	0	0	3	0	0	0	0	0	0	0	0	0	3	0
A	0	0	3	2	0	0	0	0	0	0	0	0	3	2
U	0	0	0	2	1	0	3	0	0	0	3	3	0	2
G	0	0	0	**3**	1	0	0	2	3	0	0	2	2	3
C	0	3	3	0	**6**	4	1	3	1	6	2	1	1	1
C	0	3	2	0	3	**9**	5	4	3	4	5	1	0	0
A	0	0	6	2	1	**5**	8	4	3	2	3	4	4	0
U	0	0	2	5	1	4	**8**	7	3	2	5	6	3	3
U	0	0	1	1	4	3	7	**7**	6	2	5	8	5	3
G	0	0	0	4	0	3	3	6	**10**	6	5	4	7	8
A	0	0	3	0	3	1	2	2	6	9	5	4	7	6
C	0	3	0	2	3	6	2	5	5	9	8	4	3	6
G	0	0	2	3	1	2	5	1	8	5	8	7	3	6
G	0	0	0	5	2	1	1	4	4	7	4	7	6	6

random. This notion of a random sequence preserves the frequency of the bases and merely allows the order to vary. For each random sequence a similarity score is calculated using, for example, the Needleman and Wunsch algorithm. The process is repeated many times to build up a reference set of scores, i.e. the sampling distribution of the similarity score for random sequences. The probability of a score greater than that observed can then be estimated. Some computer packages for sequence comparison have options for applying the method.

Non-randomness in DNA sequences and the limitations of random shuffling

The random shuffling method is a useful preliminary procedure for evaluating sequence similarities which provides a check on possible overinterpretation in a particular case by screening out similarities which could simply be chance events. However, it is important to stress that apparent statistical significance may be due to the limitations of random shuffling as a plausible model of randomness in the DNA sequence. Lipman *et al*. (1984) have shown how the nearest neighbour properties of a sequence and local base fluctuations can affect the statistical significance and, in particular, how random shuffling can overestimate the statistical significance of a similarity. The effect can be appreciated by considering the significance of the number of matches of the sequence ATCG with itself. Under random shuffling the probability of four matches is 1/24 but if A & T are always neighbours and similarly for G & C then the chances are 1/8. One approach is to generate random sequences which preserve some of the properties of the sequence such as nearest neighbour frequencies or other groupings of the bases. A different method when searching a databank is to use the distribution of scores within the library sequences as a reference set for evaluating a particular score (Section 13.4.6).

13.4.6 Searching for sequence similarities in a large databank

So far we have dealt with methods for comparing a pair of sequences, i.e. sequence alignment, dot-matrix plots and dynamic programming algorithms for finding optimal alignments. However, with the development of large databanks of DNA and protein sequences there is the different but related problem of searching a bank for sequences which are similar to a given 'query' sequence. In principle, the search is a series of pairwise comparisons of the query sequence with those in the databank. In practice, a problem arises because of the large amount of material to be processed so that computing time becomes prohibitive. For example, in the European Molecular Biology Laboratory DNA sequence databank (Release number 18, February 1989) there were 22 938 sequences typically about 1000 bases long (Kirkwood, 1989). This has led to the development of algorithms for rapid similarity searches of nucleic acid and protein databanks. Two programs currently in widespread use are FASTP (Lipman & Pearson, 1985) and FASTA (Pearson & Lipman, 1988). A detailed description of the algorithms used in these packages is beyond the scope of this book. However, both methods involve a modification of a basic algorithm proposed by Wilbur and Lipman (1983) which is described briefly below.

Wilbur and Lipman algorithm

The method of Wilbur and Lipman is motivated by the observation that the dot-matrix plot often shows diagonals with relatively many matches and that such diagonals remain when the matrix is filtered to show only matches of length k or more. The method uses only matching k-tuples or k-words (i.e. strings of matching bases of length k appearing in the dot-matrix as a diagonal of length k) ignoring shorter matches. The value of k is selected by the user. A skeleton outline of the algorithm is as follows.

(1) The first step is finding all the matching k-tuples. High speed is achieved by working with a *look-up table* for the query sequence (Dumas & Ninio, 1982). This contains a list of the positions of each base element so that matches can be located without having to compare every element of one sequence with all the elements of the other.
(2) The k-tuples are restricted to those which fall on a subset of the diagonals in the dot-matrix plot. This region is referred to as the *window space* and is defined as those diagonals which fall within a specified distance w of a *significant* diagonal, i.e. one which contains a certain number of k-tuple matches.
(3) Matches are scored provided that they are part of a k-tuple match and that they fall within the window space. Then an alignment is found by applying a modification of the Needleman and Wunsch algorithm to matches within the window space.

In practice the user must specify values for the parameters k and w. Wilbur and Lipman illustrated their method using 28 pairs of nucleic acid sequences selected from the Los Alamos Nucleic Data Bank using $k = 4$, a window $w = 10$ with gap penalty of 6.

```
Score  Number of
       sequences
 <2      1 :=
  4      0 :
  6      1 :=
  8      4 :==
 10     10 :=====
 12     35 :==================
 14     76 :======================================
 16    215 :==================================================
 18    283 :==================================================
 20    390 :==================================================
 22    362 :==================================================
 24    313 :==================================================
 26    307 :==================================================
 28    246 :==================================================
 30    157 :==================================================
 32     87 :============================================
 34     65 :=================================
 36     36 :==================
 38     24 :============
 40     21 :===========
 42     11 :======
 44      8 :====
 46      3 :==
 48      1 :=
 50      2 :=
 52      2 :=
 54      0 :
 56      0 :
 58      0 :
 60      0 :
 62      2 :=
 64      0 :
 66      0 :
 68      0 :
 70      0 :
 72      0 :
 74      0 :
 76      3 :==
 78      1 :=
 80      1 :=
>80     10 :=====
526898 residues in 2677 sequences, mean score; 22.8 (5.75)
ktup = 2, scan time 2:26.76
```

Figure 13.24 Histogram of initial similarity scores produced by FASTP in a comparison of the protein sequence of bovine cyclic AMP-dependent kinase, OKBO2C with 2677 sequences in the NBRF protein sequence library. Note truncation of the histogram heights for frequencies greater than 100. After Lipman & Pearson (1985).

The results were in excellent agreement with those produced by the standard Needleman and Wunsch algorithm which effectively uses $k = 1$.

Output from a databank search using FASTP/FASTA

Typical output from a databank search gives a histogram of the similarity scores between the query sequence and those in the library together with alignments for those sequences with high similarity scores. The histogram helps to assess values relative to the variation in the databank. Figure 13.24 shows a histogram of similarity scores obtained by comparing a sequence for bovine cyclic AMP-dependent kinase, OKBO2C with the 2677 protein sequences in the National Biomedical Research Foundation (NBRF) library (Lipman & Pearson, 1985). This highlights 15 rather extreme scores greater than 75 (including the self comparison). The corresponding alignments are then identified for further analysis.

Assessing statistical significance

Lipman and Pearson (1985) stress the importance of applying biological information in evaluating a potential similarity but they note that when the biological context is not clear an estimate of the statistical significance may be useful. They propose a random shuffling method in which the query sequence is compared with versions of the library sequence whose elements have been randomly permuted. From a number of random permutations a score is calculated as

$$z = \frac{\text{similarity score} - \text{mean of random scores}}{\text{standard deviation of random scores}}.$$

Based on experience of a large number of database searches, Lipman and Pearson suggest the following guidelines for significance: $z > 3$—possibly significant; $z > 6$—probably significant; $z > 10$—significant. An alternative approach is to use the values from the database as a reference set for calculating the mean and standard deviation. This makes some allowance for the possible non-random structure of the sequences but does not allow for sequences being of different lengths. Another problem of interpretation is that we are selecting extreme scores from some distribution so that the more comparisons being made the more likely an extreme value will occur simply by chance.

EXERCISES

Most of the analyses described in the chapter require a computer to carry them out. In some cases appropriate software is available so that the calculations can be performed with a few lines of code. For example, there are packages for time series analysis with commands for calculating autocorrelation coefficients, partial autocorrelation coefficients and for fitting autoregressive models. In other situations it may be necessary to use a package or a general purpose programming language to develop appropriate programs. For example, at the present time a versatile package suitable for non-specialists for simulating point patterns is not available. The examples given below are intended to encourage readers to explore and use existing software and to write their own programs if needs be. We start by describing some example sets of data which are referred to in the various exercises. A useful first step would be to plot out the observations and to look for interesting patterns. Indeed, this simple approach will often be sufficient to grasp the main features of the data.

Canadian lynx data

A famous time series in ecology is the annual record of the numbers of the Canadian lynx trapped in the Mackenzie River district in north west Canada from 1821 to 1934 (Elton & Nicholson, 1942). A remarkable feature of the data are the regular quasi-cyclical fluctuations with peaks every 10–11 years.

Year	N	Year	N	Year	N	Year	N	Year	N	Year	N
1821	269	1841	151	1861	236	1881	469	1901	758	1921	229
1822	321	1842	45	1862	245	1882	736	1902	1307	1922	399
1823	585	1843	68	1863	552	1883	2042	1903	3465	1923	1132
1824	871	1844	213	1864	1623	1884	2811	1904	6991	1924	2432
1825	1475	1845	546	1865	3311	1885	4431	1905	6313	1925	3574
1826	2821	1846	1033	1866	6721	1886	2511	1906	3794	1926	2935
1827	3928	1847	2129	1867	4245	1887	389	1907	1836	1927	1537
1828	5943	1848	2536	1868	687	1888	73	1908	345	1928	529
1829	4950	1849	957	1869	255	1889	39	1909	382	1929	485
1830	2577	1850	361	1870	473	1890	49	1910	808	1930	662
1831	523	1851	377	1871	358	1891	59	1911	1388	1931	1000
1832	98	1852	225	1872	784	1892	188	1912	2713	1932	1590
1833	184	1853	360	1873	1594	1893	377	1913	3800	1933	2657
1834	279	1854	731	1874	1676	1894	1292	1914	3091	1934	3396
1835	409	1855	1638	1875	2251	1895	4031	1915	2985		
1836	2285	1856	2725	1876	1426	1896	3495	1916	3790		
1837	2685	1857	2871	1877	756	1897	587	1917	674		
1838	3409	1858	2119	1878	299	1898	105	1918	81		
1839	1824	1859	684	1879	201	1899	153	1919	80		
1840	409	1860	299	1880	229	1890	387	1920	108		

Redwood seedlings data

These data on the locations of 62 redwood seedlings were extracted by B.D. Ripley (1977) from a larger set of data in Strauss (1975). They were taken in an approximately square plot of side 23 m but they are given below as the (x, y) coordinates in a unit square (from Diggle, 1983). Various analyses of these data can be found in these references.

x	0.364	0.898	0.864	0.966	0.864	0.686	0.500	0.483	0.339	0.483	0.186	0.203	0.186
y	0.082	0.082	0.180	0.541	0.902	0.328	0.598	0.672	0.836	0.820	0.402	0.525	0.541
x	0.483	0.898	0.780	0.898	0.746	0.644	0.525	0.220	0.381	0.483	0.203	0.102	0.186
y	0.082	0.098	0.123	0.754	0.902	0.279	0.574	0.795	0.836	0.770	0.426	0.574	0.557
x	0.441	0.839	0.780	0.898	0.678	0.610	0.585	0.220	0.407	0.508	0.220	0.119	0.500
y	0.098	0.082	0.164	0.779	0.246	0.344	0.574	0.836	0.852	0.754	0.459	0.574	0.098
x	0.898	0.763	1.000	0.703	0.627	0.559	0.263	0.263	0.441	0.237	0.136	0.483	0.898
y	0.164	0.148	0.836	0.279	0.344	0.639	0.852	0.697	0.754	0.475	0.574	0.148	0.189
x	0.949	0.966	0.729	0.644	0.525	0.288	0.441	0.186	0.203	0.119			
y	0.525	0.959	0.262	0.361	0.656	0.852	0.820	0.377	0.500	0.623			

13.1 In time series analysis an essential preliminary is to graph the data as a *time plot*. This will often reveal features of the series such as trends, periodic effects and provide a check on whether the series is stationary. The appearance of the time plot can, however, be affected by the shape of the graph as measured by the aspect ratio (i.e. the ratio of length to breadth). For example, differences between steep slopes are difficult to discern by eye and long thin graphs (high aspect ratio) which compress the y-axis relative to the x-axis can help to compare slopes.

Examine time plots of the lynx data using aspect ratios of 2 : 1, 4 : 1 and 8 : 1 both for the original counts and their logarithms. Note how increasing the aspect ratio makes the asymmetry of the increasing and decreasing phase of each cycle more apparent.

13.2 For the Canadian lynx data carry out the following analyses.

(a) Calculate the sample autocorrelation function and the partial autocorrelation function using: (i) the raw counts; (ii) log transformed data.

(b) Confirm that the second-order autoregressive scheme fitted to the logarithm of the count using Minitab is given by

$$Y_{t+1} - 6.69 = 1.39(Y_t - 6.69) - 0.753(Y_{t-1} - 6.68) + Z_t$$

where the Z_t are independent random variables sampled from a Normal distribution with mean zero and standard deviation equal to 0.277.

(c) Simulate data using the model and compare it with the original series using: (i) a time plot; (ii) the sample autocorrelation function; (iii) the sample partial autocorrelation function.

(d) Discuss the limitations of the model for describing changes in the size of the lynx population.

13.3 *Runs test.* A runs test is used to test for non-randomness in the order of a series of observations. It can be applied to a series of binary responses (0/1), or a series of continuous responses scored as + or − depending on whether the observation lies above or below the mean, or median, of the series. The runs test is also sometimes applied to the residuals from a fitted regression model (Draper & Smith, 1981). To illustrate the definition of a run consider the artificial series $(+++)(-)(++)(----)(+)(--)$ in which there are 6 runs indicated by parentheses. To calculate the statistical significance of the observed number of runs we calculate its sampling distribution over random permutations of the observations in the series. In this case the mean and variance are given by

$$\text{mean}[U] = \frac{2n_1n_2}{(n_1+n_2)} + 1; \qquad \text{var}[U] = \frac{2n_1n_2(2n_1n_2 - n_1 - n_2)}{(n_1+n_2)^2(n_1+n_2-1)}$$

where n_1 and n_2 are the number of + and − signs in the series. For n_1 and $n_2 > 10$, the distribution of U is approximately Normal.

The series below were derived from simulated data of: (i) white noise; (ii) AR(1) scheme, $\alpha = 0.5$; (iii) AR(1) scheme $\alpha = -0.5$. A plus sign indicates an observation above the series mean, a minus sign below the mean. In each case apply the runs test and comment on your findings.

a: −−−−−−++−+−++−+−+−+−−−+++−−−−++++−−++−−+−+−+−+−−−−
b: −++++−−−−+++−−−+++−++++++++++−+−−−+−−−−−−−−−−−−−++
c: −++−+−−+−+−−−++++−+−+−+++−++−++−−−+−+−+−+−−+−−+−−+

Apply the runs test to the series of Canadian lynx data.
(Note: Some statistical packages, e.g. Minitab, have a procedure to perform the runs test.)

13.4 Consider an autocorrelated series in which each observation has the same mean and variance.

(a) Show that the variance of the mean $(\bar{Y})$ in a series of n observations is given by

$$\text{var}[\bar{Y}] = \frac{\sigma^2}{n}\{1 + (n-1)\bar{\rho}\}$$

where σ^2 is the variance of an observation and $\bar{\rho}$ is the average of all the $n(n-1)/2$ pairwise correlations between the observations.

(Hint: Use the fact that the variance of a sum of variables is equal to the sum of the individual variances plus twice the sum of the covariances between each pair of observations. To see the general form of the result consider the particular case $n = 2$ and $n = 3$.)

(b) Demonstrate that for the autocorrelated series the sample variance is a biased estimate of the population variance by showing that its mean, or expected, value is given by

$$\text{mean}\,[s^2] = \sigma^2(1 - \bar{\rho}).$$

(Hint: Expand the formula for the sample variance and use the result in (a) to obtain the expected value of $\bar{Y}^2$.)

(c) Treating a serially correlated sample as an independent one results in doubly biased results. Discuss this using the above results.

13.5 In the two-dimensional Poisson process point events occur at random independently and with uniform rate in a plane region of infinite extent.

(a) Show that, for a two-dimensional Poisson process with rate equal to λ events per unit area, the distribution of the *squared* nearest neighbour distance from any event is exponential with mean $1/\pi\lambda$.

(b) Carry out a simulation exercise to confirm the theoretical results that the mean and variance of the nearest neighbour distance (D) from any event are given by

$$\text{mean}\,[D] = \frac{1}{2\sqrt{\lambda}}; \qquad \text{var}\,[D] = \frac{(4 - \pi)}{4\pi\lambda}.$$

(c) Discuss the problems of applying these results in testing for non-randomness in a mapped point pattern.

13.6 Consider the following model for the firing pattern in a cell. When the cell is active firing occurs at random with a constant rate measured in spikes per second. Immediately after firing there is a refractory or 'rest' period of fixed length in which the cell is inactive.

(a) Simulate a series of data with 1000 spikes for a cell with an active firing rate of 10 spikes per second and a refractory period of 20 msec.

(b) Consider the general case in which the rest period is T s and the active firing rate is λ spikes per second. Show that the mean inter-spike interval is equal to $T + 1/\lambda$ and hence that the mean overall firing rate is given by

$$\text{mofr} = \frac{\lambda}{(1 + \lambda T)}.$$

(c) For the simulated data in (a), calculate the histogram and the cumulative empirical distribution function of the inter-spike intervals and compare them with the corresponding probability density function and cumulative distribution function for a Poisson process with the same mean overall firing rate.

(d) Use the simulated data to confirm the result that the distribution of the interval between spikes has mean $T + 1/\lambda$ and standard deviation $1/\lambda$.

13.7 *Testing for spatial randomness using count data.* One way to test for non-randomness in a mapped point pattern is to divide the area into sub-regions and to compare the number of points in each sub-region with the expected count for a random distribution. If there are m sub-regions of equal size with counts $n_1, n_2, \ldots, n_m$ then the expected number in each sub-region for a random distribution with the same total count is the mean count per area ($\bar{n}$) averaged over the whole area. A possible measure of discrepancy between the observed and expected counts is the Pearson chi-square statistic given by

$$X^2 = \sum_{i=1}^{m} \frac{(n_i - \bar{n})^2}{\bar{n}}.$$

On the null hypothesis of spatial randomness, the distribution of the statistic is *approximately* chi-squared with $(m - 1)$ degrees of freedom provided that the mean count is not too small (a rule of thumb is that it should be greater than 5).

For the redwood seedling data:

(a) Apply the above test for spatial randomness by dividing the whole area into a 3×3 grid of square sub-regions; interpret the results and suggest two contrasting explanations for the observed pattern;

(b) Repeat (a) using a 2×2 grid of square sub-regions and comment on the difference in findings.

13.8 Apply the following techniques using distance methods to demonstrate departures from randomness in the redwood seeding data.

(a) Calculate the empirical cumulative distribution function (ECDF) of the nearest neighbour distances between seedlings and compare it with 10 ECDFs for sets of 62 points located at random. Interpret the result.

(b) Superimpose an 8×8 grid of points symmetrically placed on the area and calculate the ECDF of the distance from each point of the grid to the location of the nearest seedling. Compare this point-event ECDF with that obtained by measuring distances from the grid to a set of 62 randomly located events and interpret the result.

13.9 Consider the following form of a Poisson cluster process to describe the pattern of redwood seedlings (Section 13.3.7).

Step 1. Locate the positions of n parents at random within the square.
Step 2. Allocate the 62 seedlings at random to the parents.
Step 3. Position each seedling relative to the location of its parent with independent x and y displacements sampled from a Normal distribution with mean zero and standard deviation (σ).

(a) Simulate a single set of data for each of the nine possible parameter combinations from $n = (15, 25, 40)$ and $\sigma = (0.02, 0.04, 0.06)$. Plot the data for comparison with the observed pattern.
(Hint: To make sure that each simulated set of data contains 62 points, think of the square area as being embedded in the centre of 9 squares each with a copy of the pattern. So, for example, an x coordinate of $1 + \Delta x$ is located at $x = \Delta x$ and a value of $-\Delta x$ is located at $x = 1 - \Delta x$. Similar rules apply to the y coordinates.)

(b) For each set of simulated data calculate the empirical cumulative distribution function of the nearest neighbour distances and compare it with the observed ECDF.

13.10 Apply the Smith and Waterman algorithm to find the optimal local alignment of the two short sequences of RNA given below using scores of 3 for a match, -1 for a mismatch and a penalty for a gap of length k equal to $W_k = 3 + k$.

Sequence x GCCUCG
Sequence y GCCAUUG

13.11 Wilbur and Lipman (1983) illustrated their algorithm for rapid similarity searches using the DNA sequences from two nucleotide fragments given below (read from left to right). Use a computer program for DNA sequence analysis (e.g. FASTA) to compare the two sequences and interpret the results.

Sequence x: AACGTCAAGGCCGCCTGGGGTAAGGTCGGCGCGCACGCTG
GCGAGTATGGTGCGGAGGCCCTGGAGAGGATGTTCCTGTC
CTTCCCCACCACCAAGACCTACTTCCCGCACTTCGACCTG
AGCCACGGCTCTG
Sequence y: TGCTGTCTCTTGCCTGTGGGGAAAGGTGAACTCCGATGAA
GTTGGTGGTGAGGCCCTGGGCAGGCTGCTGGTTGTCTACC
CTTGGACCCAGCGGTACTTTGATAGCTTTGGAGACCTATC
CTCTGCCTCTGCTA

14 Descriptive Models of Complex Relationships: Multiple Regression and Response Surface Models

14.1 INTRODUCTION

This chapter deals with models for describing a complex relationship between a response variable and several explanatory variables. An example discussed in this chapter concerns predicting the weight of timber in a tree from easily-made measurements of its height and diameter. Such a relationship is useful for estimating the amount of timber in a stand without felling any trees. The weight of timber is affected by many factors which may vary from one tree to another so there is no simple relationship with height and diameter. We will see, however, that much of the complexity can be subsumed in a simple statistical model which gives useful predictions (Section 14.2).

Another frequently occurring problem is to determine conditions under which a response is optimum. An example we discuss is how heat loss in pigs is affected by environmental temperature and energy intake. For economy and welfare reasons when rearing pigs it is important to know the temperature at which heat loss is least and how this varies with energy intake. One approach is to measure heat loss over a range of values of temperature and energy intake. Then we can estimate the underlying *response surface* by using flexible empirical models to describe its main features (Section 14.3).

A third problem which arises in analysing relationships involving several variables is to examine the effects of one variable while allowing for the effects of others. In some situations this may be achieved by doing controlled experiments but in others this may not be feasible. For example, in epidemiology we are often interested in relating the incidence of a disease to factors such as weight, diet and smoking habit in a population of individuals. Typically, we are dealing with a sample of individuals in which there is an uneven distribution of the various possible combinations of factors. In this case models can be useful for trying to tease out effects of some variables while controlling for the effects of others. We present two rather different examples of this approach. The first involves fitness in humans and measurements of oxygen uptake in relation to the variables age, weight, pulse rates and run time in a sample of men completing a 1.5 mile test run (Section 14.4). The second concerns survival in house sparrows in relation to size measurements of a sample of birds grounded following a severe storm (Section 14.5). The analysis of observational data cannot establish causal links but it may suggest which variables are important. Sometimes the finding can be tested experimentally, but in other cases this might not be possible: for example, morphometric measurements are specific to an individual and cannot be varied experimentally.

Sometimes there may be many explanatory variables and the problem is to decide which ones are important. We may wish to test whether there is any relationship at all and if so to assess the magnitude of the contribution of a particular variable, perhaps to see if it can be effectively ignored. This is a difficult problem which can require fitting and comparison of a large number of models. Moreover, there may be many different models all apparently 'equally good', in some sense. In Section 14.4.2 we present a general framework for analysing this type of problem and, in particular, give a graphical method which is useful for comparing a large number of models.

This chapter is largely about situations in which there is no body of theory to predict the form of the underlying systematic relationship. The emphasis therefore is on *empirical descriptions* rather than on *mechanistic models* based on postulates about the biological processes. Despite their descriptive nature, however, such models can be very useful for the wide range of problems illustrated above. A rather different situation is when a mechanistic model is available but it is very complicated and difficult to fit. In this case, it can sometimes be profitable to *approximate* the behaviour of the complex model by a simpler descriptive model which is easier to fit and to analyse.

14.2 TWO EXPLANATORY VARIABLES

In this section we deal with the problem of describing the relationship between a response variable and two explanatory variables. In the example described below we are interested in relating the volume of timber in an individual tree to its diameter and height for the purposes of predicting timber volume. We consider the typical situation where the response variable is subject to random variation which obscures the underlying relationship. For example, two trees with the same diameter and height will in general differ in volume due to differences in other characteristics, conditions of growth, genetic make-up, etc.

14.2.1 Two variable linear-Normal model

In Chapter 4 we dealt with the straight line-Normal model for relating a response variable to a single explanatory variable in which the observed response consists of an underlying straight line relationship plus a random disturbance, i.e.

$$Y = \alpha + \beta x + \varepsilon,$$

where $\varepsilon \sim N(0, \sigma^2)$. For a given value of x the values of Y follow a Normal distribution with mean $\alpha + \beta x$ and constant variance σ^2.

We extend this basic model to two explanatory variables x_1 and x_2 as follows. First, suppose that for a fixed value of x_1, the relationship between Y and x_2 follows the straight line-Normal model. Then, assume that the intercept in the model varies linearly with x_1. This gives the two variable linear-Normal model as

$$\begin{aligned} Y &= \begin{pmatrix} \text{intercept which} \\ \text{varies linearly} \\ \text{with } x_1 \end{pmatrix} + \beta_2 x_2 + \varepsilon \\ &= (\alpha + \beta_1 x_1) + \beta_2 x_2 + \varepsilon \end{aligned}$$

where $\varepsilon \sim N(0, \sigma^2)$. In this case the observations at specific values of x_1 and x_2 follow a Normal distribution with mean equal to a linear function of two explanatory variables, i.e. mean Y at given x_1 and x_2, or mean $(Y | x_1, x_2) = \alpha + \beta_1 x_1 + \beta_2 x_2$. This *underlying* relationship can be visualised as a series of parallel lines showing:

(1) mean $(Y | x_1, x_2)$ against x_1 for different values of x_2;
(2) mean $(Y | x_1, x_2)$ against x_2 for different values of x_1.

Figure 14.1 shows some simulated data for the particular model

$$Y = 10 + 2x_1 - x_2 + \varepsilon$$

where the variance of the random component $\sigma^2 = 1$. In this case the random variability is relatively small so that the underlying series of parallel lines is apparent. The model can also be represented in 3-D as a plane representing the underlying relationship mean $(Y | x_1, x_2)$, with observations being subject to vertical deviations from the surface.

The two variable linear-Normal model is sometimes referred to as a *multiple regression model* with two predictors or explanatory variables. The parameters β_1 and β_2 are called the *regression coefficients* or sometimes the *partial regression coefficients*. The coefficient β_1 is the change in the mean value of Y for a unit change in x_1 for any fixed value of x_2. Similarly, β_2 measures the effect of changing x_2 by one unit. The model is linear because the *parameters* occur as linear combinations. For fitting the model to a set of data we identify each observation with a subscript i and write the model as follows

Two variable linear-Normal model

$$Y_i = \alpha + \beta_1 x_{1i} + \beta_2 x_{2i} + \varepsilon_i, \qquad i = 1, 2, \ldots, n,$$

where $\varepsilon_i \sim N(0, \sigma^2)$.

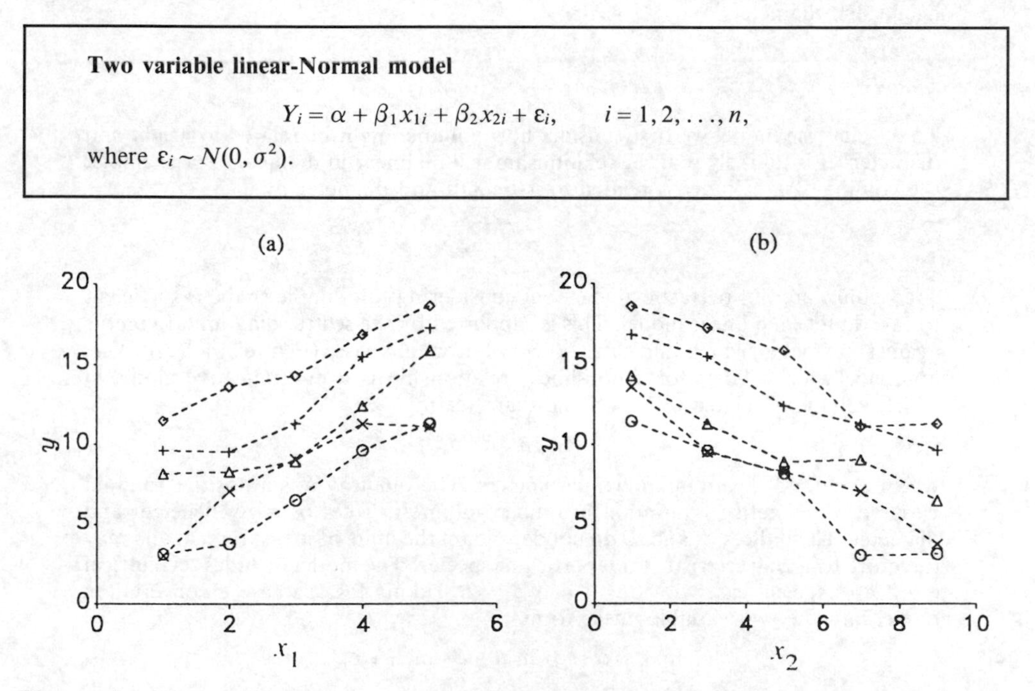

Figure 14.1 Illustration of the two variable linear-Normal model using simulated data: (a) y against x_1 for values of $x_2 = 1$ (top), 3, 5, 7, 9 (bottom); (b) y against x_2 for values of $x_1 = 1$ (bottom), 2, 3, 4, 5 (top). See text for details of model.

14.2.2 Fitting the two variable linear-Normal model

To fit the two variable linear-Normal model to a set of data we use the method of least squares and find parameter estimates $\hat{\alpha}$, $\hat{\beta}_1$ and $\hat{\beta}_2$ for which the sum of squares of the deviations of the observations from the fitted values is least, i.e. we minimise

$$S = \sum_{i=1}^{n} (y_i - \hat{\alpha} - \hat{\beta}_1 x_{i1} - \hat{\beta}_2 x_{i2})^2.$$

Since the model is linear this is a straightforward exercise and nowadays there are many statistical packages with facilities for fitting multiple regression models with two or more explanatory variables.

Example 14.1 Application of the two variable linear-Normal model to the Minitab tree data

A problem in forestry is to estimate the volume of timber in a given area of forest. Since it is difficult to measure the volume of a tree directly, foresters have developed indirect methods which relate the volume to other, more easily made, measurements such as height and diameter. In practice, a sample of trees of various heights and diameters is felled to provide the data for a predictive equation. Ryan *et al.* (1985) present measurements on volume, height, and diameter at 4 ft 6 in above the ground, in a sample of 31 black cherry trees felled in Allegheny National Forest, Pennsylvania. Figure 14.2 depicts the data and shows that tree volume tends to increase with diameter and height and also that taller trees tend to have larger diameters. Here we build a simple model to describe how tree volume varies with height and diameter.

Model

To develop the model we first consider how volume might be related to height and diameter. It is unlikely that the relationship will be linear in diameter: for example, the volume of a cylinder is related to its length and diameter by

$$V = \frac{\pi l d^2}{4}.$$

Tree trunks are not perfect cylinders but considering this simple shape is sufficient to cast doubt on a linear model. This is supported by the scatter diagram of volume against diameter which indicates a curvilinear increase (Figure 14.2(b)). One approach which allows for a non-linear relationship is a multiplicative model in which volume is related to length and diameter by

$$V = c h^{\beta_1} d^{\beta_2} \delta,$$

where c, β_1 and β_2 are unknown parameters. The quantity δ is a positive random variable which reflects variation in timber volume in trees of a given height and diameter. The indices β_1 and β_2 do not depend on the units of measurement and may therefore be a characteristic for a particular species. The model includes 'cylindrical trees' as a special case with $\beta_1 = 1$ and $\beta_2 = 2$. Taking logs to base e converts the model into the two variable linear form

$$\ln V = \alpha + \beta_1 \ln h + \beta_2 \ln d + \varepsilon,$$

where $\alpha = \ln(c)$ and $\varepsilon = \ln \delta$. The transformation linearises the model but it also affects the variability of the response variable (Chapter 11). A constant variance in the transformed volumes corresponds to larger variation in larger trees with

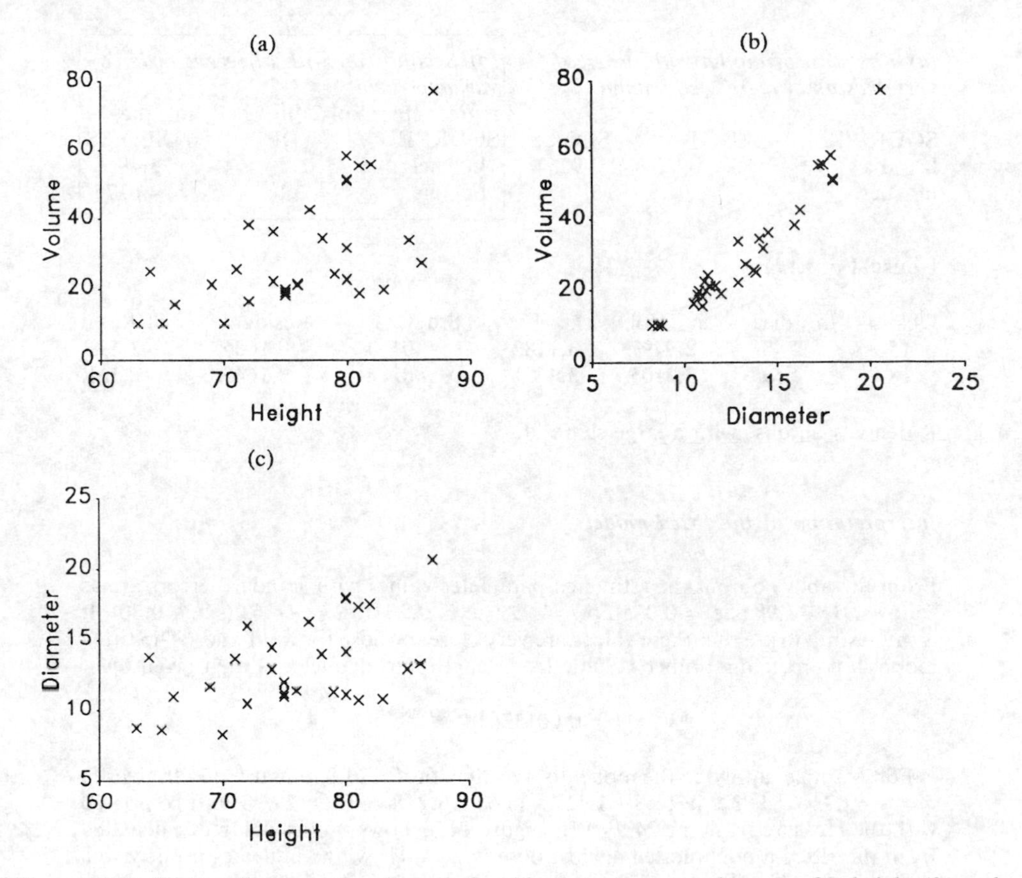

Figure 14.2 Scatter diagrams showing relationships between tree volume (cu ft), height (ft) and diameter (in) in a sample of black cherry trees.

standard deviation proportional to the mean of the untransformed volumes. Figure 14.2(a) indicates that timber volumes are more variable in larger trees so a plausible first step is to fit the model by least squares using log transformed measurements.

Fitting the model by least squares using Minitab

MTB> regress 'ln_Vol' 2 'ln_dia' 'ln_hei'
The regression equation is

ln_Vol = −6.63 + 1.98 ln_dia + 1.12 ln_hei

Predictor	Coef	Stdev	t-ratio	p
Constant	−6.6316	0.7998	−8.29	0.000
ln_dia	1.98265	0.07501	26.43	0.000
ln_hei	1.1171	0.2044	5.46	0.000

s = 0.08139 R-sq = 97.8% R-sq(adj) = 97.6%

Analysis of variance

SOURCE	DF	SS	MS	F	p
Regression	2	8.1232	4.0616	613.19	0.000
Error	28	0.1855	0.0066		
Total	30	8.3087			

(a) First run of model with order of variables as in regress command

SOURCE	DF	SEQ SS
ln_dia	1	7.9254
ln_hei	1	0.1978

(b) Second run with different order of variables using
regress 'ln_Vol'2 'ln_hei' 'ln_dia'

SOURCE	DF	SEQ SS
ln_hei	1	3.4957
ln_dia	1	4.6275

Unusual observations

Obs.	ln_dia	ln_Vol	Fit	Stdev.Fit	Residual	St.Resid
15	2.48	2.9497	3.1182	0.0154	−0.1686	−2.11R
18	2.59	3.3105	3.4751	0.0288	−0.1645	−2.16R

R denotes an obs. with a large st. resid.

Interpretation of the fitted model

From the above output the estimated parameter values with standard errors are as follows: $\hat{\beta}_1 = 1.98$ (s.e. = 0.075); $\hat{\beta}_2 = 1.12$ (s.e. = 0.204); $\hat{\alpha} = -6.63$ (s.e. = 0.80). It is interesting to see that these values are very close to those for a cylinder. The fitted model for predicting timber volume from length and diameter is then given by

$$V = 0.00132h^{1.12}d^{1.98}.$$

For example, applying the model to tree 20 with $d = 13.8$ in and $l = 64$ ft gives ln $V = -6.632 + 1.983 \times \ln 13.8 + 1.117 \times \ln 64 = 3.218$ and $V = 24.98$ cu ft compared with the measured volume 24.9 cu ft. Figure 14.3 shows timber volumes calculated from the fitted model plotted against observed values which indicates a fairly well fitting model. Finally, note that the output also contains a reference to two unusual observations for trees with relatively low timber volumes (see Section 14.6.3).

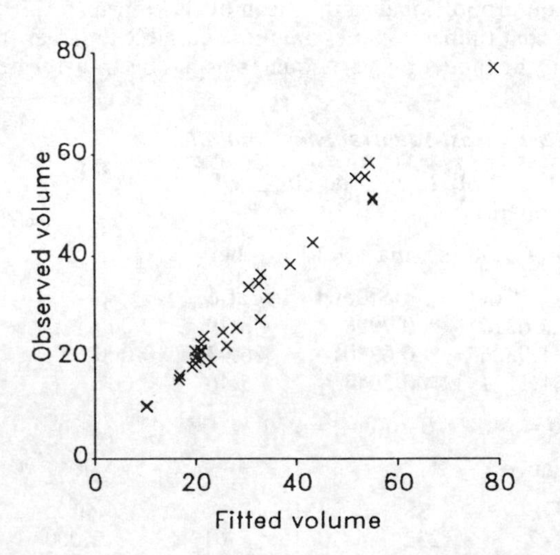

Figure 14.3 Observed timber volumes plotted against values calculated from the two variable linear-Normal model fitted to the log measurements of the Minitab tree data.

14.2.3 Assessing the importance of the variables in the two variable linear model

In the application of the two variable linear model to the tree data the emphasis was on fitting a model to estimate timber volume from height and diameter. In other situations we may be interested in assessing the relative contribution of the explanatory variables in the fitted model to see whether one variable is more important than the other in accounting for the variation in the response variable. In some cases we might not know *a priori* that a relationship exists at all and then we would want to test for one. In this section we deal with assessing the contribution of the two explanatory variables in the fitted model. To illustrate the approach we use the output from the model fitted to the tree data (Example 14.1). We begin with a brief description of the analysis of variance which forms the basis of the method.

Analysis of variance table

The analysis of variance table for the two variable linear-Normal model is similar to that for the straight line-Normal model (Chapter 5) with modifications to the degrees of freedom for residual and regression sums of squares to allow for fitting an extra parameter. The balance sheet of variation is then:

(1) *Total sum of squares of the observations about the overall mean* = *Residual, or Error, sum of squares* + *Regression, or Model, sum of squares*;
(2) A corresponding breakdown of the degrees of freedom as *Total* $(n-1)$ = *Residual* $(n-3)$ + Regression (2);
(3) Mean squares obtained by dividing the sums of squares by their degrees of freedom.

The residual mean square is used to estimate the variance of the response about the underlying relationship. For the tree data the estimate is 0.0066 with its square root $s = 0.081$, which is the estimated standard deviation of the response. Two further quantities derived from the analysis of variance table are: (a) the coefficient of determination, R^2, which is the ratio of the regression sum of squares to the total sum of squares, expressed as a percentage and sometimes referred to as the percentage variation accounted for by the model $- R^2$ is equivalent to the square of the correlation coefficient between the observations and the fitted values; (b) a value of R^2 adjusted to allow for the number of parameters in the model and calculated as 1 minus the ratio of the residual mean square to the total mean square, expressed as a percentage.

Partitioning the regression sum of squares and comparing models

The regression or model sum of squares in the analysis of variance table shows the variation accounted for by fitting both explanatory variables but it does not show the relative contribution of each one. There are different ways of doing this and we deal with two methods: (a) a sequential partition of the regression sum of squares as shown on the Minitab output; (b) listing all possible models.

Sequential partition of the regression sum of squares

The Minitab output shows a sequential partition of the regression sum of squares into sums of squares (SEQ SS) corresponding to each variable *in the order in which it is entered into the model.* In the example, log diameter is fitted first followed by log height. The sequential partition is then: (a) regression sum of squares fitting log diameter *alone* (7.9254); (b) sum of squares due to fitting log height *after* fitting log diameter, calculated by subtracting (a) from the regression sum of squares fitting both diameter and height, i.e. $8.1232 - 7.9254 = 0.1978$. By definition, the sums of squares in the sequential partition add up to the regression sum of squares for fitting both variables. The corresponding partition fitting log height first is then: (a) regression sum of squares fitting height *alone* (3.4957); (b) sum of squares fitting log diameter *after* fitting log height (4.6275).

In the sequential partition of the regression sum of squares the contribution of each variable depends on the order in which it is fitted. In the above example, log height *alone* accounts for 42% of the total variation but when fitted *after* log diameter it accounts for only 2%. By contrast, log diameter *alone* accounts for 95% of the total variation and 56% even *after* fitting height. This occurs because the explanatory variables are not independent of each other, in this case taller trees tend to have larger diameters (Figure 14.2). So, by fitting diameter first most of the effect of height is already allowed for. The effect of the correlation between the explanatory variables on the analysis of the two variable linear model is explored further in Section 14.2.4.

All possible models

Another way to summarise the contributions of the explanatory variables is to tabulate the residual sum of squares, degrees of freedom and corresponding mean squares for models with *all* combinations of the explanatory variables. We could also include the regression sum of squares but since sums of squares for residual and regression add up to the total sum of squares the regression sum of squares would be redundant. For the tree data example there are only three models (apart from the model just fitting a constant) to consider comprising log diameter alone, log height alone and both variables together. Table 14.1 shows that adding log height to the model makes only a relatively small reduction to the residual mean square. Note that information in this table can be related to the sequential partition given in Example 14.1. The residual sum of squares for a model with log diameter alone minus the residual sum of squares fitting both variables is equal to $0.3833 - 0.1855 = 0.1978$ which is the sum of squares due to fitting log height after log diameter.

Table 14.1 Summary of residual variation for different models relating log volume to log diameter (x_1) and log height (x_2) using the tree data

Variables in model		Residual sum of squares	Degrees of freedom	Residual mean square
—	—	8.3087	30	0.2770
x_1	—	0.3833	29	0.0132
—	x_2	4.8130	29	0.1660
x_1	x_2	0.1855	28	0.0066

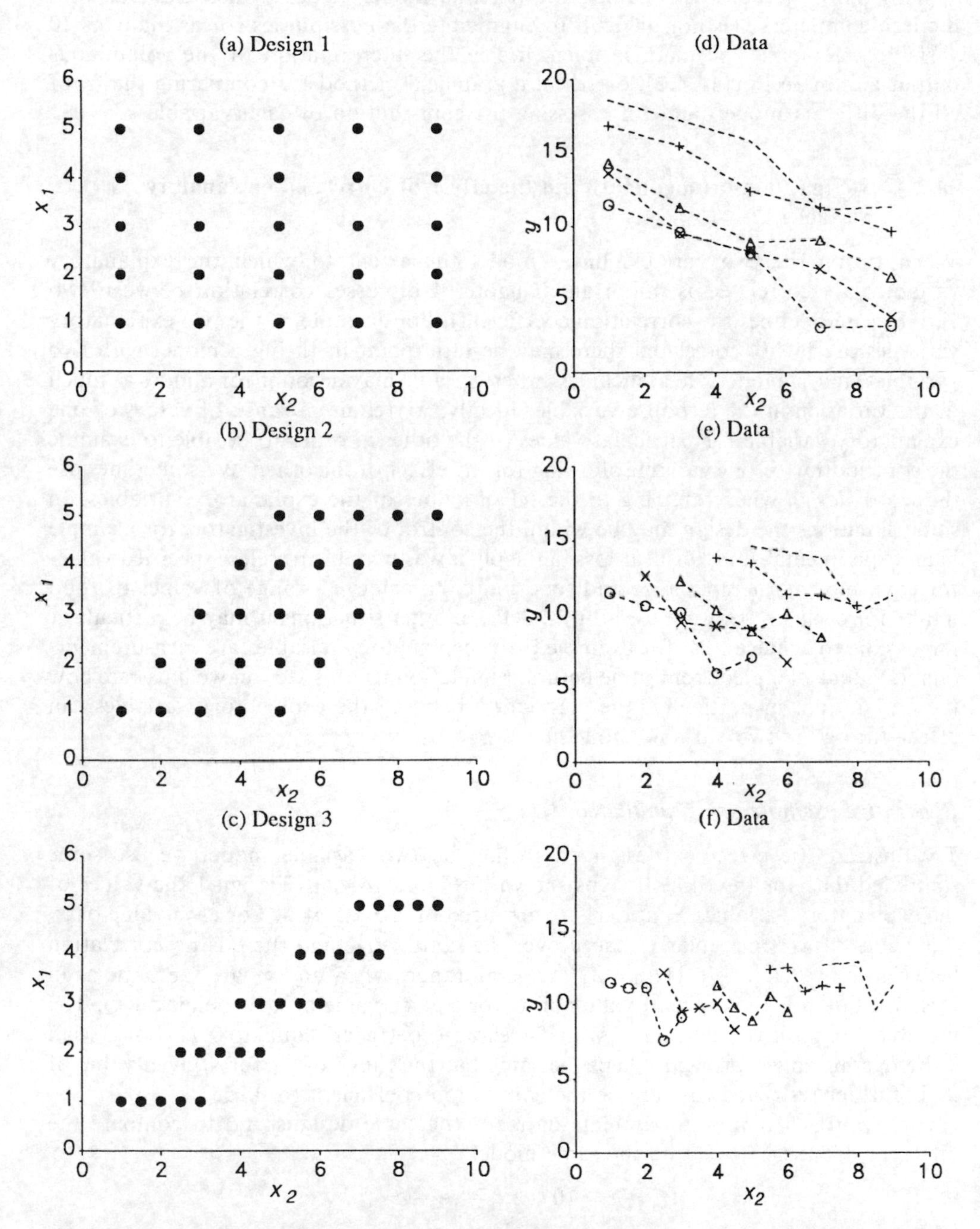

Figure 14.4 Examples to illustrate the effect of design on fitting the two variable linear-Normal model. (a)–(c) Different designs; (d)–(f) simulated data from a two variable linear-Normal model. See Section 14.2.4 for details of model. Key: ○– – –○ $x_1 = 1$; ×– – –× $x_1 = 2$; △– – –△ $x_1 = 3$; +– – –+ $x_1 = 4$; – – – – $x_1 = 5$.

With more variables the number of possible models quickly increases but with modern computers it is not difficult to enumerate the possibilities for as many as 10 variables or more. The main problem lies in the interpretation of the voluminous output and in Section 14.4.2 we present a graphical method for comparing the fit of all the different models and for assessing the contribution of each variable.

14.2.4 Design, non-orthogonality and the effect of correlated explanatory variables

When fitting the two variable linear model the extent to which the explanatory variables are correlated is important (Chapter 4 discusses correlation between two variables and defines the correlation coefficient). For example, if the two explanatory variables are highly correlated there may be little point in fitting a model with two variables since a model which includes either variable may account for almost as much of the variation in the response variable. Ideally, we require a range of values of one explanatory variable for particular values of the other in order to be able to examine the contribution of one variable allowing for the effect of the other. We sometimes use the word *design* when referring to the set of values of the explanatory variables. In some situations the design may be within the control of the investigator, for example in an experimental study of heat loss in the pig it was possible to select specified values for environmental temperature and food intake to achieve a range of values of food intake for each temperature (Section 14.3.1). In other situations it may be difficult or impossible to achieve this, for example if the explanatory variables are measurements on individuals sampled from some natural population. In this section we illustrate how the design and, in particular, the correlation between the explanatory variables can affect fitting the two variable model.

Illustrative example using simulated data

To illustrate the effect of design on fitting the two variable model we use some simulated data for the three designs shown in Figure 14.4. In Design 1 the values of the explanatory variables x_1 and x_2 are arranged on a 5×5 grid. For each value of x_2 the values of x_1 are regularly spaced over the same range and there is no correlation between the variables. In Design 2 the overall ranges of x_1 and x_2 are the same as in Design 1 but now the range of values of x_1 for a given value of x_2 depends on x_2. The variables are positively correlated with correlation coefficient equal to 0.71. For Design 3 the overall ranges are again kept the same but the range of x_1 for a given value of x_2 is further restricted to increase the correlation coefficient to 0.95.

Our approach is now to simulate data for the three designs and to compare the different data sets for fitting the same model

$$Y = 10 + 2x_1 - x_2 + \varepsilon,$$

where $\varepsilon \sim N(0, 1)$. For each design we used the *same* set of random disturbances to avoid differences in results due to random variations. Table 14.2 gives the three sets of data which are depicted in Figure 14.4(d)–(f).

In each case the model was fitted by least squares and the results summarised in terms of: (a) sequential partitions of the regression sum of squares (Table 14.3); (b)

Table 14.2 Simulated data to illustrate effect of design on fitting the two variable linear-Normal model

Random disturbances (ε_i)	Values of explanatory variables and simulated data								
	Design 1			Design 2			Design 3		
	x_{1i}	x_{2i}	y_i	x_{1i}	x_{2i}	y_i	x_{1i}	x_{2i}	y_i
0.484	1	1	11.484	1	1	11.484	1	1.0	11.484
0.611	1	3	9.611	1	2	10.611	1	1.5	11.111
1.160	1	5	8.160	1	3	10.160	1	2.0	11.160
−1.945	1	7	3.055	1	4	6.055	1	2.5	7.555
0.153	1	9	3.153	1	5	7.153	1	3.0	9.153
0.622	2	1	13.622	2	2	12.622	2	2.5	12.122
−1.487	2	3	9.513	2	3	9.513	2	3.0	9.513
−0.739	2	5	8.261	2	4	9.261	2	3.5	9.761
0.047	2	7	7.047	2	5	9.047	2	4.0	10.047
−1.218	2	9	3.782	2	6	6.782	2	4.5	8.282
−0.711	3	1	14.289	3	3	12.289	3	4.0	11.289
−1.709	3	3	11.291	3	4	10.291	3	4.5	9.791
−2.088	3	5	8.912	3	5	8.912	3	5.0	8.912
0.055	3	7	9.055	3	6	10.055	3	5.5	10.555
−0.508	3	9	6.492	3	7	8.492	3	6.0	9.492
−0.143	4	1	16.857	4	4	13.857	4	5.5	12.357
0.472	4	3	15.472	4	5	13.472	4	6.0	12.472
−0.617	4	5	12.383	4	6	11.383	4	6.5	10.883
0.284	4	7	11.284	4	7	11.284	4	7.0	11.284
0.614	4	9	9.614	4	8	10.614	4	7.5	11.114
−0.378	5	1	18.622	5	5	14.622	5	7.0	12.622
0.231	5	3	17.231	5	6	14.231	5	7.5	12.731
0.835	5	5	15.835	5	7	13.835	5	8.0	12.835
−1.877	5	7	11.123	5	8	10.123	5	8.5	9.623
0.284	5	9	11.284	5	9	11.284	5	9.0	11.284

Table 14.3 Illustration of effect of design on fitting the two variable linear-Normal model: partitioning the regression sum of squares

Design	Sequential partitions of the regression sum of squares			
	Fitting x_1 first		Fitting x_2 first	
	Source	SS	Source	SS
	x_1 alone	202.60	x_1 after x_2	202.60
1	x_2 after x_1	210.75	x_2 alone	210.75
$R^2 = 95\%$	x_1 and x_2	413.35	x_1 and x_2	413.35
	x_1 alone	51.31	x_1 after x_2	106.71
2	x_2 after x_1	55.44	x_2 alone	0.04
$R^2 = 83\%$	x_1 and x_2	106.75	x_1 and x_2	106.75
	x_1 alone	13.16	x_1 after x_2	23.59
3	x_2 after x_1	15.29	x_2 alone	4.86
$R^2 = 57\%$	x_1 and x_2	28.45	x_1 and x_2	28.45

the estimated regression coefficients and their standard errors (Table 14.4). The main points are as follows.

(1) For Design 1, the contribution of each variable to the regression sum of squares is the *same* whatever the order in which the variables are added when fitting the model. As a consequence the regression sum of squares can be partitioned into separate contributions for each variable. For Designs 2 and 3 the contributions depend on the order of fitting. The effect is particularly striking for Design 2 where, for this realisation, only a very small amount of variation ($SS = 0.04$) is accounted for by fitting x_2 alone; after fitting x_1 this increases to 55.44. However, a similar effect is observed in Design 3.
(2) When fitting x_1 and x_2 the standard error of the estimated regression coefficient for x_1 is smallest for Design 1, intermediate for Design 2 and largest for Design 3, being about three times that for Design 1 (Table 14.4, left hand side). The reduction in precision is because of the restricted range of values of x_1 for given values of x_2 (Figure 14.4(b) and (c)). In other words the effect of correlation between the explanatory variables is to reduce the precision of the estimated regression coefficients. For the variable x_1 we can attribute the reduction in precision to the effect of the correlation because we have deliberately kept the values of x_1 the same in each design since this also affects precision. The effect occurs with x_2 although in this case the interpretation is less straightforward because the designs differ in the distribution of their x_2 values.
(3) Table 14.4 also shows estimated regression coefficients fitting the variables singly. A remarkable property of Design 1 that the estimated regression coefficient of each variable is the same whether or not the other variable is included in the fitted model. For Designs 2 and 3 this is not so and failing to allow for variation in both variables can give very misleading estimated coefficients.

Orthogonality and non-orthogonality

Design 1 illustrates the importance of orthogonality between two explanatory variables. Orthogonal literally means consisting of right-angles, or perpendicular; Figure 14.4(a) conveys this idea in design. In a statistical sense, two variables are orthogonal if they are uncorrelated. This is so for the grid of points in Design 1 but

Table 14.4 Illustration of effect of design for fitting the two variable linear-Normal model: estimated regression coefficients

Design	Estimated regression coefficients (s.e.)			
	Fitting both variables		Fitting single variables	
	x_1	x_2	x_1	x_2
1	2.01 (0.14)	−1.03 (0.07)	2.01 (0.45)	−1.03 (0.22)
2	2.07 (0.20)	−1.05 (0.14)	1.01 (0.26)	−0.02 (0.24)
3	2.17 (0.44)	−1.11 (0.28)	0.51 (0.18)	0.20 (0.13)

orthogonality need not imply a rectangular arrangement; two explanatory variables with a more irregular distribution could be uncorrelated. The implications for fitting the two variable model are illustrated above and summarised as follows.

(1) If two variables are orthogonal their contribution to the regression sum of squares is the same whatever the order of fitting, i.e. there is a single partition of the regression sum of squares with separate contributions due to each variable.
(2) For orthogonal variables the estimated regression coefficient for one variable is unaffected by including the other variable in the fitted model.
(3) For non-orthogonal variables the contribution of one variable to the regression sum of squares and its estimated regression coefficient will depend on whether the other variable is included in the fitted model.

Non-orthogonality is a problem for fitting models with several explanatory variables. Assessing the contribution of a particular variable can be difficult because of the different combinations of the other variables which could be included in the fitted model. In Sections 14.4 and 14.5 we deal with problems of several explanatory variables and illustrate some effects of non-orthogonality in fitting more complex models.

14.3 EXPLORING A RESPONSE SURFACE WITH TWO VARIABLES

14.3.1 Introduction

The two variable linear model described in Section 14.2 is the simplest model for describing the relationship between a response variable and two explanatory variables. The systematic part of the model can be visualised as a series of parallel straight lines, each relating the response to one of the variables for a fixed value of the other, and we can also represent the underlying relationship as a plane in 3-D. Often, however, response surfaces are more complicated than this. Figure 14.5(a) shows an example using some data on heat loss in the pig where for a given food intake there is a curvilinear relationship with temperature indicating a minimum heat loss. In this section we present models for describing more complex response surfaces. These models are descriptive since they are not based on postulates about the biological processes or mechanisms. Nevertheless they can be useful for estimating biologically significant quantities such as the position of a minimum or for predicting the value of the change in response for given changes in the explanatory variable, provided that this does not involve extrapolation from the range over which the model was fitted. In this section we shall be mainly concerned with formulating and fitting models. In Chapter 7 we deal with aspects of design and, in particular, the choice of values of the explanatory variables.

Example 14.2 Heat loss in the pig

Plane of nutrition and environmental temperature markedly affect the rate of heat loss from the growing pig. Close & Mount (1978) report the results of an experiment in which heat loss was measured continuously on individually housed male pigs for

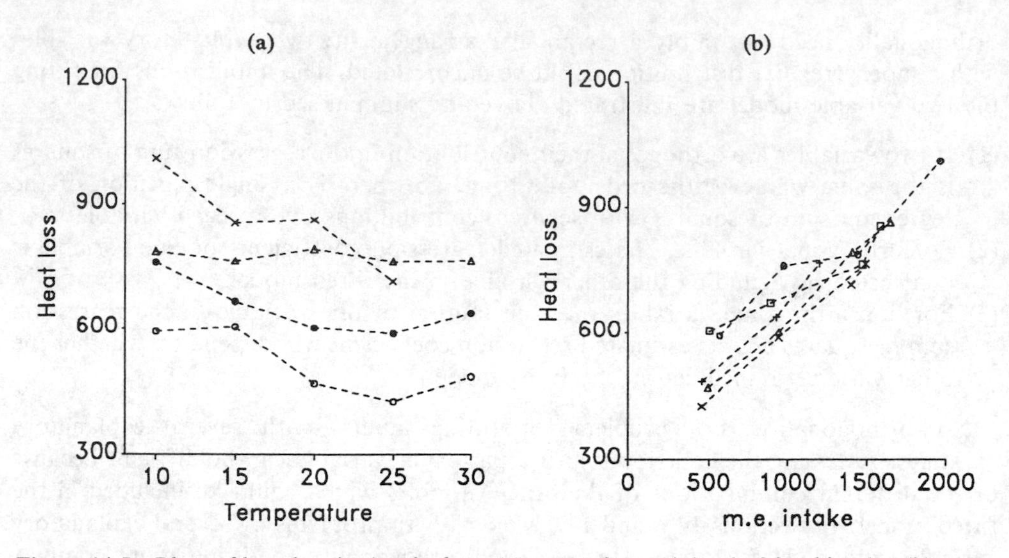

Figure 14.5 Plots of heat loss in the pig data: (a) means for each level of food intake; (b) means for each temperature. See Table 14.5.

14 days. The environmental temperature was maintained at 10, 15, 20, 25 and 30 °C, and each animal received one of four levels of feeding. These represented intakes of approximately once, twice and three times the maintenance energy and an *ad lib* level. Some pigs lost their appetite in the hot conditions so that one treatment combination was missing. For each of the 19 combinations of temperature and feeding level there were two pigs. Table 14.5 gives the heat loss of each pig together with its metabolisable energy intake.

Figure 14.5(a) shows a plot of heat loss against temperature for each level of food intake using an average for each pair of pigs at the same temperature and same level of food intake. This indicates that heat loss is related to temperature but in a way which depends on food intake. Figure 14.5(b) shows heat loss against m.e. intake for each temperature.

Purpose of the model

One of the aims of the experiment was to find a model to describe how heat loss varies with temperature and metabolisable energy intake. The model can be used to estimate the temperature at which heat loss is least and to examine how this critical temperature varies with plane of nutrition. The problem has two aspects. First, we need to describe the underlying systematic part of the relationship. This will require a mathematical function involving the two explanatory variables and some unknown parameters. Then, we need to fit the model, i.e. estimate the unknown parameters. To do this we fit a surface to a scatter of points in 3-D (Figure 14.6).

14.3.2 Building up a complex response surface: fitting models in one variable for fixed values of the others

When responses are measured over a range of values of one variable for fixed values of the other it is often useful to begin by fitting separate models in one variable, and

Table 14.5 Heat loss (h) and metabolisable energy intake (m) for 38 experimental pigs at different environmental temperatures (t)

Pig	t (°C)	Food level	m (Kj/kg$^{0.75}$/d)	h	Pig	t (°C)	Food level	m (Kj/kg$^{0.75}$/d)	h
1	10	1	612	521	20	20	2	935	548
2	10	1	529	665	21	20	3	1464	788
3	10	2	1008	689	22	20	3	1360	795
4	10	2	945	830	23	20	4	1706	834
5	10	3	1423	675	24	20	4	1603	891
6	10	3	1469	896	25	25	1	471	478
7	10	4	1949	981	26	25	1	440	372
8	10	4	1981	1037	27	25	2	893	593
9	15	1	496	625	28	25	2	990	584
10	15	1	513	583	29	25	3	1474	728
11	15	2	875	656	30	25	3	1514	798
12	15	2	911	687	31	25	4	1157	638
13	15	3	1442	743	32	25	4	1654	791
14	15	3	1510	782	33	30	1	469	505
15	15	4	1783	907	34	30	1	441	464
16	15	4	1398	800	35	30	2	876	559
17	20	1	484	454	36	30	2	980	716
18	20	1	506	484	37	30	3	1213	752
19	20	2	954	656	38	30	3	1191	778

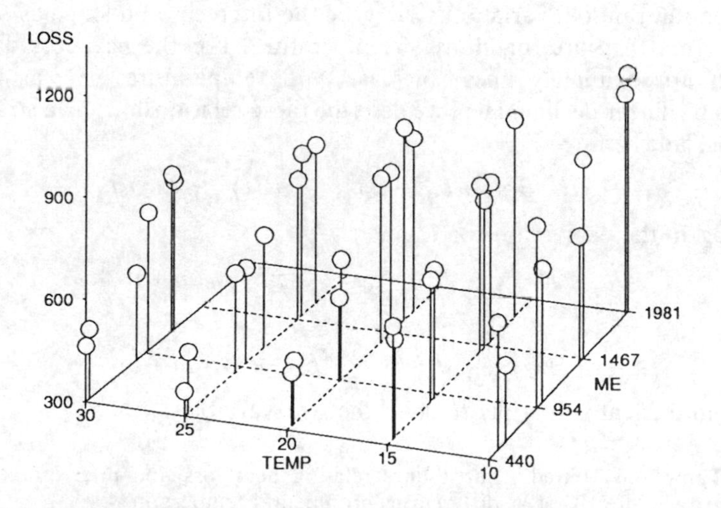

Figure 14.6 3-D plot showing heat loss in pigs in relation to environmental temperature and metabolisable energy intake.

then to see how the estimated parameters are related to values of the other variable. For example, with the pig heat loss data we could relate heat loss to temperature for given levels of m.e. intake, or relate heat loss to m.e. intake for fixed values of temperature. Here we use the latter to avoid the slight complication of variation in values of m.e. intake due to differences between individual pigs. We shall see that this is also simpler.

Plots of heat loss against m.e. intake for each temperature shows that the relationship appears linear apart from the random variation due to differences between individual pigs (Figure 14.5(b)). Table 14.6 gives slopes and intercepts of lines fitted by least squares together with the residual mean square to estimate the variance about the line. This suggests:

(1) An approximately linear increase of the slope with temperature;
(2) A curvilinear decrease of the intercept with temperature which flattens off at higher temperatures;
(3) More variability about the line at 10 °C.

In the next section we combine these results into a model relating heat loss to m.e. intake and temperature.

14.3.3 Two variable quadratic response surface model

In the previous section we showed that the relationship between heat loss and m.e. intake of pigs can be approximated by a straight line with slope and intercept which vary with environmental temperature. We now show how to combine these separate straight lines into a single model relating heat loss to two variables, m.e. intake (m) and temperature (t).

For a given temperature we have

$$H = \alpha(t) + \beta(t)m + \varepsilon,$$

where ε denotes random variation. We write the intercept and slope as $\alpha(t)$ and $\beta(t)$ to indicate that they are functions of temperature. For the heat loss data the slope showed an approximately linear increase with temperature whereas the intercept showed a curvilinear decline. Here we describe these relationships by a straight line and a quadratic, and write

$$\alpha(t) = a_\alpha + b_\alpha t + c_\alpha t^2; \qquad \beta(t) = a_\beta + b_\beta t.$$

Substituting in the expression for H gives

$$H = a_\alpha + b_\alpha t + c_\alpha t^2 + a_\beta m + b_\beta tm + \varepsilon$$

or

$$H = \alpha + \beta_1 t + \beta_{11} t^2 + \beta_2 m + \beta_{12} tm + \varepsilon,$$

where the numerical subscripts refer to the two variables.

Table 14.6 Fitted straight lines relating heat loss and m.e. intake in growing pigs kept at different environmental temperatures

Temperature (°C)	Estimated intercept (s.e.)	Estimated slope (s.e.)	Residual mean square	Residual degrees of freedom
10	450 (92.1)	0.271 (0.068)	10211	6
15	489 (31.5)	0.210 (0.026)	1115	6
20	293 (42.0)	0.344 (0.035)	1913	6
25	285 (33.7)	0.315 (0.029)	1274	6
30	301 (56.5)	0.381 (0.062)	2200	4

The product term *tm* which is present because the *slope* of heat loss on m.e. intake varies linearly with temperature is called the *linear by linear interaction term*. The underlying mean response surface is a special case of the two variable quadratic response surface which includes quadratic terms in both variables. For the pig data this would correspond to an underlying quadratic relationship between heat loss and m.e. intake for a given temperature.

In general, for two explanatory variables x_1 and x_2 and a response Y with constant variance we have

Two variable quadratic-Normal response surface model

$$Y = \alpha + \beta_1 x_1 + \beta_2 x_2 + \beta_{11} x_1^2 + \beta_{22} x_2^2 + \beta_{12} x_1 x_2 + \varepsilon$$

where $\varepsilon \sim N(0, \sigma^2)$.

14.3.4 Fitting the two variable quadratic response surface model

The two variable quadratic response surface model is a linear model (in the parameters) which can be fitted by least squares. This can be done using a statistical package with a facility for fitting multiple regression models with several explanatory variables discussed in Section 14.4. The model involves only two measured explanatory variables but by treating the quadratic and interaction terms as extra variables it can be fitted as a multiple regression model with five explanatory variables.

Example 14.3 Application of the quadratic response surface model to the heat loss in the pig data

Fitting the model

The model developed for the pig heat loss data is a special case of the quadratic response surface model in which the coefficient for a quadratic term in m.e. intake is zero. To fit the model we recall that the variance in heat loss was greater at 10 °C than at the other temperatures so that ordinary least squares which assumes a constant variance is not appropriate. Instead, we fit by weighted least squares with weights inversely proportional to the variance of the response. Here we assume a constant variance for temperatures above 10 °C and use the residual mean squares from Table 14.6 to estimate the weights. The ratio of the residual mean square at 10 °C to the average of the values at 15, 20, 25 and 30 °C is $10211/1626 = 6.28$ giving weights of 0.16, 1, 1, 1 and 1 for the five temperatures. The output below was obtained using Minitab to fit a multiple regression model by weighted least squares.

Minitab output

Heat loss, temperature and m.e. intake in columns c1, c2 and c3.

```
MTB > name c1 'H' c2 't' c3 'm'     # Name columns containing data
MTB > let c4 = 't' * 't'            # Quadratic in temperature
MTB > let c5 = 't' * 'm'            # t × m interaction
MTB > name c4 't_sq' c5 'tm'        # Name derived variables
```

```
MTB > #
MTB > set c6                          # Set up a column for the weights
DATA > 8(0.16) 30(1)
DATA > end
MTB > name c6 'weight'
MTB > #
MTB > regress 'H' 4 't' 'm' 't_sq' 'tm';
SUBC > weight = 'weight'.

The regression equation is

H = 973 - 49.8 t + 0.138 m + 0.908 t_sq + 0.00767 tm

Predictor        Coef          Stdev     t-ratio       p
Constant        973.1          149.3        6.52   0.000
t              -49.79          12.39       -4.02   0.000
m             0.13770        0.07147        1.93   0.063
t_sq           0.9081         0.2548        3.56   0.001
tm           0.007668       0.003397        2.26   0.031

Analysis of variance

SOURCE       DF          SS         MS        F        p
Regression    4      566908     141727    71.42    0.000
Error        33       65487       1984
Total        37      632395
SOURCE       DF      SEQ SS
t             1       60825
t_sq          1      477515
m             1       18456
tm            1       10111
```

Interpreting and using the fitted model

The fitted response surface is

$$\hat{H} = 973.1 - 49.79t + 0.1377m + 0.9081t^2 + 0.007668tm.$$

The quadratic term in temperature is statistically significantly different from zero ($p < 0.001$) and so is the interaction between temperature and m.e. intake ($p < 0.05$). Surprisingly the evidence from the analysis for a linear term in m.e. intake is not very strong ($p = 0.06$).

One of the aims of the study was to estimate the *critical temperature* i.e. the temperature at which heat loss is least. We can use the fitted model to do this by setting the partial derivative with respect to t equal to zero, i.e.

$$\frac{\partial \hat{H}}{\partial t} = 0 = -49.79 + 1.8162t + 0.007668m,$$

i.e.

$$t = 27.4 - 0.00422m.$$

This gives a minimum because the second partial derivative is positive. So, the model predicts a critical temperature which decreases with increasing m.e. intake. These estimated critical temperatures calculated at the mean m.e. intake in the four levels of food intake are as follows: (496, 25.3); (937, 23.4); (1406, 21.5); (1654, 20.4). Figure 14.7 illustrates the fitted quadratic relationship with temperature for the above values of m.e. intake levels.

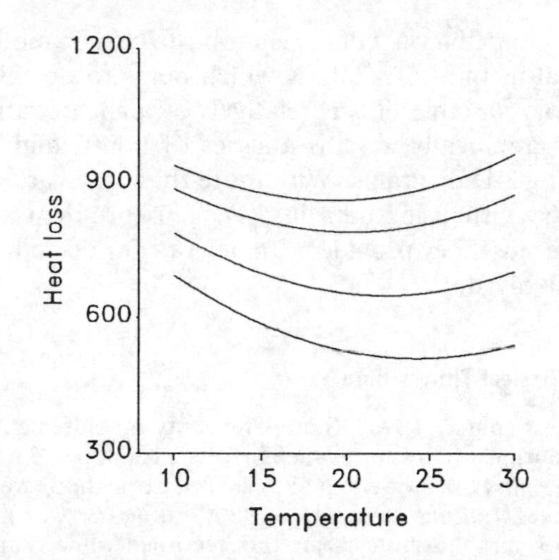

Figure 14.7 Fitted quadratic response surface model relating heat loss in pigs to m.e. intake and environmental temperature shown for different levels of m.e. intake: 496 (bottom), 937, 1406 and 1654 (top).

14.4 MODELS WITH SEVERAL EXPLANATORY VARIABLES

14.4.1 Aspects of analysing relationships with three or more variables

So far we have dealt with models for describing the relationship between a response variable and two explanatory variables. These include the two variable linear model and the two variable quadratic response surface model. For fitting a two variable response surface model the quadratic and interaction terms can be treated as additional explanatory variables but the basic relationship is a surface which can be visualised in three dimensions. Here we consider models with three or more explanatory variables.

An important aspect of analysing a relationship with several explanatory variables is assessing the importance of the different variables in the model and of comparing models involving various combinations of variables. This problem arises in different situations. For example, in predicting values of the response variable we might look for a model with a relatively small number of explanatory variables chosen from observations involving a larger set. The aim is to find a simpler model which predicts almost as well as a model including all the variables. This could have advantages for reducing costs. In other situations we might not know in advance which variables are related to the response and part of the analysis will be to test for a real effect, as well as assessing the magnitude of the contribution to the model if an effect is detected.

In Section 14.2.4 we showed that in the two variable case when there is non-orthogonality arising from correlation between the explanatory variables, the contribution of a particular variable will depend on whether or not the other variable is included in the fitted model. With several explanatory variables a particular variable can be included in many possible fitted models comprising different combinations of the other variables. Assessing the importance of the variable then becomes more complicated since the contribution will depend on the chosen model. In Section 14.4.4 we present a graphical method of analysis which helps to do this.

With two explanatory variables it was relatively straightforward to display the data and the fitted model graphically, e.g. as a series of points and lines on a graph, or representing them using 3-D diagrams. With more than two variables it becomes more difficult to do this: try visualising data in 5-D. This problem actually increases the value of the models because they provide a framework for detecting relationships when many variables are involved.

Example 14.4 Physical fitness data

In a physical fitness course at N.C. State University measurements were made on a sample of men during a fitness test which involved running 1.5 miles (SAS, 1985). These were *age*, *weight* (kg), *oxygen* uptake rate (ml per kg body weight per minute), time to run 1.5 miles (*runtime*), heart rate while resting (*rstpulse*), heart rate while running (*runpulse*), and maximum heart rate recorded while running (*maxpulse*). The problem posed was how oxygen uptake is related to the other variables. Figure 14.8 shows scatter diagrams of oxygen uptake against each of the six explanatory variables. There is a marked decrease with *runtime*—men who ran faster have a higher uptake rate. The diagrams also suggest decreasing, but more variable, relationships between oxygen uptake and the other variables. This presentation raises two questions. First, are the apparent decreases with all the variables real or could they be due to chance random variation? There is no doubt about the decrease with *runtime* but the effects are less obvious for the other variables. Second, if the trends are real, could they be due to an indirect effect of *runtime*? For example, oxygen uptake appears higher in the young men but this may be because they were fitter and ran the race more quickly. We cannot answer these questions by looking at variables one at time. What we must do is analyse oxygen uptake in relation to two or more of the explanatory variables. In Section 14.4.4 we present one approach which involves comparing simple descriptive models containing different combinations of the explanatory variables. The aim is to identify the variables which affect oxygen uptake and to assess the contribution of these variables to the relationship.

14.4.2 *p*-Variable linear-Normal model

The p-variable linear-Normal model for describing the relationship between a response variable Y and p explanatory variables $x_1, x_2, \ldots, x_p$ is given by

$$Y = \alpha + \beta_1 x_1 + \beta_2 x_2 + \cdots + \beta_p x_p + \varepsilon,$$

where $\varepsilon \sim N(0, \sigma^2)$ denotes random variation in the responses. The parameters $\beta_1, \beta_2, \ldots, \beta_p$ are the *regression coefficients* or the *partial regression coefficients*. For example, β_1 is the change in the mean value of Y for a unit change in x_1 for any fixed values of the other variables. The model is sometimes referred to as a *multiple regression model with p explanatory variables*. The explanatory variables may correspond to different biological variables or some could be derived as functions of the others—for example, as quadratic or interaction terms. A model which is linear in all the different variables is the simplest possible case and in many applications we would add to it additional terms for quadratic and interaction effects as part of the model building exercise and also as a check on the adequacy of the simple model.

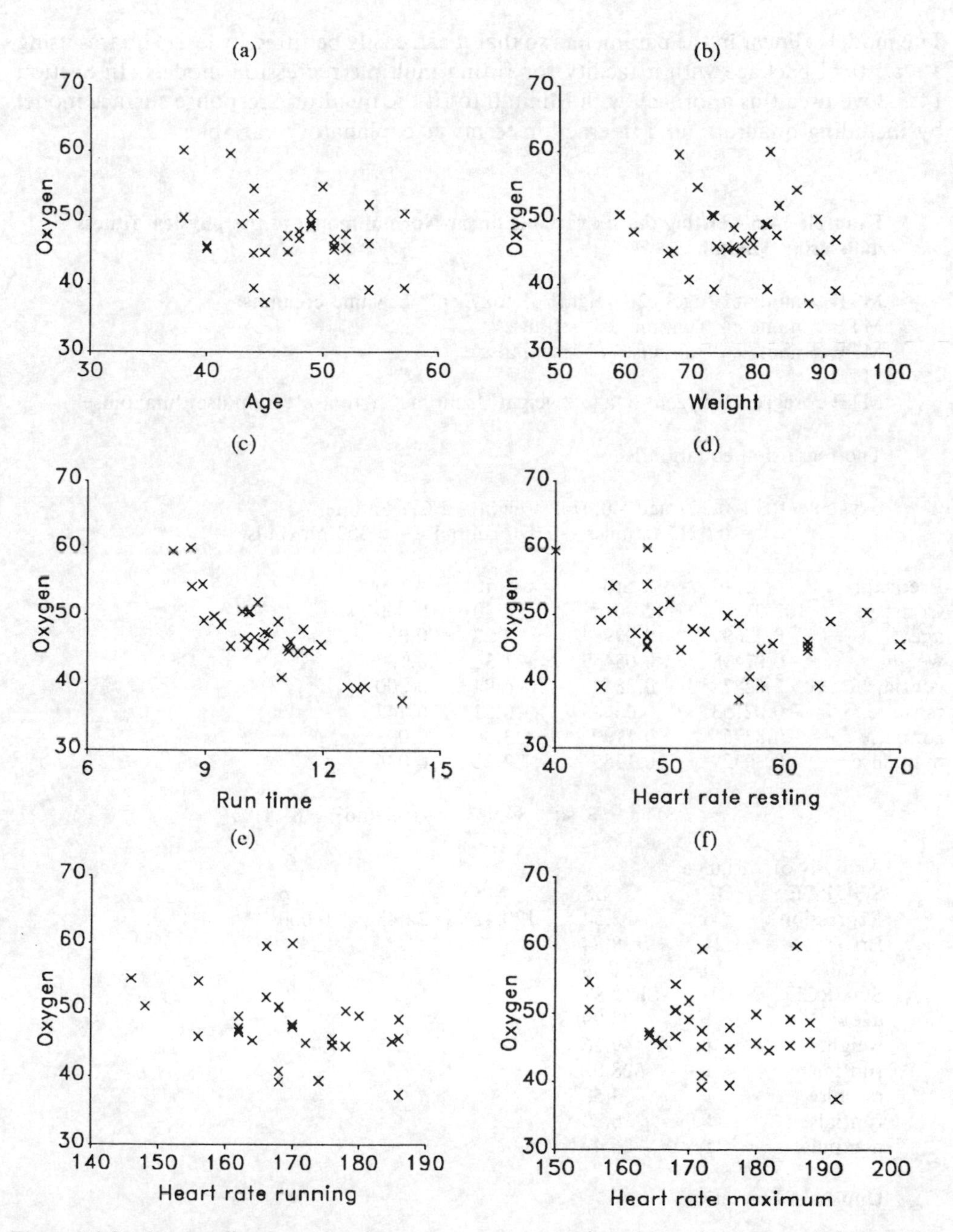

Figure 14.8 Scatter diagrams showing oxygen uptake against the explanatory variables in the physical fitness data.

14.4.3 Fitting the *p*-variable linear-Normal model

The model is linear in the parameters so that it can easily be fitted by least squares using a statistical package with a facility for fitting multiple regression models. In Section 14.3.3 we used this approach with Minitab to fit the quadratic response surface model by including quadratic and interaction terms as explanatory variables.

Example 14.5 Fitting the six variable linear-Normal model to the physical fitness data using Minitab

```
MTB > name c1 'age' c2 'weight' c3 'oxygen' # Name columns
MTB > name c4 'runtime' c5 'rstpulse'
MTB > name c6 'runpulse' c7 'maxpulse'
MTB > #
MTB > regress 'oxygen' 6 'age' 'weight' 'runtime' 'rstpulse' 'runpulse' 'maxpulse'
```

The regression equation is

oxygen = 103 – 0.227 age – 0.0742 weight – 2.63 runtime
– 0.0215 rstpulse – 0.370 runpulse + 0.303 maxpulse

Predictor	Coef	Stdev	t-ratio	p
Constant	102.93	12.40	8.30	0.000
age	– 0.22697	0.09984	– 2.27	0.032
weight	– 0.07418	0.05459	– 1.36	0.187
runtime	– 2.6287	0.3846	– 6.84	0.000
rstpulse	– 0.02153	0.06605	– 0.33	0.747
runpulse	– 0.3696	0.1199	– 3.08	0.005
maxpulse	0.3032	0.1365	2.22	0.036

s = 2.317   R-sq = 84.9%   R-sq(adj) = 81.1.%

Analysis of variance

SOURCE	DF	SS	MS	F	p
Regression	6	722.54	120.42	22.43	0.000
Error	24	128.84	5.37		
Total	30	851.38			

SOURCE	DF	SEQ SS
age	1	78.99
weight	1	49.26
runtime	1	528.02
rstpulse	1	3.58
runpulse	1	36.20
maxpulse	1	26.49

Unusual observations

Obs.	age	oxygen	Fit	Stdev. Fit	Residual	St. Resid
8	54.0	51.855	46.470	0.832	5.385	2.49R
9	51.0	40.836	46.239	0.664	– 5.403	– 2.43R

R denotes an obs. with a large st.resid.

Interpretation of output

The output gives the fitted model with the estimated regression coefficients for each of the six variables together with their standard errors. The corresponding t-ratios are obtained by dividing the estimated coefficients by their standard errors and are used to test the null hypothesis that a particular coefficient is zero. For example, the estimated coefficient for *weight* is -0.0742 (s.e. $= 0.0546$) with t-ratio $-0.0742/0.0546 = -1.36$. The probability of observing a more extreme t-ratio on the null hypothesis that the true coefficient is zero is obtained from tables of Student's t-distribution with degrees of freedom equal to the residual (or error) degrees of freedom of 24. The output gives the p-value as 0.187 so that in this model *weight* is not detected as statistically significant at the 5% level. Similarly, the estimated coefficient for *rstpulse* of -0.0215 (*s.e.* $= 0.0661$) is not statistically significant at 5% (t-ratio $= -0.33$, $p = 0.747$). The t-ratios for the remaining variables are all statistically significant at 5%. The fitted model accounts for $R^2 = 84.9\%$ of the total variation with an estimated residual standard deviation of $s = 2.32$. The output shows a sequential partition of the regression sum of squares corresponding to the order of fitting variables: *age*, *weight*, *runtime*, *rstpulse*, *runpulse* and *maxpulse*. Changing the order would give a different partition in which the contribution of a given variable would change due to the effect of non-orthogonality, and in Section 14.4.2 we present a more detailed analysis of the contribution of each variable. Finally, the output contains a list of 'unusual observations', in this case with relatively large deviations from the fitted model. These so-called *regression diagnostics* are dealt with in Section 14.6.

14.4.4 Comparing models and model selection

When relating a response variable to several explanatory variables we often want to assess the importance of each variable in the relationship. For example, the problem arises in deciding which variables to include in a predictive model. Can some variables be ignored with little loss for prediction? In other situations we might wish to test whether the contribution of a variable is real and not simply a result of random variation. Both aspects arise in fitting the six variable model to the physical fitness data where the coefficients for two of the variables are not significantly different from zero. This suggests fitting a simpler model with fewer explanatory variables. One approach to this situation is to fit models with selected combinations of the explanatory variables and to compare how well they fit. However, as the number of explanatory variables increases there are many possible combinations of variables and therefore many models to choose from. In this section we present some systematic methods of analysis which can be useful for comparing models involving several explanatory variables.

Correlations amongst the explanatory variables

In Section 14.2.4 we saw how non-orthogonality arising from correlation between two explanatory variables can result in the contribution of a particular variable being enhanced or diluted by including the other variable. This problem applies to models with several explanatory variables but the effects can be much more complicated because of the many models involving different combinations of variables. For comparing such models it is often useful to examine the relationships amongst the explanatory variables. One approach is to plot scatter diagrams and to calculate correlation coefficients. If two explanatory variables are highly correlated then there may be little to be gained by including them both in the same model, but this is not a hard and fast rule. Table 14.7 shows correlation coefficients for the physical fitness data in typical matrix form. There is a large positive correlation between *runpulse* and *maxpulse* (Figure 14.9), and a large negative correlation between *oxygen* and *runtime* (Figure 14.8(c)).

Significance tests of variables in the model

In comparing models a question which often arises is whether the addition of one or more variables to a model leads to a significant improvement in fit. By adding more variables we reduce the residual sum of squares but could this reduction be a chance effect due to the random element in the model? Similarly, we might consider the reduction in fit by removing some variable or variables from a specified model. Here

Table 14.7 Matrix of correlation coefficients for the variables in the physical fitness data

	oxygen	*age*	*weight*	*runtime*	*rstpulse*	*runpulse*
age	−0.31					
weight	−0.16	−0.23				
runtime	−0.86	0.19	0.14			
rstpulse	−0.40	−0.16	0.04	0.45		
runpulse	−0.40	−0.34	0.18	0.31	0.35	
maxpulse	−0.24	−0.43	0.25	0.23	0.30	0.93

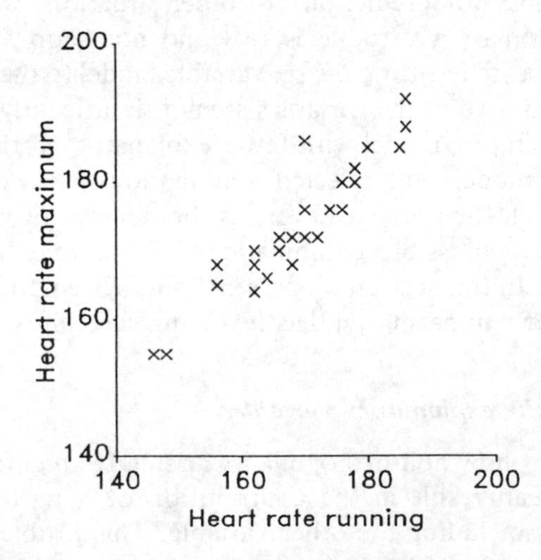

Figure 14.9 Scatter plot of *maxpulse* against *runpulse* for physical fitness data showing high positive correlation.

we give a formal test procedure for the statistical significance of a specified set of variables in a model. In practice the method is often used as a criterion for deciding which variables to add to a model or in screening out redundant variables. These applications require special consideration because of the problem of multiple comparisons which arise when carrying out a number of significance tests. This consideration is discussed briefly below, but first we describe the method for a single comparison.

Suppose we wish to test whether adding a set of variables to a given model leads to a greater improvement in fit than could be expected by chance alone. To do this we see whether the reduction in the residual sum of squares by including the extra variables is significantly different from zero. Suppose that the residual sum of squares for the two models are SS_1 and SS_2 with corresponding degrees of freedom df_1 and df_2, where Model 2 contains the variables in Model 1 as well as the extra variables. Then the test-statistic is the ratio

$$F = \frac{(SS_1 - SS_2)/(df_1 - df_2)}{SS_2/df_2}.$$

The numerator is the reduction in the residual sum of squares due to fitting the extra variables divided by the number of extra variables. The denominator is the residual mean square for Model 2 which contains the extra variables. The method tests the null hypothesis that Model 1 is the true model. When the null hypothesis is true the sampling distribution of the test-statistic follows an F-distribution with degrees of freedom $(df_1 - df_2)$ and df_2 (assuming that the observations are Normally distributed with constant variance).

To illustrate the method we use the physical fitness data and test to see whether adding the five extra variables gives a significant improvement in fit to a model in *runtime*, i.e. we compare the residual sum of squares for the *full* model with that for the single variable model in *runtime*. For the full model we have $SS_2 = 128.84$, $df_2 = 24$, $MS_2 = 128.84/24 = 5.368$ (Example 14.5) while fitting a model in *runtime* alone gives $SS_1 = 218.48$, $df_1 = 29$. The F-ratio then is $(218.48 - 128.84)/(5.368 \times 5) = 3.34$ $(p = 0.02)$. We therefore reject the hypothesis that a linear model in *runtime* is sufficient to explain the variation in oxygen uptake. This could mean that one or more of the other variables is related to oxygen uptake after allowing for *runtime*. The result could alternatively imply that the underlying relationship with *runtime* is not linear and that the departures from linearity are correlated with one or more of the other variables.

For the special case when there is only a single extra variable the value of the F-ratio is equal to the square of the t-ratio for the estimated regression coefficient. Example 14.5 illustrates this for the variable *maxpulse* in the physical fitness data. The sequential partition of the regression sum of squares shows that the reduction in the residual sum of squares by adding *maxpulse* to the other five variables is 26.49. The F-ratio is $26.49/5.368 = 4.935$. The computer output gives a t-ratio equal to 2.22 which is equal to $\sqrt{4.935}$, illustrating that when there is only a single extra variable, $F = t^2$.

Use of residual mean square for full model

In the above illustration of the F-ratio test we compared a *reduced* model with only a single variable with the *full* model including all the explanatory variables. In other situations we may wish to compare two reduced models both of which involve subsets of the explanatory variables. In this case a sensible modification to the test is to use the residual mean square for the full model in the denominator of the calculated F-ratio. The idea of this is to increase the power of the test by using an unbiased estimator of the residual variance rather than using the residual mean square for Model 2 which would be inflated if some important variables were not included in Model 2. Example 14.6 illustrates this approach using the physical fitness data.

Forward selection and backward elimination procedures

Forward selection and backward elimination procedures are two methods for model selection which can be routinely carried out using a range of statistical packages. In forward selection, variables are successively added to the model and retained if the fit is significantly improved, whereas in backward elimination variables are removed if their contribution is not significant. The aim is to find a model which is 'best fitting' in some sense. We mention these methods only briefly because we prefer the approach which examines all possible models for a given set of explanatory variables. With modern computing power this approach is feasible even for as many as 10 variables. Moreover, the output from the models can be conveniently summarised in the form of a graphical analysis of deviance (see below) which greatly helps to compare the models. A deficiency of forward selection and backward elimination procedures is that although they can find models which fit well, other models which fit almost as well, and in some cases better, may be missed. By examining all possible models this problem does not arise.

In forward selection algorithms, variables are added depending on an 'F-to-enter' value calculated as the F-ratio for that variable. If the 'F-to-enter' value exceeds some threshold level then the variable is included. In backward elimination variables are dropped if their 'F-to-remove' value is too small. Often the threshold values are obtained as the percentage points of the F-distribution for some specified significance level: if we set $\alpha = 0.05$ then the 'F-to-enter' value has to exceed the upper 5% point of the F-distribution for the variable to be included in the model. A particular application could well involve many such tests so that there is a problem of multiple comparisons and a tendency to include variables in the model even when there is no effect as 'significant' values arise by chance. For a more detailed discussion of forward and backward selection procedures see Draper & Smith (1966).

All possible models

The most satisfactory way to compare models involving different combinations of the explanatory variables is to examine all the possible models. For example, in the fitness data there are six explanatory variables so that if we consider models which are linear in the variables we have: 6 single-variable models; 15 (6 choose 2) two-variable models; 20 three-variable models; 15 four-variable models; 6 five-variable models; 1 six-variable model; making a total of 63 possible models. The model with no variables is called the *null model* which corresponds to fitting a single mean. The model which includes all the variables is called the *full model*. The number of possible models increases rapidly with the number of explanatory variables; for $n = 10$ there are 1024 possible models, including the null model. In practice, to examine the fit of all possible models it is best to restrict the number of explanatory variables to less than say six or seven. For a moderate number of explanatory variables fitting the set of all possible models is relatively straightforward and some statistical packages have procedures for doing this (SAS and Genstat 5). The more difficult problem is summarising the voluminous computer output and interpreting the results.

Some statistical criteria for comparing different models and for assessing the contribution of a particular variable are: (a) R^2—the percentage of the variation accounted for; (b) the residual mean square; (c) the statistical significance of the estimated regression coefficients; (d) consistency of the coefficients in different models. Biological considerations are also important and may dictate that a variable be included although in purely numerical terms its contribution may appear to be small. In this section we shall deal with p-variable linear-Normal models but similar considerations apply to other models and Section 14.5 presents an illustration using a regression model for binary responses.

A useful first step is to tabulate the fitted models showing the variables included in each model, percentage of variation accounted for (R^2), the residual mean square and some indication of the statistical significance of each variable.

Example 14.6 All possible models for physical fitness data

Table 14.8 shows the various linear models fitted to the physical fitness data. These are grouped according to the number of explanatory variables and sorted within each group by increasing values of R^2. In this presentation the statistical significance is based on the F-ratio test which uses the residual mean square for the full model in the denominator. Note that the table shows the results of many such tests so that there is the problem of multiple comparisons: on average, 5% of tests will detect significant results when none are present and this needs to be kept in mind when interpreting the results.

(1) *runtime* alone (Model 6) accounts for 74% of the total variation compared with 85% accounted for by the model with all six variables. In every model in which it occurs the coefficient for *runtime* is statistically significant ($p < 0.01$). In the above illustration of the F-ratio test we showed that the additional 11% of the variation accounted for by adding all the variables to *runtime* is statistically significant.
(2) Model 57 which includes all variables except *runtime* accounts for 55% of the total variation and each variable is statistically significant.
(3) No single variable when added to *runtime* is statistically significant (Models 17–21). However, when *runpulse* is added with either *age* or *maxpulse* it is significant (Models 40 and 41).
(4) The only time that *weight* is significant is when *runtime* is not in the model and when either *age* or *runpulse* is present.
(5) The sign of the estimated coefficient for *maxpulse* changes from negative to positive when included with *runpulse*. In the full model the coefficient is statistically significant but in Section 14.6.3 we shall see that this depends on the inclusion of a single individual.
(6) The table illustrates how by using the residual mean square for the full model increases the power of the test for detecting the effects of the variables. For example, if we test for an effect of *age* in Model 3 using the residual mean square of 26.6 it is not detected. However, by using the residual mean square of 5.4 for the full model the F-ratio is increased by a factor of about five and is highly significant.

Table 14.8 Summary of fitted models involving all the 64 possible linear combinations of the six physical fitness variables: *w—weight*; *x—maxpulse*; *a—age*; *r—runpulse*; *s—rstpulse*; *t—runtime*.

Model		Variables in model						R^2 (%)	Residual *MS*
	p	*w*	*x*	*a*	*r*	*s*	*t*		
0	0	.	.	.	.	.	.	0	28.4
1	1	−*w*	.	.	.	.	.	3	28.5
2	1	.	−**X**	.	.	.	.	6	27.7
3	1	.	.	−**A**	.	.	.	9	26.6
4	1	.	.	.	−**R**	.	.	16	24.7
5	1	.	.	.	.	−**S**	.	16	24.7
6	1	.	.	.	.	.	−**T**	74	7.5
7	2	−*w*	−**X**	.	.	.	.	7	28.3
8	2	−**W**	.	−**A**	.	.	.	15	25.8
9	2	−*w*	.	.	−**R**	.	.	17	25.3
10	2	.	−*x*	.	.	−**S**	.	17	25.1
11	2	−*w*	.	.	.	−**S**	.	18	24.9
12	2	.	.	.	−**R**	−**S**	.	24	23.3
13	2	.	−**X**	−**A**	.	.	.	26	22.5
14	2	.	+**X**	.	−**R**	.	.	29	21.6
15	2	.	.	−**A**	.	−**S**	.	30	21.3
16	2	.	.	−**A**	−**R**	.	.	38	19.0
17	2	.	.	.	.	−*s*	−**T**	74	7.8
18	2	−*w*	.	.	.	.	−**T**	74	7.8
19	2	.	−*x*	.	.	.	−**T**	75	7.7
20	2	.	.	.	−*r*	.	−**T**	76	7.3
21	2	.	.	−*a*	.	.	−**T**	76	7.2
22	3	−*w*	−*x*	.	.	−**S**	.	19	25.6
23	3	−*w*	.	.	−**R**	−**S**	.	24	23.8
24	3	−**W**	−**X**	−**A**	.	.	.	29	22.4
25	3	−**W**	+**X**	.	−**R**	.	.	32	21.4
26	3	.	+**X**	.	−**R**	−**S**	.	35	20.3
27	3	−**W**	.	−**A**	.	−**S**	.	36	20.2
28	3	.	−**X**	−**A**	.	−**S**	.	39	19.2
29	3	−**W**	.	−**A**	−**R**	.	.	41	18.6
30	3	.	+**X**	−**A**	−**R**	.	.	42	18.2
31	3	.	.	−**A**	−**R**	−**S**	.	47	16.8
32	3	−*w*	.	.	.	−*s*	−**T**	75	8.0
33	3	.	−*x*	.	.	−*s*	−**T**	75	8.0
34	3	−*w*	−*x*	.	.	.	−**T**	75	8.0
35	3	−*w*	.	.	−*r*	.	−**T**	76	7.5
36	3	.	.	.	−*r*	−*s*	−**T**	76	7.5
37	3	.	.	−*a*	.	−*s*	−**T**	77	7.3
38	3	−*w*	.	−*a*	.	.	−**T**	77	7.2
39	3	.	−*x*	−**A**	.	.	−**T**	78	6.9
40	3	.	+**X**	.	−**R**	.	−**T**	81	6.0

Table 14.8 (*continued*).

Model		Variables in model						R^2 (%)	Residual MS
	p	w	x	a	r	s	t		
41	3	.	.	$-\underline{\mathbf{A}}$	$-\mathbf{R}$	.	$-\underline{\mathbf{T}}$	81	6.0
42	4	$-\mathbf{W}$	$+\underline{\mathbf{X}}$	.	$-\underline{\mathbf{R}}$	$-\underline{\mathbf{S}}$	.	39	20.1
43	4	$-\mathbf{W}$	$-\underline{\mathbf{X}}$	$-\underline{\mathbf{A}}$	.	$-\underline{\mathbf{S}}$	.	43	18.8
44	4	$-\underline{\mathbf{W}}$	$+\underline{\mathbf{X}}$	$-\underline{\mathbf{A}}$	$-\underline{\mathbf{R}}$	.	.	47	17.3
45	4	.	$+\mathbf{X}$	$-\underline{\mathbf{A}}$	$-\underline{\mathbf{R}}$	$-\underline{\mathbf{S}}$	.	50	16.3
46	4	$-\mathbf{W}$	.	$-\underline{\mathbf{A}}$	$-\underline{\mathbf{R}}$	$-\underline{\mathbf{S}}$	.	50	16.3
47	4	$-w$	$-x$	.	.	$-s$	$-\underline{\mathbf{T}}$	75	8.3
48	4	$-w$	.	.	$-r$	$-s$	$-\underline{\mathbf{T}}$	76	7.8
49	4	$-w$	.	$-\mathbf{A}$	.	$-s$	$-\underline{\mathbf{T}}$	78	7.4
50	4	.	$-x$	$-\mathbf{A}$	.	$-s$	$-\underline{\mathbf{T}}$	78	7.1
51	4	$-w$	$-x$	$-\mathbf{A}$	.	.	$-\underline{\mathbf{T}}$	79	7.0
52	4	.	$+\mathbf{X}$	.	$-\underline{\mathbf{R}}$	$-s$	$-\underline{\mathbf{T}}$	81	6.2
53	4	.	.	$-\underline{\mathbf{A}}$	$-\mathbf{R}$	$-s$	$-\underline{\mathbf{T}}$	81	6.2
54	4	$-w$	$+\underline{\mathbf{X}}$	.	$-\underline{\mathbf{R}}$	.	$-\underline{\mathbf{T}}$	82	6.0
55	4	$-w$	.	$-\underline{\mathbf{A}}$	$-\mathbf{R}$	.	$-\underline{\mathbf{T}}$	82	6.0
56	4	.	$+x$	$-a$	$-\underline{\mathbf{R}}$	.	$-\underline{\mathbf{T}}$	84	5.3
57	5	$-\underline{\mathbf{W}}$	$-\underline{\mathbf{X}}$	$-\underline{\mathbf{A}}$	$-\underline{\mathbf{R}}$	$-\underline{\mathbf{S}}$	.	55	5.2
58	5	$-w$	$-x$	$-\mathbf{A}$	.	$-s$	$-\underline{\mathbf{T}}$	79	7.2
59	5	$-w$	$+\underline{\mathbf{X}}$	.	$-\underline{\mathbf{R}}$	$+s$	$-\underline{\mathbf{T}}$	82	6.3
60	5	$-w$	.	$-\underline{\mathbf{A}}$	$-\underline{\mathbf{R}}$	$-s$	$-\underline{\mathbf{T}}$	82	6.2
61	5	.	$+x$	$-a$	$-\underline{\mathbf{R}}$	$-s$	$-\underline{\mathbf{T}}$	84	5.5
62	5	$-w$	$+\mathbf{X}$	$-\underline{\mathbf{A}}$	$-\underline{\mathbf{R}}$	.	$-\underline{\mathbf{T}}$	85	5.2
63	6	$-w$	$+\mathbf{X}$	$-\mathbf{A}$	$-\underline{\mathbf{R}}$	$-s$	$-\underline{\mathbf{T}}$	85	5.4

Lower case *italic*—$p > 0.05$; Upper case **bold**—$p < 0.05$; Upper case **bold underlined**—$p < 0.01$. F-tests use residual mean square for the full model with all six variables. The sign given is that of the regression coefficient.

The analysis shows the effects of non-orthogonality whereby the contribution of different variables depends on which other variables are included in the model. It also shows how there can be different models which account for almost the same amount of variation. So, an automated search for a single best fit model will, in general, provide an incomplete picture.

Graphical analysis of variance

The above table summarising the fit of all the possible models contains a lot of information which can be rather tedious to extract. A *graphical analysis of variance* (Brown, 1992) displays the same information but in a way which helps to show more clearly the contribution of the different variables and how they are affected by other variables in the fitted model. It is helpful for highlighting the non-orthogonality in the data and for interpreting the results.

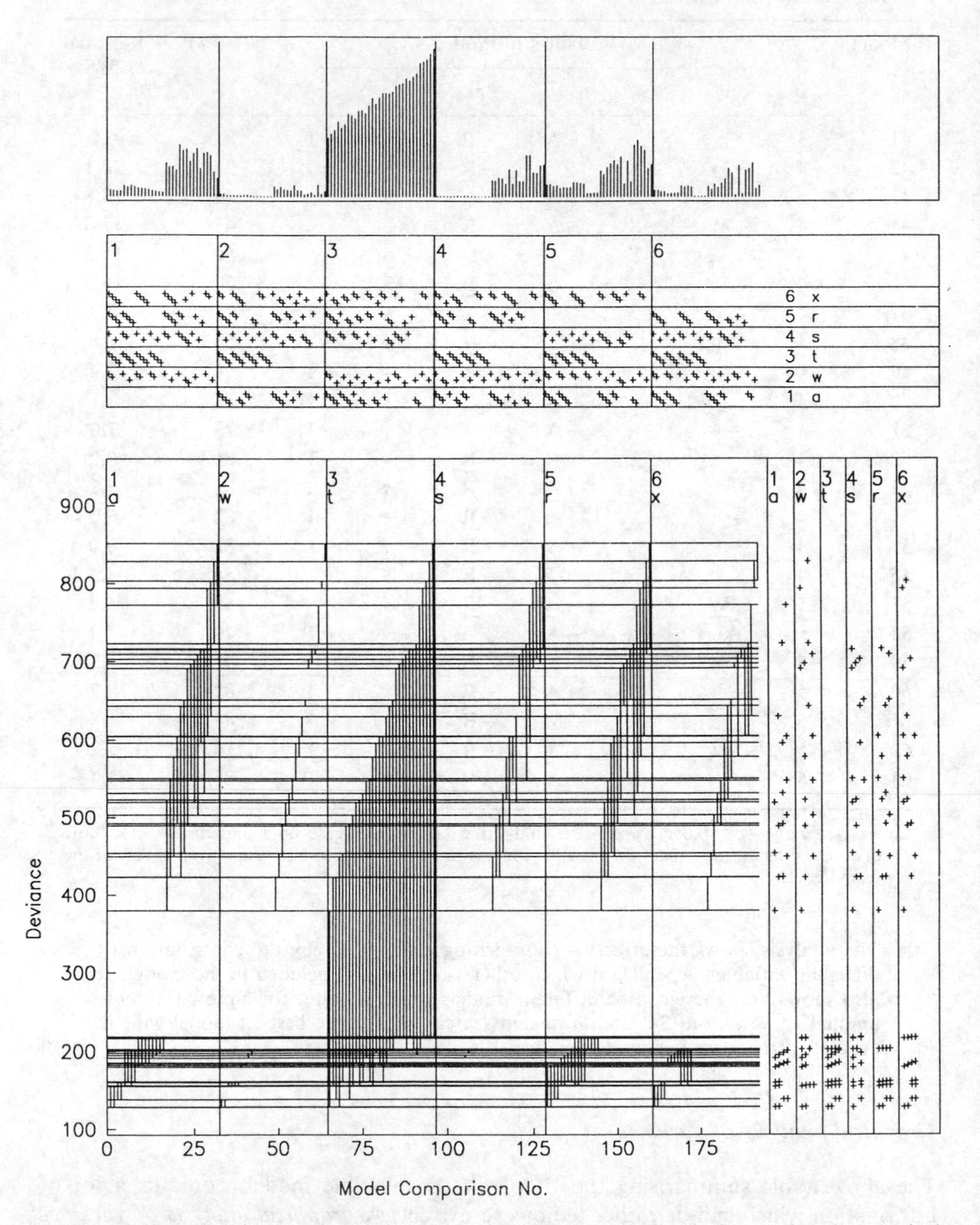

Figure 14.10 Graphical analysis of variance (or more generally deviance) for the physical fitness data (see text for method of reading graph). Key: a–*age*, w–*weight*, t–*runtime*, s–*rstpulse*, r–*runpulse*, x–*maxpulse*.

Figure 14.10 illustrates the graphical analysis of deviance using the physical fitness data. The elements in the diagram are as follows:

(1) Horizontal lines marking the residual sum of squares for the models with the various combinations of explanatory variables shown in the key on the right hand side. The presence of a + in a column on the right indicates that the term is in the model. The pluses are staggered for successive models, so that models with very similar residual sums of squares or deviances can be discriminated. The top line corresponds to the *null model* with no variables; the bottom line to the *full* model with all six variables.

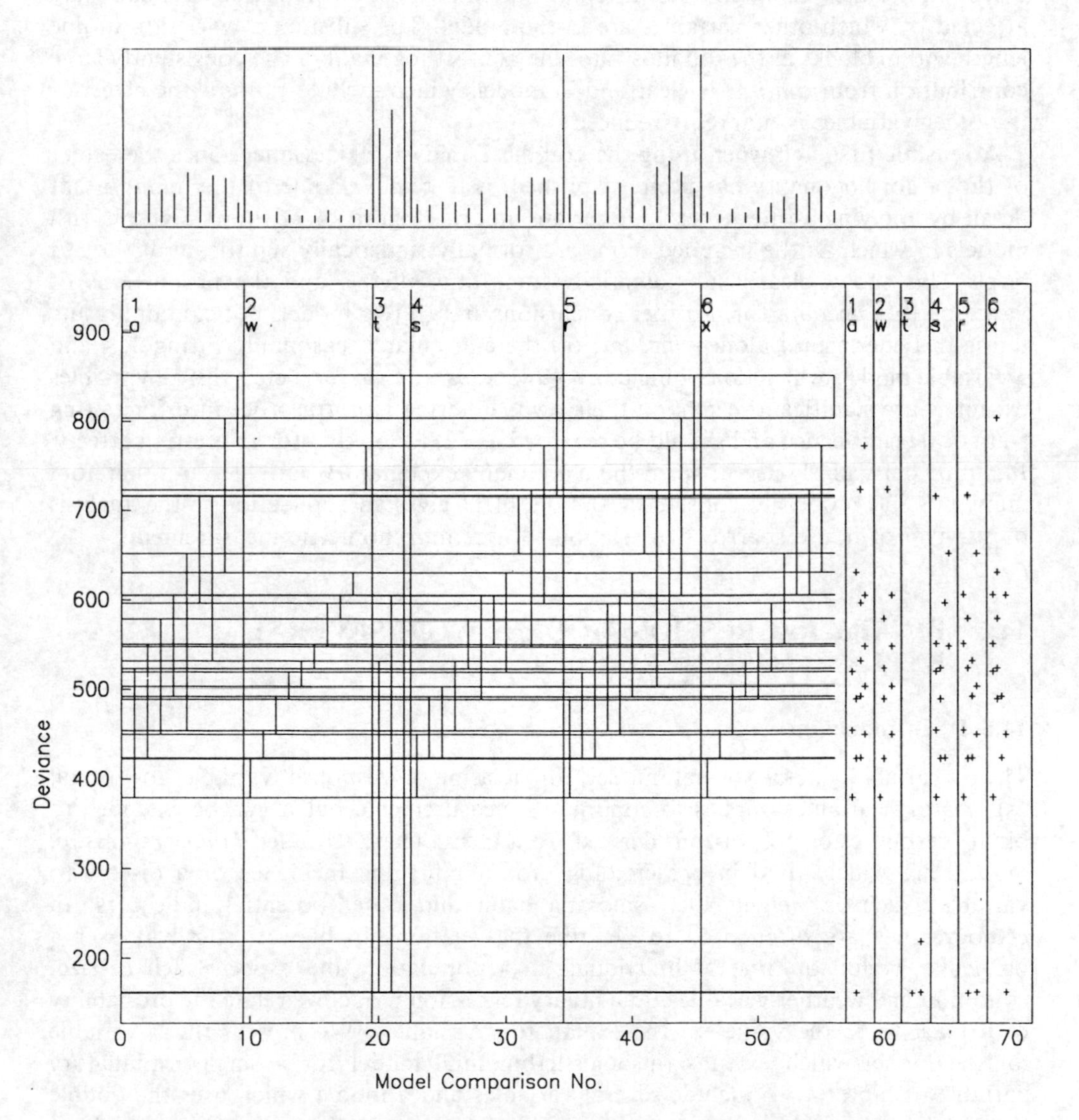

Figure 14.11 Filtered graphical analysis of deviance for the physical fitness data showing only those models in which each term is statistically significant at 5%.

(2) Vertical lines linking the horizontal lines to show the effect of adding a given variable to a particular model grouped according to the variable added (named at top of block); the length of the joining line is the reduction in residual sum of squares due to adding the variable.
(3) A top register showing the vertical lines in the body of the diagram but rescaled and drawn with a common baseline for comparing the lengths of the lines and hence the contributions to the models. The register below gives a key showing for each vertical line the variables included in the model before addition of the term under examination.

It is immediately apparent that the contribution of most of the variables is markedly affected by which other variables are in the model. The substantial variation in line length within blocks at the top illustrates the non-orthogonality. The consistently large contribution from *runtime* is clear and in models which include *runtime* the effect of the other variables is markedly reduced.

A sensible first step when using the graphical analysis of deviance, once the extent of the nonorthogonality has been established, is to remove some of the unimportant detail by applying a filter to exclude certain models. Figure 14.11 shows a graph with models in which all the included terms are formally statistically significant at the 5% level. This shows clearly all potentially important models and the proportions of variation they account for. Further conclusions are: (i) two models not containing any terms fit better than t alone—*atr*, *trx*; (ii) there are many reasonably fitting 3, 4 and 5 variable models which do not include t—*awsrx*, *asrx*, *awrx*, *asr*, etc.; (iii) all variables except w are significant at 5% on their own. Filtering is *not* a procedure for testing statistical significance and should be regarded as aid to looking for interesting effects. It can be particularly useful when there are many explanatory variables and therefore many possible models to consider. Example 14.11 gives an application to the analysis of survival of house sparrows in relation to nine morphometric measurements.

14.5 BINARY REGRESSION MODELS WITH SEVERAL EXPLANATORY VARIABLES

14.5.1 Introduction

The p-variable linear-Normal model for relating a response variable to several explanatory variables applies to continuous measurements but it can be extended to binary responses or proportions derived from them. These so-called *binary regression models* are widely used in epidemiology for relating the incidence of a disease to variables such as weight, diet, smoking habit and other potential indicators. In ecology, they are often used to describe the relationship between survival over a particular period of time of individuals in a population and aspects such as size, condition and weather variables. In a binary regression model we relate the probability of a reaction to the values of the explanatory variables. We present the p-variable logistic model which extends the logistic-binomial model for a single explanatory variable (Chapter 4) to include several variables and a model which uses the double exponential function to relate the probability of reaction to the explanatory variables. Both models are special cases of the generalised linear models discussed in Chapter 11 and the ideas discussed here apply to this wider class of models.

Example 14.7 Survival in house sparrows

In 1898 a number of exhausted house sparrows were taken to the laboratory of Hermon C. Bumpus after a severe storm. Bumpus saw this as an opportunity to illustrate the process of natural selection and measured nine characters on each bird to relate them to survival (Bumpus, 1898). Table 14.9 summarises his data on males showing average values of each character for the 51 birds which survived and 36 which perished. Bumpus concluded that birds which survived '... are shorter and weigh less (i.e. are of smaller body), they have longer wing bones, longer legs, longer sternums and greater brain capacity'. Table 14.9 illustrates these findings. Differences for length and weight are highly statistically significant; values for the remaining characters are consistently larger in survivors but the differences are not statistically significant at the 5% level. The analysis of each variable is sufficient to demonstrate that birds which survived differed fundamentally from those which perished. It says little, however, about the effect on survival of different *combinations* of the characters, or whether some characters may emerge as significant when considered in concert with others. To address these questions we use a binary regression model to relate the probability of survival to the morphometric measurements.

14.5.2 *p*-Variable logistic and double exponential binary regression models

One of the most widely used binary regression models is the p-variable logistic model in which the probability of a reaction is related to the values of the explanatory variables, $x_1, x_2, \ldots, x_p$ by

$$P(\text{reaction}) = p = \frac{e^{\alpha + \beta_1 x_1 + \beta_2 x_2 + \cdots + \beta_p x_p}}{1 + e^{\alpha + \beta_1 x_1 + \beta_2 x_2 + \cdots + \beta_p x_p}},$$

i.e. a logistic function (symmetrical S-shaped curve) of a linear combination of the explanatory variables. The model is a generalised linear model with logit link function, i.e. $\ln[p/(1-p)]$ is linear in the parameters (Chapter 11).

Another important model is the double exponential in which the probability of a reaction is given by

$$P(\text{reaction}) = p = \exp[-e^{\alpha + \beta_1 x_1 + \beta_2 x_2 + \cdots + \beta_p x_p}].$$

Table 14.9 Average morphometric measurements of male house sparrows for those which survived and those which perished

Variable	Code	Mean		Difference (s.e.)
		Survive ($n = 51$)	Perish ($n = 36$)	
Length (mm)	l	159.25	162.08	−2.83 (0.40)
Alar extent (mm)	a	247.41	247.56	−0.15 (1.00)
Weight (g)	w	25.48	26.27	−0.793 (0.309)
Head length (mm)	b	31.70	31.56	0.138 (0.140)
Humerus length (in)	h	0.739	0.730	0.0089 (0.0052)
Femur length (in)	f	0.717	0.709	0.0075 (0.0051)
Tibio-tarsus length (in)	t	1.136	1.126	0.0093 (0.0088)
Skull width (in)	s	0.604	0.602	0.0022 (0.0032)
Sternum length (in)	k	0.856	0.844	0.0109 (0.0076)

This is an asymmetrical sigmoidal curve. It corresponds to the generalised linear model with *complementary log-log link* (or *double log link*) $\ln[-\ln(1-\mu)]$ applied to the probability of no reaction. The model is sometimes used in dilution assay where the probability of infection in a sample from a well-mixed solution is related to the density of bacteria by $p = 1 - e^{-\lambda}$. The model also has some advantages for analysing survival data because it guarantees a positive instantaneous mortality rate m given by $p = e^{-mt}$ (Manly, 1976).

Both the p-variable logistic and the double exponential model can be fitted to a set of independent binary responses by using the algorithm for fitting generalised linear models. Example 14.8 shows an application of the double exponential model to the data on survival in sparrows (Manly, 1976).

Significance tests of variables in model

In generalised linear models significance tests of the variables in the model are calculated by comparing *residual deviances* of models with and without the specified variables. The procedure is analogous to using the residual sums of squares in linear-Normal models (Section 14.5.2). Consider first a test of significance of one variable, say x_p in the p-variable model. Let D_p denote the residual deviance for the model which includes x_p and let D_{p-1} be the corresponding deviance for the *reduced* model in which x_p is removed. The test-statistic uses the change in deviance due to removing the variable, i.e.

$$G = X_1^2 = D_{p-1} - D_p,$$

a large change indicating a significant effect. This tests the null hypothesis that the model with fewer parameters is true. On the null hypothesis the test-statistic is distributed *approximately* as chi-square with 1 degree of freedom and the p-value can be looked up in tables. The approach extends to testing the significance of a set of variables. If there are r variables to be tested then on the null hypothesis the test-statistic is approximately distributed as chi-squared with r degrees of freedom. Example 14.8 gives an illustration.

Example 14.8 Application of the *p*-variable double exponential model to survival in house sparrows: fitting the model using Genstat 5

Response variable ydie: 0 – bird survives; 1 – bird perishes.

```
MODEL    [distribution = BINOMIAL;    link = COMPLEMENTARY]    ydie;
NBINOMIAL = n
FIT 1,a,w,b,h,f,t,s,k

*****  Regression Analysis  *****
   Response variable:   ydie
     Binomial totals:   n
        Distribution:   Binomial
       Link function:   Complementary log-log
        Fitted terms:   Constant, 1,a,w,b,h,f,t,s,k
```

```
*** Summary of analysis ***
    Dispersion parameter is 1
                                          mean      deviance
               d.f.     deviance      deviance         ratio
Regression        9        54.42        6.0462          7.32
Residual         77        63.59        0.8259
Total            86       118.01        1.3722

* MESSAGE: The following units have high leverage:
                                  2      0.29
                                 27      0.29
                                 39      0.30
                                 45      0.39
                                 59      0.30
                                 60      0.32
                                 81      0.54
                                 83      0.30

*** Estimates of regression coefficients ***
                 estimate         s.e.          t
Constant         -16.0          18.2        -0.88
l                  0.574         0.144       3.99
a                 -0.0599        0.0883     -0.86
w                  0.668         0.228       2.93
b                 -0.483         0.421      -1.15
h                -24.8          21.2        -1.17
f                -16.0          23.6        -0.68
t                 -0.65          9.13       -0.07
s                -32.3          19.7        -1.64
k                -16.77          8.62       -1.95

* MESSAGE: s.e.s are based on dispersion parameter with value 1
```

Interpretation of fitted model

The output gives an analysis of deviance table which shows a breakdown of the total deviance as the sum of deviances for regression and residual. The total deviance is the residual deviance for the *null model* obtained by fitting an overall mean. To test the statistical significance of the fitted model we use the method based on differences of deviances (Section 14.5.2). A portmanteau test (of the hypothesis that the coefficients of the explanatory variables are zero) uses the difference between the deviances for total and residual, i.e. the regression deviance, which is distributed *approximately* as chi-square with 9 degrees of freedom. The value of 54.42 indicates that *at least one* of the coefficients is non-zero ($p < 0.001$). The statistical significance of any particular variable could be tested in a similar way from the reduction in the residual deviance on removing the variable from the model.

On the interpretation of the signs of the coefficients in the fitted model, a positive value means a reduction in survival with that variable, a negative value an increase in survival. Thus, longer heavier birds survived less well whereas survival tended to increase with depth of sternum. We shall see below, however, that further analysis shows an effect of other variables which is not apparent here.

Example 14.9 Application of p-variable double exponential model to survival in house sparrows: II comparing models

In the fitted full model containing nine variables only the deviances for length, weight and sternum length were statistically significant. However, it is possible that other coefficients are significantly different from zero in a reduced model with fewer variables because of non-orthogonality. In this case the correlations amongst the

Table 14.10 Summary of fit of selected models relating survival in house sparrows to morphometric variables: *l*—length; *w*—weight; *h*—humerus; *f*—femur; *k*—sternum (keel); *t*—tibio-tarsus; *b*—beak and head; *s*—skull; *a*—alar

	Variables in model									Residual deviance
	a	*s*	*b*	*t*	*k*	*f*	*h*	*w*	*l*	
Null model	.	.	.	.	.	.	.	.	.	118.0
Single	− *a*	.	.	.	.	.	.	.	.	118.0
variable	.	− *s*	.	.	.	.	.	.	.	117.5
models	.	.	− *b*	.	.	.	.	.	.	117.1
.	.	.	.	− *t*	.	.	.	.	.	117.0
.	.	.	.	.	− *k*	.	.	.	.	116.1
.	.	.	.	.	.	− *f*	.	.	.	116.0
.	.	.	.	.	.	.	− *h*	.	.	114.8
.	.	.	.	.	.	.	.	$+\underline{\mathbf{W}}$	.	110.8
.	.	.	.	.	.	.	.	.	$+\underline{\mathbf{L}}$	99.62
2-variable	.	.	.	.	.	.	.	*w*	$+\underline{\mathbf{L}}$	99.18
models	.	− **S**	.	.	.	.	.	.	$+\underline{\mathbf{L}}$	93.11
with	.	.	$-\underline{\mathbf{B}}$	.	.	.	.	.	$+\underline{\mathbf{L}}$	92.38
length	.	.	.	.	$-\underline{\mathbf{K}}$	.	.	.	$+\underline{\mathbf{L}}$	89.94
.	$-\underline{\mathbf{A}}$	.	.	.	.	.	.	.	$+\underline{\mathbf{L}}$	89.78
.	.	.	.	$-\underline{\mathbf{T}}$	.	.	.	.	$+\underline{\mathbf{L}}$	89.34
.	.	.	.	.	.	$-\underline{\mathbf{F}}$	.	.	$+\underline{\mathbf{L}}$	85.21
.	.	.	.	.	.	.	$-\underline{\mathbf{H}}$	.	$+\underline{\mathbf{L}}$	82.36
3-variable	− *a*	.	.	.	.	.	$-\underline{\mathbf{H}}$	.	$+\underline{\mathbf{L}}$	82.35
models	.	.	.	− *t*	.	.	$-\underline{\mathbf{H}}$	.	$+\underline{\mathbf{L}}$	82.33
with	.	.	.	.	.	− *f*	− *h*	.	$+\underline{\mathbf{L}}$	82.06
length and	.	.	− *b*	.	.	.	$-\underline{\mathbf{H}}$	.	$+\underline{\mathbf{L}}$	81.07
humerus	.	− *s*	.	.	.	.	$-\underline{\mathbf{H}}$	.	$+\underline{\mathbf{L}}$	79.66
.	.	.	.	.	− *k*	.	$-\underline{\mathbf{H}}$	.	$+\underline{\mathbf{L}}$	79.34
.	.	.	.	.	.	.	$-\underline{\mathbf{H}}$	$+\underline{\mathbf{W}}$	$+\underline{\mathbf{L}}$	74.62
4-variable	− *a*	.	.	.	.	.	$-\underline{\mathbf{H}}$	$+\underline{\mathbf{W}}$	$+\underline{\mathbf{L}}$	74.52
models	.	.	.	− *t*	.	.	$-\underline{\mathbf{H}}$	$+\underline{\mathbf{W}}$	$+\underline{\mathbf{L}}$	74.07
with	.	.	.	.	.	− *f*	$-\underline{\mathbf{H}}$	$+\underline{\mathbf{W}}$	$+\underline{\mathbf{L}}$	73.78
length,	.	− *s*	.	.	.	.	$-\underline{\mathbf{H}}$	$+\underline{\mathbf{W}}$	$+\underline{\mathbf{L}}$	71.67
humerus and	.	.	− *b*	.	.	.	$-\underline{\mathbf{H}}$	$+\underline{\mathbf{W}}$	$+\underline{\mathbf{L}}$	71.54
weight	.	.	.	.	− **K**	.	$-\underline{\mathbf{H}}$	$+\underline{\mathbf{W}}$	$+\underline{\mathbf{L}}$	70.77
9-variable model	− *a*	− *s*	− *b*	− *t*	− **K**	− *f*	− *h*	$+\underline{\mathbf{W}}$	$+\underline{\mathbf{L}}$	63.59

Lower case *italic*—$p > 0.05$; Upper case **bold**—$p < 0.05$; Upper case **bold underlined**—$p < 0.01$. Tests based on change in deviance due to adding variable.

explanatory variables are all positive, being largest between: humerus and femur ($r = 0.88$); humerus and tibio-tarsus ($r = 0.77$); femur and tibio-tarsus ($r = 0.81$). To examine this further we can fit and compare all possible $2^9 = 512$ models using the methods of Section 14.5.2 and, in particular, the graphical analysis of deviance. Before we do this we present a forward selection procedure which starts with single variable models and adds those variables in turn for which the reduction in deviance is least. Table 14.10 gives the results. The main points are as follows.

(1) Fitting single variable models shows length with the smallest residual deviance followed by weight. In each case the effects are statistically significant ($p < 0.01$).
(2) When the variables are added in turn to a basic model in length all except weight give a significant reduction in the residual deviance with the largest reduction for humerus. So allowing for length *helps* to detect an effect of variables which are not significant in single variable models. Figure 14.12 shows how this can arise in an artificial example and illustrates how survival varies with combinations of length and humerus. By contrast, when weight is added in the deviance in length alone there is only a very small reduction to the model. In this case including both variables in the model *dilutes* the effect of weight. This is because both variables are similarly related to survival and correlated with each other: the effect of weight is largely subsumed by including length.
(3) Adding single variables to a basic model in length and humerus shows, rather surprisingly, that the largest reduction is due to adding weight. So including *both* length and humerus helps to detect weight which was not significant when added to length.
(4) When the remaining variables are added singly to a model with length, humerus and weight the addition of sternum length is just significant with a reduction in deviance of $(74.62 - 70.75) = 3.86$.

Example 14.10 Application of *p*-variable double exponential model to survival in house sparrows: III interpretation of fitted models

In the fitted 3-variable model with length, humerus and weight the probability of survival is given by

$$p = \exp[-\exp(-32.7 + 0.373l - 55.8h + 0.506w)]$$

i.e. a double exponential function of a linear predictor in length, weight and length of humerus. The signs of the coefficients correspond to greater survival in relatively short (+), light (+) long-limbed (−) birds. Figure 14.13 shows the fitted model and the observed proportion of birds surviving in relation to the value of the linear predictor grouped in bands of unit width. The fitted model describes the pattern of survival rather well. It is interesting to see that the range of predicted survival rates for the sample varied from close to zero to almost unity. This indicates strong selection pressure caused by the storm and the greater relative fitness of short birds with long limbs. The effect of weight is probably of less evolutionary significance since it is likely to reflect recent and ephemeral events.

The analysis confirms the effects of length and weight which were apparent from comparisons based on the means of each variable and provides more compelling evidence for greater survival of longer-limbed birds. The above model is, however, one of many other possible models which provide a reasonably good fit to this set of data (see graphical analysis of deviance below).

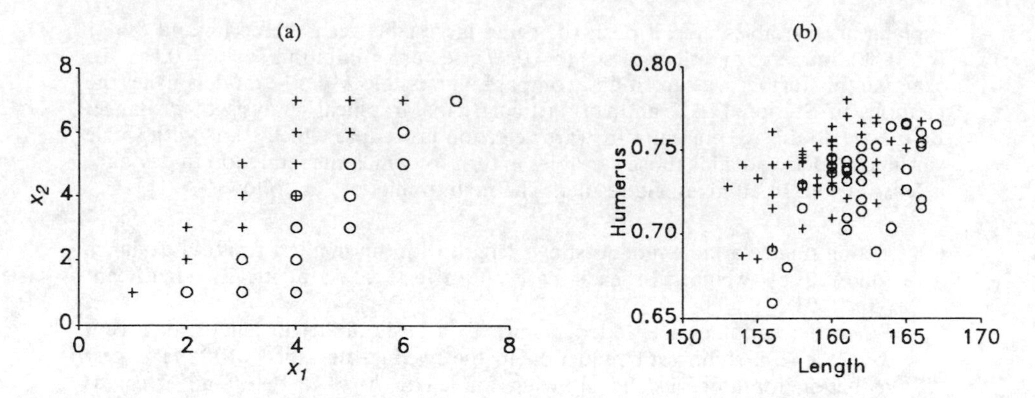

Figure 14.12 Illustration of how allowing for the effect of one variable can help to detect the effect of another. Symbols denote values of the binary response variable—survive (+), die (○). For survival in relation to a single variable there is a relatively large overlap between groups but when both variables are considered the separation becomes apparent. (a) Simulated data; (b) Bumpus data.

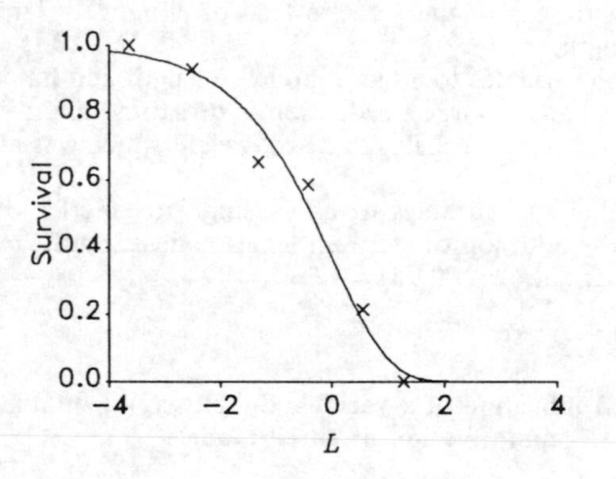

Figure 14.13 Fitted double exponential model relating the probability of survival in house sparrows to a linear predictor L in total length, weight and length of humerus; small values of L correspond to higher survival in relatively short, light and long-limbed birds.

Graphical analysis of deviance for the p-variable double exponential model fitted to survival in house sparrows

In the above analysis we considered a simple forward selection procedure adding variables in turn to a basic model in length. This approach detected a significant effect of several variables which was not apparent in the full model, in particular humerus. Here we summarise the results of fitting all the possible $2^9 = 512$ models using the graphical analysis of deviance (rather than of variance since we are now dealing with a generalised linear model). Figure 14.14 gives a filtered picture of models in which all the terms are statistically significant at the 1% level. This shows up several models which were not apparent with the stepwise procedure. These are: (a) models with *w* and other single variables, i.e. *wh*, *wf*, *wt* and *wk*; (b) models with *lw* and the addition of more than one variable, i.e. *lwab* and *lwtk*. These models are missed by the stepwise procedure which starts with *l* and then adds *h*. This illustrates the general point that

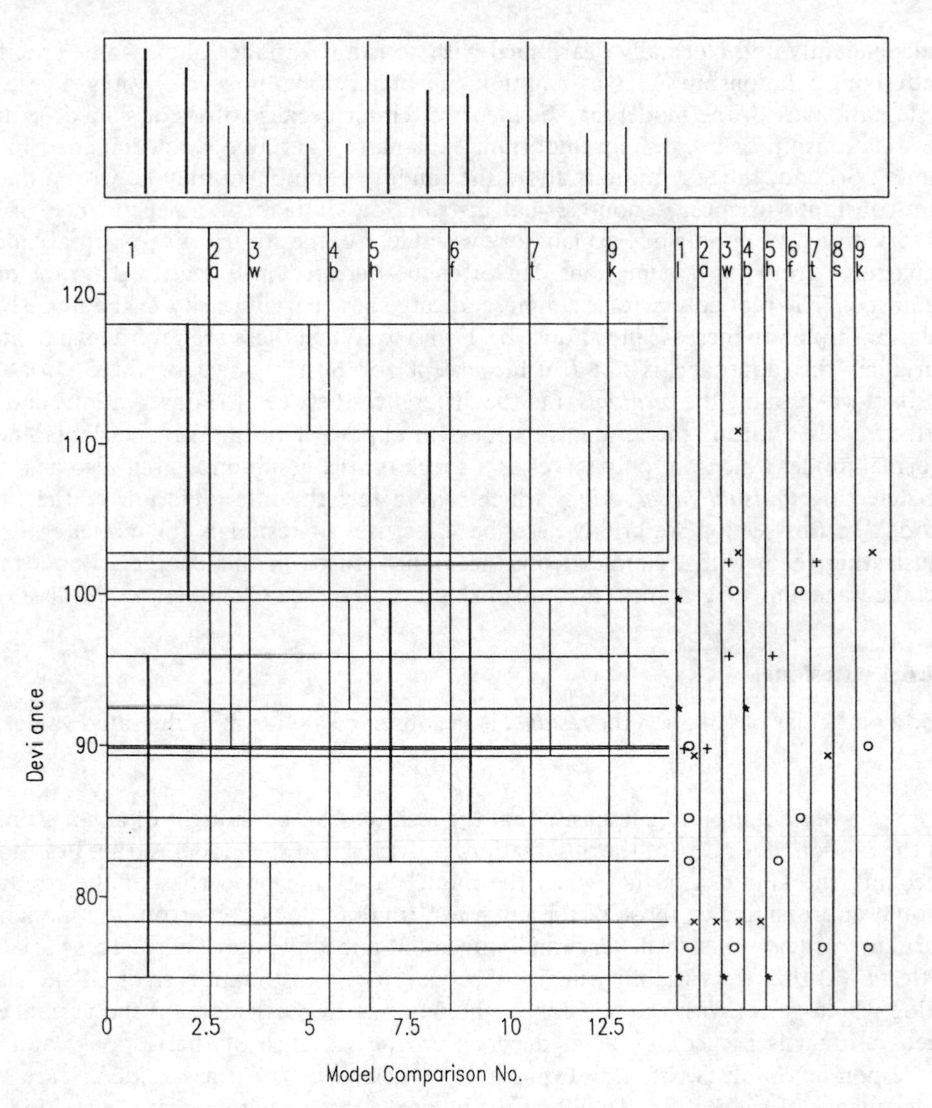

Figure 14.14 Filtered graphical analysis of deviance for the double exponential models fitted to the house sparrow survival data showing all models in which all the variables are statistically significant at the 1% level. This graph employs different symbols for each model in the right hand columns, to aid discrimination between models. A further new feature is that the vertical lines are coded according to the formal significance level of the term: thin line, $p < 0.01$, thick line, $p < 0.001$.

building up one at a time does not show up the cases where the addition of two or three variables as a group *does* improve the fit significantly, but where the addition of the best of the variables *does not*.

14.6 MODEL CHECKING: EXAMINATION OF RESIDUALS

In the p-variable linear-Normal model a basic assumption is that the observations are independently and Normally distributed with constant variance about some specified underlying relationship. This assumption can break down in several ways. First, the systematic part of the model may be incorrect either because of the assumed form of the relationship or because an important explanatory variable is not included in the model. Second, the assumptions about the random component may be wrong due to non-constant variance, a non-Normal distribution or lack of independence of the observations. With only one explanatory variable, i.e. the straight line-Normal model, departures from the assumptions can often be detected in a scatter diagram of Y against x. The plot could, for example, indicate some curvilinearity in the underlying relationship or an increase in variance of Y with x. When there are several explanatory variables this approach is not feasible except for possibly two or three variables. Instead we examine the *residuals*, i.e. the differences between the observations and the fitted values. Plotting the residuals is an essential part of fitting the p-variable linear-Normal model which not only serves as a check on the assumptions but also as a way of detecting *unusual observations* which may affect the interpretation of the fitted model. In this section we briefly describe some uses of residuals for model checking illustrating them with examples from the models fitted in this chapter. For a more detailed account of graphical methods using residuals see Chambers *et al.* (1983).

14.6.1 Residuals

For a particular observation the *residual* is the observed value minus the fitted value, i.e.

$$\hat{\varepsilon}_i = Y_i - \hat{Y}_i.$$

We use a hat notation to emphasise that the residual can be thought of as an estimate of the random deviation of the observation, provided that the model is true. For model checking the key idea is that when the model holds the properties of the residuals should approximate to those of the unknown random deviations from the model. In particular, the residuals will vary randomly about a mean value of zero. This variation reflects: (a) the real variability in the observations; (b) estimation error of the fitted value which varies with the position in the dataset. In particular: (a) the variance of each residual is not constant but depends on the variance of the response and the corresponding values of the explanatory variables; (b) the residuals are not independent of each other. To allow for the non-constant variance of the residual we sometimes work with a *standardised residual* obtained as the residual divided by its estimated standard error, i.e.

$$\hat{\varepsilon}_i' = \frac{\hat{\varepsilon}_i}{s\sqrt{1-h_i}} = \frac{Y_i - \hat{Y}_i}{s\sqrt{1-h_i}},$$

where s is the estimated standard deviation (square root of the residual mean square) and h_i is called the *leverage* of the observation. The idea of leverage is discussed below (Section 14.6.3) but for now it is enough to know that the leverage of an observation is a measure of the remoteness of the values of its explanatory variables from the centre of the observations. The advantage of the standardised residual is that when the model is true the residuals all have variance approximately equal to one so that they are particularly appropriate in checking for non-constant variance. Also, we expect that *approximately* 95% of the values should lie in the range $(-2, 2)$. Statistical packages often calculate standardised residuals when fitting the p-variable linear-Normal model and draw attention to values outside of this range. In the plotting methods below we use standardised residuals produced by the statistical package Minitab. However, ordinary residuals are often effective for detecting departures from the model and are easy to calculate if standardised residuals are not available.

14.6.2 Residual plots

Residuals are used for model checking by plotting them out in various ways. Such plots can detect lack of fit in either the systematic or random part of the model. Residual plots can also reveal observations which are unusual in some way and many statistical packages record *rogue* observations or *outliers* in *regression diagnostics* (Section 14.6.3). Here we are more concerned with lack of fit in either the systematic or random part of the model rather than with outlying data values.

Residuals vs explanatory variables

A plot of the standardised residuals against the values of an explanatory variable can

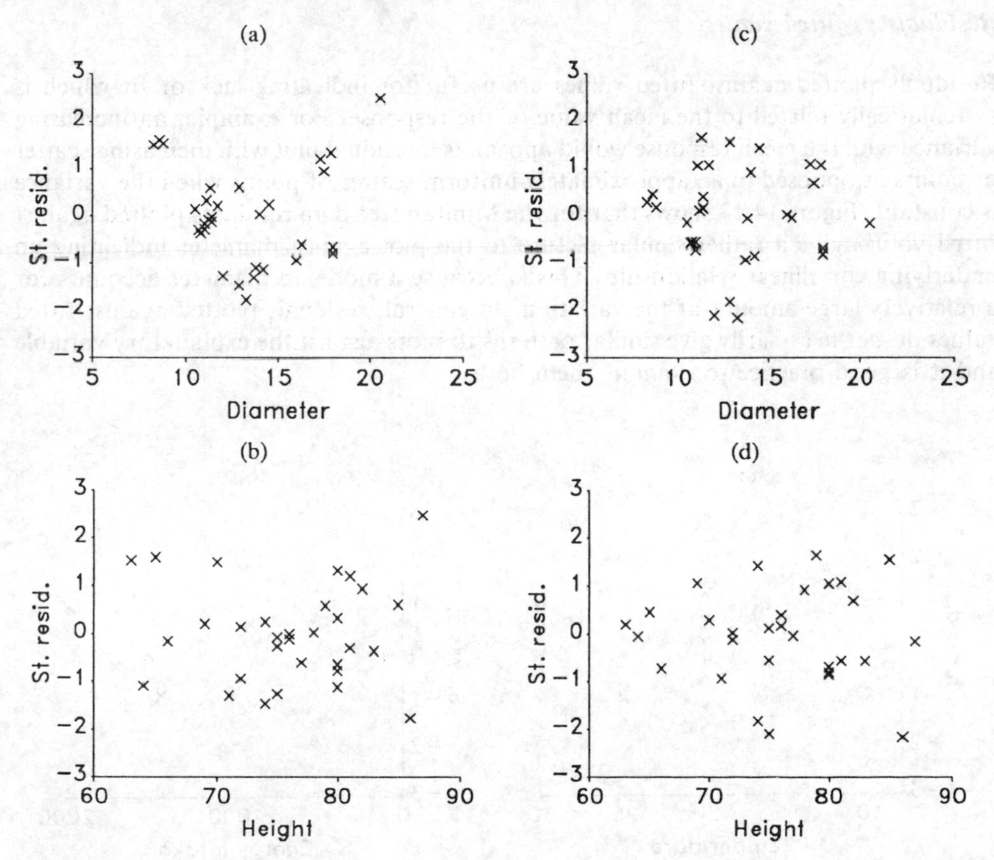

Figure 14.15 Standardised residuals plotted against height and diameter for the two variable linear model fitted to the Minitab tree data. (a) and (b) Raw data; (c) and (d) log transformed data.

detect systematic lack of fit or a non-constant variance associated with that variable. When the model is true all we should see is a random scatter of points about the value zero with approximately 95% of values in the range $(-2, 2)$.

Figure 14.15 shows plots for the Minitab tree data using residuals from the two variable linear model relating tree volume to diameter and length for both the raw data and log transformed values. For the untransformed data, there is a tendency for negative residuals at intermediate values of diameter which indicates an underlying curvilinear relationship. The effect is not apparent in the plot based on log transformed values which suggests that the data conform to the power function model (Example 14.1).

Figure 14.16 shows a plot of the standardised residuals against temperature and metabolisable energy intake for the quadratic response surface model fitted to the heat loss in the pig data *by ordinary least squares*. Recall that in Example 14.3 we fitted the model by weighted least squares, to allow for the greater variability in the relationship with energy intake at 10 °C. The effect is apparent in the plot against temperature as a greater scatter of points at 10 °C.

Residuals vs fitted values

Residuals plotted against fitted values are useful for indicating lack of fit which is systematically related to the mean value of the response. For example, an increasing variance with the mean response would appear as a residual plot with increasing scatter of points as opposed to an approximately uniform scatter of points when the variance is constant. Figure 14.17 shows that for the Minitab tree data residuals plotted against fitted values give a rather similar picture to the plot against diameter indicating an underlying curvilinear relationship. This is because a model in diameter accounts for a relatively large amount of the variation. In general, residuals plotted against fitted values do not necessarily give similar patterns to plots against the explanatory variable and it is good practice to examine them both.

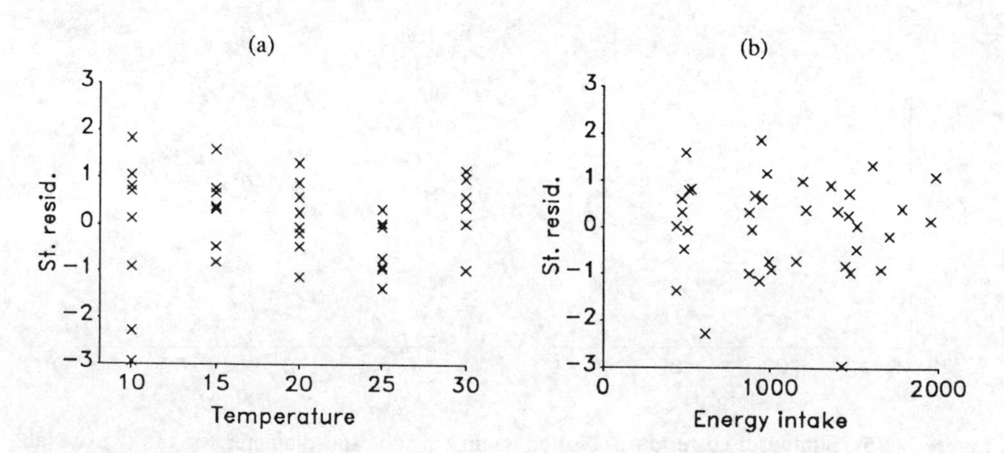

Figure 14.16 Standardised residuals against temperature and energy intake for the quadratic response surface model fitted to the heat loss in the pig data by ordinary least squares.

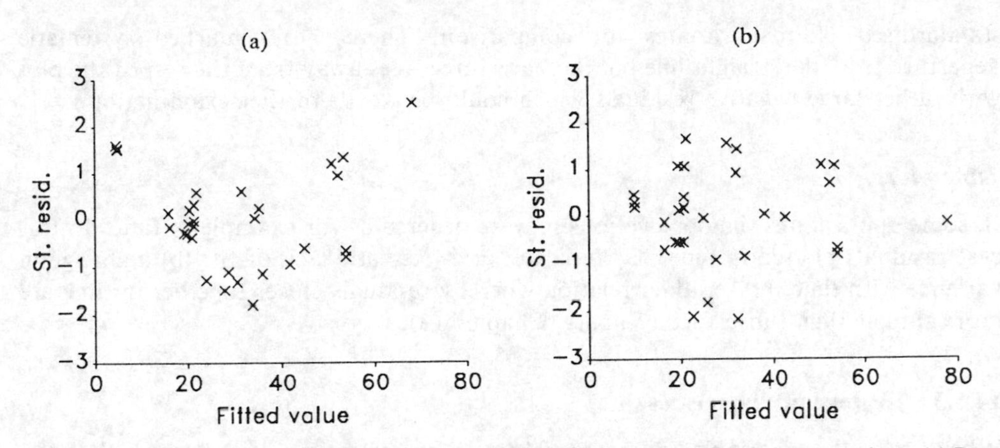

Figure 14.17 Standardised residuals plotted against fitted values for Minitab tree data. (a) Raw data; (b) log transformed data.

Residuals vs Normal scores

Residuals plotted against Normal scores, i.e. as a Normal probability plot, serve as a check on a Normal distribution of responses (Chambers *et al.*, 1983). When the assumption holds the residuals fall on or near a straight line. There are, however, no simple rules for deciding on what constitutes deviation from the line and the interpretation is made more difficult because the residuals are not independent. In practice, generating plots with the same number of points by simulating standardised Normal variates can be useful for assessing the pattern in a particular residual plot.

Figure 14.18(a) shows standardised residuals plotted against Normal scores for the Minitab tree data using log transformed data and Figure 14.18(b) shows a plot using

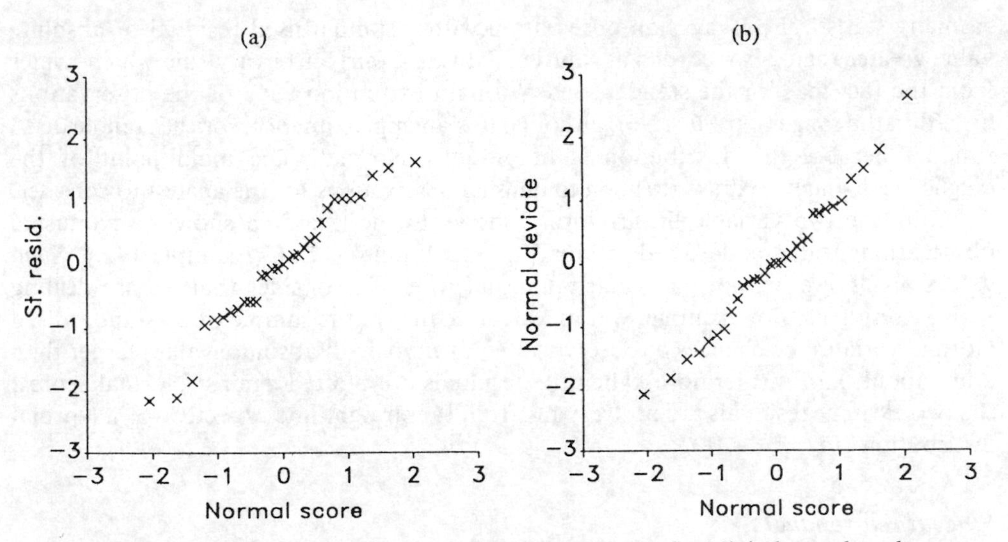

Figure 14.18 (a) Standardised residuals against Normal scores for Minitab tree data, log transformed; (b) plot for simulated standardised Normal variates. Normal scores are approximations to expected values for a sample of size n of ordered values from a $N(0, 1)$ distribution.

standardised Normal variates for comparison. There is no marked systematic departure from the straight line but there are three trees away from the rest of the plot with rather large negative residuals which could be worth further examination.

Time plots

In some applications the observations may be ordered as for example in time. In this case residuals plotted in time may indicate: (a) systematic lack of fit; (b) a change in variance with time; (c) serial correlation whereby residuals closer together in time are more similar than those further apart (Chapter 13).

14.6.3 Regression diagnostics

When fitting the p-variable linear-Normal model many statistical packages will record *unusual observations* in the output as *regression diagnostics*. Broadly speaking, an observation may be unusual in the response, in the explanatory variables or in both. Sometimes unusual observations can be explained in a satisfactory way, for example as errors in the data or a contaminated sample. In other situations there may be no obvious explanation and it is then wise to examine its effect when fitting the model by removing suspect observations from the analysis and reporting the effects. Observations which have a relatively large effect on the fitted model are sometimes referred to as *influential observations*. Diagnostic regression analysis, which is about detecting influential and outlying observations and quantifying their effects, is a large area of statistical analysis (Atkinson, 1985). Here we deal very briefly with two types of regression diagnostics.

Unusual responses: large residuals

In many statistical packages an observation with a standardised residual of absolute value greater than 2 is recorded as unusual. This is a fairly arbitrary rule which comes from the fact that for the standardised Normal distribution 95% of the observations lie within the range $(-1.96, 1.96)$, although it is an approximation for the standardised residual because the distribution is not exactly Normal. The main point of the diagnostic is simply to draw attention to unusual observations for the analyst to consider.

Fitting the two variable linear-Normal model to the tree data shows two unusual observations with standardised residuals of -2.11 and -2.16 (Example 14.1). Such values are, however, not particularly unusual when we consider that we are dealing with 31 residuals. For example, we can calculate that in the sample of 31 standardised Normal variates the chances of observing two or more with absolute values larger than 2 are about 0.5. Furthermore, when the residuals are plotted against Normal scores, the two extreme residuals are not very far from the straight line expected for a Normal distribution (Figure 14.18).

Studentised residual

The Studentised residual for a particular observation is obtained in a similar way to the standardised residual except that the observation is removed from the fitted model to estimate the residual mean square. If there is a single outlier in the data then Studentised residuals will be more likely to detect it than the standardised residuals because the residual mean square is not inflated by the unusual observation. However, with more than one outlier the method breaks down. In general, the only reliable way of checking for outliers is to fit the model by using a so-called *robust* fitting method which gives less weight to observations with large residuals. The problem with ordinary least squares is that it tends to produce residuals which have the properties that the least squares method requires, i.e. there is a kind of assumption-fulfilling tendency. Robust estimation is a large and complex topic which is outside of the scope of this book. For further details, see Tukey (1977) and Mosteller & Tukey (1977).

Unusual explanatory variables: high leverage

The leverage of an observation is a measure of how remote the values of its explanatory variables are from those of the other observations. The leverage is calculated from the values of the explanatory variables but we shall skip over the technical details of this. To illustrate the idea, however, we quote the result for the straight line model for which the leverage of the ith observation is given by

$$h_i = \frac{1}{n} + \frac{(x_i - \bar{x})^2}{\sum_{i=1}^{n} (x_i - \bar{x})^2},$$

where n is the number of observations and $\bar{x}$ is the mean of the x-values, i.e. high leverage occurs when x_i is far from the mean $\bar{x}$. It can be shown that the variance of the residual for an observation is equal to $\sigma^2(1 - h_i)$, where σ^2 is the variance of a single observation. So for an observation with leverage close to 1 the variance of the residual will be small. In other words, observations with high leverage have a relatively large effect in determining the fitted model. For this reason statistical packages often record observations with high leverage as part of the output for the fitted model. For example, in fitting the p-variable linear model Minitab records observations as having a large influence for which $h_i > 3p/n$.

In the course of fitting all the possible linear models to the physical fitness data, using Minitab, an observation with high leverage was detected for the three-variable model with *runtime*, *runpulse* and *maxpulse* (Example 14.6, Table 14.8, Model 40). For this individual (number 21): *runtime* = 8.63 min; *runpulse* = 170 per min; *maxpulse* = 186 per min. The high leverage occurs because the individual has a relatively high *maxpulse* for his *runpulse*. It is interesting to compare the fitted models with and without the observation. For the full data the estimated slope coefficient for *runpulse* = −0.375 (s.e. = 0.124, $t = -3.03$) and for *maxpulse* = 0.354 (s.e. = 0.135, $t = 2.63$). Without individual 21 the corresponding values are for *runpulse* = −0.240 (s.e. 0.149, $t = -1.61$) and for *maxpulse* = 0.185 (s.e. = 0.171, $t = 1.08$). For both variables the effect of removing the observations is to dilute the statistical significance of the estimated coefficients so that neither is significant at the 5% level. However, further analysis shows that when the individual is excluded the effect of *runpulse* is still significant when included in combination with *age*, as it was using the full dataset.

EXERCISES

14.1 The data below are measurements of the inorganic phosphorus (x_1), organic phosphorus (x_2) and estimated plant-available phosphorus (p.p.m.) in samples of 18 Iowa soils at 20 °C (Snedecor & Cochran, 1967). Fit a two variable linear-Normal model to analyse the relationship between the amount of phosphorus available to the plant and the organic and inorganic components.

Sample	x_1	x_2	y
1	0.4	53	64
2	0.4	23	60
3	3.1	19	71
4	0.6	34	61
5	4.7	24	54
6	1.7	65	77
7	9.4	44	81
8	10.1	31	93
9	11.6	29	93
10	12.6	58	51
11	10.9	37	76
12	23.1	46	96
13	23.1	50	77
14	21.6	44	93
15	23.1	56	95
16	1.9	36	54
17	26.8	58	168
18	29.9	51	99

14.2 The two datasets below were simulated using a two variable linear-Normal model for two different designs but with the same set of random deviations. For each case, fit both single variable straight line-Normal models and the two variable model, and discuss your findings. Relate the results to the theory given in Exercise 14.10.

Dataset 1			Dataset 2			Dataset 1			Dataset 2		
x_1	x_2	y	x_1	x_2	y	x_1	x_2	y	x_1	x_2	y
1	1	3.3	1	1.0	3.3	3	7	10.3	3	5.5	9.5
1	3	6.9	1	1.5	6.1	3	9	12.5	3	6.0	11.0
1	5	8.2	1	2.0	6.7	4	1	10.2	4	5.5	12.4
1	7	8.6	1	2.5	6.4	4	3	8.7	4	6.0	10.2
1	9	8.5	1	3.0	5.5	4	5	11.8	4	6.5	12.5
2	1	5.8	2	2.5	6.6	4	7	16.3	4	7.0	16.3
2	3	11.1	2	3.0	11.1	4	9	14.5	4	7.5	13.8
2	5	11.1	2	3.5	10.4	5	1	10.6	5	7.0	13.6
2	7	11.0	2	4.0	9.5	5	3	14.4	5	7.5	16.6
2	9	10.1	2	4.5	7.9	5	5	13.8	5	8.0	15.3
3	1	6.8	3	4.0	8.3	5	7	13.8	5	8.5	14.6
3	3	7.8	3	4.5	8.5	5	9	10.9	5	9.0	10.9
3	5	9.1	3	5.0	9.1						

14.3 In the two variable regression problem a surprising situation can arise in which there is almost no correlation between y and x_1 and y and x_2, and yet a multiple regression model involving x_1 and x_2 accounts for almost all, or even all, of the variation in the y values. The data in the table below are an extreme example (Hamilton, 1987). Analyse them to illustrate the effect.

Use the example to comment on some potential problems of: (i) discarding correlated x variables from a regression analysis; (ii) $y - x$ scatter plots and simple correlations; (iii) forward selection techniques with correlated explanatory variables.

x_1	x_2	y
2.23	9.66	12.37
2.57	8.94	12.66
3.87	4.40	12.00
3.10	6.64	11.93
3.39	4.91	11.06
2.83	8.52	13.03
3.02	8.04	13.13
2.14	9.05	11.44
3.04	7.71	12.86
3.26	5.11	10.84
3.39	5.05	11.20
2.35	8.51	11.56
2.76	6.59	10.83
3.90	4.90	12.63
3.16	6.96	12.46

14.4 To estimate the effect of a second variable x_2 in the two variable linear-Normal model, a colleague suggests the following *stagewise* approach. First fit a simple linear regression in x_1 followed by a simple linear regression of the residuals from that regression $(y - \hat{y})$ on x_2. The approach is unsatisfactory because theory shows that

$$\hat{\beta}_{2\text{stage}} = \hat{\beta}_2(1 - r^2_{x_1 x_2}),$$

where $\hat{\beta}_2$ is the estimate of the regression coefficient for x_2 fitting the two variable model by least squares and $r_{x_1 x_2}$ is the sample correlation coefficient between the values of the explanatory variables x_1 and x_2.

Illustrate the above result for the two datasets in Exercise 14.2.

14.5 For the heat loss in the pig data (Example 14.2, Table 14.5), fit the two variable quadratic response surface model using centred explanatory variables and compare your results with those given in Example 14.3.

How could you test for the presence of a quadratic term in metabolisable energy intake?

14.6 Fur seals wean their pups in between foraging trips to sea. The data below are from a study of pup growth rate and show measurements on 37 pups for the variables: growth rate (*growth*, kg per day), birth date of the pup (*bdate*, days from 31 October), number of foraging trips (*ntrip*), mean foraging trip duration (*tripd*) and mean attendance time ashore (*atten*) in days (data provided by Dr N. Lunn).

Use the p-variable Normal model to analyse the *growth* in relation to the explanatory variables *bdate*, *ntrip*, *tripd*, *atten*.

Pup	*growth*	*bdate*	*ntrip*	*tripd*	*atten*	Pup	*growth*	*bdate*	*ntrip*	*tripd*	*atten*
1	0.082	69	11	4.2	2.3	20	0.075	41	12	7.2	1.5
2	0.045	53	9	9.1	2.1	21	0.042	27	15	6.4	1.9
3	0.107	67	16	4.4	1.6	22	0.056	34	12	7.5	2.2
4	0.054	23	12	7.8	1.7	23	0.082	40	12	7.0	2.1
5	0.077	35	10	8.3	2.7	24	0.070	42	9	7.7	1.9
6	0.076	36	13	6.7	2.1	25	0.053	29	13	7.7	2.2
7	0.049	36	10	9.1	2.6	26	0.049	36	12	7.9	2.7
8	0.062	43	14	6.1	1.7	27	0.039	36	13	7.2	1.5
9	0.057	26	9	10.8	2.3	28	0.049	50	10	8.8	1.6
10	0.049	27	10	9.4	3.0	29	0.039	55	10	9.2	2.1
11	0.054	30	17	5.4	2.0	30	0.093	41	10	8.7	2.0
12	0.070	40	12	6.8	1.8	31	0.059	42	12	5.8	2.0
13	0.063	47	9	6.8	2.1	32	0.054	26	12	8.2	1.8
14	0.042	51	10	8.0	1.6	33	0.080	28	14	7.2	1.4
15	0.084	51	14	6.0	1.4	34	0.067	39	11	8.1	2.3
16	0.067	27	14	6.8	1.6	35	0.079	43	11	8.2	1.8
17	0.058	30	11	9.0	1.8	36	0.051	18	11	8.8	2.2
18	0.061	38	15	6.1	1.9	37	0.052	26	10	9.2	1.9
19	0.090	39	17	5.4	1.5						

14.7 Male and female puffins cannot be distinguished on plumage but male tend to have deeper and longer bills. The table below shows measurements on a random sample of 21 males and 20 females for curved bill length (*clen*), straight bill length (*slen*) and bill depth (*depth*), measured in mm (data provided by Dr M.P. Harris).

(a) Develop a logistic regression model relating the probability of a bird being a male to its bill measurements.

(b) Suppose that a bird is sexed as a male if the probability corresponding to its bill measurements is greater than half. Apply your model to the sample to calculate the proportion of males misclassified as females and *vice versa*.

(c) What problems might there be in applying the model to sex a further sample of birds in the field?

Males				Females			
Bird	*clen*	*slen*	*depth*	Bird	*clen*	*slen*	*depth*
1	44.1	28.8	33.5	1	43.0	28.6	32.3
2	44.0	29.7	35.9	2	42.0	29.3	31.4
3	44.1	29.0	34.9	3	43.2	29.1	34.0
4	42.6	28.2	34.6	4	42.9	29.2	29.8
5	46.4	28.8	39.0	5	40.0	27.2	30.6
6	42.2	27.1	34.3	6	40.6	27.5	29.1
7	44.5	28.0	36.9	7	42.1	28.0	28.6
8	44.2	28.8	34.3	8	42.0	27.6	32.0
9	43.7	28.0	37.0	9	43.5	28.0	34.0
10	44.0	27.6	36.5	10	41.4	27.5	32.5
11	45.3	30.0	36.1	11	41.6	28.5	32.6
12	44.3	29.9	35.8	12	44.7	30.4	32.8
13	46.0	30.5	34.1	13	41.3	27.6	33.2
14	45.2	29.5	34.5	14	40.7	28.0	33.8
15	43.2	27.7	34.0	15	41.2	27.0	32.9
16	45.1	28.2	35.5	16	42.3	27.2	32.6
17	44.6	27.8	37.6	17	44.0	28.5	30.7
18	42.2	29.0	34.0	18	39.6	26.6	31.9
19	42.7	26.6	34.4	19	43.8	29.0	34.7
20	44.9	29.2	34.8	20	41.9	27.8	33.8
21	44.8	29.1	34.5				

14.8 For each of the three datasets below, analyse the residuals from the fitted two variable linear-Normal model and discuss your findings.

x_1	x_2	y (Dataset 1)	y (Dataset 2)	y (Dataset 3)
1	1	11.5	11.5	11.1
2	1	12.1	11.4	13.4
3	1	13.0	13.8	16.5
4	1	11.9	14.2	16.6
5	1	10.8	19.3	17.8
1	3	10.0	10.7	10.2
2	3	9.2	13.4	12.8
3	3	10.8	8.3	14.9
4	3	12.3	17.8	15.4
5	3	7.8	17.8	16.9
1	5	7.2	7.7	4.9
2	5	8.1	8.6	10.7
3	5	8.3	10.9	18.0
4	5	8.3	11.3	13.1
5	5	8.4	14.8	15.7
1	7	6.1	4.1	4.8
2	7	6.7	6.3	7.1
3	7	8.5	6.0	9.5
4	7	6.1	10.7	12.9
5	7	7.1	13.3	14.0
1	9	1.9	3.3	0.9
2	9	5.4	5.5	5.5
3	9	3.7	7.5	7.1
4	9	4.3	10.2	9.3
5	9	2.7	10.1	11.0

14.9 In many applications of multiple regression, the subset of variables to include in the model is not known and the data are used to decide which variables to include. This can lead to selection bias from 'significant' relationships which occur by chance, especially when there are many explanatory variables. The effect is illustrated by the following simulation exercise. For the values of x_1 given in the table below, 10 sets of data were simulated using the straight line-Normal model

$$Y_i = 1.0 + 0.5x_{1i} + \varepsilon_i, \qquad i = 1, \ldots, 20$$

where $\varepsilon_i \sim N(0, 1)$. Then, the set of explanatory variables was augmented by adding the nine variables $x_2, \ldots, x_{10}$ whose values are given in the table below. Each dataset was analysed by stepwise multiple regression using Minitab with the command STEPWISE and default values of 'F-to-enter' and 'F-to-remove' equal to 4. The table below shows the variables selected in each analysis and that in four cases the models contain at least one redundant variable.

(a) Carry out a further 10 simulations to reinforce the results.
(b) Are you surprised to find that x_1 was always correctly selected?

x_{1i}	1	2	3	4	5	6	7	8	9	10	11	12	13	14	15	16	17	18	19	20
x_{2i}	8	20	11	7	16	13	4	2	6	3	17	14	19	10	5	15	12	1	9	18
x_{3i}	9	3	20	10	6	14	8	19	7	1	15	2	13	12	11	5	18	16	4	17
x_{4i}	9	13	11	12	6	16	20	10	18	17	3	14	2	8	15	1	19	7	4	5
x_{5i}	19	13	6	20	16	10	12	15	3	11	5	2	18	4	8	9	17	14	1	7
x_{6i}	11	10	3	13	8	16	5	1	15	19	4	9	7	18	17	14	20	12	6	2
x_{7i}	18	14	10	1	19	3	11	4	8	2	16	20	6	7	13	5	12	17	9	15
x_{8i}	9	6	16	11	15	12	8	4	1	2	20	14	5	18	17	19	3	10	13	7
x_{9i}	6	3	7	15	10	16	14	9	5	12	2	4	20	11	8	17	1	13	19	18
x_{10i}	20	7	15	17	4	10	5	12	13	14	2	19	6	16	1	8	11	9	18	3

Data set	Variables selected	Data set	Variables selected
1	x_1	6	x_1, x_3, x_{10}
2	x_1, x_{10}	7	x_1
3	x_1	8	x_1, x_9
4	x_1, x_6	9	x_1
5	x_1	10	x_1

14.10 The two variable linear-Normal model with explanatory variables centred about their means is written as

$$Y_i = \alpha + \beta_1(x_{1i} - \bar{x}_1) + \beta_2(x_{2i} - \bar{x}_2) + \varepsilon.$$

(a) Show that the least squares estimates of the slope coefficients are given by

$$\hat{\beta}_1 = \frac{S_{Yx_1}S_{x_2x_2} - S_{Yx_2}S_{x_1x_2}}{S_{x_1x_1}S_{x_2x_2} - S^2_{x_1x_2}}; \qquad \hat{\beta}_2 = \frac{S_{Yx_2}S_{x_1x_1} - S_{Yx_1}S_{x_1x_2}}{S_{x_1x_1}S_{x_2x_2} - S^2_{x_1x_2}},$$

(b) By expressing $\hat{\beta}_1$ as a linear combination of the responses, show that its mean and variance are given by

$$\text{mean}\,[\hat{\beta}_1] = \beta_1; \qquad \text{var}\,[\hat{\beta}_1] = \frac{\sigma^2}{S_{x_1x_1}(1 - r^2_{x_1x_2})},$$

where $r_{x_1x_2}$ is the sample correlation coefficient between the values of the explanatory variables. Illustrate these results using 30 sets of simulated data from the two variable model fitted to dataset 2 in Exercise 14.2.

(c) Comment on the relevance of this result, and a similar one for $\hat{\beta}_2$, to the design and the analysis in an investigation of the relationship between Y and x_1 and x_2.

15 Models of the Brain: Neural Networks

15.1 INTRODUCTION

The study of neural networks had its origins in a paper published almost fifty years ago by Warren McCullough and Walter Pitts (1943). Their suggestion was that many aspects of brain function could be modelled using a directed network of neurones which co-ordinate their firing in a very simple way: any neurone fires if a weighted sum of the inputs to it from other neurones exceeds its threshold; if the threshold is not reached it does not fire. The synaptic weight between any two neurones indicates the contribution which the firing of the first neurone makes to the total input of the second neurone. The use of different sets of weights renders these networks very flexible, and McCullough and Pitts showed how networks composed of varying numbers of neurones with different sets of weights, and different connection architectures, could represent a wide range of logical functions. Their work demonstrated that neural networks had considerable potential for modelling some mental activities. The subject developed slowly—with one or two landmark discoveries which are discussed below—over the next thirty years or so, but, in the 1980s, two things happened which resulted in a renaissance of the subject. First, the introduction of parallel processing computers and other new computing techniques meant that sufficient computer power was available for substantial investigations of real problems. Secondly, some important discoveries were made of interesting new models and learning techniques. The subject has since undergone a very rapid development.

15.1.1 The reasons for studying neural networks

Today the term neural network has different meanings for different people. For the computer scientist or control engineer, it means a network—of simple processors showing some of the characteristics of neurones in real organisms—that is important because of its special computing capabilities. Usually, at one end of the network, data is fed in from measuring instruments in the external world, and at the other end the network produces outputs which might be read directly by an observer or used to control a device. In between, the network consists of a series of layers of 'neurones' which fire or not depending on their current state, their inputs from earlier layers and their own properties. Those that fire send impulses to connected neurones in subsequent layers. The networks are not programmed in the way that other computers are, by analysing the task required and giving the computer specific instructions to perform it. Surprisingly, they can be trained to perform complex tasks by *random* adjustment of the connection strengths between neurones. Those patterns of connections which produce a better approximation to the desired relationship between input and output are reinforced; those which do not are discouraged by weakening the connections. By this means, complicated tasks can be carried out without the costs involved in a careful analysis of the problem and the coding of the instructions in the computer language. Also if the task changes, the network can usually be modified by a further training session. The networks are particularly suitable for implementation on parallel processing computers. Then they can often respond extremely quickly, and are useful for real-time applications which require a short response time.

The biologist is more interested in networks of neurones in real organisms, exhibiting a wide range of properties and subject to a plethora of influences. Real neurones have a cell body, to which are attached dendrites which receive inputs from the terminals of other neurones (see Figure 15.1). From the cell body projects the axon, which is very long in some cases, and, at the end of the axon, the cell branches out to a number of terminals, which then establish contact with, and send impulses on to, the dendrites of other neurones. The interior of the resting cell is at a different potential from the outside fluid, and consequently there is a potential difference across the cell membrane, called the membrane potential. Any inputs via its dendrites in the form of pulses of increased voltage are quickly transmitted by the cell along the dendrites to the cell body. When these integrate to a value at the cell body which is greater than the cell threshold, the cell fires: for a very brief time, the membrane potential reverses its sign, after which the cell reverts back to its previous equilibrium. But before the cell body returns to equilibrium, an impulse, in the form of the brief potential reversal, is transmitted along the axon to the terminals of the cell, which can then be transmitted to other cells further on in the network. The point at which a terminal of one cell lies adjacent to a dendrite of another and hence the route by which impulses are transmitted from cell to cell is called a synapse.

For neurophysiologists, this is a very crude description of a neurone. They might go on to study the biophysics of the channels in the membrane through which the potential-reversing ions pass in both directions, and whether these channels open in response to differences in voltage or chemical concentration. What proteins are part of the membrane and what are their functions? Do they act as enzymes or are they involved in cell-to-cell recognition? What neurotransmitters are released when a nerve impulse reaches the end of the axon and how does this result in transmission of the impulse to the dendrites of an adjacent neurone? These are important questions, and have implications for many practical questions, but even with a knowledge of how these building blocks of the brain function, there still remains the problem of its higher level functioning.

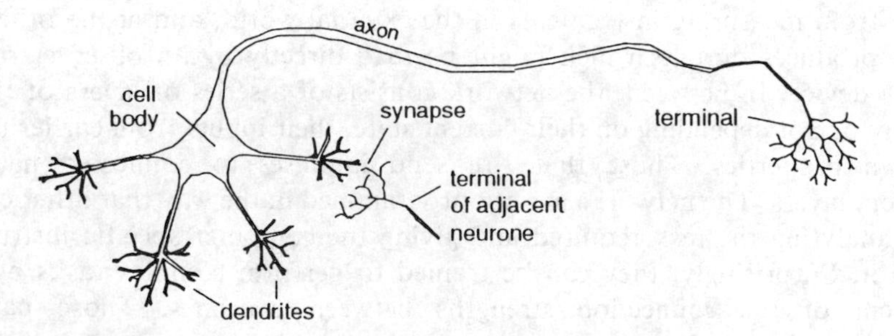

Figure 15.1 The components of a neurone.

In between the computer scientists and those physiologists whose main concerns are the detailed mechanisms of brain function, are psychologists and neurobiologists who are interested in how the brain functions as a whole, and in theories of the mind. They ask questions such as: how is it that even though individual neurones function relatively slowly, some mental activities are very fast? Since individually neurones are not very reliable, how is it that most tasks are done so well? These considerations have led some biologists and computer scientists to the idea that the brain functions by parallel distributed processing. The speed of response is high, because most of the processing occurs simultaneously (i.e. in parallel) in different parts of the network. There is also some redundancy which can be used to correct errors. In this chapter, we outline a few basic ideas and models within this broad framework, which are of particular interest to psychologists and neurobiologists studying overall brain function, although some of the models have been developed for practical applications outside biology.

15.1.2 A simplified model of a neurone

As we said above, real neurones have a very complex structure, but those in neural network models are usually very simple. In the simplest model neural networks, following McCullough and Pitts (1943) model, the neurone is represented by a node as in Figure 15.2, which has inputs from the terminals of other neurones, $x_1, x_2, \ldots, x_n$. x_i takes the value 1 or 0 depending on whether the ith neurone is firing or not. The neurone forms a weighted sum of the inputs

$$w_1x_1 + w_2x_2 + \cdots + w_nx_n$$

and if this is greater than the threshold for the neurone, T, the neurone fires. Then the output of the neurone, y' takes the value 1, and this can then be transmitted to become the input of any neurones further on in the network. If

$$w_1x_1 + w_2x_2 + \cdots + w_nx_n \leqslant T$$

the cell does not fire and its output, $y' = 0$. The quantity w_i is thus measuring the strength of the synapse between the terminal of the ith neurone in the previous layer

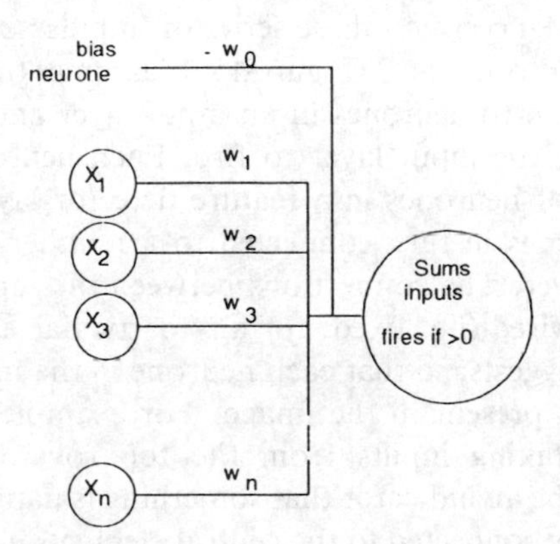

Figure 15.2 The McCullough–Pitts model of a neurone.

and the dendrite of the neurone under consideration. In many neural network models, the neurones are arranged in layers (based on the known structure of the cerebral cortex) with input being to the first layer, and then each layer communicating to the next in turn until the output layer is reached. Other models which depart from this simple plan have interesting features such as feedback loops.

We mention one small point of notation before continuing. The thresholds, T, can be adjusted in most neural network training procedures, and so they are usually written so as to emphasise their similarity to ordinary synaptic weights. We follow the convention used in most neural network software: an extra neurone is included (in the layer before the neurone whose synaptic weights are being considered) called a *bias neurone* which always fires with a value 1. First rewrite the above inequalities by taking the T term to the other side of the inequality, and replace it by $-w_0x_0$, where x_0 always takes the value 1. For example, the condition for the output neurone to fire

$$w_1x_1 + w_2x_2 + \cdots + w_nx_n > T$$

can be written

$$w_0x_0 + w_1x_1 + w_2x_2 + \cdots + w_nx_n > 0.$$

$T = -w_0$ is just like another weight which can be adjusted according to the same rules as the remainder, but which is attached to a neurone which is always firing.

15.2 THE PERCEPTRON

About fifteen years after the paper of McCullough and Pitts, substantial progress was made in the application of their ideas to perception when Rosenblatt (1958, 1962) devised the *Perceptron*. He showed how a McCullough–Pitts network with modifiable weights could be trained to perform simple perceptual tasks, and demonstrated its abilities in concrete implementations.

For the purpose of discussion of the idea of the Perceptron, we could consider the eye to be like a camera in which the image of the outside world is projected and focused on to the retina. The image on the retina is converted to electrical signals via the light sensitive cells (rods and cones) which send impulses to the brain when the cells are illuminated. The brain then converts these series of impulses into an assessment of the viewed object. The Perceptron model (Figure 15.3) assumes that the light sensitive cells in the retina are connected to neurones in an input layer and that impulses from the image cause neurones in the input layer to fire. Each neurone in the input layer is connected to a number of neurones in a feature detector layer, and each neurone in the feature detector layer is in turn connected to a number of neurones in the final output or Perceptron layer. The connections between the input layer and the feature detector layer are hard-wired (i.e. fixed) for any particular application, and could be devised, as their name suggests, so that each neurone in the feature detector layer fires if a particular feature is present in the image. For example, in Figure 15.3, feature detector neurone a is taking inputs from the top row of the elementary retina illustrated, and so could be an indicator that something is happening at the top; feature detector neurone c is only connected to the central element of the retina, and therefore could be an indicator of events in the centre of the retina. The weights and thresholds of the feature detector neurones also impact on this behaviour, but for the moment it is

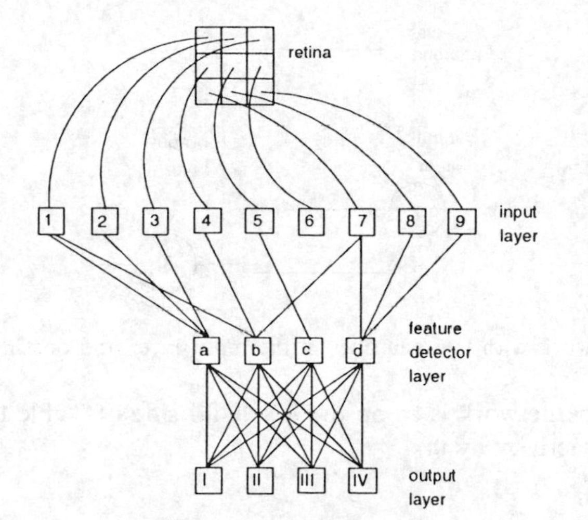

Figure 15.3 The Perceptron model.

sufficient to know that the weights could be set so that the feature detector neurones detect features in the original retina array. The connections between the feature detector layer and the final Perceptron layer are adjustable so as to give the network some flexibility of performance.

The process of adjusting these weights to enable the network to perform a range of tasks is called training the network. The network is also said to learn. For this purpose, a training dataset is used which consists of a list of inputs together with their desired associated outputs. During the training process, these inputs are presented in some order, usually at random. After each presentation, the network weights are adjusted using a specified algorithm so as to improve the performance for the input presented. After sufficient presentations, the network is usually trained, and it can then be used for predicting new cases.

15.2.1 A two layer neural network

There are two distinct phases of operation of an artificial neural network: the learning or training stage and the operation stage. To illustrate some of these ideas, we start with a very simple network which has just two layers, an input layer with two neurones, 1 and 2, and an output layer with one neurone, 3 as in Figure 15.4. This could be considered as a Perceptron with no middle (feature detector) layer, i.e. direct connections between the two neurones in the input layer and the single output neurone. The connections between the two input neurones and the output neurone have variable weights which can be trained to perform the very elementary task that neurone 3 in the output layer should fire if both 1 and 2 are on, but otherwise should be silent. However, we first illustrate the operation of the network, assuming it has already been trained. We demonstrate in Section 15.2.2 how the network can learn how to perform this task, using a performance learning rule.

If we restrict the states of neurones to be either firing or not (i.e. taking values 1 or 0 only), there are only four possible cases in the input layer. Thus the desired

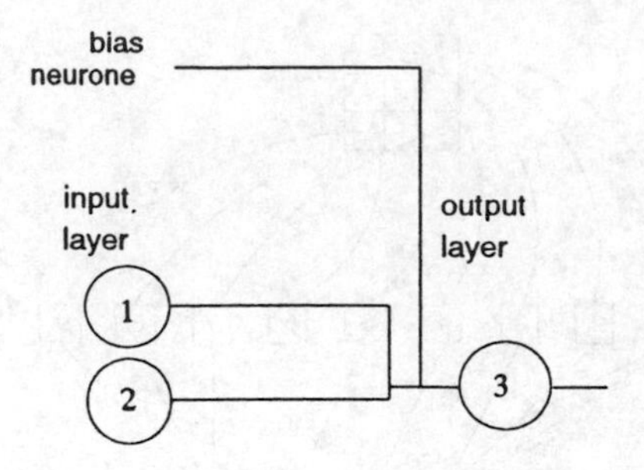

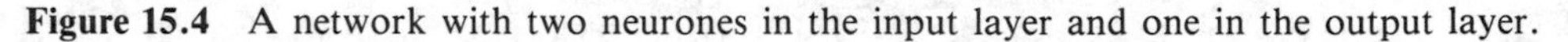

Figure 15.4 A network with two neurones in the input layer and one in the output layer.

performance of the network is as on the left hand side of Table 15.1, if we indicate firing by 1 and inactivity by 0.

15.2.2 Learning rules

Supervised, unsupervised and reinforcement learning

One useful categorisation of learning rules is related to how the learning is guided. *Supervised learning*, or *learning with a teacher*, is done by direct comparison of the actual output of the network with the desired output for a set of training input/output relationships. The correct answers for the output at each output neurone are fed in during training, and the network adjusts in response to the errors it has made. A related form of learning is *reinforcement learning* in which all that the network is told is whether the response is correct or not. *Unsupervised learning* is not guided at all, and in this case often all that the network can do is to recognise correlations in the incoming patterns, to create categories from these, and to output for a given input pattern which category it has assigned. In this case, the network often functions in a manner similar to the statistical technique of cluster analysis.

Table 15.1 The operation of the simple two layer network discussed in Section 15.2.1. On the left hand side the four possible different cases for the inputs and the resulting output are listed. The weights of the synapses of the trained network are given by $w_1 = 2$, $w_2 = 1$ and $w_0 = -T = -2$. On the right is illustrated how the output neurone 'calculates' the weighted sum of the inputs including the bias neurone representing the threshold. When this is above zero the output neurone fires, and when equal to or below zero it does not fire

The desired relationship between the inputs and the output				Operation of the network	
	Input layer		Output layer		
	State of 1	State of 2	State of 3	$w_0(1) + w_1x_1 + w_2x_2$	Therefore state of 3
Case 1	0	0	0	$-2 + (2)(0) + (1)(0) = -2$	0
Case 2	0	1	0	$-2 + (2)(0) + (1)(1) = -1$	0
Case 3	1	0	0	$-2 + (2)(1) + (1)(0) = 0$	0
Case 4	1	1	1	$-2 + (2)(1) + (1)(1) = 1$	1

Coincidence learning: Hebb's learning rule

One method of training networks was suggested by work of Donald Hebb reported in 1949. When discussing real neurones, Hebb suggested that if the output of a neurone and one of its input neurones fire together consistently, then the strength of the connection in the synapse involved will tend to increase, i.e. in our terms, the weight for that synapse will increase. Coincidences between input activity and desired output activity tend to be emphasised and hence the term *coincidence learning* is used. Hebbian learning can be either supervised or unsupervised. In the supervised version, during the learning stage, the correct response (i.e. that which the network should have after training) for the particular input is fed into the output layer, and the weights—measuring the strengths of the synaptic connections—reflecting connecting neurones which are both firing are increased. The weights of neurones which are not both firing are not adjusted. Note that this uses the desired output in the training procedure, but ignores the existing performance.

Training the simple two layer network

The Hebb learning rule does not take the existing performance into account, but most learning rules adjust the weights in response to the performance of the neurone or the network. We illustrate one such algorithm, associated with the Perceptron, and devised by Rosenblatt. We know from the above definition that the output neurone, neurone 3, fires if the weighted sum of the inputs (i.e. the outputs from neurones 1 and 2) exceeds the threshold T. Thus we need to determine the values w_1, w_2 and $T = -w_0$ in the following criterion for the firing of neurone 3

$$\text{neurone 3 fires, i.e.} \quad y' = 1 \quad \text{if} \quad w_0(1) + w_1x_1 + w_2x_2 > 0.$$

which will ensure that the network has the properties given in Table 15.1. The problem is very simple to solve in this particular case, but what we need here is some *general* method of adjusting the weights and the threshold which is likely to be successful in this specific case and in a range of other more complex cases.

Let us try the Perceptron Learning Rule. During the learning phase, the possible inputs are presented in some order to the network and the correct output for the network is also fed into the network at the appropriate position so that the neurone at that position, neurone 3, can make some adjustment to its synaptic weights. In this example, the order of presenting the training cases is systematic, but this is not necessary. Quite often the cases are presented in a random order, or if there are very many cases, they are sampled at random. The threshold T (or $-w_0$) can also be adjusted during the learning process. After each presentation of an input and a desired output the following rule is applied.

Perceptron learning rule (simplest version)

(1) If the output is the correct one for the particular input combination, make no adjustment to the weights;
(2) If the output should be 1, but is 0, *increment* the weights on the active lines only by a constant c_1;
(3) If the output should be 0, but is 1, *decrement* the weights on the active lines only by the same constant c_1.

Table 15.2 The learning process in the simple 2-layer network in Section 15.2.2. During the learning process the sequence of inputs at 1 and 2 and correct outputs at 3 given in the left hand side of Table 15.1 are applied. The stages in the learning process are described on the right hand side of the table. The changes in weights ($c_1 = 1$ in this example) and resulting new weights are given in italics in the appropriate columns in the same rows as the comments on the right

Trial no.	Input layer				Bias neurone 0		Linear combination	Output layer		Error in output	*Action*
	Neurone 1		Neurone 2		0			State of 3 actual	State of 3 desired		
	x_1	w_1	x_2	w_2	x_0	w_0	$\Sigma\, w_i x_i$				
		0		*0*		*0*					*Initial wts.*
1	0		0		1		0	0	0	n	
		—		—		—					*No adjustment*
		0		*0*		*0*					*Same wts.*
2	1		1		1		0	0	1	y	
		+1		*+1*		*+1*					*Increment 1, 2, 0*
		1		*1*		*1*					*New wts.*
3	1		0		1		2	1	0	y	
		−1				*−1*					*Decrement 1, 0*
		0		*1*		*0*					*New wts.*
4	0		1		1		1	1	0	y	
				−1		*−1*					*Decrement 2, 0*
		0		*0*		*−1*					*New wts.*
5	0		0		1		−1	0	0	n	
		—		—		—					*No adjustment*
		0		*0*		*−1*					*Same wts.*
6	1		1		1		−1	0	1	y	
		+1		*+1*		*+1*					*Increment 1, 2, 0*
		1		*1*		*0*					*New wts.*
7	1		0		1		1	1	0	y	
		−1				*−1*					*Decrement 1, 0*
		0		*1*		*−1*					*New wts.*
8	0		1		1		0	0	0	n	
		—		—		—					*No adjustment*
		0		*1*		*−1*					*Same wts.*
9	0		0		1		−1	0	0	n	
		—		—							*No adjustment*
		0		*1*		*−1*					*Same wts.*
10	1		1		1		0	0	1	y	
		+1		*+1*		*+1*					*Increment 1, 2, 0*
		1		*2*		*0*					*New wts.*
11	1		0		1		1	1	0	y	
		−1				*−1*					*Decrement 1, 0*
		0		*2*		*−1*					*New wts.*
12	0		1		1		1	1	0	y	
				−1		*−1*					*Decrement 2, 0*

Table 15.2 (*Continued*)

Trial no.	Input layer				Bias neurone 0		Linear combination	Output layer		Error in output	*Action*
	Neurone 1		Neurone 2					State of 3 actual	State of 3 desired		
	x_1	w_1	x_2	w_2	x_0	w_0	$\Sigma\ w_i x_i$				
		0		*1*		*−2*					*New wts.*
13	0		0		1		−2	0	0	n	
		—		—		—					*No adjustment*
		0		*1*		*−2*					*Same wts.*
14	1		1		1		−1	0	1	y	
		+1		*+1*		*+1*					*Increment 1, 2, 0*
		1		*2*		*−1*					*New wts.*
15	1		0		1		0	0	0	n	
											No adjustment
		1		*2*		*−1*					*Same wts.*
16	0		1		1		1	1	0	y	
				−1		*−1*					*Decrement 2, 0*
		1		*1*		*−2*					*New wts.*
17	0		0		1		−2	0	0	n	
		—		—		—					*No adjustment*
		1		*1*		*−2*					*Same wts.*
18	1		1		1		0	0	1	y	
		+1		*+1*		*+1*					*Increment 1, 2, 0*
		2		*2*		*−1*					*New wts.*
19	1		0		1		1	1	0	y	
		−1				*−1*					*Decrement 1, 0*
		1		*2*		*−2*					*New wts.*
20	0		1		1		0	0	0	n	
		—		—		—					*No adjustment*
		1		*2*		*−2*					*Same wts.*
21	0		0		1		−2	0	0	n	
		—		—		—					*No adjustment*
		1		*2*		*−2*					*Same wts.*
22	1		1		1		1	1	1	y	
		—		—		—					*No adjustment*
		1		*2*		*−2*					*Same wts.*
23	1		0		1		−1	0	0	y	
		—		—		—					*No adjustment*

The example in Table 15.2 trains the network from the state in which all the weights are zero to the fully trained state. The reader might initially only examine the first few and the last few training steps. All the steps function in the same way, but the whole sequence is presented here to show that it is possible to go from arbitrary weights to a set of weights which has the desired performance.

At the beginning of the training process, the network weights are all zero. The first input is $(x_1, x_2) = (0, 0)$ and the desired output is 0. $w_0x_0 + w_1x_1 + w_2x_2 = (0)(1) + (0)(0) + (0)(0) = 0$ and so the calculated output is 0 whereas the desired output is 0. Therefore there is no adjustment to the weights. At the second trial, $(x_1, x_2) = (1, 1)$ is presented, and the desired output for this input is 1. $w_0x_0 + w_1x_1 + w_2x_2 = (0)(1) + (0)(1) + (0)(1) = 0$ resulting in an output of 0. So there is an error. In this case the weights on the active lines are incremented by 1, making new weights (1, 1, 1). And so the process continues, until a point arrives when the weights do not change after all the presentations in the training dataset have been made. Then the training is complete. At the end, the weights are essentially the same as those used in the demonstration of the operation of the trained network in Section 15.2.

In a real case, we would not be able to present all possible inputs in a sequence like this, so in general input vectors and the corresponding desired output vectors would be sampled at random from the whole set. When the whole set is known, it is common to randomise the order of presentation of the set rather than repeatedly presenting them in a fixed order as above. When using a performance learning algorithm, learning is assumed to be complete after many vectors have been presented with no further learning occurring.

15.2.3 Limitations of the Perceptron

Although Rosenblatt's theories proved that the Perceptron could be trained to perform certain tasks (i.e. that his automatic training algorithms such as that given above would always converge to a network which could perform the task under consideration), it turned out that the range of tasks that the simple Perceptron could perform was rather limited. One task that the Perceptron cannot perform is the EXCLUSIVE OR (*XOR*) function, which we can illustrate by reference to Figure 15.5. The desired network is one in which the output neurone 3 fires if *either* 1 or 2 are on *but* not both. Since the standard training rules for this simple two neurone case correspond to moving the straight line $w_1x_1 + w_2x_2 - T = 0$ about the plane, we can easily see that it is not possible to separate the plane using one straight line into two regions corresponding to these two situations: (i) either 1 or 2 on, but not both: $(x_1, x_2) = (1, 0)$ or $(0, 1)$; (ii) otherwise: $(x_1, x_2) = (0, 0)$ or $(1, 1)$.

Many other similar functions are outside the reach of the Perceptron. The inability of the Perceptron to deal with this problem is a specific example of a more general restriction of the Perceptron: it cannot be trained to determine the parity of a two-dimensional black and white diagram (the number of distinct black regions into which the plane is split). More than one layer of adjustable weights—adding what are known as hidden layers—are necessary for these so called 'hard learning' problems, and this is how they are dealt with in current neural networks. One disadvantage of networks with hidden layers in the 1960s was that effective automatic methods of training them were not then available.

Partially as a result of a strong attack by Minsky and Papert (1969) embodying these criticisms, the development of neural networks, at least in the USA, again slowed down. It was not until the 1980s that the pace of development began to increase again, when improvements in computing power, especially using parallel processors, enormously increased the potential of the neural network technique. Further reasons

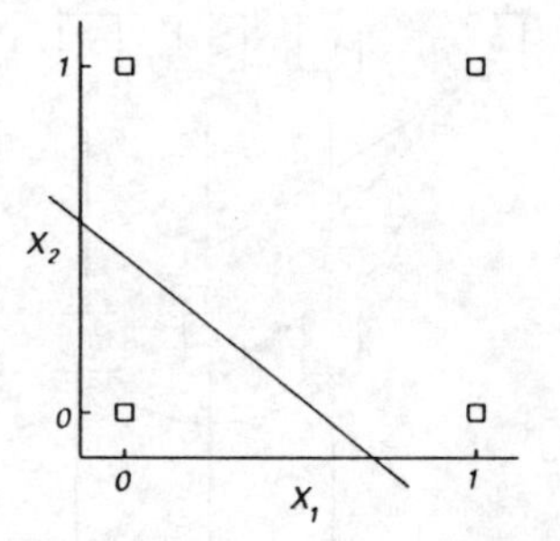

Figure 15.5 An illustration of the learning process in the simple two layer network discussed in Section 15.2.1. The state of the neurone 1 (either 0 or 1) is plotted along the x_1-axis, and of neurone 2 along the x_2-axis. The rule by which neurone 3 decides its output is given by the line across the diagram

$$w_1x_1 + w_2x_2 - T = 0,$$

$$\text{i.e.} \quad x_2 = -\frac{w_1}{w_2}x_1 + \frac{T}{w_2}.$$

Points on one side of this line satisfy

$$w_1x_1 + w_2x_2 - T > 0 \quad \text{i.e.} \quad \text{neurone 3 fires}$$

and those on the other side satisfy

$$w_1x_1 + w_2x_2 - T < 0 \quad \text{i.e.} \quad \text{neurone 3 does not fire.}$$

Thus this line divides the plane into two regions: that including points (i.e. states of the inputs 1 and 2) for which neurone 3 fires, and that including points for which neurone 3 does not fire. The process of training the network consists of moving this line about the plane so as to separate the plane into the two regions appropriately.

for the increased interest were some important discoveries. The first major discovery was the development by John Hopfield in 1982 to 1985 of what have come to be known as Hopfield networks with their interesting properties. A second which substantially increased the utility of neural networks was the discovery of backpropagation, a powerful automatic method of training networks with hidden layers.

15.3 THE HOPFIELD NETWORK

Stimulated by work on invertebrate neurophysiology, John Hopfield, in an important paper in 1982, made advances that turned out to be of theoretical and practical importance for both computing and neural modelling. He worked with a class of *auto-associative neural networks*. These are networks rather like the Perceptron but with the added feature that the neurones in the middle layer have outputs which in turn act as inputs to other neurones in that layer (Figure 15.6), as well as normal outputs to the final layer. An auto-associative network must therefore operate in a more complex way than the Perceptron discussed above. In the simple Perceptron, the neurones in the middle layer are set to fire or not, depending on the weighted sum of the inputs from the initial layer, and then immediately feed their output for further processing in the final layer. The neurones which fire in any one layer thus fire synchronously. For the

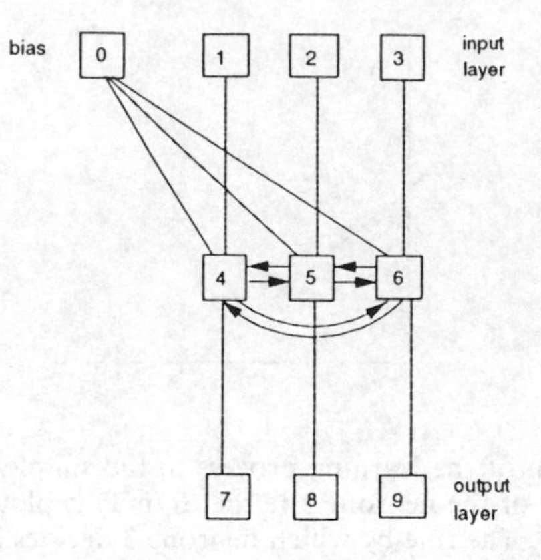

Figure 15.6 The Hopfield auto-associative network.

Hopfield network, the neurones in the input layer which fire do so simultaneously. They transfer this pattern of firing without change to the middle layer, and then become inactive. The middle layer then goes through a series of updates as described below until it settles down to a constant pattern of activity, and only then transfers its pattern of firing again without change to the output layer. The initial transfer between the input layer and the middle layer and the eventual output from the middle layer to the output layer occurs without any calculations involving weights and thresholds. In between the middle layer functions independently, and it is only here that calculations involving weights and thresholds take place. In the next few paragraphs, we consider an example with three input and three output neurones and three neurones in the middle layer.

In Figure 15.6, depending on what the input pattern is, some of the neurones 1 to 3 in the input layer initially fire and output their activity to neurones 4 to 6 in the middle layer. Immediately after this they become and subsequently remain silent. The firing pattern of the neurones in the middle layer at this time merely reflects the initial activation of the input layer. As we said above, the novel feature of an auto-associative network is that each of them is simultaneously feeding its output to the other neurones in the layer. In what time sequence do they perform and respond to this feedback? The rule that is generally used in practice is asynchronous, random updating. Each neurone establishes its initial state (firing or not depending on what is initially fed into it from the input layer) and continues in that state until its turn comes to update its state. The neurones take their turns in a random sequence. Each is updated according to its own Poisson process with the same rate on average for all three neurones. These neurones thus behave unlike real neurones which fire only for a few milliseconds: any neurone in the Hopfield network which starts to fire at a particular update continues to fire until it decides to change its state at a later update. The rule determining whether any neurone—when its turn comes to update—will fire or not is the same as for the Perceptron: if the weighted sum of the inputs (which consists only of the current outputs from other members of the layer) is greater than the threshold for the neurone, the neurone will fire; otherwise it is silent. Note that it is not dependent on its own state at the time.

Consider the example in Table 15.3. Neurones 1 and 3 in the input layer fire initially and then stop, whereas 2 does not. They feed these states into the neurones in the

Table 15.3 The time course of a typical individual run of a 3-neurone Hopfield network with the following synaptic weights:

From the bias neurone (0) to each neurone in the middle layer:

$$w_{40} = 1, \quad w_{50} = 0, \quad w_{60} = 0;$$

Between the neurones in the middle layer:

$$w_{45} = w_{54} = -2, \quad w_{46} = w_{64} = -4, \quad w_{56} = w_{65} = 3.$$

The neurone updated at each stage is decided by a random process in which each neurone of the middle layer has the same chance of being selected

Neurone firing patterns							Neurone updated	$\Sigma\, w_{ij}x_j$	New state	Energy
Bias	Input layer			Middle layer						
0	1	2	3	4	5	6				
1	1	0	1							
1				1	0	1				3
							4	−3	0	
1				0	0	1				0
							4	−3	0	
1				0	0	1				0
							6	0	0	
1				0	0	0				0
							5	0	0	
1				0	0	0				0
							4	1	1	
1				1	0	0				−1
							5	−2	0	
1				1	0	0				−1
							6	−4	0	
1				1	0	0				−1
							4	1	1	
1				1	0	0				−1

middle layer to which they are connected and then remain silent. At this stage the only neurones firing are 4 and 6. Then the random, asynchronous updating starts. Suppose that the order in which the neurones are updated is 4, 4, 6, 5, 4, 5, 6, 4. Neurone 4 is updated first, and as a result it stops firing (since $w_{45}x_5 + w_{46}x_6 + w_{40}x_0 = (-2)(0) + (-4)(1) + (1)(1) = -3$, which is less than zero). Neurone 4 is updated again and it remains in the same state. Neurone 6 is next, and taking into account the inputs at that time from 4 and 5, it stops firing; and so on as in Table 15.3.

Table 15.4 The behaviour of a 3-neurone Hopfield network. This network has three neurones in the middle layer, with synaptic weights as follows:

From the bias neurone (0) to each neurone in the middle layer: $w_{40} = 1$, $w_{50} = 0$, $w_{60} = 0$
Between the neurones in the middle layer: $w_{45} = -2$, $w_{46} = -4$, $w_{56} = 3$.

The table gives the following.

(1) All the possible states of the network (there are $2^3 = 8$ of them, since there are two possibilities—firing or not—for each of three neurones).
(2) All the possible transitions by considering the possibility that each neurone of the three will be the next to be updated, and what the result of that updating will be. Each such updating is equally likely, and as there are three of them, each has a probability of 1/3.
(3) The energy level of each state.

	Prob.	Original network state neurone			$w_{j0} + \Sigma\ w_{ji}x_i$ for updated neurone	New network state neurone			Energy
		4	5	6		4	5	6	
From network state 1		0	0	0					0
(a) update 4	1/3				1	1	0	0	−1
(b) update 5	1/3				0	0	0	0	0
(c) update 6	1/3				0	0	0	0	0
From network state 2		0	0	1					0
(a) update 4	1/3				−3	0	0	1	0
(b) update 5	1/3				3	0	1	1	−3
(c) update 6	1/3				0	0	0	0	0
From network state 3		0	1	0					0
(a) update 4	1/3				−1	0	1	0	0
(b) update 5	1/3				0	0	0	0	0
(c) update 6	1/3				3	0	1	1	−3
From network state 4		0	1	1					−3
(a) update 4	1/3				−5	0	1	1	−3
(b) update 5	1/3				3	0	1	1	−3
(c) update 6	1/3				3	0	1	1	−3
From network state 5		1	0	0					−1
(a) update 4	1/3				1	1	0	0	−1
(b) update 5	1/3				−2	1	0	0	−1
(c) update 6	1/3				−4	1	0	0	−1
From network state 6		1	0	1					3
(a) update 4	1/3				−3	0	0	1	0
(b) update 5	1/3				1	1	1	1	2
(c) update 6	1/3				−4	1	0	0	−1
From network state 7		1	1	0					1
(a) update 4	1/3				−1	0	1	0	0
(b) update 5	1/3				−2	1	0	0	−1
(c) update 6	1/3				−1	1	1	0	1
From network state 8		1	1	1					2
(a) update 4	1/3				−5	0	1	1	−3
(b) update 5	1/3				1	1	1	1	2
(c) update 6	1/3				−1	1	1	0	1

Thus the network eventually settles into a pattern of activity which does not change. At this time or after a fixed number of iterations depending on how the network is programmed, the middle layer feeds its output to the output layer.

Hopfield showed that an equilibrium pattern of firing will always occur eventually, provided that certain conditions hold. He demonstrated (i) that an energy function could be defined for each possible state of the network, and (ii) that the firing pattern of the network develops in such a way as to reduce the value of this energy function, eventually settling into stable configurations of firing which correspond to local minima of the energy function. The conditions are that the synaptic weights connecting the ith and jth neurones in the middle layer satisfy $w_{ij} = w_{ji}$ and $w_{ii} = 0$; i.e. the weights are symmetric, and there is no connection from a neurone to itself. The energy function (note that $T_i = -w_{io}$) he defined as

$$E = -\frac{1}{2}\sum_i \sum_{j \neq i} w_{ij}x_i x_j + \sum_i x_i T_i$$

which in this case is

$$-\tfrac{1}{2}(w_{45}x_4x_5 + w_{46}x_4x_6 + w_{54}x_5x_4 + w_{56}x_5x_6 + w_{64}x_6x_4 + w_{65}x_6x_5) + (x_4T_4 + x_5T_5 + x_6T_6).$$

Using the symmetry relations of the w_{ij}, this can also be written as

$$-(w_{45}x_4x_5 + w_{46}x_4x_6 + w_{56}x_5x_6) + (x_4T_4 + x_5T_5 + x_6T_6).$$

The last column in Table 15.3 demonstrates that this particular trial in this network conforms to the general rule, i.e. the energy does decline or at least stays the same as the iterations proceed.

It is possible to subject any network of this form to an analysis listing (a) the possible paths and (b) the probabilities associated with the possible next steps at each stage of every path. This is done for a Hopfield network with three neurones in the middle layer in Table 15.4. From this, all possible sequences of states of the network can be obtained. It will be observed that any changes of state of the network are to a lower or at least the same energy level. The energy level never rises. The sequences can also be plotted in a figure demonstrating this reduction in energy (see Figure 15.7).

There are two equilibria which, once entered, are never left: (1, 0, 0) and (0, 1, 1). The network if left long enough always ends up in one of these states. By tracing paths through Figure 15.7, the probabilities of ending up in state (1, 0, 0) starting from every state can be calculated. They are given in Table 15.5.

From almost all of the states the eventual equilibrium cannot be predicted with any certainty. An exception (apart from the equilibrium states themselves) is (0, 0, 0) which always moves on to (1, 0, 0). From states (1, 1, 0) and (1, 0, 1) there is a greater likelihood of eventual arrival at (1, 0, 0). Thus there are basins of attraction around each equilibrium. (0, 0, 0) and (1, 0, 0) are in the basin of attraction of equilibrium (1, 0, 0), and from states (1, 1, 0) and (1, 0, 1) the network is quite likely to move into this basin of attraction. From (1, 1, 1) on the other hand, there is a greater probability of eventual arrival at (0, 1, 1).

Larger networks potentially have a richer structure in the sense of having more equilibrium states. They also frequently have larger regions of the space of possible

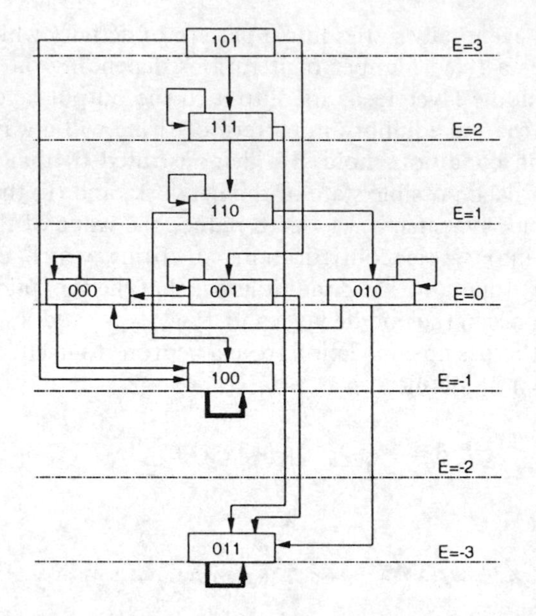

Figure 15.7 The possible network states and transitions between states for the middle layer of the Hopfield net defined in Table 15.4 (after Aleksander and Morton (1990)). The lines between the states give the transitions, the arrow indicating the direction of change. Line thickness is proportional to the probability that the corresponding change of network state will occur from the starting state. In all but three cases this probability is 1/3. That from state (0, 0, 0) to itself has a probability of 2/3 and those from (0, 1, 1) to (0, 1, 1) and (1, 0, 0) to (1, 0, 0) have probability 1. The energy level of each state is indicated by E in the right hand column.

Table 15.5 The probabilities of eventual equilibrium in state (1, 0, 0) starting from the states given

Initial state	1, 0, 0	0, 0, 0	1, 1, 0	1, 0, 1	0, 0, 1	0, 1, 0	1, 1, 1	0, 1, 1
Probability	1	1	3/4	15/24	1/2	1/2	3/8	0

network states which are basins of attraction of the equilibria. Once the network has entered one of the catchment areas, it always, or with high probability, eventually arrives at the corresponding equilibrium. The network thus has the property that certain firing patterns are very likely or certain eventually to evoke one of these equilibria. We have not discussed how these networks might learn, but they can be trained using various training algorithms, some of which are related to the Hebb coincidence learning algorithms.

There are two other reasons why the discovery of Hopfield nets generated a great deal of interest, one with implications in psychology and the other to do with computing. The first is that this could be a simple model explaining how associations occur between ideas; how a partial notion evokes a more complete one. The partial notions are the states in the basin of attraction of the equilibrium which acts as a label for the complete notion. These networks could also provide a simple model for the uncertainty of associations: some states evoke one equilibrium sometimes, and a second equilibrium at other times. To make the discussion more concrete, consider a simple perceptual assessment: the categorisation of a shape as rectangular, triangular or circular. The state corresponding to the notion 'the shape has no vertices' would belong in the basin of attraction of the 'circle' equilibrium state, whereas 'the shape has vertices' has a less certain outcome: either the 'triangle' or 'rectangle' equilibrium state. The Hopfield net could also be used in computer engineering for producing a content-addressable memory: a memory which is not indexed by some number but by part of the contents of the memory (see Exercises 15.4 and 15.5).

15.4 BACK-PROPAGATION

Another key advance of the 1980s was the discovery of an automatic way of training classes of network with hidden layers. Hidden layers are those which have variable synaptic weights which can be adjusted during the training process, but whose performance cannot be compared with a target performance because another layer with variable weights lies between them and the desired output. So far we have only dealt with simpler networks. In Section 15.2 we considered a two layer Perceptron. In Section 15.3 we discussed Hopfield networks with one middle layer of neurones which fed their output without further modification into the final output layer. The input and output layers here are only channels for the input and output and do not take part in the dynamics of the network. Both these networks can be trained by performance learning algorithms which compare the observed output of the layer with the desired output.

We saw in the discussion of the limitations of the Perceptron that other hidden layers are necessary for networks to perform certain tasks. There is a problem with performance learning for networks which have hidden layers, as we only know what the desired output of the output layer is, or the final layer before the output layer in cases where the output layer is merely acting as a passive transmitter. We cannot infer from this what the desired output for the earlier layers should be. Various workers had come close to finding a solution to this problem but it was only in 1986 that the back-propagation algorithm was perfected.

The method of training a back-propagation network is to start with some synaptic weights (almost always determined by a random process), and to apply a given input and desired output selected from the training dataset, usually at random. The input is first propagated forward until the eventual output is obtained which is compared with the desired output. This difference is used to adjust the weights in the final layer. Then the error is propagated back to the previous layer, and this error used to adjust the weights in this layer. This process is continued until the first layer is reached. Then a further input and desired output are presented and the process repeated. The method by which the error is back-propagated through the network was discovered by D.E. Rumelhart, G.E. Hinton and R.J. Williams in 1986 and several others at about the same time (Parker, 1985, and Le Cun, 1985). The training rule is known as the Generalised Delta Rule. The details of back-propagation methods are beyond the scope of this book and the interested reader is referred to Rumelhart *et al.* (1986).

This discovery was very important because networks with hidden layers are essential for many processing tasks, and it enabled such networks to be trained automatically. However, there are many practical problems with the training procedures. The error sum of squares calculated using a validation dataset and expressed as a function of all the variable weights can be used as a measure of network performance. This function in most applications has many local minima, and also rather flat regions, in weight space. Consequently it is often difficult to find the best performing set of weights, which would be situated at the global minimum in the weight space. Nevertheless, the method has been used with considerable success in some large networks, which have been able to perform in ways that are both interesting and useful.

15.5 OTHER WORK

This chapter does no more than briefly introduce a few important concepts and methods in neural networks, and in this section we list some more substantial treatments of the subject, and briefly set this work in the context of some of the more interesting known complexities of real brain function.

15.5.1 Connectionist models in context

Invertebrates have a much simpler nervous system than vertebrates, and a rather more limited repertoire of behaviour. The snail *Aplysia* has a nervous system consisting of tens of thousands of cells, compared with more than 10^9 in higher animals. Moreover the neurones are collected together into discrete groups called ganglia, and each ganglion usually contains of the order of a thousand neurones. The first finding which conflicted with the usual assumptions of the modellers was that neurones are by no means all the same. In the abdominal ganglion of *Aplysia*, neurones vary in many ways: size, shape, pigmentation, the neurotransmitters they utilise, and their firing patterns. Using these differences, investigators have been able to identify the same series of neurones in many different individuals, thus showing a partial invariance from individual to individual in the cells found in particular positions. A further interesting finding has been that the firing of a single identified cell seems to be the trigger responsible for initiating some responses, for example for speeding up the heart rate, or causing a flight response to some danger. Networks of these neurones in other invertebrates are involved in coordinated oscillations, which also relate to particular physiological functions (Selverston & Moulins, 1985). Studies have also shown different neural mechanisms involved in different kinds of learning—in habituation and sensitisation for example (Kandel, 1989; Sahley, 1990).

Some parts of the vertebrate brain also contain quite heterogeneous collections of cells serving a variety of functions. Communication in some areas of the brain is not so specific as neurone–neurone contact, nor is it so fast: some neuropeptides diffuse over larger areas on longer time scales, and appear to function more as broadcasters, possibly bringing groups of neurones into synchrony. A massive literature exists in this general area, and we here suggest two works which discuss mechanisms: Leng (1988) reviews pulsatile phenomena in neuroendocrine systems, and Coen (1985) provides a readable introduction to a wide area of brain function, both vertebrate and invertebrate. For general background, Kandel *et al.* (1991) impressively survey most areas of neural science.

15.5.2 Further reading on neural networks

From a biological point of view, the publication of the text *Parallel Distributed Processing: Explorations in the Microstructure of Cognition* edited and partly written by David Rumelhart and Jay McClelland (1986) was an important landmark. It contains a detailed discussion of possible models for many mental processes within a parallel distributed processing (PDP) framework, and on its publication clarified many previously obscure issues. The books by Hertz *et al.* (1991), Aleksander & Morton (1990) and Hecht-Nielsen (1990) contain good reviews of neural networks as computational techniques. Koch & Segev (1989) cover a number of aspects of neuronal modelling including single neurone models. More biologically and psychologically orientated works include the volume edited by Hanson & Olson (1990), and the books by Arbib (1989) and Grossberg (1982).

EXERCISES

15.1 For a neural network with three input neurones connected with synaptic weights w_1, w_2, w_3 to a single output neurone with threshold T, can values of the weights and the threshold be determined so that the network performs in the following ways:
(a) if one or more input neurones fire, the output neurone fires;
(b) if two or less input neurones fire, the output neurone fires;
(c) if three neurones are firing, the output neurone fires;
(d) if any two neurones fire, but the third does not, the output neurone fires.
In each case justify your answer.

15.2 What are Hebbian and Performance Learning?
Can Hebbian learning be utilised to train the network given in Tables 15.2 to perform the task given there?

15.3 Document the performance of a Hopfield network like that given in Figure 15.6, consisting of three neurones in the middle layer, but with weights as follows: bias weights $w_{40} = 0$, $w_{50} = 1$, $w_{60} = 0$; inter-neuronal weights $w_{45} = w_{54} = 1$, $w_{46} = w_{64} = 1$, $w_{56} = w_{65} = -2$.
(a) Draw up a table like that in Table 15.4 which lists the probability distribution of new network states reached in one updating from all possible initial network states.
(b) Using this table, draw up a flow diagram showing the possible network states, with associated energy levels, and transitions between states, indicating for each transition its probability.
(c) List the equilibrium network states and their basins of attraction.

15.4 *Use of a Hopfield network as a content-addressable memory.* Consider the network in Figure 15.7. This network has the potential for being a memory consisting of the two equilibrium states $Eq_1 = (x_4, x_5, x_6) = (1, 0, 0)$ and $Eq_2 = (x_4, x_5, x_6) = (0, 1, 1)$, which can be addressed by specifying the value of any one or more elements, since the two equilibrium states have no neuronal states in common. This is done in the network by fixing any neurone (i.e. clamping it) or pair of neurones at the specified value.
(a) Show that if neurone 4 is clamped at 1, the equilibrium eventually resulting would always be Eq_1; and similarly for clamping neurone 5 at 0 or neurone 6 at 0.
(b) Show that clamping neurone 5 or 6 to 1 results eventually in Eq_2.
(c) Show that clamping neurone 4 to 0 does not necessarily result in arrival at equilibrium Eq_2. What are the eventual equilibria in this clamped state, and how likely are they to occur?
(d) Interpret these results as demonstrations of the value and limitations of Hopfield networks as content-addressable memories.

15.5 *Simulated annealing and the Boltzmann machine*. Annealing is a process used by blacksmiths for tempering steel, in which a metal bar is repeatedly heated to a progression of temperatures—starting with very high temperatures and lowering them in steps—and alternately cooled in order to strengthen it. Simulated annealing is an iterative search technique for finding a minimum of a function which involves a random element at each stage (equivalent to the shaking up of the molecules involved in the blacksmith's heating) so as to prevent the system falling into a local minimum and staying there. At the start of the process, the random element perturbing the direction of optimum search is high and the temperature is then correspondingly said to be high. As the search proceeds, the temperature is lowered, the magnitude of the random element is reduced, until eventually a minimum is reached, and the temperature is so low that no further random element occurs in the search procedure. Then the system remains in this equilibrium. This can be applied to the Hopfield network to correct for the occurrence of false local minima which were demonstrated in Exercise 15.4.

In the Hopfield network as given in the previous exercise, once a neurone has been selected for updating, the outcome is determined by the value of what is known as the *activation function* which equals $A_j = w_{j0} + \Sigma_i w_{ij}x_i$ for the jth neurone. If $A_j > 0$, the new state of the jth neurone $= x_j^*$ is 1, otherwise it is zero.

(a) The simulated annealing procedure replaces this decision by a probabilistic outcome involving the logistic function, where

$$P(x_j^* = 1) = \frac{1}{1 + e^{-A_j/T}}.$$

Show that the maximum slope of this logistic function is dependent on T, and that, as T gets smaller and smaller, the slope gets steeper; until, as $T \to 0$, the function approaches a step function, and hence approaches the deterministic decision rule of the unmodified Hopfield network.

(b) Discuss how such a rule could overcome the problems of false minima, encountered in the solution to Exercise 15.4. (A variant of this technique is used in the Boltzmann machine, devised by Hinton and Sejnowksi (1986).)

Part V

GLOSSARY, TABLES, REFERENCES AND INDEX

Mathematical Glossary

A1 ELEMENTARY ALGEBRA AND ARITHMETIC

Modulus or absolute value of a real number x is the magnitude of x regardless of sign and is denoted by $|x|$.
Factorial of a whole number n is the product of all the integers $1, 2, \ldots, n$ and is denoted by $n!$

Examples

$|-5| = |5| = 5$

$5! = 5 \times 4 \times 3 \times 2 \times 1 = 120$

$n! = n \times (n-1) \times (n-2) \times \cdots \times 1$

$0! = 1$ (by definition)

Rules for indices

$x^a . x^b = x^{a+b}$ — $2^2 \times 2^3 = 4 \times 8 = 32 = 2^5 = 2^{2+3}$

$x^a / x^b = x^{a-b}$ — $3^3/3^2 = 27/9 = 3 = 3^{3-2}$

$(x^a)^b = x^{ab}$ — $(2^3)^2 = 8^2 = 64 = 2^6 = 2^{2\times 3}$

$x^0 = 1, \ x \neq 0$

$x^{-a} = 1/x^a$

Logarithms

If $x = a^y$ then $y = \log_a x$.
y is the logarithm to base a of x.

Base 10

If $x = 10^y$ then $y = \log_{10} x$.

$x = 100 = 10^2$

i.e. logarithm to base 10 of 100 is 2 or $\log_{10} 100 = 2$.

$\log_{10} e = 0.4343$

Base e

Logs to base e = 2.71828 are called natural or Naperian logarithms, written

$\log_e x$ or $\ln x$.

$\log_e 10 = 2.3026.$

Rules of logarithms

$\log_a(xy) = \log_a x + \log_a y$ — $\log_{10}(2 \times 3) = \log_{10} 2 + \log_{10} 3$

$\log_a(x/y) = \log_a x - \log_a y$ — $\log_e(3/4) = \log_e 3 - \log_e 4$

$\log_a x^k = k \log_a x$ — $\log_{10} 2^3 = 3 \log_{10} 2$

These rules follow from the rules for indices.

Change of base

$\log_b x = \log_b a \times \log_a x$

Examples

$\log_e 100 = \log_e 10 \times \log_{10} 100$

$= 2.3025 \times 2 = 4.605$

Progressions and series

Arithmetic progression (AP)

$a, a+d, a+2d, \ldots, a+(n-1)d$

Sum of first n terms

$$S_n = a + (a+d) + (a+2d) + \cdots + \{a+(n-1)d\} = \frac{n}{2}\{2a+(n-1)d\}.$$

$1, 3, 5, 7$

$a = 1, d = 2, n = 4.$

$S_4 = 1 + (1 + 1 \times 2) + (1 + 2 \times 2) + (1 + 3 \times 2) = 16.$

$S_4 = \frac{4}{2}\{2 \times 1 + 3 \times 2\} = 2 \times 8 = 16.$

Sum of first n natural numbers

$$S_n = 1 + 2 + 3 + \cdots + n = \frac{n}{2}\{2 + (n-1)\} = \frac{n(n+1)}{2}.$$

Geometric progression (GP)

$a, ar, ar^2, \ldots, ar^{n-1}$

$1, 1/2, 1/2^2, 1/2^3$

$(a = 1, r = 1/2, n = 4)$

Sum of first n terms

$$S_n = a + ar + ar^2 + \cdots + ar^{n-1} = \frac{a(1-r^n)}{(1-r)}.$$

For $|r| < 1$, sum to infinity

$$S_\infty = \frac{a}{(1-r)}$$

$S_4 = 1 + \frac{1}{2} + (\frac{1}{2})^2 + (\frac{1}{2})^3 = 1.875$

$$S_4 = \frac{(1-(1/2)^4)}{(1-1/2)} = 1.875.$$

$$S_\infty = \frac{1}{1-0.5} = 2.$$

A2 TRIGONOMETRIC FORMULAE

In any right-angle triangle,

$$\sin\theta = \frac{\text{Opposite}}{\text{Hypotenuse}}$$

$$\cos\theta = \frac{\text{Adjacent}}{\text{Hypotenuse}}$$

$$\tan\theta = \frac{\text{Opposite}}{\text{Adjacent}} = \frac{\sin\theta}{\cos\theta}.$$

$\operatorname{cosec}\theta = 1/\sin\theta$

$\sec\theta = 1/\cos\theta$

$\cot\theta = 1/\tan\theta$

	Angle				
Degrees	$0°$	$30°$	$45°$	$60°$	$90°$
Radians	0	$\pi/6$	$\pi/4$	$\pi/3$	$\pi/2$
sin	0	$1/2$	$1/\sqrt{2}$	$\sqrt{3}/2$	1
cos	1	$\sqrt{3}/2$	$1/\sqrt{2}$	$1/2$	0
tan	0	$1/\sqrt{3}$	1	$\sqrt{3}$	∞

$\sin(-\theta) = -\sin(\theta)$

$\cos(-\theta) = \cos(\theta)$

$\sin(A + B) = \sin A \cos B + \cos A \sin B$

$\cos(A + B) = \cos A \cos B - \sin A \sin B$

Examples

$\sin 2A = 2 \sin A \cos A$

$$\cos 2A = \cos^2 A - \sin^2 A = 2\cos^2 A - 1 = 1 - 2\sin^2 A$$

A3 COMPLEX NUMBERS

$$z = x + iy \text{ where}$$

x = real part(z), y = imaginary part(z) provided x and y are both real.

$$z = r\,e^{i\theta} = r(\cos\theta + i\sin\theta)$$

where r is called the *modulus* of z, and θ the *argument* of z (for r, θ both real).

$z = 3 + 4i$

real part(z) = 3, imaginary part(z) = 4. Then

$z = 5\,e^{i\theta}$ where $\theta = \tan^{-1}(4/3)$

De Moivre's theorem

$(\cos\theta + i\sin\theta)^k = \cos k\theta + i\sin k\theta$

A4 TABLE OF STANDARD DERIVATIVES

$y = f(x)$	$\frac{dy}{dx}$
x^n	nx^{n-1}
$\sqrt{x}$	$\frac{1}{2\sqrt{x}}$
$\frac{1}{x}$	$-\frac{1}{x^2}$
e^{ax}	$a\,e^{ax}$
$\log_e x$	$\frac{1}{x}$
a^x	$a^x \log_e a$

$y = f(x)$	$\frac{dy}{dx}$
$\sin x$	$\cos x$
$\cos x$	$-\sin x$
$\tan x$	$\sec^2 x$
$\sin^{-1}(x/a)$	$\frac{1}{\sqrt{a^2 - x^2}}$
$\cos^{-1}(x/a)$	$-\frac{1}{\sqrt{a^2 - x^2}}$
$\tan^{-1}(x/a)$	$\frac{a}{(a^2 + x^2)}$

Differentiation formulae

Product

$$\frac{\mathrm{d}}{\mathrm{d}x}\{f(x)g(x)\} = f\frac{\mathrm{d}g}{\mathrm{d}x} + g\frac{\mathrm{d}f}{\mathrm{d}x} = fg' + gf'$$

Examples

$f(x) = x;\ g(x) = \mathrm{e}^{ax}$

$$\frac{\mathrm{d}}{\mathrm{d}x} x\,\mathrm{e}^{ax} = \mathrm{e}^{ax} + ax\,\mathrm{e}^{ax}$$

Ratio

$$\frac{\mathrm{d}}{\mathrm{d}x}\left\{\frac{f(x)}{g(x)}\right\} = \frac{\frac{\mathrm{d}f}{\mathrm{d}x}g - f\frac{\mathrm{d}g}{\mathrm{d}x}}{g^2} = \frac{f'g - fg'}{g^2}$$

$f(x) = ax;\ g(x) = b + x$

$$\frac{\mathrm{d}}{\mathrm{d}x}\left\{\frac{ax}{b+x}\right\} = \frac{a(b+x) - ax}{(b+x)^2} = \frac{ab}{(b+x)^2}$$

Chain rule

$$\frac{\mathrm{d}}{\mathrm{d}x}\{f(g(x))\} = \frac{\mathrm{d}f}{\mathrm{d}g}\frac{\mathrm{d}g}{\mathrm{d}x}$$

$f(g) = \mathrm{e}^{g};\ g(x) = -x^2.$

$$\frac{\mathrm{d}f}{\mathrm{d}g} = \mathrm{e}^{g};\ \frac{\mathrm{d}g}{\mathrm{d}x} = -2x$$

$$\frac{\mathrm{d}f}{\mathrm{d}x} = \frac{\mathrm{d}f}{\mathrm{d}g}\frac{\mathrm{d}g}{\mathrm{d}x} = \mathrm{e}^{g}(-2x) = -2x\,\mathrm{e}^{-x^2}$$

Inverse functions

$$\frac{\mathrm{d}y}{\mathrm{d}x} = \frac{1}{\mathrm{d}x/\mathrm{d}y}$$

$y = \sin^{-1} x,\ -1 \leqslant x \leqslant 1.$

so that $x = \sin y$ and $\frac{\mathrm{d}x}{\mathrm{d}y} = \cos y$

Hence

$$\frac{\mathrm{d}y}{\mathrm{d}x} = \frac{1}{\mathrm{d}x/\mathrm{d}y} = \frac{1}{\cos y} = \frac{1}{\sqrt{1-x^2}}$$

A5 TAYLOR SERIES

In a Taylor series the value of a function $f(x)$ is expressed in terms of the value of the function and its derivatives at some origin $x = x_0$ as follows

$$f(x) = f(x_0) + (x - x_0)\frac{\mathrm{d}f}{\mathrm{d}x} + \frac{(x-x_0)^2}{2!}\frac{\mathrm{d}^2 f}{\mathrm{d}x^2} + \frac{(x-x_0)^3}{3!}\frac{\mathrm{d}^3 f}{\mathrm{d}x^3} + \cdots + \frac{(x-x_0)^n}{n!}\frac{\mathrm{d}^n f}{\mathrm{d}x^n} + \cdots$$

where the derivatives are evaluated at $x = x_0$.
An important application is the *approximation* of a function by using the first two terms of the series, i.e.

$$f(x) \sim f(x_0) + (x - x_0)\left.\frac{\mathrm{d}f}{\mathrm{d}x}\right|_{x=x_0}$$

Maclaurin series (special case when $x_0 = 0$)

$$f(x) = f(0) + xf'(0) + \frac{x^2}{2!} f''(0) + \frac{x^3}{3!} f^3(0) + \cdots + \frac{x^n}{n!} f^n(0) + \cdots$$

Some useful series expansions

Exponential

$$e^x = 1 + x + \frac{x^2}{2!} + \cdots + \frac{x^n}{n!} + \cdots$$

Sine

$$\sin x = x - \frac{x^3}{3!} + \frac{x^5}{5!} + \cdots + (-1)^n \frac{x^{2n+1}}{(2n+1)!} + \cdots$$

Cosine

$$\cos x = 1 - \frac{x^2}{2!} + \frac{x^4}{4!} + \cdots + (-1)^n \frac{x^{2n}}{(2n!)} + \cdots$$

Logarithmic

$$\log_e(1 + x) = x - \frac{x^2}{2} + \frac{x^3}{3} - \cdots + (-1)^{n+1} \frac{x^n}{n} + \cdots \qquad (-1 < x \leqslant 1)$$

A6 SOLUTION OF EQUATIONS—NEWTON'S METHOD

To find a solution of the equation

$$f(x) = 0.$$

Start with an approximate solution x_0.
Find a new approximate solution by expanding function as first two terms of a Taylor series expansion about x_0:

$$f(x_1) \sim f(x_0) + (x_1 - x_0)f'(x_0)$$

so that

$$x_1 = x_0 - f(x_0)/f'(x_0).$$

Repeat the procedure applying the general formula

$$x_{n+1} = x_n - f(x_n)/f'(x_n)$$

until successive values are the same to the required accuracy.

For a simple epidemic model the proportion of susceptibles which become infected satisfies the equation

$$x = 1 - e^{-R_0 x}$$

where R_0 is the reproductive rate.
Find x when $R_0 = 1.25$.
Write as

$$f(x) = 1 - x - e^{-R_0 x}$$

$$f'(x) = -1 + R_0 e^{-R_0 x}.$$

Start with $x_0 = 0.5$, and results are as in table below.

n	x_n	$f(x_n)$	$f'(x_n)$	x_{n+1}
0	0.5000	−0.0353	−0.3309	0.3933
1	0.3933	−0.0049	−0.2355	0.3725
2	0.3725	−0.0002	−0.2153	0.3716
3	0.3716	−0.00005	−0.2144	0.3714
4	0.3714	−0.000006	−0.2142	0.3714

A7 TABLE OF STANDARD INTEGRALS

$y = f(x)$	$\int f(x)\,dx$
a	$ax + C$
x^a $(a \neq -1)$	$\frac{x^{a+1}}{(a+1)} + C$
$\sqrt{x}$	$\frac{2x^{3/2}}{3} + C$
$\frac{1}{x}$	$\log_e \|x\| + C$
$\frac{1}{(a+x)}$	$\log_e \|(a+x\| + C$
e^{ax}	$\frac{e^{ax}}{a} + C$

$y = f(x)$	$\int f(x)\,dx$
$\sin x$	$-\cos x + C$
$\cos x$	$\sin x + C$
$\tan x$	$-\log_e \|\cos x\| + C$
$\frac{1}{x^2 + a^2}$	$\frac{1}{a}\tan^{-1}\left(\frac{x}{a}\right) + C$
$\frac{1}{\sqrt{a^2 - x^2}}$	$\sin^{-1}\left(\frac{x}{a}\right) + C$
$\log_e x$	$x \log_e \|x\| - x + C$

Integration techniques

Integration by change of variable

$$x = g(z)$$

$$\int f(x) = \int f(g(z)) \frac{dx}{dz}\,dz.$$

Examples

$$\int x\,e^{-x^2}\,dx$$

Let $x = \sqrt{z}$ so that $\frac{dx}{dz} = \frac{1}{2\sqrt{z}}$.

This gives

$$\int x\,e^{-x^2}\,dx = \int \sqrt{z}\,e^{-z}\frac{1}{2\sqrt{z}}\,dz = \int \tfrac{1}{2}e^{-z}\,dz$$

$$= -\tfrac{1}{2}e^{-z} = -\tfrac{1}{2}e^{-x^2}.$$

Integration by parts

$$\int f(x)\frac{dg}{dx}\,dx = f(x)g(x) - \int g(x)\frac{df}{dx}\,dx$$

$$\int x\,e^{-x}\,dx = -x\,e^{-x} + \int e^{-x}\,dx$$

$$= -x\,e^{-x} - e^{-x} + C.$$

A8 PARTIAL DIFFERENTIATION

Examples

Partial derivative of f with respect to x, keeping y constant

$$\frac{\partial f}{\partial x} = \lim_{\delta x \to 0} \frac{f(x+\delta x, y) - f(x, y)}{\delta x}$$

$\partial f/\partial y$ is the partial derivative of f with respect to y, keeping x constant.

If $R = f(t, x) = 10 - 3t + 4x$,

(the rate at which R changes if t is increased but x is kept constant) $= \dfrac{\partial R}{\partial t} = -3$

(the rate at which R changes if x is increased but t is kept constant) $= \dfrac{\partial R}{\partial x} = 4$

Chain rule for changing variables

(a) If $f(x, y)$ is a function of x and y which are each both functions of another pair of variables, u and v, written $x = x(u, v)$, $y = y(u, v)$. Then

$$\frac{\partial f}{\partial u} = \frac{\partial f}{\partial x}\frac{\partial x}{\partial u} + \frac{\partial f}{\partial y}\frac{\partial y}{\partial u}$$

$$\frac{\partial f}{\partial v} = \frac{\partial f}{\partial x}\frac{\partial x}{\partial v} + \frac{\partial f}{\partial y}\frac{\partial y}{\partial v}$$

(Note: variables involved at each stage are

$$f \leftarrow \begin{Bmatrix} x \\ y \end{Bmatrix} \leftarrow \begin{Bmatrix} u \\ v \end{Bmatrix}.)$$

$f(x, y) = \dfrac{3x}{y}$ $\quad x = 2u + v \quad y = u - v$

$\dfrac{\partial f}{\partial x} = \dfrac{3}{y} \quad \dfrac{\partial x}{\partial u} = 2 \quad \dfrac{\partial y}{\partial u} = 1$

$\dfrac{\partial f}{\partial y} = \dfrac{-3x}{y^2} \quad \dfrac{\partial x}{\partial v} = 1 \quad \dfrac{\partial y}{\partial v} = -1$

$$\frac{\partial f}{\partial u} = \frac{\partial f}{\partial x}\frac{\partial x}{\partial u} + \frac{\partial f}{\partial y}\frac{\partial y}{\partial u}$$

$$= \left(\frac{3}{y}\right)2 + \left(\frac{-3x}{y^2}\right)(1) = \frac{6y - 3x}{y^2}$$

$$= \frac{6(u-v) - 3(2u+v)}{(u-v)^2} = \frac{-9v}{(u-v)^2}$$

(b) If $f(x, y)$ is a function of x and y which are each both functions of the single variable, t, written $x = x(t)$, $y = y(t)$.
Then

$$\frac{df}{dt} = \frac{\partial f}{\partial x}\frac{dx}{dt} + \frac{\partial f}{\partial y}\frac{dy}{dt}$$

(Note as there is only one variable involved at the second stage,

$$f \leftarrow \begin{Bmatrix} x \\ y \end{Bmatrix} \leftarrow \{t\}$$

there is no need for partial derivatives, ordinary derivatives will do at this stage.)

$f(x, y) = \dfrac{3x}{y} \quad x = \sin t \quad y = \cos t$

$\dfrac{\partial f}{\partial x} = \dfrac{3}{y} \quad \dfrac{dx}{dt} = \cos t \quad \dfrac{dy}{dt} = -\sin t$

$\dfrac{\partial f}{\partial y} = \dfrac{-3x}{y^2}$

$$\frac{df}{dt} = \frac{\partial f}{\partial x}\frac{dx}{dt} + \frac{\partial f}{\partial y}\frac{dy}{dt}$$

$$= \left(\frac{3}{y}\right)\cos t + \left(\frac{-3x}{y^2}\right)(-\sin t)$$

$$= 3 + 3\tan^2 t = 3\sec^2 t.$$

Second partial derivatives

Since $\partial f/\partial x, \partial f/\partial y$ are themselves functions of x and y, they too can be differentiated with respect to x and y:

$$\frac{\partial^2 f}{\partial x^2}=\frac{\partial}{\partial x}\left(\frac{\partial f}{\partial x}\right),\quad \frac{\partial^2 f}{\partial y \partial x}=\frac{\partial}{\partial y}\left(\frac{\partial f}{\partial x}\right),$$

$$\frac{\partial^2 f}{\partial x \partial y}=\frac{\partial}{\partial x}\left(\frac{\partial f}{\partial y}\right),\quad \frac{\partial^2 f}{\partial y^2}=\frac{\partial}{\partial y}\left(\frac{\partial f}{\partial y}\right)$$

Note that

$$\frac{\partial^2 f}{\partial y \partial x}=\frac{\partial^2 f}{\partial x \partial y}.$$

Examples

$$f(x, y)=\frac{3x}{y}$$

$$\frac{\partial f}{\partial x}=\frac{3}{y} \qquad \frac{\partial f}{\partial y}=\frac{-3x}{y^2}$$

$$\frac{\partial^2 f}{\partial x^2}=0 \qquad \frac{\partial^2 f}{\partial x \partial y}=-\frac{3}{y^2} \qquad \frac{\partial^2 f}{\partial y^2}=\frac{6x}{y^3}$$

Maxima, minima and other turning points of functions of two variables

Define

$$c_{10}=\frac{\partial f}{\partial x} \qquad c_{01}=\frac{\partial f}{\partial y}$$

$$c_{11}=\frac{\partial^2 f}{\partial x \partial y}$$

$$c_{20}=\frac{\partial^2 f}{\partial x^2} \qquad c_{02}=\frac{\partial^2 f}{\partial y^2}.$$

A point (x_0, y_0) is a *stationary point* of $f(x, y)$ if

$$\frac{\partial f}{\partial x}=\frac{\partial f}{\partial y}=0 \text{ at } (x_0, y_0).$$

It is

- a *local maximum* if $c_{20}<0$ and $c_{20}c_{02}-c_{11}^2>0$
- a *local minimum* if $c_{20}>0$ and $c_{20}c_{02}-c_{11}^2>0$
- a *saddle point* if $c_{20}c_{02}-c_{11}^2<0$
- *undetermined* if $c_{20}c_{02}-c_{11}^2=0$.

The nature of the turning point at (0, 0) of the function

$$f(x, y)=x^3+3xy^2-3x^2-3y^2$$

can be determined as follows.

$$\frac{\partial f}{\partial x}=c_{10}=3x^2+3y^2-6x$$

$$\frac{\partial f}{\partial y}=c_{01}=6xy-6y=6y(x-1)$$

$$c_{20}=6x-6=6(x-1),$$

$$c_{02}=6x-6=6(x-1), \quad c_{11}=6y$$

At $(0,0)$, $c_{20}=-6$, $c_{02}=-6$, $c_{11}=0$, $c_{20}c_{02}-c_{11}^2=36$, and hence $(0, 0)$ is a maximum.

A9 DIFFERENTIAL EQUATIONS

First order differential equations

Variables separable

Form

$$\frac{\mathrm{d}y}{\mathrm{d}x}=f(x)g(y)$$

$$(2+3x)\frac{\mathrm{d}y}{\mathrm{d}x}=\mathrm{e}^y$$

Solution

(1) Separate variables:

$$\frac{dy}{g(y)} = f(x)\,dx$$

$$e^{-y}\,dy = \frac{dx}{2+3x}$$

(2) Integrate both sides separately

$$\int \frac{dy}{g(y)} = \int f(x)\,dx + C$$

$$\int e^{-y}\,dy = \int \frac{dx}{2+3x}$$

$$-e^{-y} = \tfrac{1}{3}\log(2+3x) + C$$

Linear differential equations

Form

$$a(x)\frac{dy}{dx} + b(x)y = c(x)$$

$$x\frac{dy}{dx} + 2x^2y = 5x^2$$

Solution

(1) Rewrite in form

$$\frac{dy}{dx} + P(x)y = Q(x)$$

(by dividing throughout by $a(x)$).

$$\frac{dy}{dx} + 2xy = 5x$$

(2) Multiply by *integrating factor*

$$\mu(x) = \exp\left[\int P(x)\,dx\right]$$

$$\mu(x) = \exp\left[\int P(x)\,dx\right] = \exp\left[\int 2x\,dx\right] = \exp(x^2).$$

to obtain

$$\mu(x)\frac{dy}{dx} + \mu(x)P(x)y = \mu(x)Q(x)$$

$$e^{x^2}\frac{dy}{dx} + 2x\,e^{x^2}y = 5x\,e^{x^2}$$

(3) Rewrite left hand side as

$$\frac{d}{dx}[\mu(x)y].$$

$$\frac{d}{dx}[y\,e^{x^2}] = \frac{5}{2}\frac{d}{dx}(e^{x^2})$$

(4) Solution then is

$$y = \frac{\int \mu(x)Q(x)\,dx + C}{\mu(x)}.$$

$$y\,e^{x^2} = \tfrac{5}{2}e^{x^2} + C$$

or $y = \tfrac{5}{2} + C\,e^{-x^2}$

A10 DIFFERENCE EQUATIONS

In general, a difference equation relates the value of x_{n+k} to values of $n, x_n, x_{n+1}, \ldots, x_{n+k-1}$, where k is the *order* of the difference equation.
A difference equation is *linear* if the response variable appears only in a linear combination, otherwise it is *non-linear*.

Linear first order homogeneous

$x_{n+1} = 2x_n.$

Linear second order homogeneous

$x_{n+2} = x_{n+1} + x_n.$

Non-linear first order homogeneous

$x_{n+1} = 3x_n(1 - x_n).$

A difference equation is *homogeneous* if the equation is satisfied when the response variable is set to zero, otherwise it is *non-homogeneous*.

Examples

Linear first order non-homogeneous

$$x_{n+1} = 2x_n + 1/n.$$

Linear first order difference equations

General form

$$x_{n+1} = a(n)x_n + b(n)$$

where $a(n)$ and $b(n)$ are *known* functions of n.

General solution homogeneous case

$$x_{n+1} = a(n)x_n.$$

Solve by successive substitution:

$$x_1 = a(0)x_0,$$
$$x_2 = a(1)x_1 = a(1)a(0)x_0,$$
$$x_3 = a(2)x_2 = a(2)a(1)a(0)x_0,$$

etc.
In general,

$$x_n = a(n-1)a(n-2)\ldots a(1)a(0)x_0.$$

Constant coefficient—important special case in which $x_{n+1} = ax_n$ where a is constant. The solution is

$$x_n = a^n x_0.$$

Linear second order difference equations with constant coefficients

General form

$$x_{n+2} = a_1 x_{n+1} + a_0 x_n + b(n)$$

where a_0 and a_1 are constants and $b(n)$ is a known function of n.

General solution homogeneous case

General form

$$x_{n+2} = a_1 x_{n+1} + a_0 x_n.$$

Fibonacci sequence

$$x_{n+2} = x_{n+1} + x_n$$

Principle of solution

Look for solutions of the form $x_n = A\lambda^n$ where A and λ are constants.
Substitution gives

$$A\lambda^{n+2} = a_1 A\lambda^{n+1} + a_0 A\lambda^n.$$

Rearranging and dividing by $A\lambda^n$ shows that λ satisfies the *auxiliary equation*

$$\lambda^2 - a_1\lambda - a_0 = 0,$$

i.e.

$$\lambda_1, \lambda_2 = \frac{a_1 \pm \sqrt{a_1^2 + 4a_0}}{2}.$$

For Fibonacci sequence

$$\lambda^2 - \lambda - 1 = 0.$$

i.e.

$$\lambda_1, \lambda_2 = \frac{1 \pm \sqrt{5}}{2}.$$

Cases

(i) Roots real and different ($\lambda_1 \neq \lambda_2$)

$$x_n = A_1\lambda_1^n + A_2\lambda_2^n$$

where A_1 and A_2 are constants determined by initial conditions.

Examples

Fibonacci sequence

$$x_n = A_1\left(\frac{1+\sqrt{5}}{2}\right)^n + A_2\left(\frac{1-\sqrt{5}}{2}\right)^n,$$

Initial conditions $x_0 = 0$ and $x_1 = 1$ gives

$$A_1 + A_2 = 0$$

$$A_1 + A_2 + \sqrt{5}(A_1 - A_2) = 2.$$

i.e.

$$A_1 = -A_2 = \frac{1}{\sqrt{5}}.$$

(ii) Roots real and equal ($\lambda_1 = \lambda_2 = \lambda = a_1/2$)

$$x_n = (A_1 + nA_2)\lambda^n.$$

For this special case $x_n = An\lambda^n$ is also a solution of the equations. Check by substitution.

$$x_{n+2} - a_1x_{n+1} - a_0x_n = 0$$

gives

$$(n+2)\lambda^{n+2} - a_1(n+1)\lambda^{n+1} - a_0n\lambda^n = 0$$

so that $(n+2)\lambda^2 - a_1(n+1)\lambda - a_0n = 0$ and $n(\lambda^2 - a_1\lambda - a_0) + 2\lambda^2 - a_1\lambda = 0$, i.e. $\lambda = a_1/2$.

(iii) Complex roots ($\lambda_1 = r\,e^{i\theta}, \lambda_2 = r\,e^{-i\theta}$)

$$x_n = r(A_1\,e^{in\theta} + A_2\,e^{-in\theta}).$$

For a *real* solution $x_0 = r(A_1 + A_2)$ is real so that A_1 is the complex conjugate of A_2. In polar form $A_1 = A\,e^{i\beta}$ and $A_2 = A\,e^{-i\beta}$ with solution

$$x_n = Ar^n\{e^{i(n\theta+\beta)} + e^{-i(n\theta+\beta)}\}$$

$$\text{or} \quad x_n = Cr^n\cos(n\theta + \beta)$$

where C and β are constants determined by initial conditions.

Example based on analysis of grouse bag data, Chapter 13.

$$x_{n+2} = x_{n+1} - 0.50x_n.$$

Auxiliary equation is $\lambda^2 - \lambda + 0.50 = 0$ with roots

$$\lambda_1, \lambda_2 = \frac{1 \pm \sqrt{1-2}}{2} = \frac{1 \pm i}{2}.$$

In polar form

$$r = \sqrt{2}/2 = 0.707, \quad \theta = \tan^{-1} 1 = \pi/4.$$

Solution

$$x_n = C(0.707)^n\cos(n\pi/4 + \beta),$$

i.e. a damped cosine wave with period 8 time steps.

Equilibrium and local stability

Equilibrium value, or *stationary value* occurs when

$$x_{n+1} = x_n = x_e, \quad \text{i.e.} \quad x_e = f(x_e).$$

Local stability analysis: consider small displacement Δ_n from equilibrium and write

$$x_e + \Delta_{n+1} = f(x_e + \Delta_n) \approx f(x_e) + \Delta_n \left.\frac{df}{dx}\right|_{x_e}$$

approximately

$$\Delta_{n+1} = k\Delta_n \text{ and } \Delta_n = k^n\Delta_0.$$

Quadratic map

$$x_{n+1} = Rx_n(1 - x_n) \qquad 0 \leqslant R \leqslant 4$$

$$x_e = Rx_e(1 - x_e)$$

i.e. $x_e = (R-1)/R$

$$f(x_n) = Rx_n(1 - x_n)$$

$$k = f'(x_e) = R(1 - x_e) - Rx_e = 2 - R.$$

where $k = f'(x_e)$, slope of function at equilibrium point. This gives:

Examples

$k < -1$	*Unstable* divergent oscillation	$3 < R < 4$
$-1 < k < 0$	*Stable* convergent oscillation	$2 < R < 3$
$0 < k < 1$	*Stable* convergent exponential	$1 < R < 2$
$1 < k$	*Unstable* divergent exponential	$0 < R < 1$

i.e. stable when $|k| < 1$. — $1 < R < 3$

Global dynamic behaviour

For unstable equilibria, deterministic difference equations display fascinating dynamical global behaviour ranging from 2-point, 4-point, 8-point, ..., 2^n-point limit cycles and eventually chaos.

Quadratic map
$3 < R < 3.449$—stable limit cycles period 2;
$3.449 < R < 3.57$—stable limit cycles period $4, 8, \ldots, 2^n$;

$3.570 < R < 4$—chaos.

Chaos is characterised by *sensitivity to the initial conditions*—close initial values become widely separated after relatively few iterations of the model.

See Chapter 1.

A11 MATRICES

Definition and notation

A matrix is an array of symbols, which may be numbers, variables or functions, denoted by

$$\mathbf{A} = \begin{pmatrix} a_{11} & a_{12} & a_{13} & \ldots & a_{1n} \\ a_{21} & a_{22} & a_{23} & \ldots & a_{2n} \\ a_{31} & a_{32} & a_{33} & \ldots & a_{3n} \\ \vdots & \vdots & \vdots & & \vdots \\ a_{m1} & a_{m2} & a_{m3} & \ldots & a_{mn} \end{pmatrix}$$

A (2×3) matrix

$$\mathbf{A} = \begin{pmatrix} 1 & 3 & 4 \\ 2 & 6 & 5 \end{pmatrix}$$

A (3×4) matrix

$$\mathbf{A} = \begin{pmatrix} 1 & 4 & 2 & 5 \\ 0 & 1 & 1 & 0 \\ 6 & 3 & 1 & 2 \end{pmatrix}$$

The *dimension* (also *size* or *order*) of the matrix is given by the number of rows and columns: a matrix with m rows and n columns is called an $(m \times n)$ matrix. A *square* $(n \times n)$ matrix has equal number of rows and columns.

A (3×3) square matrix

$$\mathbf{A} = \begin{pmatrix} 0 & 1 & 3 \\ 2 & 1 & 1 \\ 6 & 0 & 2 \end{pmatrix}$$

A matrix is sometimes written as (a_{ij}) or $[a_{ij}]$ with a_{ij} is the *element* in row i column j.

A *diagonal matrix* is a square matrix with non-zero elements down the leading diagonal and zeros elsewhere.

The *trace* of a square matrix is the sum of the elements in the leading diagonal, denoted by $\mathrm{Tr}(\mathbf{A})$.

$\mathrm{Tr}(\mathbf{A}) = 0 + 1 + 2 = 3.$

Examples

A *column vector* is an $(m \times 1)$ matrix. A *row vector* is an $(1 \times n)$ matrix. Vectors are usually denoted by bold lower case letters, i.e.

$$\mathbf{c} = \begin{pmatrix} c_1 \\ c_2 \\ \vdots \\ c_m \end{pmatrix}; \quad \mathbf{r} = (r_1, r_2, \ldots, r_n)$$

A (3×1) column vector

$$\mathbf{c} = \begin{pmatrix} 1 \\ 4 \\ 2 \end{pmatrix}$$

A (1×2) row vector

$$\mathbf{r} = (1, 3)$$

A *scalar* is a number, i.e. a (1×1) matrix.

Manipulating matrices

Addition

The sum of two matrices is defined only if they have the same dimension and is obtained by adding the corresponding elements.

$$\begin{pmatrix} a_{11} & a_{12} \\ a_{21} & a_{22} \\ a_{31} & a_{32} \end{pmatrix} + \begin{pmatrix} b_{11} & b_{12} \\ b_{21} & b_{22} \\ b_{31} & b_{32} \end{pmatrix} = \begin{pmatrix} a_{11} + b_{11} & a_{12} + b_{12} \\ a_{21} + b_{21} & a_{22} + b_{22} \\ a_{31} + b_{31} & a_{32} + b_{32} \end{pmatrix}$$

Scalar multiplication

To multiply a matrix by a scalar, each element is multiplied by the scalar.

$$k \begin{pmatrix} a_{11} & a_{12} & a_{13} \\ a_{21} & a_{22} & a_{23} \end{pmatrix} = \begin{pmatrix} ka_{11} & ka_{12} & ka_{13} \\ ka_{21} & ka_{22} & ka_{23} \end{pmatrix}$$

Scalar product

The scalar product of two vectors of equal length is a scalar obtained as the sum of the products of the corresponding elements and denoted by

$$\langle a, b \rangle = a_1 b_1 + a_2 b_2 + \cdots + a_n b_n$$

$$\mathbf{a} = (1, 2, 3); \quad \mathbf{b} = (0, -1, 4)$$

$$\langle \mathbf{a}, \mathbf{b} \rangle = 1 \times 0 - 1 \times 2 + 3 \times 4$$

$$= 10$$

Multiplying two matrices

Two matrices $\mathbf{A}$ and $\mathbf{B}$ can be multiplied together, to form $\mathbf{C} = \mathbf{AB}$, only if the number of columns of $\mathbf{A}$ is equal to the number of rows of $\mathbf{B}$. If $\mathbf{A}$ is an $(m \times r)$ matrix and $\mathbf{B}$ is a $(r \times n)$ matrix the product $\mathbf{C} = \mathbf{AB}$ is an $(m \times n)$ with elements given by

$$c_{ij} = a_{i1} b_{1j} + a_{i2} b_{2j} + \cdots + a_{ir} b_{rj}$$

$$= \sum_{s=1}^{r} a_{is} b_{sj}.$$

This is the sum of the products of the elements in row i of $\mathbf{A}$ with the corresponding elements in column j of $\mathbf{B}$, i.e. the scalar product of the vector formed from row i with the vector formed from column j.

$$\mathbf{A} = \begin{pmatrix} 1 & 1 & 2 \\ 2 & 3 & 1 \\ 1 & 0 & 1 \\ 1 & 2 & 1 \end{pmatrix} \qquad \mathbf{B} = \begin{pmatrix} 1 & 3 \\ 2 & 1 \\ 0 & 5 \end{pmatrix}$$

$$\mathbf{AB} = \begin{pmatrix} 1 \times 1 + 1 \times 2 + 2 \times 0 & 1 \times 3 + 1 \times 1 + 2 \times 5 \\ 2 \times 1 + 3 \times 2 + 1 \times 0 & 2 \times 3 + 3 \times 1 + 1 \times 5 \\ 1 \times 1 + 0 \times 2 + 1 \times 0 & 1 \times 3 + 0 \times 1 + 1 \times 5 \\ 1 \times 1 + 2 \times 2 + 1 \times 0 & 1 \times 3 + 2 \times 1 + 1 \times 5 \end{pmatrix}$$

$$= \begin{pmatrix} 3 & 14 \\ 8 & 14 \\ 1 & 8 \\ 5 & 10 \end{pmatrix}$$

Special matrices

Identity matrix (I)

This is a square matrix with ones down the diagonal and zeros elsewhere, i.e.

$$\mathbf{I} = \begin{pmatrix} 1 & 0 & 0 & \dots & 0 \\ 0 & 1 & 0 & \dots & 0 \\ 0 & 0 & 1 & \dots & 0 \\ \vdots & \vdots & \vdots & & \vdots \\ 0 & 0 & 0 & \dots & 1 \end{pmatrix}$$

For any matrix A, $\mathbf{IA} = \mathbf{A}$.

Examples

$$\mathbf{I} = \begin{pmatrix} 1 & 0 & 0 \\ 0 & 1 & 0 \\ 0 & 0 & 1 \end{pmatrix}$$

Transposed matrix

The transpose of an $(m \times n)$ matrix $\mathbf{A}$ is an $(n \times m)$ matrix obtained by interchanging the rows and columns of $\mathbf{A}$ and denoted by

$$\mathbf{A}^{\mathrm{T}} = \begin{pmatrix} a_{11} & a_{21} & a_{31} & \dots & a_{m1} \\ a_{12} & a_{22} & a_{32} & \dots & a_{m2} \\ a_{13} & a_{23} & a_{33} & \dots & a_{m3} \\ \vdots & \vdots & \vdots & & \vdots \\ a_{1n} & a_{2n} & a_{3n} & \dots & a_{mn} \end{pmatrix}$$

or $\mathbf{A}'$. If $\mathbf{A}$ is square and $\mathbf{A}^{\mathrm{T}} = \mathbf{A}$ then $\mathbf{A}$ is symmetric.

$$\mathbf{A} = \begin{pmatrix} 3 & 1 & 2 \\ 4 & 0 & 5 \end{pmatrix}$$

$$\mathbf{A}^{\mathrm{T}} = \begin{pmatrix} 3 & 4 \\ 1 & 0 \\ 2 & 5 \end{pmatrix}$$

Determinant of a matrix

The *determinant* of a square matrix $\mathbf{A}$ is a scalar quantity denoted by det $\mathbf{A}$, $|\mathbf{A}|$ or Δ.

For a 2×2 matrix

$$\det \mathbf{A} = \begin{vmatrix} a_{11} & a_{12} \\ a_{21} & a_{22} \end{vmatrix} = a_{11}a_{22} - a_{12}a_{21}.$$

$$\begin{vmatrix} 2 & 1 \\ 3 & 4 \end{vmatrix} = 2 \times 4 - 1 \times 3 = 5$$

For a 3×3 matrix

$$\begin{vmatrix} a_{11} & a_{12} & a_{13} \\ a_{21} & a_{22} & a_{23} \\ a_{31} & a_{32} & a_{33} \end{vmatrix} = a_{11} \begin{vmatrix} a_{22} & a_{23} \\ a_{32} & a_{33} \end{vmatrix} - a_{12} \begin{vmatrix} a_{21} & a_{23} \\ a_{31} & a_{33} \end{vmatrix} + a_{13} \begin{vmatrix} a_{21} & a_{22} \\ a_{31} & a_{32} \end{vmatrix}.$$

$$\begin{vmatrix} 2 & 4 & 1 \\ 0 & 6 & 3 \\ 1 & 2 & 1 \end{vmatrix} = 2 \begin{vmatrix} 6 & 3 \\ 2 & 1 \end{vmatrix} - 4 \begin{vmatrix} 0 & 3 \\ 1 & 1 \end{vmatrix} + 1 \begin{vmatrix} 0 & 6 \\ 1 & 2 \end{vmatrix}$$

$$= 2(6 \times 1 - 3 \times 2) - 4(0 \times 1 - 1 \times 3) + 1(0 \times 2 - 6 \times 1)$$

$$= 0 + 12 - 6$$

$$= 6.$$

The determinant of an $(n \times n)$ matrix is obtained as

$$\det \mathbf{A} = a_{11}C_{11} + a_{12}C_{12} + \dots + a_{1n}C_{1n},$$

where C_{1j} is the determinant of the matrix formed by omitting row 1 column j, multiplied by $(-1)^{j+1}$. There are computer algorithms for evaluating determinants.

If the determinant is non-zero the matrix is *non-singular*, i.e. has an inverse (see below).

Inverse matrix (A^{-1})

For an $(n \times n)$ matrix $\mathbf{A}$ there *may* be an inverse $(n \times n)$ matrix denoted by $\mathbf{A}^{-1}$ such that

$$\mathbf{A}\mathbf{A}^{-1} = \mathbf{I} = \mathbf{A}^{-1}\mathbf{A}.$$

where $\mathbf{I}$ is the identity matrix.

A special case is the (2×2) matrix for which

$$\mathbf{A}^{-1} = \begin{pmatrix} a_{11} & a_{12} \\ a_{21} & a_{22} \end{pmatrix}^{-1}$$

$$= \frac{1}{(a_{11}a_{22} - a_{12}a_{21})} \begin{pmatrix} a_{22} & -a_{12} \\ -a_{21} & a_{11} \end{pmatrix}$$

$$= \frac{1}{\det \mathbf{A}} \begin{pmatrix} a_{22} & -a_{12} \\ -a_{21} & a_{11} \end{pmatrix}.$$

If the inverse exists, the matrix $\mathbf{A}$ is *non-singular*, if not it is *singular*. A necessary and sufficient condition for an inverse of $\mathbf{A}$ to exist is that the determinant of $\mathbf{A}$ is non-zero.

Computer routines for calculating the inverse of a matrix are widely available.

Examples

For

$$\mathbf{A} = \begin{pmatrix} 2 & 5 \\ 1 & 3 \end{pmatrix}; \quad \det \mathbf{A} = 1$$

and

$$\begin{pmatrix} 2 & 5 \\ 1 & 3 \end{pmatrix}^{-1} = \begin{pmatrix} 3 & -5 \\ -1 & 2 \end{pmatrix}.$$

Check by matrix multiplication:

$$\begin{pmatrix} 2 & 5 \\ 1 & 3 \end{pmatrix}\begin{pmatrix} 3 & -5 \\ -1 & 2 \end{pmatrix} = \begin{pmatrix} 1 & 0 \\ 0 & 1 \end{pmatrix}.$$

Systems of linear equations

The system of linear equations

$$a_{11}x_1 + a_{12}x_2 + \cdots + a_{1n}x_n = b_1$$
$$a_{21}x_1 + a_{22}x_2 + \cdots + a_{2n}x_n = b_2$$
$$\cdots\cdots\cdots\cdots\cdots\cdots\cdots\cdots$$
$$a_{n1}x_1 + a_{n2}x_2 + \cdots + a_{nn}x_n = b_n$$

can be represented in matrix form

$$\mathbf{Ax} = \mathbf{b}$$

where $\mathbf{A}$ is an $(n \times n)$ matrix and $\mathbf{x}$ and $\mathbf{b}$ are column vectors of length n.
The solution exists only if $\det \mathbf{A}$ is non-zero, and is given by $\mathbf{x} = \mathbf{A}^{-1}\mathbf{b}$.

$$2x_1 + 5x_2 = 1$$
$$x_1 + 3x_2 = 2$$

i.e.

$$\begin{pmatrix} 2 & 5 \\ 1 & 3 \end{pmatrix}\begin{pmatrix} x_1 \\ x_2 \end{pmatrix} = \begin{pmatrix} 1 \\ 2 \end{pmatrix}.$$

Solution:

$$\begin{pmatrix} x_1 \\ x_2 \end{pmatrix} = \begin{pmatrix} 3 & -5 \\ -1 & 2 \end{pmatrix}\begin{pmatrix} 1 \\ 2 \end{pmatrix} = \begin{pmatrix} -7 \\ 3 \end{pmatrix},$$

i.e. $x_1 = -7, x_2 = 3$.

Eigenvalues and eigenvectors

If $\mathbf{A}$ is an $(n \times n)$ matrix, $\mathbf{x}$ a column vector and λ a scalar such that $\mathbf{Ax} = \lambda\mathbf{x}$ then $\mathbf{x}$ is called an *eigenvector* of $\mathbf{A}$ with *eigenvalue* λ.

$$\begin{pmatrix} 2 & -1 \\ 4 & -3 \end{pmatrix}\begin{pmatrix} x_1 \\ x_2 \end{pmatrix} = \lambda\begin{pmatrix} x_1 \\ x_2 \end{pmatrix}$$

Characteristic or auxiliary equation

The eigenvector and eigenvalue satisfy the equation

$$(\mathbf{A} - \lambda\mathbf{I})\mathbf{x} = 0$$

$$\begin{pmatrix} 2-\lambda & -1 \\ 4 & -3-\lambda \end{pmatrix}\begin{pmatrix} x_1 \\ x_2 \end{pmatrix} = 0$$

where **I** is the $(n \times n)$ identity matrix. For a non-zero solution the matrix $\mathbf{A} - \lambda\mathbf{I}$ must be singular, i.e.

$$\det(\mathbf{A} - \lambda\mathbf{I}) = 0.$$

This equation for λ is called the *characteristic* or *auxiliary equation* of **A**. For an $(n \times n)$ matrix it is a polynomial of degree n.
An important special case is the 2×2 matrix

$$\begin{vmatrix} a_{11} - \lambda & a_{12} \\ a_{21} & a_{22} - \lambda \end{vmatrix} = (a_{11} - \lambda)(a_{22} - \lambda) - a_{12}a_{22}$$

with characteristic equation

$$\lambda^2 - \lambda(a_{11} + a_{22}) + (a_{11}a_{22} - a_{12}a_{21}) = 0.$$

This is a quadratic equation with coefficients given by the trace and the determinant of the matrix. Eigenvalues can be real or complex occurring in conjugate pairs.

Examples

$$\begin{vmatrix} 2 - \lambda & -1 \\ 4 & -3 - \lambda \end{vmatrix} = 0$$

i.e.

$$\lambda^2 + \lambda - 2 = 0$$

$$(\lambda + 2)(\lambda - 1) = 0$$

with eigenvalues of 1 and -2.

Finding eigenvectors

The calculations for finding eigenvalues and eigenvectors are complicated since they involve evaluating determinants, finding roots of polynomials and solving systems of equations with singular matrices. Computer algorithms are available to do the computations. The numerical example opposite illustrates the idea. Eigenvectors are determined only up to a constant value since if $\mathbf{Ax} = \lambda\mathbf{x}$ then $\mathbf{A}k\mathbf{x} = \lambda k\mathbf{x}$ and $k\mathbf{x}$ is also an eigenvector corresponding to λ.

The eigenvalue with the largest absolute value is called the *dominant eigenvalue* with corresponding *dominant eigenvector*.

For the above example the eigenvector corresponding to the eigenvalue of 1 satisfies

$$\begin{pmatrix} 2 & -1 \\ 4 & -3 \end{pmatrix}\begin{pmatrix} x_1 \\ x_2 \end{pmatrix} = \begin{pmatrix} x_1 \\ x_2 \end{pmatrix},$$

or

$$2x_1 - 1x_2 = x_1$$

$$4x_1 - 3x_2 = x_2.$$

Both equations give $x_1 = x_2$ and an eigenvector proportional to $x = (1, 1)$. Note that there is no unique solution of the equation.
Similarly, the eigenvector corresponding to the eigenvalue of -2 is proportional to $(1, 4)$.

A12 PROPERTIES OF RANDOM VARIABLES

Random variables: X, Y, Z

Constants: a, b, c

Let $\mathrm{E}[X] = 2, \mathrm{E}[Y] = 3$,
and $\mathrm{var}[X] = 4, \mathrm{var}[Y] = 9$.

Expectations or means

$$\mathrm{E}[a + bY] = a + b\mathrm{E}[Y]$$

$\mathrm{E}[1 + 2Y] = 1 + 2\mathrm{E}[Y] = 1 + (2)(3) = 7$

$\mathrm{E}[4Y] = 4\mathrm{E}[Y] = (4)(3) = 12$

$$\mathrm{E}[X + Y] = \mathrm{E}[X] + \mathrm{E}[Y]$$

$\mathrm{E}[X + Y] = 2 + 3 = 5$

$$\mathrm{E}[XY] = \mathrm{E}[X]\mathrm{E}[Y] + \mathrm{cov}[X, Y]$$

i.e. $\mathrm{cov}[X, Y] = \mathrm{E}[XY] - \mathrm{E}[X]\mathrm{E}[Y]$

If $\mathrm{cov}[X, Y] = -5$

$\mathrm{E}[XY] = 2 \times 3 + (-5) = 1$

If X, Y independent,

$$E[XY] = E[X]E[Y]$$

i.e. $\text{cov}[X, Y] = 0$.

Examples

If X, Y independent.

$E[XY] = (2)(3) = 6$

Variances

$$\text{var}[a + bY] = b^2 \text{ var}[Y]$$

$\text{var}[1 + 2Y] = 2^2 \text{ var}[Y]$

$= 4 \text{ var}[Y] = (4)(9) = 36$

$\text{var}[4Y] = 4^2 \text{ var}[Y]$

$= 16 \text{ var}[Y] = (16)(9) = 144.$

$$\text{var}[X + Y] = \text{var}[X] + \text{var}[Y] + 2 \text{ cov}[X, Y]$$

If $\text{cov}[X, Y] = 6$, i.e. $\text{correlation}[X, Y] = 1$

$\text{var}[X + Y] = 4 + 9 + 12 = 25$

If $\text{cov}[X, Y] = -6$, i.e.
$\text{correlation}[X, Y] = -1$

$\text{var}[X + Y] = 4 + 9 - 12 = 1$

If X and Y are independent

$$\text{var}[X + Y] = \text{var}[X] + \text{var}[Y]$$

If X, Y independent,

$\text{var}[X + Y] = 4 + 9 = 13$

If X and Y are independent

$$\text{var}[XY] = \text{var}[X] (E[Y])^2 + (E[X])^2 \text{ var}[Y] + \text{var}[X] \text{var}[Y]$$

If X, Y are independent,

$\text{var}[XY] = (4)(3^2) + (2^2)(9) + (4)(9) = 108$

Statistical Tables

Table S1 Cumulative probability function of the standardised **Normal distribution**, N(0,1). (Reproduced, with permission, from *New Cambridge Elementary Statistical Tables*, by D.V. Lindley and W.F. Scott, 1984, published by Cambridge University Press.)

$$\Phi(x) = \int_{-\infty}^{x} f(x)\,dx \qquad \text{where} \quad f(x) = \frac{1}{\sqrt{2\pi}} \exp\left(-\frac{1}{2}x^2\right) dx$$

$\Phi(x)$ in body of table.

	last digit of x									
x	0	1	2	3	4	5	6	7	8	9
0.0	0.5000	0.5040	0.5080	0.5120	0.5160	0.5199	0.5239	0.5279	0.5319	0.5359
0.1	0.5398	0.5438	0.5478	0.5517	0.5557	0.5596	0.5636	0.5675	0.5714	0.5753
0.2	0.5793	0.5832	0.5871	0.5910	0.5948	0.5987	0.6026	0.6064	0.6103	0.6141
0.3	0.6179	0.6217	0.6255	0.6293	0.6331	0.6368	0.6406	0.6443	0.6480	0.6517
0.4	0.6554	0.6591	0.6628	0.6664	0.6700	0.6736	0.6772	0.6808	0.6844	0.6879
0.5	0.6915	0.6950	0.6985	0.7019	0.7054	0.7088	0.7123	0.7157	0.7190	0.7224
0.6	0.7257	0.7291	0.7324	0.7357	0.7389	0.7422	0.7454	0.7486	0.7517	0.7549
0.7	0.7580	0.7611	0.7642	0.7673	0.7704	0.7734	0.7764	0.7794	0.7823	0.7852
0.8	0.7881	0.7910	0.7939	0.7967	0.7995	0.8023	0.8051	0.8078	0.8106	0.8133
0.9	0.8159	0.8186	0.8212	0.8238	0.8264	0.8289	0.8315	0.8340	0.8365	0.8389
1.0	0.8413	0.8438	0.8461	0.8485	0.8508	0.8531	0.8554	0.8577	0.8599	0.8621
1.1	0.8643	0.8665	0.8686	0.8708	0.8729	0.8749	0.8770	0.8790	0.8810	0.8830
1.2	0.8849	0.8869	0.8888	0.8907	0.8925	0.8944	0.8962	0.8980	0.8997	0.9015
1.3	0.9032	0.9049	0.9066	0.9082	0.9099	0.9115	0.9131	0.9147	0.9162	0.9177
1.4	0.9192	0.9207	0.9222	0.9236	0.9251	0.9265	0.9279	0.9292	0.9306	0.9319
1.5	0.9332	0.9345	0.9357	0.9370	0.9382	0.9394	0.9406	0.9418	0.9429	0.9441
1.6	0.9452	0.9463	0.9474	0.9484	0.9495	0.9505	0.9515	0.9525	0.9535	0.9545
1.7	0.9554	0.9564	0.9573	0.9582	0.9591	0.9599	0.9608	0.9616	0.9625	0.9633
1.8	0.9641	0.9649	0.9656	0.9664	0.9671	0.9678	0.9686	0.9693	0.9699	0.9706
1.9	0.9713	0.9719	0.9726	0.9732	0.9738	0.9744	0.9750	0.9756	0.9761	0.9767
2.0	0.97725	0.97778	0.97831	0.97882	0.97932	0.97982	0.98030	0.98077	0.98124	0.98169
2.1	0.98214	0.98257	0.98300	0.98341	0.98382	0.98422	0.98461	0.98500	0.98537	0.98574
2.2	0.98610	0.98645	0.98679	0.98713	0.98745	0.98778	0.98809	0.98840	0.98870	0.98899
2.3	0.98928	0.98956	0.98983	0.99010	0.99036	0.99061	0.99086	0.99111	0.99134	0.99158
2.4	0.99180	0.99202	0.99224	0.99245	0.99266	0.99286	0.99305	0.99324	0.99343	0.99361
2.5	0.99379	0.99396	0.99413	0.99430	0.99446	0.99461	0.99477	0.99492	0.99506	0.99520
2.6	0.99534	0.99547	0.99560	0.99573	0.99585	0.99598	0.99609	0.99621	0.99632	0.99643
2.7	0.99653	0.99664	0.99674	0.99683	0.99693	0.99702	0.99711	0.99720	0.99728	0.99736
2.8	0.99744	0.99752	0.99760	0.99767	0.99774	0.99781	0.99788	0.99795	0.99801	0.99807
2.9	0.99813	0.99819	0.99825	0.99831	0.99836	0.99841	0.99846	0.99851	0.99856	0.99861
3.0	0.99865	0.99869	0.99874	0.99878	0.99882	0.99886	0.99889	0.99893	0.99896	0.99900
3.1	0.99903	0.99906	0.99910	0.99913	0.99916	0.99918	0.99921	0.99924	0.99926	0.99929
3.2	0.99931	0.99934	0.99936	0.99938	0.99940	0.99942	0.99944	0.99946	0.99948	0.99950

Upper percentage points of the standardised Normal distribution

Percentage	10%	5%	2.5%	1%	0.5%	0.1%	0.05%
Normal deviate	1.28	1.64	1.96	2.33	2.58	3.09	3.29

Table S2 Two-sided percentage points of **Student's *t*-distribution**. (Reproduced, with permission, from *New Cambridge Elementary Statistical Tables*, by D.V. Lindley and W.F. Scott, 1984, published by Cambridge University Press.)
The table gives *two-sided* percentage points of Student's *t*-distribution on the degrees of freedom stated. These are points which cut off the total percentage stated in both tails together (*i.e.* half of the percentage given in each tail).

d.f.	10%	5%	1%	0.1%	*d.f.*	10%	5%	1%	0.1%
1	6.31	12.71	63.66	636.6	12	1.78	2.18	3.06	4.32
2	2.92	4.30	9.93	31.60	14	1.76	2.15	2.98	4.14
3	2.35	3.18	5.84	12.92	16	1.75	2.12	2.92	4.02
4	2.13	2.78	4.60	8.61	18	1.73	2.10	2.88	3.92
5	2.02	2.57	4.03	6.87	20	1.73	2.09	2.85	3.85
6	1.94	2.45	3.71	5.96	30	1.70	2.04	2.75	3.65
7	1.90	2.37	3.50	5.41	40	1.68	2.02	2.70	3.55
8	1.86	2.31	3.36	5.04	60	1.67	2.00	2.66	3.46
9	1.83	2.26	3.25	4.78	120	1.66	1.98	2.62	3.37
10	1.81	2.23	3.17	4.59	∞	1.65	1.96	2.58	3.29

Table S3 One-sided percentage points of the χ^2 **distribution**. (Reproduced, with permission, from *New Cambridge Elementary Statistical Tables*, by D.V. Lindley and W.F. Scott, 1984, published by Cambridge University Press.)
The table gives *one-sided* percentage points of chi-squared distribution on the degrees of freedom stated. These are points which cut off the percentage stated in the upper tail of the distribution.

d.f.	10%	5%	1%	0.1%	*d.f.*	10%	5%	1%	0.1%
1	2.71	3.84	6.64	10.83	12	18.55	21.03	26.22	32.91
2	4.61	5.99	9.21	13.82	14	21.06	23.68	29.14	36.12
3	6.25	7.82	11.34	16.27	16	23.54	26.30	32.00	39.25
4	7.78	9.49	13.28	18.47	18	25.99	28.87	34.81	42.31
5	9.24	11.07	15.09	20.52	20	28.41	31.41	37.57	45.31
6	10.64	12.59	16.81	22.46	30	40.26	43.77	50.89	59.70
7	12.02	14.07	18.48	24.32	40	51.81	55.76	63.69	73.40
8	13.36	15.51	20.09	26.12	60	74.40	79.08	88.38	99.61
9	14.68	16.92	21.67	27.88					
10	15.99	18.31	23.21	29.59					

Table S4 One-sided percentage points of ***F*-distribution**. (Reproduced, with permission, from *New Cambridge Elementary Statistical Tables*, by D.V. Lindley and W.F. Scott, 1984, published by Cambridge University Press.)

The table gives *one-sided* percentage points of the *F*-distribution on the two sets of degrees of freedom stated, i.e. of F_{v_1,v_2} where v_1 is the value given along the top margin of the table, and v_2 along the side margin. These are points which cut off the percentage stated in the upper tail of the distribution.

5% points of F

v_2	$v_1 = 1$	2	3	4	5	6	8	12	24	∞
1	161.4	199.5	215.7	224.6	230.2	234.0	238.9	243.9	249.1	254.3
2	18.51	19.00	19.16	19.25	19.30	19.33	19.37	19.41	19.45	19.50
3	10.13	9.55	9.28	9.12	9.01	8.94	8.85	8.75	8.64	8.53
4	7.71	6.94	6.59	6.39	6.26	6.16	6.04	5.91	5.77	5.63
5	6.61	5.79	5.41	5.19	5.05	4.95	4.82	4.68	4.53	4.37
6	5.99	5.14	4.76	4.53	4.39	4.28	4.15	4.00	3.84	3.67
7	5.59	4.74	4.35	4.12	3.97	3.87	3.73	3.57	3.41	3.23
8	5.32	4.46	4.07	3.84	3.69	3.58	3.44	3.28	3.12	2.93
9	5.12	4.26	3.86	3.63	3.48	3.37	3.23	3.07	2.90	2.71
10	4.97	4.10	3.71	3.48	3.33	3.22	3.07	2.91	2.74	2.54
12	4.75	3.89	3.49	3.26	3.11	3.00	2.85	2.69	2.51	2.30
14	4.60	3.74	3.34	3.11	2.96	2.85	2.70	2.53	2.35	2.13
16	4.49	3.63	3.24	3.01	2.85	2.74	2.59	2.43	2.24	2.01
18	4.41	3.56	3.16	2.93	2.77	2.66	2.51	2.34	2.15	1.92
20	4.35	3.49	3.10	2.87	2.71	2.60	2.45	2.28	2.08	1.84
30	4.17	3.32	2.92	2.69	2.53	2.42	2.27	2.09	1.89	1.62
40	4.09	3.23	2.84	2.61	2.45	2.34	2.18	2.00	1.79	1.51
60	4.00	3.15	2.76	2.53	2.37	2.25	2.10	1.92	1.70	1.39
120	3.92	3.07	2.68	2.45	2.29	2.18	2.02	1.83	1.61	1.25
∞	3.84	3.00	2.61	2.37	2.21	2.10	1.94	1.75	1.52	1.00

1% points of F

v_2	$v_1 = 1$	2	3	4	5	6	8	12	24	∞
1	4052	4999	5403	5625	5764	5859	5981	6106	6235	6366
2	98.50	99.00	99.17	99.25	99.30	99.33	99.37	99.42	99.46	99.50
3	34.12	30.82	29.46	28.71	28.24	27.91	27.49	27.05	26.60	26.13
4	21.20	18.00	16.69	15.98	15.52	15.21	14.80	14.37	13.93	13.46
5	16.26	13.27	12.06	11.39	10.97	10.67	10.29	9.89	9.47	9.02
6	13.75	10.92	9.78	9.15	8.75	8.47	8.10	7.72	7.31	6.88
7	12.25	9.55	8.45	7.85	7.46	7.19	6.84	6.47	6.07	5.65
8	11.26	8.65	7.59	7.01	6.63	6.37	6.03	5.67	5.28	4.86
9	10.56	8.02	6.99	6.42	6.06	5.80	5.47	5.11	4.73	4.31
10	10.04	7.56	6.55	5.99	5.64	5.39	5.06	4.71	4.33	3.91
12	9.33	6.93	5.95	5.41	5.06	4.82	4.50	4.16	3.78	3.36
14	8.86	6.52	5.56	5.04	4.70	4.46	4.14	3.80	3.43	3.00
16	8.53	6.23	5.29	4.77	4.44	4.20	3.89	3.55	3.18	2.75
18	8.29	6.01	5.09	4.58	4.25	4.02	3.71	3.37	3.00	2.57
20	8.10	5.85	4.94	4.43	4.10	3.87	3.56	3.23	2.86	2.42
30	7.56	5.39	4.51	4.02	3.70	3.47	3.17	2.84	2.47	2.01
40	7.31	5.18	4.31	3.83	3.51	3.29	2.99	2.67	2.29	1.81
60	7.08	4.98	4.13	3.65	3.34	3.12	2.82	2.50	2.12	1.60
120	6.85	4.79	3.95	3.48	3.17	2.96	2.66	2.34	1.95	1.38
∞	6.64	4.61	3.78	3.32	3.02	2.80	2.51	2.19	1.79	1.00

0.1% points of F

ν_2	$\nu_1 = 1$	2	3	4	5	6	8	12	24	∞
1	405280	500000	540380	562500	576400	585940	598140	610670	623500	636620
2	998.5	999.0	999.2	999.2	999.3	999.3	999.4	999.4	999.5	999.5
3	167.0	148.5	141.1	137.1	134.6	132.8	130.6	128.3	125.9	123.5
4	74.14	61.25	56.18	53.44	51.71	50.53	49.00	47.41	45.77	44.05
5	47.18	37.12	33.20	31.09	29.75	28.83	27.65	26.42	25.13	23.79
6	35.51	27.00	23.70	21.92	20.80	20.03	19.03	17.99	16.90	15.75
7	29.25	21.69	18.77	17.20	16.21	15.52	14.63	13.71	12.73	11.70
8	25.41	18.49	15.83	14.39	13.48	12.86	12.05	11.19	10.30	9.33
9	22.86	16.39	13.90	12.56	11.71	11.13	10.37	9.57	8.72	7.81
10	21.04	14.91	12.55	11.28	10.48	9.93	9.20	8.45	7.64	6.76
12	18.64	12.97	10.80	9.63	8.89	8.38	7.71	7.01	6.25	5.42
14	17.14	11.78	9.73	8.62	7.92	7.44	6.80	6.13	5.41	4.60
16	16.12	10.97	9.01	7.94	7.27	6.81	6.20	5.55	4.85	4.06
18	15.38	10.39	8.49	7.46	6.81	6.36	5.76	5.13	4.45	3.67
20	14.82	9.95	8.10	7.10	6.46	6.02	5.44	4.82	4.15	3.38
30	13.29	8.77	7.05	6.13	5.53	5.12	4.58	4.00	3.36	2.59
40	12.61	8.25	6.60	5.70	5.13	4.73	4.21	3.64	3.01	2.23
60	11.97	7.77	6.17	5.31	4.76	4.37	3.87	3.32	2.69	1.89
120	11.38	7.32	5.78	4.95	4.42	4.04	3.55	3.02	2.40	1.54
∞	10.83	6.91	5.42	4.62	4.10	3.74	3.27	2.74	2.13	1.00

References

ALBERTS, B., BRAY, D., LEWIS, J., RAFF, M., ROBERTS, K. & WATSON, J.D. (1983) *Molecular Biology of the Cell*. Garland, New York.

ALEKSANDER, I. & MORTON, H. (1990) *An Introduction to Neural Computing*. Chapman & Hall, London.

ANDERSON, R.M. (ed.) (1982) *The Population Dynamics of Infectious Diseases: Theory and Applications*. Chapman & Hall, London.

ARBIB, M.A. (1989) *The Metaphorical Brain 2: Neural Networks and Beyond*. Wiley, New York.

ATKINSON, A.C. (1985) *Plots, Transformations, and Regression*. Clarendon Press, Oxford.

BACON, P.J. (ed.) (1985) *Population Dynamics of Rabies in Wildlife*. Academic Press, London.

BARNETT, V.D. (1962) The Monte Carlo solution of a competing species problem. *Biometrics*, **18**, 76–103.

BEDDINGTON, J.R., FREE, C.A. & LAWTON, J.H. (1975) Dynamic complexity in predator–prey models framed in difference equations. *Nature*, **255**, 58–60.

BERRIDGE, M. J. & RAPP, P. E. (1979) A comparative survey of the function, mechanism and control of cellular oscillators. *Journal of Experimental Biology*, **81**, 217–279.

BERTALANFFY, L. von (1938) A quantitative theory of organic growth (inquiries on growth laws II). *Human Biology*, **10**(2), 181–213.

BEST, E.N. (1979) Null space in the Hodgkin–Huxley equations: a critical test. *Biophys. J.*, **27**, 87–104.

BJERKNES, R. & IVERSEN, O.H. (1968) In DRISCHEL, H. (ed.) *Proc. First Intern. Symp. Bio-kybernetik*, Karl Marx Universitat, 253.

BLANKSMA, J.J. (1902) *Rec. trav. chim.*, **21**, 366.

BLISS, C.I. (1935) The calculation of the dosage–mortality curve. *Annals of Applied Biology*, **22**, 134–167.

BOX, G.E.P. & JENKINS, G.M. (1970) *Time Series Analysis, Forecasting and Control*. Holden-Day, San Francisco.

BRADLEY, D.W. & BRADLEY, R.A. (1983) Application of Sequence Comparison to the Study of Bird Songs. In SANKOFF, D. & KRUSKAL, J.B. (eds) *Time Warps, String Edits, and Macromolecules: the Theory and Practice of Sequence Comparison*. Addison-Wesley, Reading, Massachusetts, Chapter 6.

BRIGGS, G.E. & HALDANE. J.B.S. (1925) *Biochem. J.*, **19**, 338–339.

BROADBENT, S.R. & KENDALL, D.G. (1953) The random walk of *Trichostrongylus retortaeformis*. *Biometrics*, **9**, 460–466.

BRODY, S. & PROCTOR, R.C. (1932) Growth and development with special reference to domestic animals. Further investigations of surface area in energy metabolism. *Mo. Res. Bull.*, **116**.

BROWN, D. (1975) A test of randomness of nest spacing. *Wildfowl*, **26**, 102–103.

BROWN, D. (1992) A graphical analysis of deviance. *Applied Statistics*, **41**, 55–62.

BUMPUS, H.C. (1898) The elimination of the unfit as illustrated by the introduced sparrow *Passer domesticus*. *Biological Lectures*, Marine Biol. Lab., Woods Hole, 209–226.

CAIN, A.J., KING, J.M.B. & SHEPPARD, P.M. (1960) New data on the genetics of polymorphism in the snail *Cepea nemoralis L. Genetics*, **45**, 393–411.

CARLSON, T. (1913) Über Geschwindigkeit und Grösse der Hefevermehrung in Würze. *Biochem. Z.*, **57**, 313–334.

CASWELL, H. (1989) *Matrix Population Models*. Sinauer Associates, Inc., Massachusetts.

CAUGHLEY, G. (1976) Plant–Herbivore Systems. In R.M. MAY (ed.) *Theoretical Ecology*. Blackwell Scientific Publications, Oxford, 94–113.

CHAMBERS, J.M., CLEVELAND, W.S., KLEINER, B. & TUKEY, P.A. (1983) *Graphical Methods for Data Analysis*. Duxbury Press, Boston.

CHATFIELD, C. (1984) *The Analysis of Time Series: An Introduction*. Chapman & Hall, London.

CLARKE, M.F., LECHYCKA, M. & COOK, C.A. (1940) The biological assay of riboflavin. *J. Nutr.*, **20**, 133–144.

CLOSE, W.H. & MOUNT, L.E. (1978) The effects of plane of nutrition and environmental temperature on the energy metabolism of the growing pig. 1. Heat loss and critical temperature. *Br. J. Nutr.*, **40**, 413–421.

COCHRAN, W.G. & COX, G.M. (1957) *Experimental Designs*. Wiley, New York.

COEN, C. (ed.) (1985) *Functions of the Brain*. Clarendon Press, Oxford.

COX, D.R. (1958) *Planning of Experiments*. Wiley, New York.

COX, D.R. & SNELL, E.J. (1981) *Applied Statistics*. Chapman & Hall, London.

CROW, J.F. (1986) *Basic Concepts in Population, Quantitative, and Evolutionary Genetics*. Freeman, New York.

CROWDER, M.J. & HAND, D.J. (1990) *The Analysis of Repeated Measures*. Chapman & Hall, London.

DARWIN, C.R. (1859) *Origin of Species*. Dent, London.

DAUNCEY, M.J., BROWN, D., HAYASHI, M. & INGRAM, D.L. (1988) Thyroid hormone nuclear receptors in skeletal muscle as influenced by environmental temperature and energy intake. *Quarterly Journal of Experimental Physiology*, **73**, 183–191.

DIGBY, P., GALWEY, N. & LANE, P. (1989) *Genstat 5: A Second Course*. Clarendon Press, Oxford.

DIGGLE, P.J. (1983) *Statistical Analysis of Spatial Point Patterns*. Academic Press, London.

DIGGLE, P.J. (1990) *Time Series: A Biostatistical Introduction*. Clarendon Press, Oxford.

DRAPER, N. R. & SMITH, H. (1981) *Applied Regression Analysis*. 2nd edn. Wiley, New York.

DUMAS, J.P. & NINIO, J. (1982) Efficient algorithms for folding and comparing nucleic acid sequences. *Nucleic Acids Res.*, **10**, 197–206.

EDELSTEIN-KESHET, L. (1988) *Mathematical Models in Biology*. Random House, New York, Section 8.8.

ELTON, C.S. & NICHOLSON, M. (1942) The ten year cycle in numbers of lynx in Canada. *J. Anim. Ecol.*, **11**, 215–244.

FALCONER, D.S. (1977) *Introduction to Quantitative Genetics*. Longman, London.

FEIGL, P. & ZELEN, M. (1965) Estimation of exponential survival probabilities with concomitant information. *Biometrics*, **21**, 826–838.

FIRBANK, L.G. & WATKINSON, A.R. (1985) On the analysis of competition within two-species mixtures of plants. *Journal of Applied Ecology*, **22**, 503–517.

FISHER, R.A. (1947) The analysis of covariance method for the relation between a part and the whole. *Biometrics,* **3**, 65–68.

FITCH, W.M. & SMITH, T.F. (1983) Optimal sequence alignments. *Proc. Natl. Acad. Sci. U.S.A.*, **80**, 1382–1386.

FITZHUGH, R. (1961) Impulses and physiological states in theoretical models of nerve membrane. *Biophys. J.*, **1**, 445–466.

FORD, E.D. & DIGGLE, P.J. (1981) Competition for light in a plant monoculture modelled as a spatial stochastic process. *Annals of Botany*, **48**, 481–500.

GALTON, F. (1889) *Natural Inheritance*. Macmillan, London. p. 83.

GARDNER, M. (1971) Mathematical games. *Scientific American*, **224**, February, 112.

GARDNER, M. (1983) *Wheels, Life and Other Mathematical Amusements*. Freeman, New York.

GAUSE, G.F. (1934) *The Struggle for Existence*. Hafner, New York.

GENSTAT 5 COMMITTEE (1987) *Genstat 5 Reference Manual*. Clarendon Press, Oxford.

GILPIN, M.E. (1973) Do hares eat lynx? *American Naturalist*, **107,** 727–730.

GILPIN, M.E. (1975) Limit cycles in competition communities. *The American Naturalist*, **109**, 51–60.

GLASS, L. & MACKEY, M. C. (1978) Pathological conditions resulting from instabilities in physiological control systems. *Annals of the New York Academy of Sciences*, **316**, 214–235.

GODFREY, K. (1983) *Compartment Models and Their Application*. Academic Press, London.

GOLDBETER, A. (1980) Models for oscillations and excitability in biochemical systems. In SEGEL, L.A. (ed.) *Mathematical Models in Molecular and Cellular Biology*. Cambridge University Press, Cambridge, 248–291.

GREIG-SMITH, P. (1952) The use of random and contiguous quadrats in the study of the structure of plant communities. *Annals of Botany*, **16**, 293–316.

GROSSBERG, S. (1982) *Studies of Mind and Brain: Neural Principles of Learning, Perception, Development, Cognition and Motor Control*. Reidel Press, Boston.

HALDANE, J.B.S. (1964) A defence of beanbag genetics. *Perspectives in Biology & Medicine*. 343–359.

HAMILTON, D. (1987) Sometimes $R^2 > r^2_{yx_1} + r^2_{yx_2}$: correlated variables are not always redundant. *American Statistician*, **41**, 129–132.

HANES, C.S. (1932) Studies on plant amylases. I. *Biochem. J.*, **26**, 1406–1421.

HANSON, F.B. & TUCKWELL, H.C. (1981) Logistic growth with random density independent disasters. *Theoretical Population Biology*, **19**, 1–18.

HANSON, S.J. & OLSON, C.R. (eds) (1990) *Connectionist Modeling and Brain Function: the Developing Interface*. MIT Press, Cambridge, Massachusetts.

HARRISON, F.A., HILLS, F., PATERSON, J.Y.F. & SAUNDERS, R.C. (1986) The measurement of liver blood flow in conscious calves. *Quarterly Journal of Experimental Physiology*, **71**, 235–247.

HARWOOD, J. (1981) Managing gray seal populations for optimal stability. In FOWLER, C.W. & SMITH, T.D. (eds) *Dynamics of Large Mammal Populations*. Wiley, New York, 159–172.

HASSELL, M.P., COMINS, H.N. & MAY, R.M. (1991) Spatial structure and chaos in insect population dynamics. *Nature*, **353**, 255–258.

HASSELL, M.P. (1975) Density-dependence in single-species populations. *Journal of Animal Ecology*, **44**, 283–295.

HASSELL, M.P. (1978) *The Dynamics of Arthropod Predator–Prey Systems*. Princeton University Press, Princeton.

HASTINGS, N.A.J. & PEACOCK, J.B. (1974) *Statistical Distributions: A Handbook for Students and Practitioners*. Butterworths, London.

HEALEY, M.J.R. (1972) Statistical analysis of radioimmunoassay data. *Biochem. J.*, **130**, 207–210.

HEBB, D.O. (1949) *The Organization of Behaviour*. Wiley, New York.

HECHT-NIELSEN, R. (1990) *Neurocomputing*. Addison-Wesley, Reading, Massachusetts.

HEGYI, F. (1974) A simulation model for managing Jack pine stands. In FRIES, J. (ed.) *Growth Models for Tree and Forest Simulation*, 74–90, Research Notes 30, Department of Forest Yield Research, Royal College of Forestry, Stockholm.

HENRI, V. (1902) *C. r. hebd. Acad. Sci., Paris*, **135**, 916–919.

HERTZ, J., KROGH, A. & PALMER, R.G. (1991) *Introduction to the Theory of Neural Computation*. Addison-Wesley, Redwood City, CA.

HINTON, G.E. & SEJNOWSKI, T.J. (1986) Learning and relearning in Boltzmann machines. In RUMELHART, D.E. & McCLELLAND, J.L. (eds) *Parallel Distributed Processing*. MIT Press, Massachusetts.

HODGKIN, A.L. & HUXLEY, A.F. (1952) A quantitative description of membrane current and its application to conduction and excitation in nerve. *J. Physiol.*, **117**, 500–544.

HOFSTETTER, F.B. (1954) Untersuchungen an einer Population der Turkentaube. *J. Orn.*, **95**, 348–410.

HOLLING, C.S. (1959) Some characteristics of simple types of predation and parasitism. *Canadian Entomologist*, **91**, 385–398.

HOPFIELD, J. J. (1982) Neural networks and physical systems with emergent collective computational abilities. *Proceedings of the National Academy of Sciences, USA*, **79**, 2554–2558.

HOSMER, D.W. Jr. & LEMESHOW, S. (1989) *Applied Logistic Regression*. Wiley, New York.

HUDSON, R. (1965) The spread of the Collared Dove in Britain and Ireland. *British Birds*, **58**(4), 105–139.

HUFFAKER, C.B. (1958) Experimental studies on predation: dispersion factors and predator–prey interactions. *Hilgardia*, **27**, 343–383.

HUNT, M. J., LENNIG, M. & MERMELSTEIN, P. (1983) Use of dynamic programming in a syllable-based continuous speech recognition system. In SANKOFF, D. & KRUSKAL, J.B. (eds) *Time Warps, String Edits, and Macromolecules: The Theory and Practice of Sequence Comparison*, Addison-Wesley, Reading, Massachusetts, Chapter 5.

HUTCHINSON, G.E. (1948) Circular causal systems in ecology. *Ann. N.Y. Acad. Sci.*, **50**, 221–246.

HUXLEY, J.S. (1932) *Problems of Relative Growth*. Methuen & Co. Ltd., London.

INGHE, O. (1989) Genet and ramet survivorship under different mortality regimes—a cellular automata model. *J. Theoret. Biol.*, **138**, 257–270.

ISHAM, V. (1988) Mathematical modelling of the transmission dynamics of HIV infection and AIDS: a review. *Journal of the Royal Statistical Society A*, **151**, 5–30.

KANDEL, E. (1989) Small systems of neurones. In LLINAS, R.R. (ed.) *The Biology of the Brain: From Neurons to Networks* (Readings from Scientific American), 70–86.

KANDEL, E.R., SCHWARTZ, J.H. & JESSELL, T.M. (1991) *Principles of Neural Science*. Elsevier, New York.

KEMPTHORNE, O. (1952) *The Design and Analysis of Experiments*. Wiley, New York.

KEMPTON, R.A. & HOWES, C.W. (1981) The use of neighbouring plot values in the analysis of variety trials. *Applied Statistics*, **30**, No. 1, 59–70.

KENDALL, D.G. (1974) Pole-seeking Brownian motion and bird navigation. *J. Roy. Statist. Soc.*, **B**, **36**, 365–417.

KENDALL, M.G. (1976) *Time-Series*. Griffin, London.

KERMACK, W.O. & McKENDRICK, A.G. (1927) Contributions to the mathematical theory of epidemics. *Proc. R. Soc. (A)*, **115**, 700–721.

KIRKWOOD, T.B.L. (1989) Methods for comparing DNA sequences. *Biometric Bulletin*, **6**(4), 20–21.

KOCH, C. & SEGEV, I. (1989) *Methods in Neuronal Modeling*. MIT Press, Cambridge, Massachusetts.

KOCH, H. von (1904) Sur une courbe continue sans tangent, obtenue par une construction géometrique élémentaire. *Arkiv för Matematik*, **1**, 681–704.

KUHN, R. (1923) Über Spezifität der Enzyme. II Saccharase- und Raffinasewirkung des Invertins. *Hoppe-Seyler's Zeitschr. f. Physiol. Chemie*, **125**, 28–92.

LACK, D. (1965) *The Life of the Robin*. Collins.

LAIRD, A.K. (1965) Dynamics of relative growth. *Growth*, **29**, 249–263.

LANGTON, C.G. (1986) Studying artificial life with cellular automata. *Physica*, **22D**, 120–149.

LE CUN, Y. (1985) Une procedure d'apprentissage pour reseau à seuil assymetrique (A learning procedure for asymmetric threshold network). *Proceedings of Cognitiva*, **85**, 599–604. Paris.

LENG, G. (ed.) (1988) *Pulsatility in Neuroendocrine Systems*. CRC Press, Boca Raton, Florida.

LESLIE, P.H. (1945) On the use of matrices in certain population mathematics. *Biometrika*, **33**, 182–212.

LESLIE, P.H. (1948) Some further notes on the use of matrices in population mathematics. *Biometrika*, **35**, 213–245.

LINDLEY, D.V. & SCOTT, F.W. (1984) *New Cambridge Elementary Statistical Tables*. Cambridge University Press, Cambridge.

LIGGETT, T.M. (1985) *Interacting Particle Systems*. Springer-Verlag, New York.

LIN, C.C. & SEGEL, L.A. (1973) *Mathematics Applied to Deterministic Problems in the Natural Sciences*. Macmillan, New York.

LINDENMAYER, A. (1968) Mathematical models for cellular interaction in development, Parts I and II. *Journal of Theoretical Biology*, **18**, 280–315.

LINDENMAYER, A. (1975) Developmental systems and languages in their biological context. In HERMAN, G.T. & ROZENBERG, G. *Developmental Systems and Languages*. North-Holland Publishing Company, Amsterdam.

LINDENMAYER, A. (1975) Developmental algorithms for multicellular organisms: a survey of L-systems. *Journal of Theoretical Biology*, **54**, 3–22.

LINEWEAVER, H. & BURK, D. (1934) The determination of enzyme dissociation constants. *J. Amer. Chem. Soc.*, **56**, 658–666.

LIPMAN, D.J. & PEARSON, W.R. (1985) Rapid and sensitive protein similarity searches. *Science*, **227**, 1435–1441.

LIPMAN, D.J., WILBUR, W.J., SMITH, T.F. & WATERMAN, M.S. (1984) On the statistical significance of nucleic acid similarities. *Nucleic Acids Research*, **12**, 215–226.

LOETSCH, F. & HALLER, K.E. (1964) *Forestry Inventory*, **1**, BLV Verlag, Munchen.

LOTKA, A.J. (1925) *Elements of Physical Biology*. Williams and Wilkins, Baltimore.

MacKENZIE, J.M.D. (1952) Fluctuations in the numbers of British tetraonids. *Journal of Animal Ecology*, **21**, 128–153.

MANLY, B.F.J. (1976) Some examples of the double exponential fitness function. *Heredity*, **36**(2), 229–234.

MANLY, B.F.J. (1990) *Stage-Structured Populations: Sampling, Analysis and Simulation*. Chapman & Hall, London.

MATTHEWS, G.V.T. (1955) *Bird Navigation*. Cambridge University Press.

MATTHEWS, G.V.T. (1974) On bird navigation, with some statistical undertones. *J. Roy. Statist. Soc.*, **B**, **36**, 349–364.

MAY, R.M. & ANDERSON, R.M. (1987) Transmission dynamics of HIV infection. *Nature*, **326**, 137–142.

MAY, R.M. (1967a) Models for single populations. In R.M. MAY (ed.) *Theoretical Ecology*, Blackwell Scientific Publications, Oxford, 4–25.

MAY, R.M. (1976b) Models for Two Interacting Populations. In R.M. MAY (ed.) *Theoretical Ecology*, Blackwell Scientific Publications, Oxford, 49–70.

MAY, R.M. (1987) Chaos and the dynamics of biological populations. *Proc. R. Soc. Lond.*, **A**, **413**, 27–44.

MAYNARD SMITH, J. (1965) The spatial and temporal organization of cells. In *Mathematics and Computer Science in Biology and Medicine*. M.R.C., HMSO, 247–254.

MAYNARD SMITH, J. (1974) *Models in Ecology*. Cambridge University Press, Cambridge.

McCULLOUGH, W.S. & PITTS, W. (1943) A logical calculus of the ideas immanent in nervous activity. *Bulletin of Mathematical Biophysics*, **5**, 115–133.

MEDLEY, G.F., ANDERSON, R.M., COX, D.R. & BILLARD, L. (1987) Incubation period of AIDS in patients infected via blood transfusion. *Nature*, **328**, 719–721.

MEINERT, C.L. & McHUGH, R.B. (1968) The biometry of an isotope displacement microassay. *Mathematical Biosciences*, **2**, 319–338.

MEINHARDT, H. (1982) *Models of Biological Pattern Formation*. Academic Press, London.

MICHAELIS, L. & MENTEN, M.L. (1913) Die Kinetik der Invertinwirkung. *Biochem. Z.*, **49**, 333–369.

MINSKY, M. & PAPERT, S. (1968) *Perceptrons: an Introduction to Computational Geometry*. MIT Press, Cambridge, Massachusetts.

MOLLISON, D. & KUULASMAA, K. (1985) Spatial Epidemic Models: Theory and Simulations. In P.J. BACON (ed.) *Population Dynamics of Rabies in Wildlife*. Academic Press, London.

MOLLISON, D. (1977) Spatial contact models for ecological and epidemiological spread. *J. R. Statist. Soc.*, B, **39**, 283–326.

MOLLISON, D. (1981) The Importance of Demographic Stochasticity in Population Dynamics. In HIORNS, R.W. & COOKE, D. (eds) *The Mathematical Theory of the Dynamics of Biological Populations II*. Academic Press, New York, 99–107.

MORGAN, B.J.T. (1984) *Elements of Simulation*. Chapman & Hall, London.

MOSTELLER, F. & TUKEY, J.W. (1977) *Data Analysis and Regression*. Addison-Wesley, London.

MOUNTFORD, M.D., WATSON, A., MOSS, R., PARR, R. & ROTHERY, P. (1989) Land inheritance and population cycles of red grouse. In LANCE, A.N. & LAWTON, J.H. (eds) *Red Grouse Population Processes*. Royal Society for the Protection of Birds, Sandy, 78–83.

MURRAY, J.D. (1989) *Mathematical Biology*. Springer-Verlag, Berlin.

NAGUMO, J., ARIMOTO, S. & YOSHIZAWA, S. (1962) An active pulse transmission line simulating nerve axon. *Proc. IRE.*, **50**, 2061–2070.

NEEDLEMAN, S.B. & WUNSCH, C.D. (1970) A general method applicable to the search for similarities in the amino acid sequences of two proteins. *J. Mol. Biol.*, **48**, 44–453.

NEWTON, I. & CAMPBELL, C.R.G. (1975) Breeding of ducks at Loch Leven, Kinross. *Wildlife*, **26**, 83–103.

NICHOLSON, A.J. & BAILEY, V.A. (1935) The balance of animal populations. *Proceedings of the Zoological Society of London*, **3**, 551–598.

NISONOFF, A. & PRESSMAN, D. (1958) Heterogeneity of antibody sites in their relative combining affinities for structurally related haptens. *J. Immunol.*, **81**, 126–135.

NUMATA, M. (1961) Forest vegetation in the vicinity of Choshi. Coastal flora vegetation at Choshi, Chiba Prefecture IV. *Bull. Choshi Marine Lab. Chiba Univ.*, **3**, 28–48.

PACALA, S.W., HASSELL, M.P. & MAY, R.M. (1990) Host–parasitoid association in patchy environments. *Nature*, **344**, 150–153.

PANDEY, R.B. (1989) Computer simulation of a cellular automata model for the immune response in a retrovirus system. *Journal of Statistical Physics*, **54**, 997–1010.

PAPADAKIS, J.S. (1937) Methode statistique pour des experiences sur champ. *Bulletin de l'Institut Amel. Plantes a Salonique*, **23**.

PARK, T. (1954) Experimental studies of interspecific competition. II. Temperature, humidity and competition in two species of *Tribolium*. *Physiological Zoology*, **27**, 177–238.

PARKER, D.B. (1985) *Learning Logic*. (TR–47). Massachusetts Institute of Technology, Centre for Computational Research in Economics and Management Science, Cambridge, Massachusetts.

PEARL, R., EDWARDS, T.I. & MINER, J.R. (1934) The growth of *Cucumis melo* seedlings at different temperatures. *J. Gen. Physiol.*, **17**, 687–700.

PEARL, R., REED, L.J. & KISH, J.F. (1940) The logistic curve and the census count of 1940. *Science*, **92**, 486–488.

PEARSON, W.R. & LIPMAN, D.J. (1988) Improved tools for biological sequence comparison. *Proc. Natl. Acad. Sci. USA*, **85**, 2444–2448.

PEITGEN, H., JÜRGENS, H. & SAUPE, D. (1992) *Fractals for the Classroom*. Springer-Verlag, New York.

PETO, J. (1986) AIDS and promiscuity. *Lancet*, 1986 Volume II, 979.

POULAIN, D.A., BROWN, D. & WAKERLEY, J.B. (1988) Statistical analysis of patterns of electrical activity in vasopressin and oxytocin-secreting neurones. In LENG, G. (ed.) *Pulsatility in Neuroendocrine Systems*, 119–154.

PREECE, M.A. & BAINES, M.J. (1978) A new family of mathematical models describing the human growth curve. *Annals of Human Biology*, **5**, 1–24.

PRESS, W.H., FLANNERY, B.P., TEUKOLSKY, S.A. & VETTERLING, W.T. (1989) *Numerical Recipes: the Art of Scientific Computing (FORTRAN version)*. Cambridge University Press, Cambridge.

PRUSINKIEWICZ, P. & LINDENMAYER, A. (1990) *The Algorithmic Beauty of Plants*. Springer-Verlag, New York.

RACE, R.R. & SANGER, R. (1954) *Blood Groups in Man*. Blackwell, Oxford.

RAGGETT, G.F. (1982) Modelling the Eyam plague. *The Institute of Mathematics and its Applications*, **18**, 221–226.

RAPP, P. E. (1980) Metabolic regulation as a control system. In SEGEL, L.A. (ed.) *Mathematical Models in Molecular and Cellular Biology*. Cambridge University Press. pp. 146–155.

RENSHAW, E. (1986) A survey of stepping-stone models in population dynamics. *Adv. Appl. Prob.*, **18**, 581–627.

RENSHAW, E. (1991) *Modelling Biological Populations in Space and Time*. Cambridge University Press, Cambridge.

RICHARDS, F.J. (1959). A flexible growth function for empirical use. *Journal of Experimental Botany*, **10**, 290–300.

RICKLEFS, R.E. (1968). A graphical method for fitting equations to growth curves. *Ecology*, **48**, 978–983.

RIPLEY, B.D. (1977) Modelling spatial patterns (with discussion). *Journal of the Royal Statistical Society*, Series B, **39**, 172–212.

RIPLEY, B.D. (1981) *Spatial Statistics*. Wiley, New York.

ROBINSON, D.L. (1987) Estimation and use of variance components. *The Statistician*, **36**, 3–14.

RODBARD, D., RAYFORD, P.L., COOPER, J. & ROSS, G.T. (1968) Statistical quality control of radioimmunoassays. *J. Clin. Endocrinol. Metab.*, **28**, 412ff.
RODBARD, D. (1974) Statistical quality control and routine data processing for radioimmunoassays and immunoradiometric assays. *Clinical Chemistry*, **20**, 1255–1270.
ROSENBLATT, F. (1958) The Perceptron, a probabilistic model for information storage and organization in the brain. *Psych. Rev.*, **65**, 386–408.
ROSENBLATT, F. (1962) *Principles of Neurodynamics: Perceptrons and the Theory of Brain Mechanisms*. Spartan Books, Washington DC.
ROSENZWEIG, M.L. & MacARTHUR, R.H. (1963) Graphical representation and stability conditions of predator–prey interactions. *American Naturalist*, **97**, 209–223.
ROSENZWEIG, M.L. (1969) Why the prey curve has a hump. *American Naturalist*, **103**, 81–87.
ROSS, G.J.S. (1990) *Nonlinear Estimation*. Springer-Verlag, New York.
ROSS, R. (1910) *The Prevention of Malaria*. Murray, London.
ROTHERY, P. (1974) The number of pins in a point quadrat frame. *Journal of Applied Ecology*, **11**, 745–754.
ROWELL, J.G. & WALTERS, D.E. (1976) Analysing data with repeated observations on each experimental unit. *Journal of Agricultural Science, Cambridge*, **87**, 423–432.
RUMELHART, D.E., HINTON, G.E. & WILLIAMS, R.J. (1986) Learning internal representations by error propagation. In RUMELHART, D.E. & McCLELLAND, J.L. (eds) *Parallel Distributed Processing. Explorations in the Microstructure of Cognition*, Vol. 1, 318–362.
RUMELHART, D.E. & McCLELLAND, J.L. (eds) (1986) *Parallel Distributed Processing: Explorations in the Microstructure of Cognition, I & II.* MIT Press, Cambridge, Massachusetts.
RYAN, B.F., JOINER, B.L. & RYAN, T.A. Jr. (1985) *Minitab Handbook*. Duxbury Press, Boston.
SAHLEY, C.L. (1990) The behavioural analysis of associative learning in the terrestrial mollusc *Limax maximus*: the importance of inter-event relationships. In HANSON, S.J. & OLSON, C.R. (eds) *Connectionist Modelling and Brain Function: the Developing Interface*. MIT Press, Cambridge, Massachusetts.
SAS Institute (1985) *SAS/STAT User's Guide*, Release 6.03 Edition. SAS Institute, Cary.
SCAMMON, R.E. (1927) The first seriatim study of human growth. *American Journal of Physical Anthropology*, **10**, 329–336.
SIEGEL, S. (1956) *Nonparametric Statistics for the Behavioral Sciences*. McGraw-Hill, New York.
SELLERS, P. (1974) An algorithm for the distance between two finite sequences. *Combin. Theor.*, **16**, 253–258.
SELVERSTON, A.I. & MOULINS, M. (1985) Oscillatory neural networks. *Ann. Rev. Physiol.*, **47**, 29–48.
SETLOW, R.B. & POLLARD, E.C. (1962) *Molecular Biophysics*. Addison-Wesley, Reading, Massachusetts.
SHIPLEY, R.A. & CLARK, R.E. (1972) *Tracer Methods for in vivo Kinetics*. Academic Press, New York.
SKELLAM, J.G. (1951) Random dispersal in theoretical populations. *Biometrika*, **38**, 196–218.
SKELLAM, J.G. (1972) Some philosophical aspects of mathematical modelling in empirical science with special reference to ecology. In J.N.R. JEFFERS (ed.) *Mathematical Models in Ecology*. Blackwell Scientific Publications, Oxford, 13–28.
SKELLAM, J.G. (1973) The formulation and interpretation of mathematical models of diffusionary processes in population biology. In BARTLETT, M.S. & HIORNS, R.W. (eds) *The Mathematical Theory of the Dynamics of Biological Populations*. Academic Press, New York, 63–85.
SMITH, F.E. (1963) Population dynamics in *Daphnia magna* and a new model for population growth. *Ecology*, **44**, 651–663.
SMITH, T.F. & WATERMAN, M.S. (1981) Identification of common molecular subsequences. *J. Mol. Biol.*, **147**, 195–197.
SNEDECOR, G.W. & COCHRAN, W.G. (1967) *Statistical Methods* (sixth edition). Iowa University Press, Iowa.

SOKAL, R.R. & ROHLF, F.J. (1981) *Biometry*. W.H. Freeman, New York.
STENT, G.S. (1963) *Molecular Biology of Bacterial Viruses*. W.H. Freeman, Folkstone.
STEWART, I. (1990) *Does God Play Dice*? Penguin Books, London.
STRAUSS, D.J. (1975) A model for clustering. *Biometrika*, **62**, 467–475.
TANNER, J.T. (1975) The stability and the intrinsic growth rates of prey and predator populations. *Ecology*, **56**, 855–867.
THOMPSON, D.J. (1975) Towards a predator–prey model incorporating age-structure: the effects of predator and prey size on the predation of *Daphnia magna* by *Ischnura elegans*. *Journal of Animal Ecology*, **44**, 907–916.
TOFFOLI, T. & MARGOLUS, N. (1987) *Cellular Automata Machines: a New Environment for Modelling*. MIT Press, Cambridge, Massachusetts.
TONG, H. (1990) *Non-linear Time Series: A Dynamical System Approach*. Clarendon Press, Oxford.
TUFTE, E.R. (1983) *The Visual Display of Quantitative Information*. Graphics Press, Cheshire, Connecticut.
TUKEY, J.W. (1977) *Exploratory Data Analysis*. Addison-Wesley, London.
TURING, A.M. (1952) The chemical basis of morphogenesis. *Phil. Trans. Roy. Soc. London*. **B237**, 37–72.
ULBRICH, J. (1930) *Die Bisamratte*. Heinrich, Dresden.
UTIDA, S. (1957) Population fluctuation, an experimental and theoretical approach. *Cold Spring Harbor Symp. Quant. Biol.*, **22**, 139–151.
VARIOT, G. & CHAUMET, M. (1906) Tables de croissance dressées en 1905 d'après les mensurations de 44,000 enfants parisiens de la 15 ans. *C. R. Acad. Sc. Paris*, T. 142, 299–301.
VARLEY, G.C. & GRADWELL, G.R. (1968) Population models for the winter moth. *Symposium of the Royal Entomological Society of London*, **4**, 132–142.
VOLTERRA, V. (1926) Fluctuations in the abundance of a species considered mathematically. *Nature*, **118**, 558–560.
VOLTERRA, V. (1926) Variazioni e fluttuazioni del numero d'individui in specie animali conviventi. *Mem. Acad. Lincei*. **2**, 31–113. Translation in appendix to CHAPMAN, R.N. (1931) *Animal Ecology*. McGraw-Hill, New York.
VOLTERRA, V. (1931) *Leçons sur la Théorie Mathématique de la Lutte pour la Vie*. Gauthier-Villars, Paris.
VON MISES, R. (1931) *Wahrscheinlichkeitsrechnung*. Leipzig and Wien.
VON NEUMANN, J. (1966) *Theory of Self-Reproducing Automata*. (Edited and completed by A. Burks). University of Illinois Press.
WATKINSON, A.R. (1980) Density-dependence in single species populations of plants. *Journal of Theoretical Biology*, **83**, 345–357.
WILBUR, W.J. & LIPMAN, D.J. (1983) Rapid similarity searches of nucleic acid and protein data banks. *Proc. Natl. Acad. Sci. USA*, **80**, 726–730.
WILLIAMS, T. & BJERKNES, R. (1972) Stochastic model for abnormal clone spread through epithelial basal layer. *Nature*, **236**, 19–21.
WINFREE, A.T. (1980) *The Geometry of Biological Time*. Springer-Verlag, New York.
WINFREE, A.T. (1987) *When Time Breaks Down*. Princeton University Press, Princeton.
WOLFRAM, S. (1984) Universality and complexity in cellular automata. *Physica* **10D**, 1–35.
WOLFRAM, S. (1985) Twenty problems in the theory of cellular automata. *Physica Scripta*, 1985, 170–183.
WOLFRAM, S. (1986) *Theory and Applications of Cellular Automata*. World Scientific.
WOOLF, B. (1932) cited by J.B.S. HALDANE & K.G. STERN, *Allgemeine Chemie der Enzyme*. Steinkopff, Dresden & Leipzig, 119–120.
YULE, G.U. (1902) *The New Phytologist*, **1**, 222–238.
ZULLINGER, E.M., RICKLEFS, R.E., REDFORD, K.H. & MACE, G.M. (1984) Fitting sigmoidal equations to mammalian growth curves. *J. Mamm.*, **65**(4), 607–636.

Index

NOTE: Figures and Tables are indicated by *italic page numbers*; footnotes by suffix 'n'

Index compiled by Paul Nash